Modulation-Doped Field-Effect Transistors

APPLICATIONS AND CIRCUITS

Modulation-Doped Field-Effect Transistors

APPLICATIONS AND CIRCUITS

EDITED BY

Heinrich Daembkes

HEAD OF MICROWAVE COMPONENTS DEPARTMENT
DAIMLER-BENZ AG RESEARCH CENTER—ULM

**Published under the sponsorship of the
IEEE Microwave Theory and Techniques Society**

IEEE
PRESS

The Institute of Electrical and Electronics Engineers, Inc., New York

Library of Congress Cataloging-in-Publication Data

Modulation-doped field-effect transistors : applications and circuits / edited by
Heinrich Daembkes.
    p.   cm.
    ''Published under the sponsorship of the IEEE Microwave Theory and
Techniques Society.''
    ''IEEE order number: PC0262-6.''
    Includes index.
    ISBN 0-87942-256-4
    1. Modulation-doped field-effect transistors. I. Daembkes, Heinrich, 1951–
II. IEEE Microwave Theory and Techniques Society.
TK7871.95.M63   1990
621.381′5284—dc20   90-37557

Contents

Section 1.1.3: Low Noise Amplifiers

Preface

THIS is the second of two related books. The aim of these books is to provide a comprehensive collection of basic contributions both for device and circuit engineers as well as graduate students in the field of microelectronics.

The first volume is intended to be an extension, or supplement, to the IEEE Press Reprint Book by H. Fukui, entitled *Low-Noise Microwave Transistors & Amplifiers* (1981). The second volume, concentrating on applications, should be seen in conjunction with the more general reprint book by R. A. Pucel (*Monolithic Microwave Integrated Circuits,* IEEE Press, 1985). The contributions start with the physics of the relevant semiconductor heterojunctions and end with the presentation of state-of-the-art (autumn 1988) circuits in the digital and in the microwave and millimeterwave areas.

With the development of new methods of epitaxial crystal growth—especially the molecular beam epitaxy (MBE) and metal organic vaporphase epitaxy (MOVPE)—the growth of atomically abrupt heterojunctions became possible. Today, layer sequences containing heterojunctions with transitions of atomic scale are routinely grown in numerous labs. At the abrupt transition between GaAs and AlGaAs, the formation of a potential well inside the GaAs is observed. If the AlGaAs layer is n-type doped and the GaAs layer is undoped, the free electrons of the thin AlGaAs layer are transferred into this potential well, as this position is more favorable for energetical reasons. By this effect the free carriers (here, electrons) are spatially separated from their parent donor atoms, thus reducing or even eliminating the mechanism of impurity scattering. This allows the carriers to attain extremely high mobilities, especially at lower temperatures, but also at room temperature. This is also the origin of one name of the new devices:

High Electron Mobility Transistors = HEMT.

The electrons are confined inside a very narrow potential well, which reduces their nature from a three-dimensional system to a two-dimensional one. The thin electron channel is formed only along the abrupt heterointerface, and one obtains a quasi-two-dimensional electron gas:

Two-dimensional Electron-Gas Field-Effect Transistor

= TEGFET.

The fact that the doping is introduced only into a selected part of the layer sequence is expressed in the names

MOdulation-Doped FET = MODFET

and

Selectively Doped Hetero-FET = SDHT.

The most general name for these new devices is

Hetero-Field-Effect Transistor = HFET.

The existence of the two-dimensional electron gas (2DEG) with extraordinary carrier transport properties is the key element of the new hetero-field-effect transistors, which leads to very high transconductances, high transit frequencies, and low noise properties. Thus the new hetero-FETs are the ideal candidates for the next generation's digital and analog integrated circuits.

The editor is grateful to members of the IEEE Microwave Theory and Techniques Society's Administrative and Publication Committees, in particular to Dr. M. Schneider and Prof. T. Itoh for suggesting that he undertake these books and for their support. Finally, he is indebted to the members of the IEEE Press staff, and to Mrs. Weber, for their efforts and support.

Heinrich Daembkes
Editor

Part 1
MODFETs in Analog Systems

Noise Modeling and Measurement Techniques

ALAIN CAPPY

(*Invited Paper*)

Abstract —The HEMT noise behavior is presented from theoretical and experimental points of view. The general method used in the high-frequency noise analysis is described and the different approximations commonly used in the derivation of the noise parameter expressions are discussed. A comparison between the noise performance of both MESFET's and HEMT's is carried out. The measurement techniques providing the noise figure and the other noise parameters are then described and compared.

I. INTRODUCTION

DURING THE PAST five years, the high electron mobility transistors (HEMT, also called TEGFET, MODFET, SDHT ···) employing modulation-doped AlGaAs/GaAs heterostructures has demonstrated excellent performance in the field of microwave amplifiers [1], [3]. Recently, a small-signal gain of 3.6 dB and an output power of 3.4 mW with 2 dB gain have been reported at 94 GHz for a single-stage amplifier [4]. Moreover HEMT devices capable of a noise figure as low as 2.7 dB at 62 GHz with 3.8 dB associated gain have been successfully produced. More generally, Fig. 1 shows the reported HEMT noise performance at room temperature from different laboratories. These performances are superior to those of conventional MESFET's, and the reasons of this superiority are interesting.

In the next section, the general method used in noise analysis is presented, including the calculation of the gate and drain noise sources and their correlation coefficient, as well as the calculation of the noise figure and the other noise parameters. The different approximations and assumptions commonly used in the noise figure derivation are discussed and the specific influence of the gate noise and of the correlation coefficient is emphasized. A comparison between the HEMT and MESFET noise performance is then carried out in order to illustrate the main reasons of the HEMT superiority in the fields of low-noise amplifiers. Results on the HEMT low-frequency noise behavior, an important parameter in oscillator spectral purity, are then presented. Different structures with reduced low-frequency noise are proposed. The last section reviews the high-frequency noise figure measurement techniques and gives their respective advantages and drawbacks, especially in terms of measurement accuracy.

Manuscript received May 4, 1987; revised July 27, 1987.

The author is with the Centre Hyperfréquences et Semiconducteurs, Université des Sciences et Techniques de Lille, 59655 Villeneuve d'Ascq Cedex, France

IEEE Log Number 8717580.

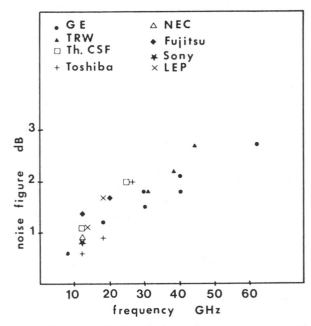

Fig. 1. Reported HEMT noise performance at room temperature.

II. HIGH-FREQUENCY NOISE MODELING IN HEMT'S

From the point of view of the circuit designer, the noise performance of the HEMT is characterized by three noise parameters: the minimum noise figure $F_{\min}$, the noise conductance G_n (or noise resistance R_n), and the optimum source impedance Z_{opt} (or optimum source admittance Y_{opt}). These noise parameters can easily be calculated from the equivalent circuit of the noisy HEMT presented in Fig. 2(a). This circuit comprises the well-known small-signal equivalent circuit and the four noise sources $\overline{e_g^2}, \overline{e_s^2}, \overline{i_g^2}, \overline{i_d^2}$. The two noise sources $\overline{e_g^2}$ and $\overline{e_s^2}$ represent the noisy behavior of access resistances R_g and R_s and are simply given by the Nyquist formula

$$\overline{e_i^2} = 4kTR_i \Delta f \tag{1}$$

where k is the Boltzmann constant, T the absolute temperature, R_i the resistance value, and Δf the frequency bandwidth. The two current noise sources $\overline{i_g^2}$ and $\overline{i_d^2}$ represent the internal noise sources of the intrinsic HEMT. These noise sources are correlated.

By means of a simple circuit manipulation, these four noise sources are transformed in two correlated noise sources $\overline{v^2}$ and $\overline{i^2}$ preceding the extrinsic HEMT, which is

Reprinted from *IEEE Transactions on Microwave Theory and Techniques*, vol. 36, no. 1, pp. 1–10, January 1988.

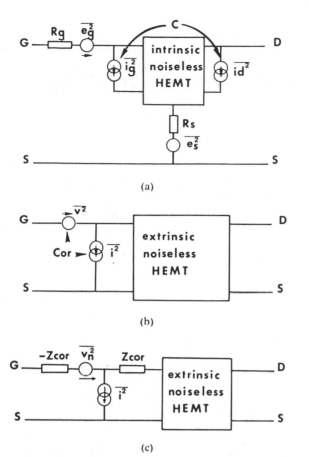

Fig. 2.   Circuit transformations for the noise figure determination.

now considered noiseless (Fig. 2(b)). Lastly, $\overline{v^2}$ and $\overline{i^2}$ are decorrelated by introduction of two (noiseless) correlation impedances (Fig. 2(c)). At this step, the calculation of F_{min} and Z_{opt} is straightforward and gives

$$F_{min} = 1 + 2g_n(R_{cor} + R_{opt}) \qquad (2)$$

where

$$Z_{cor} = R_{cor} + jX_{cor} \qquad g_n = \overline{i^2}/4kT\Delta f$$

$$r_n = \overline{v_n^2}/4kT\Delta f$$

and

$$Z_{opt} = R_{opt} + jX_{opt} = \sqrt{R_{cor}^2 + \frac{r_n}{g_n}} - jX_{cor}.$$

More details of this noise figure calculation technique can be found in [1] and [2].

This synopsis of the noise analysis shows that the determination of the HEMT's noise performance requires (i) knowledge of the small-signal equivalent circuit and (ii) the gate and drain noise sources $\overline{i_g^2}$ and $\overline{i_d^2}$, as well as their correlation coefficient, defined by

$$C = \frac{\overline{i_g i_d^*}}{\sqrt{\overline{i_g^2} \cdot \overline{i_d^2}}}. \qquad (3)$$

Therefore, a noise modeling will be divided into two steps: the calculation of the small-signal equivalent circuit and the calculation of the noise sources $\overline{i_g^2}, \overline{i_d^2}$ and the

correlation coefficient C. In fact, the method used for the noise source calculation [3]–[5] is usually derived from the Shockley impedance field method and it may be appropriate to recall briefly the principles of this method.

In the microwave frequency range, the noise arises from the fluctuation of carrier velocity due to the scatterings with phonons or ionized impurities. In a section of the HEMT channel (length dx, width Z, sheet carrier density $N(x)$), the local noise current can be expressed as [6]

$$\overline{i_d^2(x)} = q^2 Z N(x) \overline{\Delta v_\parallel^2}/dx. \qquad (4)$$

In this expression $\overline{\Delta v_\parallel^2}$ represents the mean square of the velocity fluctuations. In the microwave frequency range, the spectrum of $\overline{\Delta v_\parallel^2}$ is white, leading to a local noise current spectral density given by

$$Si_d(x) = 4q^2 Z N(x) D_\parallel(x)/dx \qquad (5)$$

where $D_\parallel(x)$ is the diffusion coefficient parallel to the electrical field direction. This expression is quite general and can be applied for a resistance at thermal equilibrium (thermal noise) and/or for the high field region of the HEMT channel (diffusion noise).

The purpose of the impedance field method is to determine the effects of the *local* noise sources $Si_d(x) \cdot \Delta f$ on the drain and gate electrodes. The device noise behavior will then be entirely characterized by two correlated equivalent gate and drain noise sources. Let $Z(x, \omega)$ be the small-signal impedance between the abscissa x and the drain electrode. The drain voltage fluctuation $\overline{v_d^2(x)}$ arising from the local noise source located between x and $x + dx$ is given by

$$\overline{v_d^2(x)} = 4q^2 Z N(x) D_\parallel(x) \left| \frac{dZ(x, \omega)}{dx} \right|^2 \Delta f \cdot dx. \qquad (6)$$

Assuming that the noise current sources located at two different abscissas x and x' are uncorrelated, the open-circuit noise voltage $\overline{v_d^2}$ is given by a summation performed over the whole length of the active channel.

$$\overline{v_d^2} = \int 4q^2 Z N(x) D_\parallel(x) \Delta f \left| \frac{dZ(x, \omega)}{dx} \right|^2 dx. \qquad (7)$$

It is then convenient to define the short-circuit noise current $\overline{i_d^2}$ by $\overline{i_d^2} = |y_{22}|^2 \overline{v_d^2}$, where y_{22} is an intrinsic admittance matrix parameter defined by $y_{22} = g_d + j\omega C_{gd}$.

The gate noise current source $\overline{i_g^2}$ and the correlation coefficient C are calculated in a similar way [5], [6], and can be expressed as

$$\overline{i_g^2} = 4q^2 Z \Delta f \omega^2 \int N(x) D_\parallel(x) \left| \frac{d}{dx}\left(\frac{\partial Q(x)}{\partial I} \right) \right|^2 dx \qquad (8)$$

$$\overline{i_g i_d^*} = j\omega 4q^2 Z \Delta f \int N(x) D_\parallel(x) \left(\frac{d}{dx} Z(x, \omega) \right)$$

$$\cdot \left(\frac{d}{dx}\left(\frac{\partial Q(x)}{\partial I} \right) \right) dx. \qquad (9)$$

In these expressions, Q is the total amount of stored

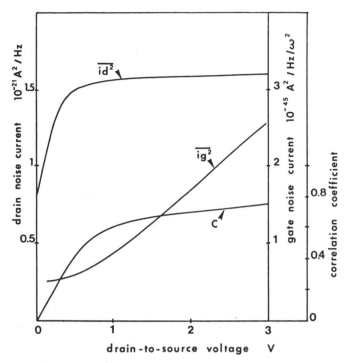

Fig. 3. Evolution of the drain and gate noise current sources and the correlation coefficient versus V_{ds}. $V_{gs} = -0.5$ V, $L_g = 0.5$ μm, $A = 500$ Å, $N_d = 10^{18}$ cm^{-3}.

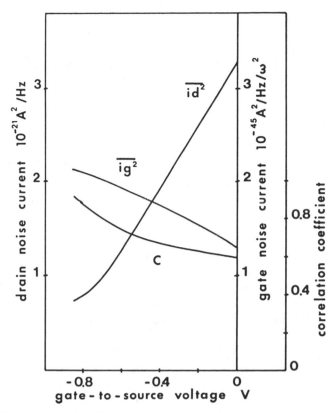

Fig. 4. Evolution of the drain and gate noise current sources and of the correlation coefficient versus V_{gs}. $V_{ds} = 2$ V, $L_g = 0.5$ μm, $A = 500$ Å, $N_d = 10^{18}$ cm^{-3}

charge in the device and therefore $\partial Q(x)/\partial I$ represents the stored charge fluctuation induced by a local noise source. Expressions (7), (8), and (9) show that the calculation of the HEMT noise performance requires, for each section of the channel, the sheet carrier density $N(x)$, the diffusion coefficient $D_{\parallel}(x)$, and the impedance and "charge" fields dZ/dx and $d/dx(\partial Q/\partial I)$. Since $N(x)$ is given by the dc drain current $I_{ds} = qZN(x)v(x)$, and $D_{\parallel}(x)$ is assumed to be dependent on the local electrical field [4] or local average energy [5], the main problem encountered in noise modeling is to calculate the impedance and "charge" fields. According to the type of modeling, this can be done in an analytical [3], [4] or a numerical form [6].

It should be emphasized that the impedance field method applies only to 1-D modeling, at least in its classical form. In most cases, the device noise properties are thus calculated *below* the onset of saturation [8]–[10], which obviously constitutes an important approximation.

It should be also noted that other techniques can be used for the noise analysis. As a matter of fact, statistical procedures, such as 2-D Monte Carlo modeling, provide the instantaneous current $I(t)$ and consequently the mean square current fluctuations $\overline{\Delta I^2(t)} = \overline{(I(t) - \bar{I})^2}$. The noise current spectral density can be easily deduced from $\overline{\Delta I^2(t)}$ by a Fourier transform [10]. However, since very long time periods cannot be studied by a 2-D Monte Carlo procedure, the results of such a method concern mainly the device noise behavior in the millimeter-wave range. Unfortunately, this method has not been applied for the HEMT case yet.

Returning to the noise source calculation problem it seems appropriate at this step to discuss the evolution as a function of the gate and drain bias voltages. For this purpose, Fig. 3 shows the evolutions of both the noise current sources $\overline{i_d^2}$ and $\overline{i_g^2}$, and the correlation coefficient C versus V_{ds} for a given V_{gs}. These results have been obtained using the numerical noise modeling of Cappy et al. [5], which takes the nonstationary electron dynamics and the carrier injection into the buffer into account. This figure shows that $\overline{i_d^2}$ and C increase in the ohmic region and tend to saturate at high drain voltage, while $\overline{i_g^2}$ increases with a near constant slope versus V_{ds}. This indicates that the hot electron effects, which are more pronounced at high V_{ds}, affect mainly the gate noise and, to a lesser extent, the correlation coefficient.

As a function of V_{gs}, Fig. 4 shows a strong decrease of $\overline{i_d^2}$ and an increase of both $\overline{i_g^2}$ and C as the dc drain current diminishes.

When the noise sources are known, the problem is to calculate the noise figure and the other noise parameters. Due to the large number of different parameters influencing the noise figure, a physical analysis of the noise properties is rather difficult in the most general case. For this reason, it is convenient to introduce the different elements step by step in order to make their influence easier to understand.

III. THE NOISE PARAMETERS

In a first approximation, the gate noise source $\overline{i_g^2}$ and feedback capacitance C_{gd} can be neglected. Following van der Ziel [7], [8], the drain noise source $\overline{i_d^2}$ can be expressed as

$$\overline{i_d^2} = 4kTg_mP\Delta f \qquad (10)$$

where g_m is the transconductance and P is a dimensionless parameter close to 1–3, depending upon the technological parameters and biasing conditions. Introducing expression (10) in the noise figure calculation yields

$$F_{min} = 1 + 2\sqrt{P} \cdot \frac{f}{f_c} \cdot \sqrt{g_m(R_s + R_g)} \qquad (11)$$

$$g_n = Pg_m(f/f_c)^2 \qquad (12)$$

$$Z_{opt} = \sqrt{\frac{g_m(R_s + R_g)}{P}} \cdot \frac{1}{C_{gs}\omega} + \frac{1}{jC_{gs}\omega} \qquad (13)$$

where F_{min} is the minimum noise figure, g_n the noise conductance, Z_{opt} the optimum input impedance, and f_c is the cutoff frequency $g_m/2\pi C_{gs}$. It can be noted that expression (11) is similar to the well-known and widely used Fukui formula, where the so-called fitting factor kf is given by $2\sqrt{P}$. Classically, expression (11) shows that a low F_{min} value requires a high f_c and small values of both the sum $R_s + R_g$ and the coefficient P. In fact, the P value greatly influences F_{min}, g_n, and Z_{opt} and its determination is obviously of primary importance.

According to Delagebeaudeuf [9], the parameter P can be approximated for operating points below the onset of saturation by

$$P = \frac{I_{ds}}{E_c L_g g_m}. \qquad (14)$$

In this expression I_{ds} is the dc drain current, E_c the critical field of an idealized $v - E$ relationship, L_g the gate length, and g_m the transconductance.

Usually the transconductance of HEMT's is higher than that of conventional MESFET's. Therefore the coefficient P and, of course, the Fukui fitting factor $kf = 2\sqrt{P}$ is lower for HEMT's. Introducing (14) in (11) yields

$$F_{min} = 1 + 2\sqrt{\frac{I_{DS}}{E_c L_g} \cdot \frac{f}{f_c}} \cdot \sqrt{R_s + R_g}. \qquad (15)$$

It should be emphasized that, as far as the design of low-noise devices is concerned, expressions (15) and (11) yield quite different conclusions. Indeed, following (11), for a given gate length and therefore a near constant cutoff frequency f_c, the noise figure can be reduced by reducing the transconductance g_m. In other words, the noise figure can be improved using a large epilayer thickness. On the contrary, the transconductance is no longer present in expression (15) and the preceding conclusion does not hold. From an experimental point of view, it is rather difficult to separate these two approaches. For this purpose, Table I presents different reported results. These results show that very high transconductance can provide a low noise figure (device I). Furthermore, with similar gate length, access resistance value, and cutoff frequency, device V [large epilayer thickness (570 Å) and rather low g_m

(220 mS/mm)] and device VI [400 Å epilayer thickness and higher g_m (275 mS/mm)] exhibit similar noise performance. It should be also noted that the kf factor deduced from experimental findings varies significantly from one device to another. For the data of Table I, kf varies from 1.2 to 2.5 without any obvious correlation with the other device parameters.

The important problem encountered in the comparison between the expressions (11) and (15) of the noise figure probably arises from the assumptions used to deduce these formulas. In particular, it seems important to consider the influence of the gate noise source and the correlation coefficient. According to the pioneering work of van der Ziel [8], the gate noise current source $\overline{i_g^2}$ can be expressed as

$$\overline{i_g^2} = 4kT\Delta f C_{gs}^2 \omega^2 R/g_m \qquad (16)$$

where R is a dimensionless multiplication parameter depending upon biasing conditions and device parameters. Neglecting the influence of C_{gd}, the calculation of F_{min} and other noise parameters can be carried out analytically and yields

$$F_{min} = 1 + 2\sqrt{P + R - 2C\sqrt{PR}} \cdot \frac{f}{f_c}$$
$$\cdot \sqrt{g_m(R_s + R_g) + \frac{PR(1 - C^2)}{R + P - 2C\sqrt{RP}}} \qquad (17)$$

$$g_n = g_m\left(\frac{f}{f_c}\right)^2 \cdot \sqrt{P + R - 2C\sqrt{RP}} \qquad (18)$$

$$Z_{opt} = \sqrt{\frac{g_m(R_s + R_g) + \frac{PR(1 - C^2)}{P + R - 2C\sqrt{RP}}}{P + R - 2C\sqrt{PR}}} \cdot \frac{1}{C_{gs}\omega}$$
$$+ \frac{1}{jC_{gs}\omega} \cdot \left(\frac{P - C\sqrt{RP}}{P + R - 2C\sqrt{PR}}\right). \qquad (19)$$

These expressions, like those obtained by Pucel [2], show different fundamental effects:

(i) The gate noise influences the noise figure *even at low frequency*.

(ii) The noise figure (in linear scale, not in dB) keeps a linear variation versus frequency even if the gate noise is taken into account.

(iii) If $R_s + R_g$ tends to zero, the noise figure is no longer close to unity if the gate noise is taken into account and the device can be characterized by an intrinsic noise figure:

$$F_{int} = 1 + 2\frac{f}{f_c} \cdot \sqrt{PR(1 - C^2)} \qquad (20)$$

(iv) Due to the correlation between the drain and gate noise current sources, the gate noise is partially subtracted from the drain noise. This important effect is expressed in the noise parameter expressions by the terms $P + R - 2\sqrt{PR}$ and $PR(1 - C^2)$. This reduction of the drain noise

TABLE I

device	I	II	III	IV	V	VI	VII	VIII
ref.	[14]	[39]	[40]	[41]	[42]	[43]	[44]	[45]
geometry	.0,25x150	0,35x65	0,35x62	0,4x200	0,5x200	0,5x200	0,5x200	0,55x200
g_m mS/mm	570	230	330	250	220	275	235	290
g_d mS/mm	32	23	17	18	15	20	15	20
C_{gs} pF/mm	1.8	0.95	2.05	1.0	1.2	1.25	1.3	1.2
C_{gd} fF/mm	100	230	78	100	50	100	50	165
R_g Ω	0,9	2.4	1.6	4	2.1	1	1	2
R_s Ω.mm	0,5	0,38	0,19	0,7	0,65	0,68	0,8	0,7
F_c GHz	50	38	26	40	29	35	29	38
f_o,GHz	50	16	35	28	48	35	48	20
F_{min}-dB 12 GHz	0,83	1.05	1,1	0,95	0,95	0,85	1.4	1.2
Gass dB	–	–	13	11,8	10,3	12,5	11	11
k_f	1.43	2.5	2	1.33	1.2	1.28	1.9	1.8
Fcalc P=1,R=0.5 C=0.9	0.8	1.2	1.2	1.0	1.2	1.0	1.2	1.0
Fint calc P=1,R=0.5 C=0.9	0.6	0.77	1.08	0.73	0.98	0.83	0.98	0.77

is the basic reason why the field effect transistor (conventional or not) is a low-noise device. Therefore, it is not obvious to neglect the gate noise and the correlation coefficient in FET noise analysis.

From a theoretical point of view, the problem is to estimate the multiplication coefficient R and the correlation coefficient C. In fact, several papers [8], [11], [12] have shown that R (in MESFET's) is close to 0.2–0.4 below the onset of saturation. This is not negligible compared with the P value (close to 0.6–2) given by the same modeling. Independently and using a completely different approach, Cappy and coworkers have found R close to 0.5 and P close to 1–1.5 for the low-noise operating conditions (I_{ds} #100 mA/mm, $V_{ds} = 3$V). More precisely, Figs. 5 and 6 show the evolution of P, R, and C as a function of V_{ds} and V_{gs} for the same device parameters as for Figs. 3 and 4.

Concerning the correlation coefficient, which is one of the main parameters influencing the intrinsic noise figure, Fig. 4 shows that this parameter is strongly underestimated when the calculation is performed below the onset of saturation. Under low-noise conditions and for an operating frequency up to 20 GHz for a half-micron gate length

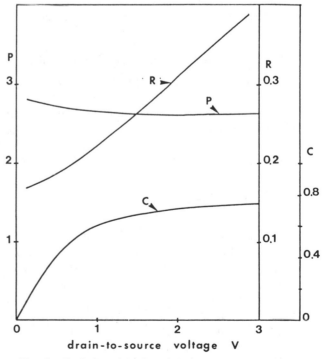

Fig. 5. Evolution of R, P, and C versus V_{ds}. $V_g = -0.5$ V.

7

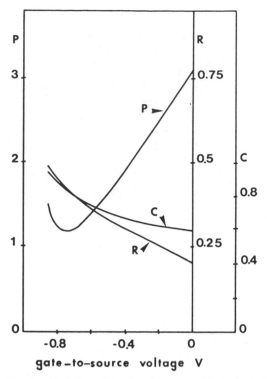

Fig. 6. Evolution of R, P, and C versus V_{gs}. $V_{ds} = 2$ V.

HEMT, C is practically purely imaginary and is close to 0.8–0.95. In addition, it seems that this parameter mainly depends on the aspect ratio L_g/A [5].

In order to show the validity of expression (17), both the extrinsic and intrinsic noise figures have been calculated for the different devices described in Table I. For this comparison, the same typical values of P, R, and C have been introduced in each case: $P = 1$, $R = 0.5$, $C = 0.9$. It should be noted that theoretical and experimental noise figures are now in good agreement. In addition, the important contribution of the intrinsic noise figure to the extrinsic noise figure should be pointed out.

IV. INFLUENCE OF THE FEEDBACK CAPACITANCE C_{gd}

In the preceding theoretical approaches, the feedback capacitance C_{gd} is neglected, which constitutes a strong approximation, especially for the noise figure derivation in the millimeter-wave range. The main influence of the gate-to-drain coupling is to make the drain noise current source frequency dependent, following the expression [5]

$$\overline{i_d^2} = \overline{i_{d_0}^2}\left(1 + \left(\frac{f}{f_0}\right)^2\right) \tag{21}$$

where $f_0 = g_d/2\pi C_{gd}$ and $\overline{i_{d_0}^2}$ is the drain noise source when C_{gd} is neglected. As a consequence, the noise figure rises steeper than the f law for frequencies beyond f_0. Furthermore, due to the evolution of g_d and C_{gd} versus V_{gs}, f_0 decreases for decreasing drain currents; consequently the optimum dc drain current increases with increasing operating frequency. These two effects have been found experimentally in the case of conventional MESFET's [13] and HEMT's [14].

The cutoff frequency f_0 constitutes an important parameter of the noise behavior of HEMT's; therefore, since a low output conductance is required for achieving high power gain, a reduction of C_{gd} is necessary in order to provide f_0 values as high as possible. From the experimental datas of Table I, f_0 is shown to be close to or greater than f_c except for devices II and VIII.

V. COMPARISON BETWEEN MESFET's AND HEMT's NOISE PERFORMANCE

A. The Noise Figure

The preceding noise analysis shows that the main parameters influencing the noise figure are the cutoff frequency f_c, the three noise parameters P, R, and C, and the sum of access resistances $R_s + R_g$.

Usually the cutoff frequency of HEMT's is larger than that of MESFET's for two main reasons:

(i) The high carrier mobility provides a more important overshoot effect, resulting in a higher average velocity and therefore a higher transconductance.

(ii) The small epilayer thickness yields high g_m, high intrinsic gate capacitance, and therefore a relatively less important effect of the parasitic capacitances.

Thus, the cutoff frequency of a typical 0.4–0.5 μm gate length HEMT is only slightly lower than that of the best 0.2–0.25 μm gate length conventional MESFET's [15], [16], which constitutes an important element of the HEMT superiority.

Concerning the noise coefficient P, R, and C, the situation is less clear since these parameters cannot be accurately provided by experiments and are mainly the results of theoretical considerations. Fortunately, for a given drain current, these parameters do not vary in a large extent, R is always close to 0.5–0.7 under low noise conditions while C mainly depends on the aspect ratio L_g/A. For short-gate-length MESFET's, C is close to 0.7–0.8 while a higher L_g/A ratio in the case of HEMT's provides a higher C value, close to 0.8–0.95.

In the case of the drain noise coefficient P, the difference between MESFET's and HEMT's is not important because of antagonistic effects, namely slight increases of P, firstly with increasing mobility and secondly with increasing epilayer thickness. The first effect arises from the relation between the low field diffusion coefficient and the carrier mobility (Einstein relation):

$$D_0 = \frac{kT}{q}\mu_0. \tag{22}$$

This relation holds for HEMT's and MESFET's at low electrical field strengths. For higher fields, theoretical works [7] have shown that the diffusion coefficient in a 2-D electron gas is similar to the bulk diffusion coefficient. Therefore, since the greater part of drain noise arises from the low field region [8], an increase of both $\overline{i_d^2}$ and P with increasing mobility is not surprising. Nevertheless, this effect is compensated in HEMT's by the decrease of P with decreasing epilayer thickness, and finally no important superiority of HEMT's over MESFET's arising from the noise coefficient P can be expected.

The last parameters determining the noise figure involve the sum of access resistances $R_s + R_g$. Since the gate resistance is dependent only on the gate fabrication process and device layout, the main problem arises from the source resistance. Furthermore, the choice of smaller gate width for increasing frequency operation involves a reduction of R_g, which becomes far less significant than the increasing value of access resistance R_s.

In the case of MESFET's, the source resistance R_s can be properly calculated from a knowledge of the active layer structure [18]. In the case of HEMT's, the source-to-gate access region is more complex and the source resistance value results from the conduction in several layers: the highly doped cap layer, the highly doped AlGaAs layer, and the 2-D gas [19], [20]. The complexity of the source–gate access involves several effects:

(i) The conventional transmission line method underestimates the parasitic source resistance [19].

(ii) Due to the capacitive effect of the depleted region at the AlGaAs/GaAs interface, the source resistance is frequency dependent [21].

Therefore, a precise determination of R_s is not obvious, especially at a high frequency of operation, yielding some difficulties in the correlation between the noise figure and the source resistance value. Nevertheless, very low values of both contact resistance ($R_c = 0.03 \ \Omega \cdot mm$) and total access resistance ($R_s = 0.5 \ \Omega \cdot mm$) have been achieved [22]. These results are comparable to the state-of-the-art of MESFET source resistances, and these two devices can be considered as equivalent as far as the parasitic resistances are concerned.

B. The Noise Conductance

In several applications such as broad-band amplifiers, the device is not matched for the minimum noise figure and the mismatch effect can be expressed as

$$F = F_{\min} + \frac{g_n}{R_0}|Z_0 - Z_{opt}|^2 \qquad (23)$$

where $F_{\min}$ is the minimum noise figure, g_n the noise conductance, $Z_0 = R_0 + jX_0$ the input termination, or source, impedance, and Z_{opt} the optimum source impedance. This expression shows that the mismatch effect is less important for low values of the noise conductance g_n. As shown in expressions (12) and (18), the noise conductance is inversely proportional to the square of the cutoff frequency f_c. Since HEMT's exhibit a higher f_c value than MESFET's, HEMT's have lower noise conductance, which results in reduced sensitivity of the noise figure to changes in source impedance and therefore permits low-noise performance over a wider bandwidth. This effect has been experimentally shown by Pospieszalski et al. [23].

To summarize, the superiority of HEMT noise performance can be related to the higher cutoff frequency f_c and to the higher correlation coefficient reducing the intrinsic noise figure F_{int}, which represents, at least for the good

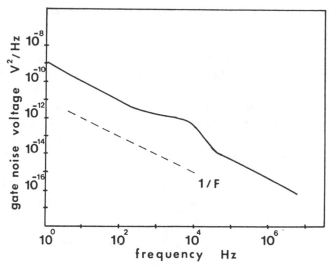

Fig. 7.   Typical HEMT equivalent gate LF noise spectra.

devices having low access resistance values, an important part of the extrinsic noise figure $F_{\min}$.

VI. LOW-FREQUENCY NOISE IN HEMT'S

In the preceding section, our attention was focused on the HEMT noise properties in the microwave frequency range. However, low-frequency (LF) noise is an important parameter in some applications, e.g. local oscillators, because the LF noise is up-converted in the microwave frequency range [24]. Below 100 MHz, the two dominant noise sources are the $G - R$ noise and the $1/f$ noise, while the diffusion noise becomes significant when $G - R$ and $1/f$ noise vanish at higher frequencies. The $G - R$ noise is caused by fluctuaions in the number of free carriers and can be related to the presence of trap centers in the forbidden gap, while the physical origin of $1/f$ noise in semiconductor devices remains unknown as yet. The LF noise, usually measured as an equivalent noise voltage referred to the input gate, can be expressed as [25]

$$\overline{v_n^2} = 4kT\Delta f \left(\rho_0 \frac{f_0}{f} + \sum_{r=1}^{n} \frac{\rho_r(\tau_r/\tau_0)}{1 + (2\pi f\tau_r)^2} \right) \qquad (24)$$

where ρ_0 is the $1/f$ noise equivalent gate resistance when $f = f_0$, ρ_r is a similar resistance for the rth component of the GR noise, τ_0 is a reference time constant, and τ_r the time constant for the trap causing GR noise. A typical evolution of $\overline{v_n^2}$ in HEMT's is presented in Fig. 7, which shows $G - R$ noise components superimposed on a background of $1/f$ noise. This is similar to the LF noise behavior of MESFET's.

The experimental study of the HEMT LF noise spectra [25]–[27] has clearly indicated several effects:

(i) The GR noise originates from (AlGa)As only and the noise magnitude is higher than that of the MESFET one.

(ii) A higher Al mode fraction enhances the trap concentration that causes the $G - R$ bulge near 10 kHz.

(iii) A close correlation exists between the amplitude of the room-temperature LF noise and the device properties at low temperature.

Therefore, the complex band structure of (Al,Ga)As and the multiple donor levels introduced by the dopant atoms [28] are the main causes of the observed strong LF noise. The use of low Al mole fraction (< 0.2) can reduce the amplitude of the GR noise, but is not well suited to provide high sheet carrier concentration and therefore high performance devices. However, it has been shown that the low-temperature parasitic effects (persistent photoconductivity, collapse, etc.) can be greatly reduced by spatially separating Al and the dopant atoms. This can be done using a superlattice [29] or an atomic planar doped layer [30]. These structures are then likely to present low LF noise and can be used for the realization of high-frequency oscillators with good spectral purity.

VII. High-Frequency Noise Measurement Techniques

A precise determination of the device noise performance is of primary importance in comparing various devices, validating the design of a low-noise device, and validating noise modeling. For this purpose, two different methods can be used: the conventional method and the least-squares fit of measured noise figures as a function of the input termination.

A. The Conventional Method

Fig. 8 shows a typical microwave transistor noise characterization system. The automatic noise-figure indicator provides the noise figure and gain of all the amplifier stages constituted by the bias networks, the tuners, and the device in the fixture. The main problem is then to deduce the noise figure F and associated G of the device using the measured F_m, G_m values.

In fact the noise figure of a cascade of noisy two-ports is given by the Friis formula:

$$F_m = F_1 + \frac{F-1}{G_1} + \frac{F_2-1}{GG_1}. \qquad (25)$$

In our case, F_1 and G_1 are the noise figure and available gain of the input matching two-port constituted by the input bias, the input tuner, and half of the test fixture, while F_2 refers to similar components for the output (Fig. 9). Since the input and output matching two-ports are passive networks, $F_1 = 1/G_1$ and $F_2 = 1/G_2$ and (25) can be written as

$$F_m = \frac{F}{G_1} + \frac{1-G_2}{GG_1G_2} \qquad (26)$$

or

$$F = G_1\left(F_m - \frac{1-G_2}{G_m}\right) \qquad (27)$$

where $G_m = GG_1G_2$.

Expression (27) shows that an accurate determination of the device noise figure F needs an accurate determination of G_1. As shown by Strid [31], the popular back-to-back method for determining G_1 can be very inaccurate; therefore, it is preferable to calculate G_1 from S parameters

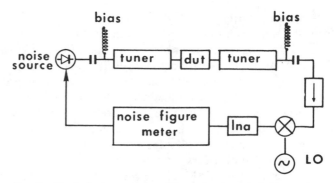

Fig. 8. Typical microwave transistor noise characterization system.

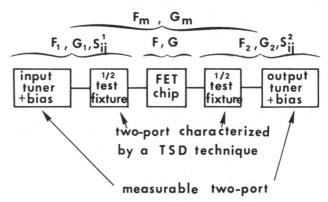

Fig. 9. Extraction of the device noise figure F and associated gain G from the measured values F_m and G_m.

that can be measured with a high accuracy. The S parameters (S_{ij}^1) of the whole input network are easily calculated from the S parameters of the input bias + tuner, which can be measured, and from the S parameters of the input half test fixture, which can be provided by a TSD technique [38]. Assuming an ideal generator impedance, we have

$$G_1 = \frac{|S_{21}^1|^2}{1-|S_{22}^1|^2}. \qquad (28)$$

The determination of G_2 can be carried out in a similar way.

This method has several drawbacks:

(i) A long measurement time is required for seeking the minimum noise figure as well as for the S parameter measurements.

(ii) A minimum observed value F_m does not always provide the device minimum noise figure after the correction for the losses.

(iii) An additional measurement is needed for the noise resistance R_n (or noise conductance g_n) determination.

For these reasons, a more systematic method has been proposed [32].

B. The Noise Parameter Determination by Least-Squares Fit

Instead of randomly searching for the real minimum noise figure, another method is to measure the noise figure for different input reflection coefficients. Although four measurements are needed to determine the four unknowns

$(F_{min}, R_n, \Gamma_{opt})$ of the relation (29)

$$F = F_{min} + \frac{4R_n}{|1 + \Gamma_{opt}|^2} \frac{|\Gamma_0 - \Gamma_{opt}|^2}{1 - |\Gamma_0|^2} \qquad (29)$$

where Γ_0 is the input reflection coefficient, it is better to perform seven or more measurements and then to determine the four noise coefficients by a least-squares fit of expression (29) [32], [33]. This technique reduces the derivation of the noise parameters to the solution of a four-linear-equation system [32]. A successive approximation technique has also been proposed to take into account the errors in the input reflection coefficient evaluation [34].

The major advantage of this measurement technique is the possibility of a fully automatic noise and gain characterization of a device [37], while its main drawback is the possibility of erroneous results or even results without physical meaning [36], especially at high microwave frequencies. These problem arise from the following causes:

(i) The computed results are highly sensitive to measurement errors in the case of a large noise resistance R_n [32].

(ii) The matrix of the four-linear-equation system which is to be solved can become singular for some values of the input termination [36].

(iii) The noise measurements are very sensitive to oscillations that can occur at low frequency, for which HEMT's are always potentially unstable. This major problem can be reduced if no tuner is used at the output of the DUT [37].

(iv) In the case of very low noise devices, the problem of input network losses is reduced but not suppressed.

For these different reasons, this method seems to be very well suited for systematic measurements in industrial laboratories rather than for measurements of high-performance devices at high frequency. Moreover, the published results concern mainly frequencies of operation lower than 12 GHz and devices providing rather high noise figures. Therefore, in the case of high-frequency (>18 GHz) noise figure determination, the conventional searching of F_{min} seems presently to be more accurate and more suitable, even if it is tedious. The recent possibility of accurate S parameter measurements up to 40 GHz, yielding accurate losses determination, confirms this assertion.

VIII. Conclusions

The general method yielding the HEMT noise parameters F_{min}, g_n, and Z_{opt} has been described. In order to consider the influence of the various parameters, different approximations have been carried out, and their results have been compared with experimental findings concerning high-performance devices. This analysis has shown that the gate noise and the correlation coefficient plays a prominant part in the noise parameter value. A comparison of the noise performance of both HEMT's and conventional MESFET's has shown that the HEMT superiority can be mainly related to the higher cutoff frequency and correlation coefficient.

Some particular aspects of the HEMT LF noise have then been discussed and structures with reduced LF noise have been proposed. Lastly, the two main experimental methods for determining the noise figure have been presented and their advantages and drawbacks have been discussed.

References

[1] H. Rothe and W. Dahlke, "Theory of noisy fourpoles," Proc. IRE, vol. 44, pp. 811–818, 1956.

[2] R. A. Pucel, H. A. Haus, and H. Statz, "Signal and noise properties of gallium arsenide field effect transistors," advances in electronics and electron physics, vol. 38, pp. 195–265, 1974.

[3] T. M. Brookes, "The noise properties of high electron mobility transistors," IEEE Trans. Electron Devices, vol. ED-33, no. 1, pp. 52–57, 1986.

[4] C. F. Whiteside, G. Bosman, H. Morkoc, and W. F. Kopp, "The dc, ac, and noise properties of the GaAs:AlGaAs modulation-doped field effect transistor channel," IEEE Trans. Electron Devices, vol. ED 33 no. 10, pp. 1439–1445, 1986.

[5] A. Cappy, A. Vanoverschelde, M. Schortgen, C. Versnaeyen, and G. Salmer, "Noise modeling in submicrometer-gate two dimensional electron-gas field effect transistor," IEEE Trans. Electron Devices, vol. ED 32, no. 12, pp. 2787–2795, 1985.

[6] B. Carnez, A. Cappy, R. Fauquembergue, E. Constant, and G. Salmer, "Noise modeling in sub-micrometer-gate FET's," IEEE Trans. Electron Devices, vol. ED 28, no. 7, pp. 784–789, 1981.

[7] A. van der Ziel, "Thermal noise in field effect transistor," Proc. IRE, vol. 50 pp. 1808–1812, 1962.

[8] A. van der Ziel, "Gate noise in field effect transistors at moderately high frequencies," Proc. IRE, vol. 51 pp. 461–467, 1963.

[9] D. Delagebeaudeuf, I. Chevrier, M. Laviron, and P. Delescluse, "A new relationship between the Fukui coefficient and optimal current value for low noise operation of field effect transistors," IEEE Electron Device Lett., vol. EDL-6 no. 9, pp. 444–445, 1985.

[10] C. Moglestue, "A Monte Carlo particle study of the intrinsic noise figure in GaAs MESFET," IEEE Trans. Electron Devices, vol. ED-32 no. 10, pp. 2092–2096, 1985.

[11] F. M. Klaasen, "High-frequency noise of the function field-effect transistor," IEEE Trans. Electron Devices, vol. ED-14, no. 7, pp. 368–373, 1967.

[12] W. Beachhold, "Noise behavior of GaAs field-effect transistor with short gate length," IEEE Trans. Electron Devices, vol. ED-19 no. 5, pp. 674–680, 1972.

[13] C. H. Oxley and A. J. Holden, "Simple models for high-frequency MESFET's and comparison with experimental results," in Proc. IEEE MOA, June 1986.

[14] P. C. Chao, S. C. Palmateer, P. M. Smith, V. K. Mishra, K. H. G. Duh, and J. C. M. Hwang, "Millimeter-wave low noise high electron mobility transistors," IEEE Electron. Device Lett., vol. EDL-6, no. 10, pp. 531–533, 1985.

[15] B. Kim, H. Q. Tserng, and H. D. Shih, "Millimeter-wave GaAs FET prepared by MBE," IEEE Electron. Device Lett., vol. EDL-6 pp. 1–2, 1985.

[16] P. W. Chye and C. Huang, "Quarter-Micron low noise GaAs FET's," IEEE Electron Device Lett., vol. EDL-3 pp. 401–403, 1982.

[17] J. Zimmermann, and Wu Yen, "Etude de la dynamique des electrons à deux dimensions dans les hétérojonctions," Rev. Phys. Appl., to be published.

[18] H. Fukui, "Design of microwave GaAs MESFET's for broad-band low noise amplifiers," IEEE Trans. Microwave Theory Tech., vol. MTT-27, pp. 643–650, 1979.

[19] S. J. Lee and C. R. Crowell, "Parasitic source and drain resistance in high-electron-mobility transistors," Solid-State Electron., vol. 28 no. 7, pp. 659–668, 1985.

[20] M. D. Feuer, "Two-layer model for source resistance in selectively doped heterojonction transistors," IEEE Trans. Electron Devices, vol. ED-32, no. 1, pp. 7–11, 1985.

[21] C. Versnayen, A. Vanoverschelde, A. Cappy, G. Salmer, and M. Schortgen, "Frequency dependence of source access resistance of heterojonction field-effect transistor," Electron. Lett., vol. 21, no. 12 pp. 539–540, 1985.

[22] K. H. G. Duh, P. C. Chao, P. M. Smith, L. F. Lester, and B. R. Lee, "60 GHz low noise high-electron mobility transistors," Electron Lett., vol. 22, no. 12, pp. 547–549, 1986.

[23] M. W. Pospieszalski, S. Weinreb, P. C. Chao, U. K. Mishra, S. C. Palmateer, P. M. Smith, and J. C. M. Huang, "Noise parameters and light sensitivity of low noise high electron mobility transistors at 300 and 12.5 K," *IEEE Trans. Electron Devices*, vol. ED-33 pp. 218–233, 1986.

[24] K. Kurokawa, "Noise in synchronized oscillators," *IEEE Trans. Microwave Theory Tech.*, vol. MTT-16, no. 4, pp. 234–240, 1968.

[25] S. M. Liu, M. B. Das, W. Kopp, and H. Morkoc, "Noise behaviour of 1 μm gate-length modulation-doped FET's from 10^{-2} to 10^8 Hz," *IEEE Electron Device Lett.*, vol. EDL-6 no. 9, pp. 453–455, 1985.

[26] J. M. Dieudonne, M. Pouysegur, J. Graffeuil, and J. L. Cazaux, "Correlation between low-frequency noise and low temperature performance of two-dimensional electron gas FET's," *IEEE Trans. Electron Devices*, vol. ED-33 no. 5, pp. 572–575, 1986.

[27] L. Loreck, H. Dambkes, K. Heime, K. Ploog, and G. Weimann, "Deep level analysis in (AlGa)As-GaAs 2D electron gas devices by means of low frequency noise measurements," *IEEE Electron Device Lett.*, vol. ED-5 no. 1, pp. 9–11, 1984.

[28] E. F. Schubert and K. Ploog, "Shallow and deep donors in direct gap n type $Al_xGa_{1-x}As$:Si grown by molecular beam epitaxy" *Phy. Rev. B*, vol. 30, no. 12, pp. 7021–7029, 1984.

[29] T. Baba, T. Mizutani, and M. Ogawa, "Elimination of persistent photoconductivily and improvement in Si activation coefficient by Al spatial separation from Ga and Si in Al-Ga-As:Si solid system—A novel short period AlAs/n-GaAs superlattice," *Jap. J. Appl. Phys.*, vol. 22 no. 10, pp. L627–L629, 1983.

[30] S. Hiyamizu, S. Sasa, T. Ishikawa, K. Kondo, and H. Ishikawa, "A new hetero structure for 2 DEG system with a Si atomic-planar-doped AlAs-GaAs-AlAs quantum well structure grown by MBE," *Jap. J. Appl. Phys.*, vol. 24, pp. L431–L433, 1985.

[31] E. Strid, "Measurement of losses in noise-matching networks," *IEEE Trans. Microwave Theory Tech.*, vol. MTT-29, no. 3, pp. 247–252, 1981.

[32] R. Q. Lane, "The determination of device noise parameters," *Proc. IEEE*, vol. 57 pp. 1461–1462, 1969.

[33] G. Garuso and M. Sannino, "Computer aided determination of microwave two port noise parameters," *IEEE Trans. Microwave Theory Tech.*, vol. MTT 26, no. 9, pp. 639–643, 1978.

[34] M. Mitama and H. Katoh, "An improved computational method for noise parameter measurement," *IEEE Trans. Microwave Theory Tech.*, vol. MTT-27, no. 6, pp. 612–615, 1979.

[35] M. Sannino, "Computer aided simultaneous determination of noise and gain parameters of microwave transistors," in *Proc. European Microwave Conf.*, (Brighton, U.K.), pp. 692–695, 1979.

[36] M. Sannino, "On the determination of device noise and gain parameters," *Proc. IEEE*, vol. 67 no. 9, pp. 1364–1367, 1979.

[37] E. F. Calandra, G. Martines, and M. Sannino, "Characterization of GaAs FET's in terms of noise, gain and scattering parameters through a noise parameters test set," *IEEE Trans. Microwave Theory Tech.*, vol. MTT-32, no. 3, pp. 231–237, 1984.

[38] R. A. Speciale, "A generalization of the TSD network calibration procedure, coupling n port scattering parameter measurements affected by leakage errors," *IEEE Trans. Microwave Theory Tech.*, vol. MTT-25, no. 12 pp. 1100–1115, 1977.

[39] J. J. Berenz, K. Nakano, and K. P. Weller, "Low noise high electron mobility transistors," in *IEEE MTT-S Symp. Dig.*, pp. 83–86, 1984.

[40] P. R. Jay, H. Derewonko, D. Adam, P. Briere, D. Delagebeaudeuf, P. Delescluse, and J. F. Rochette, "Design of TEGFET devices for optimum low noise high frequency operation," *IEEE Trans. Electron Devices*, vol. ED-33, no. 5, pp. 590–594, 1986.

[41] K. Kamei, S. Hori, H. Kawasaki, K. Shibata, H. Mashita, Y. Ashizawa, "Low noise high electron mobility transistor," in *Proc. 11th GaAs and Related Compounds Conf.*, (Biarritz, France), 1984.

[42] H. Hida, K. Ohata, Y. Suzuki, and H. Toyoshima, "A new low noise AlGaAs:GaAs 2 DEG with a surface undoped layer," *IEEE Trans. Electron Devices*, vol. ED-33 no. 5, pp. 601–607, 1986.

[43] K. Tanaka, M. Ogawa, K. Togashi, H. Takakuwa, H. Ohke, M. Kanazuwa, U. Kato, and S. Watanabe, "Low-noise HEMT using MOCVD," *IEEE Trans. Electron Devices*, vol. ED-33 no. 12, pp. 2053–2058, 1986.

[44] K. Joshin, T. Mimura, M. Niori, Y. Yamashita, K. Kosmura, and J. Saito, "Noise performance of microwave HEMT," in *IEEE MTT-S Symp. Dig.*, pp. 563–565, 1983.

[45] M. Wolny, P. Chambery, A. Briere, and J. P. Andre, "Low noise high electron mobility transistor grown by MOVPE," in *Proc. High Speed Electronics Conf.* (Stockholm, Sweden), 1986, pp. 148–151.

Microwave Characterization of (Al,Ga)As/GaAs Modulation-Doped FET's: Bias Dependence of Small-Signal Parameters

DOUGLAS J. ARNOLD, RUSS FISCHER, WILLIAM F. KOPP, TIMOTHY S. HENDERSON, AND HADIS MORKOÇ, SENIOR MEMBER, IEEE

Abstract—Microwave characterization on 1-μm gate (Al, Ga)As/GaAs modulation-doped field-effect transistors (MODFET's) was done using a network analyzer. Equivalent circuit parameters were computed and compared to those of GaAs MESFET's with an identical geometry. The MODFET's had higher current-gain and power-gain cutoff frequencies (18 and 38 GHz versus 14 and 30 GHz) and the circuit parameters g_{m0}, c_{gs}, and R_{ds} displayed sharper pinchoff effects than those of the MESFET's. C_{gs} in the MODFET displayed a gate bias dependence due to widening of the potential well in the channel. This information should prove valuable in the development of MODFET computer models for circuit simulation.

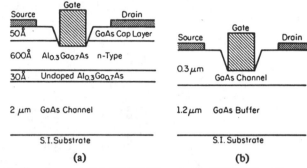

Fig. 1. Structure of (a) MODFET and (b) MESFET. The n-type $Al_xGa_{1-x}As$ was doped with Si to 2.5×10^{18} cm^{-3} in the MODFET. The MESFET channel was doped to 2×10^{17} cm^{-3}.

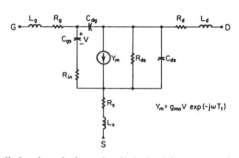

Fig. 2. Small-signal equivalent circuit derived from scattering parameter measurements. Circuit values are listed in Table I.

INTRODUCTION

(Al,Ga)As/GaAs modulation-doped field-effect transistors (MODFET's) have demonstrated their viability as microwave and high-speed switching devices. Current-gain cutoff frequencies of 30 GHz [1], [2] and 45 GHz [3] have been reported for 0.5 and 0.25-μm gate length devices, respectively. In addition, a gate delay of 12.2 ps was reported for 1.0-μm gate devices in a ring oscillator circuit at room temperature [4]. Before MODFET's can be implemented in high-speed digital and microwave circuits, an equivalent circuit suitable for computer simulations must be available. Adequate dc models exist for establishing bias conditions [5], but a bias-dependent ac model has not yet been developed. In this paper the bias dependence of the high-frequency equivalent circuit elements of MODFET's are investigated and compared to those of GaAs MESFET's.

EXPERIMENTAL

Transistor structures shown in Fig. 1 were grown by molecular beam epitaxy on semi-insulating (100) oriented substrates. The details of substrate preparation and growth can be found in [6]. The structures are indicated in Fig. 1. The MESFET structure consisted of a 1.2-μm undoped GaAs buffer grown on a Cr-doped substrate followed by a 0.3-μm GaAs channel doped with Si to 2×10^{17} cm^{-3}. The MODFET structure included a 2-μm undoped GaAs channel, a 30-Å layer of undoped Al$_{0.3}$Ga$_{0.7}$As, a 600-Å Al$_{0.3}$Ga$_{0.7}$As layer doped with Si to 2.5×10^{18} cm^{-3} and an n$^+$ 50-Å GaAs cap layer

Manuscript received January 3, 1984; revised April 13, 1984. This work was supported by the Air Force Office of Scientific Research. R. Fischer was supported by an IBM Fellowship.

The authors are with the Department of Electrical Engineering and the Coordinated Science Laboratory, University of Illinois, Urbana, IL 61801.

also doped to 2.5×10^{18} cm^{-3}. The undoped layer reduces ionized impurity scattering at the heterointerface [7].

After growth the wafers were lapped to 350 μm. Mesa isolation patterns were defined photolithographically and etched with 3:1:1 DI H$_2$O:H$_2$O$_2$:HF. Source and drain patterns were developed and AuGe/Ni/Au contacts evaporated and alloyed at 500°C in a H$_2$ atmosphere for 1 min. A 1 μm $\times$ 290 μm gate patterns was defined and recessed. Al was used as the gate metal for both the MESFET's and MODFET's. An Au overlay was added to facilitate bonding. The wafers were then thinned to 200 μm, scribed, and bonded to 50-Ω microstrip transmission lines for microwave measurements.

Scattering parameter measurements were made using a HP 8409B automatic network analyzer controlled by an HP 9845 desktop computer. Gain and admittance parameters were then computed from the measured S-parameters. The Y-parameters were then used to construct the equivalent circuit indicated in Fig. 2 [8], [9].

Reprinted from *IEEE Transactions on Electron Devices*, vol. ED-31, no. 10, pp. 1399–1402, October 1984.

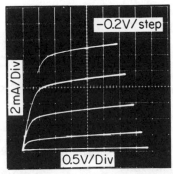

Fig. 3. Current–voltage characteristics of half a MODFET (gate width 145 μm). The scales are 2 mA/DIV vertical, 0.5 V/DIV horizontal, and −0.2 V/step.

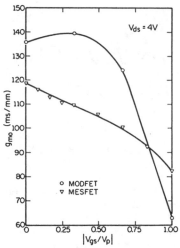

Fig. 4. Small-signal transconductance of MODFET's and MESFET's as a function of gate bias normalized to the pinchoff voltage. g_{m0} is normalized to 1 mm of gate width.

TABLE I
MODFET AND MESFET SMALL-SIGNAL EQUIVALENT CIRCUIT VALUES FOR THE CIRCUIT IN FIG. 2
(The values are normalized to 1 mm of gate width.)

	MODFET $I_{ds} = 20$mA $V_{ds} = 4$V $V_{gs} = -0.2$V	MESFET $I_{ds} = 20$mA $V_{ds} = 4$V $V_{gs} = -1.6$V
g_{mo} (mS/mm)	140	110
R_{ds} (Ωmm)	220	120
R_{in} (Ωmm)	7.0	3.6
C_{gs} (pF/mm)	1.3	1.1
C_{dg} (pF/mm)	0.089	0.11
C_{ds} (pF/mm)	0.16	0.28
τ_t (pS)	2.4	4.4
R_s (Ωmm)	1.5	0.93
R_d (Ωmm)	0.84	1.5
R_g (Ωmm)	1.2	1.2
L_s nH	0.2	0.2
L_d nH	0.5	0.5
L_g nH	0.5	0.5

RESULTS

Current–voltage characteristics for half of a MODFET ($Z = 145$ μm) are displayed in Fig. 3. The scales were 2 mA/div vertical, 0.5 V/div horizontal, and −0.2 V/step on the gate. The source resistance and the dc transconductance after subtracting the probe resistance are 1.5 $\Omega \cdot$ mm and 140 mS/mm, respectively.

Table I contains circuit element values for the equivalent circuit in Fig. 2 deduced from S-parameter measurements made in the frequency range 2–4 GHz for MODFET and MESFET devices.

Measured source and drain resistances (R_s and R_d) include contact resistance, semiconductor resistance between the gate and source, or drain contact as well as part of the distributed channel resistance under the gate. Due to variations in the positioning of the gate during fabrication, R_s and R_d values

vary among devices but $R_s + R_d$ remains approximately constant. For the bias points listed in Table I corresponding current-gain cutoff and unilateral power-gain cutoff frequencies can be estimated from the equivalent circuit by [9]

$$f_T = g_{m0}/2\pi C_{gs} \tag{1}$$

$$f_{max} = f_T/\{2[(R_{in} + R_g + R_s)/R_{ds} + 2\pi f_T R_g C_{dg}]^{1/2}\}. \tag{2}$$

The resulting values were current-gain cutoff frequencies of 18 and 14 GHz for the MODFET's and MESFET's, respectively. The unilateral gain cutoff frequency was 38 GHz for the MODFET's and 30 GHz for the MESFET's.

The discrepancy in cutoff frequencies between the two devices can be attributed to the larger transconductance in the MODFET's. The intrinsic transconductance g_m of short gate length FET's operating in the velocity saturated mode is given by [10]

$$g_m = \epsilon v_s Z/w. \tag{3}$$

Here ϵ is the permittivity of the material in the depletion region of the FET, v_s is the electron saturation velocity in the channel, and Z is the gate width. For a MESFET, w is the depletion depth. For a MODFET, $w = d + \Delta d$ where d is the thickness of the (Al,Ga)As beneath the gate and $\Delta d = 80$ Å is the average effective displacement of the electron gas in the GaAs [5]. Because MODFET channels are undoped, one expects a higher electron saturation velocity than in doped GaAs MESFET channels. Furthermore, w in MODFET's is smaller, since the permittivity of both structures is approximately equal, $\epsilon_r(Al_xGa_{1-x}As)/\epsilon_0 = 13.2 - 2.8x$, then (3) points to higher intrinsic transconductances in MODFET's. This is evidenced in Fig. 4. Here small-signal transconductance g_{m0} is plotted against gate bias normalized to the pinchoff voltage for a drain voltage of 4 V. The pinchoff voltages are 0.6 and 2.4 V for the MODFET and MESFET, respectively.

The sharper pinchoff characteristics of MODFET's is also indicated in Fig. 4. For MESFET's, w in (3) increases with bias decreasing g_m. In a MODFET, $w = d + \Delta d$ is nearly independent of bias and g_m is constant until the device is biased

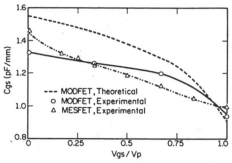

Fig. 5. The input capacitance in picofarads per millimeter of gate width versus gate voltage normalized to the pinchoff voltage.

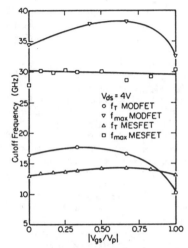

Fig. 6. Current-gain cutoff frequency f_T and unilateral power-gain frequency f_{max} for MODFET's and MESFET's in gigahertz versus gate bias normalized to pinchoff.

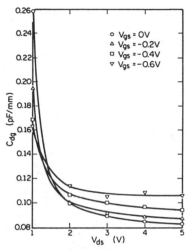

Fig. 7. Feedback capacitance for the MODFET's versus V_{ds} with V_{gs} as a parameter in picofarads per millimeter of gate width.

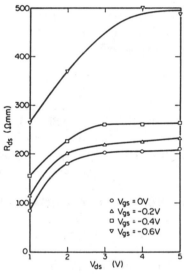

Fig. 8. Output resistance for the MODFET's versus V_{ds} with V_{gs} as a parameter in ohms times millimeters.

close to pinchoff and the approximation $\Delta d \simeq 80$ Å breaks down. The slight decrease in g_{mo} at small reverse gate voltages is a result of the parasitic MESFET which arises from parallel conduction in the $Al_x Ga_{1-x} As$.

Fig. 5 presents the input capacitance as a function of gate bias for MESFET experimental data and for MODFET experimental and theoretical data. The theoretical curve comes from a numerical simulation of a normally on MODFET [5]. The two MODFET curves show good agreement as to the functional dependence of C_{gs}. Since C_{gs} is chiefly due to the gate capacitance (C_g), the functional form of C_{gs} is explained by

$$C_g = \frac{\epsilon LZ}{W}. \tag{4}$$

Combining (3) and (4) we obtain a constant current-gain cutoff frequency

$$f_T = v_s / 2\pi L. \tag{5}$$

Fig. 6, which shows f_T and f_{max} versus gate bias for both devices, indicates that this is true for the MESFET's. For the MODFET we expected f_T to remain constant until pinchoff is approached. When the potential well containing the 2DEG opens up the MODFET acts like a highly nonoptimal MESFET, and thus f_T decreases.

The feedback capacitance C_{dg} is plotted in Fig. 7 as a func-

tion of bias. C_{dg} is mainly due to gate fringing capacitance between the gate and drain. As would be expected, C_{dg} is almost constant with increasing reverse gate bias since the gate fringing capacitance is invariant with gate bias [11], [12]. At low drain bias, C_{dg} contains part of C_g. When the device is saturated, though, C_{dg} reflects only the gate fringe capacitance. This explains the rapid decrease of C_{dg} with increasing drain bias near saturation. Similar behavior has been reported for GaAs MESFET's [11], [13], [14].

Fig. 8 shows the output resistance R_{ds} of the MODFET's versus V_{ds} with gate voltage as a parameter. At low drain voltages R_{ds} fits the theory developed for GaAs FET's [15]

$$R_{ds} \propto V_{ds}(1 - I_d/I_s). \tag{6}$$

Here I_d and I_s are the drain and drain saturation currents, respectively. This dependence upon V_{ds} can be seen in Fig. 8 for low drain voltage. In Fig. 9, which shows R_{ds} versus I_d, it is obvious that the linear dependence with drain current does not hold over the entire operating range of the device.

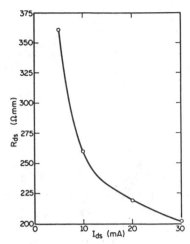

Fig. 9. Output resistance for the MODFET's as a function of drain current in ohms times millimeters.

The nonlinearities may arise from parallel resistance in the substrate. Such behavior has been observed experimentally in MESFET's [11], [13] as well.

SUMMARY

MODFET's were fabricated and the microwave performance was compared to that of GaAs MESFET's. The MODFET's had higher current-gain and power-gain cutoff frequencies, 18 and 38 GHz versus 14 and 30 GHz for the MESFET's. The frequency performance improvement was determined to be the result of larger transconductances due to the higher electron velocity in the undoped GaAs channel of the MODFET's.

The bias dependence of g_{m0}, C_{gs}, C_{dg}, and R_{ds} were investigated for the MODFET's and compared to GaAs MESFET's. These parameters exhibited sharper pinchoff behavior in MODFET's which is consistent with the dc pinchoff behavior.

ACKNOWLEDGMENT

The authors would like to acknowledge fruitful discussions with Dr. T. J. Drummond, Dr. F. Ponse, Prof. M. S. Shur, Prof. K. Heime, and K. R. Gleason, and the assistance of J. Klem and D. Perrachionne.

REFERENCES

[1] K. Joshin, T. Mimura, Y. Yamashita, K. Kosemura, and J. Saito "Noise performance of microwave HEMT," *IEEE Trans. Microwave Theory Tech.–Int. Microwave Symp. Dig.*, pp. 563–565, 1983.

[2] N. T. Linh, M. Laviron, P. Delescluse, P. N. Tung, D. Delagebeaudeuf, F. Diamond, and J. Chevrier, "Low noise performance of two-dimensional electron gas FET's," presented at the IEEE/Cornell Conf., Aug. 15–17, 1983.

[3] P. C. Chao, T. Yu, P. M. Smith, S. Wanuga, J.C.M. Hwang, W. H. Perkins, H. Lee, L. F. Eastman, and E. D. Wolf, "Quarter-micron gate length microwave high electron mobility transistor," *Electron. Lett.*, vol. 19, pp. 894–895, 1983.

[4] C. P. Lee, D. L. Miller, D. Hou, and R. J. Anderson, "Ultra high speed integrated circuits using GaAs/AlGaAs high electron mobility transistors," in *Proc. 41st Dev. Res. Conf.* (Vermont, June 1983), p. IIA-7.

[5] T. J. Drummond, H. Morkoç, K. Lee, and M. Shur, "Model for modulation doped field effect transistor," *IEEE Electron. Device Lett.*, vol. EDL-3, pp. 338–341, 1982.

[6] T. J. Drummond, H. Morkoç, and A. Y. Cho, "Molecular beam epitaxy growth of (Al, Ga)As/GaAs heterostructures," *J. Crystal Growth*, vol. 56, p. 449, 1982.

[7] L. C. Witkowski, T. J. Drummond, C. M. Stanchak, and H. Morkoç, "High mobilities in $Al_x Ga_{1-x} As$-GaAs heterojunctions," *Appl. Phys. Lett.*, vol. 37, pp. 1033–1035, Dec. 1980.

[8] W. Fischer, "Equivalent circuit and gain of MOS field effect transistors," *Solid-State Electron.*, vol. 9, pp. 71–81, 1966.

[9] P. Wolf, "Microwave properties of Schottky-barrier field-effect transistors," *IBM J. Res. Dev.*, vol. 9, p. 125.

[10] R. E. Williams and D. W. Shaw, "Graded channel FET's: Improved linearity and noise figure," *IEEE Trans. Electron Devices*, vol. ED-25, pp. 600–605, June 1978.

[11] R. W. H. Engelmann and C. A. Leichti, "Bias dependence of GaAs and InP MESFET parameters," *IEEE Trans. Electron Devices*, vol. ED-24, pp. 1288–1296, 1977.

[12] E. Wassertron and J. McKenna, "The potential due to a charged metallic strip on a semiconductor surface," *Bell Syst. Tech. J.*, vol. 49, pp. 853–877, May/June 1970.

[13] J. Sone and Y. Takayama, "A small-signal analytical theory for GaAs field-effect transistors at large drain voltages," *IEEE Trans. Electron Devices*, vol. ED-25, pp. 329–337, Mar. 1978.

[14] D. Arnold, W. Kopp, R. Fischer, T. Henderson, and H. Morkoç, "Microwave performance of GaAs MESFET's with AlGaAs layers: Effect of heterointerfaces," submitted for publication.

[15] R. A. Pucel, H. A. Hans, and H. Statz, "Signal and noise properties of gallium arsenide microwave field-effect transistors," *Advances Electron. Electron Phys.*, vol. 38, pp. 195–265, 1975.

MICROWAVE CHARACTERIZATION OF VERY HIGH TRANSCONDUCTANCE MODFET

L.H. Camnitz, P.J. Tasker, H. Lee, D. van der Merwe and L.F. Eastman

School of Electrical Engineering and National Research
and Resource Facility for Submicron Structures
420 Phillips Hall, Cornell University, Ithaca, NY 14853

ABSTRACT

Submicrometer gate length $Al_{0.3}Ga_{0.7}As$/GaAs modulation doped FET's have been fabricated using n^+ GaAs and graded $n^+Al_xGa_{1-x}As$ capping layers in a deep, narrow recess structure. Intrinsic transconductance steadily and dramatically increased from 240 mS/mm for 0.65 μm gate enhancement mode MODFET's to 570 mS/mm for 0.33 μm devices, suggesting electron velocity enhancement at short gate lengths. Maximum extrinsic transconductances of 450 mS/mm at 300K and 580 mS/mm at 77K have been measured for the 0.33 μm gate length device. An unusually high unity current gain cutoff frequency, f_T, of 55 GHz makes this device potentially superior for high speed logic. Parasitic resistances and high feedback capacitance, however, severely limit f_{max} to 70 GHz.

INTRODUCTION

The modulation-doped $Al_xGa_{1-x}As$/GaAs heterostructure field effect transistor is being intensively developed for both microwave and high speed logic applications due to its expected advantages over GaAs MESFETs.

The improved performance over conventional MESFET's stems from the higher electron mobility and higher saturation velocity in MODFET's (1). Indeed, noise performance of 0.5 μm MODFET's has equaled that of 0.25 μm MESFET's (2). In high speed logic, switching delays as low as 12.2 psec have been achieved with 1 μm gate length MODFET's (3).

Recent work has shown that transconductance in MODFET's continues to increase as gatelength decreases to 0.5 μm (4). In addition, enhanced electron velocity has been predicted for 0.25 μm gate length MESFETs using 2-dimensional Monte-Carlo models (5). It is, therefore, of great interest to evaluate the performance improvement of MODFET's as the gatelength decreases to 0.25μm. In this study, recessed gate microwave MODFET's with varying submicrometer gatelengths as low as 0.33 μm were fabricated on a conventional modulation doped structure. Their performance was evaluated using a combination of DC and microwave tests. It was found that MODFET performance continues to improve greatly as the gate length decreases to 0.33 μm.

MATERIAL GROWTH

The modulation doped structure used (see Fig. 1) employs a 200 Å n^+ $Al_{0.3}Ga_{0.7}As$ layer directly on 1.5 μm of undoped GaAs. A 300 Å n^+ $Al_xGa_{1-x}As$ layer with a graded composition forms a smooth transition to the 500 Å n^+ GaAs cap layer. This n^+ structure is employed to reduce source resistance through lower ohmic contact resistance and lower sheet resistance (6). All three layers are Si-doped at an atomic flux of 2.5×10^{18} cm^{-3}, corresponding to a doping density of 1×10^{18} cm^{-3} in the $Al_{0.3}Ga_{0.7}As$ due to the lower doping efficiency of $Al_xGA_{1-x}As$. The structure was grown in a Varian Gen II MBE machine at a substrate temperature of 600°C and a GaAs growth rate of 1 μm/hr. The sheet carrier density and 77K Hall mobility in the 2DEG were estimated to be 1×10^{12} cm^{-2} and 15,000 cm^2/V-sec, respectively, for this structure. In addition, the 300K mobility in the 2DEG was estimated to be 5000 cm^2/V-sec. The unetched layer sheet resistivity was very low (180 $\Omega/\square$) due to the conductivity of the highly doped n^+GaAs cap.

DEVICE FABRICATION

The 300 μm gate width microwave devices were fabricated by contact photolithography for all levels except for the gate, which was directly written by EBL. Mesa isolation was followed by Ni/AuGe/Ag/Au 100 Å/900 Å/1000 Å/1000 Å ohmic contact metalization. The metal was deeply alloyed at a peak temperature of 570°C, resulting in a transfer resistance of less than 0.1 Ω-mm, at both 300K and 77K (7,8).

A single level of PMMA masked the gate recess etch, and was then used to lift-off the gate metal pattern. The gate pattern was centered in the 3 μm source-drain channel. The recess trench was formed by spray etching with a solution of H_2O_2 buffered to pH 7 with NH_4OH. The resulting recess trenches were ~ 600 Å deep and uniform across the wafer. The gate metal was a sandwich of Ti/Pt/Au with a total thickness of 4000 Å.

The lengths of the gate metal after liftoff were 0.33 μm, (see Fig. 2) 0.47 μm, and 0.65 μm. Note that in the 0.33 μm device the edges of the gate metal and the trench are coincident, while in the 0.65 μm device, the trench is 0.25 μm wider than the gate footprint. This was because the angle of the PMMA resist walls varied from vertical for the 0.3 μm gate length to slightly under-

Reprinted from *30th IEEE Int. Electron Devices Meeting, Tech. Dig.*, pp. 360–363, 1984.

cut for the 0.65 μm gate length. After the gate level, the bonding pads were plated with 3 μm of Au and the wafer was thinned and scribed for testing.

DC RESULTS

The devices were enhancement mode with a transconductance (g_m) which was strongly dependent on the gatelength, L_g. Fig. 3 shows drain characteristics for a representative device with 0.33 μm gatelength. The maximum 300K extrinsic g_m of the 0.33 μm device was 135 mS (450 mS/mm). Corresponding to an intrinsic g_m of 171 mS (570 mS/mm) since the measured source resistance, R_S, was 1.7 Ω (.5 Ω·mm). At 77K, the extrinsic g_m increases to a very high value of 550 mS/mm. The large difference between this value and the intrinsic g_m of 800 mS/mm illustrates the increasing negative feedback effect of R_S as g_m increases. The source resistance was measured by the end resistance method (monitor V_{ds} while forcing a positive gate current). The values of intrinsic transconductance (Table 1) were independently confirmed by the equivalent circuit models derived from the S-parameters. The 0.47 μm and 0.63 μm devices exhibited significantly lower extrinsic g_m, especially the latter which had an 11 Ω source resistance. This excess source resistance comes from the nearly pinched-off recessed channel region just on the source side of the gate metal. The wide variation in R_S causes a wide variation in extrinsic g_m unrelated to the intrinsic MODFET. Therefore, intrinsic g_m is used to compare the performance of the devices to eliminate this potentially misleading effect.

The intrinsic g_m at 300K increases dramatically from 240 mS/mm to 570 mS/mm as L_g is decreased from 0.65 μm to 0.33 μm (see Fig. 4). This might be explained by a decrease in the spacing, d, from the gate to the 2DEG for the shorter devices, based on a variation of gate trench depth. This is unlikely, however, since the gate recess etch has no "trenching" effect. Also, the pinchoff voltage is more negative for the shortest device, opposite what would be expected if this explanation were valid.

A more plausible explanation for the g_m increase is increase in average electron velocity in the channel due to a short electrical length and higher average electric field. Unfortunately, accompanying this is a drastic decrease in output resistance R_0, from 58 Ω·mm to 22 Ω·mm. (See Fig. 5). Again, we use the intrinsic R_0 to eliminate the degenerative effect of R_S on the value of extrinsic R_0. The low output resistance is in part caused by injected electrons in the buffer layer or substrate adding excess drain current. At high drain voltage, this current becomes quite large and causes a degradation of g_m at low I_D (see Fig. 6). This effect, which is caused by the difficulty of pinching off deeply injected current, may be reduced by the incorporation of an AlGaAs buffer layer (9).

The linear source to drain resistance with the gate forward biased is known as the "ON"

resistance. A low value is important in direct coupled logic to minimize the 'low' logic voltage. An extremely low "ON" resistance value of 5 Ω (or 1.5 Ω·mm) is measured for the 0.33 μm device. This is only 3 times R_S, indicating a very highly conductive channel. The maximum I_d is a determining factor of a logic gate's switching speed. The 0.47 μm device has a maximum I_d at 300K of about 50 mA (or 167 mA/mm), while I_d of the 0.33 μm device is 90 mA (or 300 mA/mm) at 1V drain to source voltage. This is a substantial increase which should greatly increase the switching speed of a digital circuit.

RF MEASUREMENTS

S-parameter measurements were performed from 2-18 GHz. Fig. 7 shows $|h_{21}|^2$ and unilateral gain vs. frequency for the 0.33 μm device. An extrapolation of the measurement at 6 dB/octave yields a 55 GHz f_T, among the highest values reported for FET devices. In comparison, f_T values of 40 GHz are usual for optimized 0.25 μm MESFET's (10). The measured f_T value might be even higher if L_g were reduced to 0.25 μm. The high f_T value supports the view that the high transconductance comes from a higher average electron velocity. However the computation of an exact electron velocity in a short MODFET is difficult from experimental data, since it involves assumptions about fringing capacitance or electrical length which are hard to verify.

The maximum frequency of oscillation, f_{max}, is 70 GHz. Equivalent circuit models (Table 1) derived from the S-parameters show that an excessive value of gate resistance (7 Ω) degrades the unilateral gain substantially. The projected effect of reducing the gate resistance to 2 Ω is to increase f_{max} to well over 100 GHz (Fig. 8). This reduction in gate resistance can be made using a low resistance "T" shaped gate or a smaller unit gate width. The models also indicate that an equivalent increase in the available power gain (MAG) may not be realized, due to a high value of the gate to drain feedback capacitance (C_{dg}). Both the parasitic series resistances and the feedback capacitance will have to be reduced before high power gain is realized at millimeter wave frequencies.

DISCUSSION

The values of C_{dg} in the equivalent circuit model for the various gate lengths are significantly higher than those normally measured for optimized low noise MODFET's and GaAs FET's (typically 0.05 pF - 0.15 pF for a 300 μm gate width).

High values of C_{dg} were previously reported for closely spaced n^+ recessed gate structures and attributed to concentrated fringing fields on the drain side of the gate (11). In 0.25 μm GaAs FET's using a short recess structure, g_m and velocity enhancement as compared to a long recess channel devices have been reported (12). It is suggested here that in the undoped MODFET channel, the g_m enhancement is more dramatic than that in a

MESFET due to the higher electron mobility. This enhancement occurs <u>only</u> when a very short (< 0.4 μm) high field region exists between two ohmic regions. Placing a short gate in a long recess channel will allow the drain bias induced electric field to extend beyond the gate towards the ohmic drain region, lowering the electric field under the gate and resulting in lower C_{dg} but no velocity enhancement. The ohmic drain region here is defined as the n^+ contact layer or the ohmic contact itself if there is no n^+ contact layer. There appears to be a tradeoff between the enhanced electron velocity (or f_T) and the increased feedback capacitance obtained in the short electrical length structure.

In high speed logic applications, the voltage gain in a circuit is low, so that the effect of feedback capacitance on circuit performance is relatively small. This, along with the previously mentioned improvement of DC characteristics makes the short electrical length structure very attractive for these applications. For microwave amplification, the higher voltage gain increases the power gain degradation due to C_{dg} through the Miller effect. Thus, one needs to ascertain the optimum electrical length for the MODFET structure. In either case, the evidence suggests that significant improvements in high frequency performance can be made by simply reducing L_g and minimizing the parasitic resistive elements.

CONCLUSION

The performance advantages of a short electrical length MODFET with an 0.33 μm gate length include high transconductance of 450 mS/mm at 300K and 550 mS/mm at 77K. A 300 mA/mm current switching capability, a 1.5 Ω mm "ON" resistance, and 55 GHz f_T make this device extremely attractive for high speed logic applications. The measured value of f_{max}, 70 GHz, is limited in large part by parasitic gate resistance and can be increased to well over 100 GHz. However, a 0.21 pF/mm feedback capacitance is one liability of this deep, narrow recess structure. G_m and $1/R_0$ strongly increased as L_g decreased to 0.33 μm suggesting electron velocity enhancement effects.

ACKNOWLEDGEMENTS

The authors gratefully acknowledge helpful discussions with W. Jones, W. Schaff, D. Shire and D.W. van der Merwe, and technical assistance from J. Berry. We thank Dr. E. Wolf for the use of the NRRFSS electron beam microfabricator and Prof. J.P. Krusius for assistance in DC measurements. This work was supported by The National Radio Astronomy Observatory under grant 79-08925.

REFERENCES

(1) T.J. Drummond, S.L. Su, W.G. Lyons, R. Fischer, W. Kopp, H. Morkoc, K. Lee and M.S. Shur, Elect. Lett. **18** 24 p 1057-1058 Nov. 1982.

(2) N.T. Linh, M. Lauivon, P. Deleculse, P.N. Tung, D. Delagebeaudeuf, F. Diamond, and J. Cheuvier, Proc. of IEEE/Cornell Conf. on High Speed Semiconductor Devices, 1983, pp. 187-193.

(3) C.P. Lee, D.L. Miller, D. Hou, and R.J. Anderson, paper presented at 1983 Dev. Res. Conf., Burlington, VT June 20, 1983.

(4) T. Mimura, K. Nishiuchi, M. Abe, A. Shibatomi, and M. Kovayashi, 1983 IEDM Tech. Dig. pp. 99-102.

(5) Y. Awano, K. Tomizawa, N. Hashizume, and M. Kawashima, Electr. Lett. 1983 **19**, pp. 20-21.

(6) H. Dambkes, W. Brockerhoff, K. Heime, K. Plogg, G. Weimann, and W. Schapp, Elect. Lett. **20** 15 pp. 615-617 July 1984.

(7) P. Zwicknagl, S.D. Mukherjee, W.L. Jones, H. Lee, P. Capani, T. Griem, J.D. Berry, L. Rathbun and L.F. Eastman, paper presented at the GaAs and Related Compounds Symposium, Sept.1984 Biarritz, France.

(8) W.L. Jones, private communication.

(9) L.H. Camnitz, P.A. Maki, P.J. Tasker and L.F. Eastman, paper presented at the GaAs and Related Compounds Symposium Sept. 1984 Biarritz, France.

(10) P.W. Chye and C. Hung, IEEE Electron Device Lett. **EDL-3** 12 pp 401-403 Dec. 1982.

(11) R. Tyrani, Inst. Phys. Conf. Ser. No. **65**, Ch. 5, 1982.

(12) P.C. Chao, P..M. Smith, S. Wanuga, W. Perkins and E. D. Wolf, IEEE Electron Dev. Lett. **EDL-4** (9), Sept. 1983, pp. 326-328.

TABLE 1
DC Parameters

Gatelength	0.33	0.47	0.65
extr. g_m (mS)	135	72	42.5
extr. g_m (mS/mm)	450	240	142
R_s Ω	1.7	5.5	11
R_s (Ω·mm)	0.51	1.65	3.3
intr. g_m (mS)	171	112	72
intr. g_m (mS/mm)	570	375	240
extr. R_0 (Ω)	80	250	300
intr. R_0(Ω)	63	160	177
intr. R_0 (Ω·mm)	21.7	45	57.6
R_g (Ω)	7	5	4

Equivalent Circuit Model Element Values

L_g (μm)	0.33	0.47	0.65
Parameter			
L_{recess} (μm)	0.35	0.53	0.9
V_{tb} (V)	-0.1	0.1	-0.15
I_d (mA)	45	20	30
V_{ds} (V)	2.0	2.0	2.0
C_{gs} (pF)	0.34	0.44	0.5
C_{dg} (pF)	0.063	0.065	0.071
C_{ds} (pF)	0.139	0.143	0.158
R_i (Ω)	1.12	1.2	1.2
R_0 (Ω)	72.3	150	192
g_{m0} (mS)	171	120	73
τ (pS)	3.62	2.8	2.7
L_g (nH)	0.26	0.22	0.27
L_s (nH)	0.09	0.078	0.078
L_d (nH)	0.27	0.19	0.16

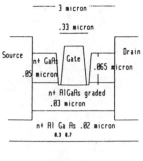

Fig. 1. Device structure.

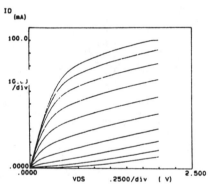

Fig. 2. SEM micrograph of gate structure (0.33μm device).

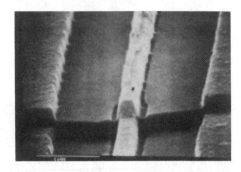

Fig. 3. $I_d(V_{ds})$ characteristics -enhancement mode (0.33μm gate,V_{gs}-0.1V/step).

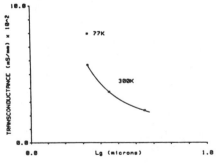

Fig. 4. Plot of intrinsic g_m as a function of L_g.

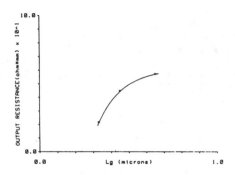

Fig. 5. Plot of intrinsic R_o as a function of L_g.

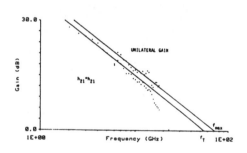

Fig. 6. $G_m(I_d)$ for 0.33μm device showing g_m degradation as V_{ds} increases.

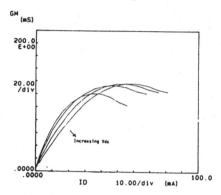

Fig. 7. Measured f_T and f_{max}.

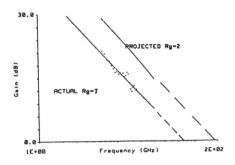

Fig. 8. Projected f_{max}.

The DC, AC, and Noise Properties of the GaAs/AlGaAs Modulation-Doped Field-Effect Transistor Channel

CHRISTOPHER F. WHITESIDE, GIJS BOSMAN, MEMBER, IEEE, HADIS MORKOÇ, SENIOR MEMBER, IEEE, AND WILLIAM F. KOPP

Abstract—The channel noise of a normally-on MODFET has been measured at $T = 300$ K as a function of bias and frequency. The results are in good agreement with theoretical predictions for velocity fluctuation noise. A new method is introduced to calculate the correct dependence of the two-dimensional carrier concentration on channel voltage. The measurements indicate that real-space-charge transfer in the bias range considered is negligible.

I. INTRODUCTION

IN RECENT YEARS much attention has been paid to GaAs-AlGaAs modulation-doped field-effect transistors (MODFET's) for potential use in high-speed logic circuits. The very high transconductance g_m and high cutoff frequencies f_T also make them of interest for low-noise microwave amplification. Excellent articles by Solomon *et al.* [1] and Linh [2] have been written reviewing the characteristics of these new transistors.

Since the first report of the noise figure of these devices in the microwave frequency range, an interest in the noise behavior has developed. The noise figures of various MODFET's have been reported recently and show improvements over conventional GaAs MESFET's of comparable gate lengths.

Up to now only noise figure measurements have been reported in the microwave frequency range. In this paper we will not focus on the noise figure, but instead report on the noise characteristics of the FET channel. At intermediate frequencies ($0.5 < f < 10$ GHz) the channel noise is due to fluctuations of the free-carrier velocity and is the major contributor to the overall device noise. Measurements of the thermal noise (i.e., velocity fluctuation noise) as a function of bias are discussed in this paper.

In Section II we will outline the theory of the impedance field method, which is used to obtain the ac and noise properties of the MODFET channel. Section III explains

Manuscript received December 27, 1985; revised May 19, 1986. This work was supported by the National Science Foundation under Grant ECS 83-00222.

C. F. Whiteside and G. Bosman are with the Department of Electrical Engineering, University of Florida, Gainesville, FL 32611.

H. Morkoç and W. Kopp are with the Electrical Engineering Department and the Coordinated Sciences Laboratory, University of Illinois, Urbana, IL 61801.

IEEE Log Number 8609953.

the methods of obtaining the charge–voltage relationship for the devices used in our experiments. Some of the methods of obtaining the charge–voltage relationship involve only low-bias data while other methods involve high-bias data. Comparing the results of the different methods can help determine the presence or absence of real-space-charge transfer. Section IV describes the MODFET structures to be considered. Measurement procedures are discussed in Section V. The experimental results will then be presented and discussed in Section VI, followed by conclusions in Section VII.

II. THEORY

In this section the procedure for obtaining the position-dependent ac channel voltage in terms of the Green's function for a MODFET channel is discussed. It will be shown how the Green's function is related to the impedance field [3], [4]. Once the impedance field is obtained, the ac and noise properties can be easily calculated. Van Vliet [5] and Nougier [6] have outlined this method for the case of the junction field-effect transistor (JFET), whereas Gummel and Blue [7] have demonstrated the method for IMPATT diodes. In this paper the impedance field for a MODFET is calculated using the proper transport equations, and the ac and noise properties of the MODFET are derived.

Application of the Impedance Field Method to the MODFET

In the following a one-dimensional collision-dominated transport model is used to obtain simple analytical expressions for the impedance and noise of the device. The advantage of this approach is that it provides physical insight into the ac and noise behavior of the channel. Clearly this treatment breaks down for very short submicrometer devices ($L < 0.5$ μm) since in that case the usual concept of mobility and diffusion needs to be generalized (see Constant [8]). Assuming no leakage current through the gate and neglecting both diffusion and displacement currents, the charge transport equation is given by

$$I = qwn_s[V(x)] \, v[E(x)] \tag{1}$$

Reprinted from *IEEE Transactions on Electron Devices*, vol. ED-33, no. 10, pp. 1439–1446, October 1986.

where w is the gate width and $v[E(x)]$ is the field-dependent carrier velocity. The contribution of displacement current to the total current in the frequency range of interest is small due to the high carrier concentration in the channel and is therefore omitted from (1). Also, since current saturation is caused by velocity saturation and not by pinchoff of the carrier concentration, the diffusion current contribution can be safely neglected. The sign convention is as follows. The source is chosen at $x = 0$, the drain at $x = L > 0$, $q > 0$, $V(x) > 0$, $E(x) < 0$, $v[E(x)] > 0$, and $I > 0$. The 2-D sheet carrier concentration $n_s[V(x)]$ is assumed to be only a function of the local electrical potential under the gate. Velocity saturation will cause accumulation and/or depletion of the sheet carrier concentration in the high-field region under the gate, making (1) invalid. For this reason the model we employ only describes the linear and triode regimes of the current–voltage characteristic. At $T = 300$ K the velocity-field characteristic of the 2-D electron gas is assumed to be identical to the one of bulk GaAs [9]. Consequently

$$v(E) = \frac{-\mu_0 E}{1 - E/E_c} \qquad (2)$$

where μ_0 is the low-field mobility taken to be 8000 cm^2/V $\cdot$ s at room temperature, and the critical field E_c is chosen to be 11.4 kV/cm. When the electric field exceeds 3.5 kV/cm, the model (see (1)) no longer holds due to the saturation effects mentioned above. The large critical field E_c is chosen to provide the proper curvature of the velocity characteristic at low electric fields. Using the bulk GaAs velocity-field characteristic as a first attempt is justified since in the high-field region under the gate the reduced sheet carrier concentration causes the quasi-triangular potential well to widen. As the well widens, the electron gas goes from quasi-two-dimensional to three-dimensional bulk. This phenomenon is confirmed by the results of Monte Carlo simulations by Cappy et al. [9]. In addition, Cappy observed that real-space-charge transfer of hot carriers from the 2-D interface back to the AlGaAs layer can be neglected in a MODFET.

The ac and dc transport equations are obtained by linearizing (1) in the following way. Introducing small-signal variations around the steady-state quantities, one obtains

$$I = I_0 + \Delta I \qquad (3)$$

$$n_s(V) = n_s(V_0) + \left(\frac{dn_s}{dV}\right)_0 \Delta V \qquad (4)$$

$$v(E) = v(E_0) + \left(\frac{dv}{dE}\right)_0 \Delta E. \qquad (5)$$

Substituting (3), (4), and (5) into (1), the expression for the dc current becomes

$$I_0 = -qwn_s[V_0(x)]\, \mu_0 E(x)/(1 - E(x)/E_c) \qquad (6)$$

and for the ac current, after neglecting second-order terms and using $\Delta E = (-d\Delta V/dx)$, one finds

$$\Delta I(x) = -qwn_s[V_0(x)]\left(\frac{dv}{dE}\right)_0 \frac{d\Delta V(x)}{dx}$$
$$+ qw\left(\frac{dn_s}{dV}\right)_0 v(E_0)\,\Delta V(x). \qquad (7)$$

Equation (7) relates the ac current and voltage at position x by a first-order differential equation. Solving for $\Delta V(x)$, one obtains with boundary condition $\Delta V(0) = 0$

$$\Delta V(x) = \frac{H(x - x')\left(1 - \dfrac{E_0(x')}{E_c}\right)\Delta I(x')}{qw\mu_0 n_s[V_0(x')] - I_0/E_c}$$

$$\cdot \exp \int_{x'}^{x} \frac{qw\mu_0 E_0(u)\left(\dfrac{dn_s}{dV}\right)_{0,u} du}{qw\mu_0 n_s[V_0(x')] - I_0/E_c} \qquad (8)$$

where $H(x - x')$ is the unit step function. Upon integration of the exponential term, (8) simplifies to Green's function

$$z(x, x', f) = \frac{H(x - x')\left(1 - \dfrac{E_0(x')}{E_c}\right)}{qw\mu_0 n_s[V_0(x)] - I_0/E_c}. \qquad (9)$$

The unit step function has the physical significance that the ac voltage at position x only depends on the ac current variations introduced between the source ($x = 0$) and position x. Hence, ac current variations only propagate a voltage response toward the drain end of the device. This arises because the conduction in a MODFET is dominated by drift. If diffusion had a significant effect on current flow and had to be included, the ac transport equation would become second order. Then the voltage response of $\Delta I(x)$ would propagate to the drain as well as to the source terminal of the device. This effect was found in short space-charge-limited diodes by Tehrani et al. [10].

To obtain the impedance field from (9), one substitutes $x = L$ (see [6]) and finds

$$z(L, x', f) = \frac{-E_0(L)}{I_0}\left(1 - \frac{E_0(x')}{E_c}\right). \qquad (10)$$

Since the device impedance measured at the terminals is equal to the integral of the impedance field over the entire length, then

$$Z(L) = \int_0^L z(L, x', f)\, dx' = \frac{-E_0(L)L}{I_0}\left(1 + \frac{V(L)}{E_c L}\right) \qquad (11)$$

where $V(L)$ is the dc voltage at L and $Z(L)$ is the device impedance.

Having obtained the impedance field and device impedance for the MODFET, the expressions for the noise can be calculated. In this paper the focus is on velocity-fluctuation noise only. It has been shown that the spectral density of velocity fluctuations is directly related to the diffusion coefficient $D(E)$ [3], which may be field dependent. Taking this field dependence into account, the spectral density of the open-circuit-voltage noise becomes

$$S_{\Delta V} = \int_0^L w4q^2 D[E_0(x')] \, n_s[V_0(x')]$$

$$\cdot \left| \frac{-E_0(L)}{I_0} \left(1 - \frac{E_0(x')}{E_c} \right) \right|^2 dx'. \quad (12)$$

The spectral density of the current noise can easily be found from

$$S_{\Delta I} = \frac{S_{\Delta V}}{|Z(L)|^2}. \quad (13)$$

Often the expressions for the noise of a device are given in terms of an ac noise temperature $T_n(E)$. With the help of the generalized Einstein relation [11]

$$D(E) = \frac{kT_n(E)}{q} \mu'(E) \quad (14)$$

where k is Boltzmann's constant, $\mu'(E)$ is the differential mobility, i.e., the derivative of the velocity-field characteristic at a given field, one gets

$$S_{\Delta V} = \left(\frac{E_0(L)}{I_0} \right)^2 4q\mu_0 wk \int_0^L T_n[E_0(x')] \, n_s[V_0(x')] \, dx' \quad (15)$$

$$S_{\Delta I} = \frac{4q\mu_0 wk}{\left(1 + \frac{V(L)}{E_c L} \right)^2 L^2} \int_0^L T_n[E_0(x')] \, n_s[V_0(x')] \, dx' \quad (16)$$

for the spectral densities of the voltage and current noise, respectively.

III. CHARGE-VOLTAGE DEPENDENCE

To successfully use the expressions for the impedance and noise in the case of a MODFET, the proper relationship between the sheet carrier concentration and the electrical potential in the channel has to be known. Several authors [12], [13] used an effective capacitance and threshold voltage to calculate $n_s(V)$ as is usually done in a MOSFET. In addition they assume a constant mobility up to saturation. Good agreement between experimental and theoretical I-V characteristics was found for some enhancement-mode devices. Also, the thermal current noise was calculated using these models [14]. However, our experimental findings could not be explained using these models.

Solving the Poisson and Schrödinger equations self-consistently, Vinter [15] shows that for the MODFET structures he considers the gate capacitance depends strongly on gate bias. This gate bias dependence of the capacitance can be observed experimentally in the transconductance measurements of Gupta et al. [16].

There are several experimental methods for obtaining the correct charge–voltage dependence. In the first method the small-signal gate-to-channel capacitance is measured as a function of gate bias. This capacitance is proportional to the derivative of the charge–voltage relationship. In addition, a measurement of the low drain bias ($V_{DS} \ll 50$ mV) total channel resistance as a function of V_G is required. This resistance is given by

$$R_{DS}(V_G) = R_{ss} + R_{dd} + \frac{L}{qw\mu_0 n_s(V_G)} \quad (17)$$

where R_{ss} and R_{dd} are the source and drain access resistances, respectively, and L is the gate length. The last term in (17) accounts for the resistance of the channel region under the gate. Taking the derivative of (17) with respect to V_G, one obtains

$$\frac{dR_{DS}(V_G)}{dV_G} = \frac{-L\left(\frac{dn_s}{dV_G} \right)}{qw\mu_0 n_s^2(V_G)}. \quad (18)$$

It is clear from (18) and the capacitance measurements that one has enough information to obtain the charge–voltage relationship without the knowledge of the access resistances R_{ss} and R_{dd}.

Direct measurements of the gate capacitance on small-geometry MODFET's ($L < 1$ μm) may prove to be difficult and/or inaccurate because of parasitic effects. If one has large-area devices fabricated on the same wafer, measurements of this sort could be done.

A second, approximate method is to use (17) and estimate the value of $R_{ss} + R_{dd}$. The charge–voltage relationship is then directly obtainable without knowledge of the capacitance. Estimates of the value of $R_{ss} + R_{dd}$ can be made with the help of calculations for heterojunction lineup in equilibrium [17] and the quasi-triangular potential-well approximation. These calculations are expected to give sheet carrier concentrations of reasonable enough accuracy to obtain the end resistances.

Finally, a third method is introduced to obtain the charge–voltage relationship. It involves the small-signal impedance of the channel for any value of drain bias for which the impedance field model outlined earlier holds.

Position L_1 (see Fig. 1) is the point taken to be the drain-side edge under the gate. At this point the maximum electric field in the channel is produced. Its value is not allowed to exceed the peak field value of 3.5 kV/cm in our impedance field model. The charge at L_1 is controlled by the gate voltage. The impedance at L_1 is given by (cf. (11))

$$Z(L_1) = \frac{-E_0(L_1) L_1}{I_0} \left(1 + \frac{V(L_1)}{E_c L_1} \right). \quad (19)$$

With reasonable knowledge of the drain resistance R_{dd}, then this impedance is

$$Z(L_1) = Z(L) - R_{dd} \quad (20)$$

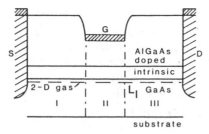

Fig. 1. Diagram of the MODFET structure used in our experiments.

and the dc voltage at L_1 is

$$V(L_1) = V_{DS} - I_0 R_{dd}. \qquad (21)$$

Then from (19) a value for $E_0(L_1)$ can be calculated using the measured current– and impedance–voltage characteristics, and the sheet carrier concentration at L_1 follows from (6). In this way the charge–voltage relationship is obtained in the presence of high electric fields under the gate. If this method gives results for $n_s(V)$ similar to the other two methods, which are based on the assumpton that the field under the gate is low, then the absence of real-space-charge transfer in MODFET operation is experimentally verified.

Note that all of the above methods are restricted to the regime of operation where the AlGaAs layer under the gate is fully depleted of carriers. Also, it is assumed that the 2-D sheet carrier mobility is not a function of gate bias.

IV. DEVICE DESCRIPTION

A diagram of the MODFET structure used in our measurements is shown in Fig. 1. The gate is recessed so that the space-charge regions of the Schottky barrier and the AlGaAs interface overlap. The gate-interface spacing determines whether the device will operate in the enhancement or depletion mode. Experiments were performed on a depletion mode MODFET with a gate length of 1 μm and a gate width of 145 μm. Spacing between the source and drain pads was 4 μm. Regions I and III represent the source and drain access resistances R_{ss} and R_{dd}, whereas Region II is the active gate region.

A GaAs buffer layer of 1-μm thickness was deposited on a GaAs semi-insulating substrate. Next a 30-Å intrinsic AlGaAs spacer layer with an aluminum mole fraction $x = 0.28$ was grown on top, followed by a 600-Å layer of AlGaAs doped to a level of 2.5×10^{18} cm^{-3}

V. MEASUREMENT PROCEDURE

Measurements of the current–voltage characteristic show that the MODFET exhibits a significant change in the current with time for fixed bias voltage. The current could be stabilized when the drain bias was pulsed with a low-duty cycle. This indicates that the current change was due to Joule heating of the channel. In our setup a pulse length of 4 ms at a 3-percent duty cycle was used. Shorter pulses could not be applied due to the rise time of the bias tee. Under these conditions the I–V characteristic of the

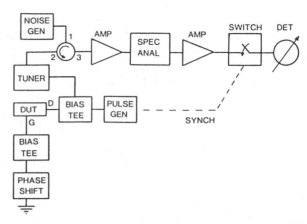

Fig. 2. Noise measurement setup for continuous or pulse bias conditions. D indicates the drain terminal of the DUT, whereas G indicates the gate terminal. Note that the source is grounded.

device was measured from the linear region up into saturation using a digital oscilloscope.

The noise measurement scheme is shown in Fig. 2 and is a modified version of the setup described by Gasquet et al. [18]. From a series of four measurements at each bias setting, the noise temperature T_n of the device under test (DUT), the power reflection coefficient Γ^2 at port 2 of the circulator, the noise temperature of the amplifier system T_A, and the gain bandwidth of the product of the system can be obtained. All losses, including any mismatch of the DUT, are accounted for in this way.

The MODFET was measured in the common source configuration with the gate high-frequency short-circuited to ground. This is done to eliminate the contribution from the equivalent current-noise source associated with the gate of the device. At a gate bias of $V_G = 0$ V a small bonding wire located very close to the gate is used to short-circuit the input. At high frequencies, when the inductive reactance of the wire becomes too large to effectively short the gate, a high-quality shorted coaxial phase shifter is placed in parallel with the gate. The length of this shorted transmission line can be adjusted to produce an electrical short circuit at the physical gate of the device. By adjusting the phase shifter a minimum in the noise temperature of the DUT at a given frequency could be observed. The minimum corresponds with the condition of a high-frequency shorted gate. This, and the measurement of a white noise spectrum indicate that the gate current-noise source has been completely canceled. When measuring the noise at gate biases other than $V_G = 0$ V, the phase shifter by itself proved to be adequate to cancel the gate noise contribution.

The output of the DUT can be matched at each bias point with the tuner to the characteristic impedance (50 Ω) of the system. The noise generator is used as a calibration noise source as well as for the purpose of matching the DUT. A circulator is used to couple both the device noise and the calibration noise generator to the amplifier stages. The contribution from the noise generator, seen by the amplifier stages, depends on the reflection coefficient seen at port 2 of the circulator. Since the impedance levels of the MODFET channel, from the linear to saturation regimes, do not cause large mismatches from the characteristic impedance, the tuner is not always needed. Measurements of T_n obtained with and without the tuner agreed well within the experimental error. If the tuner is omitted, the channel reflection coefficient and the noise can be measured simultaneously.

Using broadband circulators and amplifiers, a frequency range from 500 MHz to 3.4 GHz could be covered. A range of frequencies needs to be measured to determine whether the noise spectrum is indeed white. Frequency selection is obtained from the spectrum analyzer which shifts the high-frequency noise signal to an intermediate frequency (IF) of 21.4 MHz. At this frequency further amplification and power detection are performed. This system has the advantage that the noise of the DUT can be measured under continuous or pulse bias. If the measurement is done under pulse bias, the RF switch and the pulse bias generator are synchronized so that only noise generated during the bias pulse is measured.

Once the ac noise temperature T_n is known, the DUT equivalent current-noise spectral density is obtained from

$$S_{\Delta I}(f) = 4kT_n \, \text{Re} \, (Y) \qquad (22)$$

where Re (Y) is the real part of the small-signal admittance of the DUT. The value of Re (Y) was obtained in three different ways. The first method is to differentiate the I–V characteristic and approximate Re (Y) by dI/dV. This method is valid at low frequencies and gives only good results at high frequencies if the real part of the channel admittance does not change significantly with frequency. A second method is to calculate the absolute value of the admittance from the reflection coefficient if the tuner is omitted. Then, if parasitic susceptive elements are small in comparison with the device-channel ac conductance, the major contribution to the reflection coefficient is due to the channel conductance and again we assume Re $(Y) = |Y|$. This method can also be used under pulse bias conditions. The last method is to measure the S-parameters. However, the S-parameters can only be measured under continuous bias conditions. When possible, the data from the three methods were compared and were found to agree within the experimental error.

VI. RESULTS AND DISCUSSION

The current–voltage characteristic of the MODFET at room temperature with gate bias $V_G = 0$ V is shown in Fig. 3. The spectral densities of the open-circuit voltage and short-circuit current noise as a function of the dc current are depicted in Fig. 4. The solid lines in both figures indicate the results of the impedance field model outlined in the text. The model is valid up to 26 mA because at this current value the peak electric field in the channel reaches 3.5 kV/cm. The continued increase in the measured I–V characteristic beyond 26 mA is due to the formation of an accumulation layer under the drain side of the gate. A large voltage drop will occur across this ac-

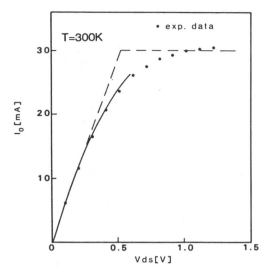

Fig. 3. Current–voltage characteristic at $V_G = 0$ V. The black dots represent the experimental data. The solid line indicates the results of our model and the dashed line indicates the results of the constant capacitance and mobility model.

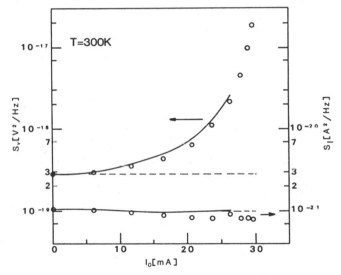

Fig. 4. Voltage and current noise spectral densities as a function of dc current I_0. The solid line indicates the results of the impedance field model outlined in the text using $D(E) = D_0$. The dashed line indicates the results of the constant capacitance and mobility model.

cumulation layer, leading to current saturation in analogy with the situation in GaAS MESFET's [19].

The charge–voltage relationship used in our calculations was derived by using the measured ac impedance and I–V characteristic (method III). A drain resistance $R_d = 6.2 \, \Omega$ was used. This value results from equilibrium calculations for heterojunction lineup.

Although calculations and measurements [20], [21] show that the diffusion coefficient begins to increase near 3.5 kV/cm for bulk GaAs, we assume that $D(E) = D_0$, since measurements of the field dependence of the diffusion coefficient on gateless MODFET structures [22] show that $D(E)$ is field independent in the field range of interest.

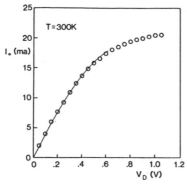

Fig. 5. Current–voltage characteristic at $V_G = 0$ V and $T = 300$ K after the cooling cycle.

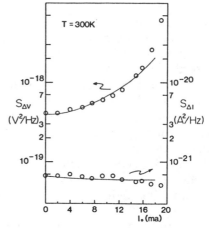

Fig. 6. Voltage and current noise spectral densities as a function of dc current I_0 at $V_G = 0$ V and $T = 300$ K after the cooling cycle.

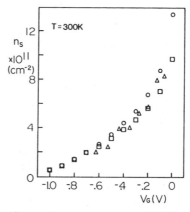

Fig. 7. Charge–voltage relationship of the MODFET. The circles and squares indicate data obtained with method II (low field) using $R_{ss} + R_{dd}$ = 19.4 and 20.9 Ω, respectively. Triangles indicate method III (high field) with $R_{dd} = 9.7$ Ω.

The dashed lines in Figs. 3 and 4 are the predicted performance curves of a MODFET using the constant capacitance and mobility models previously mentioned. Although the calculated current-noise spectral density in Fig. 4 shows good agreement with experiment, the voltage spectral density does not. This is due to the fact that these models do not correctly predict the small-signal impedance levels in this region of operation, as can be seen in the *I–V* characteristic of Fig. 3.

An attempt was made to measure this same device at liquid-nitrogen temperature. A collapse of the *I–V* characteristic was observed, similar to what other researchers found at low temperatures [23]. Therefore, measurements of the noise at these temperatures were not possible.

After returning the device to room temperature, the static *I–V* characteristic had changed, giving a lower saturation current and higher resistance in the linear regime. Since the device had not been hermetically sealed when lowered into the liquid-nitrogen bath, the number of surface states in regions I and III might have been changed, altering the width of the surface depletion region, and hence R_{ss} and R_{dd}. This would result in an increase in the access resistances and account for the lower saturation current. It is assumed that the gate metalization protected the surface in region II. This seems to be justified by the measurements of the transconductance and charge–voltage relationship both before and after the cooling cycle.

The *I–V* characteristic and noise at $T = 300$ K after the cooling cycle are shown in Figs. 5 and 6, respectively. There is little change in the behavior of the noise characteristic. Since the total channel resistance has changed, the levels of the noise are slightly different. Our model again shows good agreement with experiment.

Shown in Fig. 7 are the results for the charge–voltage relationship obtained using two different methods. The triangles indicate the results for $n_s(V_G)$ calculated using method III with an increased drain resistance of 9.7 Ω after cooling. The circles represent the values determined with the help of method II with $R_{ss} + R_{dd} = 2 \times R_{dd} = 19.4$ Ω. The charge–voltage dependence, choosing slightly different total access resistances equal to 20.9 Ω, is indicated by the squares. Note that method III involves high-field regions under the gate, while method II assumes a low-

field ohmic regime under the gate. From the fact that both methods give identical results for n_s, we conclude that real-space-charge transfer in the gate region of a MODFET is absent.

VII. CONCLUSIONS

This paper has reviewed the method for obtaining the dc, ac, and noise properties of an active semiconductor device based on the impedance field. The impedance field was then derived for the MODFET channel. Calculations using the impedance field model show excellent agreement with measurements. Methods of obtaining the correct charge–voltage relationship under the gate region were examined. The charge–voltage relationship obtained with low and high fields in the channel show excellent agreement, indicating that real-space-charge transfer is not present in the bias range up to saturation.

REFERENCES

[1] P. M. Solomon and H. Morkoç, "Modulation-doped GaAs/AlGaAs heterojunction field-effect transistors (MODFET's)—ultrahigh-speed

devices for supercomputers," *IEEE Trans. Electron Devices*, vol. ED-31, p. 1015, 1984.

[2] N. T. Linh, "Two-dimensional electron gas FETs: Microwave applications," in *Semiconductors and Semimetals*. New York: Academic, to be published.

[3] W. Shockley, J. A. Copeland, and R. P. James, "The impedance field method of noise calculation in active semiconductor devices," in *Quantum Theory of Atoms, Molecules and the Solid State*. New York: Academic Press, 1966, pp. 537–563.

[4] K. M. van Vliet, A. Friedmann, R. Zijlstra, A. Gislof, and A. van der Ziel, "Noise in single injection diodes—I. A survey of methods," *J. Appl. Phys.*, vol. 46, p. 1804, 1975.

[5] K. M. van Vliet, "The transfer-impedance method for noise in field-effect transistors," *Solid-State Electron.*, vol. 22, p. 233, 1979.

[6] J. P. Nougier, J. C. Vaissiere, and D. Gasquet, "Mathematical formulation of the impedance field method. Application to the noise of the channel of field effect transistors," in *Proc. 6th Int. Conf. Noise in Physical Systems*. Washington, DC: National Bureau of Standards, 1981, pp. 42–46.

[7] H. Gummel and J. Blue, "A small-signal theory of avalanche noise in IMPATT diodes," *IEEE Trans. Electron Devices*, vol. ED-14, p. 569, 1967.

[8] E. Constant, in *Hot-Electron Transport in Semiconductors*, L. Reggiani, Ed. Berlin: Springer-Verlag, 1985.

[9] A. Cappy, A. Vanoverschelde, J. Zimmermann, P. Philippe, C. Versnaeyen, and G. Salmer, "Etude theorique et experimentale du transistor a effet de champ a heterojunction," *Rev. Phys. Appl.*, vol. 18, p. 719, 1983.

[10] S. Tehrani, G. Bosman, C. M. Van Vliet, and L. L. Hench, "Space charge limited current noise in the presence of traps in SiC," presented at the 8th Int. Conf. on Noise in Physical Systems, Rome, Italy, Sept. 1985 (proceedings to be published by North Holland, Amsterdam).

[11] A. van der Ziel, "Equivalent temperature of hot electrons," *Solid-State Electron.*, vol. 23, p. 1035, 1980.

[12] P. Delagebeaudeuf and N. T. Linh, "Metal-(n) AlGaAs-GaAs two-dimensional electron gas FET," *IEEE Trans. Electron Devices*, vol. ED-29, p. 955, 1982.

[13] K. Lee, M. Shur, T. J. Drummond, and H. Morkoç, "Current–voltage and capacitance–voltage characteristics of modulation-doped field-effect transistors," *IEEE Trans. Electron Devices*, vol. ED-30, p. 207, 1983.

[14] A. van der Ziel and E. N. Wu, "Thermal noise in high electron mobility transistors," *Solid-State Electron.*, vol. 26, p. 383, 1983.

[15] B. Vinter, "Subbands and charge control in a two-dimensional electron gas field-effect transistor," *Appl. Phys. Lett.*, vol. 44, p. 307, 1984.

[16] A. K. Gupta, E. A. Sovero, R. L. Pierson, R. D. Stein, R. T. Chen, D. L. Miller, and J. A. Higgins, "Low-noise high electron mobility transistors for monolithic microwave integrated circuits," *IEEE Electron Device Lett.*, vol. EDL-6, p. 81, 1985.

[17] K. Lee, M. Shur, T. J. Drummond, and H. Morkoç, "Electron density of the two-dimensional electron gas in modulation-doped layers," *J. Appl. Phys.*, vol. 54, p. 2093, 1983.

[18] D. Gasquet, J. C. Vaissiere, and J. P. Nougier, "New method for wide band measurement of noise temperature of one-port networks at high pulsed bias," in *Proc. 6th Int. Conf. on Noise in Physical Systems*. Washington, DC: National Bureau of Standards, 1981, pp. 305–308.

[19] S. M. Sze, *Physics of Semiconductor Devices*, 2nd ed. New York: Wiley, 1981, ch. 6.

[20] J. G. Ruch and G. S. Kino, "Transport properties of GaAs," *Phys. Rev.*, vol. 174, p. 921, 1968.

[21] J. Andrian, G. Bosman, A. van der Ziel, and C. M. Van Vliet, "Hot electron diffusion noise associated with intervalley scattering in short GaAs devices," presented at the 8th Int. Conf. on Noise in Physical Systems, Rome, Italy, Sept. 1985 (proceedings to be published by North Holland, Amsterdam).

[22] C. Whiteside, G. Bosman, H. Morkoç, and W. Kopp, "Noise temperature and diffusion coefficient associated with the quasi-two-dimensional electron gas in an ungated MODFET structure," *IEEE Electron Device Lett.*, vol. EDL-7, p. 294, 1986.

[23] R. Fischer, T. J. Drummond, J. Klem, W. Kopp, T. S. Henderson, D. Perrachione, and H. Morkoç, "On the collapse of drain *I–V* characteristics in modulation-doped FET's at cryogenic temperatures," *IEEE Trans. Electron Devices*, vol. ED-31, p. 1028, 1984.

Simultaneous Determination of Transistor Noise, Gain, and Scattering Parameters for Amplifier Design Through Noise Figure Measurements Only

GIOVANNI MARTINES AND MARIO SANNINO

Abstract—A method for the simultaneous determination of transistor noise and gain parameters through noise figure measurements has been presented recently.

An improved version of the method is presented here which can also yield all the scattering parameters needed for designing amplifiers. By means of a proper (computer-aided) data-processing technique, s_{11}, s_{22}, $|s_{12}|$, $|s_{21}|$, and $\underline{/s_{12}s_{21}}$ are determined.

As experimental verifications, the characterization of a GaAs MESFET versus frequency (4–8 GHz) is reported.

I. INTRODUCTION

Since the standard (graphical) methods of measuring noise in linear two-ports were established by IRE in 1960, a modernized (computer-aided) technique has been introduced [1].

Following this method, the overall measuring system noise figure F_m and the available power gain G_a of the two-port under test are measured as functions of the input termination (or source) reflection coefficient Γ_s. Then the noise figure F of the two-port is derived through the well-known Friis formula

$$F_m(\Gamma_s) = F(\Gamma_s) + \frac{F' - 1}{G_a(\Gamma_s)} \qquad (1)$$

where F' is the noise figure of the "second stage" represented by all the instruments subsequent to the two-port under test.

The four noise parameters are then determined by substituting some (redundant, i.e., more than four, for accuracy reasons) values of $F(\Gamma_s)$ into the relationship which defines the device noise parameters F_o, $|\Gamma_{on}|$, $\underline{/\Gamma_{on}}$, and N_n

$$F_m(\Gamma_s) = F_o + 4N_n \frac{|\Gamma_s - \Gamma_{on}|^2}{(1 - |\Gamma_s|^2)(1 - |\Gamma_{on}|^2)}. \qquad (2)$$

The device gain parameters G_{ao}, $|\Gamma_{og}|$, $\underline{/\Gamma_{og}}$, and N_g defined by

$$\frac{1}{G_a(\Gamma_s)} = \frac{1}{G_{ao}} + 4Ng \frac{|\Gamma_s - \Gamma_{og}|^2}{(1 - |\Gamma_s|^2)(1 - |\Gamma_{og}|^2)} \qquad (3)$$

are available from the values of $G_a(\Gamma_s)$ either computed through the two-port scattering parameters determined by a network analyzer or measured by means of a signal generator and a power meter.

In a method recently proposed, both noise and gain parameter sets of transistors are simultaneously determined solely through noise figure measurements [2]. The novelty of the method lies in the use of different values of F' for each value of Γ_s. The variation of F' is easily achieved by inserting a step attenuator between the two-port and the subsequent amplifier.

In this paper, an improved version of the method is reported which can also yield information about the scattering parameters needed for amplifier design.

Manuscript received September 5, 1983; revised April 11, 1984.

G. Martines and M. Sannino are with the Dipartimento di Tripegueria Elettrica, Università di Palermo, Viale delle Scienze, 90128 Palermo, Italy.

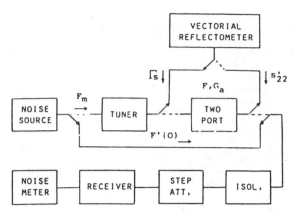

Fig. 1. Simplified block diagram of the measuring system.

The advantages of the method proposed consist in that a) different measuring systems for determining noise, gain, and scattering parameters are avoided because all the parameters are derived by processing experimental data obtained solely by means of a noise figure measuring system, and b) time requirements are reduced to those required by conventional methods for the determination of noise parameters only. Other advantages are discussed in the conclusions.

II. ANALYSIS OF THE METHOD AND MEASURING PROCEDURE

In order to show the main steps of the measuring procedure, we can refer to the simplified block diagram of Fig. 1. It differs from the conventional noise measuring system only by the use of a) a step attenuator inserted in the receiver input to change F', and b) an isolator connected to the device output in place of the matching tuner in order to match the receiver input for each value of Γ_s without seeking careful matching. Avoiding the use of the output tuner is necessary, however, when characterizing conditionally stable devices in order to prevent oscillation.

By this way, F' of (1) becomes [3]

$$F'(s'_{22}) = F'(0)/(1 - |s'_{22}|^2) \qquad (4)$$

where

$$s'_{22}(\Gamma_s) = s_{22} + \frac{s_{12}s_{21}\Gamma_s}{1 - s_{11}\Gamma_s} \qquad (5)$$

and $F'(0)$ is the value of F' in input matched condition (usually, 50 Ω for microwaves).

For each frequency and bias condition for the device under test, the measuring steps are the ones summarized as follows.

1) As for the conventional noise-measuring methods, we realize some redundant values of Γ_s by adjusting the tuner and measuring them through a vectorial reflectometer (or a network analyzer); then we measure $F_m(\Gamma_s)$. Some drawbacks can be avoided by following the criteria already suggested for selecting Γ_s [4], [5].

2) Step 1 is repeated for some (redundant, i.e., more than two) values of F' obtained through the attenuation introduced by the step attenuator.

3) The device output reflection coefficient $s'_{22}(\Gamma_s)$ for each Γ_s is measured.

4) F' for each position of the step attenuator is measured. From (1), (2), and (3), the device noise and gain parameters can now be derived [2].

5) By directly connecting the matched noise source to the

Reprinted from *IEEE Transactions on Instrumentation and Measurement*, vol. IM-34, no. 1, pp. 89–91, March 1985.

TABLE I
COMPARISON BETWEEN THE SCATTERING PARAMETERS DERIVED BY THE
METHOD PROPOSED AND THE ONES MEASURED THROUGH A
NETWORK ANALYZER [BRACKET]
(GaAs MESFET NE 24483; $V_{DS} = 3V$, $I_{DS} = 13mA = 15$ percent I_{DSS})

GHz	$\lvert s_{11} \rvert$	$\angle s_{11}$(deg.)	$\lvert s_{12} \rvert$	$\lvert s_{21} \rvert$	$\angle s_{12}s_{21}$(deg.)	$\lvert s_{22} \rvert$	$\angle s_{22}$(deg.)
4	.78 [.75]	−104.0 [−103.2]	.028 [.028]	2.52 [2.52]	84.9 [84.2]	.60 [.59]	−59.2 [−55.0]
5	.63 [.66]	−134.8 [−137.4]	.021 [.021]	2.62 [2.63]	59.7 [62.6]	.54 [.52]	−75.9 [−72.0]
6	.67 [.67]	−172.4 [−174.4]	.036 [.035]	2.61 [2.64]	41.4 [37.0]	.49 [.48]	−102.2 [−97.0]
7	.68 [.72]	153.7 [155.4]	.056 [.050]	2.24 [2.42]	−3.1 [−7.7]	.47 [.50]	−137.6 [−136.5]
8	.68 [.72]	125.9 [126.9]	.051 [.047]	1.97 [2.10]	−44.8 [43.1]	.54 [.53]	−174.0 [−167.1]

measuring system, i.e., without the tuner, $F_m(0) = F_{m50}$ for each attenuation measured.

6) With the noise source connected as above, Step 3 is repeated thus measuring the device s_{22} parameter.

From steps 5 and 6 and (1) written for $\Gamma_s = 0$, the device noise figure and available gain F_{50} and G_{50} for input matched conditions are derived with redundancy. Comparison between these measured values and the ones computed through the device noise and gain parameters already determined, provides also a test of the correctness of the results and allows one to evaluate with a good degree of approximation the loss of the tuner adopted as admittance transformer network [6].

III. DETERMINATION OF SCATTERING PARAMETERS

From the parameters determined through the above procedure, it is possible to derive also all those s-parameters needed for low-noise amplifier design.

The parameter s_{22} is measured in Step 6. The available power gain for the input matched condition is given by

$$G_a(0) = G_{as0} = \frac{\lvert s_{21} \rvert^2}{1 - \lvert s_{22} \rvert^2}. \tag{6}$$

Gain has been computed through the gain parameters and also measured in Step 5 and 6 with accuracy (redundancy). From (6), knowing s_{22}, $\lvert s_{21} \rvert$ is derived.

The relationship which relates the device available gain $G_a(\Gamma_s)$ to the scattering parameters yields

$$\lvert 1 - s_{11}\Gamma_s \rvert^2 = \frac{\lvert s_{21} \rvert^2 (1 - \lvert \Gamma_s \rvert^2)}{G_a(\Gamma_s)(1 - \lvert s'_{22}(\Gamma_s)\rvert^2)} = \gamma(\Gamma_s) \tag{7}$$

where s'_{22} is measured for all Γ_s in Step 4. From (7) and (5), $\lvert s_{12}s_{21} \rvert$ is computed.

Putting now $x = \lvert s_{11} \rvert \cos \angle s_{11}$ and $y = \lvert s_{11} \rvert \sin \angle s_{11}$, (7) yields

$$x^2 + y^2 + a_i x + b_i y + c_i = 0 \tag{8}$$

where

$$a_i = \frac{2 \cos \angle \Gamma_{si}}{\lvert \Gamma_{si} \rvert}, \quad b_i = \frac{2 \sin \angle \Gamma_{si}}{\lvert \Gamma_{si} \rvert},$$

$$c_i = \frac{\gamma(\Gamma_{si}) - 1}{\lvert \Gamma_{si} \rvert^2}, \quad \text{with } i = 1, \cdots, n. \tag{9}$$

Solving (8) by means of the least-squares method and a (computer-aided) successive approximation procedure, s_{11} is derived and $\angle s_{12}s_{21}$ is computed through (5).

IV. CONCLUDING REMARKS AND EXPERIMENTAL VERIFICATIONS

A noise-measuring method is presented which furnishes the transistor noise and gain parameters and also all those scatter-

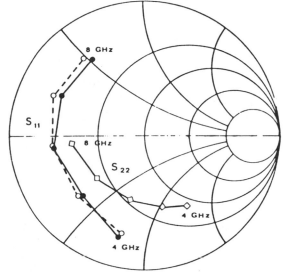

Fig. 2. The values of the parameters s_{11} and s_{22} of Table I as determined through the method presented (continous line) and as measured by a network analyzer (dashed line) are reported on the Smith chart. The values of s_{22} are nearly coincident.

ing parameters needed for designing amplifiers and for computing maximum stable gain and stability factor, i.e., s_{11}, s_{22}, $\lvert s_{12} \rvert$, $\lvert s_{21} \rvert$, and $\angle s_{12}s_{21}$. The time requirement is the same as for conventional methods for noise parameter determination only. Further, it is noteworthy that the parameters are accurately obtained because they are derived by processing redundant experimental data. Only s_{22} is obtained by conventional measurements.

The method has been examined by characterizing a low-noise GaAs MESFET NE 24483 (by NEC) versus frequency (4–8 GHz) for different bias conditions ($V_{DS} = 3$ V, $I_{DS} = 5$, 10, 15, 25, percent I_{DSS}). For each frequency, noise figure F_m has been measured for twelve values of Γ_s, realized through a coaxial slide-screw tuner, and for four values of F', realized by four steps of the attenuator, namely 0, 3, 6, and 10 dB. Since (1) is a simple relationship (a straight line), four values of F' are sufficient to accurately determine F and G_a from F_m.

Experimental results for the device scattering parameters are shown in Table I and in Fig. 2, where the parameters are measured conventionally through a network analyzer and reported for comparison.

REFERENCES

[1] R. Q. Lane, "The determination of device noise parameters," Proc. IEEE, vol. 57, pp. 1461–1462, Aug. 1969.

[2] M. Sannino, "Simultaneous determination of device noise and

gain parameters through noise measurements only," *Proc. IEEE*, vol. 68, pp. 1343–1345, Oct. 1980.

[3] K. Mishima, Y. Sawayama, and M. Sannino, "Comments on 'Simultaneous determination of device noise and gain parameters through noise measurements only,'" *Proc. IEEE*, vol. 70, pp. 100–101, Jan. 1982.

[4] M. Sannino, "On the determination of device noide and gain parameters," *Proc. IEEE* (lett.), vol. 67, pp. 1364–1366, Aug. 1969.

[5] G. Caruso and M. Sannino, "Computer-aided determination of microwave two-port noise parameters," *IEEE Trans. Microwave Theory Tech.*, vol. MTT-26, pp. 639–643, Sept. 1978.

[6] G. Martines and M. Sannino, "Determination of microwave transistor noise and gain parameters through noise-figure measurements only," *IEEE Trans. Microwave Theory Tech.*, vol. MTT-30, pp. 1255–1259, Aug. 1982.

A Method for Measurement of Losses in the Noise-Matching Microwave Network While Measuring Transistor Noise Parameters

GIOVANNI MARTINES AND MARIO SANNINO

Abstract —A new method for measuring the loss of a tuner network used as the noise-source admittance transformer in a noise parameter test set is presented. Since the method is based on noise figure measurements, the tuner losses are determined on-line while performing measurements for determining transistor noise parameters.

Experiments carried out on a coaxial slide-screw tuner by means of a computer-assisted measurement setup are reported.

I. INTRODUCTION

Modern techniques for the determination of the four noise parameters which characterize the noise behavior of microwave transistors, i.e., minimum noise figure, optimum reflection coefficient of the input termination (magnitude and phase), and noise resistance, require measurements of the noise figure $F(\Gamma_s)$ of the transistor under test for some (redundant, i.e., more than four) values of the reflection coefficients Γ_s of its input termination. These measurements are made in conjunction with a computer-aided data-processing procedure based on the least-squares method [1]. This procedure furnishes good results, provided that, in selecting the values Γ_s, well-established criteria are followed [2], [3].

The same procedure yields the four available gain parameters of the transistor under test after carrying out measurements of transistor power gain $G_a(\Gamma_s)$ for some (redundant) Γ_s through a gain-measuring setup [1]. These parameters and the scattering parameters can also be derived when determining transistor noise parameters if a proper experimental procedure for measuring $F(\Gamma_s)$ is followed [4]–[7].

The essential component of the noise-parameter-measuring setup is a coaxial or waveguide double-stub or slide-screw tuner inserted between the noise source and the device under test (DUT), in order to transform the noise source reflection coefficient Γ_{ns} to the desired value Γ_s of the DUT input termination. The DUT output noise powers are then detected, through a receiver, by a meter which measures the noises figure F_m of the whole measuring setup. This figure is given by Friis's formula for cascaded networks, i.e.

$$F_m(\Gamma_{ns}) = \alpha_{\Gamma_s}(\Gamma_{ns})\left(F(\Gamma_s) + \frac{F_r(S'_{22})-1}{G_a(\Gamma_s)} \right) \qquad (1)$$

where α_{Γ_s} represents the tuner loss, which depends on the tuner configuration; F, G_a, and S'_{22} are the noise figure, the available power gain, and the output reflection coefficient of the DUT, respectively; and F_r is the receiver noise figure.

From (1), it appears that in order to determine $F(\Gamma_s)$ from the measured noise figure F_m, previous measurements of the tuner loss $\alpha_{\Gamma_s}(\Gamma_{ns})$ are needed. Very little has appeared so far in the literature concerning this measurement.

Manuscript received December 9, 1985; revised August 7, 1986. This work was supported in part by Italian Research Council-Electronic Components, Circuits and Technologies (CCTE) Group and Ministry of Education.

The authors are with the Dipartimento di Ingegneria Elettrica, Universita di Palermo, Viale delle Scienze, Palermo, Italy.

IEEE Log Number 8611025.

Alternative measuring methods based on the use of a low-loss tuner as a reflecting input termination (e.g., sliding short) have been proposed [8], [9] and applied recently [10] to the characterization of the noise behavior of receivers. In current practice, however, the former methods are preferred for characterizing microwave transistors in terms of noise parameters, especially when there is interest in the *direct* determination of the optimum parameters using only the input-tuning procedure.

On the problem of determining the losses α_{Γ_s} of the tuner versus setting, a recent paper of Strid [11] is of interest. This suggests a unique procedure to determine the losses of a tuner by characterizing it by its scattering parameters. For any Γ_s, the loss α_{Γ_s} is then computed as the inverse of the tuner available power gain. Unfortunately, this method is very time consuming and also not convenient from an experimental viewpoint, particularly when the tuner is used in a transistor noise parameter test set. This is because it requires

a) carrying out two experimental procedures at different times by means of two different measuring systems, one to measure noise figures $F_m(\Gamma_s)$, the other to measure the tuner S-parameters; and

b) setting and resetting the tuner accurately in performing the above experimental procedures for each Γ_s.

The above drawbacks are avoided by the method proposed in this paper, which allows tuner losses to be measured with the same setup used for noise figure measurements. The tuner losses α_{Γ_s} can be determined on-line while performing the measurements of $F_m(\Gamma_s)$ intended to derive noise (gain and scattering) parameters of transistors. The method has been devised in order to be easily embodied in a system for computer-assisted measurements for transistor characterization.

II. ANALYSIS OF THE METHOD

To illustrate the principle of the method, let us first refer to a setup dedicated to measurements of tuner loss only. In Section III its application using a noise parameter test set is discussed.

In the measuring system shown in Fig. 1, the noise source is alternatively switched ON and OFF to inject two different noise power levels to the tuner under test; the noise powers are then detected by a single sideband receiver equipped with an input isolator and a noise figure indicator.

Supposing that the noise source is a well-matched one, i.e., $\Gamma_{ns} = 0$, the noise figure F_r of the tuner–receiver cascade (Fig. 1(b)) is given by

$$F_{ra}(0) = \alpha_{\Gamma_s} F_r(\Gamma_s) \qquad (2)$$

from which

$$\alpha_{\Gamma_s}|_{dB} = F_{ra}(0)|_{dB} - F_r(\Gamma_s)|_{dB} \qquad (3)$$

which furnishes the tuner loss α_{Γ_s} if $F_r(\Gamma_s)$ is known.

To compute $F_r(\Gamma_s)$, we need in general the four noise parameters of the receiver. In the case of an input-isolated receiver, however, we have, more simply, [6], [7], [12]

$$F_r(\Gamma_s) = \frac{F_r(0)}{1 - |\Gamma_s|^2}. \qquad (4)$$

The receiver noise figure $F_r(0)$ in the input-matched condition is then measured by directly connecting the source to the isolator, and $|\Gamma_s|$ is measured through a reflectometer. Relationship (4) is valid if $T_a \cong T_0$ and $\Gamma_r \cong 0$, where T_a and T_0 are the (isolator)

Reprinted from *IEEE Transactions on Microwave Theory and Techniques*, vol. MTT-35, no. 1, pp. 71–75, January 1987.

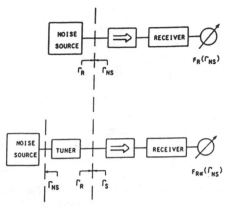

Fig. 1. The two different configurations of the measuring system for determination of tuner loss through noise figure measurements.

ambient temperature and the reference temperature (290°K), respectively, and Γ_r is the input reflection coefficient of the receiver.

If these hypotheses do not hold, we have (see the Appendix)

$$F_r(\Gamma_s) = \frac{1}{\mu'}F_r(0) + \left(\frac{1}{\mu'}-1\right)\left(\frac{T_a}{T_0}-1\right) \qquad (5)$$

where

$$\mu' = \frac{1-|\Gamma_s|^2}{|1-\Gamma_r\Gamma_s|^2}. \qquad (6)$$

Let us consider finally the case in which the noise source is not well matched and consequently $F_{r\alpha}(0)$ and $F_r(0)$ cannot be measured. If Γ_{ns1} and Γ_{ns2} are the values of the *nominal* source reflection coefficient Γ_{ns} in the OFF and ON conditions, respectively, we can compute $F_{r\alpha}(0)$ and $F_r(0)$ by means of relationships similar to (5). Thus (see also the Appendix)

$$F_{r\alpha}(0) = \mu'_2 F_{r\alpha}(\Gamma_{ns}) + (\mu'_2-1)\left(\frac{T_a}{T_0}-1\right) \qquad (7)$$

and

$$F_r(0) = \mu'_2 F_r(\Gamma_{ns}) + (\mu'_2-1)\left(\frac{T_a}{T_0}-1\right) \qquad (8)$$

with

$$\mu'_2 = \frac{1-|\Gamma_{ns2}|^2}{|1-\Gamma_{ns2}\Gamma_r|^2}. \qquad (9)$$

From (7), (8), (5), and (2), the tuner loss α_{Γ_s} is determined by computer-aided processing of measured noise figures $F_{r\alpha}(\Gamma_{ns})$ and $F_r(\Gamma_{ns})$.

Compared with gas-discharge noise sources, modern solid-state sources offer differences between mismatches in ON and OFF conditions which are usually neglected. In this case, obviously, the above relationships hold with $\Gamma_{ns1} \cong \Gamma_{ns2} \cong \Gamma_{ns}$. On the other hand, different ON–OFF mismatches, i.e., $\Gamma_{ns1} \neq \Gamma_{ns2}$, are not acceptable when characterizing transistors through a noise parameter test set, because this implies that the value Γ_s is ambiguous while measuring $F(\Gamma_s)$ of (1).

For this reason, the mismatch difference is to be reduced to a negligible value. To this end, an isolator (or attenuator) can be connected to the source output, provided that the attenuation so introduced is accurately measured in order to determine the actual noise levels furnished by the equivalent noise source so obtained [13], [14].

The method presented here is suitable for this, using the same setup shown in Fig. 1 where the two-port under test is now the isolator–(or attenuator)–tuner cascade.

III. Application of the Method while Performing Computer-Controlled Measurements of the Noise (Gain and Scattering) Parameters of Microwave Transistors

The method proposed requires that the measurements be preceded by a calibration procedure for both the receiver and the noise source employed.

The calibration procedure is (for each frequency):

a) measure the receiver input reflection coefficient Γ_r through a vectorial reflectometer;
b) measure the noise source reflection coefficient Γ_{ns}; if the hypothesis $\Gamma_{ns} \cong \Gamma_{ns1} \cong \Gamma_{ns2}$ does not hold, Γ_{ns2} must be measured; in the case of transistor characterization, however, the ambiguity must be reduced, as stated above, through an isolator or an attenuator;
c) measure the noise figure $F_r(\Gamma_{ns})$ using the setup of Fig. 1(a).

The measurements to carry out on the tuner under test for each setting and frequency are:

a) measure the noise figure $F_{r\alpha}(\Gamma_{ns})$ using the setup of Fig. 1(b);
b) measure the reflection coefficient Γ_s of the tuner output.

From this it appears that, as previously stated, the method has been conceived with the objective of conveniently solving the problem of measuring the loss of a tuner when it is used as a source admittance transformer in a transistor noise figure measuring system. Actually, we note that the measurement a) is performed through noise figure measurements and step b) is already part of the experimental procedure for the determination of the noise (and gain) parameters of the transistor. In other words, the main advantage of the method is that it requires measurements of noise figure and reflection coefficient, which can be added to those already required by experimental procedures for computer-aided determination of transistor noise and gain parameters [1], [4], [5]. No further experimental effort is needed, the tuner loss being taken into account when carrying out computer-aided data processing.

The method can be easily implemented in a system for computer-assisted measurements. To this end, the setup of Fig. 1 is assembled by employing computer-controlled instruments and computer-driven microwave switches to realize the instrument connections. This can be accomplished using a test setup for complete characterization of transistors shown in Fig. 2, where the computer-assisted version of the system described in [6] is reported.

IV. Experimental Results

The losses of the coaxial slide-screw tuner have been measured through a computer-assisted implementation of the measuring system of Fig. 1.

As instrument system controller (via HP-IB), an HP 9836 computer with a HP 6942A multiprogrammer, is used. To switch ON and OFF the solid-state noise source HP 346B and to actuate the microwave switches and the receiver programmable step attenuator HP 8494H, the multiprogrammer is equipped with relay card HP 69730A. Narrow-band (spot) noise figures are

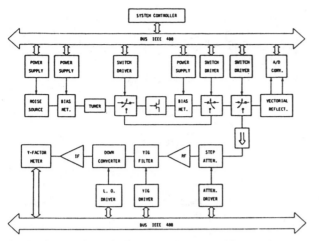

Fig. 2. Implementation of the tuner loss measuring method in the computer-controlled version of the transistor noise, gain, and scattering parameter measuring system proposed in [6].

TABLE I
APPROXIMATE VALUES OF THE VSWR OF THE TUNER OUTPUT VERSUS SETTING AND FREQUENCY

	8 GHz	10 GHz	12 GHz
P1	7.14	8.33	9.67
P2	3.76	4.31	4.74
P3	2.09	2.25	2.43
P4	1.38	1.44	1.51

measured with the known Y-factor method by a HP 436A power meter with a HP 8484A (0.1 nW–0.01 mW) probe through a 30-MHz filter (1-MHz bandwidth). Reflection coefficient measurements are carried out by using as the vectorial reflectometer a network analyzer in the reflection mode only and an A/D converter card HP 69751A of the multiprogrammer. The tuner tested is the 2–18-GHz slide double-screw Maury Microwave 2640D.

The tuner losses have been measured at 8, 10, and 12 GHz for four different spacings of one screw from the inner conductor of the tuner and for 57 positions of the carriage, as shown in Fig. 3. The other screw (the left and larger) was fully extracted. The relevant values of the VSWR are reported in Table I.

V. CONCLUSIONS

The model of the tuner tested through the method presented here is the same as that previously characterized by Strid through S-parameter measurements [11]; qualitative comparisons between the two methods show that the results are in good agreement. This does not hold, however, when measuring the loss of a tuner employed as a noise source admittance transformer in a noise parameter test set. The main advantage of the method proposed here is that it requires measurements which are part of the experimental procedure already required for characterization of noisy two-ports.

This is valid in particular for determining noise, gain, and scattering parameters with the noise parameter test set recently proposed [6], [7]. It is thus becoming an accurate and convenient method for the complete characterization of microwave transistors.

APPENDIX

COMPUTATION OF THE NOISE FIGURE $F_r(\Gamma_s)$ OF A RECEIVER WITH ISOLATED INPUT FOR ANY INPUT TERMINATION REFLECTION COEFFICIENT Γ_s AS FUNCTION OF THE NOISE FIGURE $F_r(0)$ IN INPUT MATCHED CONDITION

The noise temperature (noise power) of a receiver with an input isolator depends on the effective noise temperature of the isolator, which varies as Γ_s varies, and on the noise temperature (constant versus Γ_s) of the stages following the isolator. The isolator noise temperature is given by $T_a(\alpha_i(\Gamma_s)-1)$ [14], with $\alpha_i(\Gamma_s)=1/G_{ai}(\Gamma_s)$, where α_i and G_{ai} are the attenuation and the

available power gain, respectively. Thus, the receiver noise temperature is given by

$$
\begin{aligned}
T_{er}(\Gamma_s) &= T_a(\alpha_i(\Gamma_s)-1)+\alpha_i(\Gamma_s)\tau \\
&= \alpha_i(\Gamma_s)(T_a+\tau)-T_a
\end{aligned}
\tag{A1}
$$

where [12]

$$
\alpha_i(\Gamma_s)=\alpha_i(0)\frac{|1-\Gamma_r\Gamma_s|^2}{1-|\Gamma_s|^2}
\tag{A2}
$$

as can be derived from the relationship of the available gain

$$
G_{ai}(\Gamma_s)=\frac{|S_{21}|^2(1-|\Gamma_s|^2)}{|1-S_{11}\Gamma_s|^2(1-|S_{22}'(\Gamma_s)|^2)}.
\tag{A3}
$$

Recalling that $S_{12}=0$ for an isolator and, consequently, $S_{11}=\Gamma_r$ and $S_{22}'(\Gamma_s)=S_{22}$, the measured Y-factor is given by [13]

$$
Y=\frac{\mu_2[T_2+T_e(\Gamma_{s2})]}{\mu_1[T_1+T_e(\Gamma_{s1})]}
\tag{A4}
$$

where Γ_{s2} and Γ_{s1} are the reflection coefficients of the isolator input termination when the noise source is in the ON and OFF conditions, respectively, and Γ_2 and Γ_1 are the corresponding values of the mismatch factor given by

$$
\mu=\frac{(1-|\Gamma_s|^2)(1-|\Gamma_r|^2)}{|1-\Gamma_s\Gamma_r|^2}.
\tag{A5}
$$

From (A4) and (A5), we have

$$
Y=\frac{\mu_2(T_2-T_a)+\alpha_i(0)(T_a+\tau)(1-|\Gamma_r|^2)}{\mu_1(T_1-T_a)+\alpha_i(0)(T_a+\tau)(1-|\Gamma_r|^2)}.
\tag{A6}
$$

From (A6) and (A1), the effective noise temperature of the receiver in the input matched condition is derived in the form

$$
\begin{aligned}
T_{er}(0) &= \alpha_i(0)(T_a-\tau)-T_a \\
&= \frac{\mu_2'T_2-Y\mu_1'T_1}{Y-1}-\frac{(\mu_2'-1)-Y(\mu_1'-1)}{Y-1}T_a
\end{aligned}
\tag{A7}
$$

where μ_1' and μ_2' are given by

$$
\mu_1'=\frac{1-|\Gamma_{s1}|^2}{|1-\Gamma_{s1}\Gamma_r|^2}
\tag{A8}
$$

$$
\mu_2'=\frac{1-|\Gamma_{s2}|^2}{|1-\Gamma_{s2}\Gamma_r|^2}.
\tag{A9}
$$

Recalling that

$$
F=\frac{T_e}{T_0}-1
\tag{A10}
$$

from (A7) we obtain the receiver noise figure

$$
F_r(0)=\frac{\mu_2'T_2-Y\mu_1'T_1}{(Y-1)}-\frac{(\mu_2'-1)-Y(\mu_1'-1)}{(Y-1)T_0}T_a+1.
\tag{A11}
$$

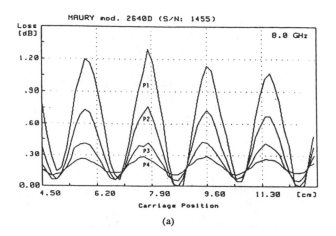

(a)

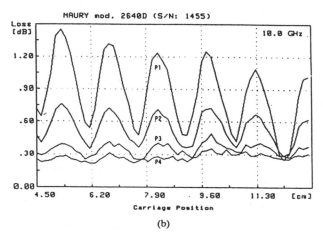

(b)

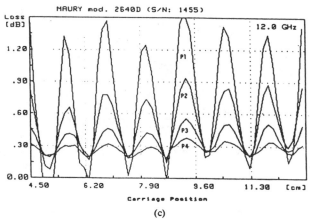

(c)

Fig. 3. Loss of the Maury Microwave 2640D tuner versus carriage position for four spacings P_1–P_4 of one screw from the inner conductor with the other screw fully extracted. here $P_1 = 0.050$, $P_2 = 0.150$, $P_3 = 0.350$ and $P_4 = 0.650$ in. The frequencies are (a) 8.0 GHz, (b) 10.0 GHz, and (c) 12.0 GHz.

If $T_1 = T_a$, (A11) becomes

$$F_r(0) = \frac{\mu_2' T_2 - (\mu_2' - 1) T_a - Y T_a}{(Y-1) T_0} + 1 \qquad (A12)$$

which, after some algebra, can be written

$$F_r(0) = \mu_2' F_r(\Gamma_s) + (\mu_2' - 1)\left(\frac{T_a}{T_0} - 1\right) \qquad (A13)$$

where

$$F_r(\Gamma_s) = \frac{T_2 - Y T_1}{(Y-1) T_0} + 1 \qquad (A14)$$

represents the receiver noise figure corresponding to the *nominal* value Γ_s of the input termination reflection coefficient, whose actual value is Γ_{s1} or Γ_{s2}.

REFERENCES

[1] R. Q. Lane, "The determination of device noise parameters," *Proc. IEEE*, vol. 57, pp. 1461–1462, Aug. 1969.

[2] M. Sannino, "On the determination of device noise and gain parameters," *Proc. IEEE (Lett.)*, vol. 67, pp. 1364–1366, Sept. 1979; also in *Low-noise Microwave Transistors and Amplifiers*, H. Fukui, Ed. New York: IEEE Press, 1981.

[3] G. Caruso and M. Sannino, "Computer-aided determination of microwave two-port noise parameters," *IEEE Trans. Microwave Theory Tech.*, vol. MTT-26, pp. 639–642, Sept. 1978.

[4] M. Sannino, "Simultaneous determination of device noise and gain parameters through noise measurements only," *Proc. IEEE (Lett.)*, vol. 68, pp. 1343–1345, Oct. 1980.

[5] G. Martines and M. Sannino, "Determination of microwave transistor noise and gain parameters through noise-figure measurements only," *IEEE Trans. Microwave Theory Tech.*, vol. MTT-30, pp. 1255–1259, Aug. 1982.

[6] E. Calandra, G. Martines, and M. Sannino, "Characterization of GaAs FET's in terms of noise, gain, and scattering parameters through a noise parameter test set," *IEEE Trans. Microwave Theory Tech.*, vol. MTT-32, pp. 231–237, Mar. 1984.

[7] G. Martines and M. Sannino, "Simultaneous determination of transistor noise, gain, and scattering parameters for amplifier design through noise figure measurements only," *IEEE Trans. Instrum. Meas.*, vol. IM-34, pp. 89–91, Mar. 1985.

[8] G. F. Engen, "A new method of characterizing amplifier noise performance," *IEEE Trans. Instrum. Meas.*, vol. IM-19, pp. 344–349, Nov. 1970.

[9] V. Adamian and A. Uhlir, "A novel procedure for receiver noise characterization," *IEEE Trans. Instrum. Meas.*, vol. IM-22, pp. 181–182, June 1973.

[10] V. Adamian and A. Uhlir, "Simplified noise evaluation of microwave receivers," *IEEE Trans. Instrum. Meas.*, vol. IM -33, pp. 136–140, June 1984.

[11] E. W. Strid, "Measurements of losses in noise matching networks," *IEEE Trans. Microwave Theory Tech.*, vol. MTT-29, pp. 247–253, Mar. 1981.

[12] M. Sannino, "Comments on 'Simultaneous determination of device noise and gain parameters through noise measurements only'," reply of the author," *Proc. IEEE (Lett.)*, vol. 70, pp. 100–101, Jan. 1982.

[13] G. Mamola and M. Sannino, "Source mismatch effects on measurements of linear two-port noise temperatures," *IEEE Trans. Instrum. Meas.*, vol. IM-24, pp. 239–242, Sept. 1975.

[14] G. Mamola and M. Sannino, "Source mismatch effects on noise measurements and their reduction," *Alta Frequenza*, vol. 44, pp. 233–239, May 1975.

An Improved Computational Method for Noise Parameter Measurement

MASATAKA MITAMA, MEMBER, IEEE, AND HIDEHIKO KATOH

Abstract—Conventional methods for noise parameter measurement for linear noisy two-ports have been improved by introducing a computational method for evaluating measured admittance errors. Derivation and comparison with a conventional method are given. Noise parameters of a packaged 0.5-μm gate-length GaAs MESFET (NE38806) were successfully measured using the proposed technique.

I. INTRODUCTION

AS IS WELL KNOWN, the noise behavior of a linear noisy two-port network can be characterized by the four noise parameters, F_0, G_0, B_0, and R_n, as

$$F = F_0 + \frac{R_n}{G_s}\left\{(G_s - G_0)^2 + (B_s - B_0)^2\right\} \qquad (1)$$

where

F noise figure,
Y_s $G_s + jB_s$ = source admittance,
F_0 minimum noise figure,
Y_0 $G_0 + jB_0$ = optimum source admittance that gives minimum noise figure,
R_n equivalent noise resistance.

The usual method [1] of obtaining noise parameters requires a special source admittance setting procedure, which is tedious.

An alternate method [2] consists of performing noise figure measurements for more than four arbitrary source admittances with a least squares method used for data processing. In this method, an estimated error (residue) ϵ'_i is defined for the ith measured data set (G_{si}, B_{si}, F_i) ($i = 1, 2, \cdots, N$) as

$$\epsilon'_i = \left| F_0 + \frac{R_n}{G_{si}}\left\{(G_{si} - G_0)^2 + (B_{si} - B_0)^2\right\} - F_i \right| \qquad (2)$$

which is shown by the dotted line ϵ'_i in Fig. 1. The weighted square sum of the estimated error is minimized. However, of the three measurement values G_{si}, B_{si}, and F_i, (2) considers the measured error in F only. That is, no consideration is made for measured G_s and B_s errors. Thus measured F errors at Y_s values away from Y_0 tend to be overvaluated, resulting in unsatisfactory computed noise parameter values, especially at higher microwave frequencies. In particular, according to Lane [2], computed results are highly sensitive to measurement errors in case of a large R_n.

Manuscript received September 7, 1978; revised January 2, 1979.
The authors are with the Central Research Laboratories, Nippon Electric Company, Ltd., 4-Miyazaki, Takatsu-ku, Kawasaki 213, Japan.

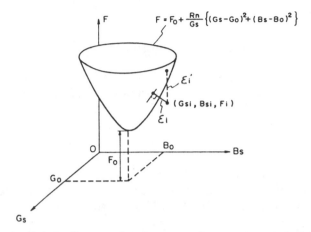

Fig. 1. Estimated error ϵ_i for the present least squares method. The dashed line designated by ϵ'_i represents the corresponding error used in the conventional method [2].

Assigning weighting values according to each measurement accuracy has been suggested [3], but is deemed impractical.

In the present paper, a computational method is introduced for evaluating measured source admittance errors, as well as the measured F error.

Another reason for unsatisfactory results with the conventional methods is the random differences between measured Y_s values and actual Y_s values imposed onto the two-port network. In the present method, a low-loss microstrip tuner is used, which helps to decrease the errors introduced by these differences.

II. LEAST SQUARES METHOD CONSIDERING Y_s ERRORS

Let $G_s = x$, $B_s = y$, $F = z$, $F_0 = a_1$, $R_n = a_2$, $G_0 = a_3$, and $B_0 = a_4$ for simplicity of notation. Let the estimated values for the measured set (X_i, Y_i, Z_i) ($i = 1, 2, \cdots, N$) in the x, y, and z coordinates be $(\hat{x}_i, \hat{y}_i, \hat{z}_i)$ and the estimated values for parameters a_1, a_2, a_3, and a_4 be $\hat{a}_1$, $\hat{a}_2$, $\hat{a}_3$, and $\hat{a}_4$. Furthermore, define the estimated errors (residues) as

$$V_{xi} = \hat{x}_i - X_i \quad V_{yi} = \hat{y}_i - Y_i \quad V_{zi} = \hat{z}_i - Z_i \qquad (3)$$

$$V_{a1} = \hat{a}_1 - a_1^0 \quad V_{a2} = \hat{a}_2 - a_2^0 \quad V_{a3} = \hat{a}_3 - a_3^0 \quad V_{a4} = \hat{a}_4 - a_4^0 \qquad (4)$$

where a_1^0, a_2^0, a_3^0, and a_4^0 are zeroth-order approximation values. These estimated values must satisfy the functional relation of

$$G(\hat{x}_i, \hat{y}_i, \hat{z}_i; \hat{a}_1, \hat{a}_2, \hat{a}_3, \hat{a}_4) = 0, \qquad i = 1, 2, \cdots, N \qquad (5)$$

Reprinted from *IEEE Transactions on Microwave Theory and Techniques*, vol. MTT-27, no. 6, pp. 612–615, June 1979.

where

$$G(x, y, z; a_1, a_2, a_3, a_4)$$
$$= -z + a_1 + \frac{a_2}{x} \left\{ (x - a_3)^2 + (y - a_4)^2 \right\}. \quad (6)$$

Equation (6) is derived by transposing the left-hand side value of (1) to the right-hand side values.

The term "estimated values" is defined such that $x_i, \cdots, a_4$ minimize

$$S = \sum_i \left(w_{xi} V_{xi}^2 + w_{yi} V_{yi}^2 + w_{zi} V_{zi}^2 \right)_{\min} \quad (7)$$

where w_{xi}, w_{yi}, and w_{zi} are weights to be determined according to measurement accuracies, and the subscript min denotes the minimum value within the parentheses.

When $w_{xi} = w_{yi} = w_{zi} = 1$, the quantity

$$\epsilon_i = \left\{ \left(w_{xi} V_{xi}^2 + w_{yi} V_{yi}^2 + w_{zi} V_{zi}^2 \right)_{\min} \right\}^{1/2} \quad (8)$$

represents the length of a segment of a line normal to the quasi-elliptic paraboloid represented by (5) projected from the measured point (X_i, Y_i, Z_i), as shown by the solid line ϵ_i in Fig. 1.

Note that the minimum value within the parentheses in (7) occurs when the line drawn from the measured point is perpendicular to the quasi-elliptic paraboloid. With (8), the evaluation of the measured Y_s error as well as that of the measured F error becomes possible, as contrasted to the conventional method [2].

Assuming that estimated errors are small, (5) can be expanded in a Taylor series to a first-order approximation as

$$G_0^i + G_x^i V_{xi} + G_y^i V_{yi} + G_z^i V_{zi}$$
$$+ G_{a1}^i V_{a1} + G_{a2}^i V_{a2} + G_{a3}^i V_{a3} + G_{a4}^i V_{a4} = 0 \quad (9)$$

where

$$G_0^i = G(X_i, Y_i, Z_i; a_1^0, a_2^0, a_3^0, a_4^0) \quad (10)$$

$$G_x^i = \left. \frac{\partial G}{\partial x} \right|_{x = X_i, y = Y_i, z = Z_i; a_1 = a_1^0, a_2 = a_2^0, a_3 = a_3^0, a_4 = a_4^0} \quad (11)$$

and $G_y^i, \cdots, G_{a4}^i$ are similar to the above. Using (9), (8) can be rewritten as

$$\epsilon_i = \sqrt{w_i} \, |d_i| \quad (12)$$

where

$$d_i = -\left(G_{a1}^i V_{a1} + G_{a2}^i V_{a2} + G_{a3}^i V_{a3} + G_{a4}^i V_{a4} + G_0^i \right) \quad (13)$$

$$1/w_i = (G_x^i)^2 / w_x + (G_y^i)^2 / w_y + (G_z^i)^2 / w_z. \quad (14)$$

The derivation of (12) is shown in the Appendix.

Then, by the least squares method, a set of linear equations is obtained as

$$\frac{\partial S}{\partial V_{a1}} = 2 \sum_i \left[w_i d_i G_{a1}^i \right] = 0$$

$$\cdot$$
$$\cdot$$
$$\cdot$$

$$\frac{\partial S}{\partial V_{a4}} = 2 \sum_i \left[w_i d_i G_{a4}^i \right] = 0. \quad (15)$$

Fig. 2. Microstrip tuner.

In order to solve (15), the initial values a_1^0, a_2^0, a_3^0, and a_4^0 were obtained by using the conventional method described in [2]. Then (15) was solved by an iteration method using (3) and (4) to obtain the V_{a1}, V_{a2}, V_{a3}, and V_{a4} that make S minimum. In the present calculation, w_{xi}, w_{yi}, and w_{zi} were set to unity.

III. Measurement

A. Microstrip Tuner

Usually, commercial stub tuners are widely used to provide Y_s values necessary for noise parameter measurements. However, good reproducibility is difficult to obtain with these stub tuners. In addition, dissipation loss must be calibrated for each Y_s value, which is tedious. To avoid these requirements, an integrated microstrip tuner, shown in Fig. 2, has been used. To a 50-Ω main line, eight open shunt stubs (lands) were constructed on a 0.8-mm-thick teflon-glass fabric board (Di Clad 522). Each stub consisted of ten 0.8-mm × 1-mm lands with 0.2-mm separation. By connecting appropriate lands with an adhesive conductive sheet, necessary Y_s values can be obtained on an arbitrary basis. Y_s values were recorded with an X-Y recorder prior to the noise figure measurement. Fig. 3 shows an example. Circles in the figure indicate the source admittance points realized. Each solid line shows the locus of discrete change of the source admittance when one particular stub length is adjusted by connecting the lands one by one with an adhesive conductive sheet. The reproducibility using this arrangement was superior to that using a stub tuner. The dissipation loss, calculated by using generalized scattering matrices [4], was 0.02 dB for the particular Y_0 circuit configuration; hence the circuit loss was neglected throughout.

B. Available Gain

In order to minimize the Y_s value inaccuracy caused by connector disconnection, a two-port network available gain was calculated using both the premeasured Y_s value and the two-port network S parameters. Since circuit losses will result in an actual available gain smaller than that computed, receiving system noise contribution will not be overevaluated. The output matching circuit con-

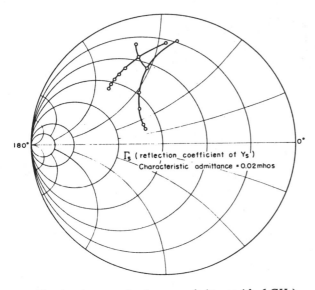

Fig. 3. An example of source admittance ($f = 6$ GHz).

sisted of two quarter-wave transformer sections, common throughout the measurement, with less than 2.5 output VSWR. In a unilateral two-port network case, as is the case of an FET, the input impedance of the two-port network can be considered as being independent of the load impedance. Output circuit loss was neglected throughout.

C. NF Measurement

Noise figure (NF) measurements were made at 6, 7, and 8 GHz with an HP 342A automatic NF meter for a packaged 0.5-μm gate-length GaAs MESFET (NE 38806). To reduce the receiving system noise contribution, a low-noise preamplifier (NF < 3 dB, Gain > 17 dB) was placed directly after the FET output matching circuit. Although the measured NF was the double-sideband (signal and image) value, it can be considered as a single-sideband value, since signal and image separation is only 60 MHz.

IV. CALCULATION RESULTS

A. Comparison with Conventional Method

In Fig. 4, the computed noise parameter standard deviation as a function of the number of measurement points N is shown both for the present and for the conventional method [2]. For each N, ten different trains of data sets were randomly selected from a total of fourteen measured points. Then, noise parameters and standard deviations were calculated for each N case. The figure shows that a smaller noise parameter deviation is obtainable with the present method for any given number of measurement points. Fig. 5 shows computed noise parameter sensitivities to an individual data set (G_{si}, B_{si}, F_i) accuracy. Noise parameters are calculated by extracting each data set one by one sequentially from fourteen data sets. It is clear that the individually measured data is less sensitive to the computed parameters with the present method. The small deviation in

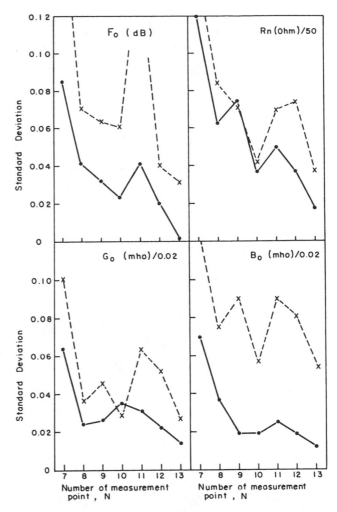

Fig. 4. Computed noise parameter standard deviation comparison with conventional method ($f = 6$ GHz) [2]. The solid line curve corresponds to the present method, while the dashed line curve corresponds to the conventional method.

TABLE I
NOISE PARAMETERS OF A 0.5-μm GATE-LENGTH GAAS MESFET
(NE 38806 OR 2SK 124). $V_p = -2.21$ V, $I_{DSS} = 51$ mA

FET	I_{DS} (mA)	V_{DS} (V)	f (GHz)	F_0 (dB)	$|\Gamma_0|$	ϕ_0 (deg.)	$R_n/50$ (ohm)
NE 38806	10	3	6	1.88	0.660	114.3	0.53
"	"	"	7	1.93	0.642	140.2	0.30
"	"	"	8	2.10	0.592	164.0	0.12

computed noise parameters and the insensitivity to individual data with the present method should be attributable to the new estimated error evaluation method described in Section II.

B. Measured Noise Parameters for NE 38806

Noise parameters for a packaged 0.5-μm gate-length GaAs MESFET (NE 38806) were measured with the present method. Results are listed in Table I. Measurement points for 6, 7, and 8 GHz were 14, 13, and 15, respectively. Because of rather small R_n values, 26.5–6 Ω, constant noise figure circles [5] indicated a mild dependence on Γ_s (reflection coefficient of Y_s) planes.

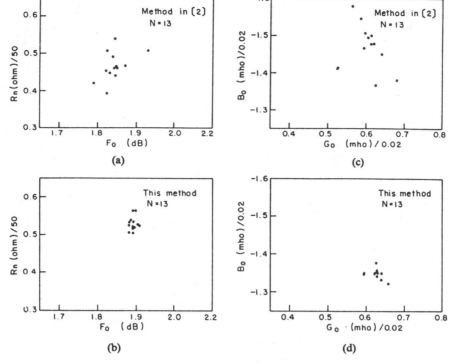

Fig. 5. Comparison of the sensitivity in computed noise parameters to individual data ($f = 6$ GHz). (a) and (c) are the conventional method [2]; (b) and (d) are the present method.

V. CONCLUSION

Conventional methods [1], [2] for noise parameter measurement have been improved by introducing a new computational method for evaluating measured source admittance errors as well as the measured noise figure error and by utilizing an integrated microstrip tuner. It was demonstrated that a smaller deviation in the computed noise parameters and insensitivity to individual data were achieved with the present method when compared with the conventional method [2]. Noise parameters of a packaged 0.5-μm gate-length GaAs MESFET (NE 38806) have been successfully measured with the new method.

APPENDIX
DERIVATION OF (12)

Consider the terms within the parentheses of (8),

$$s_{hi} = w_{xi} V_{xi}^2 + w_{yi} V_{yi}^2 + w_{zi} V_{zi}^2. \quad \text{(A.1)}$$

In the following, the superscripts and subscripts i's are dropped throughout for simplicity. Using (9) and (13), (A.1) can be rewritten as

$$s_h = w_x V_x^2 + w_y V_y^2 + \frac{w_z}{G_z^2}(d - G_x V_x - G_y V_y)^2. \quad \text{(A.2)}$$

The minimum of s_h occurs when $\partial s_h / \partial V_x = 0$ and $\partial s_h / \partial V_y = 0$, simultaneously. Thus we obtain a set of linear equations as

$$\left(w_x + w_z \frac{G_x^2}{G_z^2}\right)V_x + \left(w_z \frac{G_x G_y}{G_z^2}\right)V_y = \left(w_z \frac{G_x}{G_z^2}\right)d \quad \text{(A.3)}$$

$$\left(w_z \frac{G_x G_y}{G_z^2}\right)V_x + \left(w_y + w_z \frac{G_y^2}{G_z^2}\right)V_y = \left(w_z \frac{G_y}{G_z^2}\right)d. \quad \text{(A.4)}$$

Solving (A.3) and (A.4), we obtain

$$V_x = w\left(\frac{G_x}{w_x}\right)d \quad V_y = w\left(\frac{G_y}{w_y}\right)d \quad \text{and} \quad V_z = w\left(\frac{G_z}{w_z}\right)d. \quad \text{(A.5)}$$

Then, from (A.1) and (A.5), we have

$$\epsilon = \sqrt{(s_h)_{min}} = \sqrt{w}\,|d|.$$

ACKNOWLEDGMENT

The authors would like to thank Dr. K. Ayaki and the members of the Microwave Circuit Group, Central Research Laboratories, Nippon Electric Company, Ltd., for their encouragement and guidance.

REFERENCES

[1] IRE Subcommittee on Noise, "IRE standards on methods of measuring noise in linear twoports, 1959," Proc. IRE, vol. 48, pp. 60–68, Jan. 1960.
[2] R. Q. Lane, "The determination of device noise parameters," Proc. IEEE, vol. 57, pp. 1461–1462, Aug. 1969.
[3] M. S. Gupta, "Determination of the noise parameters of a linear 2-port," Electron. Lett., vol. 6, pp. 543–544, Aug. 20, 1970.
[4] G. E. Bodway, "Two port power flow analysis using generalized scattering parameters," Microwave J., vol. 10, pp. 61–69, May 1967.
[5] H. Fukui, "Available power gain, noise figure, and noise measure of two-ports and their graphical representations," IEEE Trans. Circuit Theory, vol. CT-13, pp. 137–142, June 1966.

The Determination of Device Noise Parameters

Abstract—A novel noise measurement technique is outlined which results in data that directly give the noise parameters of the test device when processed by a simple computer program.

The usual method[1] of obtaining device noise parameters (F_{min}, R_n, G_{opt}, B_{opt}) entails an experimental search for the "minimum" noise figure and the source admittance that results in this "minimum."

There are two reasons why this search technique is unsatisfactory in practice. First, it is tedious, and it is inaccurate in determining G_{opt}, B_{opt} since the partial derivative of noise factor, with respect to source admittance, is zero at the noise factor minimum. Second, the presence of transformation-dependent loss in the input matching network biases the obtained value of Y'_{opt}, usually in the direction of $G'_{opt} > G_{opt}$ and $B'_{opt} < B_{opt}$, since this minimizes the noise factor insertion loss product.

The methods that follow avoid these practical difficulties and by use of a simple computer program can simplify noise characterization. The noise behavior of a two-port may be expressed as

$$F = F_{min} + \frac{R_n}{G_s}|Y_s - Y_{opt}|^2 \qquad (1)$$

where $Y_{opt} = G_{opt} + jB_{opt}$ is the particular source admittance realizing F_{min}, and $Y_s = G_s + jB_s$ is the source admittance.

In principle, four (nonsingular) measurements of noise factor from different source admittances will determine the four real numbers (F_{min}, R_n, G_{opt}, B_{opt}), the noise parameters. However, since there are bound to be some experimental errors both in the measurement of the noise factor ($\sim \pm 10$ percent) and in the measurement of the admittance at which it occurs (± 10 percent), a few (three) redundant measurements will allow a statistical smoothing of the experimentally determined surface.

In order to use a readily available IBM subroutine[2] for simultaneous equation solution, (1) may be cast in a form[3] that is linear with respect to four new parameters A, B, C, and D:

$$F = A + BG_s + \frac{C + BB_s^2 + DB_s}{G_s}. \qquad (2)$$

Equation (2) represents a hyperplane having the four new parameters A, B, C, and D, where

$$F_{min} = A + \sqrt{4BC - D^2} \qquad (3)$$

$$R_n = B \qquad (4)$$

$$G_{opt} = \frac{\sqrt{4BC - D^2}}{2B} \qquad (5)$$

$$B_{opt} = \frac{-D}{2B}. \qquad (6)$$

We have usually a priori knowledge that both F_{min} and G_{opt} are positive thus allowing proper selection of the roots in (3) and (5).

A least-squares fit of the seven observed noise factors to the plane of (2) is sought; therefore, the following error criterion is established:

$$\varepsilon \equiv \frac{1}{2}\sum_{i=1}^{7} W_i\left[A + B\left(G_i + \frac{B_i^2}{G_i}\right) + \frac{C}{G_i} + \frac{DB_i}{G_i} - F_i\right]^2 \qquad (7)$$

where W_i is a weighting factor to be used if certain data are known to be less reliable than the average. Then,

$$\frac{\partial \varepsilon}{\partial A} = \sum_{i=1}^{7} W_i P = 0 \qquad (8)$$

Manuscript received May 6, 1969; revised June 5, 1969. This work was supported in part by Wright-Patterson AFB under Contract F33615-68-C-1612.

[1] "IRE standards on methods of measuring noise in linear twoports, 1959," *Proc. IRE*, vol. 48, pp. 60–68, January 1960.
[2] Subroutine Sim Q.
[3] H. Fukui, "The noise performance of microwave transistors," *IEEE Trans. Electron Devices*, vol. ED-13, pp. 329–341, March 1966.

TABLE I

Unit No.	V_c(V)	I_c(mA)	G_{opt} (mmhos)	B_{opt} (mmhos)	F_{min} (dB)	R_n (ohms)
TL01 JA155A/1 T0–18 pkg. 500 MHz CE Fixed Bias						
7	5.0	3.0	5.69	−5.32	2.77	61.96
8	5.0	3.0	4.31	−5.07	2.24	53.86
9	5.0	3.0	5.28	−4.32	2.76	66.62
10	5.0	3.0	4.84	−5.15	2.23	51.71
11	5.0	3.0	4.10	−5.27	2.26	59.69
19	5.0	3.0	5.20	−4.94	2.15	50.60
TL01 JA155A/1 T0–18 pkg. 1016 MHz CE Fixed Bias						
19	5.0	3.0	13.20	−6.18	3.11	54.19
21	5.0	3.0	11.29	−6.77	3.02	55.65
23	5.0	3.0	11.59	−6.12	3.22	53.18
24	5.0	3.0	10.49	−7.05	3.04	54.71
25	5.0	3.0	7.26	−7.12	2.85	53.58

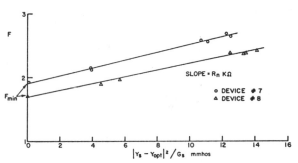

Fig. 1. Experimental data scatter.

$$\frac{\partial \varepsilon}{\partial B} = \sum_{i=1}^{7} W_i\left(G_i + \frac{B_i^2}{G_i}\right)P = 0 \qquad (9)$$

$$\frac{\partial \varepsilon}{\partial C} = \sum_{i=1}^{7} W_i\frac{P}{G_i} = 0 \qquad (10)$$

$$\frac{\partial \varepsilon}{\partial D} = \sum_{i=1}^{7} W_i\frac{B_i}{G_i}P = 0 \qquad (11)$$

where

$$P = \left[A + B\left(G_i + \frac{B_i^2}{G_i}\right) + \frac{C}{G_i} + \frac{DB_i}{G_i} - F_i\right].$$

Equations (8) through (11) are solved by an IBM 1130 computer yielding the noise parameters via (3) through (6).

A set of gain parameters[4] may be obtained by substitution of a signal generator for the noise source (or a large increase in excess noise power), viz.,

$$\frac{1}{G_{av}} = \frac{1}{G_{MA}} + \frac{R_{eq}}{G_s}|Y_s - Y_{opt,g}|^2 \qquad (12)$$

where $R_{eq} = g_{22}/|Y_{21}|^2$, G_{av} = available gain, and G_{MA} = maximum available gain.

Since the loss of the input network may be obtained for each source admittance, this loss in decibels is subtracted from the measured noise or 1/gain factor in decibels prior to processing, thus, removing the aforementioned bias error.

EXPERIMENTAL RESULTS

A developmental microwave transistor type was characterized at 0.5 GHz and 1 GHz using the program described; the results appear in Table I.

[4] ——, "Available power gain, noise figure, and noise measure of two-ports and their graphical representations," *IEEE Trans. Circuit Theory*, vol. CT-13, pp. 137–142, June 1966.

Reprinted from *Proceedings of the IEEE*, vol. 57, no. 8, pp. 1461–1462, August 1969.

Fig. 1 displays the scatter in the experimental data about the computed best straight line.

Shortcoming of the Method as Presently Used

This program was originally used for a UHF MOS structure with somewhat disappointing results due to an apparently high sensitivity to measurement error for devices with large values of R_n ($> 150 \, \Omega$). The author invites suggestions for a more sophisticated fitting routine that may result in lower error sensitivity.

Acknowledgment

The author wishes to thank R. Rohrer for suggesting the least-squares fitting program, M. Purnaiya for supplying the experimental data, A. Keet for writing the FORTRAN program, and J. Gibbons of Stanford University for general encouragement.

RICHARD Q. LANE
Res. and Develop. Lab.
Fairchild Semiconductor
Palo Alto, Calif. 94304

On the Noise Parameters of Isolator and Receiver with Isolator at the Input

MARIAN W. POSPIESZALSKI, SENIOR MEMBER, IEEE

Abstract —Noise parameters of an isolator and those of a receiver with an isolator at the input are reviewed. Some comments on recently published results are offered.

I. INTRODUCTION

Isolators are very commonly used in low-noise receivers as well as in noise measuring systems (for instance, [1]–[5]). Usually their purpose is to isolate either the noise source or the receiver from the rest of the system. In these cases, the noise properties of either the isolator alone or the receiver with the isolator at the input need to be known. This paper offers a brief discussion of the noise properties of these two-ports and gives closed-form expressions in some idealized cases for the set of noise parameters, namely minimum noise temperature T_{min}, optimum source reflection coefficient Γ_{opt}, and noise parameter N as defined in [9]. A short discussion of some of the recently published results [4], [5], [11], [15] is also given.

II. THEORY

Consider a linear, noisy system schematically presented in Fig. 1. Signal parameters of both an isolator and a receiver are represented by chain matrices $[A_I]$ and $[A_R]$ and their noise parameters by correlation matrices $[C_{AI}]$ and $[C_{AR}]$, respectively [6]. An isolator is a passive, nonreciprocal, linear two-port with thermal noise generators only and, therefore, its noise parameters can be derived from its signal parameters [7]. The appropriate equivalent networks with pertinent formulas [7], [8] are given in Fig. 2. Then the correlation matrix $[C_A]$ completely characterizing the noise parameters of the system at the input port of the isolator is [8]

$$[C_A] = [C_{AI}] + [A_I][C_{AR}][A_J]^\dagger \qquad (1)$$

where the "dagger" designates the complex conjugate of the transpose of $[A_I]$ matrix. Any desired set of noise parameters can be derived from $[C_A]$ (for instance, [6], [8]–[10]).

It should be stressed that this approach is not limited by the particular realization of an isolator as, for instance, a Faraday rotation isolator or an isolator made of a circulator with one port terminated. The noise properties of both isolators are the same if they are at the same physical temperature and their two-port signal parameters are the same.

Although the formulas presented in Figs. 1 and 2 and also (1) lend themselves easily to computer implementation (for instance, [13], [14]), and, therefore, are convenient to use in computer-aided design and/or computer-aided measurement, it is very instructive to discuss the conventional noise parameters of an ideal isolator, which is equivalent to an ideal circulator with one port terminated (Fig. 3(a)). It follows directly from Twiss's [7] general approach or from simple physical reasoning that the noise parameters of an ideal isolator are

$$T_{min} = 0, \quad \Gamma_{opt} = 0, \quad N = \frac{T_a}{4T_0} \qquad (2)$$

Manuscript received June 24, 1985; revised November 12, 1985.
M. W. Pospieszalski is with the National Radio Astronomy Observatory, Charlottesville, VA 22903. The National Radio Astronomy Observatory is operated by Associated Universities, Inc., under contract with the National Science Foundation.
IEEE Log Number 8407183.

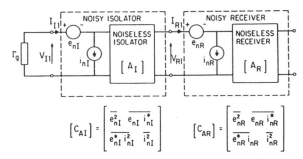

$$[C_{AI}] = \begin{bmatrix} \overline{e_{nI}^2} & \overline{e_{nI}\, i_{nI}^*} \\ \overline{e_{nI}^*\, i_{nI}^2} & \overline{i_{nI}^2} \end{bmatrix} \qquad [C_{AR}] = \begin{bmatrix} \overline{e_{nR}^2} & \overline{e_{nR}\, i_{nR}^*} \\ \overline{e_{nR}^*\, i_{nR}} & \overline{i_{nR}^2} \end{bmatrix}$$

Fig. 1. A cascade connection of isolator and receiver.

where

T_{min}	minimum noise temperature,
Γ_{opt}	optimum reflection coefficient of the source,
$T_0 = 290$ K	standard temperature,
T_a	physical temperature of a circulator termination, (or physical temperature of an isolator),
N	noise parameter defined in [9].

It is instructive to give physical interpretation of the noise parameters given by (2). An ideal isolator emits a noise wave from its input port, which is totally absorbed by the source if $\Gamma_g = \Gamma_{opt} = 0$. In this case, no noise generated by the isolator appears at its output and $T_{min} = 0$. If $\Gamma_g \neq 0$, part of the noise is reflected back and appears at the isolator output, which gives rise to parameter $N > 0$.

Small losses L of an isolator in the forward direction can be modeled accurately by a cascade connection of an ideal isolator and a matched attenuator, as shown in Fig. 3(b). In this case of a slightly lossy isolator, the noise parameters are

$$T_{min} = T_a(L-1) \quad \Gamma_{opt=0} \quad N = \frac{T_a + T_{min}}{4T_0}. \qquad (3)$$

If an isolator cannot be described by these simple models, its noise properties are best treated by the general approach outlined in Fig. 2.

Finally, if an ideal isolator is followed by a receiver as showed in Fig. 4 (small losses of an isolator can be modeled as part of a receiver), the noise parameters of this system at the input port of the isolator are

$$T_{min} = T_R, \quad \Gamma_{opt} = 0, \quad N = \frac{T_a + T_{min}}{4T_0}. \qquad (4)$$

Therefore, the noise temperature T_n of the system of Fig. 4 for arbitrary Γ_g is

$$T_n = T_{min} + (T_a + T_{min})\frac{|\Gamma_g|^2}{1 - |\Gamma_g|^2}. \qquad (5)$$

It is clear from (5) why only the magnitude of the source reflection coefficient needs to be known and also why it is advantageous to keep the termination of a circulator cold.

If the simple model of Fig. 4 does not apply, the use of (1) is recommended, which requires the knowledge of all four noise parameters of a receiver, two-port signal parameters of an isolator, and its physical temperature.

III. COMMENTS

The equivalence of noise behavior between a Faraday rotation isolator and an isolator made of a circulator with one port

Reprinted from *IEEE Transactions on Microwave Theory and Techniques*, vol. MTT-34, no. 4, pp. 451–453, April 1986.

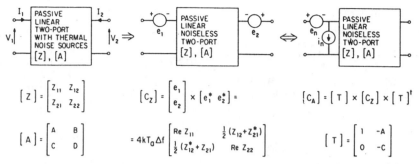

$$[Z] = \begin{bmatrix} Z_{11} & Z_{12} \\ Z_{21} & Z_{22} \end{bmatrix}$$

$$[C_Z] = \begin{bmatrix} e_1 \\ e_2 \end{bmatrix} \times \begin{bmatrix} e_1^* & e_2^* \end{bmatrix} =$$

$$[C_A] = [T] \times [C_Z] \times [T]^t$$

$$[A] = \begin{bmatrix} A & B \\ C & D \end{bmatrix} = 4kT_0\Delta f \begin{bmatrix} \operatorname{Re} Z_{11} & \frac{1}{2}(Z_{12} + Z_{21}^*) \\ \frac{1}{2}(Z_{12}^* + Z_{21}) & \operatorname{Re} Z_{22} \end{bmatrix}$$

$$[T] = \begin{bmatrix} 1 & -A \\ 0 & -C \end{bmatrix}$$

Fig. 2. An isolator as a passive, nonreciprocal, linear two-port with thermal noise sources only. The formulas given are from [7] and [8].

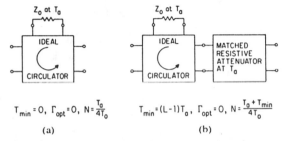

Fig. 3. (a) An ideal isolator, as an ideal circulator with one port terminated and its noise parameters. (b) An approximate model of slightly lossy isolator and its noise parameters.

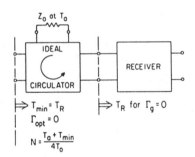

Fig. 4. An approximate model of a cascade connection of isolator and receiver and its noise parameters.

terminated has been discussed in a recent paper [11], where the approach presented by Siegman in an earlier work [12] has been reviewed. The conclusions of both papers [11], [12] on this subject follow directly from the much more general result of Twiss [7].

In [4], [5], and [15], expressions for the noise figure of a system with the isolator at the input for arbitrary Γ_g are given. It should be noted that these expressions are valid only if the physical temperature of the isolator T_a is equal to the standard temperature $T_0 = 290$ K. This condition was not clearly stated in [4], [5], and [15].

The formulas presented in this paper follow directly from results published many years ago [7], [8]. The author feels, however, in view of recently published [4], [5], [11], [15] that it is worthwhile to present these in the form congruent with that commonly used in the description of noisy two-ports.

REFERENCES

[1] S. Weinreb and A. R. Kerr, "Cryogenic cooling of mixers for millimeter and centimeter wavelength," *IEEE J. Solid-State Circuits*, vol. SC-8, pp. 58–63, Feb. 1973.

[2] S. Weinreb, "Low-noise cooled GASFET amplifiers," *IEEE Trans. Microwave Theory Tech.*, vol. MTT-28, pp. 1041–1053, Oct. 1980.

[3] A. D. Sutherland and A. van der Ziel, "Some pitfalls in millimeter-wave noise measurement utilizing a cross-correlation receiver," *IEEE Trans. Microwave Theory Tech.*, vol. MTT-30, pp. 715–718, May 1982.

[4] E. F. Calandra, G. Martines, and M. Sannino, "Characterization of GaAs FET's in terms of noise, gain, and scattering parameters through a noise parameter test set," *IEEE Trans. Microwave Theory Tech.*, vol. MTT-32, pp. 231–237, Mar. 1984.

[5] G. Martines and M. Sannino, "Simultaneous determination of transistor noise, gain, and scattering parameters for amplifier design through noise figure measurement only," *IEEE Trans. Instrum. Meas.*, vol. IM-34, pp. 89–91, Mar. 1985.

[6] H. Rothe and W. Dahlke, "Theory of noisy four poles," *Proc. IRE*, vol. 44, pp. 811–818, June 1956.

[7] R. Q. Twiss, "Nyquist's and Thevenin's theorems generalized for nonreciprocal linear networks," *J. Appl. Phys.*, vol. 26, no. 5, pp. 599–602, May 1955.

[8] H. Hillbrand and P. Russer, "An efficient method for computer-aided noise analysis of linear amplifier networks," *IEEE Trans. Circuits Syst.*, vol. CAS-23, pp. 235–238, Apr. 1976.

[9] J. Lange, "Noise characterization of linear two-ports in terms of invariant parameters," *IEEE J. Solid-State Circuits*, vol. SC-2, pp. 37–40, June 1967.

[10] P. Penfield, "Wave representation of amplifier noise," *IRE Trans. Circuit Theory*, vol. CT-9, pp. 84–86, Mar. 1962.

[11] A. D. Sutherland, "On the inherent noise of an ideal two-port isolator," *IEEE Trans. Microwave Theory Tech.*, vol. MTT-30, p. 830, pp. 830–832, May 1982.

[12] A. E. Siegman, "Thermal noise in microwave systems, Part I," *Microwave J.*, vol. 4, pp. 81–90, Mar. 1961.

[13] D. L. Fenstermacher, "A computer-aided analysis routine including optimization for microwave circuits and their noise," National Radio Astronomy Observatory Electronics Div. Internal Rep. No. 217, July 1981.

[14] J. Granlund, "FARANT on the HP9816 computer," National Radio Astronomy Observatory Electronics Div. Internal Rep. no. 250, Aug. 1984.

[15] K. Mishima and Y. Sawayama, "Comments on 'Simultaneous determination of device noise and gain parameters through noise measurement only,'" *Proc. IEEE*, vol. 70, p. 100, Jan. 1982.

Editor's Note: Corrections to this article that appeared in *IEEE Transactions on Microwave Theory and Techniques*, vol. MTT-34, no. 6, p. 746, June 1986, have been incorporated into this reprint.

A New Method For Calculating the Noise Parameters of MESFET's and TEGFET's

A. CAPPY, M. SCHORTGEN, AND G. SALMER

Abstract—Analytical formulas for the intrinsic noise sources of both MESFET's and TEGFET's are derived from a numerical noise modeling. Using these expressions the calculated noise figure is in good agreement with experimental findings. The influence of gd and Cgd on the noise figure is then pointed out and a comparison between the kf factor of MESFET's and TEGFET's is presented.

I. INTRODUCTION

DURING THE PAST three years, works on quarter micrometer gate conventional MESFET's and half micrometer (or less) TEGFET's have demonstrated the very attractive possibilities of these devices in the field of low-noise amplifiers beyond 20 GHz.

In order to study the noise properties of these devices the Fukui Formula [5] is classicaly used. Nevertheless, the theoretical derivation of this simple expression is based upon drastic assumptions [4], and this simple formula can be unreliable, especially in the millimeter-wave range. In order to overcome this difficulty, the purpose of this letter is to describe a new method for calculating the noise parameters (NF, R_n, G_n, Γ_{opt}) of both MESFET's and TEGFET's.

II. CALCULATION OF THE NOISE FIGURE

In order to calculate the noise figure and the other noise parameters of a field-effect transistor, the short circuit gate and drain noise currents $\langle ig^2 \rangle$ and $\langle id^2 \rangle$ as well as the correlation coefficient C between these noise sources must be known. The theoretical determination of these quantities has been described in a previous paper [1] in the case of conventional MESFET's. Recently a similar analysis has been used to study the noise performances of TEGFET's and to compare the possibilities of these two structures in the millimeter-wave range [2]. These noise modelings are based upon the main following assumptions:

Uncorrelated noise sources are assumed to be distributed in each section of the channel. For given dc drain current and gate-to-source voltage, the short circuits gate and drain noise currents are calculated from the knowledge of the noise source in each incremental section using a numerical FET modeling [1]. To do that, the drain–source voltage and the stored charge fluctuations $(\Delta V_{ds})_j$ and $(\Delta Q)_j$ resulting from a small current fluctuation in the jth section of the channel are computed. This calculation is repeated for each section [1].

The open-circuit voltage $\langle v^2 \rangle$, the gate noise current $\langle ig^2 \rangle$,

Manuscript received January 21, 1985; revised March 22, 1985.

The authors are with Centre Hyperfréquences et Semiconducteurs, L.A. C.N.R.S. no. 287, U.E.R. D'I.E.E.A., Bât. P4, 59655 Villeneuve D'Ascq Cedex, France.

and the correlation coefficient $C' = \langle ig \cdot v* \rangle / \sqrt{\langle ig^2 \rangle \cdot \langle v^2 \rangle}$ are then computed by a quadratic summation [1]. The short-circuit drain noise current $\langle id^2 \rangle$ and the correlation coefficient between the short-circuit gate and drain noise currents are easily calculated using the following expressions:

$$\langle id^2 \rangle = |y_{22}|^2 \cdot \langle v^2 \rangle \tag{1}$$

$$C = \frac{y_{22}*}{|y_{22}|} \cdot C' \tag{2}$$

where * denotes the complex conjugate, $y_{22} = gd + j\omega \cdot Cgd$ is the intrinsic admittance parameter, gd the output conductance and CgD the gate-to-drain capacitance.

When the intrinsic noise sources are known, the noise figure and the other noise parameters are calculated using the rigorous analysis of Rothe and Dahlke [3]. In this analysis the intrinsic ($\langle id^2 \rangle$, $\langle ig^2 \rangle$) and extrinsic ($\langle eg^2 \rangle = 4kT Rg \Delta f$, $\langle e_s^2 \rangle = 4kT R_s \Delta f$) noise sources are represented in terms of a noise resistance R_n, a noise conductance Gn and a correlation impedance Z_c.

These three quantities are calculated as follows:
Let us define

$$v_g = z_{11} i_g + z_{12} id \tag{3}$$

$$v_d = z_{21} i_g + z_{22} id \tag{4}$$

where z_{ij} are the intrinsic impedance parameters. Using these two parameters, a straightforward calculation gives

$$Z_c = Z_{11} - Z_{21} \cdot \frac{(R_s + \text{Cor } \sqrt{R_{nd} \cdot R_{ng}})}{R_s + R_{nd}} \tag{5}$$

$$G_n = \frac{1}{|Z_{21}|^2}(R_s + R_{nd}) \tag{6}$$

$$R_n = R_s + R_g + R_{ng} - \left| \frac{Z_{11} - Z_c}{Z_{21}} \right|^2 (R_s + R_{nd}) \tag{7}$$

where

$$R_{nd} = \langle v_d^2 \rangle / 4kT \ \Delta f, \ R_{ng} = \langle v_g^2 \rangle / 4kT \ \Delta f \tag{7}$$

$$\text{Cor} = \langle v_g v_d* \rangle / \sqrt{\langle v_g^2 \rangle \cdot \langle v_d^2 \rangle}$$

Z_{ij} extrinsic impedance parameters.

Thus, the noise figure related to Z_c, R_n, G_n [3], can be easily computed if the whole set of small signal parameters,

Reprinted from *IEEE Electron Device Letters*, vol. EDL-6, no. 6, pp. 270–272, June 1985.

43

but especially the intrinsic noise source $\langle id^2\rangle$, $\langle ig^2\rangle$ and the correlation coefficient are known.

The evolution of $\langle id^2\rangle$, $\langle ig^2\rangle$, and C has been systematically studied with the numerical modeling as a function of the main technological parameters and analytical formulas for these quantities has been derived

$$\langle id^2\rangle = 4kT\,\Delta f \cdot \frac{gd^2 + \omega^2 + Cgd^2}{gd^2} \cdot \frac{gm}{Cgs}$$
$$\cdot\, Lg \cdot (\alpha Z + \beta Ids) \tag{8}$$

$$\langle ig^2\rangle = \frac{2kT\Delta f\; Cgs^2\omega^2}{gm}, $$

with $\alpha = 2.10^5$ pF/cm^2
and $\beta = 1.2510^2$ pF/(mA · cm). \hfill (9)

In these expressions gm and Cgs are the transconductance and the gate-to-source capacitance, z the gate width, and Lg the gate length. It has been found that the two fitting parameters α and β do not present a significant dependence upon material properties and consequently are quite the same for both MESFET's and TEGFET's. Obviously material parameters influence $\langle id^2\rangle$ and $\langle ig^2\rangle$ values since gm and gd strongly depend upon material properties. For a typical MESFET ($z = 300\ \mu m$, $Ids = 10$ mA), αZ (6000 pF/cm) is larger than βIds (1250 pF/cm).

The derivation of the correlation coefficient C is more difficult. In fact, it is the correlation coefficient C' that has been calculated while C has been deduced from equation [2]. It has been found that C' depends especially upon the gate length over epilayer ratio Lg/a. In the case of quarter micrometer gate MESFET's Lg/a is close to 3 and the correlation coefficient C' is close to $j\,0.7$, while for half-micrometer gate MESFET's or for TEGFET's, the Lg/A ratio is greater than 5 and the correlation coefficient C' is close to $j\,0.8$.

III. RESULTS

A comparison between the theoretical noise figure and the measured ones has been carried out in the case of various devices, described in the literature [7]–[10] (Table I). In this table the value of the correlation coefficient C' which provides the best fitting is also presented. It can be noted that in all cases, the theoretical results are in good agreement with experimental ones.

A. Influence of gd and Cgd

Fig. 1) represents the evolution of the noise figure of a typical quarter micrometer MESFET as a function of the frequency. In this figure, the noise figure and frequency are plotted in linear scales.

According to Fukui formula, the noise figure should increase linearly with the frequency. This is verified below 20 GHz where the k_F factor is close to 2.5 but not at higher frequencies.

The greater increase of the noise figure with increasing frequency can be principally explained by the influence of the gate-to-drain capacitance (Fig. 2). As a matter of fact Cgd

TABLE I

	L_g (μm)	F (GHz)	NF (meas)	NF (calc)	C'	g_d (mS/mm)
MESFET AVANTEK	0.25	18	1.55	1.54		19
		21.7	1.98	1.81	j0.65	
		32	2.6	2.44		
HUGHES	0.3	18	1.69	1.69	j0.75	25.8
TRW	0.35	18	1.5	1.61	j0.8	22.8
TEGFET FUJITSU	0.5	8	1.3	1.23		22.2
		11.3	1.77	1.77	j0.8	
		20	3.1	2.82		
THOMSON-CSF	0.5	17.5	1.5	1.64	j0.8	19.4

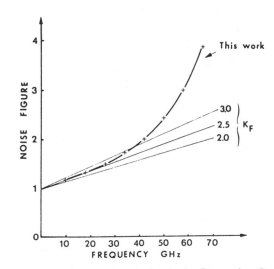

Fig. 1. Comparison between the calculated noise figure using (5)–(10) and the Fukui Formula. The noise figure is in linear scale, not in decibels.

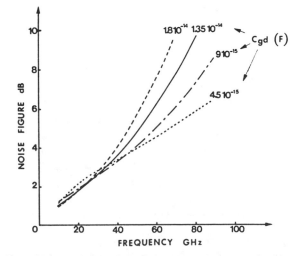

Fig. 2. Influence of the value of the gate-to-drain capacitance on the evolution of the noise figure versus frequency $Z = 150\ \mu m$; $Rs + Rg = 5\ \Omega$; $gd = 4$ mS.

44

influences the drain noise current $\langle id^2 \rangle$ since it is proportional to $\delta(\omega) = (gd^2 + \omega^2 Cgd^2)/gd^2$.

A low value of Cgd is then required to achieve a low value of both $\langle id^2 \rangle$ and noise figure in the millimeter-wave range.

In addition, for a given value of Cgd, $\langle id^2 \rangle$ will be lower for high values of the output conductance. This can explain why all the performances reported in Table I have been achieved with devices exhibiting a high output conductance as compared to the usual values achieved for both NESFET's or TEGFET's (typically 10–15 nm·S/mm.)

B. Comparison Between MESFET's and TEGFET's k_F Factors

The superiority of TEGFET's in the field of low-noise amplifiers is usually related to a lower k_F factor. This k_F factor can be written as [5]

$$k_F = 2 \cdot \sqrt{\frac{\langle id^2 \rangle}{4kTgm\Delta f}}$$

$$= 2 \cdot \sqrt{\frac{gd^2 + \omega^2 Cgd^2}{gd^2} \cdot \frac{L_g}{Cgs} (\alpha Z + \beta Ids)} \cdot \qquad (10)$$

At low-frequency $\delta(\omega) = 1$ and under low-noise conditions $Cgs = \epsilon LgZ/a + Cp$ where Cp is the fringing capacitance. Thus [10] becomes

$$k_F = 2 \quad \frac{a}{\epsilon Z} \cdot \frac{\alpha Z + \beta Ids}{1 + \dfrac{Cpa}{\epsilon ZLg}} \cdot \qquad (11)$$

This expression shows that k_F can be reduced with decreasing epilayer thickness, but k_F is less sensitive to the material properties than other parameters of the Fukui Equation such as gm. Due to very high doping levels, the AlGaAs layer of TEGFET's is always thinner than the GaAs MESFET's one, and thus for similar dc drain current and fringing capacitance under low-noise condition. The k_F factor of a TEGFET, is lower than the MESFET one.

REFERENCES

[1] B. Carnez, A. Cappy, R. Fauquembergue, E. Constant, and G. Salmer, "Noise modeling in submicrometer-gate FET's," *IEEE Trans. Electron Devices*, vol. ED-28, pp. 784–789, 1981.

[2] A. Cappy, A. Vanoverschelde, M. Schortgen, C. Versnaeyen, and G. Salmer, "Comparison between TEGFET's and FET's RF performances in the millimeter wave range," in *11th Int. Symp. Gallium Arseniure Related Compounds* (Biarritz), 1984.

[3] H. Rothe and W. Dahlke, "Theory of noise fourpoles," in *Proc. IRE*, vol. 44, p. 811, 1956.

[4] R. A. Pucel, H. A. Haus, and H. Statz, "Signal and noise properties of GaAs microwave FET," *Advan. Electron Electron. Phys.*, vol. 38, p. 195, 1975.

[5] H. Fukui, "Optimal noise figure of microwave GaAs MESFET," *IEEE Trans. Electron Devices*, vol. ED-26, pp. 1032–1037, 1979.

[6] A. Cappy, thèse d'état, to be published.

[7] P. W. Chye and C. Huang, "Quarter micron low noise GaAs FET's," *IEEE Electron Device Lett.*, vol. EDL-3, p. 401, 1982.

[8] M. Feng, H. Kanber, V. K. Eu, E. Watkins, and L. R. Hackett, "Ultrahigh frequency operation of ion-implanted GaAs metal-semiconductor field-effect transistors," *Appl. Phys. Lett.*, vol. 44, no. 2, p. 231, Jan. 1984.

[9] J. J. Berenz, K. Nakano, and K. P. Weller, "Low noise high electron mobility transistors," *IEEE Microwave Theory Tech. Symp.* San Francisco, CA, 1984, pp. 83–86.

[10] M. Niori, T. Saito, S. Joshin, and T. Mimura, "A 20 GHz low noise HEMT amplifier for satellite communication," in *ISSCC Dig. Techn. Paper*, pp. 198–199.

Comparison of GaAs MESFET Noise Figures

H. GORONKIN, SENIOR MEMBER, IEEE, AND V. NAIR, MEMBER, IEEE

Abstract—A method of comparing noise figures of GaAs MESFET's is presented. The noise measure M graphed against gate length for devices having the lowest value of M gives a figure of merit graph against which other devices may be compared. This is useful in determining the relative value of material and process improvements for a given gate length.

L OW-NOISE MESFET's have been fabricated using a variety of material choices, processes, and design. The literature contains reports of low-noise performance of epitaxial and implanted FET's, with gate lengths as small as 0.20 μm, and modulation-doped FET's. How can we compare the noise performance of these FET's with one other to predict future trends?

One method of comparison is to plot the noise measure $M = F - 1$, versus gate length. This is based on utilizing the form of either physical or empirical noise models which can be written as

$$F = 1 + fL_g \text{ [constants and device parameters]}$$
$$+ \text{ [higher order terms]} \quad (1)$$

where f is frequency, L_g is gate length, and F is the noise figure related to the measured quantity

$$NF = 10 \log_{10} F. \quad (2)$$

A procedure will be shown to provide a comparison of the noise figure for a particular device with the lowest value for a given technology.

In this letter GaAs MESFET's are compared. Data from several sources were used here (Table 1). Most available data was clustered around three frequencies: 8, 12, and 18–20 GHz. For commercial devices, where the gate length could not be determined by SEM measurements in our laboratory, the value given by the data sheet was used. For each gate length and frequency, the lowest published noise figure or best data taken in our laboratory was selected. Both epitaxial and implanted MESFET's are included.

The data in Table I is shown in Fig. 1. For all three frequencies, the data is connected by straight lines by the least-square method. If we restrict ourselves to first-order terms in (1), all three lines should go through the origin, however, these data extrapolate to a finite noise figure for zero gate length.

There are several fundamental reasons for this behavior. At submicrometer gate lengths the parasitic capacitance due

Manuscript received October 4, 1984; revised November 8, 1984.
The authors are with Semiconductor Research and Development Laboratories, Motorola, Inc., Phoenix, AZ 85008.

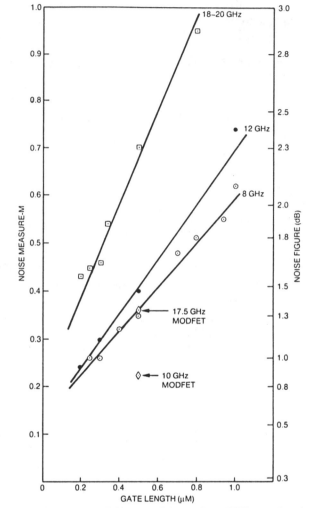

Fig. 1. Noise measure of the state-of-the-art GaAs FET's as a function of gate length.

to fringing effect and gate pads put a lower limit on the gate capacitance and hence on the minimum noise figure. The fringing capacitance, which does not depend on gate length, is over 30 percent of the source-to-gate capacitance of a half micrometer device [8]. The capacitance of a typical gate pad formed on a substrate of carrier concentration of about 10^{14} atom/cc is almost equal to the gate capacitance of a half-micrometer device [9]. This parasitic capacitance increases noise figure and decreases the gain.

The frequency dependence of the input conductance of the FET is responsible for degradation of the minimum noise figure at higher frequencies. If we consider a simplified noise model as described by Podell [10], the FET input can be viewed as a parallel RC network. The input conductance is

Reprinted from *IEEE Electron Device Letters*, vol. EDL-6, no. 1, pp. 47–49, January 1985.

IEEE ELECTRON DEVICE LETTERS, VOL. EDL-6, NO. 1, JANUARY 1985

TABLE I
MINIMUM NOISE FIGURE OF THE STATE-OF-THE-ART GaAs FET's

	Gate Length (μm)	8 GHz NF/G$_A$ (db)	12 GHz NF/G$_A$ (dB)	18-20 GHz NF/G$_A$ (db)	REFERENCE
VPE	0.2		0.95/11.5	1.55/12.3	1
I/B	0.25			1.6/8.5	2
VPE	0.25	1.0/11.5			D/S (NE673)
MBE/LRG	0.3	1.0/13.0			3
I/B	0.3		1.15/10.0	1.63/9.5	4
OMVPE	0.35			1.9/9.0	5
VPE	0.4	1.2/11.0			D/S (NE700)
VPE	0.5	1.3/12.0			D/S (MGF1403)
I/B	0.5		1.46/10.2		6
VPE	0.5			2.3/5.0	D/S (NE137)
VPE	0.7	1.7/10.5			D/S (NE218)
VPE	0.8	1.8/9.5			Motorola
I/S	0.8			2.9/6.1	7
VPE	0.94	1.9/7.3			Motorola
VPE	1.0	2.0/7.0			D/S (GAT4)
VPE	1.0		2.4/7.5		D/S (NE218)

I/B - Implanted into Buffer   D/S - Data Sheet

VPE - Vapor Phase Epitaxy   VPE/G - VPE Graded Channel

MBE - Molecular Beam Epitaxy (GaAs on GaAs)   I/S = Implant into Substrate

LRG - Low Resistance Gate

proportional to $\omega^2 C^2$. For ultra submicrometer devices the input of the FET will be dominated by the extrinsic parasitic capacitance and associated conductance. This will limit the noise figure to a finite value as the gate length approaches zero.

It is interesting to compare the slopes of Fig. 1 with the theoretical predictions of Pucel et al. [8]. According to their theory, the noise figure decreases at the rate of 2.5 dB/μm at 10 GHz for devices with gate lengths in the 1–2.5-μm region. For submicrometer devices the noise figure is expected to fall at a slower rate. From the data shown in Fig. 1, the noise figure decreases at the rate of 1.35 dB/μm at 8 GHz and 1.49 dB/μm at 12 GHz. We believe that noise figure will further flatten out as gate length approaches zero due to the parasitic effects.

In order to examine the frequency dependence of the data in Fig. 1, the slope of each line was graphed against frequency in Fig. 2. A straight line was drawn from the origin to connect the three points. The resulting fit seems satisfactory. The slope of the line is seen to represent a constant for MESFET technology

$$\frac{\partial M}{\partial L_g \partial f} = 0.054 \; \mu m^{-1} \, GHz^{-1}. \tag{3}$$

It is apparent that the diversity of device and material data contained in the brackets of (1) is washed out in this analysis.

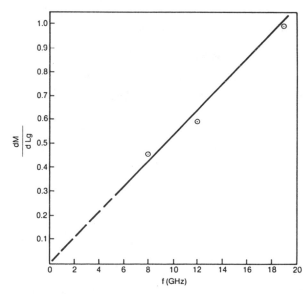

Fig. 2. Frequency dependence of noise measure/unit gate length.

It can be concluded that when all other material and device parameters have been optimized the gate length will be the significant factor determining performance. Alternately, for a particular gate length and frequency, the difference between measured noise figure and that of the curve in Fig. 1 will indicate the degree of improvement that is possible by optimizing the other parameters. For comparison let us consider the data of Feng *et al.* [11]. Their data, based on Fukui analysis [12] gives $\partial M/\partial L_g \partial f \approx 0.064 - 0.083$. We obtain a slightly lower value because of the nonzero intercept in Fig. 1, whereas Fukui has $M = 0$ at $L = 0$. Hence, Fukui equation underestimates M for short L and would overestimate M for very long gate length.

The diamonds in Fig. 1 correspond to modulation-doped MESFET's [13]. At 10 GHz there is approximately a 0.5-dB improvement in noise figure for a 0.5-μm gate and a 0.9-dB improvement at 18 GHz. By comparing the data in this manner, it can be clearly seen that low-noise MODFET's have much lower noise than conventional MESFET's. There are several possible reasons for this. The gate to source capacitance for optimized doping profiles in the AlGaAs may be lower than for MESFET's. Transconductance g_m, is higher in MODFET's due to reduced impurity scattering. The higher transconductance in MODFET's, which arises from reduced impurity scattering, reduces the ratio $C_{gs}/\sqrt{g_m}$ which, in turn, reduces the noise measure proportionally. Finally, the two-dimensional gas may be missing a component of noise that is normally found in MESFET's. This will be due to the electrons having only two degrees of freedom in the MODFET as compared to three in conventional devices.

In summary, a method of comparing GaAs FET noise figures for a variety of material and process technologies has been presented. The noise measure plot can be extrapolated to predict performance improvements. MODFET's and MESFET's fall on two different performance curves. While it is necessary to have additional data points to establish MODFET noise performance curves, the available data (shown in Fig. 1) indicate that the noise in two-dimensional electron gas devices is substantially lower than the noise in three-dimensional electron gas devices.

ACKNOWLEDGMENT

The authors wish to thank Dr. R. Vaitkus for a critical review of this work and helpful suggestions. Acknowledgment is also due to D. Scheitlin and Dr. W. Atkins for providing us with the noise figure data on Motorola FET's. Thanks are also due to M. Scott for the preparation of the manuscript.

REFERENCES

[1] P. W. Chye and C. Huang, "Quarter micron low noise GaAs FET's," *IEEE Electron Device Lett.*, vol. EDL-3, pp. 401–403, Dec. 1982.

[2] M. Feng *et al.*, "Optimization of ion implanted GaAs low noise FET," 1983 Device Res. Conf.

[3] S. G. Bandy *et al.*, "Submicron GaAs microwave FET's with low parasitic gate and source resistances," *IEEE Electron Device Lett.*, vol. EDL-4, pp. 42–44, Feb. 1983.

[4] M. Feng, H. Kanber, V. K. Eu, E. Watkins, and L. R. Hackett, "Ultrahigh frequency operation of ion implanted GaAs metal–semiconductor field-effect transistors," *Appl. Phys. Lett.*, vol. 44, no. 2, pp. 231–233, Jan. 15, 1984.

[5] V. Aebi, S. Bandy, C. Nishimoto, and G. Zdasiuk, "Low noise microwave field effect transistors using organometallic GaAs," *Appl. Phys. Lett.*, vol. 44, no. 11, pp. 1056–1058, June 1, 1984.

[6] M. Feng, V. K. Eu, T. Zielinski, H. Kanber, and W. B. Henderson, "GaAs MESFET's made by ion implantation into MOCVD buffer layers," *IEEE Electron Device Lett.*, vol. EDL-5, pp. 18–20, Jan. 1984.

[7] A. K. Gupta, D. P. Siu, K. T. Ip, and W. C. Petersen, "Low noise MESFET's for ion implanted GaAs MMIC's," in *IEEE Microwave and Millimeter-Wave Monolithic Circuits Symp.*, May 1983, pp. 27–30.

[8] R. A. Pucel, D. J. Masse, and C. F. Krumm, "Noise performance of gallium arsenide field effect transistors," *IEEE J. Solid-State Circuits*, vol. SC-11, pp. 243–255, Apr. 1976.

[9] H. A. Cooke, "Microwave field effect transistors 1978," *Microwave J.*, pp. 43–48, Apr. 1978.

[10] A. F. Podell, "A functional GaAs FET noise model," *IEEE Trans. Electron Devices*, vol. ED-28, pp. 511–517, May 1981.

[11] M. Feng, V. K. Eu, H. Kanber, E. Watkins, J. M. Schellenberg, and H. Yanasaki, "Low noise GaAs metal–semiconductor field-effect transistor made by ion implantation," *Appl. Phys. Lett.*, vol. 40, no. 9, pp. 802–804, May 1982.

[12] H. Fukui, "Optimal noise figure of microwave GaAs MESFET's," *IEEE Trans. Electron Devices*, vol. ED-26, pp. 1032–1037, July 1979.

[13] P. Delescluse, J. F. Rochette, M. Laviion, D. Delagebeauduef, P. Etienne, J. Chevrier, P. Jay, and N. T. Linh, "Ultra low noise two dimensional electron gas field effect transistor made by MBE," Abs. distributed at 1984 MBE Symp., San Francisco, CA.

A New Relationship Between the Fukui Coefficient and Optimal Current Value for Low-Noise Operation of Field-Effect Transistors

D. DELAGEBEAUDEUF, J. CHEVRIER, M. LAVIRON, AND P. DELESCLUSE

Abstract—In this letter, a new relationship is derived between the Fukui coefficient, appearing in the Fukui's equation, and the optimal current for low-noise operation of field-effect devices.

It's validity has been evidenced by considering several experimental cases including MESFET's, HEMT, and TEGFET'S.

I. INTRODUCTION

AN APPROXIMATE but useful expression is often to estimate, or interpret, the noise performance of field-effect transistors—the "Fukui equation" (1) which can be written

$$F_{\min} = 1 + K_F \frac{Cgs}{gmo} \omega \sqrt{gmo(RS + RG)} \qquad (1)$$

where

Cgs   gate–source capacitance,
gmo   intrinsic transconductance,
ω angular frequency,
RS   source resistance, and
Rg   gate resistance

and K_F a coefficient which Fukui relates to the material quality. In fact the Fukui equation is an approximation of a more general expression given by Pucel, Haus, and Statz (2) where the induced gate noise and the correlation between gate and drain noises are taken into account. The Fukui equation is then obtained by neglecting induced gate noise and quadratic terms in ω and assuming a correlation coefficient close to unity. This is a low-frequency approximation which retains, as dominant noise terms, only the drain noise current and the thermal noise arising from the gate and source resistances.

One then obtains a simplified equation which by identification with Fukui's equation gives

$$K_F = 2 \sqrt{\mathbb{P}} \qquad (2)$$

where $\mathbb{P}$ is a noise parameter defined by

$$\overline{\delta id^2} = 4KT0\, gmo\, \mathbb{P} \Delta f \qquad (3)$$

$\overline{\delta id^2}$ being the square of the drain noise current for a bandwidth Δf. This letter proposes an expression giving K_F as a function of the optimal bias current I opt together with some experimental evidence concerning the validity of the expression.

II. DERIVATION OF THE FUKUI COEFFICIENT EXPRESSION

According to Brewitt–Taylor *et al.* (3), a good estimate for the noise figure of a field-effect transistor in saturation may be obtained by limiting the calculation to the knee point of the drain I-V characteristics. The mobility in the channel may then be assumed constant since the electric field value is everywhere less than the critical value $\mathcal{E}_c$ for velocity saturation which occurs only at the drain end of the gate.

Suppose now, that a perturbing noise current δ in (xo) is injected at point xo in the channel under the gate and extracted at $xo + \Delta xo$, it is possible to calculate the drain current response δid, and it may be shown easily that (3)

$$\delta id = \delta \text{ in } (xo) \frac{\Delta xo}{L}. \qquad (4)$$

If δ in (xo) is identified with a diffusion noise source then

$$\overline{\delta \text{ in}^2} (xo) = 4q\Delta f \frac{D_{/\!/}(xo)Qo(xo)}{\Delta xo} \qquad (5)$$

where $D_{/\!/}(xo)$ is the parallel diffusion coefficient, Δf the bandwidth, and $Qo(xo)$ the total mobile electric charge per unit gate length at xo (including possibly injected charge responsible for the finite output resistance). From (4) and (5), remembering that the local diffusion noise sources are uncorrelated we have

$$\overline{\delta id^2} = \frac{4q\Delta f}{L^2} \int_0^L D(xo)Qo(xo) \, dxo. \qquad (6)$$

As suggested by Brewitt–Taylor *et al.*, it is a reasonable approximation to take for the diffusion coefficient the constant low-field value

$$D_{/\!/} \sim \frac{KT0}{q} \mu_o \qquad (7)$$

then

$$\overline{\delta id^2} \sim \frac{4KT0\mu 0 \Delta f}{L^2} \int_0^L Qo(xo) \, dxo \qquad (8)$$

Manuscript received May 15, 1985.
The authors are with Thomson-CSF, Domaine de Corbeville, BP 10, 91401 Orsay Cedex, France.

Reprinted from *IEEE Electron Device Letters,* vol. EDL-6, no. 9, pp. 444–445, September 1985.

49

which compared to (3) gives

$$\mathbb{P} \sim \frac{\mu 0}{gmo\ L^2} \int_0^L Qo(xo)\ dxo. \tag{9}$$

But any field-effect device obeys the general relationship

$$Io = vo(x)Qo(x) \tag{10}$$

Where $vo(x)$ is the carrier velocity at x and $Qo(x)$ the mobile electric charge per unit gate length at the same point.

Dividing (10) by $vo(x)$ and integrating over the gate length, one obtains

$$I_0\tau = \int_0^L Qo(x)\ dx. \tag{11}$$

Then (9) can be put in the form

$$\mathbb{P} \sim \frac{\mu 0 I 0 \tau}{gmo\ L^2} \tag{12}$$

from which according to (2)

$$KF \sim 2 \sqrt{\frac{\mu 0 I 0 \tau}{gmo\ L^2}}. \tag{13}$$

Equation (13) may be further simplified for the case of *short-gate* FET's by assuming that the minimum noise conditions are reached before the onset of pinchoff regime; then

$$\tau \sim \frac{L}{v\ max} \tag{14}$$

where $V\ max = \mu_0 \mathcal{E}_c$ is the saturated electron velocity. The proposed final form for the relation between the Fukui coefficient and the optimum current is then

$$KF \sim 2 \sqrt{\frac{I0\ opt}{\xi c L\ gmo}}. \tag{15}$$

III. Experimental Evidence for the Validity of (15)

Table I gives some useful characteristics for noise considerations of several field-effect transistors including MESFET's, HEMT, and TEGFET's. These characteristics result from static and microwave measurements and permit the

TABLE I

	Z (µ)	L (µ)	Rs+Rg (Ω)	gmo (mS)	Cgs (pF)	f (GHz)	Fmin (db)	Iopt (mA)	KF (Fmin)	KF (Iopt)	
NEC673	280	0.3	3.5	50	0.25	10	1.34	10	2.75	2.75	MESFET
MF4155	140	0.5	3.5	28	0.175	10	1.3	10	2.8	2.9	
HEMT	200	0.5	11.8	55	0.28	8	1.3	7.5	1.7	1.8	HEMT
TF3287	300	0.8	11	38	0.4	10	2.15	8	1.6	1.7	
TF3468	300	0.8	8	46	0.39	10	2.1	6	1.6	1.4	
TF3524	300	0.65	6.9	40	0.32	10	1.9	8	2.1	1.9	
TF3584	300	0.6	11.1	44	0.48	10	2.8	6	1.9	1.6	
TF3562	300	0.5	4.75	42	0.4	10	2.9	20	3.6	3.3	TEGFET
TF3588	300	0.55	8.7	55	0.31	10	1.5	8	1.65	1.75	
TF4108	300	0.5	4.8	69	0.38	10	1.1	9	1.5	1.5	
TF4164	140	0.5	5.9	34	0.16	17.5	1.4	5.8	1.65	2	
TF4262-1	140	0.5	7.9	46	0.27	17.5	2	5.9	1.5	1.7	
TF4262-2	140	0.5	6.6	19	0.13	17.5	2.12	7.4	2.4	3	
TF3569	300	0.7	9	35	0.27	10	2	8.6	2.15	2	
TF3590	300	0.6	10.4	68	0.47	10	2.4	10	2	1.7	

◆ ◆ FUJITSU
◆ THOMSON -CSF

calculation of $KF\ (F\ min)$ by use of (1) and that of $KF\ (I\ opt)$ by use of (15) (with $\mathcal{E}_c = 3.5 \times 10^5$ V/m for all cases).

The good agreement observed between the two determinations demonstrates the validity of (15) and that of the various approximations involved in its derivation.

IV. Conclusion

An expression has been derived that allows the empirical coefficient of the Fukui equation to be related to the optimum current for low-noise operation of field-effect devices.

Consideration of experimental data from a number of transistors (MESFET, HEMT, and TEGFET) demonstrates good agreement between the two possible determinations of K_F.

References

[1] H. Fukui, "Optimal noise figure of microwave GaAs MESFET's," *IEEE Trans. Electron Devices,* vol. ED-26, pp. 1032–1037, July 1979.
[2] R. A. Pucel, H. A. Haus, and H. Statz, "Signal and noise properties of gallium arsenide field effect transistors," in *Advances in Electron Physics.* New York: Academic, 1975, pp. 195–265.
[3] C. R. Brewitt–Taylor, P. N. Robson, and J. E. Sitch, "Noise figure of MESFET's," *IEE Proc.,* vol. 127, Pt. I, no. 1, Feb. 1980.

The Noise Properties of High Electron Mobility Transistors

TOM M. BROOKES, MEMBER, IEEE

Abstract—A simple analytic model for the HEMT, based on the Pucel theory for MESFET's, is developed which can be used to calculate the noise properties of the transistor. Good agreement between calculation and experiment is found. The dependence of noise temperature on gate length and channel thickness is presented.

I. INTRODUCTION

THE HIGH electron mobility transistor has been shown to be capable of very low noise temperatures and high associated gain at microwave frequencies. Several papers have been published on the small-signal properties of this device, notably by Lee *et al.* [1]. This paper describes a method for calculating the noise properties of the device by modifying the Pucel [2] noise theory so that it applies to the HEMT. An outline of the calculation of the noise parameters is presented here: a more complete derivation being given in [12].

II. THE DC AND SMALL-SIGNAL PARAMETERS OF THE HEMT

The model proposed by Grebene and Ghandi [3] for the operation of the JFET in saturation will be modified to apply to the HEMT. As will be described, approximations made in this model are more appropriate in the case of a HEMT, and therefore lead to a more accurate analysis. However, effects such as velocity overshoot and the formation of Gunn domains, which are so important in short-gate devices, cannot be included in this model, so some reservations must apply to the predictions.

A. The dc Model

The Grebene and Ghandi model divides the device into two regions: a linear region and a velocity saturated region. In region I, $0 < x < L_1$. Ohm's law applies, and the electron velocity is proportional to the electric field. In region II, $L_1 < x < L$ and the carriers are assumed to travel at a fixed (saturated) velocity v_s. The boundary between the two is determined by satisfying the current-continuity and field-continuity equations.

Manuscript received March 29, 1985; revised August 5, 1985. The National Radio Astronomy Observatory is operated by the Associated Universities, Inc., under contract with the National Science Foundation.

The author was with the National Radio Astronomy Observatory, Charlottesville, VA 22903. He is now with the University of Manchester, Nuffield Radio Astronomy Laboratories, Jodrell Bank, Macclesfield, Cheshire SK11 9DL, United Kingdom.

IEEE Log Number 8405795.

The linear region of operation has been described in the literature by Delagebeaudeuf and Linh [4] and Lee *et al.* [1]. Their models will be used to describe the operation of the HEMT in this region. In the saturated region, the model proposed by Grebene and Ghandi for the operation of the JFET in saturation will be used to calculate the potential and field distribution in the channel.

The problem will be solved as follows. Firstly, determine the current flowing in region I where the electrons obey Ohm's law and the electron density in the channel is roughly proportional to the applied gate voltage. Then, by calculating the current flowing in region II and equating this to that flowing in region I, and simultaneously matching the electric field at boundary, calculate the length of region I.

The models of Delagebeaudeuf and Linh [4] and Drummond and Morkoç [5] are used to determine the currents and potentials in region I. Following Delagebeaudeuf and Linh, the charge in the channel at a point x is given by

$$Q(x) = \frac{\epsilon_2}{(d + \Delta d)} [V_{sg} - V_{off} - V(x)] \qquad (1)$$

where $V(x)$ is the channel voltage at a point x. To simplify the later analysis, the following reduced potentials will be defined

$$s = \frac{V_{sg} - V_{off}}{V_{off}}$$

$$w = \frac{V_{sg} - V_{off} - V(x)}{V_{off}}$$

$$p = \frac{V_{sg} - V_{off} - V(L_1)}{V_{off}}$$

where L_1 is the point in the channel where the electrons reach their saturated velocity. Following the analysis of Pucel *et al.* [2], and using the relationship between applied voltage and channel charge density given by Drummond and Morkoç [5], the following equation is obtained:

$$I_d = Q(x) \, \mu_0 Z \frac{dV}{dx}$$

which, when integrated over the range $0 < x < L_1$, is equal to

$$I_d L_1 = \frac{\epsilon_2}{2(d + \Delta d)} \mu_0 Z V_{off}^2 (s^2 - p^2). \qquad (2)$$

Reprinted from *IEEE Transactions on Electron Devices*, vol. ED-33, no. 1, pp. 52–57, January 1986.

In the saturated region, the current is simply given by the product of the number of charge carriers and their velocity, so

$$I_d = Q(L_1) Z v_s = \frac{Z\epsilon_2}{(d + \Delta d)} V_{\text{off}} p v_s \qquad (3)$$

which must equal the current flowing in the unsaturated region. Equations (2) and (3) allow L_1 to be calculated.

$$L_1 = \frac{(s^2 - p^2)}{2p} \frac{V_{\text{off}}}{E_s} \qquad (4)$$

where $E_s = v_s/\mu_0$ is the field at which velocity saturation occurs.

B. The Potential Distribution in the Channel

The potential drop in the channel in the ohmic region is given by

$$V_p = -(V(L_1) - V(0)) = V_{\text{off}}(s - p).$$

In region II, the potential drop must be obtained by the integration of the longitudinal electric field present over the range $L_2 < x < L$. This field is determined only by the free electrons on the drain electrode.

Grebene and Ghandi [3] analyzed this situation for the JFET. A critical assumption in their analysis was that the thickness of the layer in which the current flowed was very much smaller than the thickness of the doped semiconductor (or alternatively, the distance from the gate electrode to the channel). This assumption is very much better satisfied in the case of the HEMT as the current flow is confined to a sheet.

Following their analysis, the potential in the channel may be approximated by the following equation:

$$\Phi(x) = \frac{2(d + \Delta d)}{\pi} E_s \sinh \frac{(\pi(x - L_1))}{2(d + \Delta d)}$$

so the voltage drop in II is

$$V_{\text{II}} = \frac{2(d + \Delta d)}{\pi} E_s \sinh \left(\frac{\pi(L_2)}{2(d + \Delta d)}\right).$$

Adding this to the drop in region I yields an expression for the source to drain voltage

$$V_{sd} = V_{\text{off}}\left(s - p + \frac{2(d + \Delta d)}{\pi L} \Psi \sinh \left(\frac{\pi L_2}{2(d + \Delta d)}\right)\right) \qquad (5)$$

where Ψ is the saturation parameter defined as

$$\Psi = \frac{E_s L}{V_{\text{off}}}.$$

Combining (4) and (5) allows L_1 to be eliminated and the reduced potentials s and p to be calculated for a given bias condition. The resulting nonlinear equations are most easily solved by computer.

C. Small-Signal Parameters

The small signal parameters r_d and g_m can be evaluated by small purturbations of the applied bias. This could be done numerically, but more physical insight is provided if analytical expressions can be derived. These parameters are defined as follows:

$$r_d = -\left.\frac{\partial V_{sd}}{\partial I_d}\right|_{V_{sg}}$$

$$g_m = -\left.\frac{\partial I_d}{\partial V_{sg}}\right|_{V_{sd}}.$$

As a similar method is used to derive both parameters, only the derivation of r_d will be described in detail.

1) *Drain Resistance:* r_d can be written as follows:

$$-r_d = +\left.\frac{\partial V_{sd}}{\partial I_d}\right|_{V_{sg}}$$

$$= \left[V_{\text{off}}\frac{dp}{dI_d} + \frac{\Psi}{L} \cosh \left(\frac{\pi L_2}{2(d + \Delta d)}\right) \frac{dL_2}{dI_d}\right]. \qquad (6)$$

From (3)

$$p = \frac{\epsilon_2}{(d + \Delta d)} \frac{I_d}{Z v_s V_{\text{off}}}$$

so that

$$\frac{dp}{dI_d} = \frac{(d + \Delta d)}{\epsilon_2 Z v_s V_{\text{off}}}.$$

The relationship $L = L_1 + L_2$ is used to determine the derivative dL_2/dI_d, and hence, by differentiating (4) one obtains

$$\frac{dL_1}{dp} = \frac{d}{dp}\left[\frac{s^2}{2p} - \frac{p}{2}\right] \frac{V_{\text{off}}}{v_s} = -\frac{V_{\text{off}}}{2E_s}\left[\frac{s^2 + p^2}{p^2}\right].$$

Substituting these values into the equation yields the result

$$r_d = \frac{(d + \Delta d)}{\epsilon_2 Z v_s}\left[1 + \frac{s^2 - p^2}{2p^2} \cosh \left(\frac{\pi L_2}{2(d + \Delta d)}\right)\right]$$

and similarly

$$g_m = \frac{\epsilon_2}{(d + \Delta d)} Z v_s \frac{\left[1 - \frac{s}{p} \cosh \frac{\pi L_2}{2(d + \Delta d)}\right]}{\left[1 - \frac{s^2 - p^2}{2p^2} \cosh \left(\frac{\pi L_2}{2(d + \Delta d)}\right)\right]}.$$

2) *The Gate to Source Capacitance:* The final small-signal parameter that needs to be determined is the gate to source capacitance C_{sg}. This is defined as the rate of change of free charge on the gate with respect to the gate voltage with the source voltage held constant. To determine C_{sg}, we need to calculate the charge stored in region I and then add the charge stored in region II. In short-gate-length microwave devices, there is an additional component to be added caused by the fringing capacitance

at the edge of the gate electrode. This component can be dominant for devices with gates shorter than 0.5 μm. A full description of the calculation of the gate to source capacitance is given in Lee et al. [1].

III. Noise Analysis of a HEMT

The dc analysis described in Section II has followed the analysis performed by Pucel, Haus, and Statz [2], [6] for the GaAs FET. In this section their noise analysis will be applied to the HEMT.

The analysis proceeded in four stages. In the first the drain circuit noise, assuming open-circuit conditions, is calculated. Then the gate circuit noise with short-circuit drain is determined and the correlation coefficient between these two noise sources calculated. Finally, using a simple equivalent circuit for the HEMT, the minimum noise figure and associated parameters are derived.

A. Drain Circuit Noise in a HEMT

The calculation of the mean square noise voltage at the drain caused by thermal noise fluctuations in region I will proceed as follows. The mean square magnitude of the noise voltage at a point x_0 in the channel will be calculated, assuming that it is Johnson noise with a field-dependent electron temperature. The noise voltage at the end of region I will then be determined by adding all these contributions together, assuming that they are incoherent. Finally, the "amplification" of this noise voltage by region II will yield the mean square drain voltage $|v_d^2|$.

Van der Ziel [7] calculated the noise voltage in the drain circuit for the JFET. Pucel et al. [2] and Statz et al. [6] enhanced this analysis to include both a field-dependent noise temperature and the effects of region II on the noise voltage development at the end of region I. Following the analysis of Pucel, modified to take account of the different mode of operation in the HEMT, it can be shown [12] that the mean square noise voltage developed at the drain by thermal noise fluctuations in the channel is equal to

$$|\overline{v_d^2}| = 4kT_0\Delta f \frac{V_{\text{off}}}{I_d} \cosh^2\left(\frac{\pi L_2}{2(d + \Delta d)}\right)\{P_0 + P_\delta\} \quad (10)$$

where

$$P_0 = \frac{t}{3p^2}(p^3 - s^3)$$

and

$$P_\delta = \frac{\delta}{p}\ln\left(\frac{p}{s}\right). \quad (10a)$$

Hence, the mean square noise voltage on the drain is given by (10). The noise current flowing may then be obtained by simple circuit analysis and is given by

$$|\overline{i_g^2}| = \frac{|\overline{v_d^2}|}{r_d^2}$$

$$= 4kT_0\Delta f \frac{V_{\text{off}}}{I_d r_d^2}\cosh^2\left(\frac{\pi L_2}{2(d + \Delta d)}\right)\{P_0 + P_\delta\}. \quad (12)$$

A different method is used to calculate the noise generated by sources in region II. Noting that the dc and small signal analysis of region II employed in this model is exactly the same as that of Pucel et al. [2] for the GaAs FET, their analysis can be followed directly. The mean square noise voltage $|v_{d_2}^2|$ at the drain caused by noise sources in region II is

$$|\overline{v_{d_2}^2}| = I_D \frac{64(d + \Delta d)^2 gD\Delta f}{\pi^5 v_s^5 \epsilon_2^2 b_p^2 Z^2}\sin^2\left(\frac{\pi b_p}{2a}\right)$$

$$\cdot \exp\left[\frac{\pi L_2}{2(d + \Delta d)} - 4\exp\frac{\pi L_2}{2(d + \Delta d)}\right.$$

$$\left. + 3 + \frac{\pi L_2}{d + \Delta d}\right].$$

Taking the limit of this expression as the thickness of the conducting layer b_p tends to zero yields a constant proportional to D, the high field diffusion constant. It has been shown [14], [15] that this quantity decreases very quickly as the electric field increases above E_s. For this reason the noise contributions from region II have been ignored. The agreement between the theory and experiment shown later justifies this assumption.

D. Gate Circuit Noise

A different procedure is followed to calculate the noise current induced on the gate by the noise fluctuations in the channel. The calculation is described in detail in [12], so only a short outline will be presented here.

The previous calculations have shown that there are noise voltage fluctuations along the channel that will induce noise charges on the gate since it is capacitively coupled to the channel. As these charges are time dependent, noise currents will flow. Pucel et al. [2] and Statz et al. [6] determined these noise currents for the GaAs FET. A similar analysis, performed for the HEMT, shows that the induced charge per unit gate width Z caused by an elementary thermal noise fluctuation at a point x_0 in the channel is [12]

$$\Delta q_r = \frac{\epsilon_2}{(d + \Delta d)}L_1\frac{\Delta i_d}{I_d}V_{\text{off}}\left[\kappa - \gamma(p - \omega) - \frac{L_2}{L_1}p\right]$$

where

$$\gamma = \frac{Z\epsilon_2\mu_0 r_d p V_{\text{off}}}{L_1(d + \Delta d)\cosh\left(\frac{\pi L_2}{2(d + \Delta d)}\right)}$$

$$\kappa = \frac{1}{(s^2 - p^2)}\left[s^2(p - s) - \frac{(p^3 - s^3)}{3}\right].$$

By integration of this expression along the channel so as to sum up the contributions from all points, and accounting for the elevated electron temperature, the mean square value of the fluctuating charge can be determined. The spectral components of the noise currents can then be obtained by multiplying the spectral components of the fluctuating charge calculated above by the angular frequency. Doing this, one arrives at the following expression for the mean square of the induced gate noise:

$$|\overline{i_g^2}| = \omega^2 \Delta [R_0 + R_\delta] \qquad (13)$$

where

$$R_0 = \frac{(p^3 - s^3)}{3p^2} [\kappa'^2 + \gamma^2 p^2 - 2\kappa'\gamma p]$$

$$+ \frac{(p^4 - s^4)}{2p^2} [\gamma^2 p - \kappa'\gamma] + \frac{(p^5 - s^5)}{5p^2} \gamma^2$$

$$R_\delta = \delta p \ln \left(\frac{p}{s}\right) [\kappa'^2 + \gamma^2 p^2 - 2\kappa' p]$$

$$+ 2p(p - s)[\gamma^2 p - \kappa'\gamma] + \frac{(p^2 - s^2)}{2} p\gamma^2$$

$$\Delta = 4kT_0 \frac{\Delta f}{I_d} \frac{V_{off}}{r_d^2} \cosh^2 \left(\frac{\pi L}{2(d + \Delta d)}\right) B$$

$$B = \left[\frac{\epsilon_2 L_1 V_{off}}{(d + \Delta d) I_d}\right]^2$$

$$\kappa' = \kappa - \frac{L_2}{L_1}.$$

The terms, R_0 and R_δ refer to the noise produced by normal thermal noise and the hot-electron noise.

Using similar arguments to those in the previous section, the noise current induced in the gate circuit by the sources in region II can be shown to be negligible. Thus, the strength of the noise sources in a HEMT have been determined.

It remains to calculate the correlation between the gate and drain circuit noise. Since they arise from the same fundamental noise mechanism, they are strongly correlated. Assuming that the correlation coefficient is purely imaginary, as it is in the case of the JFET (Klassan [8]). It can be calculated using the methods of Pucel et al. [2], Statz et al. [6], and Van der Ziel [7]. The calculation detailed in [12] shows that

$$C = \frac{\overline{i_g^* i_d}}{\sqrt{|\overline{i_g^2}||\overline{i_d^2}|}} = \frac{\frac{1}{p^2} [S_0 + S_\delta]}{\sqrt{[P_0 + P_\delta][R_0 + R_\delta]}}$$

where

$$S_0 = \frac{1}{p^2} \left[\frac{(\kappa' - \gamma p)(p^5 - s^5)}{5} + \frac{\gamma(p^6 - s^6)}{6}\right] \qquad (14)$$

and

$$S_\delta = \delta \frac{1}{p^2} \left[\frac{(\kappa' - \gamma p)(p^2 - s^2)}{2} + \frac{\gamma(p^3 - s^3)}{3}\right].$$

C. Noise Coefficients

The following set of dimensionless noise coefficients are normally defined

$$P = \frac{|\overline{i_d^2}|}{4kT_0 \Delta f g_m}$$

$$R = \frac{|\overline{i_g^2}| g_m}{2kT_0 \omega^2 C_{sg}^2}$$

and C is the correlation coefficient between them. These parameters can be defined simply in terms of the quantities just calculated. They have the following values:

$$P = \frac{2(d + \Delta d)}{r_d^2 \epsilon_2} \frac{L_1}{Z\mu_0 V_{off} g_m (s^2 - p^2)}$$

$$\cdot \cosh^2 \left(\frac{\pi L}{2(d + \Delta d)}\right) [P_0 + P_\delta]$$

$$R = Z^2 \frac{V_{off}}{I_d C_{sg}^2} \frac{B}{C_{sg}^2} g_m [R_0 + R_\delta]$$

$$C = \frac{\frac{1}{p^2} [S_0 + S_\delta]}{\sqrt{[P_0 + P_\delta][R_0 + R_\delta]}}.$$

The expressions for the noise parameters for the HEMT presented in the next section can be simplified if the following functions of P, R, and C are used:

$$K_g = P\left((1 - C)\sqrt{\frac{R}{P}}\right)^2 + (1 - C^2)\frac{R}{P}\right)$$

$$K_c = \frac{1 - C\sqrt{\frac{R}{P}}}{\left(\left(1 - C\sqrt{\frac{R}{P}}\right)^2 + (1 - C^2)\frac{R}{P}\right)}$$

$$K_r = \frac{R(1 - C^2)}{\left((1 - C)\sqrt{\frac{R}{P}}\right)^2 + (1 - C^2)\frac{R}{P}}.$$

These coefficients will be used in the following section to calculate the noise figure and associated parameters for the HEMT.

IV. RESULTS

The noise temperature predicted by this theory is proportional to frequency, so the analysis is restricted to high frequencies where the contribution to the noise from traps and other low-frequency noise generators is negligible. The validity of any theory can only be tested by comparing its predictions to experiment. This requires a detailed knowledge of the device dimensions and the associated HEMT will suffice to determine the noise temperature, and the small puturbations introduced by the other circuit elements can be introduced if necessary [2].

The equivalent circuit chosen, shown in Fig. 1, is identical that of Pucel et al. [2], so the derivation of the noise parameters follows theirs exactly. Thus, the noise param-

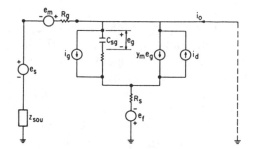

Fig. 1. The equivalent circuit for the HEMT used in the noise analysis. Note that there are no feedback elements.

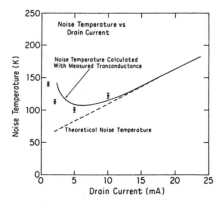

Fig. 2. The calculated and measured noise temperatures for the HEMT. The curves were obtained using measured and calculated values of transconductance, thus indicating the importance of maintaining a large transconductance at low currents if a low noise temperature is to be obtained.

eters are given by

$$r_n = (R_g + R_s) + K_r \frac{1 + \omega^2 C_{sd}^2 R_i^2}{g_m}$$

$$g_n = K_g \frac{\omega^2 C_{sg}^2}{g_m}$$

$$Z_{opt} = (R_s + R_g) + K_g R_i + \frac{1}{j\omega C_{sg}}$$

$$T_{min} = 2T_0 g_n [R_c + R_{opt}].$$

The noise temperature is then

$$T_n = T_{min} + \frac{T_0 g_n}{R_{sou}} [(R_{sou} - R_{opt})^2 + (X_{sou} - X_{opt})^2]$$

where Z_{sou} is the source impedance presented to the transistor.

In measurements made on an actual transistor, it is very difficult to measure all the noise parameters, particularly if it is packaged and the measured noise parameters have to be de-embedded to arrive at the "chip" noise parameters. For this reason, the comparison between theory and experiment will be limited to the noise temperature which is invariant under lossless transformations.

The theory has been applied to devices supplied to NRAO by Cornell University. The noise performance and the material parameters have already been reported by Camnitz et al. [9]. The parameters of the device determined from fitting to S-parameter data and to the dc parameters of the device are given in Table I. Fig. 2 shows the calculated and experimental noise temperature. Note that the calculated noise temperature increases with current even without the inclusion of diffusion noise, unlike the MESFET. This is a consequence of the different relationship between the bias and charge density in the conducting layer for the HEMT. In comparing the noise temperature, the measurements and calculations at the same drain current are used. The disagreement at low currents is probably caused by the drop in the transconductance in the transistor, which does not occur in the model. The agreement is greatly improved by using measured transconductances, shown in the second curve. This clearly shows the importance of maintaining a high transconductance at low currents if low-noise temperatures are to be achieved. Small nonuniformities in the thickness of the

TABLE I
THE QUANTUM WELL HEMT

Parameter		Value
Gate Length	L	0.35 μm
Gate Width	Z	300 μm
Undoped epithickness	e_2	0.035 μm
Doped epithickness	d_2	0.035 μm
Doping Density	N_d	3.8 x 10^{17} cm^{-3}
Mobility	μ_0	4500 cm^2 V^{-1}s^{-1}
Saturation Field	E_s	2900 V cm^{-1}
Barrier Height	ΔE_c	0.1V
Source Resistance	R_s	5Ω
Gate Resistance	R_g	5Ω
Drain Resistance	R_d	3Ω
Intrinsic Resistance	R_i	1Ω

epilayer have been shown to produce a large drop in the transconductance in the MESFET at low currents and a consequence increase in the noise temperature [11]. A similar effect could be occurring here, particularly if it is noted that the epilayer is only 0.05 μm thick about 200 atomic layers. It is important then to maintain the uniformity of the epilayer in low-noise devices.

The agreement between the theory and experiment is fair. Thus, it seems justified to use the theory to investigate dependence on the noise parameters on the various material parameters. The device whose parameters are given in Table I was used as the basis for this analysis. All the results are for a drain voltage of 2 V and 5-mA drain current. Further restrictions were placed on the analysis to maintain a gate length to epilayer thickness ratio of greater than 3.

Figs. 3 and 4 show how the minimum noise temperature changes as the gate length and doped epilayer thickness were varied. The noise temperature is very nearly proportional to the gate length in this model. Thus, reducing the gate length from 0.4 to 0.3 μm will decrease the noise

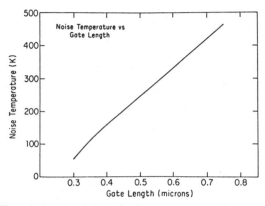

Fig. 3. The calculated variation of noise temperature with gate length for the device of Table I. The model restricts the analysis to ratios of gate length to epithickness greater than 3.

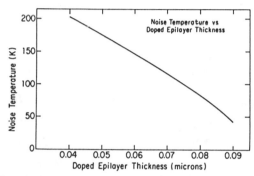

Fig. 4. The calculated variation of noise temperature with doped epithickness for the device of Table I. The analysis is restricted to ratios of gate length to epithickness greater than 3 where the model is valid. The reduction in the noise temperature is primarily caused by a decrease in the gate to source capacitance.

temperature by 25 percent. Some effects which occur in short-gate-length devices, such as velocity overshoot, may cause the rate of change of noise temperature with gate length to be different, but the trend is correct.

The noise temperature is roughly inversely proportional to the thickness of the doped epilayer. Combining these two results shows that a device with as low a ratio of gate length to epilayer thickness as possible should produce the lowest noise, providing the transconductance can be maintained at the low operating current.

The variation of the noise temperature with doping density and with the thickness of the undoped epilayer was also explored. For a fixed drain current, little variation of the noise temperature was seen as these parameters were altered over a range of approximately 10 to 1.

V. Conclusions

The simple noise model of a HEMT presented here has been shown to be in reasonable agreement with experiment. The primary noise mechanism at room temperature is thermal noise; high-field diffusion noise important in the MESFET can be ignored. Simple simulations have

shown that it is important to be able to operate the device at low currents and with a high transconductance to obtain the lowest noise temperature. Nonuniformities in the epilayer thickness have been suggested as the reason for the fall in transconductance at low currents; so, by growing more uniform layers, a better device should result. It may be possible in future to use this simple model to design a low-noise HEMT, or at least to guide the designer in which parameters are to be tightly controlled.

Acknowledgment

I am very grateful to Dr. S. Weinreb for inviting me to Charlottesville and making it possible for me to perform this work. I would also like to thank Dr. J. Granlund for checking the analysis, Dr. M. Pospieszalski for providing the results at 8 GHz, R. Harris for constructing the amplifiers, and also to Cornell, GE, and TRW for providing HEMT's. I would also like to thank all the people in Charlottesville who made my stay so pleasant and in this way contributed much to the work.

References

[1] K. Lee, M. S. Shur, T. J. Drummond, and H. Morkoç, "Current–voltage and capacitance–voltage characteristics of modulation doped field transistors," *IEEE Trans. Electron Devices*, vol. ED-30, pp. 207–212, 1983.

[2] R. A. Pucel, H. A. Haus, and H. Statz, *Advances in Electronics and Electron Physics*. New York: Academic, 1975, pp. 195–265.

[3] A. B. Grebene and S. K. Ghandi, "General theory for pinched off operation junction gate FET," *Solid-State Electron.*, vol. 12, pp. 573–589, 1969.

[4] D. Delagebeaudeuf and N. T. Linh, "Metal(n) AlGaAs-GaAs two-dimensional electron gas FET," *IEEE Trans. Electron Devices*, vol. ED-29, pp. 955–960, 1982.

[5] T. J. Drummond and H. Morkoç, "Model for modulation doped field effect transistors," *IEEE Electron Device Lett.*, vol. EDL-3, pp. 338–341, 1982.

[6] H. Statz, H. A. Haus, and R. A. Pucel, "Noise characteristics of gallium arsenide field effect transistors," *IEEE Trans. Electron Devices*, vol. ED-21, pp. 549–562, 1974.

[7] A. Van der Ziel, "Gate noise in field effect transistors at moderately high frequencies," *Proc. IEEE*, vol. 51, pp. 461–467, 1963.

[8] F. M. Klassnan, "High frequency noise of the junction field effect transistor," *IEEE Trans. Electron Devices*, vol. ED-14, pp. 368–373, 1967.

[9] L. H. Camnitz, P. A. Maki, and L. F. Eastman, "Quarter micron quantum well HEMT with an AlGaAs buffer layer," presented at the Acc. 1984 Int. Symp. on GaAs and Related Compounds, Biarritz, France.

[10] W. Baechtold, "Noise behaviour of GaAs field effect transistors with short gate lengths," *IEEE Trans. Electron Devices*, vol. ED-19, pp. 97–104, 1972.

[11] T. M. Brookes, "Noise in GaAs FET's with a non-uniform channel thickness," *IEEE Trans. Electron Devices*, vol. ED-29, pp. 1632–1634, 1982.

[12] ——, "The noise properties of high electron mobility transistors," NRAO Electronics Division Internal Report, 1984.

[13] K. Lee, M. S. Shur, T. J. Drummond, and H. Morkoç, "Parasitic MESFET in (Al,Ga)As/GaAs modulation doped FET's and MODFET characterisation," *IEEE Trans. Electron Devices*, vol. ED-31, pp. 29–35, 1984.

[14] W. Fawcett and H. D. Rees, "Calculation of the hot electron rate for GaAs," *Phys. Lett.*, p. 578, Aug. 1969.

[15] J. G. Ruch and G. Stuno, "Transport properties of GaAs," *Phys. Rev.*, pp. 921–931, Oct. 1968.

Design of TEGFET Devices for Optimum Low-Noise High-Frequency Operation

PAUL R. JAY, HENRI DEREWONKO, DIDIER ADAM, PIERRE BRIERE,
DANIEL DELAGEBEAUDEUF, PHILIPE DELESCLUSE, AND
JEAN-FRANÇOIS ROCHETTE

Abstract—In the context of applications requiring high gain and minimal noise at microwave and millimeter-wave frequencies, the optimization of a number of interrelated structural and electrical parameters is necessary.

This paper discusses the relative importance of these parameters in the design of TEGFET devices and their influence on the considerations necessary for wideband LNA design. The discussion is illustrated with recent microwave data obtained with millimeter-wave TEGFET devices.

I. INTRODUCTION

THE CONCEPT of a field-effect transistor using an AlGaAs/GaAs super lattice was first proposed by Dingle in 1978 [1]. The difficulty of obtaining good gate control with such a multilayered device led Delagebeaudeuf and Linh [2] and Mimura [3] to propose, one year later, a single heterojunction device that has become known under various acronyms (TEGFET, HEMT, MODFET, SDHT, etc.).

The primary advantages of this type of device stem from the higher mobility enjoyed by the conducting electrons, not only at room temperature, but particularly at lower temperatures, and the improved electrical characteristics of the Schottky-barrier control electrode when placed on a ternary AlGaAs layer.

The following paragraphs seek to summarize the relative contributions of the different geometrical and electrical parameters involved when considering an optimized design for devices to operate in the 18–40 GHz region as low-noise microwave amplifiers. Recent microwave results on 0.35 × 62 µm devices are included to illustrate the various points discussed.

Many of these considerations are also relevant in the design of devices for high-speed logic applications.

II. ANALYSIS OF DEVICE PERFORMANCE

While the achievement of high performance generally requires the simultaneous improvement of many technological parameters within a device, it is important to un-

Manuscript received September 30, 1985; revised December 24, 1985.
P. R. Jay was with Thomson Semiconducteurs, BP 48, 91401 Orsay Cedex, France. He is now with BNR, Ottawa, Ontario, K1Y 4H7 Canada.
H. Derewonko, D. Adam, P. Briere, D. Delagebeaudeuf, P. Delescluse, and J. F. Rochette are with Thomson Semiconducteurs, BP 48, 91401 Orsay Cedex, France.
IEEE Log Number 8607714.

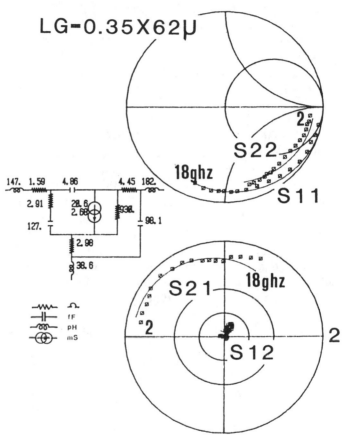

Fig. 1. Microwave *s* parameters and equivalent circuit of a typical high-frequency TEGFET.

derstand the relative contributions of the various parameters and their mutual interplay. An equivalent circuit representation of the transistor (such as shown in Fig. 1) may be obtained by optimizing the component values to fit closely to the small-signal microwave scattering parameters measured on a typical device. It is widely appreciated that such a representation of the transistor should be treated with caution and that in fact reliable values for the elements cannot be obtained using microwave measurements alone. Other determinations of certain parameters (gate resistance, for example) are very important if the equivalent circuit is to have some physical significance. Nevertheless, providing that care is taken in the

Reprinted from *IEEE Transactions on Electron Devices,* vol. ED-33, no. 5, pp. 590–594, May 1986.

determination and use of the equivalent circuit, it provides an excellent means of communication between the microwave appraisal and the device technology.

Given an approximate set of values for the equivalent circuit elements, their relative influence on the microwave behavior may be observed using simple expressions for the noise and gain

$$F = 1 + K_f \cdot 2\pi f\, C_{gs}[(R_s + R_g)/g_m]^{1/2}$$

$$\text{MAG} = \tfrac{1}{4}(f_T/f)^2/[(g_d + 2\pi f_T\, C_{gd})$$

$$\cdot\ (R_s + R_g + R_i + 2\pi f_T\, L_s)]$$

where K_f is a constant related to the optimal current for low-noise operation [4] and $f_T = g_m/2\pi C_{gs}$. The first equation is known as the Fukui equation [5] and is valid for frequencies below about $f_T/2$ while the second is an approximation that supposes unconditional stability of the device. These expressions suffice as a basis for a qualitative discussion of the effects of various parameters on the device performance.

III. Material Considerations

A multilayer epitaxial structure of the type shown in Fig. 2 is generally used and is conveniently and successfully prepared by molecular-beam epitaxy (MBE), although recently HEMT devices have been made [6] using layers grown by organometallic vapor-phase epitaxy (OMVPE). The "undoped" GaAs layer must be of high purity (to minimize impurity scattering of the conduction electrons at the heterojunction interface) and it appears to be advantageous for the residual doping of this layer to be slightly p-type (rather than n-type) in order to reduce the parallel conduction in the rather thick "buffer" layer. Fortuitously, MBE material has this property whereas such a situation is more difficult to achieve using OMVPE.

The next layer grown is usually a thin undoped AlGaAs layer to "space" the conducting electrons in the GaAs further from their host donor atoms in the doped AlGaAs in order to minimize coulombic scattering. Unfortunately this separation is associated with a reduction in the number of carriers transferred from the wide gap ternary to the GaAs. The optimum thickness of the spacer layer then becomes a compromise between protection of the carrier mobility and sheet carrier density. This optimum is also a function of the operating situation envisaged for the device, since, while at low temperatures the high mobilities involved are yet further enhanced by the presence of a spacer layer, at room temperature its improvement of the mobility is only slight and is even deleterious for the sheet carrier concentration. For room-temperature devices, the addition of a spacer layer is probably not necessary, though this may not be the case if cooled operation is required (see below).

Since a higher doping in the n-type AlGaAs results in an increased sheet concentration at the interface, it must be borne in mind that the appropriate n-type concentration in the AlGaAs is limited by two factors:

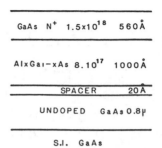

Fig. 2. Example of a TEGFET epitaxial structure.

1) the electrical behavior of the metal Schottky-barrier gate on the n-AlGaAs and
2) the presence of deep trapping "DX" centers associated with the Al content and Si donor centers in the n-type ternary layer.

High-quality Schottky barriers are now being obtained with n-type AlGaAs doped up to about 2.10^{18} cm^{-3}, and the second problem can be reduced either by diminishing the Al concentration to the lowest values possible or by incorporating the dopant in thin GaAs layers interposed within a GaAs/AlGaAs superlattice, thereby separating the Si and Al atoms [7].

A highly doped n$^+$ GaAs surface layer not only serves to facilitate the ohmic contact technology but also helps to protect the otherwise chemically reactive AlGaAs layer and additionally reduces the lateral access resistance between the source contact and the gate region [8].

IV. Geometrical Considerations

One geometrical factor that is directly related to a material aspect is that of gate length, since, as in the case of classical MESFET's, it is essential to keep as large as possible the ratio of gate length L_g to channel thickness a (or in the TEGFET case distance from the gate electrode to the heterojunction interface).

It is not often remarked that, whereas in MESFET devices it is difficult to obtain L_g/a ratios larger than 3, it is possible in a TEGFET structure to achieve more advantageous values such $L_g/a \simeq 10$ even for gate lengths as low as 0.25 μm.

This in itself allows higher values of transconductance to be obtained, albeit with an associated increase of C_{gs}.

Since both g_m and C_{gs} scale linearly with gate width, the cut-off frequency $f_T = g_m/2\pi\, C_{gs}$ should remain constant whatever the gate width involved. From the noise point of view, the factor of importance is

$$C_{gs}\, \sqrt{(R_s + R_g)/g_m}$$

or in an equivalent form

$$C_{gs}/g_m\, \sqrt{g_m(R_s + R_g)}.$$

Since C_{gs}/g_m is independent of gate width Z, g_m proportional to Z, R_s scales as Z^{-1}, and R_g scales as Z (for a constant number of gate fingers), then noise figure decreases with gate width. The choice of smaller gate width, that already follows from impedance considerations when frequency operation increases, is then beneficial to noise figure.

However, as gate widths become smaller, nonvarying contributions, such as gate pad parasitic capacitance, become more significant and when these become on the order of C_{gs} itself, a further reduction of Z is less useful. As R_g gets smaller for smaller Z, it rapidly becomes far less significant than the *increasing* value of access resistance R_s. We have demonstrated TEGFET devices using "mushroom" cross section gate electrodes to diminish R_g [9], and found that the contribution of reduced R_g did not warrant the additional process technology necessary to achieve these gates. Rather, at unit widths of 50 μm or less, it is essential to address the contribution from the series source resistance R_s, which is minimized by reducing source–gate separation as far as possible. A good clearly defined contact edge is necessary in this case.

V. Microwave Performance

Fig. 3 shows the type of geometry evolved bearing in mind the foregoing considerations; the values of the equivalent circuit in Fig. 1 correspond to a transistor of this geometry. Shown in Fig. 4 are the variations of minimum noise figure and associated gain as a function of frequency from 12 to 25 GHz measured using special off-chip matching techniques described in [9]. Many workers quote values of $f_T = g_m/2\pi C_{gs}$ to qualify their devices, though as mentioned above, these values become somewhat ambiguous if the true value of C_{gs} is complicated by a parasitic contribution. Perhaps a more useful comparison may be made using f_{max} as derived by extrapolating the frequency variation of MAG to unity gain. For the device in Fig. 4 a value of f_{max} around 100 GHz may be extrapolated assuming a 6-dB/octave fall-off.

VI. Temperature Variation

We have discussed elsewhere [10] the various factors conducive to improved TEGFET performance at lower temperatures. The obvious improvement from room temperature down to 77 K stems from an enhancement of transport properties that not only causes g_m° to almost double, but also renders the otherwise parasitic source resistance R_s vanishingly small. The reduced metallization resistance also helps by reducing R_g to about half its room-temperature value. In this way very low values of the noise figure may be obtained, making cooled TEGFET devices an economically attractive alternative to parametric amplifier or maser front ends. Even moderate cooling with simple "Peltier" cooling modules can be beneficial, and Fig. 5 shows the temperature variation of noise figure and associated gain measured on a 0.35-μm gate length $\times$ 62-μm gate width TEGFET.

It is also worth noting that while most of the resistive contributions respond more or less linearly as temperature is reduced, those mechanisms relating to electromigration vary exponentially with T, implying considerably improved reliability from devices operated at lower temperatures.

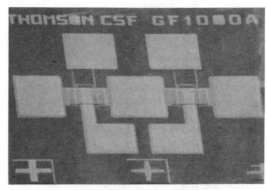

Fig. 3. Illustration showing two-cell layout of the transistor geometry (the two drain pads are in the foreground).

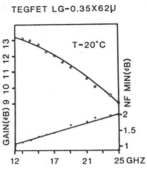

Fig. 4. Minimum noise figure and associated gain between 12 and 25 GHz.

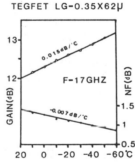

Fig. 5. Temperature variation of noise figure and associated gain.

In applications where temperature variation of performance is an undesirable effect, it is important to eliminate the spacer layer (as discussed above) and minimize the value of source resistance since the mobility changes affect this parameter more markedly than the intrinsic transconductance itself.

VII. Microwave Amplifier Design

Another major advantage of the TEGFET type of structure is that it is prepared with a fabrication technology very similar to that of classical GaAs MESFET's and, in particular, is made using semi-insulating substrates. This fact allows the TEGFET to be co-integrated with other components in an MMIC. From the point of view of matching for optimum noise performance, whether for discrete devices or for MMIC low-noise amplifiers, the sensitivity of the noise figure to the complex generator admittance is an important factor. Fig. 6 shows how this dependence for a TEGFET changes between 6 and 17

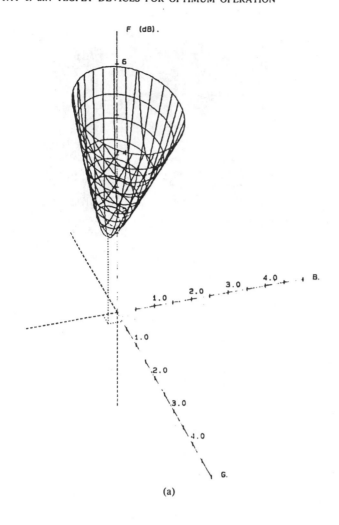

(a)

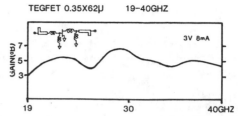

TEGFET 0.35X62µ   19–40GHZ

Fig. 7. Gain versus frequency for a 20–40-GHz wideband TEGFET amplifier.

GHz. It may be seen that a wide range of impedance variation corresponds to the noise minimum at 17 GHz indicating that TEGFET usage helps to facilitate matching conditions in a situation where "tweaking" might not be possible.

Prototype wide-band amplifiers have been assembled using a prematching technique with the 0.35-µm TEGFET chips described.

Single-stage single-ended amplifiers were measured although the layout of this transistor is such that two devices on the same chip are available for balanced amplifier configurations. Fig. 7 shows a wide-band gain of 5 dB from 20 to 40 GHz using a 62-µm gate-width TEGFET. The performance achieved with a 0.35-µm TEGFET is comparable to that obtained using smaller gate-length MESFET devices, offering high-frequency low-noise amplification with simpler fabrication technology.

VIII. CONCLUSION

We have discussed the various material and geometrical factors involved in designing a small-gate-width TEGFET for low-noise amplification above 18 GHz. The results demonstrate that such a device is potentially useful for wide-band low-noise amplification up to 40 GHz.

ACKNOWLEDGMENT

The authors gratefully acknowledge A. M. de Parscau, P. Lacroix, and L. Chusseau for their technical contributions as well as B. Carnez and C. Rumelhard for useful discussions.

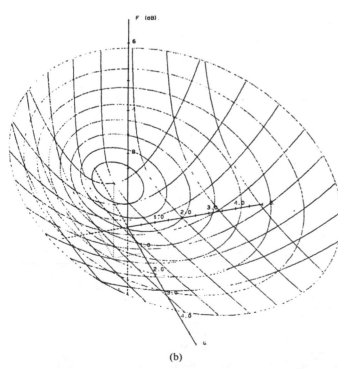

(b)

Fig. 6. Sensitivity of TEGFET noise figure to the complex generator admittance (normalized for 50 Ω). (a) At 6 GHz: $G_{opt} = 0.27$; $B_{opt} = -0.38$; $R_n = 58.7$. (b) At 17 GHz: $G_{opt} = 1.08$; $B_{opt} = -1.04$; $R_n = 13.1$.

REFERENCES

[1] R. Dingle, U.S. Patent, 4 163 237, Apr. 24, 1978.
[2] D. Delagebeaudeuf and N. T. Linh, French Patent, 79 078 03, Mar. 28, 1979.
[3] T. Mimura, Japanese Patent, 54-171026, Dec. 29, 1979.
[4] D. Delagebeaudeuf, J. Chevrier, M. Laviron, and P. Delescluse, "A new relationship between the Fukui coefficient and optimal current value for low noise operation of field effect transistors," *IEEE Electron Device Lett.*, vol. EDL-6, no. 9, Sept. 1985.
[5] H. Fukui, "Design of microwave GaAs MESFET's for broadband low noise amplifiers," *IEEE Trans. Microwave Theory Tech.*, vol. MTT-27, no. 7, July 1979.
[6] Y. Takanashi and N. Kobayashi, "AlGaAs/GaAs 2-DEG FET's fabricated from OMCVD wafers," *IEEE Electron Device Lett.*, vol. EDL-6, no. 3, Mar. 1985.
[7] T. Baba, T. Mizutani, and M. Ogawa, *Japan. J. Appl. Phys.*, vol. 22, no. 10, Oct. 1983.
[8] M. Laviron, D. Delagebeaudeuf, J. F. Rochette, P. R. Jay, P. Delescluse, J. Chevrier, and N. T. Linh, "Ultra low-noise and high-

frequency microwave operation of FET's made by MBE," in *Inst. Phys. Conf. Ser. no. 74, GaAs and Related Compounds* (Biarritz, France), p. 539, 1984.

[9] P. R. Jay, D. Adam, D. Delagebeaudeuf, H. Derewonko, and L. Chusseau, "High gain low-noise TEGFET devices for 18–40 GHz use," presented at the Cornell Conf., July 27–31, 1985.

[10] D. Delagebeaudeuf, P. Delescluse, and P. R. Jay, "Extremely low noise and low temperature TEGFET operation," presented at 15th European Micr. Conf., Paris, Sept. 9–10, 1985.

A High Aspect Ratio Design Approach to Millimeter-Wave HEMT Structures

MUKUNDA B. DAS, SENIOR MEMBER, IEEE

Abstract— In MESFET and HEMT structures as the gate length is reduced below 0.5 μm in an attempt to achieve amplification at highest possible frequencies, it is essential that the depletion depth under the gate be also reduced in order to preserve a high aspect ratio that ensures a high device voltage gain factor (g_m/g_0) and a reasonable value of stable power gain at high frequencies. Results based on this design approach indicate that an n-AlGaAs/GaAs HEMT structure with 0.25-μm gate length could provide stable power gain in excess of 6 dB at the unity current gain frequency of 92.4 GHz, and for an aspect ratio of ten it is difficult to reduce the gate length below 0.25 μm.

INTRODUCTION

IN RECENT months several research groups have reported interesting results concerning device fabrication technology and microwave performance of submicrometer gate length MESFET [1]-[3] and HEMT [4], [5] structures. In general, these studies have indicated that while a shortest possible separation between the gate and the source increases the device transconductance, and shortening of the channel length below 0.5 μm improves the carrier saturation velocity, they do not necessarily lead to the desired improvement of the device power gain performance. This situation requires a careful optimization approach to design device structures in order to achieve the highest possible power gain with acceptable stability margin. This investigation is an attempt to provide such an optimization approach concerning the HEMT structures.

The structural optimization begins with the recognition of the following facts.

1) The edge region or the fringing gate-to-drain feedback capacitance (C_{DG}) cannot be reduced beyond a practical minimum that is independent of the depletion depth under the gate.

Manuscript received February 13, 1984; revised May 23, 1984.
The author is with the Department of Electrical Engineering, Solid State Device Laboratory, and Materials Research Laboratory, Pennsylvania State University, University Park, PA 16802. He was on IPA assignment under the Air Force Systems Command with the Electronic Research Branch, AFWAL, Wright-Patterson Air Force Base, OH.

Reprinted from *IEEE Transactions on Electron Devices,* vol. ED-32, no. 1, pp. 11–17, January 1985.

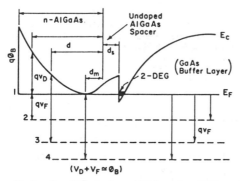

Fig. 1. The effects of n-AlGaAs layer thickness variation of the magnitude of the forward bias required to obtain the same value of n_{s0}.

2) A sizeable fraction of the source series resistance (r_{SS}) is due to the contact resistance and thus its reduction by lowering the source-to-gate separation is ultimately ineffective.

3) The magnitude of the stable power gain margin (Y_{21}/Y_{12}) at f_τ is ideally determined by the gate capacitance to the feedback capacitance ratio (C_G/C_{DG}) and in lateral structures this approximates to gate aspect ratio, i.e., the gate length to the gate depletion depth ratio (L_g/d_0). A high aspect ratio is also desirable in order to obtain a high voltage gain factor (g_m/g_0).

4) A minimum value of the depletion depth under that gate (d_0) can only be obtained by selecting the highest possible doping concentration.

THE HEMT STRUCTURE

In order to be able to reduce the total depletion depth under the gate, the behavior of the HEMT structure with varying thickness of the n-AlGaAs layer should be understood. Fig. 1 depicts a composite schematic band diagram of a Schottky-gate heterojunction structure with different n-AlGaAs thicknesses each under a different appropriate forward bias, between the metal gate and the two-dimensional electron gas (2-DEG), via the source ohmic contact, such that the sheet carrier concentration (n_{s0}) remains at its maximum value. It is clear from this diagram that for the minimum thickness (d_m) of the highly doped n-AlGaAs layer, required for maximum n_{s0}, the forward bias required is the same as the magnitude of the Schottky-barrier potential ϕ_B. The value of d_m depends on the doping concentration in the n-AlGaAs layer and to a small extent on the thickness of the AlGaAs spacer layer (d_s). Utilizing the data given in Delagebeaudeuf and Linh [6] it has been found that $d_m \simeq 70$ Å and $n_{s0} = 1.41 \times 10^{12}$ cm^{-2} when $d_s = 60$ Å and $N_D = 2 \times 10^{18}$ cm^{-3}. Under this condition the potential drops across d_m and d_s are ~ 74 and ~ 126 mV, respectively. In view of the difficulties in the calculation of these quantities when the doping in the n-AlGaAs is high and the accuracy of the various assumptions involved in the quantum well modeling is in doubt, the exact values of d_m and n_{s0} may differ from those given earlier. Nonetheless, based on these values, dimensions of an HEMT structure with 0.25-µm gate length are optimized for best possible performance according to the electrical device model calculations outlined in the following.

ELECTRICAL CHARACTERISTICS

An important parameter of the HEMT structure is the effective capacitance of the gate that can be related to the sheet carrier concentration in the 2-DEG in terms of the effective gate voltage above its threshold value (V_{TH}). This capacitance is defined by ($d_m + d_s + d_i$), where d_m and d_s are as identified earlier and d_i is the effective distance of the centroid of the 2-DEG from the interface between the undoped AlGaAs and GaAs buffer layers. Approximate value of d_i could be as much [7] as 80 A. Thus the value of the per unit area gate capacitance (C_0) can be calculated to be 5.1×10^{-7} F/cm^2. Note that the notations used here for different thicknesses differ from those given in [7].

SATURATION CURRENT AND TRANSCONDUCTANCE

The available carrier sheet concentration in the 2-DEG can be controlled by the applied gate voltage (V_{GS}) according to the relationship [6], [7]

$$qn_s = C_0(V_{GS} - V_{TH}) \tag{1}$$

where

$$V_{TH} = \left(\phi_B - \frac{qN_D d_m^2}{2\epsilon} - \Delta\phi_c\right) \tag{2}$$

and the Fermi level in n-AlGaAs is assumed almost coincident with E_c. For the minimum n-AlGaAs thickness $d_m = 70$ Å, when $N_D = 2 \times 10^{18}$ cm^{-3}, $\Delta\phi_c \simeq 320$ mV (the heterojunction discontinuity potential), and $V_{GS} \sim |\phi_B|$, the maximum effective bias voltage ($V_{GS} - V_{TH}$) becomes 400 mV. However, using $n_{s0} = 1.4 \times 10^{12}$ cm^{-2} one obtains $qn_{s0}/C_0 \simeq 442$ mV. Using a lower value of n_{s0} reported by Lee et al. [8] this bias becomes 345 mV. Thus there arises the uncertainty as to which value should be taken for calculations. For the calculations reported in this paper maximum effective bias used is 442 mV and other values used are 221 and 110.5 mV. Use of 345 mV instead of 442 would reduce the saturation current and the transconductance, as given later in Table II, by less than 25 and 2.4 percent, respectively; the series source and drain resistance would change negligibly because of the contact resistance. The HF performance, as presented later in Table III, would not be significantly affected by the possible difference in the maximum bias voltage cited.

The dc and small-signal ac parameters values were calculated using the equations [9] given in Table I. The saturation current and transconductance expressions were obtained assuming that the current saturation occurs due to carrier velocity saturation [10], [11]. The output conductance (g_0) was not determined theoretically; instead, assumed values have been used in the present calculations on the basis of experimentally observed behavior of this parameter in short-channel devices. For example, in the 0.25-µm gate length devices with 2.1- and 0.5-µm channel, g_m/g_0 values of 10 and 12.3, respectively, have been observed [2]. The saturation carrier velocities in these two channel length cases were 1.1×10^7 cm/s and 1.9×10^7 cm/s, respectively. This apparent increase in the

TABLE I
DEVICE PERFORMANCE EQUATIONS AND ASSUMPTIONS

Basic Equations	Assumptions
1. Drain saturation current vs. gate voltage $$I_{D-sat} = g_m\big|_{max} V_c(1 + U_G - \sqrt{1 + 2U_G})$$ where $U_G = V_G/V_c$ and $$g_m\big|_{max} = \frac{Z \epsilon_s v_{sat}}{d_m + d_s + d_i}$$	Z: channel width V_G: $(V_{GS} - V_{TH})$ ϵ_s: semiconductor permittivity $V_c = L_g \mathcal{E}_c$ L_g: gate length $\mathcal{E}_c = v_{sat}/\mu_o$ $\mu = \mu_o/(1 + \mathcal{E}/\mathcal{E}_c)$
2. Transconductance vs. gate voltage $$g_m = g_m\big|_{max}\left(1 - \frac{1}{\sqrt{1 + 2V_G/V_c}}\right)$$	γ_1 approaches unity with the increase of velocity saturated carriers.
3. Input gate capacitance $$C_G = \frac{\gamma_1 Z \epsilon_s L_g}{d_m + d_s + d_i}, \quad \frac{2}{3} \le \gamma_1 \le 1$$	
4. Drain-to-gate feedback capacitance $$C_{DG} = Z \epsilon_s \gamma_2, \quad \gamma_2 \approx 1$$	γ_2 depends on the location of the saturation point and additional contributions from interelectrode parasitic effects.
5. Drain-to-source output capacitance $$C_{DS} = Z\epsilon_s K(m)/2K(n)$$	K: complete elliptic integral of the first kind L_{CH}: channel length $n = L_g/L_{CH}$ $m^2 + n^2 = 1$ $$G_{MA}\big|_{up} = \frac{(Y_{21})^2}{4\,Re(Y_{11})\,Re(Y_{22})}$$ $$f_\tau = \frac{g_m}{2\pi C_G}$$
Amplifier Performance	
6. Upper-bound of maximum available power gain, $$G_{MA}\big|_{up} = \frac{\theta}{4}\left(\frac{f_\tau}{f}\right)^2\left(\frac{g_m}{g_o}\right)$$ where $$\frac{1}{\theta} = \frac{1}{n} + (r_{SS} + r_{GG})g_m/\Delta$$ $$3 \le n \le 5$$	$$\eta = \frac{1}{g_m r_{CH}}$$ $$\Delta = (1 + g_m r_{SS})$$ g_o: output conductance r_{SS}: source resistance r_{GG}: gate series resistance $g_m \gg \omega C_{DG}$
7. Upper-bound of maximum oscillation frequency, $$f_{max}\big|_{up} \approx f_\tau\sqrt{\frac{\theta}{4}\left(\frac{g_m}{g_o}\right)}$$	(see Appendix A) $$\gamma_3^2 = 1 + \left(\frac{1}{\eta} + \frac{g_m r_{SS}}{\Delta}\right)^2$$
8. Maximum stable power gain margin, $$G_{ms} \approx \left(\frac{f_\tau}{f}\right)\frac{C_G}{C_{DG}}\frac{1}{\Delta}\frac{1}{\gamma_3}$$	
9. Actual maximum available power gain, $$G_{MA} = \frac{G_{MA}\big|_{up}}{1 + \delta_F}$$	$$\delta_F = \frac{C_{DG}(1 + 2.5C_{DG}/C_G)\Delta^3}{5C_G g_o(r_{CH}\Delta + r_{SS} + r_{GG})}$$
10. Maximum frequency of oscillations $$f_{max} = \frac{f_{max}\big|_{up}}{\sqrt{1 + \delta_F}}$$	(See Appendix B)

TABLE II
PHYSICAL AND ELECTRICAL PARAMETERS (300 K)

(a) Calculated parameters vs. gate bias voltage

$(V_{GS}-V_{TH})$	I_{D-sat}	g_m	$1/g_m$	n	r_{CH}	γ_1	C_G	C_G/C_{DG}	f_τ	$\tau \approx C_G/g_m$
mV	mA/mm	mS/mm	Ω.mm	–	Ω.mm	–	pF/mm	–	GHz	psec
442	256.2	725.1	1.379	3	0.4596	0.98	1.248	8.05	92.4	1.72
221	104.5	638.8	1.575	3.5	0.4500	0.95	1.210	7.80	83.4	1.91
110.5	39.6	523.3	1.911	4.5	0.4244	0.90	1.146	7.39	71.6	2.22

(b) Assumed parameters

$d_m = 70\text{Å}$	$\mu_o = 8\times10^3\ cm^2/Vsec$	$\rho_c = 10^{-6}\ \text{-}cm^2$
$d_i = 80\text{Å}$	$\epsilon = 1.07\times10^{-12}F/cm$	$r_{GG} = 0.1\Omega$
$d_s = 60\text{Å}$	$v_{sat} = 1.9\times10^7\ cm/sec$	(for $Z = 0.1cm$)
$L_g = 0.25\mu m$	$\Delta\phi_c = 320\ mV$	$r_{DG} = 0$
$L_{CH} = 0.75\mu m$	$n_{so} = 1.41\times10^{12}cm^{-2}$	$g_m/g_o = 10$
$Z = 0.1\ cm$		
		$C_{SG} \ll C_G$

(c) Calculated fixed parameters:

$\rho_s = 555\,\Omega/square$	$g_m/max = 968\ mS/mm$
$\mathcal{E}_c = 2.375\times10^3\ V/cm$	$r_{SS} = 0.32\ \Omega.mm$
$V_c = 59.3\ mV$	$r_{DD} = 0.32\ \Omega.mm$
$V_{TH} = 500\ mV$	$C_{DG} = 0.155pF/mm.$

steady-state carrier saturation velocity with reducing channel length is caused by the reduction of collisions giving rise to what is known as the ballistic motion [12]. A recent study [13] of high field electron transport in GaAs has revealed that the peak steady-state electron velocity can approach 4×10^7 cm/s for a 0.5-μm channel under a uniform field of 10 kV/cm. It is, however, difficult to translate this high carrier velocity in an FET channel where the field is highly nonuniform and perhaps for this reason the effective saturation velocity is v_{sat}

in a 0.5-μm channel [2] is only 1.9×10^7 cm/s. The assumed manner in which the carrier mobility depends on the electric field is indicated in the mobility expression given in Table I and this has been used in the present design calculations as others have used previously [11].

The current and transconductance values calculated for the assumed 0.25-μm gate length device with 1-mm gate width at specified gate bias conditions are presented in Table II.

PARASITIC CAPACITANCE AND RESISTANCE ELEMENTS

The origins of parasitic capacitance elements are identified in the device cross-sectional schematic shown in Fig. 2(a) and an equivalent network representation that includes these elements is shown in Fig. 2(b).

The most important of the parasitic capacitances is the drain-to-gate feedback capacitance (C_{DG}). This mainly consists of the sidewall depletion in the n-AlGaAs, a small parasitic coupling between the metal gate edge and the 2-DEG, and another small parasitic component outside of the semiconductor between the metallized gate and drain sidewalls. Under a large operating drain voltage the semiconductor depletion sidewall capacitance will tend to decrease. A detailed calculation [14] of this capacitance is difficult; however, an approximate realistic value can be obtained considering that this is equivalent to the edge region fringing capacitance between a parallel plate capacitor with the lower plate (channel) extending indefinitely beyond the edge of the upper plate (gate). This is almost independent of the separation between the parallel plates as indicated in Table I. This capacitance can be

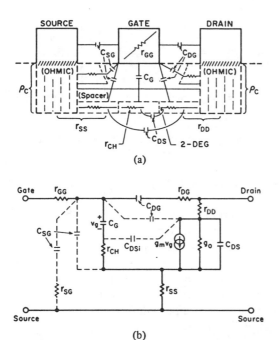

Fig. 2. (a) Cross-sectional schematic of a HEMT structure, and (b) its lumped equivalent network representation. (Circuit elements shown by dotted lines are of less significance.)

represented in two parts, one being the inner component and the other the outer component (see Fig. 2(a)). The total value of C_{DG} for a 0.25-μm drain–gate separation can have a maximum value of 0.203 pF/mm and a minimum value of 0.155 pF/mm depending on the drain bias voltage. Included in this is the capacitance between the metallization sidewalls ($\sim$0.035 pF/mm).

The source-to-gate capacitance (C_{SG}) has a similar origin as that of C_{DG} and due to a lower gate–source voltage its value would be larger than that of C_{DG}. In a device with a minimum depletion depth ($\sim$210 Å) and a gate length of 0.25 μm, the gate capacitance C_G becomes much larger than C_{SG} and for this reason an exact knowledge of the latter is unimportant.

The source-to-drain parasitic capacitance (C_{DS}) can be modeled [8] by assuming that this is equivalent to the one-sided capacitance between two conducting coplanar stripes separated by a small gap [15]. This will be distributed between the drain and source resistances. For 0.25-μm gate length and 0.75-μm channel length this capacitance becomes $\sim$0.084 pF/mm. There is another inner component of C_{DS} due to the distributed channel resistance under the gate and its coupling with the drain conducting region via the buffer layer (see Fig. 2(b)). This capacitance could be as much as the outer component of C_{DS} cited earlier.

There are three resistances that can be presented in series with the terminals of the FET structure. The most important of these is the source series resistance (r_{SS}); it consists of two parts—one arises due to the contact resistance (ρ_c) and the other is due to the ungated channel between the source and the gate. Assuming $n_{s0} \simeq 1.41 \times 10^{12}$ cm^{-2} and a mobility value of 8×10^3 cm^2/V $\cdot$ s, one obtains 0.235 $\Omega \cdot$ mm for the contact related part of r_{SS} when $\rho_c = 10^{-6}$ $\Omega \cdot$ cm^2. Assuming a source-gate separation of 0.25 μm, one obtains 0.083 $\Omega \cdot$ mm for the ungated part of r_{SS} for the same mobility and

sheet carrier concentration values. This would imply that r_{SS} can have a value of $\sim$0.32 $\Omega \cdot$ mm. This is quite in line with the reported [4] value of $\sim$1.27 $\Omega \cdot$ mm for a gate–source gap of 0.7 μm. For the determination of the drain series resistance (r_{DD}) a similar approach can be used and usually its value can be somewhat higher than r_{SS} due to a higher gate–drain separation. Too high a separation in subhalf-micrometer gate would, however, be deleterious to obtaining a high carrier saturation velocity.

The series resistance of the gate metallization is usually lowered by parallel gate pad arrangement such that only small fraction of the entire gate width (Z) is encountered in signal flow path. For millimeter-wave devices this fractional gate-width should remain below 50 μm so that lumped network element modeling concept is not invalidated. By a careful arrangement of the gate metallization, Chao et al. [4] have demonstrated that for 150-μm gate width for 0.25-μm gate length a gate resistance (r_{GG}) value of 0.8 Ω is possible. An extrapolation based on this value with multiple gate pad arrangement should give r_{GG} below 0.1 Ω for a total gate width of 1 mm. In discrete devices wirebounding and distributed gate pattern would require an appropriate representation by suitable inductive elements in series with the gate, source, and drain terminals. In integrated circuits, however, their effects can be minimized or absorbed in passive circuit elements by a careful chip layout procedure. In this work these inductances are considered to be absent.

Power Gains and Amplifier Stability

Depletion-mode high electron mobility FET structures were analyzed [16] for power gain and stability performance, based on an n-AlGaAs doping concentration of $\sim$7 $\times$ 10^{17} cm^{-3} and gate lengths of 1, 0.5, and 0.25 μm with different channel lengths, using the detailed form of the equivalent network model (see Fig. 2(b)) following the standard circuit analysis approach [17]. Due to the large values of the parasitic resistance and capacitance elements and comparatively low values of the gate capacitance, no useful simplification of the power gain expressions could be made and the calculated data indicated that although there is potential power gain in some cases, the stability margins in all cases were too small for practical use. When the parasitic capacitance values, however, are much lower than the active gate capacitance magnitude, and the associated parasitic RC time constant values are much smaller than the effective channel carrier transit time, then the power gain expressions can be simplified as given in (6), (7), and (8) of Table I. It is particularly instructive to note how a high $g_m r_{SS}$ product can reduce the upper bound of the power gain by 6 to 8 dB depending on the value of η, and also its influence on the stable power gain margin. The basis for the derivation [18] of the upper bound of the power gain is given in the Appendix, Section A.

In the case when the drain-to-gate feedback capacitance is not negligible comparable to the gate capacitance, the actual power gain can become lower than its upper-bound value. Approximate results representing this situation are also given in (9) and (10) in Table I, and the relevant analytical steps are given in the Appendix, Section B. For the purpose of numer-

TABLE III
BIAS DEPENDENCE OF UPPER-BOUND PERFORMANCE

I_{D-sat} mA/mm	f_τ GHz	G_{MA}/UP at f_τ	G_{MS} at f_τ	f_{max}/UP GHz
256.2	92.4	6.33dB	6.67dB	191
104.5	83.4	6.93dB	6.70dB	185
39.6	71.6	7.84dB	7.15dB	176

TABLE IV
COMPARISON OF ACTUAL PERFORMANCE WHEN $f_\tau = 92.4$ GHz

Performance	(Wolf)[19]	(This work)	(Curtice)[20]
f_{max}(GHz)	138.5	141	154
G_{mA} (dB) (at f_τ)	3.54	3.73	4.46

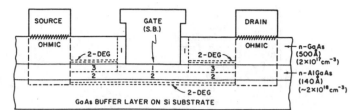

Fig. 3. Cross-sectional schematic of a proposed high-aspect ratio enhancement-mode HEMT structure.

[20] gives the highest value since he introduced a negative term under the square root sign in (3) that includes the effect of the inner part of C_{DS} as shown in Fig. 2(b).

The significance and importance of the proposed high gate aspect ratio design can be best appreciated by focusing on the ratio C_{DG}/C_G in (8) of Table I and in (4). In the former, at f_τ and beyond, the stability is seen to be directly proportional to the aspect ratio (see also (3) and (4) in Table I). In the latter, the aspect ratio should be high for the realization of the upper bound of performance.

ical determination of various power gains versus frequency behavior of a 0.25-μm gate length device with a high aspect ratio, all structural dimensions and relevant electrical parameter values are listed in Table II. The results showing the upper bound of power gain performance are given in Table III.

Prediction of realistic values of f_{max} and power gain at f_τ would be of great interest to those who are currently involved in the R and D activities concerning HEMT structures. For this reason calculations based on the equations obtained in this paper and reported by others [19], [20] are compared in Talbe IV for the selected device structure. The equation for f_{max} obtained by Wolf [19] can be recast in the following manner:

DEVICE REALIZATION

Manufacturing of high aspect ratio enhancement-mode HEMT's would require careful control of the thickness and doping concentration in the n-AlGaAs layer and its protection from the effect of free-surface potential in the gate-to-drain and gate-to-source separation regions. A structure that would achieve these objectives is schematically shown in Fig. 3. In the gate region the capping n-GaAs layer is removed by etching utilizing the differential etching properties [5] of AlGaAs and GaAs. The thickness of this layer is predetermined and grown by MBE technique. A typical combination of doping and thickness values are indicated in Fig. 3. For approximately zero threshold voltage and complete depletion of the n-AlGaAs layer, under the maximum allowed positive gate bias voltage,

$$f_{max} \simeq \frac{f_\tau}{\sqrt{4g_0(R_{CH} + r_{SS} + r_{GG}) + \frac{2C_{DG}}{C_G}\left\{\frac{C_{DG}}{C_G} + g_m(R_{CH} + r_{SS})\right\}}} . \tag{3}$$

For comparison purposes the f_{max} equation given in Table I (see (10)) can also be rearranged in the same fashion, as given in the following:

the relevant band and field diagrams are shown in Fig. 4. The extent of the total gate voltage swing is indicated by the hatched area, plus that due to d_i.

$$f_{max} = \frac{f_\tau}{\sqrt{4g_0\left(R_{CH} + \frac{r_{SS} + r_{GG}}{\Delta}\right) + \frac{4}{5}\frac{C_{DG}}{C_G}\left(1 + \frac{2.5C_{DG}}{C_G}\right)\Delta^2}} \tag{4}$$

where
$$\Delta = (1 + g_g r_{SS}).$$

There are differences in these two equations and the reason is that Wolf [19] ignored the source feedback term $g_m r_{ss}$, and his estimation of the imaginary part of y_{21} was also different (see the Appendix, Section B). Nevertheless, the closeness of results calculated from these equations is quite remarkable in the specific case considered. The equation reported by Curtice

Note that the gate metal not only forms a Schottky barrier at the n-AlGaAs interface, but the same is also formed at the n-GaAs sidewalls. This provides complete electrical isolation due to depletion and for this reason the n-GaAs layer doping should be not too high. Within the interelectrode spacing n-AlGaAs also provides 2-DEG to the n-GaAs and this should have no undesirable effects on the device performance. The free-surface potential should only affect the n-GaAs surface

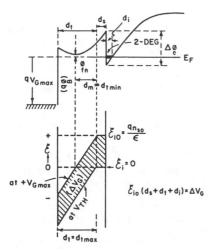

Fig. 4. Band and field diagrams for an E-HEMT showing the basis of its operation and voltage swing limitation.

and thus protect the thin active n-AlGaAs layer and this could enhance the device reliability.

For convenience the gate metallization can be achieved by a technique similar to that employed in the self-aligned gate technique [3] involving refractory metals. This would permit alloying of the source and the drain ohmic contacts after the gate is defined. The gate aspect ratio in the suggested device structure could easily exceed ten for a gate-length of 0.25 μm as required for the desired millimeter-wave performance.

DISCUSSIONS AND CONCLUSIONS

The calculated power gain performance, based on device parameter values, in line with what have been reported in the literature, clearly demonstrates the feasibility of achieving a millimeter-wave HEMT structure that would operate in enhancement mode. It is perhaps not unrealistic to extrapolate this to MESFET structures, too. For practical realization of very thin depletion depth under the gate electrode, it might even be appropriate to increase the doping concentration to 4×10^{18} cm^{-3}. The need to reduce the source series resistance is there; however, an order of magnitude reduction of the contact resistivity will be required to reduce r_{SS} by a factor of two. The parameters that have the most influence on the power gain performance are the carrier saturating velocity (through f_τ) and the voltage gain factor (g_m/g_0), assuming that the gate aspect ratio has already been fixed due to practical limitation. Since v_{sat} and g_0 both increase with the reduction of channel length, no easy compromises are possible to achieve high power gains at higher and higher frequencies. To improve the situation in subhalf-micrometer structures, reduction of g_0 by incorporation [21] of a carefully buried p$^+$-n$^+$ region below the active channel, that would naturally be fully depleted, should be considered.

The predicted upper-bound values of the maximum frequency of oscillations (191 GHz) and the power gain (6.33 dB) at 92.4 GHz are encouraging figures for future development and research. For the aspect ratio used, however, the feedback capacitance is still appreciable compared to the gate capacitance and this reduces the f_{max} and power gain at 92.4 GHz considerably as given in Table IV. It is believed that by

further increasing the aspect ratio, by reducing the spacer layer and/or the n-AlGaAs layer thickness (by increasing doping), the actual performance may be pushed towards the upper-bound values of the structure. Reduction of contact resistance would also be helpful.

In view of the imprecise values of v_{sat} and g_0, at the present state of understanding of device physical processes, it is not possible to predict exactly as to how high in frequency the device performance can be extended. Nevertheless, aided by the proposed high gate aspect ratio design approach, it should be possible to obtain reasonable millimeter-wave amplifying performance from a 0.25-μm gate HEMT structure. This being an enhancement HEMT, its usefulness in high-speed switching circuits need not be further emphasized.

APPENDIX

A. The Basis for the $G_{mA/up}$ Expression

The network model of Fig. 2(b) can be used to define Y' parameters that includes the effects of r_{SS} and then the overall Y-parameters can be calculated by including the effects of r_{GG}. Following this approach the upper bound of the maximum available power gain can be approximated to

$$\frac{1}{4} \frac{|Y_{21}|^2}{Re(Y_{11})Re(Y_{22})} \simeq \frac{|Y_{21}'|^2}{4\omega^2 C_G'^2 (r_{CH}' + r_{GG})g_0'} \quad (A1)$$

where

$$r_{CH}' = \frac{\Delta}{\eta g_m} + r_{SS}; \quad g_0' = g_0/\Delta$$

$$C_G' = C_G/\Delta; \quad \Delta = 1 + g_m r_{SS}$$

and

$$|Y_{21}'| \approx g_m/\Delta.$$

These approximations are valid when r_{GG} is low, C_{DG} is much smaller than C_G and $\omega^2 C_{DG}^2 (r_{DD}/g_0') \ll 1$, as would be the case for high aspect ratio designs. A final form of (A1) above is given in Table I, (6).

B. The Effect of C_{DG} on the Power Gain

In practice, the C_G/C_{DG} ratio may be less than 10, and this could have noticeable effect on the actual power gain. This effect can be calculated by not neglecting the $Re(Y_{12}Y_{21})$ term in the stability factor given by

$$K = \frac{2Re(Y_{11})Re(Y_{22}) - Re(Y_{12}Y_{21})}{|Y_{12}Y_{21}|} \quad (A2)$$

when $K \geqslant 2$, which is usually true above f_τ, the actual gain can be written as

$$G_{mA/up} = \frac{G_{mA/up}}{1 + \delta_F} \quad (A3)$$

where

$$\delta_F = \frac{-Re(Y_{21}Y_{22})}{2 Re(Y_{11})Re(Y_{22})}.$$

In [18], the intrinsic Y_{21}^i parameter is given by

$$Y_{21}^i \simeq \frac{g_m \exp(-k_{c0} f/f_\tau)}{1 + if/sf_\tau} \qquad (A4)$$

where $s \simeq 4$ and $k_{c0} \simeq 0.61$, similar to those for MOSFET's [22], [23]. For frequencies below $2f_\tau$, the imaginary part of Y_{21}^i as given in (A4) and the same quantity as may be obtained from the intrinsic part of the newwork model in Fig. 2(b) can be shown to become $-j\omega C_G(k_{c0} + 1)/s$ and $-j\omega C_G/\eta$, respectively. This difference arises due to the network representation of g_m with reference to the signal voltage across the gate capacitance. The magnitude $|Y_{21}|$ remains, however, nearly the same in both representations. Thus for greater accuracy, we have taken

$$Im(Y_{21}^i) \simeq -j\frac{\omega C_G}{2.5}. \qquad (A5)$$

The value of the delay factor k_{c0} in (A4) can be increased to represent additional delay that takes place in the velocity saturated region near the drain, and in that case the factor 2.5 should be appropriately reduced. Thus from the preceding we have

$$Re(Y_{12} Y_{21}) \approx -\omega^2 C_{DG}\left(C_{DG} + \frac{C_G}{2.5}\right) \qquad (A6)$$

and for the product $Re(Y_{22})Re(Y_{11})$, the approximate form in the denominator of (A1) can be used. Finally, the expression of G_{mA} and f_{max} as given in Table I can be obtained.

ACKNOWLEDGMENT

The author would like to thank Dr. P. Stover, Dr. D. Langer, and Dr. C. Litton for their encouragement and support. He is also thankful to Dr. G. Norris for useful discussions on heterostructure transport properties. He is grateful to the reviewers for their critical evaluation and helpful suggestions.

REFERENCES

[1] H. Yamasaki, E. T. Watkins, J. M. Schellenberg, L. H. Hackett, and M. Feng, "Design of optimized EHF FET's," presented at the 1983 IEEE/Cornell Conf. on High-Speed Semiconductor Devices and Circuits, Cornell Univ., Ithaca, NY, Aug. 15–17, 1983.

[2] P. C. Chao, P. M. Smith, S. Wanuga, W. H. Parkins, and E. D. Wolf, "Channel-length effects in quarter-micrometer gate-length GaAs MESFET's," IEEE Electron Device Lett., vol. EDL-4, pp. 326–328, 1983.

[3] N. Kato, Y. Matsuoka, K. Ohwada, and S. Moriya, "Influence of n^+ layer-gate gap on short-channel effects of GaAs self-aligned MESFET's (SAINT)," IEEE Electron Device Lett., vol. EDL-4, pp. 417–419, 1983.

[4] P. C. Chao, T. Yu, P. M. Smith, J. C. M. Hwang, S. Wanuga, and W. H. Parkins, "Quarter-micron gate length microwave high electron mobility transistors," presented at the 1983 IEEE/Cornell Conf. on High-Speed Semiconductor Devices and Circuits, Cornell Univ., Ithaca, NY, Aug. 15–17, 1983; also Electron. Lett., p. 894, Oct. 1983.

[5] T. Mimura, K. Nishiuchi, M. Abe, A. Shibatomi, and M. Kobayashi, "High electron mobility transistors for LSI circuits," in IEDM Tech. Dig., 1983.

[6] D. Delagebeaudeuf and N. T. Linh, "Metal-(n)AlGaAs-GaAs two-dimensional electron gas FET," IEEE Trans. Electron Devices, vol. ED-29, pp. 955–960, 1982.

[7] K. Lee, M. S. Shur, T. J. Drummond, and H. Morkoç, "Current-voltage and capacitance–voltage characteristics of modulation-doped field-effect transistors," IEEE Trans. Electron Devices, vol. ED-30, pp. 207–212, 1983.

[8] —, "Electron density of the two-dimensional electron gas in modulation doped layers," J. Appl. Phys., vol. 54, no. 4, pp. 2093–2096, 1983.

[9] M. B. Das, "DC and small signal AC characteristics of n-AlGaAs/GaAs HEMT structures," Res. Rep. 2 and "Network models of microwave HEMT structures," Res. Rep. 3, AFWAL/AADR, Wright-Patterson AFB, OH, 1983.

[10] K. Lehovec and R. Zuleeg, "Voltage–current characteristics of GaAs and J-FET's in the hot electron range," Solid-State Electron., vol. 13, pp. 1415–1426, 1970.

[11] R. Zuleeg and K. Lehovec, "Temperature-dependence of the saturation current of MOST's," IEEE Trans. Electron Devices, vol. ED-19, pp. 987–989, 1968.

[12] L. F. Eastman, "Use of molecular beam epitaxy in research and development of selected high speed compound semiconductor devices," J. Vac. Sci. Technol. B, vol. 1, pp. 131–134, 1983.

[13] B. R. Nag and M. Debroy, "Electron transport in sub-micron GaAs channels at 300 K," Appl. Phys. A, Solids and Surfaces, vol. 31, pp. 65–70, 1983.

[14] E. Wasserstrom and J. McKenna, "The potential due to a charged metallic strip on a semiconductor surface," Bell Syst. Tech. J., vol. 49, pp. 853–877, 1970.

[15] W. R. Smythe, Static and Dynamic Electricity. New York: McGraw-Hill, 1950.

[16] M. B. Das, "Power gain versus frequency behavior of HEMT structures," Res. Rep. 4, AFWAL/AADR, Wright-Patterson AFB, OH, Oct. 1983.

[17] J. M. Rollett, "Stability and power gain invariants of linear two-parts," IEEE Trans. Circuit Theory, vol. CT-9, pp. 29–32, 1962.

[18] M. B. Das and P. Schmidt, "High-frequency limitations of abrupt-junction FET's," IEEE Trans. Electron Devices, vol. ED-20, pp. 779–792, 1973.

[19] P. Wolf, "Microwave properties of Schottky-barrier field effect transistors," IBM J. Res. Dev., vol. 14, pp. 125–141, 1970.

[20] W. R. Curtice, "The performance of submicrometer gate-length GaAs MESFET's," IEEE Trans. Electron Devices, vol. ED-30, pp. 1693–1699, 1983.

[21] T. S. Tan, E. B. Stoneham, G. Patterson, and D. M. Collins, "GaAs FET channel structure investigation using MBE," in GaAs IC Symp. Dig., pp. 38–41, 1983.

[22] M. B. Das, "High frequency network properties of MOS transistors including the substrate resistivity effects," IEEE Trans. Electron Devices, vol. ED-16, pp. 1049–1069, 1969.

[23] M. B. Das, "Generalized high-frequency network theory of field-effect transistors," Proc. IEE (London), vol. 114, pp. 50–59, 1967.

Millimeter-Wave Low-Noise High Electron Mobility Transistors

P. C. CHAO, MEMBER, IEEE, S. C. PALMATEER, MEMBER, IEEE, P. M. SMITH, MEMBER, IEEE, U. K. MISHRA, MEMBER, IEEE, K. H. G. DUH, MEMBER, IEEE, AND J. C. M. HWANG, SENIOR MEMBER, IEEE

Abstract—High electron mobility transistors (HEMT's) have been fabricated which demonstrate excellent millimeter-wave performance. A maximum extrinsic transconductance as high as 430 mS/mm, corresponding to an intrinsic transconductance of 580 mS/mm, was observed in these transistors. A unity current gain cutoff frequency f_T as high as 80 GHz and a maximum frequency of oscillation f_{max} of 120 GHz were projected for these HEMT's. At 40 GHz, a minimum noise figure of 2.1 dB with an associated gain of 7.0 dB has also been measured. These are the highest f_T, f_{max}, and the best noise performance reported to date. The results clearly demonstrate the potential of HEMT's for millimeter-wave low-noise applications.

Rapid progress in material growth and device fabrication techniques involving AlGaAs/GaAs heterostructures has made the high electron mobility transistor (HEMT) a very promising microwave device, for low-noise [1], [2], high-power [3], [4], and high-speed [5] applications. In early 1985, we reported a noise figure as low as 1.3 dB with an associated gain of 8.2 dB at 18 GHz, and a unity current gain cutoff frequency f_T of 70 GHz for the 0.25-μm gate-length HEMT [6]. Considerable efforts have been made since then in optimizing both the epitaxial structure and the parasitic elements of the device. As a result, significant improvements in microwave performance have been obtained in our latest 0.25-μm HEMT's. These performance improvements (power gain in particular) allow the devices to be used in the millimeter-wave frequency range. In this letter we present the room-temperature performance of the improved HEMT's, which have demonstrated state-of-the-art noise and power gain for frequencies up to 40 GHz.

The HEMT devices were fabricated on selectively doped AlGaAs/GaAs heterostructures grown by molecular-beam epitaxy. The detailed material structure and growth conditions have been discussed elsewhere [7]. The HEMT wafer, which has a 20–40-Å spacer layer, exhibits at 77 K a sheet charge density of ~1 × 10^{12}/cm^2 and a mobility of 50 000 cm^2/V·s. Devices which have 0.25-μm-long gates were fabricated on the wafer by electron-beam lithography. The gates with a low-resistance T-shaped cross section, were delineated by the tri-layer resist technique [8]. Details of the HEMT fabrication can be found in [6].

The devices have a typical pinchoff voltage of ~ − 1.0 V. The measured source resistance of the HEMT device was 0.6 Ω·mm. A maximum normalized transconductance g_m of 430 mS/mm, which corresponds to an intrinsic g_m of 580 mS/mm,

Manuscript received July 3, 1985; revised August 6, 1985.
The authors are with the Electronics Laboratory, General Electric Company, Syracuse, NY 13221.

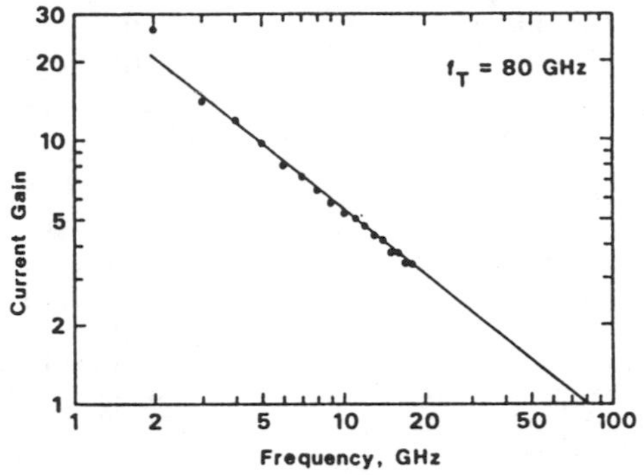

Fig. 1. Current gain as a function of frequency.

was observed at room temperature. As expected for very short gate devices, g_m in these HEMT's is observed to be strongly dependent on gate voltage and decreases rapidly with decreasing drain current, resulting in a soft pinchoff.

The RF performance of the HEMT's was measured from 2 to 40 GHz. The current gain h_{21} of the devices was determined from the S-parameters measured over a range of 2–18 GHz. As shown in Fig. 1, the f_T obtained by extrapolating these h_{21} values is 80 GHz. This is the highest f_T ever reported for any three-terminal device. The equivalent circuit, obtained by computer-fitting the measured 2–18 GHz S-parameters of the HEMT, is shown in Fig. 2. The equivalent circuit model [9], based on the physical structure of the HEMT, was found to provide an excellent fit to the measured HEMT data. From Fig. 2, significant improvement in the feedback capacitance C_{dg}, transconductance g_m, output conductance g_o, and the source and gate resistances R_s and R_g is observed in the latest HEMT when compared with our previous device [6].

The 2–40-GHz power gain results are plotted in Fig. 3, along with the gain calculated based on the equivalent circuit shown in Fig. 2. The device is conditionally stable (stability factor $K < 1$) at least to 18 GHz. The measured maximum gain of the HEMT was 11.9 dB at 30 GHz and 9.5 dB at 40 GHz. The 30- and 40-GHz results were measured using a high-frequency test fixture which exhibited an insertion loss of less than 1 dB over most of the 26–40-GHz band. The fixture includes gate and drain bias insertion networks which were designed to provide stability. No instability/oscillation problems were observed when testing these 0.25-μm HEMT devices. From Fig. 3, if the gain is extrapolated at a slope of − 6 dB per octave, a gain of 6 dB at 60 GHz and a maximum frequency of oscillation f_{max} of 120 GHz are obtained.

The device noise performance has also been measured up to

Reprinted from *IEEE Electron Device Letters*, vol. EDL-6, no. 10, pp. 531–533, October 1985.

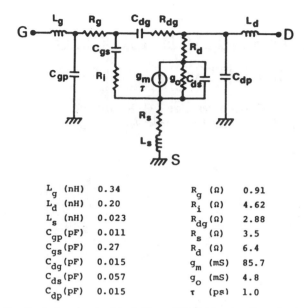

L_g (nH)	0.34	R_g (Ω)	0.91
L_d (nH)	0.20	R_i (Ω)	4.62
L_s (nH)	0.023	R_{dg} (Ω)	2.88
C_{gp} (pF)	0.011	R_s (Ω)	3.5
C_{gs} (pF)	0.27	R_d (Ω)	6.4
C_{dg} (pF)	0.015	g_m (mS)	85.7
C_{ds} (pF)	0.057	g_o (mS)	4.8
C_{dp} (pF)	0.015	τ (ps)	1.0

Fig. 2. Equivalent circuit of the 0.25-μm HEMT biased at V_{ds} = 3.5 V, I_{ds} = 37.5 mA (device width = 150 μm).

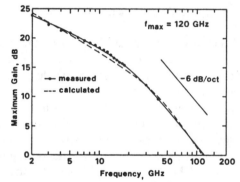

Fig. 3. Measured and calculated maximum gain of the HEMT.

40 GHz. A hot/cold manual Y-factor method was used to calibrate the solid-state noise source. The noise performance of a HEMT device at 40 GHz is shown in Fig. 4 as a function of drain current. The lowest noise figure F_{min} was obtained at a drain current of ~ 120-percent I_{dss} (i.e., with the gate slightly forward biased), where I_{dss} is the zero gate bias drain current. The relatively high value (in percent I_{dss}) of optimum bias current for this HEMT device can be related to the g_m compression that occurs at lower drain currents, as discussed previously. It is worth noting that in these HEMT's, the maximum associated gain and the lowest noise figure occur at the same drain bias current. This is in contrast to the typical transistors, in which minimum noise figure occurs at 15–25-percent I_{dss} while maximum associated gain occurs near I_{dss}. Table I summarizes the measured noise performance of the 0.25-μm HEMT's. These noise results are also the best ever reported for transistors operating at these frequencies. Fig. 5 summarizes the published HEMT noise performance from major laboratories. Extrapolating our noise results shown in Fig. 5, a noise figure of ~3 dB is expected at 60 GHz.

In conclusion, 0.25-μm HEMT's have been fabricated with significantly improved microwave performance. A cutoff frequency f_T of 80 GHz and an f_{max} of 120 GHz have been

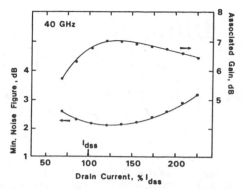

Fig. 4. Drain current dependence of the HEMT noise performance at 40 GHz.

TABLE I
SUMMARY OF HEMT ROOM-TEMPERATURE NOISE PERFORMANCE

Frequency (GHz)	Min. Noise Figure (dB)	Associated Gain (dB)
18	1.2	11.6
30	1.8	9.7
40	2.1	7.0

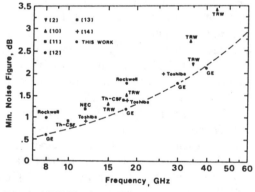

Fig. 5. Reported HEMT noise performance at room temperature from major laboratories.

obtained. These devices have also demonstrated state-of-the-art low-noise performance for frequencies up to 40 GHz. These results highlight the promise of the HEMT for millimeter-wave low-noise applications.

ACKNOWLEDGMENT

The authors wish to acknowledge G. Dimovski, L. Lester, and B. Lee for their technical assistance, and Prof. L. F. Eastman, Cornell University, for helpful discussions.

REFERENCES

[1] P. C. Chao, P. M. Smith, U. K. Mishra, S. C. Palmateer, J. C. M. Hwang, M. Pospieszalski, T. Brooks, and S. Weinreb, "Cryogenic noise performance of quarter-micron gate-length high electron mobility transistors," presented at the 43rd Annual Device Research Conf., Boulder, CO, June 1985.
[2] M. Sholley, J. Berenz, A. Nichols, K. Nakano, R. Sawires, and J. Abell, "36.0–40.0 GHz HEMT low-noise amplifier," in *IEEE MTT-S Dig.*, June 1985, pp. 555–558.

[3] P. M. Smith, U. K. Mishra, P. C. Chao, S. C. Palmateer, and J. C. M. Hwang, "Power performance of microwave high electron mobility transistors," *IEEE Electron Device Lett.,* vol. EDL-6, pp. 86–87, Feb. 1985.

[4] K. Hikosaka, Y. Hirachi, T. Mimura, and M. Abe, "Microwave high power performance of double heterojunction high electron mobility transistor," presented at the 43rd Annual Device Research Conf., Boulder, CO, June 1985.

[5] N. C. Cirillo, Jr. and J. K. Abrokwah, "8.5-picosecond ring oscillator gate delay with self-aligned gate modulation-doped n$^+$-(Al,Ga)As/ GaAs FET's," presented at the 43rd Annual Device Research Conf., Boulder, CO, June 1985.

[6] U. K. Mishra, S. C. Palmateer, P. C. Chao, P. M. Smith, and J. C. M. Hwang, "Microwave performance of 0.25-μm gate-length high electron mobility transistors," *IEEE Electron Device Lett.,* vol. EDL-6, pp. 142–145, Mar. 1985.

[7] S. C. Palmateer, P. A. Maki, W. Katz, A. R. Calawa, J. C. M. Hwang and L. F. Eastman, "The influence of V:III flux ratio on unintentional impurity incorporation during MBE growth," in *Proc. Gallium Arsenide and Related Compounds 1984* (Inst. Phys. Conf. Series 74), 1985, pp. 217–222.

[8] P. C. Chao, P. M. Smith, S. C. Palmateer, and J. C. M. Hwang, "Electron-beam fabrication of GaAs low-noise MESFET's using a new tri-layer resist technique," *IEEE Trans. Electron Devices,* vol. ED-32, pp. 1042–1046, June 1985.

[9] M. B. Das, "A high aspect ratio design approach to millimeter-wave HEMT structures," *IEEE Trans. Electron Devices,* vol. ED-32, pp. 11–17, Jan. 1985.

[10] J. J. Berenz, K. Nakano, and K. P. Weller, "Low noise high electron mobility transistors," in *IEEE MTT-S Dig.,* June 1984, pp. 83–86.

[11] K. Ohata, H. Hida, H. Miyamoto, T. Baba, T. Mizutani, and M. Ogawa, "Planar AlGaAs/GaAs selectively-doped structure FET's," in *Proc. 15th Conf. Solid State Devices* (Japan), 1984, pp. 111–117.

[12] A. K. Gupta, E. A. Sovero, R. L. Pierson, R. D. Stein, R. T. Chen, D. L. Miller, and J. A. Higgins, "Low-noise high electron mobility transistors for monolithic microwave integrated circuits," *IEEE Electron Device Lett.,* vol. EDL-6, pp. 81–82, Feb. 1985.

[13] P. Delescluse, J. F. Rochette, M. Laviron, D. Delagebeaudeuf, P. Etienne, J. Chevrier, P. Jay, and N. T. Linh, "Ultra low noise two dimensional electron gas field effect transistor made by MBE," presented at 21th WOCSMMAD, FL, Feb. 1985.

[14] K. Shibata, B. Abe, H. Kawasaki, S. Hori, and K. Kamei, "Broadband HEMT and GaAs FET amplifiers for 18–26.5 GHz," in *IEEE MTT-S Dig.,* June 1985, pp. 547–550.

NOISE PERFORMANCE OF MICROWAVE HEMT

K.Joshin,T.Mimura,M.Niori,Y.Yamashita,K.Kosemura and J.Saito

Fujitsu Laboratories Ltd.
1677,Ono,Atsugi,243-01,Japan

Abstract

Low noise HEMTs (High Electron Mobility Transistors) with 0.5μm gate have been made using direct electron beam lithography. At 12 GHz a noise figure of 1.4 dB with an associated gain of 11 dB has been obtained at room temperature. Noise figure has been reduced to 0.35 dB by decreasing ambient temperature to 100K.

Introduction

HEMTs based on a modulation-doped GaAs/n-AlGaAs single heterojunction structure offer new possibilities for high speed logic and microwave devices.[1,2] Due to its high electron mobility,it is expected that HEMTs will provide good high frequency operation,specifically higher cutoff frequency f_T for current gain than those of conventional GaAs MESFETs. For GaAs MESFETs the minimum noise figure is described by[3]

$$Fmin=1+K \frac{f}{f_T}\sqrt{Gm(Rg+Rs)} \quad , \quad (1)$$

where K is a fitting factor and f is the operating frequency.

According to (1),higher cutoff frequency leads to a lower minimum noise figure. Thus a superior noise performance is also expected of HEMTs. This paper describes the fabrication process and the noise performance of 0.5μm gate length microwave HEMTs.

Material Preparation

A cross-sectional structure of an MBE wafer is shown in Fig.1. The epitaxial layers consisting of undoped GaAs,Si-doped n-type AlGaAs and n-type GaAs were grown on a semi-insulating GaAs substrate at a temperature of 680°C by MBE. For fabrication of the prototype HEMT,the doping concentration was $2 \times 10^{18} cm^{-3}$,and the n-AlGaAs and the n-GaAs were 0.05μm in thickness. The surface n-GaAs layer was prepared to obtain improved ohmic contacts.

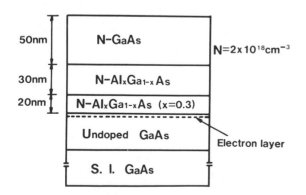

Fig.1 Schematic representation of an MBE wafer.

The hall mobilities,μ, the sheet electron concentrations,n,and the sheet resistances,ρ,were measured with and without the n-GaAs cap layer. These measurements are listed in Table 1. The n-GaAs layer was removed by selective dry etching.[4]

Due to the low mobility electrons in n-GaAs and n-AlGaAs layers,the electron mobility of the as-grown wafer was lower than that of wafer with the n-GaAs layer stripped off. The hall mobilities of 2-dimentional electron gas (2DEG) were about 6000 $cm^2/V \cdot s$ at room temperature and 26000 $cm^2/V \cdot s$ at 77K. The mobility at 77K is lower than ordinarily reported because there was no undoped AlGaAs layer in the heterojunction.[5]

Device Fabrication

Fabrication steps for the 0.5μm gate HEMT is shown in Fig.2(a). (1)To isolate the mesas,the Si-doped layers were etched by Argon ion beam. (2)Ohmic source-drain contacts were made of alloyed AuGe/Au at 450°C. (3)The gate region was wet-chemically recessed to adjust the drain current. (4)The process was completed by a 0.8μm thick aluminum gate formation.

All patterns were delineated by direct electron beam lithography. The gate length was about 0.5μm with a total gate width of 200μm,and the channel length was 2μm. Fig.2(b) is a gate cross-sectional view by SEM.

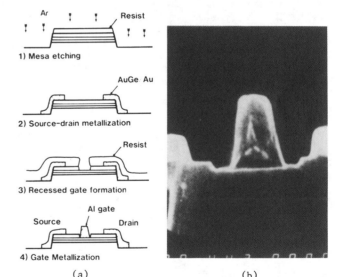

(a)   (b)

Fig.2 (a)Fabrication process and (b)SEM view of the cross-section of the microwave HEMT.

Performance

DC Characteristics

The drain I-V characteristics of the HEMT are shown in Fig.3. Low drain voltage at current saturation and high transconductance of about 60 mS are obtained.

	n (cm^{-2})	μ $(cm^2/V \cdot s)$	ρ $(\Omega/\square)$
300K with n-GaAs	1.1×10^{13}	2800	197
without n-GaAs	1.2×10^{12}	6000	894
77K with n-GaAs	3.9×10^{12}	13000	122
without n-GaAs	0.98×10^{12}	26000	248

Table 1 Hall mobilities,μ,sheet electron concentrations,n,and sheet resistances,ρ,of the wafer.

Reprinted from *1983 IEEE MTT-S Int. Microwave Symp., Tech. Dig.*, pp. 563-565, 1983.

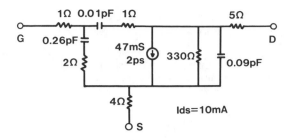

Fig.5 Equivalent circuit of the HEMT.

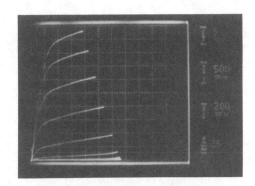

Fig.3 Drain I-V characteristics.

Because of the conduction of the low mobility electrons in the AlGaAs layer,the transconductance was low at small gate voltages. The n-AlGaAs layer in the present device may be thick enough to prevent complete depletion of the AlGaAs layer.

High Frequency Performance

For microwave evaluation,HEMT chips were mounted on carriers with 0.38 mm aluminum microstrip input and output transmission lines,which were slid into a test fixture. Noise figure measurements were made at 12 GHz. Tuning was done by gold chips along the 50 Ω input and output lines of the carrier. The results are shown in Fig.4. A minimum noise figure of 1.4 dB with 11 dB associated gain was obtained,which is comparable to the best result yet reported for GaAs MESFETs with comparable gate length. Also,the noise figure depends weakly on the drain current. It is important to note that the reported noise figure was obtained by measuring only two or three devices. Hence,the minimum noise figure of 1.4 dB may be the mean value of this wafer lot.

Current gains are plotted in Fig.6. The cutoff frequencies were 35 GHz at Ids=20mA and 30 GHz at 10mA.The relatively low noise figure of 1.4 dB is obtained with higher cutoff frequency (30GHz) than those of conventional GaAs MESFETs,in spite of the relatively high source resistance. Thus,with reduction of the source resistance to less than 2 Ω,the minimum noise figure would be reduced to 1 dB at 12 GHz.

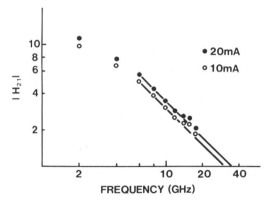

Fig.6 |h21| as a function of frequency.

Temperature Dependence of Noise Figure

Even though the mobility of 2DEG was high at room temperature,at lower temperatures a notable increase in mobility can be seen. We measured the temperature dependence of S-parameters, the noise figure,and associated gain. In Fig.7,the measurement setup is shown. The carrier of the HEMT chip,which led to the stainless steel jacketed coaxial cables with hermetically sealed connectors, was fixed on the cooling head of a cryo pump.

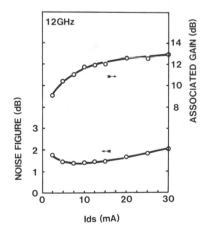

Fig.4 Noise figure and associated gain as a function of drain current.

Device Model

The S-parameters of HEMTs were measured over a 2-18GHz frequency range,using an automatic network analyzer. These S-parameters were then used to determine the maximum available gain,the current gain,and the equivalent circuit element values. The circuit model was the same as for GaAs MESFETs,as shown in Fig.5. At Ids=10mA,the gate capacitance was 0.26 pF and the source resistance was 4 Ω,both larger than those of low noise GaAs MESFETs. They seem to be the critical factors for noise performance.

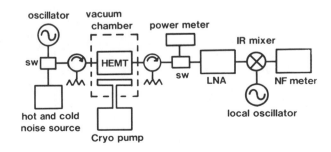

Fig.7 Setup for measuring temperature dependence of noise figure and associated gain.

Results of low-temperature measurements indicate significant improvement in the microwave performance of the HEMTs.

Fig.8 shows S-parameters between 2 and 14 GHz at temperatures of 300K,200K and 100K. Notable increase in S21 was observed with decrease of temperature.

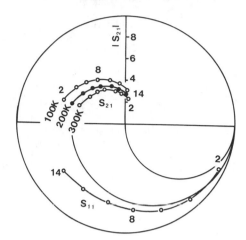

Fig.8   S11 and S21 parameters of the HEMT.

The temperature dependence of the noise figure and associated gain are shown in Fig.9. As temperature is decreased these quantities are greatly improved. At 100K the noise temperature of the HEMT was 24K (0.35 dB) with 12dB associated gain. This improvement in the noise figure and associated gain is thought to be due to the reduction of thermal noise and the increase of the electron mobility from $6000\,cm^2/V \cdot s$ to $26000\,cm^2/V \cdot s$.

Summary

High performance 0.5μm gate low noise HEMTs have been developed. The preliminary results (noise figure of 1.4 dB with 11 dB associated gain at 12 GHz) show the promise of HEMTs for low noise applications.

Acknowledgements

The authors would like to thank M.Abe,S.Yamamoto, H.Ishikawa, S.Hiyamizu,H.Komizo,Y.Tokumitsu,N.Nakayama, and T.Saito for their supports on this work and warm encouragement. They are also grateful to M.Nunokawa for his assistance in fabricating the devices.

References

1.Abe,M.,Mimura,T.,Yokoyama,N.and Ishikawa,H.,"New technology towards GaAs LSI/VLSI for computer applications," IEEE.Trans.Microwave Theory & Tech.,vol.Mtt-30,p992-998,July,1982.
2.Niori,M.,Saito,T.,Joshin,K.and Mimura,T.,"A 20 GHz high electron mobility transistor amplifier for satellite communications," ISSCC DIGEST OF TECHNICAL PAPERS,p198-199,Feb.,1983.
3.Fukui,H.,"Optimal Noise Figure of Microwave GaAs MES-FET's," IEEE.Trans.Electron Devices,vol.ED-26,p.1032-1037,July,1979..
4.Hikosaka,K.,Mimura,T.and Joshin,K.,"Selective Dry Etching of AlGaAs-GaAs Heterojunction," Jpn.J.Appl. Phys.,20,p.L847-L850,1981.
5.Hiyamizu,S.and Mimura,T.,"High Mobility Electrons in Selectively Doped GaAs/n-AlGaAs Heterostructures Grown by MBE and their Application to High-Speed Devices," J.Cryst.Growth,vol.56,p.455-463,Jan.,1982.

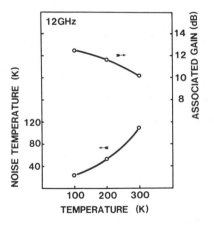

Fig.9 Noise temperature and associated gain as a function of temperature.

Low Noise HEMT with Self-Aligned Gate Structure

K. Joshin, Y. Yamashita, M. Niori, J. Saito, T. Mimura and M. Abe

Fujitsu Ltd., Atsugi

1677 Ono, Atsugi 243-01, Japan

Low noise HEMTs (High Electron Mobility Transistors) with self-aligned gate structure have been successfully developed using direct electron beam lithography and selective-dry etching technique. Their source resistances are reduced to one half of the previous result. The 0.4 μm gate length HEMT yielded 1.08 dB noise figure with 12.7 dB associated gain at 12 GHz and 1.7 dB noise figure with 8.8 dB associated gain at 20 GHz. These are comparable to the best results yet reported for quarter-micron gate GaAs MESFETs.

1. Introduction

HEMT based on a selectively doped GaAs/n-AlGaAs single heterojunction structure offers new possibilities for microwave devices [1], [2], [3]. Higher frequency operation and lower-noise performance than GaAs MESFETs are expected of HEMTs, due to high electron mobility and high saturation velocity of a two-dimensional electron gas (2DEG) at the heterojunction interface.

We have already reported on the microwave performance of HEMT with 0.5 μm gate length[1]. Its cut-off frequency is 30 GHz which is 20 % higher than GaAs MESFETs and its noise figure is 1.4 dB at 12 GHz.

To make further improvement in noise performance, it is necessary to know the relation between noise figure and device parameters of HEMT. For GaAs MESFETs, there is the well-known formula obtained by Fukui[4]. For HEMT, recently, Wu et al.[5], calculated the drain thermal noise and gate induced noise, using the simple drain current model. However their calculation did not include the parasitic resistance effect. In an actual device, parasitic resistances play an important role in determining a device noise figure. Thus, in addition to their calculation model, we take account of parasitic resistance effect. We followed their calculation to obtain insight into the HEMT noise performance relative to device parameters.

The calculation result shows that HEMT with shorter gate length and with lower source resistance has the lower noise figure. To reduce source resistance we developed the self-aligned gate technique. It uses an anisotropic dry etching technique instead of wet-chemical etching. The device with self-aligned gate structure exhibited the extremely low noise figure of 1.08 dB at 12 GHz.

This paper describes the noise figure calculation, fabrication process and noise performance of self-aligned gate HEMTs.

2. Modeling of HEMT noise figure

In the calculation of HEMT noise figure, it is assumed that electron mobility is constant throughout the channel, and that drain current saturation occurs when the electric field in the channel reaches the critical value E_c. The drain current in a HEMT device is determined by

$$I_d = \mu \, (C_g/L_g)(V_g - V_{off} - V(x))(dV/dx) \quad ,$$

where μ is the mobility, $C_g = \varepsilon W L_g/d$ is the gate capacitance, W is the channel width, L_g is the gate length, d is the AlGaAs thickness, ε is the dielectric constant of AlGaAs, and V(x) is the channel potential at x. The only noise considered in this model is the Johnson noise. The noise at the drain current saturation region is not considered, since the noise figure does not change significantly with increasing drain voltage once the knee point of drain current has been reached. The HEMT noise figure is calculated at the knee of the drain current-voltage characteristics. Also we ignored the hot electron effect. To calculate the noise figure, we used the expression of the power spectrum density of drain thermal noise and induced gate noise which are calculated by Wu, et al.,

$$S_{I_d} = 4kTg_m P \ , \qquad S_{I_g} = 4kT\omega^2 C_g^2 P/g_m$$

We also used the expression for the minimum noise figure of a GaAs MESFET obtained by Pucel, et al. [6],

$$F_{min} = 1 + 2(\omega C_g/g_m)\sqrt{K_g(K_r + g_m(R_s + R_g))} \quad (1)$$

$$K_g = P[(1-C\sqrt{R/P})^2 + (1-C^2)R/P]$$

$$K_r = PR(1-C^2)/K_g \ ,$$

where P is drain noise coefficient, R is gate noise coefficient, C is noise correlation coefficient, R_s is a source resistance and R_g is a gate resistance.

When parasitic resistances remain, the second term of the square root in Eq. (1) makes the main contribution. To first order approximation, when the K_r term is ignored, the optimum value F_o of the minimum noise figure can be described by,

$$F_o = 1 + 18 f L_g \sqrt{\varepsilon W(R_s + R_g)/(d \mu E_c)} \quad (2)$$

where f is operating frequency. Eq. (2) shows that a device with a thicker AlGaAs layer and with a smaller gate width has a lower optimum noise figure. As a result, small C_g devices and low g_m devices are expected to have a low noise figure. Since gate fringing capacitance is not considered in this model, there should be an optimal value for AlGaAs thickness and gate width.

Fig. 1 shows the dependence of the optimal value of the minimum noise figure on the gate length and the parasitic resistances. It can be seen, like the result obtained by semi-empirical

formula [4], that shortening the gate length and minimizing the parasitic gate and source resistances are essential to lower the noise figure. According to these results, HEMTs with a shorter gate length and a lower source resistance than the previous ones, were fabricated by using the self-aligned gate technique.

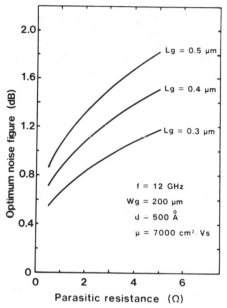

Fig. 1   Calculated optimum noise figure as a function of parasitic resistance for gate length of 0.3 μm, 0.4 μm and 0.5 μm.

3. Device fabrication

The epitaxial layers for HEMTs were grown by molecular beam epitaxy. The layers consist of undoped GaAs, Si-doped n-type AlGaAs, n-type AlGaAs with the mole fraction of AlAs gradually changed, and n-type GaAs, on 2-inch diameter semi-insulated GaAs wafer. The doping concentration is $2 \times 10^{18}/cm^3$. Growth temperature is 680 °C. To reduce the ohmic contact resistance, the n-GaAs cap layer was added. The gradually changed AlGaAs layer was grown to eliminate the discontinuity in the conduction band between n-AlGaAs and n-GaAs, and to obtain low source resistance. Hall mobility of the 2-dimensional electron gas was about 6000 $cm^2/V \cdot s$ at room temperature. Undoped AlGaAs " spacer layer " was not grown in the heterojunction, to obtain as high sheet carrier density and low sheet resistance as possible. This structure was adopted to produce a HEMT with a low noise figure at room temperature.

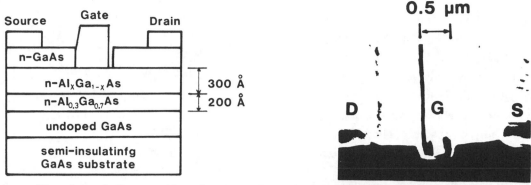

Fig. 2   Cross-sectional view of self-aligned gate HEMT.

HEMTs were fabricated with 0.4 μm gate length and 200 μm total gate width. The channel length between source and drain contacts is 1.6 μm. Fabrication process is simple. To isolate the mesa, Si-doped n-type layers were wet-chemical etched. Source-drain contact areas were defined by photolithography. Ohmic metals AuGe/Au were lifted off and alloyed at 450 °C. It is confirmed that the alloyed region reaches to the buried 2DEG. Gate areas were delineated by direct electron beam lithography.

To make the self-aligned gate structure (Fig. 2), the anisotropic dry etching technique with vertical wall character was used. The GaAs cap layer of the gate region is selectively removed by this technique. Using the mask resist of the dry etching process, evaporated Al gate metal was lifted off. Since the gate metal and the n-GaAs layer were self-aligned, the source resistance was reduced from 4 Ω to 2.5 Ω. Patterned gate length is 0.5 μm but angled evaporation of Al reduced the device gate length to about 0.4 μm. The etching gas is composed of CCl_2F_2 and He. The dry etching process is essentially a self-terminated one. The uniformity of the pinch-off voltage (-2.0 V) is excellent. We have not found any significant damage to the 2DEG.

4. Performance

The drain I-V characteristics of the HEMT are shown in Fig. 3. Because of the conduction of the low mobility electrons in the AlGaAs layer, the transconductance was low at low gate voltages. The maximum transconductance was 40 mS, that is 200 mS/mm gate width.

For microwave evaluation, HEMT chips were

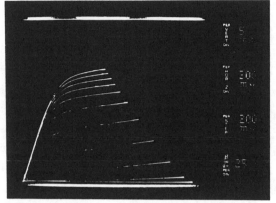

Fig. 3   Drain I-V characteristics.

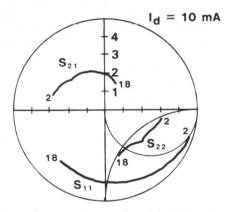

Fig. 4   S-parameters of self-aligned gate HEMT.

mounted on carriers with 0.38 mm alumina microstrip input and output transmission lines, which were slid into a test fixture. The S-parameters were measured over a 2-18 GHz frequency range, using an automatic network analyzer. Fig. 4 shows S-parameters at 10 mA drain current. The small phase rotation of S_{11} indicates small gate capacitance corresponding to the low transconductance. These S-parameters were then used to determine the maximum available gain, the current gain, and the equivalent circuit element values.

Noise figures were measured at 12 GHz and 19.6 GHz. Tuning was done by gold chips along the 50 Ω input and output lines of the carrier. Fig. 5 shows the drain current dependence of minimum noise figure and associated gain at 19.6 GHz. Extremely low noise figure and high gain was obtained at samll drain current. 1.08 dB noise figure with 12.7 dB associated gain at 12 GHz and 1.7 dB noise figure with 8.8 dB associated gain at 19.6 GHz were obtained. Fig. 6 shows the measured noise figure and the frequency dependence of the calculated noise figure for parasitic resistances of 4 Ω, 1 Ω and 0 Ω.

Table 1 summarizes the device parameters and the noise performance of the HEMTs with 0.4 μm and 0.5 μm gate length. In spite of longer gate length HEMTs exhibit noise figures comparable to the best data for quarter-micron gate GaAs MESFET. With further reduction of gate length and source resistance, further improvement of noise performance can be expected.

5. Conclusion

High electron mobility transistors with self-aligned gate structure have been developed. With the reduced gate length and source resistance, HEMT exhibited 1.08 dB noise figure with 12.7 dB associated gain at 12 GHz, and 1.7 dB noise figure with 8.8 dB associated gain at 19.6 GHz. These results show the great potential of HEMTs for low noise application.

Acknowledgment

The authors thank M. Kobayashi, A. Shibatomi, S. Yamamoto, Y. Tokumitsu, and H. Ishiwari for their support in this work. They are also grateful to M. Nunokawa for his assistance in fabricating the devices.

Reference

[1] K. Joshin, et al., IEEE. MTT-S Int. Microwave Symposium Digest p.563, 1983.

[2] J. J. Berenz, et al., IEEE. Microwave and Millimeter-wave Monolithic Circuit Symposium Digest. p. 83, 1984.

[3] K. Ohata, et al., IEEE. MTT-S Int. Microwave Symposium Digest p.434, 1984.

[4] H. Fukui, IEEE. Trans. Electron Devices, vol. ED-26, p. 1032, 1979.

[5] E. N. Wu, et al., Solid-State Electronics vol. 26, p. 639, 1983.

[6] R. A. Pucel, et al., Advances in electron physics p.195, 1975.

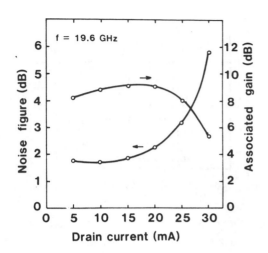

Fig. 5   Minimum noise figure and associated gain as a function of drain current.

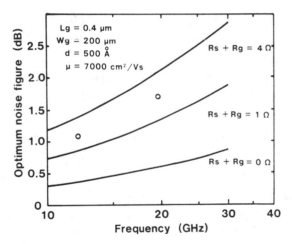

Fig. 6   Optimum noise figure as a function of frequency. ——— Calculated noise figure for parasitic resistances of 4 Ω, 1 Ω and 0 Ω. O Measured noise figure.

	Lg (μm)	Rs (Ω)	Fo (dB) (12 GHz)	Fo (dB) (19.6 GHz)
wet-recessed gate	0.5	4.0	1.4	2.2
self-aligned gate	0.4	2.5	1.08	1.7

Table 1   Summary of device parameters and noise performance of self-aligned gate HEMT.

A New Low-Noise AlGaAs/GaAs 2DEG FET with a Surface Undoped Layer

HIKARU HIDA, MEMBER, IEEE, KEIICHI OHATA, MEMBER, IEEE,
YASUYUKI SUZUKI, AND HIDEO TOYOSHIMA

Abstract—A high-performance N-AlGaAs/GaAs selectively doped two-dimensional electron gas (2DEG) FET with a surface undoped layer has been designed and demonstrated. Simple analysis based on the short-channel approximation revealed that an increase in a total layer thickness between a gate electrode and 2DEG at a hetero-interface results in a higher cutoff frequency and a lower noise figure than conventional 2DEG FET's. This is because the gate capacitance can be markedly reduced without a significant decrease in the transconductance owing to a parasitic source resistance. The surface undoped layer intentionally employed in this work can permit the total layer thickness to increase, i.e., the gate capacitance to reduce, without changes in the 2DEG density and in the source resistance. This structure also gives high gate breakdown voltage because of a small neutral region in n-(AlGa)As and a low surface electron field, which possibly yields excellent performance 2DEG FET's for practical use. Fabricated (AlGa)As/GaAs 2DEG FET's exhibited noticeable room-temperature performances of 0.95-dB noise figure with 10.3-dB associated gain at 12- and 45-GHz cutoff frequency. These are the best data ever reported for 0.5-μm gate length FET's.

I. INTRODUCTION

A GREAT DEAL of interest is attracted by two-dimensional electron gas field-effect transistors (2DEG FET's) as high-speed and high-frequency devices, based on an (AlGa)As/GaAs system [1]–[11]. In these previous reports, it is generally understood that reducing the total doped layer thickness and raising the impurity doping level in the (AlGa)As layer leads to high-performance 2DEG FET's [12]. However, it is very difficult to obtain high-quality crystals of highly doped (AlGa)As from the epitaxial growth aspect. Furthermore, even though a relatively highly doped (AlGa)As layer would be employed, the 2DEG FET would not exhibit significant enhancement in an extrinsic transconductance due to a still high source resistance, which arises from a physical limit in the sheet density of 2DEG. On the contrary, this causes an increase in both input capacitance and gate leakage current. Consequently, high-frequency performances are seriously degraded. On the other hand, concerning LSI's that require high uniformity and controllability of device characteristics, conventional FET's seem not to be suitable because of being greatly sensitive to surface fluctuations. It is therefore of great importance to realize improved-structure 2DEG FET's. Recently, the 0.8-μm gate pulse-doped

2DEG FET with a high cutoff frequency of 33 GHz has been reported, which implys the high potentiality of the FET [13]. However, a FET design to achieve high performances has not been reported in detail.

It is the purpose of this paper to report a newly developed high-performance 2DEG FET with a surface undoped layer, putting great emphasis upon lowering the input capacitance. In order to attain high microwave performances, the effective cutoff frequency and the minimum noise figure have been calculated under a short-channel approximation. Moreover, fringing capacitance, layer parameters, and gate threshold voltage are also analyzed. Finally, 2DEG FET's with a new layer structure have been fabricated and high performances of a pronounced low-noise figure and a high cutoff frequency have been demonstrated.

II. ANALYSIS

A. Approach to Higher Microwave Performance 2DEG FET's

For an actual FET, the following empirical equation given by Fukui [14] is well known as describing a noise figure NF, that is

$$NF = 1 + K \frac{f}{f_T} \sqrt{g_{m0}(R_s + R_g)} \qquad (1)$$

where f is the operating frequency, f_T the cut-off frequency, K the fitting factor, g_{m0} the intrinsic transconductance, R_s the source resistance, and R_g the gate resistance.

Cappy *et al.* have recently reported newly calculated results for 2DEG FET RF performances [15]. The report said that Fukui's expression (especially $K = 1.5$) was verified below 20 GHz for the noise figure of a 2DEG FET, while not at higher frequencies. Therefore, our attention will be focused upon characteristics in the microwave range on the basis of (1).

Under the short-channel approximation, g_{m0} is given by

$$g_{m0} = \frac{C_{gs}^0 v_s}{L_g}. \qquad (2)$$

with

$$C_{gs}^0 = \frac{\epsilon_s L_g W_g}{t_i}. \qquad (3)$$

Manuscript received October 2, 1985; revised February 6, 1986.
The authors are with the Microelectronics Research Laboratories, NEC Corporation 4-1-1, Miyazaki, Miyamae-ku, Kawasaki 213, Japan.
IEEE Log Number 8608202.

Reprinted from *IEEE Transactions on Electron Devices*, vol. ED-33, no. 5, pp. 601–607, May 1986.

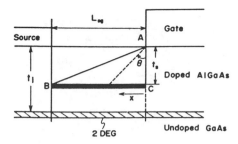

Fig. 1. A cross-sectional view of a 2DEG FET between a source and a gate.

where C_{gs}^0 is the intrinsic gate capacticance, v_s the electron saturation velocity, ϵ_s and t_i the permittivity and the thickness of the (AlGa)As layer, respectively, and W_g and L_g the gate width and the gate length, respectively. As a result, f_T is written as

$$f_T = \frac{g_{m0}}{2\pi C_{gs}^0} \approx \frac{v_s}{2\pi L_g}. \tag{4}$$

Substituting (2)–(4) into (1) yields

$$NF = 1 + 2\pi Kf \sqrt{\frac{\epsilon_s L_g^2 W_g (R_s + R_g)}{t_i v_s}}. \tag{5}$$

Equation (5) indicates that an increase in the (AlGa)As thickness t_i leads to a lower noise figure when other parameters fixed. It is also clear that to reduce the gate length and the total parasitic resistance plays an important role in it. The thickness t_i itself basically associated with C_{gs}^0 as seen in (3) and small t_i, that is large C_{gs}^0, may increase the gate-induced noise. The value of t_i, however, should be optimized in a certain range since an input capacitance C_{in} including C_{gs}^0, a fringing capacitance C_F, and another parasitic capacitance C_0 may be indeed more significant. Firstly, let us evaluate the gate fringing capacitance C_F. The C_F would be defined as the capacitance between the gate and the conducting layer in the (AlGa)As layer as shown in Fig. 1. Therefore, C_F is calculated by integrating $\epsilon_s W_g \cos\theta/t_s$ from zero to L_{sg} in x. Thus, one can obtain

$$C_F = \epsilon_s W_g \log\left[\frac{L_{sg}}{t_s} + \sqrt{\left(\frac{L_{sg}}{t_s}\right)^2 + 1}\right]. \tag{6}$$

If $L_{sg} \gg t_s$

$$C_F \approx \epsilon_s W_g \log \frac{2L_{sg}}{t_s} \tag{7}$$

where x is the distance from the source edge of the gate, θ the angle measured from the source edge of the gate to the direction of the arrow, L_{sg} the source-to-gate spacing, and t_s the depleted (AlGa)As layer thickness due to the surface potential. Note that t_s is equal to t_i if the (AlGa)As layer is fully depleted. For $L_{sg} = 0.5\ \mu m$, $W_g = 200\ \mu m$, $\epsilon_s = 12.5\ \epsilon_0$ (ϵ_0 is the permittivity in vacuum), and $t_s = 500\ \text{Å}$, $C_F \approx 0.03$ pF was obtained, whereas $C_{gs}^0 \approx 0.21$ pF. As a result, the ratio C_F/C_{sg}^0 was estimated about 15–

20 percent in the vicinity of $t_s = 500\ \text{Å}$. This ratio largely increases with an increase in t_i. This result quantitatively predicts that a large increase in t_i does not always lead to a large reduction of C_{in} because C_F and C_0 components in C_{in} can be no longer neglected compared with C_{gs}^0. In addition, the charge neutral region, that is the conducting region in the (AlGa)As layer normally ascribed to an increase in t_i, results in a crucial problem of a gate leakage current increase, and the noise figure is exceedingly degraded. Simultaneously, it becomes inevitably difficult to achieve a high power gain due to g_{m0} lowering. Through the above argument, it seems that the (AlGa)As layer thickness t_i is desirable to be raised up to a certain extent, keeping the charge neutral region beneath the gate as small as possible, so long as Fukui's formula under the short-channel approximation could be valid. However, it is, indeed difficult that the optimum value of t_i is uniquely determined. Then, the following two factors are investigated so as to obtain an aiming value of t_i. First, a low-noise first-stage amplifier incorporated into the system requires as a high gain as possible to reduce total system noise. Usually, the power gain under the matching condition for the minimum noise figure is typically about 11 dB for 2DEG FET's with typical thickness t_i of 450 Å, in accordance with our experimental data. Therefore, if the power gain is practically demanded more than 10 dB, an extreme increase in t_i is not appropriate. Second, in order to find a more quantitative value of the aiming thickness, the parasitic capacitance effect is taken in (5). Namely, C_{gs}^0 in (5) is replaced by $(C_{gs}^0 + C_F + C_0)$. Assuming that C_0 is around 0.15 pF, which includes a gate pad capacitance, the thickness t_i to give minimum noise figure is estimated to be about 600 Å for 0.5-μm gate 2DEG FET's. Note that the optimum thickness value depends upon FET geometry, and so it should be optimized case by case on the basis of the above design method.

Now let us consider the effective cutoff frequency f_{Teff}, which significantly affects high-frequency performance. f_{Teff} is conveniently defined as an extrinsic value of f_T, that is

$$f_{Teff} \equiv \frac{g_m}{2\pi C_{in}} = \frac{g_{m0}}{2\pi(C_{gs}^0 + C_p)(1 + g_{m0}R_s)} \tag{8}$$

where g_m is the extrinsic transconductance and C_p the total parasitic capacitance except for C_{gs}^0. Using (2)–(4), it is reduced to

$$f_{Teff} \approx \frac{v_s}{2\pi L_g(1 + R_s\epsilon_s W_g v_s/t_i)}, \quad \text{for} \quad C_{gs}^0 \gg C_p \tag{9}$$

$$f_{Teff} \approx \frac{\epsilon_s W_g v_s}{2\pi C_p(t_i + R_s\epsilon_s W_g v_s)}, \quad \text{for} \quad C_{gs}^0 \ll C_p. \tag{10}$$

These equations indicate that an increase in f_{Teff} requires an increase in t_i for $C_{gs}^0 \gg C_p$ when a source resistance is significantly large in particular, and by contrary, a decrease in f_{Teff} for $C_{gs}^0 \ll C_p$. Also from the analysis, it is

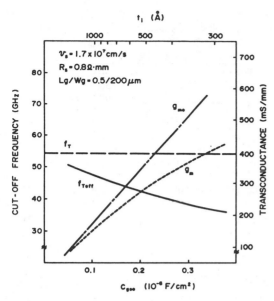

Fig. 2. Cut-off frequency and transconductance as a function of gate sheet capacitance or layer thickness between a gate electrode and 2DEG at the hetero-interface. Calculated from (2)–(4) and (8).

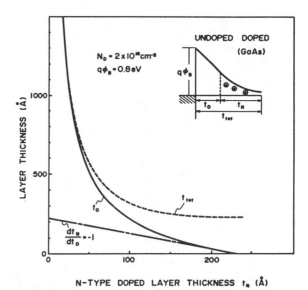

Fig. 3. n-type doped layer thickness t_N versus surface undoped layer thickness t_0 and total layer thickness t_{tot} ($\equiv t_0 + t_N$). Calculated from (11). The differential coefficient $dt_N/dt_0 = -1$ is shown as the dot-dashed line. Energy band diagram under the Schottky gate is also inserted for an undoped/doped GaAs system.

understood that an optimum value of t_i possibly exists as mentioned above. In microwave devices, $C_{gs}^0 \gg C_p$ is realistic; thus f_T, $f_{T\text{eff}}$, g_{m0}, and g_m depending on C_{gs0} ($\equiv \epsilon_s/t_i$) and/or t_i are calculated in this case as shown in Fig. 2 for a 0.5-μm gate 2DEG FET. Here $v_s = 1.7 \times 10^7$ cm/s, $R_s = 0.8$ $\Omega \cdot$ mm, and $W_g = 200$ μm were used as parameters. It should be noted that $f_{T\text{eff}}$ is considerably improved as decreasing C_{gs0}, that is, as increasing t_i. These results substantially arise from a nonlinear and/or slightly saturated dependence of g_m on C_{gs0} due to large R_s. In fact, however, the parasitic capacitance C_P cannot be neglected, and rather contributes to diminishing $f_{T\text{eff}}$. The increase in C_P, while deteriorating $f_{T\text{eff}}$, leads to weak dependence of $f_{T\text{eff}}$ on t_i for more than about 500 Å. The high power gain requires an increase both in $f_{T\text{eff}}$ and $g^{-1}d$ (g_d is the output conductance). However, the large increase in t_i results in reducing the aspect ratio, which yields an increment in g_d. Then, the aiming value of t_i is likely to be around 500 Å. Consequently, it is clearly found from Fig. 2 that $f_{T\text{eff}}$ more than 40 GHz can be expected for t_i nearly equal to 500 Å even in a 0.5 μm (not a quarter micrometer) gate 2DEG FET unless C_P is dominant. This is very promising from a simple fabrication aspect.

B. Design of a 2DEG FET with a Surface Undoped Layer

According to the above analysis, it would be expected that a FET structure with an undoped layer on the surface is extremely superior to conventional ones in many points such as a higher cutoff frequency, a lower noise figure, a larger breakdown voltage, and so forth. Briefly, this is because the surface undoped layer can allow t_i to increase without an increase in the source resistance as well as with a decrease in the gate leakage current. Here, let us determine the relationship between an undoped layer thick-

ness t_0 and an n-type doped layer thickness t_N in the depleted layer under the Schottky gate, which are schematically illustrated in the energy band diagram inserted in Fig. 3. On assumption of the Boltzmann distribution for free carriers in the bulk semiconductor, the relative formula is easily derived from solving Poisson's equation. The expression is

$$ t_0 = -\frac{\epsilon_1}{2\epsilon_2} t_N + \frac{\epsilon_1}{qN_D t_N} \left(\Phi_B - \frac{k_B T}{q} \ln \frac{N_C}{N_D} \right) \quad (11) $$

where ϵ_1 and ϵ_2 are the permittivity of the undoped layer and that of the doped layer, respectively, q the electronic charge, N_D the donor concentration, Φ_B the Schottky-barrier height, k_B the Boltzmann constant, T the temperature, and N_C the effective density of state in the conduction band. Fig. 3 shows the calculated results from (11) using $N_D = 2 \times 10^{18}$ cm^{-3} and $q\Phi_B = 0.8$ eV. The solid line and the dashed one indicate t_0 and t_{tot} ($\equiv t_0 + t_N$) depending on t_N, respectively. For instance, $t_0 = 350$ Å gives $t_N \approx 70$ Å. The differential coefficient $dt_N/dt_0 = -1$ is also plotted in Fig. 3 as the dot-dashed line. In (11), differentiating t_N with t_0 under $\epsilon_1 = \epsilon_2$ yields

$$ -\frac{1}{2} \leq \frac{dt_N}{dt_0} < 0. \quad (12) $$

These relations reveal that the threshold voltage sensitivity to surface layer thickness fluctuation for a surface undoped layer FET is less than a half of that in a conventional fully modulation doped layer FET, in which case $dt_N/dt_0 = -1$, and is further lessened as an increase in t_0. Moreover, as shown in Fig. 3, an electric field under the Schottky gate is lowered by utilizing the surface undoped layer, which results in a high gate breakdown voltage. As a result, it is obvious that the surface undoped layer en-

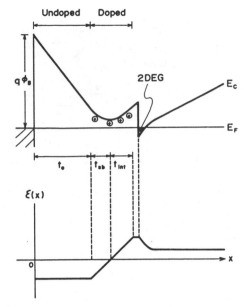

Fig. 4. Energy band diagram and electric field profile of a 2DEG FET with a surface undoped layer under the gate in thermal equilibrium. t_{int} was calculated in [5].

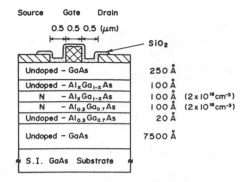

Fig. 5. Structure of the newly developed 2DEG FET.

ables us to realize highly reliable FET's with high yield and reproducibility.

It is easy to apply the surface undoped layer to 2DEG FET's. Fig. 4 shows an energy band diagram and electric field profile $\epsilon(x)$ beneath the gate electrode in thermal equilibrium for that case. A smaller electric field at the gate–undoped layer interface can be explicitly seen. As implied earlier, it is preferable to make the layers between the gate electrode and the hetero-interface depleted as fully as possible in order to maintain a large breakdown voltage and a negligibly small gate leakage current. Here the threshold gate voltage V_T can be easily obtained under the fully depleted layer approximation. Thus V_T is given by

$$V_T = \Phi_B - \frac{\Delta E_C}{q} - \frac{qN_D t_{Ntot}}{2\epsilon_s}(2t_0 + t_{Ntot}) \quad (13)$$

with

$$t_{Ntot} = t_{sb} + t_{int}. \quad (14)$$

where ΔE_c is the conduction band discontinuity at the hetero-interface, t_{sb}, which is equal to t_N in this case, is the n-doped depleted layer thickness due to the Schottky barrier, and t_{int} is the n-doped layer thickness depleted at the hetero-interface due to the conduction band discontinuity. As previously reported in [5], t_{int} is basically designed as a minimum thickness to give a maximum density of 2DEG, which is very important to give small R_s and achieve a high-performance 2DEG FET, especially for planar structure. With respect to derivation of (13), we assumed that the energy level of the conduction band E_c is equal to E_F at the hetero-interface. As expected from (12) and (13), V_T variation in the FET due to fluctuation of the surface layer thickness is noticeably small in comparison with that in a conventional one. Therefore, this is

an attractive advantage, especially for IC applications. Consequently, we can simply design surface undoped layer 2DEG FET's by means of the above equations.

III. FET STRUCTURE AND FABRICATION

The developed FET structure is shown in Fig. 5. The layer structure is designed on the basis of the above analyses, which is composed of undoped GaAs (7500 Å), undoped $Al_{0.3}Ga_{0.7}As$ (20 Å), Si-doped $Al_{0.3}Ga_{0.7}As$ (100 Å), Si-doped $Al_xGa_{1-x}As$ with graded composition of AlAs mole fraction (100 Å), undoped $Al_xGa_{1-x}As$ (100 Å), undoped GaAs (250 Å) on a semi-insulating GaAs substrate grown by molecular-beam epitaxy (MBE). In our MBE system, all the molecular-beam fluxes are automatically controlled by computer. The Si-doped $Al_{0.3}Ga_{0.7}As$ thickness was minimized and the surface GaAs layer was employed to minimize the influence of DX centers to make ohmic contact formation easy and to suppress the surface oxidation. The graded $Al_xGa_{1-x}As$ layer was used to avoid the reduction of 2DEG density due to a band offset. The fabrication process employed in this work was previously reported in detail as a closely spaced electrode structure FET process [2], [16]. In brief, a 0.5-μm long-gate and 0.4-μm-thick gate electrode was made with Al by a side-etching technique. Then Ni/Au-Ge ohmic metals were evaporated through the same photoresist mask as that used in the side-etching process and alloyed with those semiconductor layers after the liftoff process. Therefore, source and drain ohmic contacts were formed in self-alignment with the gate electrode. This simple process enables us to obtain planar short-channel FET's with a small source resistance. Finally, the device surface was passivated with a 1500-Å-thick SiO_2 film.

IV. dc PERFORMANCE

Fig. 6 shows a typical drain current–voltage characteristics of the newly developed FET at room temperature. The gate length is 0.5 μm and the gate width is 200 μm. The threshold voltage V_T of −1.3 V is in reasonable agreement with the calculated one from (13) if considering a low activation efficiency in Si-doped (AlGa)As due to deep levels, the so-called DX centers. Electrical parameters of the FET obtained by dc measurement are listed in Table I. The maximum transconductance g_{mmax} was 230 mS/mm at 300 K and 390 mS/mm at 77 K. These values,

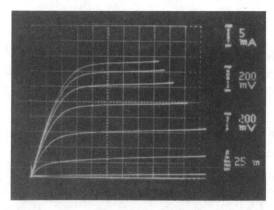

Fig. 6. Typical drain current–voltage characteristics of a 0.5-μm gate 2DEG FET with a surface undoped layer. Gate width is 200 μm. $H = 0.2$ V/div, $V = 5$ mA/div, gate voltage = -0.2 V/step.

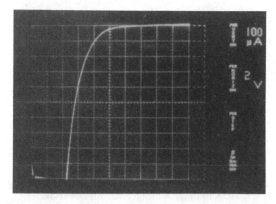

Fig. 7. Reverse gate current–voltage characteristics of the FET shown in Fig. 6. $H = 2$ V/div, $V = 100$ μA/div.

TABLE I
SUMMARY OF dc CHARACTERISTICS OF THE DEVELOPED FET AT 300 K

R_s (Ω·mm)	R_c (Ω·mm)	g_{mmax} (mS/mm)	g_{mo} (mS/mm)	g_d (mS/mm)	V_T (V)	BV_{gR} (V)	BV_{DS} (V)
1.0	0.2	230	320	5	-1.3	≥ 10	≥ 16

which are slightly smaller than the previously reported ones, are responsible for increasing the total layer thickness t_i to reduce the input capacitance. However, they can be regarded as sufficiently high. The drain conductance was as small as 5 mS/mm, which was achieved both by decreasing the buffer layer thickness and by optimizing the alloyed depth of ohmic contacts. Source resistance and contact resistance were evaluated by the transmission line method (TLM) using both ungated and gated FET test patterns with various kinds of source-to-drain spacing to avoid underestimation due to parallel conduction. The source resistance R_s was 1 Ω · mm and the contact resistance was 0.2 Ω · mm. The g_m compression at large drain current, which is clearly seen in Fig. 6, is basically due to a charge neutral region in (AlGa)As layer. However, it can be also caused by a still high source resistance, which will be reported elsewhere in detail. The V_T shift in the dark, when cooling down to 77 K, was relatively small (≈ 0.3 V) despite of the large negative V_T, which is similar to that in the optimized 2DEG FET reported in [5]. Fig. 7 shows the gate current–voltage characteristics in the reverse direction. The reverse gate breakdown voltage BV_{gR} obtained from the onset of conductance was higher than 10 V. And also the drain breakdown voltage BV_{DS} was more than 16 V near pinchoff. These results were attained by employing a comparably thick surface undoped layer with keeping the neutral region as small as possible. Therefore, the FET's also have good feasibility for high-power device applications.

V. RF PERFORMANCE

S-parameter measurement from 2 to 18 GHz was carried out for a 0.5-μm gate 2DEG FET with 200-μm gate width. The measured data under the bias condition of drain

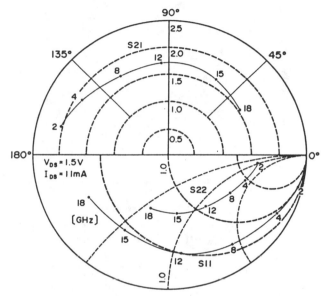

Fig. 8. Scattering parameters of the developed 0.5 μm × 200 μm FET under the bias condition of $V_{DS} = 1.5$ V and $I_{DS} = 11$ mA. The data are plotted on the Smith chart for S_{11} and S_{22} and on the polar-magnitude coordinates for S_{21}.

voltage V_{DS} of 1.5 V and drain current I_{DS} of 11 mA are shown in Fig. 8. The small phase rotations of S_{11} and S_{22} are clearly seen, possibly due to the small input and feedback capacitances. The magnitude of S_{21} is large enough. In order to determine a small-signal equivalent circuit of the 2DEG FET, a FET model in "SUPERCOMPACT" was employed. The equivalent circuit fitted to the above S-parameter data was determined as shown in Fig. 9. The small value of the input gate capacitance (0.24 pF) reasonably coincides with the calculated one (~ 0.22 pF) including the fringing capacitance using $t_i = 570$ Å. The feedback capacitance is also as small as 0.01 pF. Using S-parameters measured at $V_{DS} = 1.5$ V and $I_{DS} = 26$ mA, the dependence of the current gain $|H_{21}|$ on frequency was plotted in Fig. 10. The cutoff frequency obtained by extrapolating the data was more than 45 GHz, and this is the best data in all the previous reports for 0.5-μm gate FET's. From (4), v_s was estimated to be as large as 1.4 × 10^7 cm/s, which agrees well with the values reported so far [13].

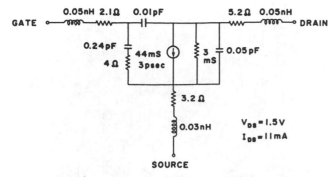

Fig. 9. Small-signal equivalent circuit derived from the scattering parameters shown in Fig. 8 using "SUPERCOMPACT."

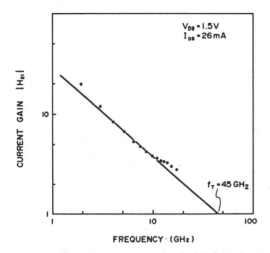

Fig. 10. Frequency dependence of current gain $|H_{21}|$ for the developed FET at $V_{DS} = 1.5$ V and $I_{DS} = 26$ mA.

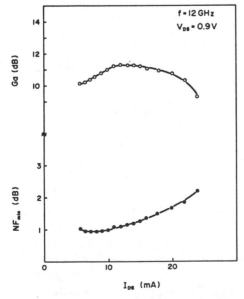

Fig. 11. Drain current dependences of minimum noise figure NF_{min} and associated gain at 12 GHz under $V_{DS} = 0.9$ V.

A 2DEG FET designed on the basis of the above calculation also exhibited a marked room temperature noise performance. Fig. 11 shows the minimum noise figure NF_{min} and the associated gain G_a as a function of source-to-drain current I_{DS} at 12 GHz, where drain bias voltage

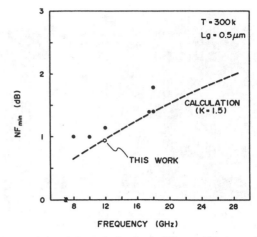

Fig. 12. Sumarized minimum noise figure NF_{min} at 300 K as a function of frequency for 0.5-μm gate 2DEG FET's. Closed circles are depicted referring to the previous reports [4]–[7] and the open circle is for this work. The dashed line shows the calculated result from (1) or (5) using a fitting factor $K = 1.5$.

is 0.9 V. A 0.95-dB minimum noise figure with 10.3-dB associated gain was obtained at 12 GHz. The associated gain reduction in a relatively large drain current is possibly due to a g_m compression.

Finally, the representative minimum noise figures at 300 K reported for 0.5-μm gate 2DEG FET's to date are plotted in Fig. 12 as a function of frequency, where the NF_{min} obtained in this work is also shown. The calculated result from (1) or (5) is also shown as the dashed line to predict the NF_{min} at other frequencies. The typical parameters used in the calculation were basically selected from the equivalent circuit shown in Fig. 10. The fitting factor K is used as 1.5. If a gate length goes down to a quarter micrometer, further improved performances are anticipated even though a fringing capacitance is taken into account.

VI. CONCLUSION

A new low-noise 2DEG FET with a surface undoped layer has been designed based on the (AlGa)As/GaAs selectively doped structure. The simple analyses of the basis of the short-channel approximation indicate that the developed FET is able to exhibit a high cutoff frequency and a superior low-noise figure owing mainly to decrease both in a gate leakage current and in an input capacitance compared to conventional FET's.

Fabricated 2DEG FET's have exhibited noticeably high performances of a 0.95-dB minimum noise figure with a 10.3-dB associated again at 12 GHz and of a 45-GHz cutoff frequency at room temperature. According to the equivalent circuit calculated from the S-parameter data, the high performances have been found to be fundamentally attained by the small input capacitance due to the surface undoped layer. The above data are the best ones ever reported for 0.5-μm gate FET's. Furthermore, the developed FET's have a gate breakdown voltage of -10 V and a high drain breakdown voltage of more than 16 V near pinchoff. Therefore, they are also promising as high-power FET's. Finally, we point out that they can play a significant role as key devices in integrated circuits because they have less sensitivity of the threshold voltage to the fluctuation of the layer thickness.

ACKNOWLEDGMENT

The authors would like to thank Dr. Y. Takayama and Dr. N. Kawamura for their helpful suggestions and encouragements during this work. They also thank M. Ogawa, Dr. T. Itoh, and H. Miyamoto for their valuable discussions.

REFERENCES

[1] K. Joshin, Y. Yamashita, M. Niori, J. Saito, T. Mimura, and M. Abe, "Low noise HEMT with self-aligned gate structure," in *Proc. 16th Int. Conf. Solid State Devices and Materials*, pp. 347–350, Aug. 1984.

[2] K. Ohata, H. Hida, H. Miyamoto, M. Ogawa, T. Baba, and T. Mizutani, "A low noise AlGaAs/GaAs FET with p^+-gate and selectively doped structure," in *IEEE MTT-S Int. Microwave Symp. Dig.*, pp. 434–436, May 1984.

[3] J. J. Berenz, K. Nakano, and K. P. Weller, "Low noise high electron mobility transistors," in *IEEE MTT-S Int. Microwave Symp. Dig.*, pp. 98–101, May 1984.

[4] K. Kamei, S. Hori, H. Kawasaki, K. Shibata, H. Mashita, and Y. Ashizawa, "Low noise high electron mobility transistor," in *Proc. 11th Int. Symp. GaAs and Related Compounds*, pp. 545–550, Sept. 1984.

[5] H. Hida, H. Miyamoto, K. Ohata, T. Itoh, T. Baba, and M. Ogawa, "Planar AlGaAs/GaAs selectively doped structure with high performances and high stabilities," in *Proc. 11th Int. Symp. GaAs and Related Compounds*, pp. 551–556, Sept. 1984.

[6] M. Laviron, D. Delagebeaudeuf, J. F. Rochette, P. R. Jay, P. Delescule, J. Chevrier, and N. Y. Linh, "Ultra low noise and high frequency microwave operation of FET's made by MBE," in *Proc. 11th Int. Symp. GaAs and Related Compounds*, pp. 539–543, Sept. 1984.

[7] A. K. Gupta, E. A. Sovero, R. L. Pierson, R. D. Stein, R. T. Chen, D. L. Miller, and J. A. Higgins, "Low noise high electron mobility transistors for monolithic microwave integrated circuits," *IEEE Electron Device Lett.*, vol. EDL-6, no. 2, pp. 81–82, Feb. 1985.

[8] U. K. Mishra, S. C. Palmateer, P. C. Chao, P. M. Smith, and J. C. M. Hwang, "Microwave performance of 0.25-μm gate length high electron mobility transitors," *IEEE Electron Device Lett.*, vol. EDL-6, no. 3, pp. 142–145, Mar. 1985.

[9] S. Kuroda, T. Mimura, M. Suzuki, N. Kobayashi, K. Nishiuchi, A. Shibatomi, and M. Abe, "New device structure for 4 Kb HEMT SRAM," in *IEEE GaAs IC Symp. Tech. Dig.*, pp. 162–165, Oct. 1983.

[10] R. H. Hendel, S. S. Pei, C. W. Tu, B. J. Roman, N. J. Shah, and R. Dingle, "Realization of sub-10 picosecond switching times in selectively doped (Al, Ga)As/GaAs heterostructure transistors," in *IEEE IEDM Tech. Dig.*, pp. 857–858, Dec. 1984.

[11] N. C. Cirillo, J. K. Abrokwah, A. M. Fraash, and P. J. Vold, "Ultrahigh-speed ring oscillators based on self-aligned-gate modulation-doped n^+-(Al, Ga)As/GaAs FET's," *Electron. Lett.*, vol. 21, no. 17, pp. 772–773, Aug. 1985.

[12] T. J. Drummond, R. Fischer, S. L. Su, W. G. Lyons, H. Morkoç, K. Lee, and M. S. Shur, "Characteristics of modulation-doped Al_xGa_{1-x}As/GaAs field-effect transistors: Effect of donor-electron seperation," *Appl. Phys. Lett.*, vol. 42, no. 3, pp. 262–264, Feb. 1983.

[13] M. Hueshen, N. Moll, E. Gowen, and J. Miller, "Pulse doped MOD-FETS," in *IEDM Tech. Dig.*, pp. 348–351, Dec. 1984.

[14] H. Fukui, "Optimal noise figure of microwave GaAs MESFET's," *IEEE Trans. Electron Devices*, vol. ED-26, no. 7, pp. 1032–1037, July 1979.

[15] A. Cappy, A. Vanoverschelde, M. Shortgen, C. Versnaeyeen, and G. Salmer, "Comparison between TEGFET's and FET's RF performances in the millimeter wave range," in *Proc. 11th Int. Symp. GaAs and Related Compounds*, pp. 533–538, Sept. 1984.

[16] T. Furutsuka, T. Tsuji, F. Katano, M. Kanamori, A. Higashisaka, and Y. Takayama, "High-speed E/D GaAs ICs with closely-spaced FET electrodes," in *Proc. 14th Int. Conf. Solid State Devices*, pp. 335–339, Aug. 1982.

Extremely low-noise 0.25-μm-gate HEMTs

K. Kamei, H. Kawasaki, S. Hori, K. Shibata, M. Higashiura, M.O. Watanabe*
and Y. Ashizawa*

Komukai Works, Research and Development Center, Toshiba Corporation,
1, Komukai Toshiba-cho, Saiwai-ku, Kawasaki 210, Japan

Abstract. Low-noise 0.25-μm-gate high electron mobility transistors
(HEMTs) have been developed, using epitaxial wafers grown by MBE. Lowest
noise figures of 0.66 and 0.85 dB with associated gains of 11.8 and 9.4
dB are obtained at 12 and 18 GHz, respectively, at room temperature.
The carrier concentration of GaAs buffer layer and the carrier profile
at $Al_{0.3}Ga_{0.7}As$/GaAs buffer layer interface have been carefully controlled.
In order to reduce parasitic source resistance and attain high transconductance, heavily doped $Al_{0.3}Ga_{0.7}As$ and GaAs cap layers have been
adopted, in addition to a narrow gap design between source and gate.

1. Introduction

Extensive work has recently been done in many laboratories on high electron
mobility transistors (HEMTs), otherwise called TEGFETs or MODFETs), in view
of their superior microwave performance as compared to conventional GaAs FETs.
At 18 GHz, noise figure of 1.4 dB for a 0.4-μm-gate HEMT (Kamei et al. 1984)
and 1.3 dB for a 0.25-μm-gate HEMT (Mishra et al. 1985) have been reported
so far. Using HEMTs, low-noise amplifiers (Mochizuki et al. 1985, Sholly et
al. 1985) or wide-band amplifiers (Shibata et al. 1985) were already built
in the last few months, showing the superiority of HEMTs to GaAs FETs in
practical applications.

Demands are, however, still strong for lower-noise transistors from system
side. For instance, in some satellite communication systems, low-noise
amplifiers using HEMTs are still operated under a cooled condition (Iwakuni
et al. 1985). In order to make such cooled operation unnecessary and, hence,
to make power consumption and amplifier size smaller, transistors such as
with less than 1-dB noise figure at 18 GHz are required.

In this paper, we report the state of the art microwave performance of
HEMTs developed using epitaxial wafers grown by MBE. We also report important
factors for material preparations and device structures for obtaining good
HEMTs, as well as their microwave performance.

2. Material Preparation

The epitaxial wafers used in this work consist of a 1-μm thick undoped GaAs
buffer layer, a 200-Å thick n-$Al_{0.3}Ga_{0.7}As$ layer and a 300-Å thick n-GaAs
cap layer, successively grown on undoped GaAs LEC substrates with 2-inch
diameter and (100) orientation. The $Al_{0.3}Ga_{0.7}As$ and GaAs cap layers were
heavily doped to 3×10^{18} cm^{-3} with Si. An undoped $Al_{0.3}Ga_{0.7}As$ spacer layer

was omitted to reduce parasitic source resistance and attain high transconductance. The arsenic source was baked before growth for a time long
enough to evaporate impurities. After several successive runs of growth to
confirm that undoped GaAs layers grown to a 5-μm thickness have a carrier
concentration less than 10^{14} cm^{-3}, epitaxial wafers for HEMTs were prepared.

The measured sheet carrier concentration n_s and the electron Hall mobility
μ_e of the epitaxial wafers for HEMT fabrication are 1.5×10^{12} cm^{-2} and
4200 cm^2 V^{-1} s^{-1} at 300 K and 1.2×10^{12} cm^{-2} and 15200 cm^2 V^{-1} s^{-1} at 77 K,
respectively. They are measured under a dark condition after removing the
n-GaAs cap layer and part of the n-$Al_{0.3}Ga_{0.7}As$ layer. The values of μ_e are
much lower than the reported ones (Heiblum et al. 1984) due to the omission
of spacer layer and the growth of heavily doped $Al_{0.3}Ga_{0.7}As$ layer in our work.

In addition to the n_s and μ_e measurements, the
carrier profile including the $Al_{0.3}Ga_{0.7}As$ and
GaAs buffer layer interface has been measured
by a C-V profiling technique. The measurements
have been performed at 77 K, because a large
leakage current occurred through the Schottky
barrier formed on the heavily doped $Al_{0.3}Ga_{0.7}As$
layer at room temperature. It has become clear
through the measurements that a carrier profile
is very sensitive to material structural
parameters (Watanabe et al. 1985). Therefore,
careful control of growth conditions has been
especially needed to obtain a steep profile.
The epitaxial wafers with a high peak carrier
concentration and a steep carrier profile have
been selectively adopted for HEMT fabrication.
Figure 1 shows an example of excellent carrier
profile. The peak carrier concentration is
4.5×10^{18} cm^{-3}. The distance in depth between
the peak and the one-order-of-magnitude less
carrier concentration is 40 Å.

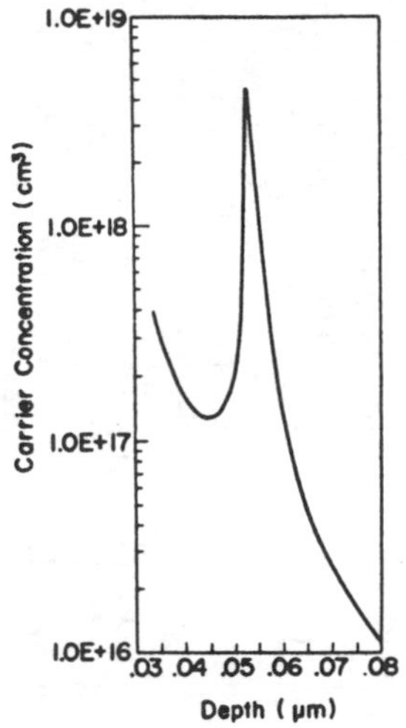

3. Device Fabrication

The cross section of designed
HEMT is schematically drawn in
Fig. 2. After 0.3-μm-high mesa
formation by wet etching, source
and drain ohmic contacts,
separated by 3 μm, were formed,
by alloying evaporated Ni/AuGe
at 450 °C for two minutes.
The gate pattern with 0.25-μm
length and 200-μm width was
delineated using a direct
electron beam lithography by
exposing a 6000-Å-thick AZ2415
resist film at a beam density of
70 μC/cm^2. The gate pattern was
defined with a 0.5-μm spacing
from the source contact edge
within 0.2-μm alignment
accuracy to reduce source

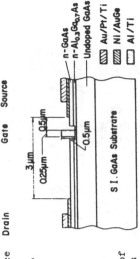

Fig. 1 Carrier profile
measured at 77 K.

Fig. 2 Schematic cross section of
0.25-μm-gate HEMT.

Reprinted with permission from 1985 Int. Symp. GaAs and Related Compounds, Inst. Phys. Conf. Ser. No. 79, Chapter 10, pp. 541–546.
©1986 Adam Hilger Ltd/Inst. of Physics Publishing.

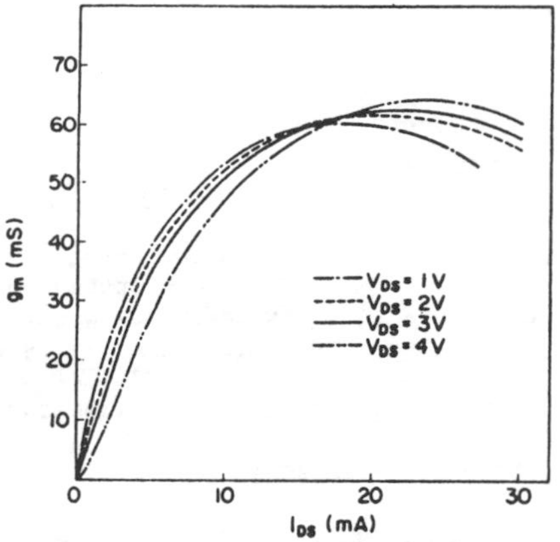

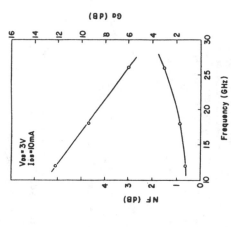

Fig. 5 Drain current versus gm curves for various drain voltages.

Fig. 6 Optimized noise figure (NF) and associated gain (Ga) versus frequency of 0.25-μm-gate HEMT.

resistance. After resist development under precisely controlled conditions as reported earlier by Kamei et al. 1980, the exposed area was recessed by wet etching while monitoring the current between the source and the drain. Al/Ti was then deposited by an electron beam evaporator to a thickness of 0.5 μm. Then, the Al/Ti other than the gate electrode was removed away together with the resist film. An Au/Pt/Ti multi-metal-layer structure was used to form bonding pads.

Figure 3 (a) shows a microphotograph of the fabricated HEMT chip. The chip size is 250 μm x 500 μm. Two gate bonding pads are provided to reduce the gate resistance. Hence, the unit gate width is 50 μm. Figure 3 (b) shows a SEM magnified photograph near the channel area. The gate electrode laid on the bottom of the recess is seen to be about 0.5 μm away from the source contact edge.

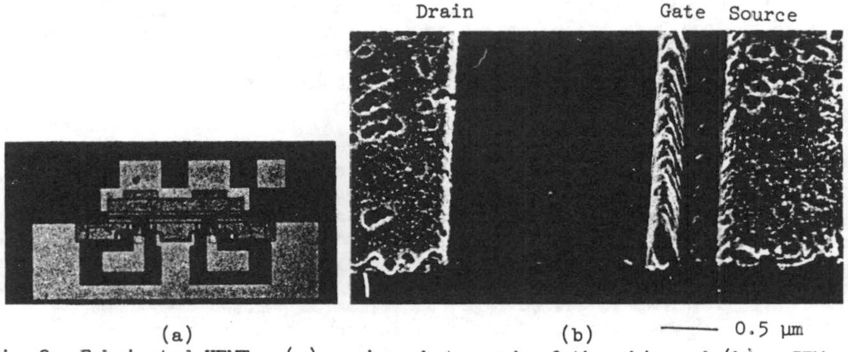

profile. Further suppression of carrier injection will be achieved by using different material structures such as insertion of an $Al_{0.3}Ga_{0.7}As$ buffer layer between GaAs substrate and GaAs buffer layer (Drummond et al. 1982).

The breakdown voltage between the gate and the drain is measured to be -6 V, when defined at a leakage current of 10 μA. This is a little lower than reported before (Kamei et al. 1984) since heavily doped epitaxial layers are used in the present work.

5. Microwave Performance

Microwave noise figure and associated gain have been measured at 12, 18 and 26 GHz on the HEMT chips mounted in microwave test fixtures. Figure 6 shows a frequency dependence of noise figure (NF) and associated gain (Ga), all measured at room temperature at a drain voltage of 3 V and a drain current of 10 mA. The measured minimum noise figures (associated gains) at 12, 18 and 26 GHz are 0.66 dB (11.8 dB), 0.85 dB (9.4 dB) and 1.5 dB (6.0 dB), respectively. These noise figures are much lower than the previously reported data of 0.95 dB at 12 GHz and 1.4 dB at 18 GHz for a 0.4-μm-gate HEMT (Kamei et al. 1984) and 1.3 dB at 18 GHz for a 0.25-μm-gate HEMT (Mishra et al. 1985). The improvement in noise figure of the developed HEMTs against a 0.25-μm-gate GaAs FET earlier developed in our laboratory is 0.5, 0.8 and 1.2 dB at 12, 18 and 26 GHz, respectively. It is expected from the comparison that HEMTs will gain a great advantage over GaAs FETs especially in millimeter-wave region.

In Fig. 7, the measured optimized noise figures (NF) and associated gains (Ga) are plotted as a function of drain current (IDS) at 12 and 18 GHz at a drain voltage of 3 V. As seen from Fig. 7, the optimized noise figures are very insensitive to the drain current and they reach a minimum at IDS≈10 mA. In the measured current range from 3 to 25 mA, the noise figure is less than 1 dB at 12 GHz. Though the dependence of 18-GHz noise figure on drain current is a little larger, the noise figure degradation from the

Drain   Gate   Source

(a)   (b)   ⎯⎯⎯⎯ 0.5 μm

Fig. 3. Fabricated HEMT : (a) a microphotograph of the chip and (b) a SEM magnified photograph near channel area.

4. Static Characteristics

Figure 4 shows the drain current/voltage characteristics of the 0.25-μm-gate HEMT observed at room temperature, where the horizontal axis, vertical axis and the gate voltage step are 1 V/div., 5 mA/div. and 200 mV, respectively. The left and the right curves are measured under dark and illuminated conditions, respectively. No difference is observed between the two dc characteristics. The transconductance (gm)/drain current (IDS) curves are plotted in Fig. 5 for various drain voltages (VDS). The maximum gm is typically 320 mS/mm and the gm at 10 mA is 250 mS/mm for VDS=3 V. It is clear from Fig. 5 that the gm is dependent on VDS and considerably decreases for smaller IDS. For IDS≤17 mA, the gm is smaller for higher VDS. Such reduction of gm for higher drain voltage and smaller drain current can be explained by the mechanism of channel-current injection into undoped GaAs buffer layer under high electric field suggested by Chandra and Eastman 1979. In our work, this type of gm reduction has been found larger for the wafers with a GaAs buffer layer having higher background carrier level, hence, with a broader carrier

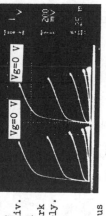

Fig. 4 Drain current/voltage characteristics of HEMT at room temperature.

source resistance R_s in the range of 1.5-1.8 Ω, which is smaller than deduced from the measured N_s. Therefore, part of the drain current is thought to flow in the heavily doped cap layer at microwave frequencies. The gate resistance R_g is found to be over 4 Ω, which is larger than calculated. Better microwave performance can be expected if we lower the gate resistance by making the gate cross sectional area larger. For this purpose, a mushroom-like gate cross section will be better than the present bullet-like one (see Fig. 3). Smaller unit gate width is also effective to reduce gate resistance.

6. Conclusion

Extremely low-noise 0.25-μm-gate HEMTs have been successfully developed by utilizing carefully controlled epitaxial wafers grown by MBE and by designing a narrow gap between source and gate electrodes. The measured noise figures at 12, 18 and 26 GHz have been found superior to hitherto reported values on HEMTs and other semiconductor devices. The developed HEMTs are being employed in our laboratory for building very low-noise amplifiers. Further improvement in device performance can be expected by reduction of gate resistance.

7. Acknowledgments

The authors would like to thank Dr. M. Ohtomo, S. Okano and Dr. T. Nakanisi for their encouragement and helpful discussions. They are also grateful to Dr. M. Kawano for preparation of the epitaxial wafers.

References

Kamei K, Hori S, Kawasaki H, Shibata K, Mashita H and Ashizawa Y, 1984 Gallium Arsenide and Related Compounds (1985 Inst. Phys. Conf. Ser 74), 545

Mishra U K, Palmateer S C, Chao P C, Smith P M and Hwang J C M, 1985 IEEE Electron Dev. Lett. EDL-6, 142

Mochizuki T, Honma K, Handa K. Akinaga W and Ohata K, 1985 IEEE MTT-s Digest, 543

Sholley M, Berenz J, Nichols A, Nakano K, Sawires R and Abell J, 1985 IEEE MTT-s Digest, 555

Shibata K, Abe B, Kawasaki H, Hori S and Kamei K, 1985 IEEE MTT-s Digest, 547

Iwakuni M, Niori M, Saito T, Hamabe T, Kurihara H, Jyoshin K and Mikuni M, 1985 IEEE MTT-s Digest, 551

Heiblum M, Mendez E E and Stern F, 1984 Appl. Phys. Lett. 44, 1064

Watanabe M O, Yoshida J, Nakanisi T, Kawano M and Kamei K, to be published in 1985 Gallium Arsenide and Related Compounds

Kamei K, Hori S, Kawasaki H, Chigira T and Kawabuchi K, 1980 IEDM Dig. Tech. Papers, 102

Chandra A and Eastman L F, 1979 Electron. Lett. 15. 90

Drummond T J, Kopp W. Thorne R E, Fisher R and Morkoc H, 1982 Appl. Phys. Lett. 40, 879

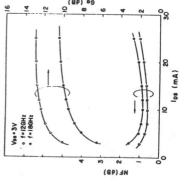

Fig. 7 Optimized noise figure (NF) and associated gain (Ga) versus drain current of 0.25-μm-gate HEMT at 12 and 18 GHz.

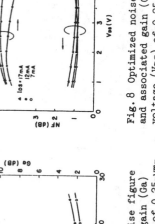

Fig. 8 Optimized noise figure (NF) and associated gain (Ga) versus drain voltage (V_{DS}) of 0.25-μm-gate HEMT at 18 GHz.

minimum value is only 0.2 dB at a drain current of 20 mA, yet with a gain increase of over 1 dB.

The noise figure is also insensitive to drain voltage as can be seen from Fig. 8 that shows measured noise figure (NF) and associated gain (Ga) versus drain voltage (V_{DS}) characteristics at 18 GHz for three levels of drain current (I_{DS}). Only for I_{DS}=7 mA and $V_{DS} \geq 5$ V, both noise figure and associated gain deteriorate noticeably. For I_{DS}=12 and 17 mA, however, noise figure and gain changes are very small. These behaviors can be explained qualitatively by the g_m dependence on drain voltage shown in Fig. 5. Anyway, the noise figure stays below 1 dB at 18 GHz in the range of measured drain voltage from 1 to 5 V, when an appropriate drain current is chosen.

In addition to very low noise figure and high gain, the observed bias dependence is very attractive for practical applications of HEMTs in microwave amplifiers. Meanwhile, when epitaxial wafers with broader profile is used for HEMT fabrication, the noise figure is found very sensitive to gate and drain bias conditions.

The values of microwave equivalent circuit elements of the HEMT have been determined from S-parameter measurements made over 2 to 18 GHz frequency range using an automatic network analyzer. The HEMT was biased at a drain voltage of 3 V and a drain current of 10 mA. Figure 9 shows the microwave equivalent circuit and its typical element values. The value of g_m is in the range of 50-55 mS (250-275 mS/mm), the gate capacitance C_{gs} in the range of 0.17-0.2 pF and the

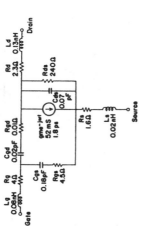

Fig. 9 Equivalent circuit of 0.25-μm-gate HEMT.

LOW NOISE HIGH ELECTRON MOBILITY TRANSISTORS

J. J. Berenz, K. Nakano, and K. P. Weller

TRW Electronic Systems Group
One Space Park
Redondo Beach, CA 90278

ABSTRACT

Sub-half-micron gate length High Electron
Mobility Transistors (HEMT) were fabricated by
direct-write electron beam lithography for low
noise EHF amplifiers. Modulation-doped epitaxial
structures were grown by molecular beam epitaxy
having 8,000 cm^2/V-sec room temperature and
77,600 cm^2/V-sec liquid nitrogen Hall mobility
for 10^{12} electrons/cm^2. Gate lengths as narrow
as 0.28 micron were defined in a recess etched
through the n$^+$ GaAs contact layer. The dc transconductance of 0.4 micron gate length depletion
mode devices exceeded 260 mS/mm. Preliminary
measurement of noise figure and associated gain
made at room temperature yielded 2.7 dB noise
figure and 5.9 dB associated gain at 34 GHz for
devices having 0.37 micron gate length. Enhancement mode devices were also fabricated having
240 mS/mm dc transconductance. These devices
yielded 1.5 dB noise figure and 10.5 dB associated gain at 18 GHz for 0.35 micron gate length.
These results are comparable to the best quarter-
micron gate length GaAs MESFET noise figures yet
reported.[1,2]

Introduction

Increasing demands on the performance required
of millimeter wave receivers has led us to the
development of discrete ultra low noise, low power
dissipation High Electron Mobility Transistors for
use in hybrid MIC pre-amplifiers. These pre-amplifiers can be used ahead of existing diode mixer and
associated local oscillator circuitry to establish
the noise figure of the receivers. High Electron
Mobility Transistors have been fabricated which
demonstrate the potential for noise performance
superior to GaAs MESFETs. In this paper we describe the design, fabrication, evaluation, and
performance of these HEMT devices.

Design

The principal factors governing device noise
figure are described by the Fukui noise figure
equation in Figure 1. The equivalent circuit
elements used in the equation are related to the
physical regions of the HEMT device structure as
shown in Figure 2. The gate-source capacitance,
C_{GS}, is the principal factor governing noise
figure; it is dominated by the channel capacitance, C_c. This capacitance can be minimized
by reducing the gate length and optimizing the
vertical channel doping profile. The intrinsic

transconductance G_m is also primarily determined
by the details of the epitaxial structure. Thus,
the gate length and epitaxial structure primarily
determine the noise figure. Minimizing the ratio
of C_{GS}/G_m minimizes the noise figure. The source
resistance, R_s, and gate metal resistance, R_G, are
parasitic elements which also contribute to the
noise figure. Their contribution can be significant, particularly at high frequency. On the
other hand, the transistor gain is primarily
determined by the cutoff frequency, f_T, which is
given by the following expression:

$$f_T = \frac{G_m}{2\pi C_{GS}} \quad . \tag{1}$$

To first order, the power gain depends on the cutoff frequency according to the following
expression:

$$G = \left(\frac{f_T}{f}\right)^2 \quad , \tag{2}$$

where f is the operating frequency. Thus, to maximize the power gain, it is necessary to minimize
the ratio C_{GS}/G_m. Although the functional dependence of gain on transconductance is greater than
for the noise figure, both noise figure and gain
are affected in the same manner by the ratio of
these quantities.

Gate length, gate width, and electrode topology selections were made in conjunction with the
design of the epitaxial structure and the selection of the fabrication technology needed to
achieve low parasitic elements and high cutoff
frequency. These key factors determine the noise
figure and gain of the transistor.

Fabrication

Sub-half-micron gate length High Electron
Mobility Transistors were fabricated by direct-
write electron beam lithography on modulation-
doped epitaxial structures grown by molecular beam
epitaxy. The epitaxial structures were grown in
a Varian MBE/Gen II system on 2-inch diameter undoped LEC substrates. To achieve high device
transconductance, while maintaining high electron
mobility, a thin 20 Å undoped Al$_{0.3}$Ga$_{0.7}$As "spacer
layer" was grown to separate the two-dimensional
electrons from their ionized "parent" donors, as

Reprinted from *1984 IEEE MTT-S Int. Microwave Symp., Tech. Dig.*, pp. 98–101, 1984.

shown in Figure 3. Typical room temperature Hall mobilities were 7700 cm^2/V-sec for 10^{12} electrons/cm^2, while liquid nitrogen temperature Hall mobilities were typically 69,000 cm^2/V-sec. The best values measured are given in the abstract.

Discrete low noise transistors were fabricated with $\leq$ 0.4 micron gate length and nominal 75 micron total gate width. Three different transistor topologies were evaluated having multiple gate feeds and short gate finger widths to minimize losses at high frequency. Device active areas were isolated by etching mesas to a depth of 2500 Å. Source and drain ohmic contact areas were defined by photolithography. Ni/AuGe/Ni/Au contact metals were evaporated, "lifted-off," and alloyed to form source and drain ohmic contacts. Note that nickel was deposited first to assist in driving the Ge through the AlGaAs layer to make a low resistance contact to the buried two-dimensional electron gas layer.

The definition of gates for HEMT devices was performed by e-beam lithography using TRW's Cambridge EBMF-2 e-beam exposure system. We have demonstrated satisfactory PMMA resist images for lift-off of thin < 0.25 μm long aluminum gates. Best results were obtained with low electron beam currents (1 nA) and with a single pass of the electron beam (i.e., a single line of exposure points), as demonstrated by the SEM photograph of a 0.4-micron long gate shown in Figure 4. The dark area surrounding the light colored gate stripe is the channel recess etched through the n$^+$ GaAs contact layer. The recessed gate is situated closer to the source contact in a 1.8-micron channel between source and drain contacts. Either depletion-mode or enhancement-mode transistor operation was achieved, depending on the depth of the gate recess. The recess depth was gauged by monitoring the drain-source saturation current during etching.

Evaluation

The dc and rf characteristics of the High Electron Mobility Transistors have been measured and analyzed to determine device parameters. DC I-V characteristics were analyzed using the University of Illinois analytical device model.[3] A typical fit to the measured DC I-V characteristics of an enhancement mode transistor is shown in Figure 5. The dc transconductance is 240 mS/mm. The rise in drain current with drain voltage in the saturation regime (i.e., low output resistance) is due to dynamic effects not included in the model.

Small-signal S-parameters of the transistors were measured in a 50-ohm microstrip test fixture over the 2-18 GHz frequency range. An equivalent circuit was determined using COMPACT* optimization to fit the elements of the equivalent circuit shown in Figure 6 to the measured S-parameter data. Table 1 lists a typical result. The inferred source resistance is relatively low (0.38 ohm-mm),

*Trademark of Compact Engineering, Palo Alto, CA.

but the gate resistance is high (37 ohms/mm) due to the thin gate metal used (2500 Å). The input resistance is also much higher than a comparable depletion-mode GaAs MESFET. In spite of these limitations, an excellent device noise figure of 1.5 dB was measured at 18 GHz for this HEMT device. And, it should be noted that relatively high associated gain (10.5 dB) was achieved at a much lower drain-source current level (3.7 mA) than GaAs MESFETs exhibiting comparable low noise performance. This is an extremely important advantage of HEMT devices over GaAs MESFETs.

Performance

Table 2 summarizes the best noise figure performance measured for the nominal 0.35-micron gate length High Electron Mobility Transistors. In spite of their longer gate length, higher gate metal resistance, and higher channel resistance, these HEMT devices exhibit noise figures comparable to the best reported for quarter micron gate length GaAs MESFETs. This being the case, ultimately lower noise figures than GaAs MESFETs should be achievable with HEMT devices having narrower gate lengths and improved parasitic elements.

In order to estimate the magnitude of the improvement possible, we extended the analytical device model to compute noise figure and gain from the device physical parameters. Excellent agreement has been obtained between calculated and measured noise performance in the case of the experimental devices which we analyzed. Theoretically, depletion-mode HEMT devices should exhibit lower noise figures than their enhancement-mode counterparts because of higher transconductance and lower gate-source capacitance. Figure 7 compares the calculated noise figure for the experimental depletion-mode devices with the measured noise figures. The noise performance projected for an optimized 0.25-micron gate length depletion-mode device at room temperature is also plotted in this figure to show the magnitude of the improvement theoretically possible. This projection of HEMT device noise figure is lower than the best measured noise figures for GaAs MESFETs and the magnitude of the improvement in noise figure over GaAs is greater at higher frequencies.

Conclusions

High Electron Mobility Transistors have been fabricated which demonstrate the potential for lower power dissipation, lower noise amplification than GaAs. As a result, these devices are expected to play a major role in future microwave and millimeter-wave monolithic receiver circuits.

References

1. P. W. Chye and C. Huang, "Quarter Micron Low Noise GaAs FET's," IEEE Electron Device Letters, Vol. EDL-3, No. 12, December 1982, pp. 401-403.

2. M. Feng, H. Kanber, V. K. Eu, E. Watkins, and
    L. R. Hackett, "Ultrahigh Frequency Operation
    of Ion Implanted GaAs Metal-Semiconductor
    Field-Effect Transistors," Appl. Phys. Lett.,
    Vol. 44, No. 2, 15 January 1984, pp. 231-233.

3. T. J. Drummond, H. Morkoc, K. Lee, and M. Shur,
    "Model for Modulation Doped Field Effect
    Transistor," IEEE Electron Device Letters,
    Vol. EDL-3, No. 11, November 1982, pp. 338-341.

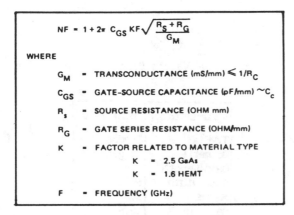

$$NF = 1 + 2\pi\, C_{GS}\, KF \sqrt{\frac{R_S + R_G}{G_M}}$$

WHERE

G_M = TRANSCONDUCTANCE (mS/mm) $\leqslant 1/R_C$

C_{GS} = GATE-SOURCE CAPACITANCE (pF/mm) $\sim C_c$

R_s = SOURCE RESISTANCE (OHM mm)

R_G = GATE SERIES RESISTANCE (OHM/mm)

K = FACTOR RELATED TO MATERIAL TYPE

  K = 2.5 GaAs

  K = 1.6 HEMT

F = FREQUENCY (GHz)

Figure 1. Fukui Noise Figure Equation

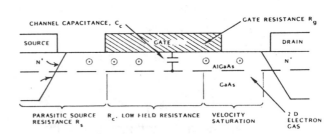

Figure 2. HEMT Cross-Section

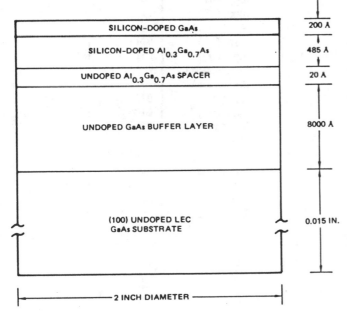

Figure 3. MBE-Grown Modulation-Doped
Epitaxial Structure

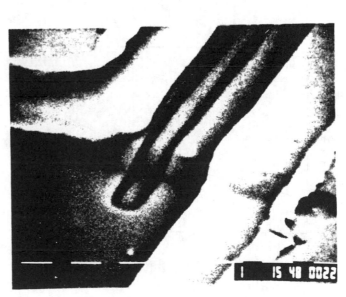

Figure 4. SEM Photo of HEMT Channel

91

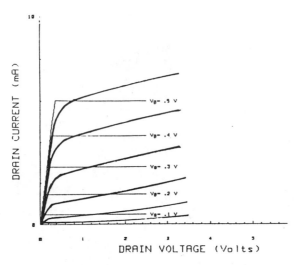

Figure 5. Comparison of Measured and Calculated I-V Characteristics of Enhancement Mode HEMT

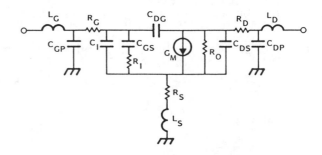

Figure 6. Equivalent Circuit

Table 2.

TRW HEMT DEVICE PERFORMANCE

NF (dB)	G_{ASS} (dB) RESULT	F (GHz)	MBE LAYER NO.	DEVICE TYPE	GATE LENGTH
1.3	9.0	15	2067	DEPLETION MODE	0.35 ᴜM
1.5	10.5	18	2078	ENHANCEMENT MODE	0.35
2.7	5.9	34	2067	DEPLETION MODE	0.35

Table 1. Equivalent Circuit Element Values

Parameter	HEMT 2078
TYPE	E-mode
Geometry	Pi-gate
Gate Length	0.35
Gate width	65
NF (dB)	1.5
Gass (dB)	10.5
Freq (GHz)	18
Id (mA)	3.7
Vd (V)	3.0
Vg (V)	0.5
ELEMENTS	
Lg (nH)	.18
Cgp (pF)	.012
Rg (ohms)	2.4
Cin (pF)	.011
Cgs (pF)	.051
Rin (ohms)	17.1
Rs (ohms)	5.9
Ls (nH)	.076
Cds (pF)	.0108
Ro (ohms)	673
Gm (mS)	15
Cdg (pF)	.0148
Rd (ohms)	6.0
Cdp (pF)	.0088
Ld (nH)	.26

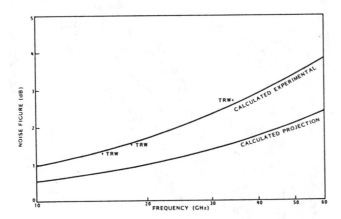

Figure 7. HEMT Noise Figure Versus Frequency

Noise Figure Characteristics of 1/2-µm Gate Single-Heterojunction High-Electron-Mobility FET's at 35 GHz

E. A. SOVERO, ADITYA K. GUPTA, MEMBER, IEEE, AND J. A. HIGGINS

Abstract—In this letter, we report room-temperature noise figure performance of Gallium Arsenide single-heterojunction high-electron-mobility transistors (HEMT's). We have measured a noise figure of 2 dB at 35 GHz with 5 dB of associated gain. The devices tested were 150 µm wide with 0.5-µm-long gates. The active layers were grown by molecular beam epitaxy (MBE). These values are the best reported results for either HEMT's or MESFET's at these frequencies, regardless of their geometry.

I. INTRODUCTION

THE UNIQUE characteristics of high-electron-mobility transistor (HEMT) devices have already demonstrated excellent device performance at microwave frequencies [1]–[3]. These devices are now fulfilling their potential of working at millimeter-wave frequencies with both high gain and low noise. In this letter we report noise figure measurements in HEMT's at 35 GHz. The devices tested utilize molecular beam epitaxy (MBE) for active layers, and electron beam lithography (EBL) for 0.5-µm-long Ti–Pt–Au gate definition. Discrete devices were mounted in specially designed chip carriers that were mounted between two WR28 waveguide transitions. The devices use via holes as low inductance connection to the ground plane on the opposite side of the substrate. All other processing steps are identical to our standard monolithic microwave integrated-circuit (MMIC) fabrication technology [4].

II. DEVICE FABRICATION

The conventional HEMT structure is based on a single GaAs/AlGaAs heterojunction with a typical sheet concentration of 10^{12} cm^{-2} in the two-dimensional electron gas (2-DEG). The epitaxial layers used for device fabrication were grown by MBE in a Varian GEN-II machine. Layer structure is grown as shown in Fig. 1, where the n-type regions are Si doped to a concentration of 10^{18} cm^{-3}. Its sheet resistance is ~190 Ω/sq. Al mole fraction is ~24 percent. Electron mobility and sheet carrier density in the 2-DEG at 77 K are 57 000 cm^2/V·s and 0.95×10^{12} cm^{-2} as determined by Hall measurements. The n-GaAs cap layer at the top is included to obtain low resistance contacts by GeAu–Ni metallization and a 450°C − 10 s alloy cycle using rapid infrared heating. A contact resistance of less than 0.2 Ω·mm is typically obtained

Manuscript received November 13, 1985; revised January 10, 1986.
The authors are with the Rockwell International Corporation, Microelectronic Research and Development Center, Thousand Oaks, CA 91360.
IEEE Log Number 8607682.

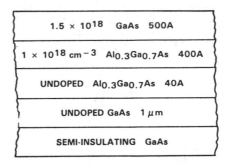

Fig. 1. Epitaxial layer structure for single heterojunction HEMT's.

at room temperature. The devices have a drain-to-source gap of 3 µm and a total gate width of 150 µm. Ti–Pt–Au gates are formed, 0.5 µm long and 6000 Å thick, by direct-write (EBL) and lift-off. Gates were recessed ~500–600 Å by wet chemical etching to adjust the drain current. Gate to source gap is ~0.5 µm. Devices are passivated by a layer of polyimide, which also serves as a crossover insulator for the gold plating step, wherein all gate feeds are brought out to a common bonding pad. The final step is to thin the wafer to 125 µm, etch via holes for source grounding, and gold plate the backside. The wafer is diced by a high-speed diamond saw.

III. MILLIMETER-WAVE MEASUREMENTS

The devices were mounted on a specially designed carrier that was inserted between two stepped waveguide transitions. The transitions match the WR28 waveguide (~300-Ω impedance) to a microstrip line (50 Ω). This is accomplished with three transmission line sections of progressively reduced waveguide heights plus a section of ridge waveguide, all one quarter-wave long at 35 GHz. The completed design is shown in Fig. 2. In the photograph the three different regions can be identified: the waveguide steps, a ridge waveguide section, and the chip carrier. The initial region made up of three quarter-wave-long steps designed to reduce the height and impedance of the waveguide (WR28) from 0.140 to 0.034 in and from 300 to 72 Ω, respectively. The second region consisted of a quarter-wave ridge waveguide section. The ridge itself was formed by a removable anodized aluminum plunger that doubles as a dc bias contact and also acts as the final transition to the microstrip line on the chip carrier. The chip carrier consisted of two 0.090-in long sections of 50 Ω on 0.10-in-thick alumina microstrip. The device was placed in an 0.020-in gap between the 50-Ω lines. Single bond wires were

Reprinted from *IEEE Electron Device Letters*, vol. EDL-7, no. 3, pp. 179–181, March 1986.

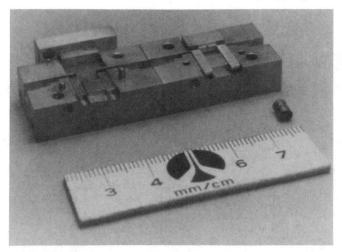

Fig. 2. Photograph of WR28 waveguide to 50-Ω microstrip transition (shown disassembled). Chip carrier can be seen in its proper position.

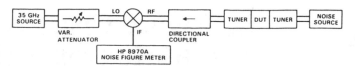

Fig. 3. Block diagram of measuring setup.

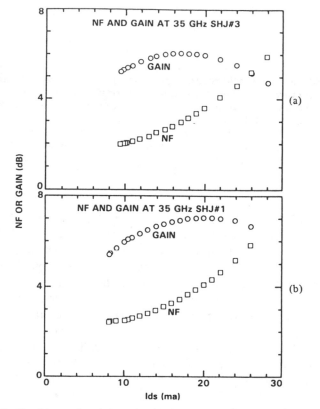

Fig. 4. Measured variation of noise figure with frequency in two single-heterojunction HEMT devices. (a) SHJ no. 3 and (b) SHJ no. 1.

used to connect the gate and drain to the input and output microstrips. Via holes connect the source directly to the ground plane of the carrier. A symmetric transition couples the output microstrip back to the waveguide.

Fig. 3 shows a block diagram of the measuring setup. We used a Hewlett-Packard Model 8970A noise figure meter. The noise source was a Micronetics, Inc., Model NSI-2640W noise diode source with an excess noise ratio (ENR) of 15.85 dB at 35 GHz. A balanced mixer was used to downconvert the signal from 35 GHz to the 10-MHz IF frequency of the HP8970A. A 35-GHz Gunn oscillator source was used as LO. The device under test was placed on the chip carrier and in the test jig described above. Input and output slide screw tuners served as matching elements. The tuners and device holder were independently tested and their insertion loss was measured at 35 GHz with a vector network analyzer. The tuners and test holder with a throughline in place of the device under test showed less than 1-dB total insertion loss. From symmetry considerations, the losses are split equally between the input and output network. For the measurement the tuners were set for minimum noise figure.

IV. DEVICE PERFORMANCE

All the measurements were taken at 35 GHz. The device was assumed to be at room temperature (293 K). Shown in Fig. 4 are plots of noise figure and associated gain as function of drain-to-source current I_{ds}. The plots show the variation of noise figure with I_{ds} in two different devices (both 0.5 μm gate by 150 μm wide). The results displayed in Fig. 4(a) give a minimum noise figure of 2 dB when measured with an I_{ds} of 9.4 mA and a corresponding associated gain of 5 dB. The maximum gain at 19 mA I_{ds} corresponds to a maximum transconductance bias. Results obtained with a second device are displayed in Fig. 4(b). This device shows a noise figure (NF) of 2.2 dB at 9.6 mA I_{ds}. When tuned for maximum gain, these devices show 8 dB of gain at 35 GHz. These results are in excellent agreement with results presented on similar devices previously [5]. The results also compare very favorably with 2.7-dB NF at 35-GHz measurements by Berenz et al. [6] on a

0.35-μm gate HEMT and 1.8-dB NF at 30 GHz with 0.25-μm gate HEMT results presented recently by Chao et al. [7].

These results are particularly significant in the fact that they were obtained with 0.5-μm gate devices. If we use as a first approximation noise model the empirical (1) as derived by Fukui [8]

$$NF(dB) = 10 \, \log_{10} \, [1 + k_f L_g f \sqrt{g_m(R_s + R_g)/(1 - g_m R_s)}]$$

(1)

where

L_g	gate length in micrometers,
f	frequency in gigahertz,
R_s	source resistance in ohms,
R_g	gate metallization resistance in ohms,
g_m	transconductance in siemens.

R_s and R_g are measured by forward biasing the gate Schottky diode. The value of g_m used was that at the bias for minimum noise (−1.5 V).

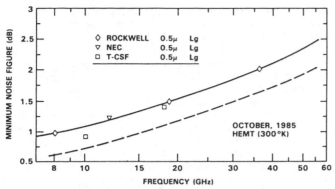

Fig. 5. Best reported noise figure performance of HEMT's at room temperature.

In light of (1), the results presented in this letter enables us to make device performance improvement assessments. It is immediately apparent that we can expect to lower the HEMT's noise figure with a reduction in gate length (say 0.25 μm). The increase in transconductance g_m should not be large enough to offset the lower noise figure due to the reduction of the gate length. Again, referring to (1) the dependence of NF on g_m is not as strong as on the gate length L_g. This discussion is best illustrated in the graph of Fig. 5. In that figure, we show the best reported noise performance of HEMT's at room temperature as function of frequency. Using published data, we find that the noise figure of the 0.5-μm devices increases monotonically with frequency (solid line). If we use the assumptions discussed above, we can estimate (dashed line) the expected performance improvement of a 0.25-μm gate-length device.

V. Conclusions

The results presented here demonstrate the excellent performance of HEMT's in the millimeter-wave region of the spectrum. We report a noise figure of 2 dB with an associated gain of 5 dB at 35 GHz with a 0.5-μm gate by 150-μm wide device. Analysis of these results indicate that further improve-

ment both in bandwidth and performance can be expected by reducing the gate length below the 0.5-μm range. Further improvements in both device geometry and layer optimization should yield devices of unsurpassed performance in both the microwave- and millimeter-wave region of the spectrum, and with a proven technology capable of a high degree of monolithic integration.

Acknowledgment

The authors would like to thank R. Pierson, Jr., for his assistance in designing and fabricating the transistor test fixture.

References

[1] N. T. Linh, M. Laviron, P. Delescluse, P. N. Tung, D. Delagebeaudeuf, F. Diamond, and J. Chevrier, "Low-noise performance of two-dimensional electron gas FET's," in *Proc. IEEE/Cornell Conf. High-Speed Semiconductor Devices and Circuits,* Aug. 1983, pp. 187–193.

[2] K. Joshin, T. Mimura, M. Nirori, Y. Yamashita, K. Kosemura, and J. Saito, "Noise performance of microwave HEMT," in *1983 IEEE MTT-S Int. Microwave Symp. Dig.,* May 1983, pp. 563–565.

[3] J. J. Berenz, K. Nakano, and K. P. Weller, "Low-noise high-electron mobility transistors," in *IEEE Microwave Millimeter-Wave Monolithic Circuits Symp.,* 1984.

[4] A. K. Gupta, G. R. Kaelin, and R. D. Stein, "Gallium Arsenide monolithic microwave-integrated circuits," in *Proc. 27th Midwest Symp. Circuits and Syst.* (Morgantown, WV), June 1984.

[5] A. K. Gupta, E. A. Sovero, R. L. Pierson, R. D. Stein, R. T. Chen, D. L. Miller, and J. A. Higgins, "Low-noise high-electron mobility transistors for monolithic microwave-integrated circuits," *IEEE Electron Dev. Lett.,* vol. EDL-6, no. 2, pp. 81–82, Feb 1985.

[6] J. J. Berenz, K. Nakano, and K. P. Weller, "Low-noise high-electron mobility transistors," *IEEE MTT-S Dig.* (San Francisco, CA), May 29–30, pp. 83–85.

[7] P. C. Chao, P. M. Smith, U. K. Mishra, S. C. Palmateer, K. H. G. Duh, and J. C. M. Hwang, "Quarter-micron low-noise high-electron mobility transistors," presented at the 10th IEEE Cornell Conf. Advanced Concepts High-Speed Semiconductor Devices and Circuits, July 1985.

[8] H. Fukui, "Determination of the basic device parameters of a GaAs MESFET," *Bell Syst. Tech. J.,* vol. 58, no. 3, pp. 771–797, Mar. 1979.

60 GHz LOW-NOISE HIGH-ELECTRON-MOBILITY TRANSISTORS

Indexing terms: Microwave devices and components, HEMTs

The noise performance of 0·25 μm-gate-length high-electron-mobility transistors at frequencies up to 62 GHz is reported. A room-temperature extrinsic transconductance g_m of 480 mS/mm and a maximum frequency of oscillation f_{max} of 135 GHz are obtained for these transistors. At 30 and 40 GHz the devices exhibit minimum noise figures of 1·5 and 1·8 dB with associated gains of 10·0 and 7·5 dB, respectively. A minimum noise figure as low as 2·7 dB with an associated gain of 3·8 dB has also been measured at 62 GHz. This is the best noise performance ever reported for HEMTs at millimetre-wave frequencies. The results clearly demonstrate the potential of short-gate-length high-electron-mobility transistors for very low-noise applications for frequencies at least up to V-band.

Introduction: High-electron-mobility transistors (HEMTs) employing GaAs/AlGaAs heterostructures have demonstrated better noise and speed performance than conventional GaAs MESFETs. The advantages of HEMTs as opposed to GaAs MESFETs are that (i) HEMTs exhibit higher electron mobility and average drift velocity in the channel, even at room temperature, (ii) a higher gate aspect ratio (i.e. gate length to gate depletion depth ratio) can be realised for extremely short-gate-length HEMTs,[1] and (iii) HEMTs have lower noise conductance and Q-value of the optimum source impedance, which result in reduced sensitivity of the noise figure to changes in source impedance and therefore broader noise bandwidth.[2] The above unique features have made HEMT devices extremely attractive for millimetre-wave (MMW) low-noise applications. Considerable progress has been made in the MMW low-noise area using HEMTs during the past couple of years. At 35 GHz, Sholley *et al.* have demonstrated a 2·1 dB noise figure for their 0·25 μm HEMTs,[3] and Sovero *et al.* recently reported a noise figure of 2·0 dB with 0·5 μm HEMTs.[4] Chao *et al.* have also reported a minimum noise figure of 2·1 dB at 40 GHz for the 0·25 μm HEMTs.[5] In this letter we report for the first time the noise performance of our most recent HEMTs at frequencies up to V-band. The devices have demonstrated the best noise performance ever reported at frequencies up to 62 GHz.

Device fabrication: The HEMT devices were fabricated on selectively doped AlGaAs/GaAs heterostructures grown by molecular-beam epitaxy. The detailed material structure and growth conditions have been discused elsewhere.[6] The HEMT wafer exhibits a sheet charge density at 77 K of approximately 1×10^{12}/cm² and a mobility of 50 000 cm²/V s. Devices having 0·25 μm-long gates were fabricated on the wafer by electron-beam lithography. The gates, having a low-resistance T-shaped cross-section, were delineated by the trilayer resist technique.[7] Details of the HEMT fabrication can be found in Reference 8.

HEMT performance: We reduced the parasitic source resistance by minimising gate-source spacing and lowering ohmic contact resistance. A transfer resistance of 0·03 Ω-mm was achieved for the HEMT ohmic contacts. This allowed us to achieve a maximum normalised transconductance g_m of 480 mS/mm at room temperature, which corresponds to an intrinsic g_m of 630 mS/mm when the source resistance is taken into account. The device small-signal gain was measured at 30, 40 and 60 GHz directly using single-stage amplifiers tuned for maximum available power gain. These measurements were performed using high-frequency test fixtures that included gate and drain bias networks which were properly designed to provide stability. For V-band measurements, E-field probes were used for the waveguide-to-microstrip transition. Microstrip circuit and adjustable backshorts were also included in the fixture to match the radiation impedance to 50 Ω optimally to achieve best transition performance. The insertion loss of the V-band test fixture from the input waveguide flange to the output flange, measured using a through line, is approximately 1·5 dB from 55 GHz to 65 GHz. At 60 GHz, a maximum available gain of 7·1 dB was measured in the HEMT. If the gain is extrapolated at a slope of −6 dB per octave, a maximum frequency of oscillation, f_{max}, of 135 GHz is obtained.

The device noise performance, measured up to 62 GHz, is listed in Table 1. Hot and cold temperature standards were

Table 1 ROOM-TEMPERATURE NOISE PERFORMANCE OF 0·25 μm HEMTs

Frequency	Minimum noise figure	Associated gain
GHz	dB	dB
30	1·5	10·0
40	1·8	7·5
62	2·7	3·8

used together with an independently calibrated solid-state noise source to verify the noise measurement accuracy. It can be seen from Table 1 that at 30 GHz the HEMT exhibits a noise figure of 1·5 dB with an associated gain G_a of 10·0 dB, and at 40 GHz a noise figure of 1·8 dB was measured with a G_a of 7·5 dB. This is the best noise performance yet reported at these frequencies.

Fig. 1a illustrates the frequency response of a V-band HEMT amplifier biased at a drain voltage V_{ds} of 3·5 V; the device power gain is 4·7 dB with an input return loss of 12 dB at 62 GHz. Note that the data is not corrected for test fixture losses. Fig. 1b also demonstrates excellent output match. Fig. 2 shows the relationship of F_{min} and F_∞ to associated gain, where F_{min} is the minimum noise figure and F_∞ represents the noise figure of an infinite chain of cascaded single-stage amplifiers (where each single-stage amplifier has the same noise figure F and associated gain G):

$$F_\infty = F + \frac{F-1}{G} + \frac{F-1}{G^2} + \cdots$$

$$= 1 + F - 1 + \frac{F-1}{G} + \frac{F-1}{G^2} + \cdots$$

$$= 1 + \frac{F-1}{1-1/G} = 1 + M$$

where M is defined as the noise measure.

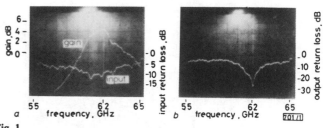

Fig. 1

a Gain and input return loss
b Output return loss for single-stage amplifier biased at $V_{ds} = 3\cdot5$ V, $I_{ds} = 10\cdot8$ mA from 55 to 65 GHz

Reprinted with permission from *Electronics Letters*, vol. 22, no. 12, pp. 647–649, June 1986.

F_∞ is a useful figure of merit for a device since it indicates the noise figure which could be achieved in a multistage amplifier using the same device. From Fig. 2, a minimum

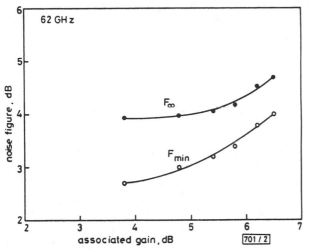

Fig. 2 *Noise figure against associated gain at 62 GHz for a 0·25 μm HEMT device*

noise figure as low as 2·7 dB was obtained in the HEMT at 62 GHz. An F_∞ of less than 4 dB was also calculated at this frequency. This is the best noise performance ever reported at such a high frequency. It is also interesting to note that F_∞ remains approximately constant when the associated gain is varied from 3·8 dB to 5·4 dB.

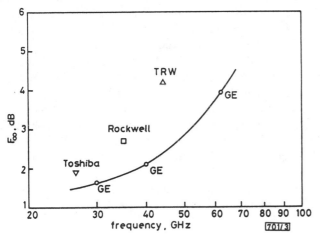

Fig. 3 *Best reported room-temperature noise measure of HEMT devices*[3,4,9]

Fig. 3 summarises the published HEMT noise figure of merit F_∞ as a function of frequency. Our devices exhibit F_∞ of

1·6, 2·1 and 3·9 dB at 30, 40 and 62 GHz, respectively. These are also the lowest noise measures ever reported for any transistors. These superior results can be attributed to the high-quality material and the advanced electron-beam fabrication technology used in this work.

Conclusions: 0·25 μm-gate-length HEMTs have been fabricated with significantly improved millimetre-wave noise performance. An f_{max} of 135 GHz has been obtained. These devices have demonstrated state-of-the-art noise performance at frequencies up to 62 GHz. These results highlight the clear superiority of HEMTs for millimetre-wave low-noise applications.

K. H. G. DUH   *17th April 1986*
P. C. CHAO
P. M. SMITH
L. F. LESTER
B. R. LEE

Electronics Laboratory
General Electric Co.
Syracuse, NY 13221, USA

References

1 DAS, M. B.: 'A high aspect ratio design approach to millimeter-wave HEMT structure', *IEEE Trans.*, 1985, **ED-32**, pp. 11–17

2 POSPIEZALSKI, M. W., WEINREB, S., CHAO, P. C., MISHRA, U. K., PALMA-TEER, S. C., SMITH, P. M., and HWANG, J. C. M.: 'Noise parameters and light sensitivity of low-noise high-electron-mobility transistors at 300 and 12·5 K', *ibid.*, 1986, **ED-33**, pp. 218–223

3 SHOLLEY, M., BERENZ, J., NAKANO, K., SAWIRES, R., NICHOLS, A., and ABELL, J.: '36·0–40·0 GHz HEMT low noise amplifier'. IEEE MTT-S int. microwave symp., St. Louis, MO, USA, pp. 555–558, 4–6 June 1985: J. BERENZ: 'HEMT technology gains on MMW-waves', *Microwaves & RF*, 1985, pp. 121–125

4 SOVERO, E. A., GUPTA, A. K., and HIGGINS, J. A.: 'Noise figure characteristics of 1/2-μm gate single-heterojunction high-electron-mobility FET's at 35 GHz', *IEEE Electron Device Lett.*, 1986, **EDL-7**, pp. 179–181

5 CHAO, P. C., PALMATEER, S. C., SMITH, P. M., MISHRA, U. K., DUH, K. H. G., and HWANG, J. C. M.: 'Millimeter-wave low-noise high electron mobility transistors', *ibid.*, 1985, **EDL-6**, p. 31

6 PALMATEER, S. C., MAKI, P. A., KATZ, W., CALAWA, A. R., HWANG, J. C. M., and EASTMAN, L. F.: 'The influence of V:III flux ratio on unintentional impurity incorporation during molecular-beam epitaxial growth'. Presented at 11th int. symp. on GaAs and related compounds, 26–28 Sept. 1984, Biarritz, France

7 CHAO, P. C., SMITH, P. M., PALMATEER, S. C., and HWANG, J. C. M.: 'Electron-beam fabrication of GaAs low-noise MESFETs using a new tri-layer resist technique', *IEEE Trans.*, 1985, **ED-32**, pp. 1042–1046

8 MISHRA, U. K., PALMATEER, S. C., CHAO, P. C., SMITH, P. M., and HWANG, J. C. M.: 'Microwave performance of 0·25 μm gate length high electron mobility transistors', *IEEE Electron Device Lett.*, **EDL-6**, pp. 142–145

9 KAMEI, K., KAWASAKI, H., HORI, S., SHIBATA, K., HIGASHIURA, M., WATANABE, M. O., and ASHIZAWA, Y.: 'Extremely low-noise 0·25 μm-gate HEMTs'. Presented at conference on gallium arsenide and related compounds, Japan, 1985

Low-Noise HEMT Using MOCVD

KUNINOBU TANAKA, MASAMICHI OGAWA, KOU TOGASHI, MEMBER, IEEE,
HIDEMI TAKAKUWA, HAJIME OHKE, MASAYOSHI KANAZAWA,
YOJI KATO, AND SEIICHI WATANABE, MEMBER, IEEE

Abstract —Low-noise HEMT AlGaAs/GaAs heterostructure devices have been developed using metal organic chemical vapor deposition (MOCVD).

The HEMT's with 0.5-μm-long and 200-μm-wide gates have shown a minimum noise figure of 0.83 dB with an associated gain of 12.5 dB at 12 GHz at room temperature. Measurements have confirmed calculations on the effect of the number of gate bonding pads on the noise figure for different gate widths. Substantial noise figure improvement was observed under low-temperature operation, especially compared to conventional GaAs MESFET's.

A two-stage amplifier designed for DBS reception using the HEMT in the first stage has displayed a noise figure under 2.0 dB from 11.7 to 12.2 GHz.

I. INTRODUCTION

TO MEET THE ever-increasing demands for low-noise, high-performance microwave circuits for satellite communications and other applications, development of very low noise devices for front-end amplifiers is being actively pursued in many laboratories [1], [2].

The limits of performance attainable using GaAs MESFET's are being approached by means of fine-pattern lithography and optimization of various device parameters. HEMT devices using AlGaAs/GaAs heterojunctions have displayed performance surpassing GaAs MESFET's within a short development period [3]. The heterojunction epitaxy for low-noise HEMT's has so far been performed using MBE (molecular beam epitaxy) by virtue of the high quality of the epitaxial interface.

We have previously reported on the microwave performance of our low-noise HEMT using MOCVD, or HIFET (hetero-interface FET), which has superior wafer throughput and surface quality (NF = 1.13 dB, PG = 10.8 dB at 12 GHz) [4].

In this paper, we will report on improvements made in the previous device performance by means of improvements in the epitaxy and device pattern and reduction of the gate length to 0.5 μm. A noise figure of 0.83 dB and an associated gain of 12.5 dB were measured at room temperature using 200-μm-wide devices.

II. DEVICE FABRICATION

The epitaxial layers necessary for the formation of the heterojunction structure are grown by MOCVD using tri-

Manuscript received May 12, 1986; revised August 3, 1986.
The authors are with the Semiconductor Group, Sony Corporation, Atsugi, Kanagawa 243, Japan.
IEEE Log Number 8610821.

methyl organometallics (TMA and TMG) and AsH_3 under atmospheric pressure. The growth temperature is 720°C and the growth rate is approximately 240 Å/min. The undoped GaAs layer is typically 5000 Å thick, while the initial thickness for the Si-doped n-AlGaAs layer is 690 Å at a doping level of 1.5×10^{18} cm^{-3}. The net background impurity concentration of the undoped GaAs layer is under 2×10^{15} cm^{-3}.

Hall mobilities of the two-dimensional electron gas created at the interface are 8030 and 148 000 cm^2/V-s at 300 and 77 K, respectively, when using an undoped $Al_{0.3}Ga_{0.7}As$ spacer layer of 100 Å.

The mobility and sheet carrier concentration of the two-dimensional electron gas (2DEG) are comparable to those reported using MBE.

A high mobility and a low sheet resistivity of the 2DEG are the most important parameters for realizing low-noise HEMT's. A high mobility of the 2DEG is desirable for low-noise operation, while a low sheet resistivity is required for reducing the source resistance.

A thin superlattice buffer (1000 to 2000 Å) of alternating undoped AlAs and GaAs layers and a 10-Å undoped AlGaAs spacer layer were introduced to satisfy the above requirements, as shown in the device cross section of Fig. 1. The 2DEG of the actual device showed a mobility of 5400 cm^2/V-s and a sheet resistivity of 800 Ω/square. When cooled to 77 K, the mobility increased to 26 000 cm^2/V-s.

The low-noise HEMT devices are fabricated on 2-in $\langle 100 \rangle$ epitaxial wafers using standard UV contact photolithography.

The gate metal is evaporated at an angle such that the gate is offset towards the source within the recess area, with part of the metal effectively covering the recess side wall. This makes possible the reduction in the actual gate length over the channel without an increase in the series gate resistance, made possible by the large gate cross section.

The modifications to the epitaxial structure, together with the gate definition processing, contribute to the reduction in the noise figure, which is mainly a function of the gate-to-source series resistance and the gate length as expressed in Fukui's equation. The gate length is nominally 0.4 to 0.5 μm and the source-to-drain spacing is 3 μm.

The chip photograph is shown in Fig. 2.

Reprinted from *IEEE Transactions on Electron Devices*, vol. ED-33, no. 12, pp. 2053–2058, December 1986.

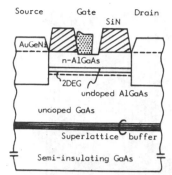

Fig. 1.   Cross section of low-noise HEMT.

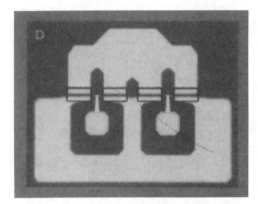

Fig. 2.   Chip photograph of low-noise HEMT.

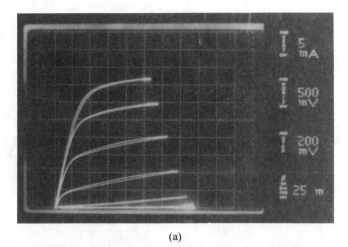

(a)

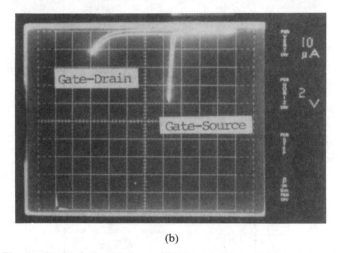

(b)

Fig. 3.   Static characteristics of HEMT. (a) Output characteristics. (b) Schottky reverse characteristics.

III. DC AND RF PERFORMANCE

The typical drain current characteristics and the gate Schottky breakdown characteristics of HEMT's having a gate length of 0.5 μm and gate width of 200 μm are shown in Fig. 3(a) and (b). At 300 K, the maximum extrinsic transconductance was 280 mS/mm at a current density of 120 mA/mm. The intrinsic transconductance calculated from the source resistance was found to have a value of 360 mS/mm. Since the source resistance mainly consists of the 2DEG element between the gate and source together with the ohmic contact resistance, reducing the gate-to-source spacing and decreasing the 2DEG sheet resistance (by increasing the sheet carrier density and mobility) will make possible an even higher extrinsic transconductance than the above value.

The gate breakdown from the Schottky barrier is an important parameter which must be taken into account under actual operating conditions. In general, a lower breakdown voltage of the Schottky gate is observed with HEMT's compared to GaAs MESFET's because of the high donor density in the AlGaAs layer. As shown in Fig. 3(b), typical breakdown voltage values of gate-to-source and gate-to-drain are > 5 V and > 10 V, respectively. An increase of the gate-to-drain breakdown voltage was accomplished by offsetting the gate metal towards the source and optimizing the gate pinchoff voltage, thereby reducing the electric field strength present across the gate and drain. These techniques are comparable to those for GaAs MESFET's.

An analytical model based on Fukui's equation [5], taking the parasitic capacitances of the gate bonding pads

into consideration, has been developed as a guide in optimizing the HEMT device performance. The number of gate bonding pads determines the equivalent gate series resistance and the input capacitance, which in turn determines the noise figure and gain.

Various combinations of gate bonding patterns and gate widths were tested for RF performance, as shown in Fig. 4. The total gate capacitance and the series gate resistance can be expressed as follows:

$$C_{gs} = W_g \cdot C_{gu} \cdot L_g + C_{\text{pad}} \cdot n$$

$$R_g = \frac{\text{Rho} \cdot W_g}{12 \cdot t_g \cdot L_g \cdot n^2}$$

where

W_g	total gate width,
C_{gu}	capacitance per unit area of gate,
L_g	gate length,
t_g	thickness of gate metal,
Rho	gate metal resistivity,
C_{pad}	capacitance per gate bonding,
n	number of bonds.

The capacitance C_{pad} includes the fringing capacitance of the bonding wires as well as the pad itself. The factor 12

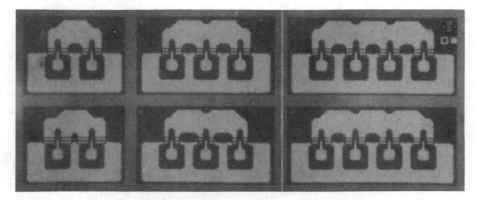

Fig. 4. Photograph of various HEMT test designs.

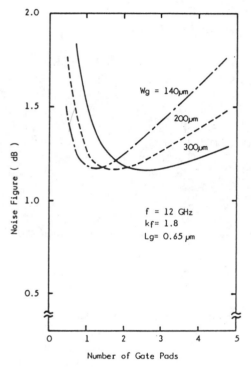

Fig. 5. Calculated noise figure for different gate widths and number of gate pads.

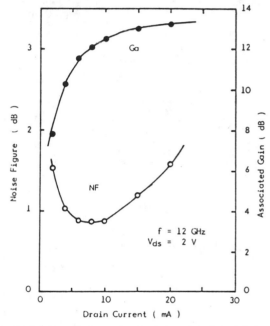

Fig. 6. Noise figure and associated gain as a function of drain current.

in the R_g equation comes from the distributed nature of current flow along the width of the FET gate.

Using the above formulas for C_{gs} and R_g in Fukui's equation, it was determined that for a device having a 200-μm gate width, two bonding pads give the minimum noise performance, whereas for a 300-μm device either two or three bonding pads will give approximately the same minimum noise figure. The calculated noise figures for 140-, 200-, and 300-μm devices as a function of number of bonds are shown in Fig. 5.

Actual noise figure measurements have confirmed this calculation. It was decided that a 200-μm, two-pad design will be used as the standard device taking into account the impedance requirements, chip area, and other factors.

All noise figure measurements were carried out by mounting the chips on 1.8-mm-square ceramic packages. Noise figures and associated gains were measured at 12 GHz using conventional sleeve tuners. The dependence of noise figure and associated gain on drain current is shown in Fig. 6. Noise figure contours as a function of drain current and drain voltage have been plotted in Fig. 7. The minimum noise figure was 0.83 dB with an associated gain of 12.5 dB at a drain current of 10 mA and drain voltage of 1 V. It can be seen that the HEMT is not sensitive to bias conditions. For 10-mA operation, the device noise figure remains under 1.0 dB from 0.5 to 3 V. With a drain voltage fixed at 1.5 V, the noise figure stays under 1.0 dB from 4 to 18 mA. These minimum noise figure values are comparable to previously reported HEMT's having 0.25-μm gate lengths fabricated by direct-write electron beam (EB) lithography on epitaxial layers grown by MBE [6], [7]. The maximum gain is observed at about 15 mA and 1.6 V. By considering circuit requirements, a bias condition of 10 mA at 2 V has been selected as the standard measurement condition.

The noise figure circles shown on the Smith chart in Fig. 8 show that the HEMT is capable of very broad band operation, having a 50-Ω unmatched noise figure of 1.77 dB. This value is about 1.0 dB better than that of a

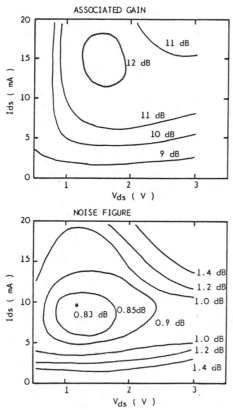

Fig. 7. Map of noise figure and associated gain as a function of drain voltage and current.

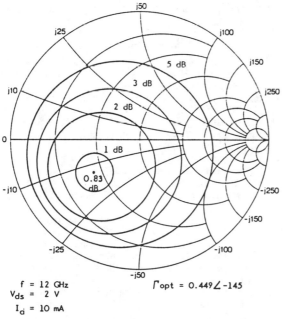

Fig. 8. Noise figure circles of low-noise HEMT.

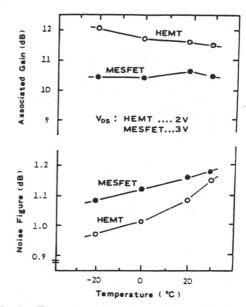

Fig. 9. Temperature dependence of low-noise HEMT.

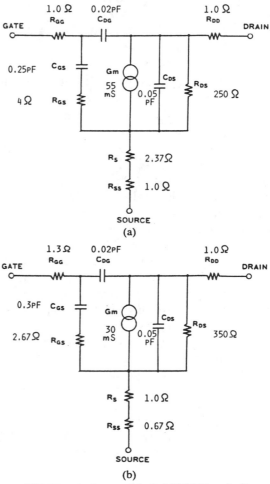

Fig. 10. (a) HEMT equivalent model. (b) MESFET equivalent model.

conventional MESFET. Thus, the HEMT is more tolerant of input mismatching, and is well suited for wide-band MMIC designs. The high transconductance of the HEMT results in an f_{max} of over 70 GHz.

In addition to 12-GHz noise performance measured at room temperature, the devices were thermoelectrically cooled to −20°C to measure the temperature dependence [8].

As shown in Fig. 9, the HEMT's have shown a larger dependence (0.1dB/20°C) than that of the MESFET's (0.05dB/20°C) at a temperature of around 10°C. (The HEMT's used in this experiment were selected for matched performance at room temperature with the MESFET's for comparison purposes.)

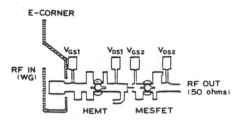

Fig. 11. Circuit pattern of two-stage amplifier using HEMT.

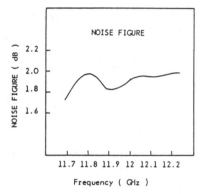

Fig. 12. Noise figure of two-stage amplifier.

The equivalent circuit of the HIFET chip has been determined from *S*-parameters taken directly from on-wafer measurements using a network analyzer. The reference planes are located on the pad surface. The nominal values of the model elements are shown in Fig. 10(a). A typical model of a 0.5-μm-long, 300-μm-wide GaAs MESFET is shown in Fig. 10(b) for comparison.

A simple two-stage low-noise amplifier for 12-GHz DBS converter systems has been built and tested to evaluate the device performance. The circuit is depicted in Fig. 11. The first stage uses a HEMT device, biased at 10 mA at a drain voltage of 2 V. The second stage uses a conventional MESFET. The input signal entering the waveguide is transferred to the microstrip using an E-corner, and the matching network transforms the 50-Ω impedance to the FET's gamma-optimum. As shown in Fig. 12, the overall noise figure of the two-stage circuit is less than 2.0 dB throughout the 500-MHz bandwidth required for DBS reception in the U.S. and Japan. The microstrip pattern used for this HEMT amplifier is only slightly different from that used for a MESFET amplifier, indicating a high degree of drop-in compatibility with conventional 0.5-μm MESFET devices currently available. The HEMT is very suitable for the first stage of DBS and other satellite receiving systems, where low-noise performance will be required at low cost and in large quantities.

IV. CONCLUSIONS

Low-noise HEMT's fabricated using AlGaAs/GaAs heterointerface structures have been successfully developed by combining standard photolithographic techniques with MOCVD technology. Excellent performance, including a noise figure of 0.83 dB and an associated gain of 12.5 dB at 12 GHz at room temperature, has been demonstrated.

A thin superlattice buffer and a 10-Å-thick undoped AlGaAs layer were introduced to increase the mobility of 2DEG at the heterointerface. The nominal gate length of 0.5 μm, formed by an angle evaporation lift-off technique, also contributed to an improvement in both RF performance and breakdown voltage.

Further refinements of the HEMT process and device technology directed at the reduction of parasitic elements will allow an even lower noise figure than those reported here.

The feasibility of high-performance HEMT devices for general-purpose microwave applications based on the high-productivity MOCVD process has been demonstrated.

ACKNOWLEDGMENT

The authors would like to thank Y. Mori, C. Kojima, and N. Watanabe for their valuable support and encouragement.

REFERENCES

[1] K. Joshin *et al.*, "Low-noise HEMT with self-aligned gate structure," in *Extended Abstracts of the 16th Conf. on Solid State Devices and Materials*, 1984, pp. 347–350.
[2] M. Laviron *et al.*, "Ultra low-noise and high frequency operation of TEGFET made by MBE," *Physics*, vol. 129b, pp. 376–379, 1985.
[3] K. Shibata, B. Abe, H. Kawasaki, S. Hori, and K. Kamei "Broad-band HEMT and GaAs FET amplifiers for 18–26.5 GHz," *IEEE MTT-S Digest*, June 1985, pp. 547–550.
[4] H. Takakuwa, *et al.*, "Low-noise microwave HEMT using MOCVD," *IEEE Trans. Electron Devices*, vol. ED-33, pp. 595–600, May 1986.
[5] H. Fukui, "Optimal noise figure of microwave GaAs MESFET's," *IEEE Trans. Electron Devices*, vol. ED-26, pp. 1032–1037, July 1979.
[6] M. Sholley *et al.*, "36.0–40.0 GHz HEMT low-noise amplifier," *IEEE MTT-S Digest, June 1985, pp.* 5455–558.
[7] U. K. Mishira, S. C. Palmateer, P. C. Chao, P. M. Smith, and J. C. M. Hwang, "Microwave performance of 0.25-micron gate length high electron mobility transistors," *IEEE Electron Device Lett.* vol. EDL-6, pp. 142–145, Mar. 1985.
[8] K. Kamei *et al.*, "Low-noise high electron mobility transistor," in *GaAs and Related Compounds Symp.*, 1984, pp. 545–550.

A LOW NOISE AlGaAs/GaAs FET WITH P+-GATE AND SELECTIVELY DOPED STRUCTURE

K. Ohata, H. Hida and H. Miyamoto

Microelectronics Research Laboratories, NEC Corporation

M. Ogawa, T. Baba and T. Mizutani

Fundamental Research Laboratories, NEC Corporation

4-1-1, Miyazaki, Miyamae-ku, Kawasaki, 213   Japan

Abstract

A low noise AlGaAs/GaAs FET with p+-gate and selectively doped structure has been developed. The FET utilizing a two dimensional electron gas has a closely spaced electrode planar structure on an MBE wafer. A 0.5 μm gate FET exhibited marked room temperature performances of 310 mS/mm transconductance and 1.2 dB noise figure with 11.7 dB associated gain at 12 GHz.

Introduction

Field effect transistors based on an N-AlGaAs/GaAs selectively doped structure have recently attracted much attention for applications to high speed ICs. Those FETs are also attractive for low noise microwave devices due to high electron mobility of the two dimensional electron gas (2DEG) in the selectively doped structure [1],[2]. A 1.4 dB noise figure at 12 GHz was reported for such FETs [2]. The FETs with the selectively doped structure reported previously usually have a recessed gate structure [2]-[4]. A surface n+-GaAs layer was used outside the gate for accumulating as much two dimensional electron gas (2DEG) density as possible and making ohmic contact formation easy. However, this structure requires precise etching of the recessed region and possibly increases the gate fringing capacitance.

In this work, a new low noise FET, based on the N-AlGaAs/GaAs selectively doped structure, has been developed. The FET has a planar p+-gate structure with small source resistance. A 0.5 μm-gate device exhibited marked room temperature performance of 1.2 dB noise figure with 11.7 dB associated gain at 12 GHz.

Device Structure

The structure of the newly developed FET is shown in Fig. 1. The gate is made with an Al electrode and an interposed highly doped p-type GaAs thin layer on n-type layers for 2DEG formation. Energy band diagrams at the gate and between the gate and the source are shown in Figs. 2(a) and (b), respectively. The surface potential of the n-type layers at the gate is raised by the p+-layer compared to that outside the gate, if the acceptor density of the p+-layer is much higher than the donor density of the n-type layers. This enables using a thicker and/or more highly doped n-layer than that used in conventional FETs with Schottky barrier gates. Therefore, much accumulation of 2 DEG and small source resistance can be attained without gate recessing. The p+-GaAs layer is so highly doped and thin that a planar gate structure and small gate resistance can also be realized. Furthermore, the gate has a large potential barrier due to the p-n

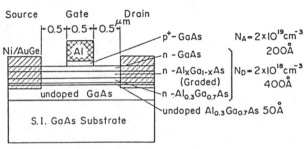

Fig. 1   Structure of the developed FET.

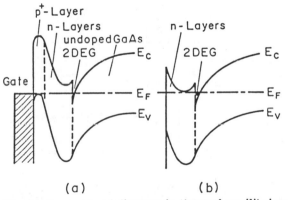

Fig. 2   Energy band diagram in thermal equilibrium.
(a) At the gate. (b) Outside the gate.

junction, so that high reliability for the gate electrode can be expected.

Other important features of the new FET shown in Fig. 1 are also related to technologies for source resistance reduction. An n-GaAs layer is used for the top layer of the n-type layers so as to make ohmic contact formation and surface protection easy. 2 DEG sheet density is limited to a relatively small value of about 1×10^{12} cm^{-2}, so that a close space structure between the source and the gate [5] is employed in order to reduce the 2 DEG layer resistance at room temperature. The gate length and the spacing between the gate and the source were designed to be 0.5 μm. Devices with four gate widths, 100 μm, 140 μm, 200 μm and 280 μm were examined. The unit gate width is 50 μm for 100 μm and 200 μm devices, and 70 μm for 140 μm and 280 μm devices. 200 μm and 280 μm devices have two gate pads.

Reprinted from *1984 IEEE MTT-S Int. Microwave Symp., Tech. Dig.*, pp. 434–436, 1984.

Consequently, advantages of this new structure are a planar gate structure with small gate fringing capacitance, small parasitic resistances, high reliability for the gate electrode and an easy fabrication process, as described in the following.

Device Fabrication

Fabrication steps for the developed FET are shown in Fig. 3. p^+-GaAs(200 Å)/n-GaAs(100 Å)/n-Al$_x$Ga$_{1-x}$As (graded composition) (200 Å) / n-Al$_{0.3}$Ga$_{0.7}$As (100 Å) /undoped Al$_{0.3}$Ga$_{0.7}$As(Spacer)(50 Å)/undoped GaAs(1 μm) layers were grown by molecular beam epitaxy (MBE) on a semi-insulating GaAs substrate: -(1). The p^+-GaAs layer was Be doped with 2×10^{19} cm^{-3} and n-type layers were Si doped with 2×10^{18} cm^{-3}. The mobilities of 2 DEG were 6500 cm^2/Vs at room temperature and 89000 cm^2/Vs at 77 K. A 0.5 μm long and 0.4 μm thick gate electrode was made with Al by side-etching technique using a photoresist mask for source and drain contacts: -(2). Ni/Au-Ge ohmic metal layers were evaporated through the photoresist mask: -(3). After the lift-off process, the Ni/Au-Ge layers were alloyed with those semiconductor layers forming source and drain ohmic contacts, so that the source and drain contacts can be formed in the self-alignment process with the gate electrode: -(4). Then, the p^+-layer outside the gate was etched using the gate, source and drain electrodes as etching masks: -(5). This p^+-layer etching is the only step in addition to the conventional fabrication process for planar Schottky barrier gate GaAs FETs [5]. Therefore, the fabrication process is simple and it is easy to form short gate electrodes. Figure 4 shows a cross sectional view of a fabricated 0.5 μm gate FET.

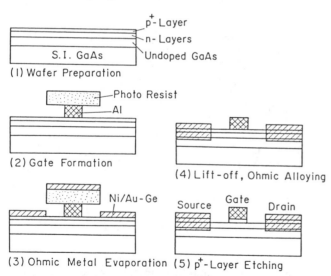

Fig. 3   Fabrication steps for the developed FET.

(1) Wafer Preparation
(2) Gate Formation
(3) Ohmic Metal Evaporation
(4) Lift-off, Ohmic Alloying
(5) p^+-Layer Etching

Fig. 4   Cross sectional view of a 0.5 μm gate FET.

Performance

The developed AlGaAs/GaAs selectively-doped-structure FET with a p^+-gate exhibited high DC and RF performances. Figure 5 shows static characteristics of a FET with a 0.5 μm long and 200 μm wide gate at room temperature. These characteristics show low saturation voltage and markedly high transconductance due to small source resistance. The maximum transconductance was 62 mS, that is 310 mS/mm gate width. The FET also exhibited enhanced transconductance of 90 mS (450 mS/mm) at 77 K. The source resistance is estimated to be 3.5 Ω at room temperature which is the sum of the contact resistance of 1 Ω and the resistance of the 2 DEG layer between the source and the gate of 2.5 Ω for 200 μm gate width. This value, that is 0.7 Ωmm, is relatively small, compared to FETs based on N-AlGaAs/GaAs selectively doped structure reported so far [1]-[4]. The drain conductance was about 15 mS/mm.

Figure 6 shows s-parameters from 2 to 12 GHz for a FET with 200 μm gate width (50 μm x 4). The magnitude of S_{21} is fairly large, corresponding to high

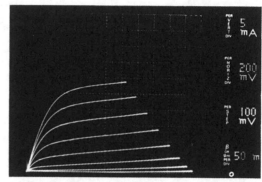

Fig. 5   Drain I-V characteristics of a 0.5 μm gate FET. Gate width is 200 μm. H:0.2 V/div, V:5 mA/div, Gate:-0.1 V/step.

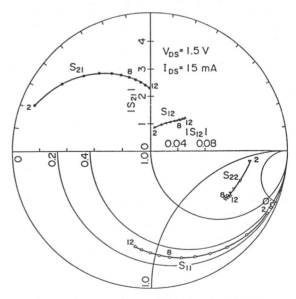

Fig. 6   S-parameters of the developed FET. Gate width is 200 μm.

transconductance. The phase rotation of S_{11} and the magnitude of S_{12} are both small. This indicates small gate capacitance and small drain gate feedback capacitance, in spite of using highly doped n-type layers. This is possibly due to small gate fringing capacitance of the planar gate structure.

The noise performance of the new FET was evaluated in the packaged form at room temperature. The FET exhibited marked noise performance as shown in Figs.7 and 8. Figure 7 shows drain current dependences of minimum noise figure NFmin and associated power gain Ga at 12 GHz for an FET with a 200 μm wide gate. Marked low noise and high gain performance was obtained, even at low drain current level. Figure 8 shows gate width dependence of the noise performance at 12 GHz. Zu represents the unit gate width. The best results were obtained at around 200 μm gate width. Room temperature performance values for 0.5 μm gate FETs were 0.34 dB noise figure with 14.8 dB associated gain and 1.2 dB noise figure with 11.7 dB associated gain at 4 GHz and 12 GHz, respectively.

Table I shows a summary of device parameters and performance for the FET with a 200 μm wide gate. Noise performances are comparable to the best data for quarter micron gate GaAs MESFETs. However, parasitic resistances and drain conductance g_d are fairly large as yet, compared to refined GaAs MESFETs. Further optimization of the device design including crystals will possibly push the noise performance improvement.

Conclusion

A new low noise FET, based on the N-AlGaAs/GaAs selectively doped structure, has been developed. The FET has a planar structure with a p+-gate and a close space structure between the gate and the source in order to have small source resistance. A 0.5 μm gate FET exhibited marked room temperature performances of 310 mS/mm transconductance , 0.34 dB noise figure with 14.8 dB associated gain at 4 GHz, and 1.2 dB noise figure with 11.7 dB associated gain at 12 GHz.

Acknowledgment

The authors wish to thank Drs. N Kawamura and Y. Takayama for their helpful suggestions and encouragements. They also thank Drs. T. Itoh, H. Terao and H. Watanabe for their valuable discussions, and Y. Ara and Y. Fujiki for microwave measurements.

References

[1] M. Laviron et al., "Low Noise Normally on and Normally off Two Dimensional Electron Gas Field-Effect Transistors", Appl. Phys. Lett., 40, pp.530-532, 1982.

[2] K. Joshin et al., "Noise Performance of Microwave HEMT," 1983 IEEE Int'l MTT-S Microwave Symp. Digest Tech. Papers, pp. 563-565, 1983.

[3] S. L. Su et al., "Modulation-Doped (Al,Ga)As/GaAs FETs with High Transconductance and Electron Velocity," Electron, Lett., 18, pp. 794-796, 1982.

[4] M. D. Feuer et al., "High-Speed Low-Voltage Ring Oscillators Based on Selectively Doped Heterojunction Transistors," IEEE Electron Device Lett., EDL-4, pp. 306-307, 1983.

[5] T. Furutsuka et al., "High-Speed E/D GaAs ICs with Closely-Spaced FET Electrodes," Proc. 14th Conf. (1982 Int'l) Solid State Devices, Tokyo, pp. 335-339, 1983.

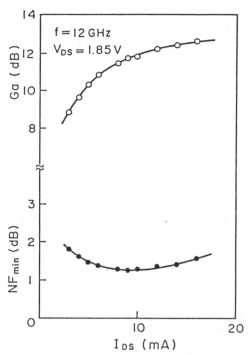

Fig. 7 Drain current dependences of minimum noise figure NFmin and accociated gain Ga.

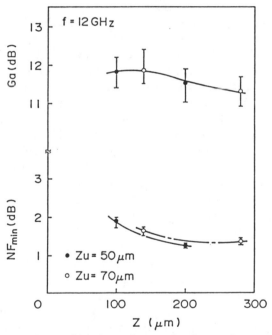

Fig. 8 Gate width dependence of noise performance.

Table I Summary of device performance for a 0.5 μm long, 200 μm wide FET.

R_s	R_g	g_m	g_d	V_p	NF (Ga)	
		(Id=10mA)	(Id=10mA)		4 GHz	12 GHz
(Ω)	(Ω)	(mS)	(mS)	(V)	(dB)	(dB)
3.5	0.9	50	3.0	-0.7~ -1.4	0.34(14.8)	1.2 (11.7)

Ultra-Low-Noise Cryogenic High-Electron-Mobility Transistors

K. H. GEORGE DUH, MEMBER, IEEE, MARIAN W. POSPIESZALSKI, SENIOR MEMBER, IEEE,
WILLIAM F. KOPP, PIN HO, MEMBER, IEEE, AMANI A. JABRA, PANE-CHANE CHAO, MEMBER, IEEE,
PHILLIP M. SMITH, MEMBER, IEEE, LUKE F. LESTER, JAMES M. BALLINGALL, MEMBER, IEEE,
AND SANDER WEINREB, FELLOW, IEEE

Abstract—Quarter-micrometer gate-length HEMT's for cryogenic low-noise application with very low light sensitivity have been developed. At room temperature, these exhibit a noise figure of 0.4 dB with associated gain of 15 dB at 8 GHz. At a temperature of 12.5 K the minimum noise temperature of 5.3 ± 1.5 K has been measured at 8.5 GHz, which is the best noise performance ever observed for any microwave transistors. The results clearly demonstrate the great potential for low-temperature low-noise applications.

I. INTRODUCTION

IN RECENT YEARS, high-electron-mobility transistors (HEMT's) have been demonstrated to show superior noise performance to conventional MESFET's. Noise performance of 0.25-μm HEMT's and MESFET's fabricated in the GE laboratory are compared in Fig. 1 over 8–60 GHz at room temperature. HEMT's not only have lower noise figures, they also have several characteristics that make them more attractive for low-noise applications than the MESFET's. The scattering parameters (S-parameters) of a HEMT in a 50-Ω system exhibit lower $|S_{22}|$ and higher $|S_{21}|$ values than those for a MESFET of the same size, providing inherently better output match and larger gain-bandwidth product. In addition, a HEMT has much lower noise conductance, g_n and usually a lower ratio $X_{g\,opt}/R_{g\,opt}$ (where $R_{g\,opt} + jX_{g\,opt}$ is the optimum source impedance) than the comparable MESFET [1] to facilitate lower noise over a broad bandwidth. Another advantage of the HEMT is that its performance improves more rapidly with cooling than does that of the MESFET. This is due to the enhancement of electron mobility by reduced ionized impurity scattering in HEMT structures resulting from spatial separation of the channel electrons and their parent ions in the highly doped AlGaAs layer.

Manuscript received September 8, 1987. The HEMT devices were developed at the General Electric Company with the support of the Jet Propulsion Laboratory, California Institute of Technology, under Contract 957352, monitored by S. Petty. The National Radio Astronomy Observatory is operated by Associated Universities, Inc. under contract with the National Science Foundation.

K. H. G. Duh, W. F. Kopp, P. Ho, A. A. Jabra, P.-C. Chao, P. M. Smith, L. F. Lester, and J. M. Ballingall are with the Electronics Laboratory, General Electric Company, Syracuse, NY 13221.

M. W. Pospieszalksi and S. Weinreb are with the National Radio Astronomy Observatory, Charlottesville, VA 22903.

IEEE Log Number 8718789.

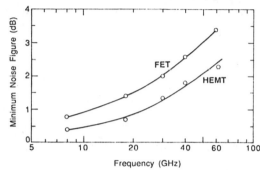

Fig. 1. Comparison of noise performance for 0.25-μm HEMT's and MESFET's at room temperature.

For good low-noise devices, one must have good pinchoff characteristics, low parasitic resistances R_g and R_s (R_g, the gate metallization resistance and R_s, the series source resistance that includes the ohmic contact resistance and the channel resistance between gate and source contacts), and high transconductance g_m. For a field-effect transistor, the minimum noise temperature can be approximately expressed by [4]

$$T_{\min} = K \cdot \omega \cdot C_{gs} \cdot \sqrt{\frac{t(Rg + Rs)}{g_m} + \frac{Kr}{g_m^2}}$$

where K and K_r are noise coefficients, $t = T_a/290$, and T_a is the physical temperature of the device. The first term represents the effective noise voltage generator in the input circuit and is dominated by the thermal noise of R_g and R_s. The second term represents that portion of the nonthermal noise coupled to the gate circuit, which is uncorrelated with the drain current. For the cryogenic operation ($T_a < 20$ K and $t < 0.07$), the second term K_r/g_m^2 becomes dominant. Therefore, for a good cryogenic low-noise device, the wafer should have a proper doping concentration and spacer thickness to achieve an enhancement of g_m at low temperature. Another important factor is the crystal quality, especially the presence of defects at the layer interfaces and in doped AlGaAs, giving rise to light sensitivity of the device dc parameters and noise performance.

Reprinted from *IEEE Transactions on Electron Devices*, vol. 35, no. 3, pp. 249–256, March 1988.

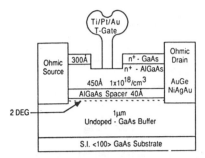

Fig. 2. Cross section of the 0.25-μm T-gate HEMT.

II. HEMT FABRICATION

The HEMT devices were fabricated on selectively doped AlGaAs/GaAs heterostructures by molecular-beam epitaxy on a 3-in substrate Varian GEN II system. The detailed material growth conditions have been discussed elsewhere [2]. Fig. 2 schematically illustrates the cross section of the HEMT devices. Two wafers were grown under identical conditions, both with 40-Å spacer layer. Both wafers, denoted here by A and B, exhibited at 77 K identical sheet charge carrier density of 8.1×10^{11} cm^{-2}, and almost the same mobities, 70 500 cm^2/V $\cdot$ s and 73 200 cm^2/V $\cdot$ s, respectively. All levels were defined by electron-beam lithography, and the T-shape gates were made using the PMMA/P(MMA-MAA)/PMMA trilayer resist technique [3]. We have developed unique processing techniques to control each of the following critical steps in the 0.25-μm fabrication: 1) formation of extremely low resistance and smooth morphology ohmic contacts, 2) precise alignment (± 0.1 μm), exposure, and development of the gate dimension 0.25 μm $\pm$ 10 percent, and 3) a uniform and reproducible gate recess etch that produces a smooth etch surface and minimizes undercut of the resist profile and the amount of oxide formed on AlGaAs.

III. DC CHARACTERISTICS

Examples of dc characteristics at different temperatures with and without light illumination for typical (0.25 $\times$ 300 μm) devices from wafer A and wafer B are shown in Figs. 3 and 4. The dependence of dc measured transconductance g_m on drain current for typical devices A and B are shown in Fig. 5. The dc characteristics of both devices are relatively independent of illumination. They do not exhibit the collapse of the family of I–V curves near the origin as reported in [5] and also observed [6] for commercial Fujitsu HEMT's. Devices from both wafer A and wafer B show some deformations of I–V characteristics in the temperature range 120–170 K, however, to a much lesser extent than the devices reported previously [1], [8]. Also, their dc characteristics remained invariant at cryogenic temperatures after illumination had been removed. The $g_m(I_D)$ characteristics show an increase of g_m upon cooling of about 70 and 75 percent at $I_D = 5$ mA and 30 and 40 percent at $I_D = 25$ mA for devices A and B, respectively. This difference between wafers is consistent

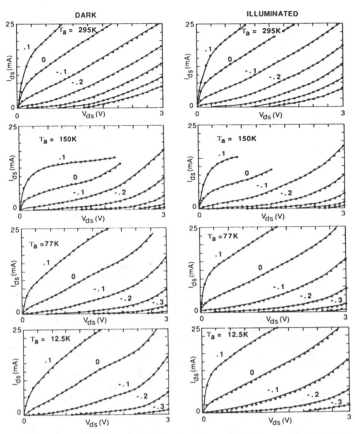

Fig. 3. Typical I–V characteristics for 0.25 μm $\times$ 300 μm wafer A HEMT at several different temperatures with and without illumination.

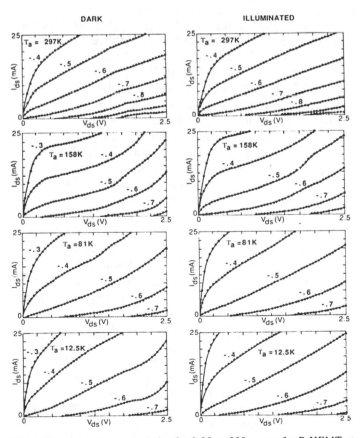

Fig. 4. Typical I–V characteristics for 0.25 $\times$ 300 μm wafer B HEMT at several different temperatures with and without illumination.

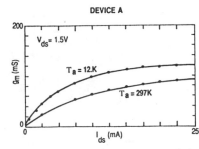
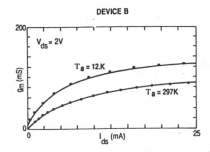

Fig. 5. The dependence of dc measured transconductance g_m on drain current for typical devices from wafer A and wafer B at room and cryogenic temperatures.

with slightly different values of electron mobility measured at 77 K for both wafers.

IV. SIGNAL AND NOISE PERFORMANCE

The room-temperature signal and noise performance was very consistent and repeatable for both wafers. The noise figure measured at 8 GHz with bias $V_D = 2$ V, $I_{DS} = 24$ mA varied between 0.4 to 0.5 dB with associated gain between 14.2 and 15.1 dB. As an example, the data for five arbitrarily selected devices from wafer B are shown in Table I and their S-parameters at low-noise bias conditions are shown in Fig. 6.

Although the room-temperature noise performance of the devices from both wafers was very similar, their cryogenic performance was different.

First, the cryogenic noise performance of the wafer A devices was not dependent on past history, i.e., conditions under which it was cooled down. Their performance was best without light illumination, typically exhibiting minimal noise temperature of 7.5 K at 8.5 GHz, which deteriorated by about 1.5 K if illuminated. Removal of light returned the noise temperature to the previous dark value. The devices from wafer B did require illumination to obtain lowest noise temperatures, typically 5.5 K at 8.5 GHz. The difference between noise performance prior to illumination (if cooled in dark) and after illumination was about 0.5 K to 1 K. Upon subsequent removal of light, the noise temperature remained at the low value or even went lower by a fraction of a Kelvin and was not observed to go up during a period of 48 h if kept cold.

Second, an important difference was repeatability of noise performance at cryogenic temperatures. Devices from wafer B showed excellent repeatability, not demonstrated by the devices from wafer A. This is illustrated by the data of Tables II and III, containing the results of cryogenic measurement for five devices of each wafer. In all cases the cryogenic noise performance could be correlated with the rate of increase of transconductance g_m with drain current, as it is illustrated in Figs. 7 and 8. The best devices of wafer A could almost match the cryogenic performance of a typical device from wafer B.

These two differences are most likely related to the difference in processing of both devices. First the devices from wafer A have a deeper gate recess etch, resulting in more enhancement-type characteristics, than the devices

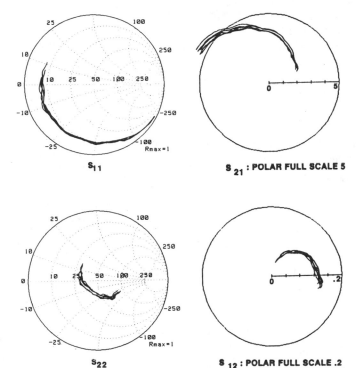

Fig. 6. The spread of S-parameters from 2 to 20 GHz of five sampled devices from wafer B at room temperature.

TABLE I
ROOM-TEMPERATURE PERFORMANCE OF WAFER B DEVICES

Device	V_{ds} (V)	V_{gs} (V)	I_{ds} (mA)	Ga (dB)	F_{min} (dB)
1	2.0	-0.591	24.0	14.6	0.45
2	2.0	-0.511	24.0	15.0	0.4
3	2.0	-0.496	24.0	14.5	0.45
4	2.0	-0.612	24.0	14.4	0.45
5	2.0	-0.483	24.0	15.1	0.4

from wafer B. This could make wafer A devices to be more sensitive to the small nonuniformity of either gate recess etch or wafer structure. Also, although both devices possess 0.25-μm gates, the gate of wafer B devices is approximately symmetrically placed between source and drain, while for wafer A devices it is offset toward the gate. In the most extensive study of low-temperature

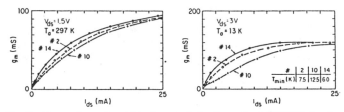

Fig. 7. Comparison of $g_m(I_D)$ characteristics for wafer A HEMT's having different cryogenic noise performance.

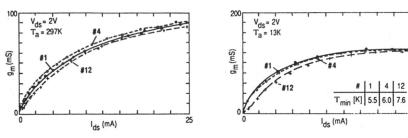

Fig. 8. Comparison of $g_m(I_D)$ characteristics for wafer B HEMT's having different cryogenic noise performance.

TABLE II
CRYOGENIC PERFORMANCE OF WAFER A HEMT's
($f = 8.5$ GHz, dark.)

HEMT #	V_d V	I_d mA	V_{gs} V	T_{min} K
2	3.0	5	-.230	7.5
7	3.0	5	-.076	12.0
10	3.0	5	-.107	12.5
14	3.0	3	-.258	6.0
15	3.0	3	-.140	7.7

TABLE III
CRYOGENIC PERFORMANCE OF WAFER B HEMT's
($f = 8.5$ GHz, illuminated.)

HEMT #	V_d V	I_d mA	V_{gs} V	T_{min} K
1	2	5	-.616	5.5
3	2	5	-.634	5.3
4	1.5	5	-.707	6.0
12	2	7.5	-.641	7.6
13	2	5	-.750	5.3

dc behavior of HEMT's [5] so far, the I-V collapse has been clearly found to be dependent on the type of the device (enchancement or depletion mode) and also on the potential distribution across the device, i.e., on its geometry. This could explain observed slight differences between the light sensitivity of two devices.

Different geometry of sample devices is also responsible for different noise parameters, which for the frequency 8.5 GHz and ambient temperature $T_a = 12.5$ K are given in Table IV. The noise parameters are for 0.25×300 μm chips mounted in standard 70-mil package and were mea-

sured using the method presented in [9]. Although the R_{opt} and g_n of two samples are different, their product $N = R_{opt} g_n$, which is invariant under lossless transformation at the input [10], is about the same. This clearly indicates that the device B layout, while not changing substantially the nature of noise mechanism involved, produced a much more useful device from the point of view of circuit applications. It is easier to realize optimal source impedance for device B, and its noise temperature is much less sensitive to changes in the source impedance.

In several studies [1], [6]–[8], it has been reported that the noise temperature versus ambient temperature characteristics for HEMT's exhibits a "hump" around the ambient temperature of 150 to 175 K, which is present for higher drain current even with light illumination. In some HEMT's the noise temperature actually increases with decreasing physical temperature. This effect could have important practical consequences as it falls in the range of temperatures attainable by radiation cooling in space applications. The HEMT's presented in this paper do not possess this effect if illuminated. If illumination is removed, then for small drain voltages and larger drain current, a "hump" on $T_n = f(T_a)$ characteristics is observed. An illustrative example is shown in Fig. 9 for wafer A HEMT's, demonstrating the dependence on $T_n = f(T_a)$ on drain current, voltage, and light illumination. Although the behavior of wafer B HEMT's was qualitatively similar, the observed differences between illuminated and not illuminated HEMT's for $V_d > 1.5$ V, $I_D \leq 15$ mA never exceeded 3 K. The light, current, and voltage dependence of this effect and its correlation with the observed deformations of drain characteristics (compare Figs. 3 and 9) together with other available evidence [5], [9] clearly link it with the existence of traps in AlGaAs, but the particular mechanisms involved are not understood.

It has been demonstrated that [4], [7], [10] the mini-

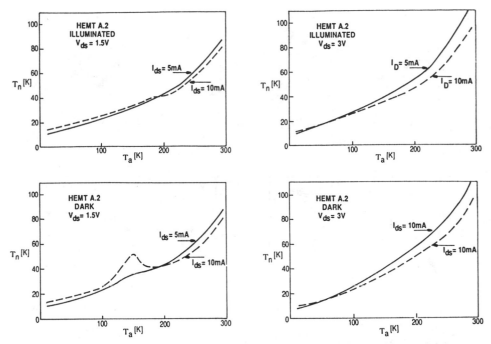

Fig. 9. The noise temperature T_n as a function of ambient temperature T_a for a single-stage 8.5-GHz amplifier with wafer A HEMT for different bias and illumination conditions. The source impedance is $(3.5 + j38)\ \Omega$.

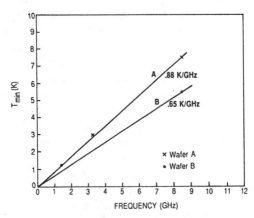

Fig. 10. Cryogenic noise performance of quarter-micrometer GE HEMT's versus frequency.

TABLE IV
NOISE PARAMETERS OF SAMPLE DEVICE AT $f = 8.5$ GHz and $T_a = 12.5$ K

Device	I_{ds} mA	V_{ds} V	T_{min} K	R_{opt} Ω	X_{opt} Ω	g_n mS	G_{AS} dB	$4NT_o$ K
A-2	5	3.0	7.5	4.3	38	2.0	13.8	10.0
B-1	5	2.0	5.5	8.2	31.5	1.0	14.2	9.5

mum noise temperature for cryogenic HEMT's and FET's from *C*- to *K*-band is proportional to frequency, which is also in agreement with studies, both theoretical and experimental, of room-temperature devices (for instance, [11], [12]). The results at *L*-band slightly deviate from the proportional dependence, most likely due to $1/f$ noise. This explanation is corroborated by room-temperature measurement of $1/f$ noise which for these two wafers exhibits a corner frequency of about 200 MHz and trails down to reach a thermal noise floor at around 2 GHz. The measured results for the devices discussed in this paper are shown in Fig. 10, yielding the best ever reported noise temperature-frequency coefficient of 0.65 K/GHz.

V. CONCLUSIONS

The performance of cryogenic HEMT's with record-breaking noise temperatures has been described. The noise performance has been qualitatively correlated with dc mea-

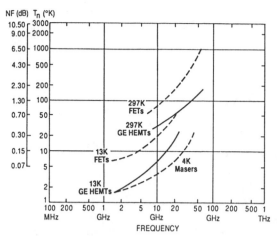

Fig. 11. State-of-the-art of low-noise cryogenic devices for frequencies below 50 GHz referred to cold input of a device. Room-temperature data for FET's and HEMT's are given for comparison. All data for cryogenic FET's and HEMT's are NRAO measured results. Maser data are for JPL and NRAO masers [13]–[15]. It should be noted that masers are inherently matched at the input and have sufficient gain to make contribution of subsequent stages insignificant. This gives them a greater advantage than presented in the figure for frequencies above X-band.

sured characteristics, light sensitivity and other technological and experimental data. While trap-related effects are a dominant factor in the performance of cryogenic HEMT's, they are far from being well understood. However, as the data presented in this paper indicates, most of the undesired effects can be avoided in conventional HEMT structure with proper wafer growth and processing.

The cryogenic noise performance of HEMT's approaches that of masers of frequencies below X-band. Fig. 11 shows the noise temperatures of various devices and natural limits. Because HEMT is very cost effective, it can replace the maser in many applications. Rapid progress in HEMT development is expected to continue. Improved performance is likely to result from a reduction in gate length and further optimization of the material structure and device fabrication process. The HEMT clearly has potential for low-temperature low-noise applications in the millimeter-wave frequencies.

ACKNOWLEDGMENT

The authors wish to thank R. Norrod and R. Harris for L-band and S-band cryogenic noise measurement and acknowledge Dr. A. W. Swanson for his strong suppot of this work. W. Lakatosh is thanked for his excellent technical assistance.

REFERENCES

[1] M. W. Pospieszalski, S. Weinreb, P. C. Chao, U. K. Mishra, S. C. Palmateer, P. M. Smith, and J. C. M. Hwang, "Noise parameters and light sensitivity of low-noise high-electron-mobility transistors," *IEEE Trans. Electron Devices*, vol. ED-33, pp. 218–223, Feb. 1986.

[2] S. C. Palmateer, P. A. Maki, W. Katz, A. R. Calawa, J. C. M. Hwang, and L. F. Eastman, "The influence of V:III flux ratio on unintentional impurity incorporation during molecular beam epitaxial growth," in *Proc. 11th Int. Symp. Gallium Arsenide and Related Compounds* (Biarritz, France), Sept. 26–28, 1984.

[3] P. C. Chao, P. M. Smith, S. C. Palmateer, and J. C. M. Hwang, "Electron-beam fabrication of GaAs low-noise MESFET's using a new tri-layer resist technique," *IEEE Trans. Electron Devices*, vol. ED-32, pp. 1042–1046, June 1985.

[4] S. Weinreb, "Low-noise, cooled GASFET amplifiers," *IEEE Trans. Microwave Theory Tech.*, vol. MTT-28, pp. 1041–1054, Oct. 1980.

[5] A. Kastalsky and R. A. Kiehl, "On the low temperature degradation of (AlGa)As/GaAs modulation doped field-effect transistors," *IEEE Trans. Electron Devices*, vol. ED-33, pp. 414–423, Mar. 1986.

[6] M. W. Pospieszalski and S. Weinreb, "FET's and HEMT's at cryogenic temperatures—Their properties and use in low-noise amplifiers," in *Proc. IEEE MTT-S Int. Microwave Symp.* (Las Vegas, NV), pp. 955–958, June 1987.

[7] M. W. Pospieszalski, S. Weinreb, R. D. Norrod, and R. Harris, "FET's and HEMT's at cryogenic temperatures—Their properties and use in low-noise amplifiers," *IEEE Trans. Microwave Theory Tech.*, to be published.

[8] M. W. Pospieszalski, "X-band noise performance of commercially-available GaAsFET's at room and cryogenic temperatures," Nat. Radio Astronomy Observatory Electron. Division Internal Rep. 260, May 1986.

[9] J. Y. Chi, R. P. Holstrom, and J. P. Salerno, "Effect of traps on low temperature high-electron-mobility transistor characteristics," *IEEE Electron Device Lett.*, vol. EDL-5, pp. 381–394, Sept. 1984.

[10] P. C. Chao, P. M. Smith, U. K. Mishra, S. C. Palmateer, J. C. M. Hwang, M. W. Pospieszalski, T. Brooks, and S. Weinreb, "Cryogenic noise performance of quarter-micrometer gate-length high-electron-mobility transistors," *IEEE Trans. Electron Devices*, vol. ED-32, p. 2528 (abstract 43rd Ann. Device Res. Conf.), Nov. 1985.

[11] R. A. Pucel, H. A. Haus, and H. Statz, "Signal and noise properties of GaAs microwave FET," in *Advances in Electronics & Electron Physics*, vol. 38, L. Morton, Ed. New York: Academic, 1975.

[12] A. Cappy, A. Vanoverschelde, M. Schortgen, C. Versnaeyen, and G. Salmer, "Noise modeling in submicrometer-gate two-dimensional electron-gas field-effect transistors," *IEEE Trans. Electron Devices*, vol. ED-32, pp. 2787–2795, Dec. 1985.

[13] S. M. Petty and D. L. Trowbridge, "Low-noise amplifiers," in *Deep Space Network—Radio Communications Instrument for Deep Space Exploration*, JPL Publ. 82-104, Jet Propulsion Lab., California Institute of Technol., Pasadena, CA, July 1983.

[14] C. R. Moore and R. C. Clauss, "A reflected wave ruby maser with K-band tuning range and large instantaneous bandwidth," *IEEE Trans. Microwave Theory Tech.*, vol. MTT-27, pp. 249–255, Mar. 1978.

[15] C. R. Moore and R. D. Norrod, unpublished data on NRAO 43-GHz ruby maser.

InGaAs/AlInAs HEMT Technology for
Millimeter Wave Applications.

Umesh K. Mishra and April S. Brown

Hughes Research Laboratories
Malibu, CA 90265

Abstract:
    We review the progress on AlInAs-GaInAs HEMTs
at the Hughes Research Laboratories. Materials
grown by MBE have yielded sheet charge densities
of 5 x 10^{18} cm^{-3} with a mobility of 9000cm^{2}V^{-1}
S^{-1} at 300K. 0.1 m long gate devices have yielded
an f$_T$ of ~170GHz. Ring oscillators made with
0.2μm long gate devices have demonstrated a
gate delay of 6pS at 300K and 5pS at 77K.
Static dividers have operated at 26.7 GHz. A
minimum noise figugure of 0.8dB with an associated gain of 8.7 dB has been obtained from a
single stage amplifier at V-band.

Introduction:
    The Al$_{.48}$In$_{.52}$As-Ga$_{.47}$In$_{.53}$As modulation
doped system is extremely attractive for both
mm-wave digital and analog applications. There
are several reasons why this system is superior
to the more conventional lattice matched AlGaAs-
GaAs and pseudomorphic AlGaAs-GaInAs material
systems. The Al$_{.48}$In$_{.52}$As- Ga$_{.47}$In$_{.53}$As
heterojunction has a higher conduction band
offset (0.5eV) which, coupled with the high Si
doping concentration possible in Al$_{.48}$In$_{.52}$As
(N$_D$~n$_s$~1 x 10^{19} cm^{-3}), allows for a larger two-
dimensional electron gas (2DEG) concentration.
The GaInAs channel has a higher electron mobility
and peak velocity which leads to superior
transport properties. Excellent optical and
electronic materials properties have been
experimentally demonstrated in high quality
materials grown by MBE. Discrete devices built
using these materials have demonstrated record
high frequency and low noise performance and
discrete circuits such as ring oscillators and
dividers have demonstrated record switching
speeds. This paper will review the work done
at Hughes Research Laboratories in the materials,
devices and circuits areas using the AlInAs -
GaInAs modulation doped system.
Materials Characteristics:
    All the modulation doped structures were grown
on (100) Fe-doped InP substrates in a Riber 2300
system equipped with a 3 inch rotating substrate
holder. The substrate temperature was held
constant during growth at approximately 500°C.
Growth rates varied from 0.75μm/hr to 1.0μm/hr.
The typical device structure is shown in Figure
1. First a 250 nm thick AlInAs buffer was grown,
followed by a 40 nm thick AlInAs - GaInAs
superlattice. The AlInAs layer served the dual

purpose of separating the active layer from the InP
surface and also provided a high band-gap layer
which confined the electrons in the active GaInAs
channel which improves the device output
conductance. The superlattice also serves two
important purposes. It smoothes the AlInAs growth
front, minimizing interface roughness scattering
resulting in higher electron mobilities in the
channel. The superlattice can also getter
outdiffusing impurities from the substrate which
further improves the materials properties of the
active channel. Next a 40 nm thick GaInAs channel
was grown, followed by a 2nm undoped AlInAs spacer.
The AlInAs donor layer was then grown, 12.5 nm
thick, and doped with Si at 4 x 10^{18} cm^{-3}. This
was followed by a 20nm thick undoped AlInAs layer
which serves as a Schottky barrier enhancing layer
and the structure was capped by a 5 nm thick n$^+$
GaInAs contact layer.
    The device structures show extremely high values
of sheet charge at (n$_s$, 3.5 x 10^{12} cm^{-2}) and
mobility (μ= 9500 cm^{2}V^{-1}s^{-1})@300 K. The
resistivity values obtained (ρ~150Ω/□) are approximately three times lower than that which can be
obtained in an AlGaAs-GaAs HEMT. Further details
of the materials properties as a function of
channel thichness, spacer layer thickness and the
nature of Si doping are available in Reference 1.
Device Fabrication and Performance:
    To determine the impact of the superior
materials properties on device performance, HEMTs
were fabricated with 1.0μm, 0.2μm and 0.1μm gate
lengths. Device fabrication followed closely the
steps described in a previous publication [2].
The I-V characteristics of a typical 1.0μm gate
length device is shown in figure 2. The extrinsic
transconductance of the device is 600 mS/mm. This
impressive performance at such a long gate length
is primarily due to the low parasitic resistances
in the structure resulting from the high two
dimensional electron gas density and mobility.
Figure 3a shows the I-V characteristics of a 25 m
wide device with 0.2μm gate length and Figure 3b
of the device with 0.1μm gate length. The
extrinsic transconductance of the two devices are
800 and 1080 mS/mm respectively. Both 50μm and
200μm wide devices are characterized from 45 GHz
to 26.5 GHz on a Cascade RF probe station. The
current gain, h$_{21}$, of the devices were extracted
from the S-parameters and the unity current gain
cut-off frequency, f$_T$, calculated by extrapolating
at 6 dB/octave. The data is presented in Figure 4.
50μm wide devices with gate lengths of 0.2μm

Reprinted from *Rec. of the IEEE GaAs Integrated Circuits Symp.*, pp. 97-100, 1988.

(A 50) and 0.1μm (B 50) exhibited an f of 120 GHz and 135 GHz respectively. The 200μm wide device with 0.1μm gate length (B 200) exhibited an f_T of ~170 GHz. Both these numbers represent world record performance for any three terminal semiconductor device. Figure 5 shows the minimum noise figure and associated gain data for a single stage amplifier measured from 60 GHz to 63.5 GHz with no corrections. The minimum aplifier noise figure was 0.8 dB with an associated gain 8.7 dB at 63 GHz [3]. This represents a substantial advance in the state-of-the-art of low noise mm-wave transistors.

Circuit Fabrication and Performance:

Ring oscillators and static frequency dividers were implemented in buffered FET logic (BFL) and capacitively enhanced logic (CEL) schemes utilizing HEMTs with 0.2-μm gate lengths. BFL produces the fastest logic gates at high fanouts, but it also dissipates the most power per logic rate because of the large output current that flows through the source follower and level-shifting diodes. The CEL logic family is unbuffered and dissipates much less power than BFL. Figures 6a and 6b show the circuit diagrams of the two logic families. The fifteen-stage ring oscillators were implemented with either BFL or CEL inverters (FI=FO=1 and FI=FO=4). Binary static dividers were also implemented in both logic families using biphase two-gate-delay flip-flops, as described in references [4] and [5].

The shortest gate delay measured for the BFL ring oscillator at 300K was 9.3 ps (fanout=1) with a power dissipation of 66.7 mW per gate. The average BFL gate delay measured across a wafer was 10ps, and the fanout sensitivity was 1.5ps per fanout. The shortest gate delay measured for the CEL ring oscillator at 300K was 6.0ps (fanout = 1) with a power dissipation of 23.8 mW per gate. The average CEL gate delay across a wafer was 7.4ps with a fanout sensitivity of 2.7 ps per fanout. The CEL gate delay reduced to less than 5.0ps with 11.35-mW power dissipation when measured at 77K. This performance surpasses the best previously reported results for a HEMT technology [6]. The dividers were tested on-wafer using a Cascade Microtech probe station at room temperature (300K). The highest operating frequency of 26.7 GHz was achieved with a CEL divider. At this frequency the divider dissipated 73.1 mW total. Figure 7 shows the input and output waveforms also measured at an input frequency of 26.7 GHz. The fastest BFL divider operated at 25.2 GHz with a 450 mW power dissipation. These results surpasses the best previously reported divider results of 20.1 GHz [7] and 22 GHz [8], which were achieved with heterojunction bipolar transistors (HBTs).

Conclusions:

AlInAs-GaInAs modulation doped structures grown by MBE on InP have demonstrated excellent electronic and optical properties. Extremely high sheet charge densities ($n_s \sim 5 \times 10^{18}$ cm^{-3}) and room temperature mobilities ($\mu \sim 9500$ cm^2 V^{-1} s^{-1}) have been achieved.0.1 m gate length HEMTs have exhibited an f_T of ~170 GHz whereas single stage amplifiers using 0.2μm gate HEMTs have

demonstrated a minimum noise figure of 0.8dB and an associated gain of 8.7 dB. Ring oscillators have demonstrated 6ps switching speeds and static frequency dividers operated at 26.7GHz at room temperature. These results confirm the tremendous impact that AlInAs-GaInAs HEMTs will have on mm-wave electronics.

Acknowledgements:

We thank J.F.Jensen, S.E. Rosenbaum, K. White and S. Vaughn for their help in circuit design and measurements. M. Thompson, P. Janke, M. Pierce, C. Hooper and L.M. Jelloian are acknowledged for their technical help.

References:

1) A.S. Brown, U.K. Mishra, J.A. Henige and M.J. Delaney, " The impact of epitaxial layer design and quality on AlInAs/GaInAs high electron mobility transistors," J. Vac. Sci. Technol. B6(2) Mar/Apr 1988

2) U.K. Mishra, A.S. Brown, L.M. Jelloian, L.H. Hackett, and M.J. Delaney, " High-Performance Submicrometer AlInAs-GaInAs HEMTs," IEEE EDL, Vol. 9, No. 1, (1988)

3) U.K. Mishra, A.S. Brown, S.E. Rosenbaum, M.J. Delaney, S. Vaughn and K. White, " Noise performance of sub-micron AlInAs-GaInAs HEMTs," Device Research Conference, Boulder CO. (1988)

4) J.F. Jensen, L.G. Salmon, D.S. Deakin, M.J. Delaney, " Ultrahigh-speed GaAs Static Frequency Divider", IEDM technical digest, pp. 476-479; Dec. 1986

5) J.F. Jensen, U.K. Mishra, A.S. Brown, R.S. Beaubien, M.A. Thompson, L.M. Jelloian, " 25 GHz Static Frequency Dividers in AlInAs-GaInAs HEMT Technology," ISSCC Technical Digest, pp. 268-269, Feb. 1988

6) Y. Awano, M. Kosugi, T. Mimura, and M. Abe; " Performance of Quarter-Micrometer-Gate Ballistic Electron HEMT," IEEE Electron Device Letters, Vol. EDL-8, No. 10, pp. 451-453, Oct. 1987

7) K.C. Wang, P.M. Asbeck, M.F. Chang, G.J. Sullivan, D.L. Miller, " A 20 GHz Frequency Divider Implemented with Heterojunction Bipolar Transistors," IEEE Electron Device Letters, Vol. EDL-8, No. 9, pp. 383-385, Sept. 1987

8) Y. Yamauchi, K. Nagata, O. Nakajima, H. Ito, T. Nittono, T. Ishibabshi, " 22GHz Divide by Four Frequency Divider using AlGaAs/GaAs HBTs," Electronics Letters, Vol. 23, No. 17, pp. 881-882; Aug. 13 1987.

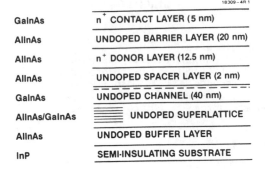

GaInAs	n⁺ CONTACT LAYER (5 nm)
AlInAs	UNDOPED BARRIER LAYER (20 nm)
AlInAs	n⁺ DONOR LAYER (12.5 nm)
AlInAs	UNDOPED SPACER LAYER (2 nm)
GaInAs	UNDOPED CHANNEL (40 nm)
AlInAs/GaInAs	UNDOPED SUPERLATTICE
AlInAs	UNDOPED BUFFER LAYER
InP	SEMI-INSULATING SUBSTRATE

Figure 1.   Schematic cross-section of
a typical AlInAs-GaInAs
modulation doped structure

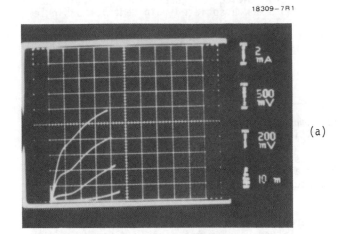

(a)

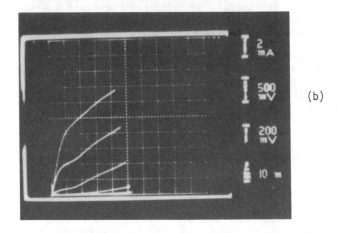

(b)

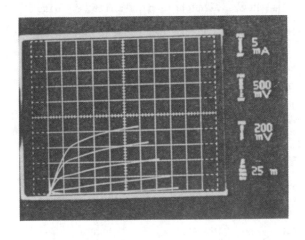

Figure 2.   I-V characteristic of a HEMT
with a 1 μm long gate

Figure 3.   I-V characteristics of HEMTs
with (a) 0.2 μm and (b) 0.1 μm
gate length

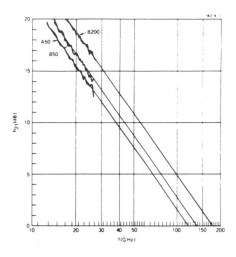

Figure 4. Current gain vs. frequency
of 0.2 μm and 0.1 μm gate
length HEMTs for 50 μm and
200 μm wide devices

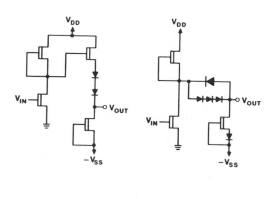

Figure 6. Circuit diagram for (a) BFL
and (b) CEL schemes

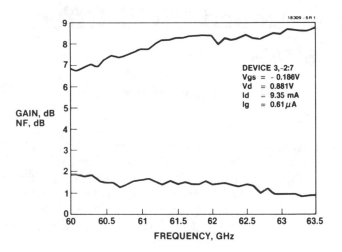

Figure 5. V-band noise figure and
associated gain data for
a single stage amplifier

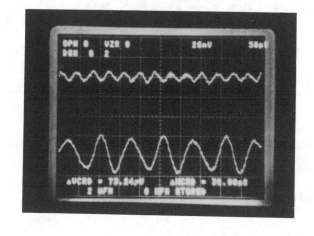

Figure 7. Input and output waveforms
of static divider operating
at 26.7 GHz

1/f Noise in Modulation-Doped Field Effect Transistors

K. H. DUH, A. van der ZIEL, fellow, ieee, and H. MORKOC, senior member, ieee

Abstract—Low frequency noise measurements are reported in modulation-doped GaAs field effect transistors. The noise spectrum is 1/f and is relatively large. At a given frequency the equivalent saturated diode current I_{eq} varies as V^2, as expected for a fluctuating resistor, and saturates when the characteristic saturates.

WE REPORT here low frequency noise measurements in "normally on" modulation-doped GaAs transistors (HEMT's) [1-3]. The noise was found to be of the 1/f type and was relatively large. The devices had a 1 μm gate length and a 4 μm distance between source and drain.

Figure 1 shows the equivalent saturated diode current, I_{eg}, vs. frequency for $V_d = 0.1$ V, $V_g = -0.2$ V. The background equivalent saturated diode current of the circuit was estimated to be 5.4 X 10^{-4} amps so that the device noise is orders of magnitude larger than the background noise. We evaluated

Manuscript received Oct. 20, 1982.

K. H. Duh and A. van der Ziel are with the Electrical Engineering Department, University of Minnesota, Minneapolis, MN 55455.

H. Morkoç is with the Electrical Engineering Department and Co-ordinated Science Laboratory, University of Illinois, Urbana, IL 61801.

Hooge's parameter α and found a value of about 3 X 10^{-4}, comparable to relatively noisy MOSFET's.

Figure 2 shows I_{eg} vs. the drain voltage, V_d, at a frequency of 400 Hz. The noise varies as V_d^2 at low V_d and saturates when V_d saturates; this agrees also with what is found for MOSFET's. The V_d^2-dependence of the noise holds for any fluctuating resistor.

As said before, the 1/f noise resembles that of a MOSFET. The top n$^+$ AlGaAs layer under the gate is depleted of electrons, the electrons are in the GaAs channel, where they can interact with traps in the AlGaAs layer near the interface by tunneling. The 1/f noise should therefore be of the number fluctuation type. The results of Fig. 2 fit very well with this picture [4].

We calculated the trap density (N_T)eff from our data and found a value of about 10^{14}/cm^3, assuming a tunneling parameter ϵ of 10^7 cm^{-1}. Since the magnitude of the noise [4] is proportional to (N_T)eff/ϵ, 10^{14}/cm^3 is comparable to p-channel MOSFET's (n-channel MOSFET's usually have the value of about 10^{15}/cm^3).

When one extrapolates the noise data at saturation to high frequencies, one comes to the conclusion that the device

Reprinted from *IEEE Electron Device Letters*, vol. EDL-4, no. 1, pp. 12–13, January 1983.

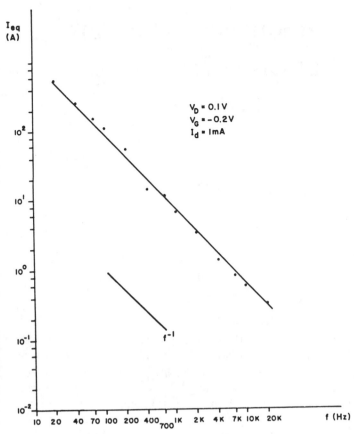

Fig. 1. I_{eq} vs. frequency for mudulation-doped GaAs field-effect transistor. $V_d = 0.1$ V, $V_g = -0.2$ V, and $I_d = 1$ mA.

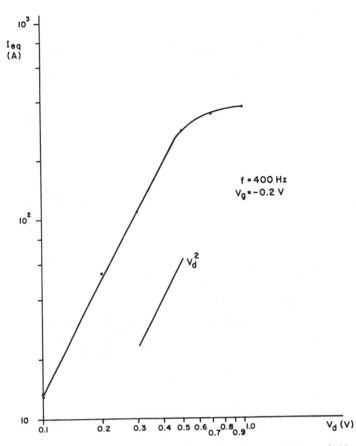

Fig. 2. I_{eq} vs. drain voltage, V_d, for modulation-doped GaAs field-effect transistor. $f = 400$ Hz, $V_g = -0.2$ V.

should show thermal noise in the 10–30 MHz frequency range. We are looking into that possibility and hope to give a report on the thermal noise of these devices at a later date.

Acknowledgment

The research at the University of Minnesota and the University of Illinois was funded by NSF and AFOSR, respectively.

References

[1] D. Delagebeaudeuf and N. T. Linh, *IEEE Trans. Electron Dev.*, ED-29, p. 955, 1982.
[2] T. Mimura, S. Hiyamizu, K. Joshen, and K. Hikosaka, *Japan J. Appl. Phys.*, vol. 20, p. L317, 1981.
[3] T. J. Drummond, W. Kopp, R. E. Thorne, R. Fisher, and H. Morkoç, *Appl. Phys. Lett.*, vol. 40, p. 879, 1982.
[4] H. S. Park, A. van der Ziel, and S. T. Liu, *Solid State Electronics*, vol. 25, no. 3, pp. 213–217, 1982.

Correlation Between Low-Frequency Noise and Low-Temperature Performance of Two-Dimensional Electron Gas FET's

JEAN-MARIE DIEUDONNE, MICHEL POUYSEGUR, JACQUES GRAFFEUIL,
AND JEAN-LOUIS CAZAUX

Abstract—We report here on low-frequency (LF) noise of GaAs/GaAlAs TEGFET's. Present investigations show that this noise is not inherently lower in TEGFET's than in MESFET's. Moreover, the bias and frequency dependence of the noise was found to indicate that traps in GaAlAs have a fundamental influence on LF noise. A close correlation is subsequently observed between the noise level at ambient temperature and the TEGFET static and dynamic performances at low temperature (130 K).

I. INTRODUCTION

LOW-FREQUENCY (LF) noise is a major drawback in GaAs MESFET's, since it limits the performances not only of ultrawide-frequency bandwidth circuits but also of nonlinear circuits where such noise is up-converted in the microwave frequency range. It is therefore of prime interest to investigate whether or not such a large LF noise also exists in two-dimensional electron gas field-effect transistors referred to in this paper as TEGFET's although other designations (i.e., HEMT, MODFET, SDHT) could also be used.

It has already been reported [1] that LF noise in TEG-FET's has a 1/f behavior and that it has the same order of magnitude as LF noise in MESFET's [2].

Also found was the close correlation between LF noise and the existence of DX centers in the GaAlAs layer, as revealed by DLTS [3]. It is also well known that at cryogenic temperatures the existence of DX centers severely disturbs the operation of TEGFET's. Therefore, a correlation should exist between the electrical performances of TEGFET's at low temperature and the amplitude of the LF noise at ambient temperature. This paper presents the data used to investigate such an expected correlation. In addition, LF noise variations with bias conditions are also

Manuscript received October 1, 1985; revised December 5, 1985.
J. M. Dieudonne, M. Pouysegur, and J. L. Cazaux are with the Laboratoire d'Automatique et d'Analyse des Systemes du Centre National de la Recherche Scientifique, 7, avenue du Colonel Roche, 31077 Toulouse Cedex, France.
J. Grauffeuil is with the Laboratoire d'Automatique et d'Analyse des Systemes du Centre National de la Recherche Scientifique, 7, avenue du Colonel Roche, 31077 Toulouse Cedex, France, and with the Universite Paul Sabatier et Centre National d'Etudes Spatiales, Toulouse, France.
IEEE Log Number 8607889.

reported, and these results can be readily used for the optimizing of the bias conditions of circuits where LF noise has to be minimized.

II. EXPERIMENTAL PROCEDURE

We believe that the most significant quantity for a practical noise evaluation of two-port devices is the noise referred back to the input [4] (i.e., the ratio of the output noise by the two-port gain). Therefore, the input voltage noise generator, the input current noise generator, and their correlation should be investigated simultaneously. Practically, in TEGFET's, as well as in MESFET's, we have verified that only the input voltage noise generator contributes significantly to the overall noise as long as the source impedance remains below the megaohm range. Consequently, this paper deals only with the input voltage noise generator.

In the noise measurement system a Takeda–Riken two-channel digital spectrum analyzer is used to collect time domain records, perform fast Fourier transform, and average the related data. The transfer function measurement capability of the spectrum analyzer is also used to determine the voltage gain of the two-port device. Great care has been taken to ensure convenient shielding and grounding of the entire system. The complete measurement procedure is remotely controlled by an HP 85 computer. The latter computer also corrects the data for extra noise sources contributed, for example, by the low-noise amplifier, and computes the resulting input voltage noise spectral intensity in squared volts per hertz in the frequency range 10 Hz–100 kHz.

In this paper, three different MBE-grown TEGFET's referred to as *A*, *B*, and *C* have been investigated. A cross sectional view of the different layers is shown in Fig. 1. The main characteristics of each of the three devices are given in Table I.

III. NOISE VARIATIONS WITH BIAS CONDITIONS

In Fig. 2, the input voltage noise spectra of device *A* at a constant low drain voltage (100 mV) are given for two different gate biases. The same figure also shows the input

Reprinted from *IEEE Transactions on Electron Devices*, vol. ED-33, no. 5, pp. 572–575, May 1986.

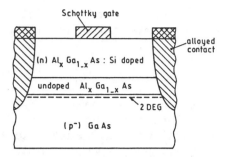

Fig. 1. Schematic cross sectional view of the TEGFET structure.

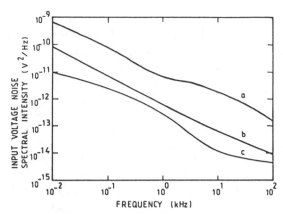

Fig. 2. Noise spectra at low drain voltage (100 mV). Curve *a*: device *A* at $V_{gs} = -0.5$ V. Curve *b*: device *A* at $V_{gs} = -2$ V. Curve *c*: a MESFET with dimensions similar to device *A*.

TABLE I
Main Physical and Electrical Characteristics of the Three Devices
(Shift of the threshold voltage and variations of drain current at $V_{ds} = 3$ V, maximum transconductance (g_m max), and unilateral power gain when the device is cooled down to 130 K. The noise voltage at ambient temperature is also given.)

DEVICE	A	B	C
Gate length	0.9 μm	0.5 μm	0.8 μm
Gate width	260 μm	150 μm	300 μm
Al composition ratio X	0.3	0.24	0.17
Idss (drain current at 300°K and Vds=3V)	(Vgs= 0V) 42 mA	(Vgs= 0V) 13 mA	(Vgs= 0.5V) 25 mA
Threshold voltage V_T at 300°K	-2.1V	-0.44V	-0.09V
Unilateral power gain at 11 GHz and 300°K	6.7 dB	16.1 dB	9.9 dB
Shift of V_T	+0.63 V	+0.24 V	+0.19 V
Variation of Idss at Vds= 3V	(Vgs= 0V) -50 %	(Vgs= 0V) -38 %	(Vgs= 0.5V) +8 %
Variation of gm max at Vds= 3V	+9 %	-15 %	+35 %
Shift of unilateral power gain at 11 GHz	+0.75 dB	+2 dB	+3.2 dB
Noise voltage in V²/Hz at 10 KHz and Vds= 200 mV	(Vgs=-1.8V) $1.04 \cdot 10^{-13}$	(Vgs=-0.3V) $2.56 \cdot 10^{-14}$	(Vgs= 0V) $5.47 \cdot 10^{-15}$

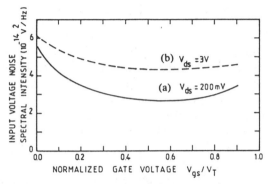

Fig. 3. Noise variations for device *B* versus gate voltage at a constant drain bias and at a frequency equal to 10 kHz. (a) $V_{ds} = 200$ mV (ohmic region). (b) $V_{ds} = 3$ V (saturation region).

voltage noise of a MESFET with similar dimensions operated in the ohmic region of the current–voltage characteristic. The MESFET's observed noise has a magnitude about 20 dB lower than that of TEGFET but has a classical $1/f^\alpha$ spectrum with $1 < \alpha < 1.5$. On the contrary, at a gate bias near threshold, the noise spectrum (curve *b* of Fig. 2) of the TEGFET is almost an ideal $1/f$ spectrum and the noise magnitude is only about 10 dB worse than that of the MESFET. Moreover, at a near-zero gate bias the presence of a quasi-constant noise magnitude in the frequency range 1–10 Khz denotes the occurrence of some single-level generation–recombination noise [5]. One recalls that for such a device at a zero gate bias a large fraction of the channel current flows through the GaAlAs layer where many traps are present. It follows that the "plateau" can be related with one of these traps. Con-

versely, at near threshold bias conditions, where the current flows through the two-dimensional electron gas, the $1/f$ spectrum indicates that a single-level single-time constant GR process cannot account for the observed noise.

In Fig. 3, the noise variations of device *B* with gate bias are displayed for two different drain voltages (200 mV: ohmic region, 3 V: normal operating conditions). In both cases a noise minimum occurs at $V_{gs}/V_T = 0.55$. The relative variation of the noise with gate bias is smaller at a large drain voltage than at a small drain voltage. This is attributed to the fact that at $V_{ds} = 3$ V most of the current flows through the two-dimensional electron gas whatever the gate voltage. Therefore, it cannot undergo any fluctuations due to trapping-detrapping on GR centers in GaAlAs. Finally, it must be stated that whatever the drain bias, the input noise keeps the same order of magnitude as reported in [6]. The observed variations are within 1 dB from $V_{ds} = 0.1$ V to $V_{ds} = 3$ V.

IV. Correlation Between LF Noise and Low-Temperature Electrical Characteristics

The major improvement in performances of the TEGFET over the MESFET is obtained at low temperature where the increase of the electron mobility in the two-dimensional electron gas of the TEGFET is much larger than the increase of the electron mobility in the doped

GaAs layer of a MESFET. However, it has been reported that in some cases the advantage of the high–low temperature mobility in the 2-D gas cannot be fully exploited if some parasitic phenomena such as a collapse in the $I(V)$ characteristics and a shift in the threshold voltage occur. Such a collapse or shift is attributed to trapping on DX centers that are activated at low temperature. Therefore, if the same traps are also responsible for the LF noise at ambient temperature, a correlation can be expected between the LF noise at room temperature and the low-temperature electrical performances. In order to investigate this point, the devices have been cooled down to 130 K and the following electrical characteristics have been measured:

current–voltage characteristics;
transconductance;
S-parameters up to 12 GHz and related performances such as the power gain at 11 GHz.

The results are reported in Table I. Firstly, it has been apparent from the observed drain current variations that there is no low-temperature current increase in devices A and B; they exhibit one collapse in the I_{ds} (V_{ds}) characteristics [7]. On the contrary, a slight increase of the low-temperature drain current is observed in device C. It also has a smaller shift in the threshold voltage. In Table I it also appears that only device C exhibits a significant gm transconductance increase at 130 K over the value at 300 K. Similarly, only device C has a correct low-temperature microwave power-gain increase. Finally, we have found that the LF noise of device C at room temperature was respectively 20 and 5 times smaller than the LF noise of devices A and B. It can therefore be concluded that the DX centers sometimes responsible for the degradation of the TEGFET performances at low temperature also account for an important fraction of the LF noise observed at room temperature. It may equally be stated that the device with both the lowest noise and the best low-temperature performances was also the one with the smaller aluminum mole concentration (0.17) in GaAlAs: this confirms the fact that the DX center concentration is dependent on the Al concentration.

Finally, we suggest that LF noise measurements at ambient temperature be systematically performed at the end of the device processing since, with the help of a modern digital analyzer, these measurements are easier to obtain at room temperature than static or dynamic measurements at cryogenic temperatures, and can nevertheless be used to predict whether the device could operate satisfactorily at low temperature.

V. CONCLUSION

We have found that the LF noise in TEGFET's has many features similar to those already observed in MESFET's: the magnitude is comparable and the relative variations with the drain voltage between 0.1 and 3 V are less than 50 percent. On the other hand, the noise variations with gate voltage exhibit a minimum. In MESFET's such a noise minimum is not systematically observed [4].

It has also been found that there exists a close correlation between static and dynamic low-temperature performances and the amplitude of the LF noise at ambient temperature. This correlation is attributed to the existence of DX centers in the GaAlAs layer. It has finally been suggested that at ambient temperature, systematic LF noise measurements could be successfully performed to predict the expected performances of TEGFET's at low temperature.

ACKNOWLEDGMENT

The authors are indebted to P. Briere, P. Jay, D. Delagebeaudeuf, N. T. Linh, H. Derowenko, and some other engineers of Thomson-Semiconductors (DAG) for providing the TEGFET's.

REFERENCES

[1] K. H. Duh, A. van der Ziel, and H. Morkoç, "1/f noise in modulation-doped field effect transistors," IEEE Electron Device Lett., vol. EDL-4, pp. 12–13, Jan. 1983.
[2] K. H. Duh and A. van der Ziel, "Hooge parameters for various FET structures," IEEE Trans. Electron Devices, vol. ED-32, pp. 662–666, Mar. 1985.
[3] L. Loreck, H. Daembkes, K. Heime, K. Ploog, and G. Weimann, "Deep level analysis in (AlGa)As-GaAs 2-D electron gas devices by means of low frequency noise measurements," IEEE Electron Device Lett., vol. EDL-5, pp. 9–11, Jan. 1984.
[4] J. Graffeuil, K. Tantrarongroj, and J. F. Sautereau, "Low frequency noise analysis for the improvement of the spectral purity of GaAs FET's oscillators," Solid-State Electron., vol. 25, pp. 367–374, 1982.
[5] J. A. Copeland, "Semiconductor impurity analysis from low-frequency noise spectra," IEEE Trans. Electron Devices, vol. ED-18, pp. 50–53, Jan. 1971.
[6] S. M. Liu, M. B. Das, W. Kopp, and H. Morkoç, "Noise behavior of 1-μm gate length modulation-doped FET's from 10^{-2} to 10^8 Hz," IEEE Electron Device Lett., vol. EDL-6, pp. 453–455, Sept. 1985.
[7] R. Fischer, T. J. Drummond, J. Klem, W. Kopp, T. S. Henderson, S. Perrachione, and H. Morkoç, "On the collapse of drain I–V characteristics in modulation-doped FET's at cryogenic temperatures," IEEE Trans. Electron Devices, vol. ED-31, pp. 1028–1032, Aug. 1985.

Deep-Level Analysis in (AlGa)As–GaAs 2-D Electron Gas Devices by Means of Low-Frequency Noise Measurements

L. LORECK, H. DÄMBKES, MEMBER, IEEE, K. HEIME, SENIOR MEMBER, IEEE, K. PLOOG, AND G. WEIMANN

Abstract—Low-frequency noise of (AlGa)As–GaAs heterostructures grown by molecular-beam epitaxy was investigated. The temperature of the samples was varied between 100 and 400 K. In the frequency range from 1 Hz to 25 kHz noise spectra can be described as superposition of several generation-recombination (GR) noise components. Four deep levels (E = 0.40, 0.42, 0.54, 0.60 eV) were detected, three of which are in agreement with those measured independently by deep-level transient spectroscopy (DLTS).

I. INTRODUCTION

AN IMPORTANT PROGRESS in the development of microwave and high-speed electron devices was the discovery of a quasi two-dimensional electron gas (TEG) located at the interface of certain semiconductor heterostructures [1]. The system (AlGa)As–GaAs is used for fabricating field-effect transistors (FET's) which are called two-dimensional electron gas FET's (TEGFET's) [2], high electron mobility transistors (HEMT's) [3] or modulation-doped FET's. (MODFET's) [4]. Electrons are transferred from the n-doped (AlGa)As into the undoped GaAs and form the TEG near the interface. Therefore ionized donors and free electrons are spatially separated and the Coulomb scattering is reduced. Very high mobilities result especially at lower temperatures.

Manuscript received September 23, 1983. This work was supported by Stiftung Volkswagenwerk.

L. Loreck, H. Dämbkes, and K. Heime are with the Universitat Duisburg, D-4100 Duisburg 1, FRG.

K. Ploog is with the Max-Planck-Institut für Festkörperforschung, D-7000 Stuttgart, FRG.

G. Weimann is with Forschungsinstitut der Deutschen Bundespost, D-6100 Darmstadt, FRG.

The deep-level characteristics of the heterostructure are important, since electron traps reduce carrier concentration and mobility and increase noise. They can be characterized by low-frequency noise measurements. Generation-recombination (GR) noise is caused by fluctuation in the number of free electrons. Discrete energy levels in the forbidden gap are able to trap electrons or act as recombination centers. Theory explains that these fluctuations cause Lorentzian shaped GR noise contributions in the specturm of ac open-circuit voltage noise [5], [6]. Each trap has a characteristic spectrum defined by its low-frequency plateau value (amplitude) and the corner frequency, from the variation of which, with temperature, the activation energy of the deep level is deduced. The trap concentration is related to the amplitude. Only levels not more than a few kT away from the Fermi level contribute to noise. By changing the temperature the Fermi level is shifted across the forbidden gap.

II. EXPERIMENTAL PROCEDURE

Two types of heterostructures grown by MBE were investigated (Table I). Their main difference is the spacer layer of undoped (AlGa)As in sample D1 which enhances the mobility of the two-dimensional electron gas.

A test pattern allowed the measurement of electron Hall mobility and contact resistance from a transmission-line structure. For the noise measurements the samples consisted of two ohmic contacts with 60-μm separation and 100-μm contact width. Ohmic AuGe/Ni contacts with a specific contact resistance of about 10^{-4} $\Omega \cdot cm^2$ were used.

Samples were mounted in a cryogenic system allowing

Reprinted from *IEEE Electron Device Letters,* vol. EDL-5, no. 1, pp. 9–11, January 1984.

TABLE I
n_s SHEET CONCENTRATION OF THE TWO-DIMENSIONAL
ELECTRON GAS

sample type		S-3	D-1
top layer GaAs		22nm	20nm
		$p=10^{15}cm^{-3}$	$p \leqslant 10^{14}cm^{-3}$
$Al_xGa_{1-x}As$		60nm	70nm
(Si-doped)		$n=1*10^{18}cm^{-3}$	$n=6*10^{17}cm^{-3}$
		x=0.25	x=0.3
spacer		–	6nm
undoped GaAs		1μm	1.1μm
		$p \geqslant 10^{15}cm^{-3}$	$p \geqslant 10^{14}cm^{-3}$
n_s/cm^2	300K	$1.2*10^{12}$	$5*10^{11}$
	77K	$7*10^{11}$	$6.5*10^{11}$
$\mu_{Hall}/cm^2/Vs$	300K	3600	7900
	77K	55 000	130 000

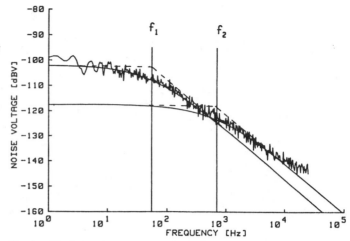

Fig. 1. Typical noise spectrum of sample $S3$ at 220 K and Lorentzian-shaped theoretical curves with corner frequencies f_1 = 55 Hz and f_2 = 700 Hz.

measurements from 20 to 400 K with 1-K accuracy. The equipment was situated in a shielded cabin. A desktop computer HP 9845 B was connected to the measurement apparatus inside the cabin via fiber-optic links.

Variable bias current was supplied by a BURSTER Digistant 6425 T current source with high impedance and low output capacitance, so that the sample noise could not be short circuited by the dc bias supply. A low-noise preamplifier ITHACO 1201 was connected in series with the HP 3582 A spectrum analyzer in order to achieve a higher input sensitivity. Smallest detectable noise voltages were about 10 nV (rms). The frequency range was 1 to 25 kHz. The computer program controlled frequency range and temperature automatically. Regular interferences such as line frequency and their multiples were suppressed by the computer evaluation.

III. MEASUREMENTS

Ohmic behavior of the samples was observed over the temperature range 200–400 K as long as the bias current was less than 5 mA. The spectrum analyzer bandwidth was 726 mHz over the whole frequency range, so that the measurement frequency was larger than the bandwidth.

No special calibration measurements were necessary, because the white noise of a metal film resistor with the same resistance as the sample was far less than that of the sample. Noise voltages were measured in decibels times volts, where 0 dB·V corresponds to 1-V (rms) single tone. Measurements were made at different temperatures with 10-K steps and stored on a magnetic tape.

IV. RESULTS AND DISCUSSION

A characteristic spectrum is shown in Fig. 1. The curves are evaluated by a superposition of several Lorentzian-type GR noise contributions. The amplitudes and characteristic corner frequencies varied with temperature. Corner frequencies are defined as $f_i = 1/2\pi\tau_i$. The activation energies ΔE_i are related to τ_i by

$$1/\tau_i \sim T^2 \exp(-\Delta E_i/kT)$$

and can be determined from an Arrhenius plot (Fig. 2(a) and (b)).

For comparison, results obtained by Hikosaka et al. [7], from deep-level transient spectroscopy (DLTS) with similar (AlGa)As-heterostructures are included in Fig. 2(a) and (b) as straight lines. Hikosaka et al., prove that the deep levels are located in the (AlGa)As layer. The good agreement between the results from noise and DLTS measurements leads us to the conclusion that the deep levels in our samples are located in the (AlGa)As layer, too.

No noise contribution from the TEG could be detected. This behavior is similar to the high-frequency noise of TEG-FET's which is considerably smaller than that of "normal" GaAs-MESFET's [9]. TEGFET's, therefore, are potentially very low noise devices.

Table II compares the levels detected by Hikosaka et al. [7], by Künzel et al. [10], and by the present authors [8]. Levels $E1$ and $E2$ could not be detected by the noise measurements, because in the available temperature range their corner frequencies are beyond the upper limit of the spectrum analyzer.

Level $E7$ was detected in sample $D1$ only (Fig. 2(a)). Since samples $S3$ and $D1$ are grown at different temperatures with different starting materials and in different MBE systems it is difficult to conclude whether $E7$ is an intrinsic or extrinsic level. The levels $E3–E5$ are common to all (AlGa)As samples. Therefore, they are not specific for the growth process or growth system, but could be either intrinsic defects or impurities common to present starting materials.

The deep-level concentrations cannot be deduced from the present noise spectra. In principle, the amplitude of the plateau value in homogeneous samples is definitely related to the trap concentration [5]. However, in the present samples a considerable transfer of electrons from the (AlGa)As layer into the TEG occurs. Transferred electrons do not contribute directly to the noise spectrum. Additional experiments on thicker (AlGa)As samples without TEG are necessary.

V. CONCLUSION

It was shown experimentally: 1) that low-frequency noise measurements are a useful method for deep-level analysis; 2)

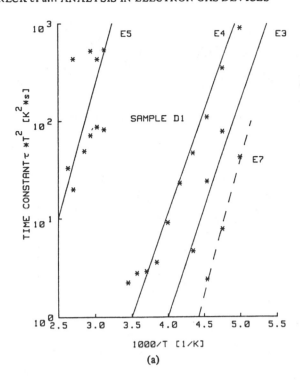

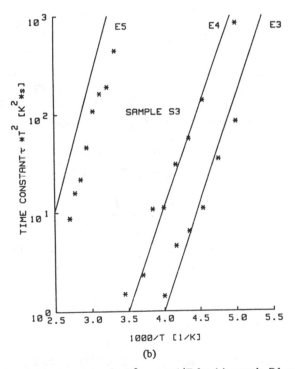

Fig. 2. Arrhenius plots of $\tau_i T^2$ versus $1/T$ for (a) sample $D1$ and (b) sample $S3$. * experimental values, this work; − − − best fit to experimental values of level $E7$; — results from [7], [10] by DLTS.

TABLE II

Level	ΔE	Hikosaka et al./7/ (DLTS)	Künzel et al./10/ (DLTS)	this work (noise)
E 1	0.26	+	+	*
E 2	0.28	+	−	*
E 3	0.42	+	+	+
E 4	0.40	+	+	+
E 5	0.60	+	+	+
E 6	0.78	+	−	−
E 7	0.54	−	−	+

+ deep level present
− deep level not present
* deep level not detectable

that GR noise in (AlGa)As–GaAs heterostructures originates from the (AlGa)As only; and 3) that the levels correspond to those detected by DLTS. Deep-level concentrations and capture cross sections can be evaluated only if either an electron transfer into the TEG and a recombination back into the (AlGa)As is excluded experimentally or included into existing theories.

REFERENCES

[1] R. Dingle, H. L. Stormer, A. C. Gossard, and W. Wiegmann, "Electron mobilities in modulation doped semiconductor heterojunction superlattices," *Appl. Phys. Lett.*, vol. 33, pp. 665–667, Oct. 1978.
[2] D. Delagebeaudeuf and N. T. Linh, "Metal-(n) AlGaAs/GaAs two-dimensional electron gas FET," *IEEE Trans. Electron Devices*, vol. ED-29, pp. 955–960, June 1982.
[3] K. Joshin, T. Mimura, N. Niori, Y. Yamashita, K. Kosemara, and J. Saito, "Noise performance of microwave HEMT," in *IEEE Microwave Technology Theory Symp. Dig.* (Boston, MA), 1983, pp. 563–565.
[4] K. H. Duh, A. van der Ziel, and H. Morkoç, "1/f noise in modulation-doped field effect transistors," *IEEE Electron Device Lett.*, vol. EDL-4, pp. 12–13, Jan. 1983.
[5] G. Bosman and R. J. J. Zijlstra, "Generation-recombination noise in p-type silicon," *Solid State Electron.*, vol. 25, no. 4, pp. 273–280, 1982.
[6] A. van der Ziel, *Noise in Measurements.* New York: Wiley, 1976.
[7] K. Hikosaka, T. Mimura, and S. Hyamizu, "Deep electron traps in MBE-grown AlGaAs ternary alloy for heterojunction devices," *Inst. Phys. Conf. Ser.*, no. 63, pp. 233–238.
[8] L. Loreck, H. Dämbkes, K. Heime, K. Ploog, and G. Weimann, "Low frequency noise in AlGaAs-GaAs 2-D electron gas devices and its correlation to deep levels," presented at 7th Int. Conf. on Noise in Physical Systems and 3rd Int. Conf. on 1/f Noise, Montpellier, May 1983.
[9] N. T. Linh, "Microwave performance of two-dimensional electron gas FET," presented at 8th European Specialist Workshop of Active Microwave Semiconductor Devices, Maidenhead, 1983.
[10] H. Künzel, K. Ploog, K. Wünstel, and B. L. Zhou, "Influence of alloy composition, substrate temperature, and doping concentration on electrical properties of Si-doped n-Al$_x$Ga$_{1-x}$As grown by molecular beam epitaxy," to be published.

Noise Behavior of 1-μm Gate-Length Modulation-Doped FET's From 10^{-2} to 10^{8} Hz

S. M. LIU, M. B. DAS, SENIOR MEMBER, IEEE, W. KOPP, AND H. MORKOÇ, SENIOR MEMBER, IEEE

Abstract—Measured equivalent gate noise voltage spectra of 1-μm gate-length MODFET's, for the frequency range of 0.01 Hz–100 MHz, are presented. They indicate that the noise consists of several high-intensity trap-related generation-recombination (g–r) noise components superimposed on a background $1/f$ noise. The g-r noise is reduced when the Al mole-fraction is lowered. The same occurs when the gate reverse bias is increased. At 100 K the g-r noise bulge moves towards subaudio frequencies clearly revealing the $1/f$ noise.

I. INTRODUCTION

THE MODULATION-DOPED FET (MODFET) has been recognized [1]–[3] for its superior low-noise performance compared with that of GaAs MESFET's at comparable microwave frequencies. In this letter we report the LF noise behavior of MODFET's with gate length of 1 μm and three different aluminum mole-fraction values ($x = 0.3$, 0.23, and 0.2). The primary objective of this work has been to determine the relative importance of the $1/f$ noise and the trap related generation–recombination (g–r) noise components in the MODFET's, as reported [4]–[5] earlier for the MESFET's. Other objectives include the determination of: 1) the effect of the Al mole-fraction variation on the noise; 2) the noise behavior at subaudio frequencies; and 3) the same at a low temperature near 77 K.

II. LF NOISE GENERATORS

There are three dominant noise current generators that become effective at low frequencies (0.01 Hz to 100 MHz) in MESFET's and MODFET's and can be considered to be present across the drain–source output terminals. These noise generators are: 1) the g-r noise; 2) the $1/f$ noise; and 3) the channel thermal noise. The thermal noise becomes significant when the g-r noise and $1/f$ noise vanish at higher frequencies.

In this work, the FET's have been operated under drain current saturation condition, since this is the common mode of their operation for amplification purposes. In this condition the LF noise, if measured as an equivalent noise voltage referred to the input gate, appears as being relatively independent of the drain current (achieved at different V_{GS}) except for some

Manuscript received April 26, 1985; revised June 19, 1983. The work done at The Pennsylvania State University was supported in part by NSF under Grant ECS-8401182 and that done at University of Illinois was supported by the Air Force Office of Scientific Research.

S. M. Liu and M. B. Das are with the Electrical Engineering Department, Solid State Device Laboratory and Materials Research Laboratory, The Pennsylvania State University, University Park, PA 16802.

W. Kopp and H. Morkoç are with the Department of Electrical Engineering and Coordinated Science Laboratory, University of Illinois, Urbana, IL 61801.

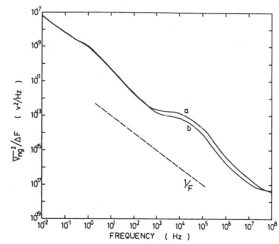

Fig. 1. Equivalent gate noise spectra of device 1483-7 ($X = 0.30$) at 300 K. a: When $V_{GS} = 0$ V. b: When $V_{GS} = -0.20$ V.

variation at frequencies where the bulges occur, as may be seen from Fig. 1. We also observed negligible variation of this noise voltage with drain bias voltage when the current is saturated. This noise voltage can be expressed as

$$\overline{v_{ng}}^2 = 4kT\Delta f \left(\frac{\rho_0 f_0}{f} + \sum_{r=1}^{n} \frac{\rho_r(\tau_r/\tau_0)}{1+(\omega\tau_r)^2} \right)$$

where ρ_0 is the $1/f$ noise equivalent gate resistance when $f = f_0$, ρ_r is a similar resistance for the rth component of the g-r noise, τ_0 is a reference time constant, and τ_r is the time constant for deep-level trap causing the g-r noise [6]. Both ρ_0 and ρ_r are usually inversely proportional to gate area.

III. EXPERIMENTAL PROCEDURE

A. Test Device Fabrication

The test MODFET devices used in this work were fabricated at the University of Illinois using single-period AlGaAs/GaAs modulation-doped heterostructures. The transistor structure consisted of a 1-μm undoped GaAs buffer layer, a thin layer of undoped (Al, Ga)As (20–60 Å), and 600 Å of Si doped (Al, Ga)As grown on a Cr-doped GaAs substrate. The V/III flux ratio was kept as low as possible while maintaining an As stabilized surface. Details of fabrication can be found elsewhere [7]. The devices were diced and good devices were selected and mounted on alumina microstrip subassembly with AuSn eutectic preformed and bonded with a thermocompression ball bonder.

Reprinted from *IEEE Electron Device Letters*, vol. EDL-6, no. 9, pp. 453–455, September 1985.

B. Noise Measurement Systems

Due to the wide frequency range covered in these measurements, three different systems were employed. As described earlier [5], an HP3581A wave analyzer was used for the range of 10 Hz–50 kHz. Additional amplification and wave mixing in combination with the same wave analyzer covered the frequency range up to 100 MHz. The subaudio range of 0.01–10 Hz was investigated using digital signal processing techniques. The signal was sampled and held for the digital voltmeter (DVM) to convert and transferred to a computer. A set of 512 precisely timed and sampled data was stored as a batch and processed using Fast Fourier Transformation (FFT). Ten batches of transformed data were arithmetically averaged to smooth out fluctuations and then plotted. Calibration was performed with known noise signal injection at the input. A digitally generated random noise source was used for the subaudio range.

For measurements at cryogenic temperatures, the device and the test fixture was mounted in a vacuum chamber. A liquid nitrogen chilled copper plate was used to cool the fixture down to 100 K, and ambient heat absorbance was then allowed to slowly bring the temperature back to room temperature. For temperature-dependent measurements, in order to eliminate the commonly observed collapse in the I–V characteristics of MODFET's in the dark [8] at temperatures below 160 K, an LED light source was used to illuminate the device during measurements.

The device noise amplified by the measurement system was plotted as a function of frequency with the input gate short-circuited. The voltage gain of the system was also plotted in the same manner. The equivalent gate input noise voltage was obtained by dividing the noise output voltage by the system voltage gain. This direct approach eliminates the need to know the actual device transconductance.

C. Experimental Results

The noise spectra showing the frequency and bias dependence of the MODFET LF noise behavior at 300 K with $x = 0.23$ are given in Fig. 1. The unusually large g-r noise bulge near 10 kHz and its bias dependence, in an opposite sense than that occurs in the case of MESFET's [5], represent a significant difference in the two cases. In the MODFET a large negative gate bias means a deeper location of the 2-DEG peak concentration compared to that at low negative or zero gate bias, and this depth variation occurs within the GaAs buffer layer. Since the deep traps causing the g-r noise bulge are present in the n-AlGaAs the effective trap concentration could be regarded to have decreased, thus lowering the g-r noise bulge. In MESFET's the total trap centers increases with reverse bias causing an increase in the g-r bulge size.

At very low frequencies the noise curves seem to point towards another g-r noise bulge at a higher level (due to a deeper trap). In this particular device the g-r noise bulge is so large that nowhere a clear $1/f$ noise spectral slope is evident except in the vicinity of 100 Hz between two g-r bulges.

The noise spectra of MODFET's with $x = 0.3, 0.23$, and 0.2 are presented in Fig. 2. These results clearly indicate that a higher mole fraction enhances the trap concentration in the n-

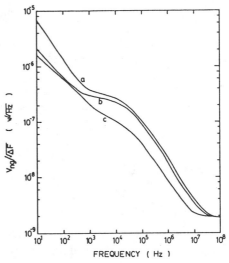

Fig. 2. Gate noise voltage density of test MODFET's with various mole fraction and comparable saturated drain current. $T = 300$ K. a: 1483-7, $X = 0.30$, $I_{DSAT} = 35$ mA, $V_{GS} = -0.20$ V. b: 1885A-10, $X = 0.23$, $I_{DSAT} = 36$ mA, $V_{GS} = 0$ V. c: 1870-3, $X = 0.20$, $I_{DSAT} = 36$ mA, $V_{GS} = -0.36$ V.

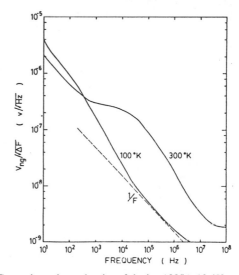

Fig. 3. Gate noise voltage density of device 1885A-10 ($X = 0.23$) at 100 and 300 K when $V_{GS} = 0$ V.

AlGaAs that causes the large bulge near 10 kHz. In the case of $x = 0.2$ the g-r noise is reduced but not sufficient enough to show the $1/f$ noise component.

The noise spectra obtained at 100 and 300 K for a MODFET with $x = 0.23$ are compared in Fig. 3. The most revealing aspects of these results are that at 100 K the g-r noise bulge moves towards a lower frequency compared to its position near 10 kHz at 300 K. This makes the underlying $1/f$ noise clearly visible at higher frequencies.

By performing noise measurements at selected frequencies when the temperature is varied from 100 to 300 K, we have been able to observe the movement of the g-r noise bulges with temperature. These results have been used to calculate the trap activation energy to be at 0.40 eV below the conduction b .d [6].

IV. Conclusion

The LF noise data obtained for the MODFET's in the frequency range of 0.01 Hz to 100 MHz have clearly indicated the existence of both g-r noise and $1/f$ noise in a superimposed manner. This is similar to the LF noise behavior of MESFET's, although the MODFET g-r noise level is much higher in the devices tested in this work. As the mole-fraction of Al is reduced in the n-AlGaAs layer the g-r noise bulge is also seen to be reduced. The dominant trap level associated with the observed g-r noise bulge has the activation energy of 0.40 eV below the conduction band. The existence of a true $1/f$ noise component observed at 100 K should provide the impetus for further investigations into the physical origin [9] of $1/f$ noise in MESFET's and MODFET's which remains unknown as yet.

ACKNOWLEDGMENT

The authors would like to thank Dr. S. Weinreb for helpful discussions on low-temperature measurements.

REFERENCES

[1] U. K. Mishra, S. C. Palmateer, P. C. Chao, P. M. Smith, and J. C. M. Hwang, "Microwave performance of 0.25 μm gate-length high electron mobility transistors," *IEEE Electron Device Lett.,* vol. EDL-6, pp. 142–145, 1985.

[2] J. J. Berenz, K. Nakano, and K. P. Weller, "Low noise high electron mobility transistors," in *IEEE MTT-S Int. Symp. Dig.,* June 1984, p. 98.

[3] N. T. Linh, M. Laviron, P. Delescluse, P. Tung, D. Delagebeaudeuf, F. Diamand, and J. Chevrier, "Low noise performance of two-dimensional electron gas FET's," in *Proc. IEEE/Cornell Conf. High-Speed Semicon. Devices Circuits,* Aug. 1983, pp. 187–193.

[4] J. Graffeuil, K. Tantrarongroj, and J. F. Sautereau, "Low frequency noise physical analysis for the improvement of the spectral purity of GaAs FET's oscillators," *Solid State Electron.,* vol. 25, pp. 367–374, 1982.

[5] S. M. Liu and M. B. Das, "Characterization of extended $1/f$ noise in microwave GaAs MESFET's," in *Proc. UGIM Symp.* (A&M Univ., TX), May 25–27, 1983, pp. 169–172.

[6] L. Loreck, H. Dumbkes, K. Heime, and K. Ploog, "Deep level analysis in (AlGa)As-GaAs MODFET's by low-frequency noise measurements," in *IEDM Tech. Dig.,* 1983, pp. 107–110.

[7] T. J. Drumond, R. Fischer, W. Kopp, H. Morkoç, K. Lee, and M. S. Shur, "Bias dependence and light sensitivity of AlGaAs/GaAs MODFET's at 77K," *IEEE Trans. Electron Devices,* vol. ED-30, pp. 1806–1811, 1983.

[8] R. Fischer, T. J. Drummond, J. Klem, W. Kopp, T. S. Henderson, D. Perrchione, and H. Morkoç, "On the collapse of drain I–V characteristics in modulation-doped FET's at cryogenic temperatures," *IEEE Trans. Electron Devices,* vol. ED-31, pp. 1028–1032, 1984.

[9] A. van der Ziel, P. H. Handel, S. C. Zhu, and K. H. Duh, "A theory of the Hooge parameters of solid state devices," *IEEE Trans. Electron Devices,* vol. ED-32, pp. 667–671, 1985.

Low-Noise High Electron Mobility Transistors For Monolithic Microwave Integrated Circuits

A. K. GUPTA, E. A. SOVERO, R. L. PIERSON, R. D. STEIN, R. T. CHEN, D. L. MILLER,

AND J. A. HIGGINS, SENIOR MEMBER, IEEE

Abstract—High electron mobility transistors (HEMT's) for monolithic microwave integrated circuits have been fabricated that have demonstrated excellent performance. External transconductance of 300 mS/mm is observed and noise figures of 1 and 1.8 dB with associated gains of 16.1 and 11.3 dB at 8 and 18 GHz, respectively, have been measured. These are comparable to the best reported noise figures for either HEMT's or MESFET's and are the highest associated gains reported for such low-noise figures. Analysis of these devices indicates that further improvements in these results is possible through optimization of HEMT layers and fabrication technology to reduce gate–source parasitic resistance.

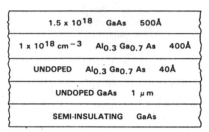

Fig. 1. Structure of MBE layer for a low-noise HEMT.

I. INTRODUCTION

WE REPORT here on the performance obtained from low-noise high electron mobility transistors (HEMT's) fabricated in our laboratory. These HEMT's utilize molecular beam epitaxy (MBE) for active layers, electron beam lithography (EBL) for 0.5-μm-long Ti-Pt–Au gates and other processing steps identical to our standard monolithic microwave integrated circuit (MMIC) fabrication technology [1]. Discrete devices having widths of 150 and 200 μm, respectively, were mounted in commercially available 70-mil microwave transistor packages for noise figure (NF) measurements. All devices had via-holes for source grounding. The devices exhibited external transconductances (g_m) as high as 300 mS/mm and excellent microwave performance. A noise figure of 1 dB with associated gain of 16.1 dB at 8 GHz and a noise figure of 1.8 dB with associated gain of 11.3 dB at 18 GHz were measured. The noise figures obtained are comparable to the best 0.5-μm gate HEMT results [2], [3], and the associated gains are significantly higher than those reported in the literature thus far.

II. DEVICE FABRICATION

Devices were fabricated on MBE grown active layers of the type shown in Fig. 1 having a sheet resistance of approximately 190 $\Omega/\square$. The n^+ GaAs cap layer at the top was included to obtain low resistance ohmic contacts by AuGe–Ni metallization and a 450°C − 10-s alloy cycle using rapid infrared heating. A contact resistance of $\leqslant 0.2\ \Omega \cdot$mm is typically obtained at room temperature. This method of ohmic contact formation results in good surface morphology and extremely smooth edges which are necessary to place the gate close to the source ($\leqslant 0.5\ \mu$m) without shorting. Ti-Pt–Au gates were formed 0.5 μm long, and 6000 Å thick

by direct-write EBL and liftoff. Gates were recessed approximately 500–600 Å. Devices were passivated by a layer of polyimide which also serves as a crossover insulator for the gold-plating step wherein all gate feeds are brought out to a common bonding pad (Fig. 2). The final step is to thin the wafer to 125 μm, etch via holes for source grounding, and plate the back side. The last few steps are standard to our MMIC process but are not strictly necessary for the evaluation of discrete devices. They were included here because, ultimately, we plan to use these HEMT's in monolithic millimeter-wave IC's and it is of interest to evaluate the devices that would be obtained after all the processing steps.

III. DEVICE PERFORMANCE

Discrete devices were mounted in 70-mil microwave transistor packages for dc and NF measurements. DC I–V characteristics of a 150–μm-wide HEMT are shown in Fig. 3. Conduction in the low mobility Al$_{0.3}$Ga$_{0.7}$As layer is responsible for the low g_m near $V_{gs} = 0$. The highest external g_m obtained is 300 mS/mm as shown in Fig. 4 where gate bias dependence of g_m is plotted for a fixed drain-source voltage. Noise figure was measured at 8 and 18 GHz (using a Hewlett-Packard 346B noise source and 8970A noise figure meter[1]) by inserting the packaged device in a coaxial test fixture with double slug tuners at input and output and tuning for minimum noise. Measured results are shown in Table I. Two types of devices with gate widths of 150 and 200 μm and unit gate widths of 38 and 50 μm, respectively, were tested. No significant difference in performance was observed. The measured noise figure was found to scale with frequency in the expected manner of (1) for both types of devices. Also included in Table I are measurements on two

---

Manuscript received October 22, 1984; revised November 26, 1984.

The authors are with Rockwell International Microelectronics Research and Development Center, Thousand Oaks, CA 91360.

[1] The 8970A has a noise figure uncertainty of ± 0.1 dB and a gain uncertainty of ± 0.2 dB. The 346B noise source has a reflection coefficient of < 0.11 in both states.

Reprinted from *IEEE Electron Device Letters,* vol. EDL-6, no. 2, pp. 81–82, February 1985.

TABLE I
MEASURED NOISE FIGURE, GAIN AND OTHER DEVICE PARAMETERS

Number	Type	L_g (μm)	W_g (μm)	f (GHz)	NF_{min} (dB)	G_a (dB)	R_s (Ω)	R_g (Ω)	k_f
1	HEMT	0.5	150	8	1.0	16.1	5.8	1.0	0.1
2	HEMT	0.5	200	18	1.8	11.3	4.1	1.1	0.1
NE13783*	MESFET	0.5	250	18	2.1	8.6	2.2	1.3	0.2
NE67383*	MESFET	0.25	250	18	2.0	7.4	2.3	1.4	0.3

*Commercially available GaAs MESFET.

Fig. 2. Photograph of low-noise HEMT. Note gold-plated interconnect for gate feeds. There are two via holes (not shown) for source grounding.

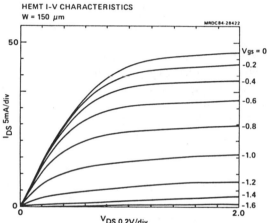

Fig. 3. DC I-V characteristics of a 150-μm HEMT.

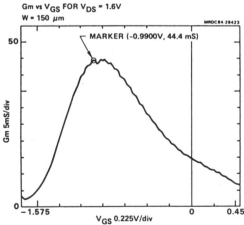

Fig. 4. Transconductance (g_m) versus gate voltage for the device of Fig. 3. Drain voltage is 1.6 V. Maximum g_m is 300 mS/mm.

commercially available low-noise devices. Measurements on these devices compare well with the manufacturer's specifications. Tuner losses were established by measurements on a network analyzer. It is very important to measure known devices in any NF test set since large errors (compared to device noise figure) can be caused by ENR calibration uncertainty and mismatches. In addition to noise figure and associated gain, we have also included in Table I measured

gate and source resistances and the fitting factor k_f in the following empirical equation by Fukui [4]:

$$NF(dB) = 10 \log_{10} \left[1 + k_f L_g f \sqrt{\frac{g_m(R_s + R_g)}{1 - g_m R_s}} \right] \quad (1)$$

where

L_g gate length (μm),
f frequency (GHz),
R_s source resistance (Ω),
R_g gate metallization resistance (Ω),
g_m transconductance (s).

R_s and R_g are measured by forward biasing the gate Schottky diode. In calculating k_f, the value of g_m used was that at the bias for minimum noise. As Table I shows, the 0.25-μm MESFET, NE67383, is only marginally better in noise figure and significantly worse in associated gain than the 0.5-μm device, NE13783. This pattern has been observed with several devices of both kinds. Three important conclusions can be drawn from the data in Table I. These are:

1) The noise performance of 0.5-μm gate-length HEMT's is comparable to the best reported 0.25-μm MESFET result of Yamasaki et al. [5].

2) The associated gain of these HEMT's is larger than previously reported for either HEMT's or MESFET's.

3) The fitting factor k_f for HEMT's is significantly lower than for MESFET's implying that for equivalent source resistances better performance is expected from HEMT's. Further reduction in R_s will require optimization of the HEMT layer.

IV. CONCLUSIONS

High electron mobility transistors (HEMT's) have been fabricated that demonstrate excellent performance. External transconductance of 300 mS/mm is observed and noise figures of 1 and 1.8 dB with associated gains of 16.1 and 11.3 dB at 8 and 18 GHz, respectively, have been measured. These are comparable to the best reported noise figures for either HEMT's or MESFET's and are the highest associated gains reported for such low-noise figures. Analysis of these devices indicates that further improvements in these results is possible through optimization of HEMT layers and fabrication technology to reduce gate source parasitic resistance.

REFERENCES

[1] A. K. Gupta, G. R. Kaelin, and R. D. Stein, "Gallium arsenide monolithic microwave integrated circuits," in *Proc. 27th Midwest Symp. Circuits and Systems* (Morgantown, WV), June 1984.

[2] N. T. Linh, M. Laviron, P. Delescluse, P. N. Tung, D. Delagebeaudeuf, F. Diamond, and J. Chevrier, "Low noise performance of two-dimensional electron gas FET's," in *Proc. IEEE/Cornell Conference on High-Speed Semiconductor Devices and Circuits*, Aug. 1983, pp. 187–193.

[3] K. Joshin, T. Mimura, M. Nirori, Y. Yamashita, K. Kosemura, and J. Saito, "Noise performance of microwave HEMT," in *1983 IEEE MTT-S Int. Microwave Symp. Dig.*, May 1983, pp. 563–565.

[4] H. Fukui, "Determination of the basic device parameters of a GaAs MESFET," *Bell Syst. Tech. J.*, vol. 58, no. 3, pp. 771–797, Mar. 1979.

[5] H. Yamasaki, L. H. Hackett, E. T. Watkins, J. M. Schellenberg, and M. Feng, "Design of optimized EHF FET's," in *Proc. IEEE/Cornell Conf. on High-Speed Semiconductor Devices and Circuits*, Aug. 1983, pp. 277–286.

LOW NOISE AMPLIFIERS USING TWO DIMENSIONAL ELECTRON GAS FETS

T. Mochizuki, K. Honma, K. Handa,

W. Akinaga and K. Ohata*

Microwave & Satellite Communications Division, NEC Corporation

4035, Ikebe-cho, Midori-ku, Yokohama, 226 Japan

* Microelectronics Research Laboratories, NEC Corporation

4-1-1, Miyazaki, Miyamae-ku, Kawasaki, 213 Japan

Abstract

Recently developed LNAs incorporating two dimensional electron gas (2 DEG) FETs for satellite communications earth stations are disclosed, which give epoch-making low noise as FET LNAs to operate in the 2, 4, 12, and 20 GHz bands at room temperature, especially under cooled state. Typically detailed further is newly developed 4 GHz band LNA with 55 K max. noise temperature at room temperature, noise temperatures of the order of 30 K across 800 MHz bandwidth (3.4 to 4.2 GHz) under thermoelectrically (TE-) cooled state (about-45°C), which has been adopted in the new earth station conducted by KDD.

Introduction

Present-day satellite communications LNAs fall in two categories: the parametric LNA and the FET LNA. As a recent trend, much lower noise temperature and broader frequency band (to meet INTELSAT X'pole compensation and WARC 79 recommendations) have become essential for INTELSAT application, while maintenance-free benefits have been called for DOMSAT application with increasing number of small earth stations operated worldwide. The former is essentially superior in noise performance, but requires complex maintenance service and is expensive. In contrast, the latter features ease of handling, compactness, and lower cost [1].

This is the reason why the realization of noise performance of the FET LNAs well comparable with the parametric LNAs has been eagerly pursued. To meet these requirements, our research and developmental efforts have been directed to the 2 DEG FET to have high potentiality as an ultra-low-noise device as well as to the low-loss and broadband noise matching technologies for an input circuit to be installed ahead of the 2 DEG FET. As a result, 2 DEG FET LNAs having ultra-low-noise characteristics in the four frequency bands (2, 4, 12, and 20 GHz) were successfully developed.

Device Structure and Performance

By 2 DEG FET is meant a device utilizing a hetero-junction between Si-doped AlGaAs and undoped GaAs to advantage, whose basic structure is shown in Fig.1 [2]. With this structure, the AlGaAs side donor region and the GaAs side two dimensional electron channel can be spatially separated each other.

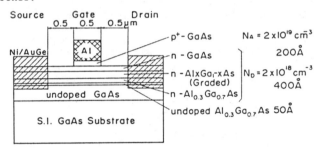

Fig. 1 Structure of the 2 DEG FET.

As a result, the two dimensional electron gas flowing along the channel formed in the high-purity GaAs region will be less susceptible to the scattering effect caused by impurities, or donors, to attain electron mobility as high as 1.5 times the values of conventional GaAs FETs at room temperature. Since it can be experimentally verified that improvements in electron mobility are pronounced under cooling state, practical application and advantages of 2 DEG FET as an ultra-low-noise device had been eagerly sought.

At present, the excellent microwave characteristics as will be described have been experimentally achieved with 2 DEG FET alone.

The noise temperature (NT) and gain vs frequency characteristics, of 2 DEG FET are shown in Fig.2. This data is calculated from the NT and gain data measured at various frequencies under room temperature conditions using the circuit of Fig.3, account being taken of losses in various circuit portions and the gain of 2 DEG FET. As obvious from Fig.2, ultra-low-noise and high-gain properties of the 2 DEG FET far surpassing the conventional GaAs FETs are secured even at room temperature.

Further, the manner in which the NT is improved by cooling 2 DEG FET with the gaseous helium refrigerator suitable for tests in a wide cooling range and the cooling circuit structure shown in Fig.4 will be described. Fig.5 illustrates the ambient-temperature-dependent NT characteristics of the 2, 4, 12, and 20 GHz 2 DEG FETs per se, respectively. These data are calculated from the

Reprinted from *1985 IEEE MTT-S Int. Microwave Symp., Tech. Dig.*, pp. 543–546, 1985.

experimental NT data obtained at 2, 4, 12, and 20 GHz under cooled state, account being taken of cooling circuit losses and the gain of the 2 DEG FET.

The NT vs ambient temperature relationships of the FETs [1], [3] can be generally expressed as

$$\frac{Te2}{Te1} = \left(\frac{Td2}{Td1}\right)^x \quad \cdots\cdots (1)$$

where

Te1 = NT of 2 DEG FET at room temperature,
Te2 = NT of 2 DEG FET under cooled state,
Td1 = room temperature,
Td2 = cooling temperature,
x = noise reduction factor.

An analysis of data taken from the graph of Fig.5 demonstrates that noise reduction factor x due to cooling of 2 DEG FET becomes more than about 1.7, apparently higher than x = 1.5 of the conventional GaAs FETs. That is, the degree of improvement in noise performance caused by cooling is more pronounced over the conventional GaAs FETs.

The 2 DEG FET featuring ultra-low-noise and high gain properties has been adopted this time as the preamplifier stage of the LNA, which plays a dominant role for reduction of overall noise. That the preamplifier stage has such properties should be extremely advantageous in noise reduction especially when the LNA constitutes a multi-stage amplifier.

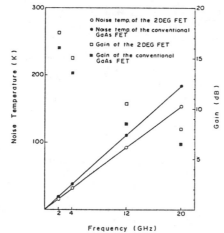

Fig.2 Noise temperature and gain of the 2 DEG FET proper as a function of frequency.

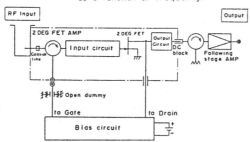

Fig.3 Circuit structure used in measuring performance data of the 2 DEG FET at room temperature.

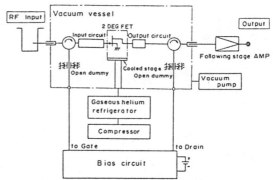

Fig.4 Circuit configuration used for experiment of cooled temperature dependence of the 2 DEG FET,s noise temperature.

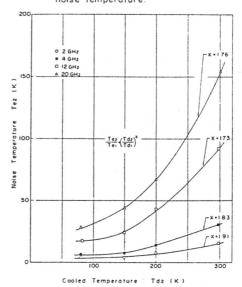

Fig 5 Cooled temperature dependence of noise temperature of the 2 DEG FET at 2, 4, 12 and 20 GHz.

LNA's Circuit Design

Design knowledge for the preamplifier stage incorporating 2 DEG FET will now be analyzed. Where a lossless impedance matching circuit is connected to the input of an FET, the noise figure (NF) of the amplifier is dependent upon the output impedance of the input circuit [4]   and is given by:

$$F = Fmin + \frac{Rn}{Gs}\{(Gs - Go)^2 + (Bs - Bo)^2\} \cdots (2)$$

where

F = NF determined by input circuit,
Fmin = optimum NF,
Rn = equivalent input noise resistance.

Here, the admittance (Yo) to give optimum NF and the source admittance (Ys) of the input circuit can be respectively expressed as

Yo = Go + jBo ..... (3)
Ys = Gs + jBs ..... (4)

As obvious from eq. (1), to construct the FET preamplifier stage to give optimum NT, it is desirable that the input side impedance be coincident with the optimum impedance for noise reduction as decided by the 2 DEG FET (Yo = Ys). Incidentally, the locus of calculated noise optimum impedances as plotted on the smith chart will invariably be inside of S_{11}^* as in Fig.6. Need arises, therefore, for inserting an isolator on the input side of the 2 DEG FET, because the degradation in the input VSWR of the FET is inevitable to secure an optimum impedance match for noise reduction. Where a suitable circuit is placed on the input side of an FET, the loss in the circuit will cause the degradation in noise performance of an amplifier. Consequently, in order for potentiality of the 2 DEG FET to be maximally displayed, it is essential that an extremely low-loss circuit capable of maintaining noise optimum impedance matching over a wide frequency band be connected between the input isolator and the 2 DEG FET as the preamplifier stage input circuit configuration.

Furthermore, an impedance-matching circuit, as the output circuit, for transforming the input impedance S_{22}^* into 50 ohms at the output port must be installed in order to lessen the effect of NT of the subsequent stages by maximizing the gain of the preamplifier stage.

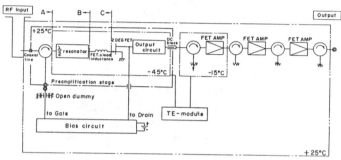

Fig.7. 4GHz TE-cooled 2 DEG FET LNA and the block diagram.

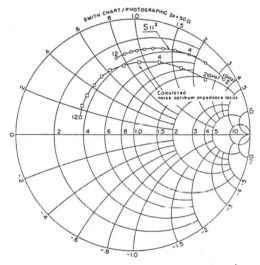

Fig.6 Noise optimum impedance locus and S_{11}^* of the 2 DEG FET.

Fig.7 shows a photo and a block diagram of the newly developed 4 GHz TE-cooled 2 DEG FET LNA as a typical example of a circuit structure involving the low-loss and noise matched circuit and the 2 DEG FET. Among the innovative concepts adopted in the design of the newly developed preamplifier stage input circuit are:

First, transformation of the impedance at the output port of the isolator's stripline is effected, for taking noise optimum impedance matching, by use only of the triplate type λ/2-line-length resonator and the FET's lead inductance. This has brought about a marked reduction of the insertion loss over conventional matching circuits.

Second, favorable noise matching can be taken with ease over as wide a bandwidth as 800 MHz due to the λ/2 resonator's effect, as opposed to conventional preamplifier stage.

The manner in which the loci formed by plotting impedances at various points in the input matching circuit on the smith chart for optimum noise matching vary from one another by the impedance transformation process is illustrated in Fig.8. A comparison between locus C for the impedance of the input circuit ahead of the 2 DEG FET as looked into the RF input side at C in Fig.7 and noise optimum impedance locus D reveals at once that they are arranged in the same H-L direction and approximately the same in arc length. Thus, combining newly developed input matching circuit and 2 DEG FET has enabled feasibility of the preamplifier stage with excessively low noise over a wide frequency band.

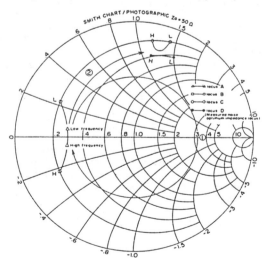

Fig. 8 Impedance transformation for taking optimum noise matching.

131

As to the bias circuit, two design precautions must be taken:

(1) Since the optimum bias voltage for the 2 DEG FET is extremely low, the bias circuit must be designed to meet this condition.

(2) The bias circuit must incorporate a protective circuit against incoming surge voltages.

Fig.9 shows a block diagram used for computation of overall NT of the 4 GHz LNA incorporating the newly developed input matching circuit at room temperature and under TE-cooled state (-45°C), and various formula used for noise budget computations. Incidentally, NT of 2 DEG FET alone under TE-cooled state is calculated using noise reduction factor x = 1.83 at 4 GHz derived by an analysis of the ambient-temperature-dependant graph of Fig.5. The noise budget in Fig.9 theoretically predicts feasibility of an LNA with NT of 47 K at room temperature and NT of 34 K under TE-cooled state.

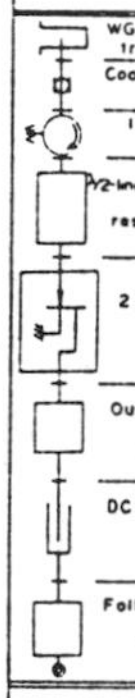

Block diagram	Noise temp gain/loss	Formula at input	at room temp (+25°C) Te/G/Ta/L	at room temp (+25°C) Noise temp. at input	under TE-cooled state (-45°C) Te/G/Ta/L	under TE-cooled state (-45°C) Noise temp. at input
WG/Coax transition	Loss L_1 (T_{a1})	$(L_1-1)T_{a1}$	L_1=0.05dB T_{a1}=298K	3.45K	L_1=0.05dB T_{a1}=298K	3.45K
Coaxial line	Loss L_2 (T_{a2})	$L_1L_2T_{a2}(1-\frac{1}{L_2})$	L_2=0.01dB T_{a2}=298K	0.69K	L_2=0.01dB T_{a2}=298K	0.69K
Isolator	Loss L_3 (T_{a3})	$L_1L_2L_3T_{a3}(1-\frac{1}{L_3})$	L_3=0.09dB T_{a3}=298K	6.33K	L_3=0.09dB T_{a3}=298K	6.33K
λ/2-line-length resonator	Loss L_4 (T_{a4})	$L_1L_2L_3L_4T_{a4}(1-\frac{1}{L_4})$	L_4=0.03dB T_{a4}=298K	2.14K	L_4=0.03dB T_{a4}=228K	1.64K
2 DEG FET	Noise: T_{e1} Gain: G_1 (T_{a5})	$L_1L_2L_3L_4T_{e1}$	T_{e1}=30.76K G_1=15dB T_{a5}=298K	32.06K	T_{e1}=18.84K G_1=15dB T_{a5}=228K	19.64K
Output circuit	Loss L_5 (T_{a6})	$\frac{L_1L_2L_3L_4L_5T_{a6}}{G_1}(1-\frac{1}{L_5})$	L_5=0.05dB T_{a6}=298	0.11K	L_5=0.05dB T_{a6}=228K	0.09K
DC block	Loss L_6 (T_{a7})	$\frac{L_1L_2L_3L_4L_5L_6T_{a7}}{G_1}(1-\frac{1}{L_6})$	L_6=0.1dB T_{a7}=298K	0.23K	L_6=0.1dB T_{a7}=228K	0.18K
Following stage	Noise: T_{e2}	$\frac{L_1L_2L_3L_4L_5L_6T_{e2}}{G_1}$	T_{e2}=60K	2.05K	T_{e2}=51K	1.74K
Overall noise temp: Te		$T_e=(L_1-1)T_{a1}+L_1L_2T_{a2}(1-\frac{1}{L_2})+L_1L_2L_3T_{a3}(1-\frac{1}{L_3})+L_1L_2L_3L_4T_{a4}(1-\frac{1}{L_4})+L_1L_2L_3L_4T_{e1}+\frac{L_1L_2L_3L_4L_5T_{a6}}{G_1}(1-\frac{1}{L_5})+\frac{L_1L_2L_3L_4L_5L_6T_{a7}}{G_1}(1-\frac{1}{L_6})+\frac{L_1L_2L_3L_4L_5L_6T_{e2}}{G_1}$		47.06K		33.76K

Fig.9 Noise budget of 4 GHz TE-cooled 2 DEG FET LNA.

Experimental Data Summary

Typical data measured at room temperature and under TE-cooled state of the overall noise characteristics of the newly developed 4 GHz band 2 DEG FET LNA shown in Fig.7 are illustrated in Fig.10. From these measured data, the noise temperature at 4 GHz is known to be 48.7 K at room temperature and 36 K under TE-cooled state. These values compare favorably with the theoretically predicted values indicated in Fig.9.

Furthermore, noise temperatures on the order of 30 K, 34 to 39.6 to be exact, unprecedentedly low values which could never be attained so far with conventional TE-cooled FET LNAs, were obtained with the LNA shown in Fig.7 over a wide frequency band of 800 MHz under TE-cooled state. This data is undoubtedly the lowest of all values reported so far of the conventional TE-cooled type FET LNAs.

Conclusion

Integrating the two approaches—ultra-low-noise property of 2 DEG FET and LNA technology, has brought about a success with the development of 2 DEG FET LNAs featuring superior noise characteristics in the 2, 4, 12, and 20 GHz bands. Incidentally, the newly developed 4 GHz TE-cooled 2 DEG FET LNA detailed as a typical example is designed to meet the specification of INTELSAT Standard-A earth station across 800 MHz bandwidth sufficiently.

The trend toward wider bandwidth and lower noise temperature will be further accelerated hereafter with the advancement of circuit, ferrite device, and cooling technologies and the improvement of the characteristics of the 2 DEG FET per se. The authors are convinced that the advent of the satellite communications FET LNAs superior in performance to the parametric LNAs can be expected before long.

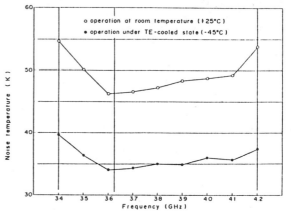

Fig 10 Typical noise performance data taken from 4GHz band TE-cooled 2 DEG FET LNA.

Acknowledgement

The authors would like to express their sincere thanks to Mr. M. Waki of KDD for his helpful encouragement.

The authors also wish to express their thanks to Dr. Y. Takayama, Manager, Microelectronics Research Laboratories, NEC, and Mr. M. Ogawa, Research Manager, Fundamental Research Laboratories, NEC, for their valuable suggestions. Thanks are also due to Messrs. S. Yokoyama, H. Shimayama and all the staff concerned at the Microwave and Satellite Communications Division, NEC, for their kind advice and cooperation.

References

(1) W. Akinaga et al., "The Present Status and Future Development of Low Noise Amplifiers for Satellite Communications," 10th AIAA, pp.125-133: Mar. 1984.

(2) K. Ohata et al., "A Low Noise AlGaAs/GaAs FET With P+-Gate and Selectively Doped Structure," 1984 IEEE Int'l MTT-S Microwave Symp. Digest Tech. Papers, pp.434-436, 1985.

(3) P. Bura, "Operation of 6 GHz FET Amplifier at Reduced Ambient Temperature," Electronics Letters, Vol.10, No.10, pp.181-182, 16 May 1974.

(4) H. Rothe and W. Dahlke, "Theory of Noisy Fourpoles," Proc. IRE, Vol.44, pp.811-818, June 1956.

Monolithic Integrated Circuit Applications of InGaAs/InAlAs HEMT's

M. Tutt, D. Pavlidis, G.I. Ng, M. Weiss, and J.L. Cazaux†

Center for High Frequency Microelectronics
Solid State Electronics Laboratory
Department of Electrical Engineering and Computer Science
The University of Michigan
Ann Arbor, Michigan 48109

ABSTRACT

The features of InGaAs/InAlAs HEMT technology are discussed in view of its potential for MMIC applications. Small and large signal characteristics benefit from the introduction of a strained layer in the channel. InGaAs HEMT's have a 3rd order intermodulation distortion point 4 dB higher than GaAs/AlGaAs, and are therefore very suitable for full multi-stage amplifier integration. A first demonstration of integration capability is shown with monolithic InGaAs attenuators which show 1.5 dB insertion loss and 18 dB dynamic range from 545MHz to 25GHz.

INTRODUCTION

High gain and low-noise circuit characteristics have been recently reported using AlGaAs/GaAs High Electron Mobility Transistors (HEMT's) in hybrid and monolithic configuration [1,2,3]. Planar-Doped AlGaAs/InGaAs pseudomorphic HEMT results at 59GHz and 94GHz [4] suggest the possibility of further performance improvement due to carrier confinement and higher electron velocity.

The ultimate in low-noise characteristics is, however, expected using InGaAs/InAlAs devices due to their inherent material and bandstructure advantages. Devices made on such heterostructures showed excellent electrical characteristics (g_m= 1080mS/mm, f_T= 165GHz with 0.1μm gate length) [5] which can be improved even further by strained channels (increased Indium composition of the $In_{0.53+x}Ga_{0.47-x}As$ layer) [6,7].

This paper describes the properties of InGaAs/InAlAs HEMT's as applied to monolithic integrated circuits and reports for the first time a working attenuator on such a material system.

PROPERTIES OF InGaAs/InAlAs HEMT's

InGaAs/InAlAs HEMT's offer distinct advantages over their GaAs/AlGaAs counterparts but present some weak points too. This section describes their features in view of their MMIC applications.

As a result of the lower electron effective mass of InGaAs, the mobility μ_n, is typically 11,400cm²/V-s instead of 8000cm²/V-s for GaAs. Their peak velocity is also higher(V_p(InGaAs) $\approx$ 1.2V_p(GaAs)) due to the larger Γ-L intervalley separation. Furthermore, the higher conduction band discontinuity (ΔE_c= 0.5eV instead of 0.28eV for GaAs/AlGaAs) results in larger sheet carrier density, better carrier confinement, and increased current density.

InGaAs/InAlAs shows also reduced trapping (DX centers) and is therefore good for cryogenic temperatures having reduced drift in device characteristics and reduced Persistent Photo-Conductivity (PPC) and I-V collapse problems.

[0]This work is supported by U.S. Army office grant No: DAAL03-87-U-0007 and Wright Patterson Air Force Base (Contract No. F33615-87-C-1406)
† Alcatel Espace, Toulouse, France.

Several parameters contribute also towards the good noise performance of InGaAs HEMT's. For example their low sheet resistance ($< 150\Omega/\square$) reduces thermal noise and the large Γ-L separation results in lower intervalley noise. Their high velocity is another factor since it contributes towards high transconductance.

There are several disadvantages of InGaAs HEMT's. The low Schottky barrier height (∼0.5eV) on N-InAlAs donor layers, the high background doping of InGaAs layers (5.0×10^{15}-$1.0\times10^{16}cm^{-3}$), and alloy clustering in highly doped InAlAs are supposed to be responsible for excessive gate leakage and poor pinch-off. Furthermore, the low Schottky barrier can make these devices more temperature sensitive. Conduction through Non-Intentionally-Doped (NID) InGaAs can result in high device conductance (G_{ds}). Another consideration is the device growth on InP substrates which are not as readily available as GaAs.

To reduce some of these effects, Schottky contacts are made on NID rather than doped InAlAs and Molecular Beam Epitaxy growth is optimized. The lowest background density values reported for InGaAs are $2\times10^{15}cm^{-3}$ [8] with excellent InGaAs/InAlAs heterostructure characteristics.

The properties of lattice matched (53% In in the InGaAs channel) HEMT's can be improved with the addition of a strained $In_{0.53+x}Ga_{0.47-x}As$ layer. A cross section of the design used in this work is shown in Fig. 1. The 2DEG is now formed in the $In_{0.53+x}Ga_{0.47-x}As$ layer (x= 0.07 and 0.12) and shows a higher mobility due to reduced carrier scattering. This is a result of larger first excited to ground state separation (E_1-E_0) and higher E_0 occupation [9]. The effective barrier to the donor region is also increased. The cross section of Fig. 1 shows also an i-$In_{0.52}Al_{0.48}As$ layer for reduced gate leakage and and n⁺-InGaAs cap for better ohmic contacts.

TECHNOLOGY AND DC CHARACTERISTICS

InGaAs/InAlAs heterostructure fabrication for devices and MMIC's follows similar steps as other III-V technologies.

For the studies reported in this paper, wet etching with H_3PO_4 : H_2O_2 : H_2O or H_2SO_4 : H_2O_2 : H_2O for PMMA compatibilty is used for mesa isolation and gate recess. Ohmic contacts are made with Ge/Au/Ni/Ti/Au and rapid thermal annealing (RTA) giving contact resistances of 0.2 to 0.3Ω-mm. Ti/Au of different thicknesses is finally used for gates and interconnects. Depending on the resistor value required for the MMIC, a seperate recess to the N InAlAs layer can be used for kΩ realization, or the entire heterostructure can be used for low value resistors.

To investigate the merits of strained versus lattice matched technology we mapped the characteristics of MBE grown wafers with discrete HEMT's and integrated attenuator and phase-shift functions. Fig. 2 shows the transconductance histogram of a wafer with 65% InGaAs channels. Very high average extrinsic and intrin-

Reprinted from the *Rec. of the IEEE GaAs Integrated Circuits Symp.*, pp. 293–296, 1988.

sic g_m's are obtained (520 and 700 mS/mm, respectively) for $1\mu m$ long gates. The recess non-uniformity caused by the wet etchant can produce threshold voltage (V_T) and drain-source current (I_{ds}) variations. A scatter plot of g_m-I_{ds} for the same wafer is shown in Fig. 3. These data, obtained over a number of devices, proved a better performance of 65% strained over other devices with smaller indium composition. Up to 40% g_m improvement could for example be obtained by changing from 60% to 65% indium channels.

The I_{ds} value at V_{gs}= 0.0V, V_{ds}=3.0V is 550mS/mm for the devices of this work (65% indium). Their DC output conductance is relatively high (39mS/mm) and is the primary limitation of these heterostructures for high power gain. However, we have recently shown that the G_{ds} can be reduced by a factor of 3-4 by the use of two adjacent channels (ie. a double channel device) [10] which provides very good carrier confinement.

MICROWAVE CHARACTERISTICS

The devices were characterized from 0.045GHz to 26.5GHz using a CASCADE probe station. From their bias dependent S-parameters, their equivalent circuits were extracted by fitting. Typical results for g_m and G_{ds} are shown in figures 4a, and 4b. g_m increases with V_{gs} due to the higher sheet carrier density until saturation of the 2 DEG is reached. Any further V_g changes result in g_m decreases due to the I_{ds} saturation. A g_m reduction with V_{ds} is also observed due to hot electron transfer to the buffer layer.

For a given V_{gs}, Fig. 4b shows an R_{ds} increase due to I_{ds} approaching saturation. An opposite tendency is, however, observed at low V_{gs} where electrons are easier to inject into the substrate. These g_m, G_{ds} nonlinearities have been used for the intermodulation distortion analysis in Section 5.

The devices showed excellent cutoff characteristics, with f_T exceeding 45 GHz. The maximum stable gain is 12.3 dB at 10 GHz and 7.5 dB at 18 GHz.

INTERMODULATION PERFORMANCE

Single stage AlGaAs/GaAs amplifiers have been demonstrated recently [3]. However, the full potential of monolithic amplifiers cannot be realized in this technology because of their third order intermodulation distortion point (IP3) which is typically 3dB smaller than that of equivalent MESFET's [11]. InGaAs/InAlAs HEMT's offer an alternative solution with their higher IP3 values. We investigated $1\mu m$ x $150\mu m$ devices of different strain and compared them to AlGaAs/GaAs HEMT's. The bias conditions for all devices were V_{ds}=2.5V and V_{gs} as required for maximum g_m.

The g_m-V_g, G_{ds}-V_d characteristics analyzed in Section 4 can be expressed in polynomial form [12]: For a given V_{ds}:

$$g_m(V) = g_{m1} + g_{m2}V_{gs} + g_{m3}V_{gs}^2 + g_{m4}V_{gs}^3 + \cdots \quad (1)$$

For a given V_{gs}:

$$G_{ds}(V) = G_{ds1} + G_{ds2}V_{ds} + G_{ds3}V_{ds}^2 + G_{ds4}V_{ds}^2 + \cdots$$

For a two-tone intermodulation with $a_1 cos\omega_1 t$ and $a_2 cos\omega_2 t$ as the input signals, the output current at the frequency $2f_1$-f_2 is:

$$I_1 = 0.25g_{m3}a_1^2a_2 + 0.25g_{m5}a_1^4a_2 + 0.375g_{m5}a_1^5a_2^3 \quad (2)$$

$$I_2 = 0.25G_{ds3}a_1^2a_2 + 0.25G_{ds5}a_1^4a_2 + 0.375G_{ds5}a_1^5a_2^3$$

The devices studied with the above technique had small signal matched or 50Ω loads at their output. The results discussed here

were not influenced by the termination condition. A 3rd order polynomial was used for Eq. 1. Fig. 5 shows the output versus input power results at the fundamental (f_1) and intermodulation ($2f_1$-f_2) frequency. The AlGaAs HEMT has an IP3 point which is $\sim$4dB below the InGaAs. For given nomimal output power (P_o) the intermodulation distortion at $2f_1$-f_2 will therfore be smaller for InGaAs (Point A) than for AlGaAs (Point B). This behavior primarily arises from the higher output power and gain of InGaAs HEMT's. Note also that the InGaAs HEMT has a larger absolute value of intermodulation power. This arises from its higher gain which results in larger output voltage swings and therefore larger influence of the G_{ds} nonlinearity. The latter is of course directly related to the intermodulation content.

Fig. 6 shows the influence of strain on Carrier to Intermodulation (C/I) ratio for different output power levels. Compared to AlGaAs/GaAs HEMT's, latticed matched InGaAs devices show higher C/I values. Best characteristics are obtained with 60% indium and the effect of strain on C/I seems to tail off beyond this value.

APPLICATION OF InGaAs TECHNOLOGY TO MONOLITHIC ATTENUATORS

As a first demonstration of integration feasibility for InGaAs/InAlAs HEMT's we have designed, fabricated, and tested a monolithic π-attenuator [13,14] with potential use in gain or signal level command. A photograph of the investigated circuit is shown in Fig. 7. It consists of three $1\mu m$ x $300\mu m$ HEMT's connected in a π configuration. The channel resistances are controlled by separate DC potentials applied to the gates via 3kΩ resistors which serve as RF-chokes. These resistors are integrated on the chip. The chip size is approximately 1mm by 1mm.

Since an attenuator is usually required to provide a wide dynamic range along with low insertion loss, the devices used must be able to provide both high- and low-impedances. Furthermore, the devices should have good isolation characteristics in the OFF state in order to avoid the degradation of attenuator characteristics. In addition to the high resistance value of the OFF FET, its capacitance is responsible for displacement current leakage which results in isolation reduction by the series FET or minimum insertion loss increase by the shunt FET's. The $300\mu m$ FET design used here gives the best possible compromise for ON and OFF attenuator characteristics.

The heterostructure design for the attenuators is of the type shown in Fig. 1 with 65% indium composition. In addition to this design, the double channel approach was also employed in order to explore its inherent features. The two channel paths of the latter design allow larger overall sheet carrier concentration and reduction of the ON-resistance. Furthermore, carriers are better confined in the channel leading to a higher output resistance.

Discrete $1\mu m$ x $150\mu m$ HEMT's of single and double channel design were tested from 500 MHz to 12.5 GHz under variable resistor conditions ($V_{ds} = 0.0V$, V_{gs} = variable). The equivalent circuit used for S-parameter fitting is shown in inset of Figure 8. The results, in Fig. 8, show an ON-OFF resistance variation of 2.85 to 103.98 Ωmm for the single channel HEMT and 1.93 to 613.6 Ωmm for the double channel HEMT. The OFF to ON resistance ratio is therefore improved by 8.7 times. Better results are expected by optimizing the material growth of the inverted heterointerface which limits the maximum achievable device mobility. As a result we used double channel HEMT's so that we may minimize G_{ds} thereby maximizing our OFF impedance.

Attenuators were measured from 545MHz to 25GHz. The best attenuator results were obtained with the double channel design. Fig. 9 shows the measured attenuation characteristics of this system. An 18dB dynamic range was obtained over the measurement band. The minimum insertion loss is 1.5 dB. The measured return loss is less than -9dB as shown in Fig. 10.

CONCLUSIONS

The characteristics of InGaAs/InAlAs HEMT's have been discussed as applied to monolithic technologies. Higher frequency performance can be obtained with the introduction of a strained layer in the channel. For a given output power level, InGaAs devices have smaller intermodulation power than their AlGaAs counterparts. Fully monoloithic multistage low noise amplifiers can therefore be realized with them in contrast to AlGaAs technology where their use is limited to their first stage only.

A monolithic attenuator operating with almost constant dynamic range from 545MHz to 25GHz has been demonstrated using this technology.

ACKNOWLEDGMENT

We are grateful to Professor P.K. Bhattacharya, Dr. J. Oh, and J. Pamalupati for layer growth.

References

[1] W. Yau, E.T. Watkins, S.K. Wang, K. Wang, B. Kiatskin, "A Four Stage V-band MOCVD HEMT Amplifier," *1987 IEEE MTT-S Digest*, Las Vegas, NV., June 1987, pp.1015-1018.

[2] J. Berenz, H.C. Yen, R. Esfandiari, K. Nakano, T. Sato, J. Velebir, K.Ip, "44 GHz Monolithic Low Noise Amplifier," *IEEE 1987 Microwave and Millimeter-Wave Monolithic Circuits Symposium*, Las Vegas, NV., June 1987, pp.15-21.

[3] Y. Yuen, C. Nishimoto, M. Glenn, Y.C. Pao, S. Bandy, G. Zdasiuk, "A Monolithic Ku-Band HEMT Low Noise Amplifier," *IEEE 1988 Microwave and Millimeter-Wave Monolithic Circuits Symposium*, New York, NY., May 1988, pp.139-42.

[4] P.C. Chao, P.M. Smith, K.H.G. Duh, J.M. Ballingall, L.F. Lester, B.R. Lee, A.A. Jebra, "High Performance 0.1μm Gate-Length Planar-Doped HEMT's," *International Electron Device Meeting Digest*, Washington DC., Dec. 1978, pp.410-13.

[5] V.K. Mishra, A.S. Brown, S.E. Rosenbaum, M.J. Delaney, S. Vaughn, K.White, "Noise Performance of Submicron AlInAs-GaInAs HEMT's," *46th Annual Device Research Conference*, Boulder, CO., June 1988, Paper IIIB-4.

[6] Y.J. Chan, D. Pavlidis, G.I. Ng, M. Jaffe, J. Singh, M. Quillec, "Effect of Channel Strain on the Electrical Characteristics of InGaAs/InAlAs HEMT's," *International Electron Device Meeting Digest*, Washington DC., Dec. 1987, pp.427-30.

[7] G.I. Ng, W.P. Hong, D. Pavlidis, M. Tutt, P.K. Bhattacharya, "Characteristics of Strained $In_{0.65}Ga_{0.35}As/In_{0.52}Al_{0.48}As$ HEMT with Optimized Transport Parameters," *IEEE Electron Device Letters*, Sept. 1988.

[8] A.S. Brown, V.K. Mishra, "The Impact of Epitaxial Layer Design and Quality on GaInAs/AlInAs HEMT Performance", *High Electron Mobility Transistor Symposium ETDL*, Fort Mommouth, NJ. Jan. 1988.

[9] W.P. Hong, G.I. Ng, P.K. Bhattacharya, D. Pavlidis, S. Willing, B. Das, " Low and High-Field Transport Properties of Pseudomorphic $In_xGa_{1-x}As/In_{0.52}Al_{0.48}As$ Modulation-Doped Heterostructures," *Journal of Applied Physics*, Aug. 1988.

[10] G. I. Ng, D. Pavlidis, M. Tutt, J. Pamalupati, J. Oh, P. K. Bhattacharya, " Strained InAlAs/InGaAs HEMT with Double Channel Design ", *Submitted for publication*

[11] J.J. Berenz, "Analog Application of HEMT's", *High Electron Mobility Transistor Symposium ETDL*, Fort Mommouth, NJ. Jan. 1988.

[12] J. A. Higgins, R. L. Kuvas " Analysis and Improvement of Intermodulation Distortion in GaAs Power FET's," *IEEE Trans. on MTT, vol. MTT-28, no. 1 pp 9-17*

[13] D. Pavlidis, Y. Archambault, J. Magarshack "Impedance Controlled Cell and its Control Circuit" Thomson-CSF Patent 84-00281. "Simple Voltage Controlled Phase Shifeters and Attenuators for MMIC's," *IEEE GaAs IC Symposium* San Diego, CA. 1981

[14] Y. Tajima, T. Tsukii, R. Mozzi, E. Tong, L. Hanes, B. Wrona, "GaAs Monolithic Wideband (2-18GHz) Variable Attenuators," *1982 IEEE MTT-S Digest*, Dallas, TX., pp.479-481.

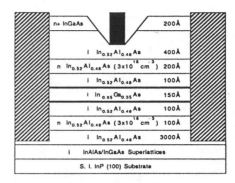

Figure 1: Cross-section of InGaAs/InAlAs HEMT's.

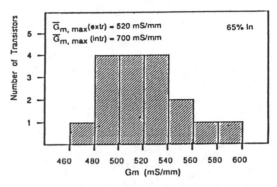

Figure 2: Transconductance of a 65% strained InGaAs/InAlAs HEMT.

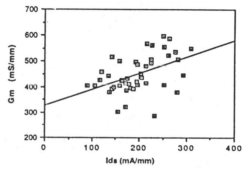

Figure 3: Extrinsic Transconductance (g_m) as a function of Drain-Source Current (I_{ds} @ g_{mMax}) scatter plot of a 65% strained InGaAs/InGaAs HEMT.

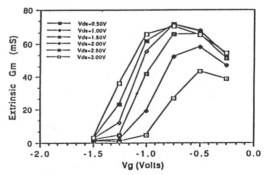

Figure 4: Small Signal bias dependence of transconductance (a) and output resistance (b) of 65% strained HEMT (1μm x 150μm).

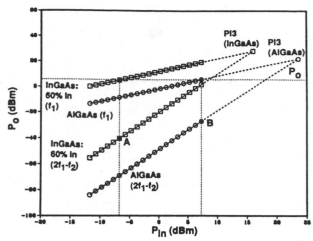

Figure 5: Intercept point (IP3) determination for AlGaAs and InGaAs HEMT's and comparison of their intermodulation characteristics for a given output power level.

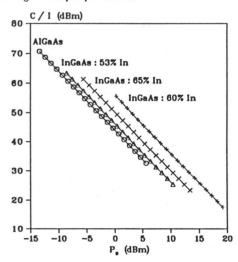

Figure 6: Carrier to Intermodulation Ratio as a function of output power level for AlGaAs and InGaAs (53%, 60%, 65% indium) HEMT's.

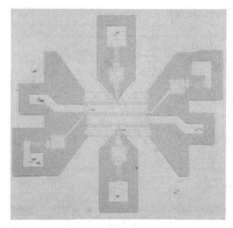

Figure 7: Photograph of InGaAs/InAlAs π-Attenuator.

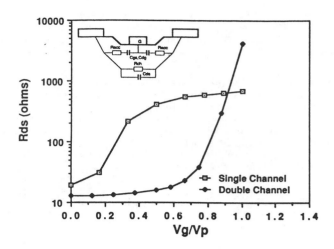

Figure 8: R_{ds} as a function of normalized gate voltage (V_g/V_p) for Double and Single Channel HEMT's.

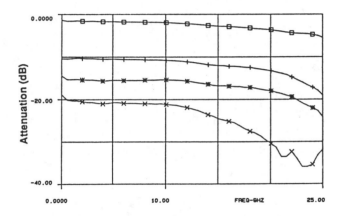

Figure 9: Attenuation (S_{21}) characteristics of the InGaAs/InAlAs π-Attenuator for various gate biases.

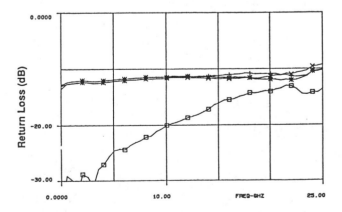

Figure 10: Return Loss characteristics of the π-Attenuator for different attenuation conditions.

136

A 12GHz-Band Monolithic HEMT Low-Noise Amplifier

N.Ayaki, A.Inoue, T.Katoh, M.Komaru, M.Noda, M.Kobiki, K.Nagahama and N.Tanino

Tel. 0727-82-5131

LSI Research & Development Laboratory
Mitsubishi Electric Corporation
4-1 Mizuhara, Itami, Hyogo, 664, Japan

ABSTRACT

A 12GHz-band two-stage monolithic HEMT (High Electron Mobility Transistor) low-noise amplifier has been developed. The HEMT used in the amplifier has a gate length of 0.5μm and shows a typical noise figure of 1.0dB at 12GHz. A noise figure of the amplifier is less than 1.7dB with an associated gain over 15.0dB in the frequency range from 11.7 to 12.7GHz.

INTRODUCTION

Monolithic microwave ICs (MMICs) have evolved into a high yielding, cost effective, high reliable technology suitable for the high volume production. The discrete HEMT has been already verified to be superior to the MESFET in noise figure. In the application of the HEMT to MMICs, however, there have been only a few works on the distributed amplifier [1][2] and on the low noise amplifier [3].

This report describes a monolithic HEMT low noise amplifier using the super low noise HEMT technology which gives as a discrete HEMT a typical noise figure of 1.0dB at 12GHz.

DESIGN

Figure 1 shows the HEMT structure. An epitaxial wafer for the HEMT is grown by molecular beam epitaxy (MBE) technique. An AlGaAs/GaAs superlattice is grown on a 3-inch diameter semi-insulating LEC GaAs substrate, followed by an about 1um thick undoped GaAs buffer layer, a 400Å thick n-AlGaAs layer and a 1500Å thick n-GaAs cap layer. The AlGaAs/GaAs superlattice is employed to reduce the dislocations. The thick cap layer is adopted in order to reduce the source-to-gate parasitic resistance and to obtain high reliability [4]. The n-GaAs and n-AlGaAs layers are Si-doped with a carrier concentration of $2 \times 10^{18}/cm^3$ and $1.5 \times 10^{18}/cm^3$, respectively.

The MMIC design is based on an equivalent circuit derived from accurate S-parameters measured by an on-wafer probing system. Figure 2 shows the S-parameters (S11,S22) of the HEMT, and Fig.3 shows an equivalent circuit and its parameters derived from the S-parameters.

Noise parameters are measured for the HEMT mounted on a chip carrier. Figure 4 shows the noise figure circles of the HEMT. The minimum noise figure (NFmin) is 1.03dB.

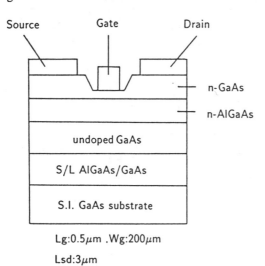

Lg:0.5μm .Wg:200μm

Lsd:3μm

Fig.1 HEMT Structure

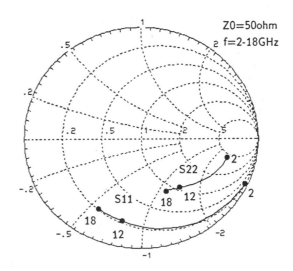

Fig.2 S-parameter of the HEMT

Reprinted from the *Rec. of the IEEE GaAs Integrated Circuits Symp.*, pp. 101–104, 1988.

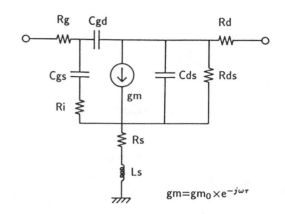

$$gm=gm_0 \times e^{-j\omega\tau}$$

Lg :0.5μm   Wg :200μm
Ls :0.012nH Cgs:0.019pF Cgd:0.041pF
Cds:0.016pF Rg :4.3ohm   Ri :1.9ohm
Rs :1.6ohm   Rd :4.0ohm   Rds:365ohm
gm_0:43mS τ :2.6ps

Fig.3 Equivalent circuit of the HEMT

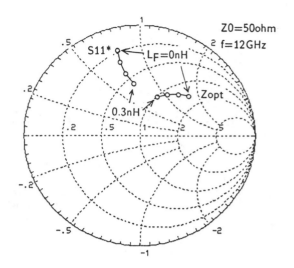

Fig.6 S11* and Zopt at 12GHz

Figure 5 shows a schematic diagram of the two-stage HEMT MMIC. This IC consists of an input matching circuit, a first HEMT, an inter-stage matching circuit, a second HEMT and an output matching circuit. The matching circuits include short stubs and bias circuits. RC filters in the bias circuits act as preventing the oscillation.

A noise figure (NF) of the amplifier depends on that of the first stage. In general, an NF matching point does not agree with the impedance matching one. Therefore, first and second stages are designed with NF matching and impedance matching, respectively. However, NF matching results in lower gain and higher input VSWR. To solve this problem, a series feedback inductor (L_F) is inserted between the source electrode of the HEMT and the via-hole ground at the first stage of the amplifier.

Figure 6 shows the effect of this series feedback inductance on S11* and Zopt at 12GHz. The start point of Zopt in Fig.6 is calculated from the data shown in Fig.4 by subtracting the length of an input transmission line. As shown in this figure, S11* approaches Zopt with increase of the series feedback inductance (L_F). However, the excess increase of this inductance degrades the gain and

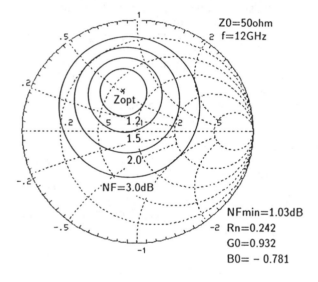

NFmin=1.03dB
Rn=0.242
G0=0.932
B0=−0.781

Fig.4 Noise figure circles of the HEMT

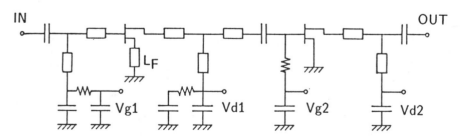

Fig.5 Schematic diagram of the two-stage HEMT MMIC

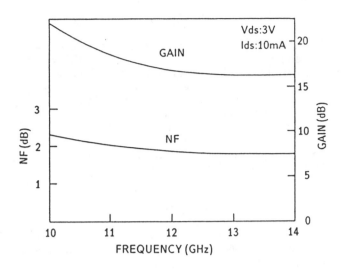

(a) Noise figure and associated gain

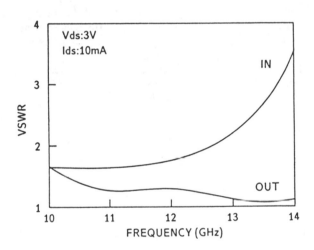

(b) Input and output VSWRs

Fig.7 Simulated results of the two-stage HEMT MMIC

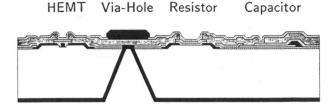

Fig.8 Schematic cross section of the HEMT MMIC

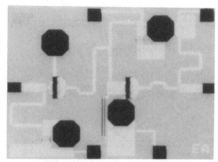

(Chip size:2.3mmx1.7mm)

Fig.9 Chip photograph of the HEMT MMIC

FABRICATION

Figure 8 shows the schematic cross section of the HEMT MMIC. The HEMT and resistors are isolated by mesa-etching. The ohmic electrodes are formed by AuGe/Ni/Au evaporation. The gate electrodes are fabricated by recess etching and lift-off. The capacitors have the structure of Metal/Insulator /Metal (MIM) and include 1500Å thick SiN. Micro-strip lines are formed by 2µm thick Ti/Au evaporation. The via-holes are opened at each ground of the HEMT and of the short stubs. Figure 9 shows a chip photograph of the HEMT MMIC. The total gate width and the gate length of each HEMT are 200µm and 0.5µm, respectively. The chip size is 2.3mm x 1.7mm, and the chip thickness is 125µm.

PERFORMANCE

Figure 10 shows the measured results together with simulated results shown in Fig.7. The drain voltage (Vds) is 3V and drain to source current (Ids) is 10mA for each HEMT. In the frequency range from 11.7 to 12.7GHz, the noise figure is less than 1.7dB with an associated gain over 15.0dB. The input VSWR is less than 1.9 and the output VSWR is less than 1.5. These results coincide considerably with the simulated data. The lowest noise figure obtained in this work is 1.57dB with an associated gain of 13.8dB at 12GHz.

the isolation. Therefore, the inductance of 0.15nH is chosen in the design. The inter-stage circuit is designed using conjugate matching to reduce the chip size. The output circuit is designed using impedance matching. SUPER-COMPACT is used to optimize the matching circuit parameters.

Figure 7 shows the simulation results. In the frequency range from 11.7 to 12.7GHz, the noise figure is less than 1.9dB with an associated gain over 16.5dB. The input VSWR is less than 2.0 and the output VSWR is less than 1.3.

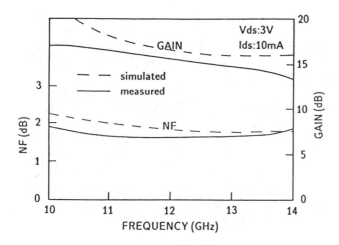

(a) Noise figure and associated gain

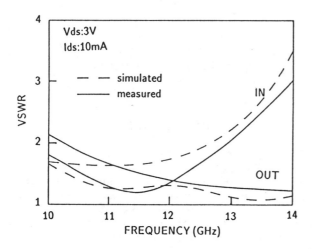

(b) Input and output VSWRs

Fig.10 Measured results of the two-stage HEMT MMIC

CONCLUSION

A 12GHz-band two-stage monolithic HEMT low-noise amplifier has been developed. The HEMT used in the amplifier has a gate length of 0.5μm and shows a typical noise figure of 1.0dB at 12GHz. In the frequency range from 11.7 to 12.7GHz, the noise figure is less than 1.7dB with an associated gain over 15.0dB. The input VSWR is less than 1.9 and the output VSWR is less than 1.5. These results suggest that the HEMT MMIC has a promising applicability for the microwave low-noise amplification.

REFERENCES

[1] Y.Yoshii, et al., "A monolithic HEMT distributed amplifier," 1987 Nat'l. Conv. Rec., IECIE, Japan, 752, March, 1987

[2] C.Nishimoto, et al., "A 2-20 GHz High-Gain, Monolithic HEMT Distributed Amplifier," 1987 IEEE MTT-S Int'l. Microwave Symp., pp.155-158, June, 1987

[3] J.Yonaki, et al., "A Q-Band Monolithic Three-Stage Amplifier," IEEE 1988 Microwave and Millimeter-wave Monolithic Circuit Symposium Digest of Papers, pp.91-98, May 1988

[4] K.Hayashi, et al., "Reliability of super low noise HEMTs," Proc. 1987 IEEE MTT-S Int'l. Microwave Symp., pp.1023-1026, June, 1987

A 20GHz High Electron Mobility Transistor Amplifier for Satellite Communications

Megumu Niori, Toshiyuki Saito

Fujitsu Radio/Satellite Communications Systems Laboratory

Kawasaki, Japan

Kazukiyo Joshin, Takasi Mimura

Fujitsu Semiconductor Devices Laboratory

Kawasaki, Japan

A RECENT REPORT[1] disclosed the development of a High Electron Mobility Transistor (HEMT) ring oscillator with switching propagation delays of 16.8ps at room temperature and 12.8ps at 77°K.

The design suggests that the HEMT will provide a lower noise figure and higher gain compared with that of conventional GaAs FETs. This paper will describe the electrical characteristics of the HEMT in the microwave region and the experimental results of a 20GHz low-noise four-stage amplifier using HEMTs.

The epitaxial structure of the HEMT for low noise application is essentially the same as that of the HEMT for high speed logic application previously cited[2]. Its electron mobility is about 8000 $cm^2V^{-1}s^{-1}$ at room temperature. The gate length and width are 0.5 and 200μ. Its typical dc transconductance is about 250mS/mm.

The measured S-parameters of the HEMT chip from 2 to 18GHz are shown in Figure 1. Figure 2 shows the equivalent circuit parameters calculated from the S-parameters. The equivalent circuit valid for GaAs FETs was adopted because of the similarities between the HEMT and GaAs FETs in device structure and S-parameters. Table 1 shows the measured minimum noise figure and the associated gain at 8, 11.3, and 20GHz, respectively. Applying the measured minimum noise figure and the equivalent circuit parameters to Fukui's relation valid for GaAs FETs[3],

$$F_{min} = 1 + 2\pi K_f f C_{gs} \sqrt{\frac{R_g + R_s}{g_m}}. \qquad (1)$$

where K_f is a fitting factor, f is a frequency, the fitting factor K_f is 1.6. It is significantly small compared with 2.5 (K_f) of the conventional GaAs FETs. Figure 3 gives the measured minimum noise figure and the predicted one calculated from equation (1). It also shows that the calculated minimum noise figure of GaAs FETs with all parameters is the same as for the HEMT in Figure 2. This shows that a minimum noise figure of below 1dB at 12GHz is obtainable, reducing the summation of R_s and R_g to less than 4Ω.

Figure 4 shows a photograph of a developed amplifier. It consists of 2 unit amplifiers, an MIC isolator, waveguide to stripline transitions and waveguide isolators. The unit amplifier is a 2-stage amplifier. The matching circuits are fabricated on 0.38mm thick alumina substrates. These substrates are mounted on a metal carrier, and HEMT chips are die-bonded on the same carrier. The performance of the 20GHz low-noise HEMT amplifier is shown in Figure 5. A noise figure of 3.9dB has been achieved with a power gain of 30dB.

Figure 6 gives the temperature dependence of noise figures and gains of the HEMT and a 0.5μ gate GaAs FET at 11.3GHz. The HEMT exhibits significant improvements both for noise figure and gain at low temperatures. This results from the drastic increase in electron mobility of the HEMT as compared with that of GaAs FETs.

Acknowledgments

The authors would like to thank K. Kosemura and Y. Yamashita for device processing, and H. Sugawara for helpful suggestions on the amplifier designs. The authors also express their thanks to Y. Tokumitsu, M. Abe, S. Yamamoto, H. Komizo and H. Ishikawa for guiding this work.

[1] Abe, M., Mimura, T., Yokoyama, N. and Ishikawa, H., "New Technology Toward GaAs LSI/VLSI for Computer Applications", *IEEE Trans. Microwave Theory Tech*, p. 992-998; July, 1982.

[2] Mimura, T., Hiyamizu, S. and Hikosaka, S., "Enhancement-Mode High Electron Mobility Transistors for Logic Applications", *Jap. J. Appl. Phys.*, p. L317-L319; May, 1981.

[3] Fukui, H., "Optimal Noise Figure of Microwave GaAs MESFETs", *IEEE Trans. Electron Devices*, p. 1032-1037; July, 1979.

Frequency (GHz)	Minimum Noise Figure (dB)	Associated Gain (dB)
8	1.3	13.0
11.3	1.7	11.2
20	3.1	7.5

TABLE 1—Measured minimum noise figure and gain of HEMTs.

Reprinted from the *Rec. of the 1983 IEEE Int. Solid-State Circuits Conf.*, pp. 198–199, 1983.

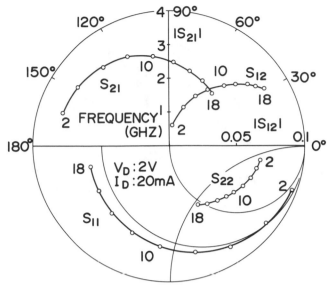

FIGURE 1—Measured S-parameters of the HEMT chip.

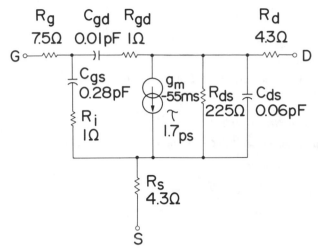

FIGURE 2—Equivalent circuit of the HEMT.

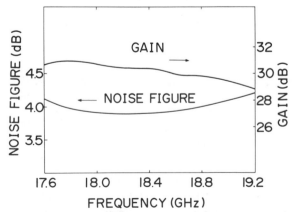

FIGURE 3—Measured and calculated values of minimum noise figure of the HEMT.

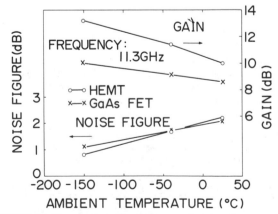

FIGURE 4—A 20GHz 4-stage HEMT amplifier.

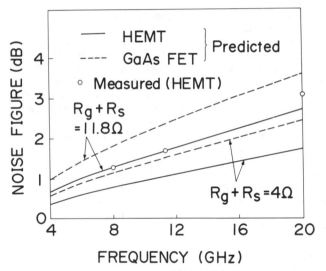

FIGURE 5—Performance of the 20GHz HEMT amplifier.

FIGURE 6—Temperature dependence of noise figure and gain of the HEMT and the GaAs FET at 11.3GHz.

20 GHz-BAND LOW-NOISE HEMT AMPLIFIER

Kiyoyasu Shibata, Kumiko Nakayama, Masao Ohtsubo, Hisao Kawasaki
Shigekazu Hori and Kiyoho Kamei

Microwave Solid-State Department
Komukai Works, Toshiba Corporation
1 Komukai-Toshiba-Cho, Saiwai-Ku
Kawasaki, 210 JAPAN

Abstract

A 20 GHz-band low-noise amplifier has been developed by using newly developed 0.25-μm gate HEMTs. The amplifier has been fabricated by cascading six single-ended unit amplifiers without any isolators at the interstages.

The HEMT amplifier exhibits a noise temperature of 170 K (NF = 2.0 dB) and a gain of 47 dB over 17.5 to 19.5 GHz in an uncooled operation. Noise temperatures of 130 K (NF = 1.6 dB) and 110 K (NF = 1.4 dB) have been obtained at -20 °C and -50 °C, respectively.

Introduction

A progress of GaAs FET performance has made it possible to build low-noise amplifiers up to millimeter-wave region[1]. However, performance of low-noise GaAs FETs is reaching its limitation because of the finest gate geometry[2], and almost little improvement can be expected for the low-noise GaAs FET amplifiers.

Recently, extensive work on HEMT (High Electron Mobility Transistor) has been done in many laboratories[3], and the noise temperature of 135 K at -55 °C in 20 GHz band has been reported from the Peltier-cooled HEMT amplifier[4]. We have also demonstrated an 18 - 26.5 GHz-band amplifier using 0.4-μm gate HEMT which shows superior performance to the 0.25-μm gate GaAs FET amplifier[5].

This paper reports on a 20 GHz-band low-noise HEMT amplifier for the satellite communication earth station by using newly developed 0.25-um gate HEMT. The developed amplifier exhibits a noise temperature of < 170 K and a gain of > 47 dB over 17.5 to 19.5 GHz in an uncooled condition. The following sections describe the HEMT employed, the unit amplifier design and the RF performance of the low-noise amplifier.

Device

Fig. 1 shows the schematic cross section of the HEMT. The HEMT wafer, grown by MBE, is made of a 1-μm thick undoped GaAs buffer layer, a 20-nm thick n-type $Al_{0.3}Ga_{0.7}As$ layer and a 30-nm thick n-type GaAs cap layer on an undoped GaAs substrate.

The n-type $Al_{0.3}Ga_{0.7}As$ and GaAs layer are heavily doped to 3×10^{18} cm^{-3} with Si. The measured sheet carrier concentration and the electron Hall mobility of the HEMT wafer are 1.5×10^{12} cm^{-2} and 4200 cm^{2}V^{-1}s^{-1} at room temperature.

A recess structure is formed to control a drain current by etching an n-type GaAs layer and a part of n-type $Al_{0.3}Ga_{0.7}As$ layer. The gate electrode with a length of 0.25 μm and a width of 200 μm is formed by Al/Ti to a thickness of 0.5 μm. The gate electrode delineated by a direct electron-beam lithography is placed with a 0.5 μm spacing from the source contact edge to reduce the source resistance. The source and drain ohmic contacts separated by 3 μm are formed by alloying the evaporated Ni/AuGe.

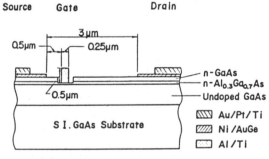

Fig. 1 Schematic cross section of HEMT chip.

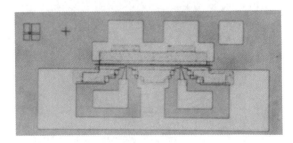

Fig. 2 Microphotograph of HEMT chip.

Reprinted from *1986 IEEE MTT-S Int. Microwave Symp., Tech. Dig.*, pp. 75–78, 1986.

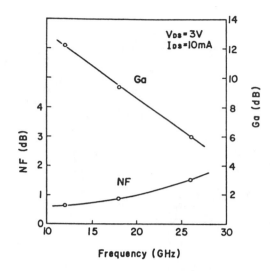

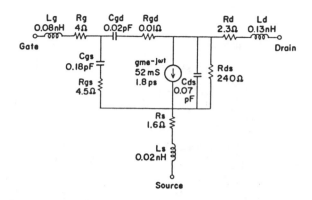

Fig. 5 Equivalent circuit with element values of 0.25-μm gate HEMT.

Fig. 3 Minimum noise figure (NF) and associated gain (Ga) vs. frequency of HEMT.

Fig. 2 shows a microphotograph of the HEMT chip. The chip size is 250 x 500 μm. Two gate bonding pads are provided to reduce the gate resistance.

Static characteristic of the HEMT has been measured and the transconductance g_m is 250 mS/mm for a drain current of 10 mA.

Microwave minimum noise figure (NF) and associated gain (Ga) have been measured on the HEMT chip mounted on a microstrip test fixture. Fig. 3 shows a frequency dependence of NF and Ga measured at room temperature at a drain current of 10 mA. The HEMT has shown typical noise figures (NF) and gains (Ga) of NF = 0.7 dB / Ga = 11.8 dB, NF = 0.9 dB / Ga = 9.4 dB and NF = 1.5 dB / Ga = 6.0 dB at 12 , 18 and 26 GHz ,respectively.

The noise parameters and S-parameters of the HEMT have been measured for the sake of amplifier design. The noise

parameters were obtained by Adamian and Uhlir's method, in which the output noise power was measured by connecting a 50 ohm load and off-set shorts at the input port[6]. Fig. 4 shows the constant noise figure and gain loci of the HEMT at 18 GHz. The minimum noise figure (NF_{min}) of 1.24 dB, the equivalent noise resistance (Rn) of 7.63 ohm and the optimum source reflection coefficient (Γ_{opt}) of 0.543 ∠126° have been obtained at 18 GHz. The S-parameters up to 24 GHz have been calculated by the equivalent circuit analysis using measured S-parameters over 2 to 18 GHz. Fig. 5 shows the equivalent circuit with element values for the drain current of 10mA. The cut-off frequency f_T is calculated to be 46 GHz.

Unit Amplifier Design

For the optimum design of six-stage low-noise amplifier, we have designed noise figure optimized (NFopt) and gain optimized (Gopt) unit amplifiers. The NFopt unit amplifiers are used for the first and the second stages and the Gopt unit amplifiers are used from the third to the sixth stages. Fig. 6 shows the equivalent circuit of the unit amplifier. The single-ended unit amplifier has been designed to give a good stability for the cascade connection. The gate bias circuit, which consists of the lumped elements with bonding wires, chip capacitors and a resistor, makes the amplifier stable in lower frequency range. A quarter-

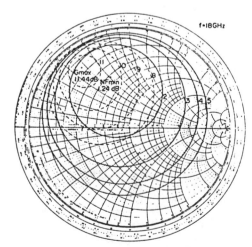

Fig. 4 Constant NF and gain loci of 0.25-μm gate HEMT.

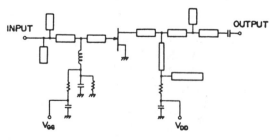

Fig. 6 Equivalent circuit of unit amplifier.

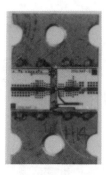

Fig. 7 Top view of unit amplifier.

wavelength open stub terminated via a resistor and a capacitor in the drain bias circuit are used to stabilize the amplifier near operating frequency. The circuit parameters for both unit amplifiers have been optimized by a computer simulation using measured noise parameters and S-parameters. The predicted noise figure (NF) and the gain (Ga) over 17.5 to 19.5 GHz are NF < 1.4 dB and Ga > 8.7 dB for the NFopt unit amplifier, and NF < 2.1 dB and Ga > 10 dB for the Gopt unit amplifier.

Fig. 7 shows the top view of the unit amplifier. The input and output matching circuits have been fabricated on 0.38-mm thick alumina substrates with Ti/Pt/Au microstrip lines and Ta_3N_4 thin film resistors. Each substrate for input and output circuits has a size of 4 x 3.6 mm and is soldered to a 0.7-mm thick Kovar carrier.

Fig. 8 shows the measured noise figure and gain performance for the NFopt and Gopt unit amplifiers under the drain voltage of 3 V and the current of 10 mA. The typical NFopt unit amplifier shows

NF < 1.5 dB and Ga > 8dB over 17.5 to 19.5 GHz, and the minimum noise figure of 1.35 dB are obtained at 18GHz. The Gopt unit amplifier shows NF < 2.5 dB and Ga > 9 dB in the same frequencies. These measured values show a good agreement to the predicted performance. Since the gain of the NFopt unit amplifier decreases in higher frequency, the Gopt unit amplifier has been designed to increase the gain at higher frequency of our interest.

Low-noise Amplifier

In order to obtain the system required gain of > 45 dB over 17.5 to 19.5 GHz, the amplifier has been built by cascading six unit amplifiers. Fig. 9 shows the configuration of the amplifier. In cascading six single-ended unit amplifiers, one or two isolators are commonly installed at the interstages in order to avoid an unexpected oscillation and an excess inband gain ripple. But the use of isolators increases the size and the cost of the amplifier, so we have designed the amplifier without any isolators at the interstages.

Fig. 10 shows the inside view of the HEMT amplifier measuring 33 x 33 x 110 mm. Six unit amplifiers are installed into the hermetically sealed housing. Waveguide to microstrip transitions with 50 ohm coaxial feedthroughs are prepared at the input and output ports. One end of the coaxial feedthrough is connected to the microstrip line of the unit amplifier by a gold ribbon, and the other end is inserted into

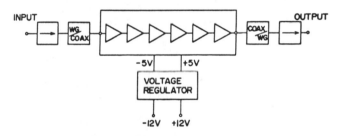

Fig. 9 Configuration of low-noise amplifier.

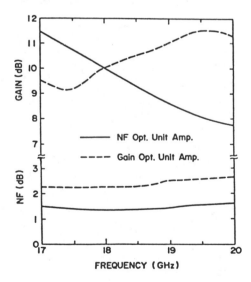

Fig. 8 Measured noise figure and gain of NFopt and Gopt unit amplifiers.

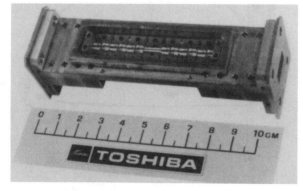

Fig. 10 Inside view of low-noise HEMT amplifier.

the waveguide. The waveguide isolators are used at the input and output ports of the amplifier to achive good VSWRs. The measured insertion loss of the input isolator is 0.2 dB. A DC voltage regulator is installed in the backside of the housing, and the amplifier can operate for an external supply voltage of 12 V with a current of 50 mA and -12 V with a current of 8 mA.

Fig. 11 shows the measured noise figure and gain of the low-noise amplifier. A noise figure of < 2.0 dB (noise temperature T_N < 170 K) and a gain of > 47 dB are obtained over the frequency range of 17.5 to 19.5 GHz. In the measured frequency band, the minimum noise figure is 1.85 dB (T_N = 154 K) at 18.5 GHz. The input and output VSWRs of the amplifier are < 1.2 over 17.5 to 19.5 GHz.

Fig. 12 shows the temperature dependence of the noise figure and the gain at 18.5 GHz. At the temperature of 50 °C, the amplifier shows NF = 2.2 dB (T_N = 190 K) and the gain of 47 dB. Furthermore, it shows the NF = 1.6 dB (T_N = 130 K) and NF = 1.4 dB (T_N = 110 K) at the temperature of -20 °C and -50 °C, respectively. It has been found that the noise figure change against temperature at a rate of 0.008 dB/°C. Since the gain of the amplifier changes at a rate of -0.012 dB/°C, the gain change of the unit amplifier is calculated to be -0.002 dB/°C.

The output power of the amplifier at 1-dB gain compression is 8.6 dBm. The obtained output power is sufficient for the low-noise amplifier for the satellite communication receivers.

Conclusion

A 20 GHz-band low-noise amplifier has been developed by using newly developed 0.25-μm gate HEMT. A noise temperature of 170 K (NF = 2.0 dB) and a gain of 47 dB

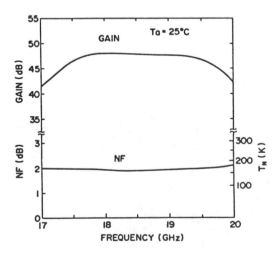

Fig. 11 Measured noise figure and gain of low-noise amplifier.

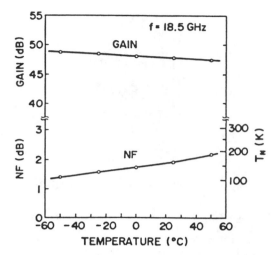

Fig. 12 Temperature dependence of noise figure and gain of low-noise amplifier.

have been obtained over 17.5 to 19.5 GHz at a room temperature.

Through this work, it has been found that the low-noise receiver with the system required noise temperature of < 200 K can be realized without a Peltier-cooling system. The uncooled HEMT amplifier will replace the Peltier-cooled amplifier, since it will be very effective to reduce the size, cost and power consumption of the satellite communication receivers.

Acknowledgement

The authors would like to thank Dr.M.Ohtomo, S.Okano, Dr.T.Nakanisi, and K.Mishima for their encouragement and helpful discussions. They are also grateful to M.Higashiura, Y.Ashizawa and Dr.M.Kawano for supplying the MBE wafers.

References
[1] E.T.Watkins,et al., "A 60 GHz GaAs FET amplifier," IEEE MTT-S Int. Microwave Symp. Digest, pp.145-147,1983.
[2] K.Kamei,et al., "Quarter micron low noise GaAs FET's operable up to 30 GHz," IEDM Tech. Papers, pp.102-105,1980.
[3] K.Kamei,et al., "Extremely low noise 0.25 μm gate HEMTs," 12th International Symposium on GaAs and Related Compounds, Karuizawa,Japan, Sept. 1985.
[4] M.Iwakuni,et al., "A 20 GHz peltier-cooled low noise HEMT amplifier," IEEE MTT-S Int. Microwave Symp. Digest, pp.551-553,1985.
[5] K.Shibata,et al., "Broadband HEMT and GaAs FET amplifier for 18-26.5 GHz," IEEE MTT-S Int. Microwave Symp. Digest, pp.547-550, 1985.
[6] V.Adamian and A.Uhlir,"Simplified noise evaluation of microwave receivers," IEEE Trans. Instrum. Meas., vol.IM-33,no.2,pp.136-140,June 1984.

36.0 - 40.0 GHz HEMT LOW NOISE AMPLIFIER

Michael Sholley, John Berenz, Arthur Nichols,
Kenichi Nakano, Ramzi Sawires, James Abell

T.R.W. Electronic Systems Group
Communications Systems Development Department
Redondo Beach, California

Abstract

This paper describes the design and development of a multistage low noise High Electron Mobility Transistor (HEMT) amplifier that exhibits state-of-the-art performance over the frequency range of 36-40 GHz. The amplifier utilizes a series of three single ended stages that are each designed around TRW'S HEMT device. Typical performance to date has been 15-17 dB gain with an associated noise figure of 4.0 to 4.6 dB.

Introduction

Until 1982 the conventional GaAs MESFET technology was rapidly approaching the limits of photolithography and fabrication techniques. Performance parameters at upper Ka-band (40 GHz) were inadequate for many receiver applications where preamplifier stages required low noise figures and high gains. With the advent of (HEMT) technology, there now exists an active device that has the potential to replace the MESFET as the device that is used in the low noise amplifiers.

High Electron Mobility Transistors (HEMT) have been fabricated which exhibit lower noise figure and higher gain at millimeter-wave frequencies than previously reported (1). Although similar in construction to the devices reported earlier, considerable progress has been made in optimizing both the epitaxial structure and the parasitic elements of the devices used in this work. Quarter-micron gate length transistors were fabricated by direct-write electron beam lithography on epitaxial layers grown by molecular beam epitaxy. The gain and noise performance achieved with these improved HEMT devices rivals the best measured for "state-of-the-art" GaAs MESFET's having the same length.

DEVICE FABRICATION

For device fabrication, single heterojunctions were grown by molecular beam epitaxy with a graded aluminum composition contact as shown in the device cross-section of Figure 1. By grading the aluminum composition over several hundred angstroms, it is possible to eliminate the barrier to current flow in the top contact layer. This modification of the epitaxial structure reduces contact resistance and increases the sheet electron concentration. The saturated drain-source current of the resulting devices is double that previously obtained with an abrupt n+ GaAs contact. Due to the greater current handling capability, these devices exhibit higher extrinsic transconductance, higher gain, and greater power output than conventional GaAs FET'S. Reduction of the source-gate resistance has lowered the noise figure. Direct-write electron beam lighography was utilized to define quarter-micron gate lengths. Reduction of the gate-source capacitance is also responsible for lowering noise figure and increasing gain.

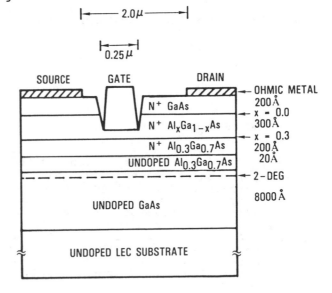

Figure 1. Cross section of HEMT device.

Reprinted from *1985 IEEE MTT-S Int. Microwave Symp., Tech. Dig.*, pp. 555-558, 1985.

To provide guidance in optimizing device performance, an analytical model has been developed to relate the dc and small-signal characteristics of the device to the physical structure. The lowest and first energy subbands of the quantum well have been incorporated in the model to improve its accuracy in describing measured device properties as a function of bias. Excellent agreement has been achieved between calculated and measured noise figure as a function of frequency using the Fukui equation [2]. Figure 2 compares the calculated noise figure and gain of a quarter-micron HEMT device made using the current technology with the measured performance of our experimental devices is close to the theoretical prediction.

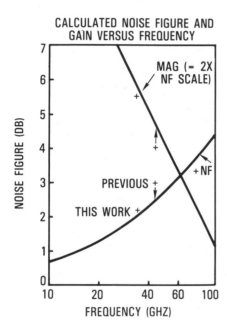

Figure II. Noise figure and gain vs.

frequency.

CIRCUIT DESIGN & CONSTRUCTION

The amplifier was constructed utilizing microstrip techniques, but due to the high frequency of operation a direct waveguide to microstrip transition was required to facilitate the assembly and testing of the individual amplifier states. As shown in Figure III the transitions were of the finline configuration [3] that utilizes a titanium tungsten, gold metallization,

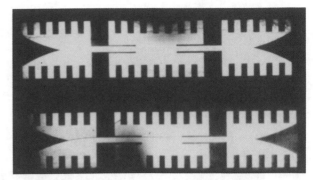

Figure III. Waveguide to microstrip

transition.

with the pattern photoetched onto 0.010" thick highly polished quartz substrate. This type of transition was chosen over the more common E-field probe which utilizes an adjustable backshort for tuning because of its greater bandwidth, lower insertion losses and an excellent VSWR. This transition requires no mechanical connection to the amplifier circuit as would a ridged waveguide transition, and allows construction of planar type amplifiers on carriers that are easily tuned and can be assembled into multistage amplifier chains. For a single stage amplifier there is no need for D.C. blocking capacitors, since there is no direct contact of the circiut metallization with the housing. The transition represents a combination of computer design and optimization. It is easily reproducible by photo-lithographic techniques. By using a (CAD) Computer Aided Design system, the turn-around-time for successive iterations and modifications was greatly reduced. Insertion loss and return loss for the transitions are shown in Figure IV. All

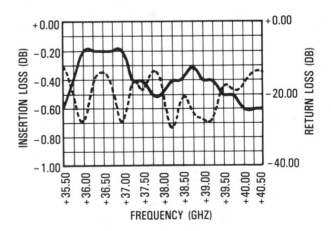

Figure IV. Waveguide to microstrip transition insertion loss and return
loss.

data is for two transitions placed back-to-back. The transitions are broadband and exhibit excellent performance over the band of interest.

A single stage amplifier was constructed using distributed matching elements as shown in Figure V.

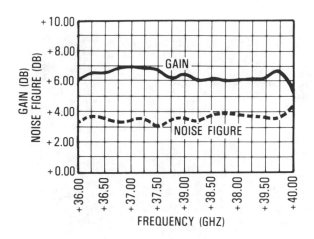

Figure VI. Noise figure and gain of single stage amplifier.

Figure V. Single stage amplifier.

The bias networks were computer optimized to insure out-of-band stability. The circuitry was photo-etched onto 0.010" thick quartz substrate using the same processes and metallization as the transitions. The amplifier housing was designed to eliminate any possible moding problems and to minimize radiation losses. Invar was chosen as the material for the carrier because its coefficient of expansion matches that of the quartz substrate.

The amplifier's noise figure was optimized by a selective mismatch of the input circuit consistent with adequate gain. The output was then tuned to compensate for the gain ripple and gain rolloff introduced by the noise tuning. Single stage amplifier performance is shown in Figures VI and VII. The 1 dB compression point at 38 GHz was +6 dBm, and the third order intercept point was typically +14 dBm.

A three stage amplifier was designed as a series of single ended cascaded stages as shown in Figure VIII. In order to optimize the amplifier's noise figure and gain an input isolator was not used. This same rationale led to the selection of a single-ended amplifier configuration so that quadrature hybrid losses associated with balanced stages could be avoided. Each stage was constructed and tuned seperately, and then mounted onto an aluminum supercarrier for installation into a hermetic receiver housing. Only moderate retuning of the first and second stage input circuitry was required after the pre-tuned stages were introduced into the multistage amplifier assembly. The output circuit of the third stage was retuned slightly to produce a flatter passband gain ripple.

Gain and noise figure of the multi-stage amplifier is shown in Figure IX. The gain was optimized and measured utilizing the standard hot/cold Y-factor method.

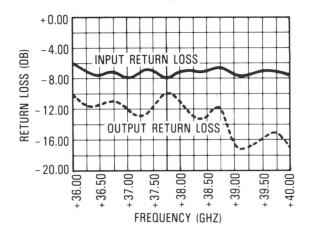

Figure VII. Input and output return losses for single stage amplifier.

149

Figure VIII. Three stage amplifier.

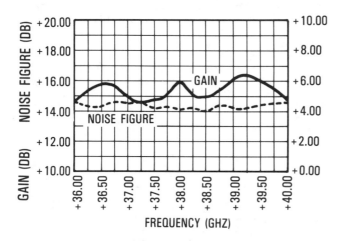

Figure IX. Noise figure and gain of three stage amplifier.

CONCLUSIONS

HEMT technology now makes possible receiver performance comparable to that currently achievable at Ku and K band frequencies. Combining low noise HEMT preamplifiers with planar construction techniques makes it possible to build high frequency high performance receivers that are low in cost, easily reproducible, light in weight, small size and with low power requirements.

Utilizing HEMT devices, the authors have constructed the first reported single and three stage amplifiers, at 36-40 GHz that surpasses that of GaAs FET technology.

Further refinements of HEMT technology directed at reducing device parasitcs will eventually allow even better sensitivity and higher gains than described in this paper. Reduction of gate metal resistance and source-gate resistance will result in HEMT device and amplifier noise figures that are lower than those reported here. In addition, higher frequencies of operation will be achieved with HEMT devices than GaAs MESFET'S because of its superior electron mobility and saturation velocity.

ACKNOWLEDGMENT

This work was performed at TRW in the Communications System Development Department under an Independent research and development program. The authors would like to thank Mr. Edward A. Freitas for the mechanical design of the carriers, housings and layouts.

REFERENCES

[1] J.J. Berenz, K. Nakano, K.P. Weller, "Low Noise High Electron Mobility Transistors," IEEE MTT-S Digest, pp83-86, San Fransisco, Ca. May 29-30, 1984.

[2] H. Fukui, "Low Noise Microwave Transistors and Amplifiers," IEEE Press pp 184-193, John Wiley and Sons, 1981.

[3] J.H.C. Van Heuven, "A New Integrated Waveguide Microstrip Transition," IEEE Transactions on Microwave Theory and Techniques, March 1976.

A 44 GHz HEMT AMPLIFIER*

Ting-Ih Hsu, Gerald Swift, Jitendra Goel, John Berenz, and Kenichi Nakano

*TRW Electronic Systems Group
One Space Park
Redondo Beach, CA 90278*

Received April 4, 1986

ABSTRACT

A 44-GHz amplifier using 0.25-μm gate length and double-heterojunction structure HEMT devices is described. Higher gain and power performance have been obtained from the amplifier using this device at millimeter-wave frequencies. A spot gain of 9.4 dB and a 1-dB gain compression point of +7.5 dBm has been achieved at 43.5 GHz.

1. Introduction

High electron mobility transistors (HEMTs) have recently achieved superior lower noise figures both in microwave and millimeter-wave frequency ranges [1-3], but have not yet shown power handling capability. A limitation in using HEMTs in medium-power amplifier applications is that relatively lower power capability of the present single heterojunction transistor has been encountered in comparison with GaAs FETs. It is known that the maximum obtainable saturated drain-source current sets an upper limit on the power output. The primary factor determining the maximum obtainable drain-source current for a single heterojunction transistor is the number of electrons in the two-dimensional electron gas.

To overcome this problem, multiple-channel structure HEMTs have been reported in the frequency range below 30 GHz [4,5]. In this paper, we will describe a 44-GHz HEMT amplifier which has achieved a 9.5-dB gain and a 1-dB gain compression point of +7.5 dBm. This improved performance results from using double heterojunction HEMTs and reducing the source-gate resistance. The HEMT device structure, amplifier development, and performance will also be described.

2. Device Description

The device using the epitaxial double heterojunction structure was fabricated to improve the current handling capability. A cross section of this double heterojunction structure is shown in Figure 1. This structure has a 300 Å undoped GaAs conductive channel which has electrons supplied from doped AlGaAs layers on both sides of the channel. As a result, approximately twice the sheet electron concentration of the quantum well with one-sided doping is achieved, as indicated by the Hall measurement results presented in Table 1.

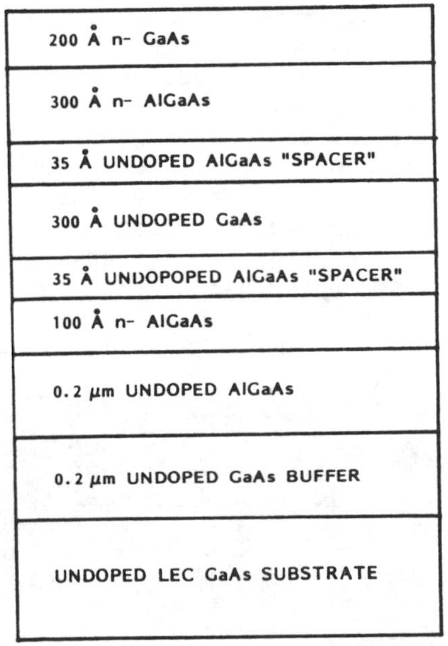

| 200 Å n- GaAs |
| 300 Å n- AlGaAs |
| 35 Å UNDOPED AlGaAs "SPACER" |
| 300 Å UNDOPED GaAs |
| 35 Å UNDOPED AlGaAs "SPACER" |
| 100 Å n- AlGaAs |
| 0.2 μm UNDOPED AlGaAs |
| 0.2 μm UNDOPED GaAs BUFFER |
| UNDOPED LEC GaAs SUBSTRATE |

Figure 1. Cross section of double heterojunction HEMT device

Table 1. Hall measurement results for different HEMT structures

RUN NO.	STRUCTURE	$\mu(300°K)$ cm2/V-SEC	$n(300°K)$ ELECTRON/cm^2	$\mu(77°K)$ cm^4/V-SEC	$n(77°K)$ ELECTRON/cm^2
2167	SINGLE HETERO-JUNCTION HEMT	7340	1.1 E12	56800	9.2 E11
2409	DOUBLE HETERO-JUNCTION HEMT	5464	2.7 E12	36440	1.6 E12

Reprinted with permission from the *International Journal of Infrared & Millimeter Waves,* vol. 7, no. 7, pp. 999–1004, 1986.

The HEMT devices were fabricated with 0.25-μm gate length and 80-μm total gate width. The same previously successful basic interdigitated device geometry was also used in this work [1,2]. The ohmic contact was alloyed using nickel, gold-germanium, nickel, and gold contact metals at temperatures above 450°C; the ohmic contact resistance was significantly reduced. Typical I-V characteristics of these HEMT devices are shown in Figure 2. Devices from wafer 2420 have a dc transconductance of 250 ms/mm and an I_{dss} of 28 mA. It has also been observed that the breakdown voltage of these double heterojunction devices is twice that of the single heterojunction devices; this is well suited for the power applications.

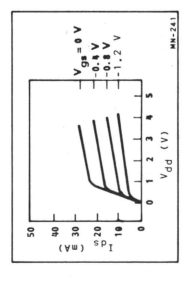

Figure 2. IV characteristics of HEMT wafer #2420

3. Amplifier Development and Performance

A 44-GHz HEMT amplifier has been constructed using MIC technology; microstrip was chosen as the transmission medium. The amplifier hardware consists of the split-block (Figure 3) and provides rectangular waveguides at both the input and output ports. The field converts from the quasi-TEM mode of the microstrip to TE_{10} mode of the rectangular waveguide by using a smooth taper-type transition on the substrate; low insertion loss, low VSWR, and wideband characteristics have been achieved.

The amplifier circuit consists of the above-mentioned waveguide-to-microstrip transition circuit, 50-ohm microstrip line, and bias circuits at both input and output ports. The HEMT device operates in the common source configuration, and

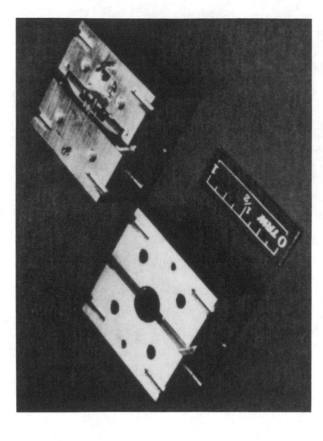

Figure 3. Photo of 44 GHz HEMT amplifier circuit

the RF input and output ports are extracted from the gate and drain of the HEMT, respectively. Bias is fed through the lowpass filter circuit etched on the substrate. Furthermore, the combination of chip resistor and capacitor is employed to provide low frequency stability. A 0.010-inch thick quartz substrate is used; the HEMT device is mounted on a machined ridge located between the ends of two microstrip sections, so the wire bonding effect on the device can be reduced. Tuning patterns and matching circuits on the substrate are obtained experimentally, permitting rapid amplifier development. HEMT amplifiers with the double heterojunction structure devices have been developed. The bias condition and gain performance of a typical HEMT amplifier are shown in Figure 4. When the amplifier was biased as V_{dd} = 4 V, V_{gg} = -0.4 V, and I_{ds} = 21.5 mA, a gain of 9.5 dB and its 1-dB gain compression point of +7.5 dBm have been achieved at 43.5 GHz. Table 2 also lists RF performance results of the HEMT with different structures.

4. Conclusions

The current handling capability of the HEMT device with the single heterojunction structure has been increased by

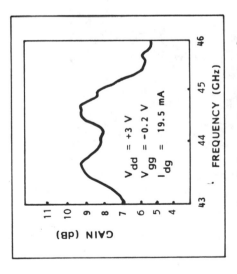

Figure 4. Gain performance of double-heterojunction HEMT amplifier

Table 2. Measured RF performance of HEMT with different structures

WAFER NO.	GAIN (dB)	P (1 dB GAIN COMPRESSION) (dBm)	η (EFFICIENCY) (%)	FREQUENCY (GHz)	STRUCTURE
2241-4	5.6	2.5	4.1	44.5	SINGLE-HETERO-JUNCTION HEMT
2420-2	9.5	7.5	5.5	43.5	DOUBLE HETERO-JUNCTION HEMT

using the double heterojunction structure. In addition, the source-gate resistance of the HEMT has been significantly reduced by alloying gate metallization at temperatures higher than 450°C. Higher gain and power have been obtained with the HEMT amplifier using this double heterojunction structure. The power capability of HEMTs can be further improved by increasing the gate periphery.

References

1. J.J. Berenz, K. Nakano, and K.P. Weller,"Low Noise High Electron Mobility Transistors," 1984 IEEE Microwave and Millimeter-Wave Circuits Symposium Digest, pp. 83-86.

2. J.J. Berenz, "High Electron Mobility Transistors," 1984 IEEE MTT-S Newsletter, Summer 1984, pp. 43-52.

3. J. Berenz, K. Nakano, T. Sato, and K. Fawcett, "Modula-tion-Doped FET DCML Comparator," Electronics Letters, vol. 21, no. 6, 14 March 1985, pp. 242-243.

4. K. Hikosaka, Y. Hirachi, T. Mimura, and M. Ake, "A Micro-wave Power Double-Heterojunction High Electron Mobility Transistor," IEEE Electron Device Letters, vol. EDL-6, no. 7, pp. 341-343, July 1985.

5. N. Sheng, C. Lee, R. Chen, D. Miller, and S. Lee, "Multiple-Channel GaAs/AlGaAs High Electron Mobility Transistor," IEEE Electron Device Letters, vol. EDL-6, no. 6, pp. 307-310, June 1985.

*This work was supported by the Air Force Space Division under Contract No. F04701-84-C-0113.

HEMT 60 GHz AMPLIFIER

Indexing terms: Microwave devices and components, Transistors

A high electron mobility transistor (HEMT) amplifier has been fabricated which exhibits 7·5 dB gain at 61 GHz. This result was obtained with a quarter-micrometre gate-length depletion-mode HEMT. Reduction of source-gate resistance and gate length are primarily responsible for this performance. The letter describes the materials and device processing technology developed for fabricating these devices.

Recent advances in high electron mobility transistor (HEMT) fabrication combined with hybrid MIC amplifier development have lead to state-of-the-art amplifier performance at 60 GHz. The gain of a quarter-micrometre gate-length depletion-mode HEMT has exceeded the gain of quarter- and subquarter-micrometre gate-length GaAs MESFETs at this frequency.[1,2] Reduction of source-gate resistance and gate length are primarily responsible for this result. This letter describes the materials and device processing technology developed for fabricating these devices.

Device description: The same basic interdigitated device geometry successfully used at frequencies up to 45 GHz was also used in this work.[3,4] It consists of two gate fingers, each 40 μm wide, sharing a common drain contact. Oxygen ion implantation was used for device active-area definition. Improvements have primarily been made in the epitaxial layer structure and ohmic contact to reduce source-gate resistance. The source-gate resistance consists of two components: the channel access resistance and the ohmic contact resistance. The channel access resistance has been effectively reduced by increasing the thickness of the n^+ GaAs cap layer to 500 Å. The ohmic contact resistance has been significantly reduced by alloying the nickel, gold-germanium, nickel and gold contact metals at temperatures greater than 450°C. DC transconductance up to 325 mS/mm and source-gate resistance as small as 0.20 ± 0.05 Ω mm has been achieved using this approach.

The drain I/V characteristics of the depletion-mode device used to obtain the 60 GHz amplifier performance is shown in Fig. 1. The DC transconductance in this case is 200 mS/mm. An undoped aluminium arsenide (AlAs) stop-etch layer was

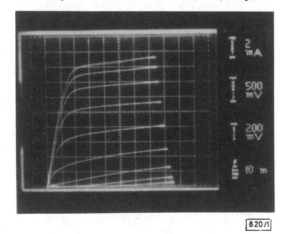

Fig. 1 *Drain I/V characteristic of HEMT*

incorporated in the epitaxial structure to control the depth of

the gate recess achieved by selective etching.[5] This stop-etch layer increases both the source-gate resistance and the output resistance in comparison to structures in which it is omitted.

Fig. 2 *HEMT 60 GHz amplifier circuit*

Amplifier performance: A single-stage amplifier has been constructed using hybrid MIC fabrication technology for device evaluation at 60 GHz. The circuit is shown in Fig. 2. It consists of low-loss, broadband finline waveguide-to-microstrip transitions fabricated on quartz substrates. The device is mounted on a ridge situated between the ends of the two microstrip-line sections. Bias is supplied through lowpass filter sections engraved on the quartz. Approximate input and output matching networks are constructed by bonding gold ribbons from the tuning pads to the microstrip lines. The response of the amplifier for device bias conditions of $V_g = -0.6$ V, $V_d = 2.5$ V and $I_d = 12.2$ mA is shown in Fig. 3. 7·5 dB amplifier gain was produced at 61 GHz. This gain includes losses in the amplifier circuit. The amplifier was stable; the device was not oscillating during these measurements.

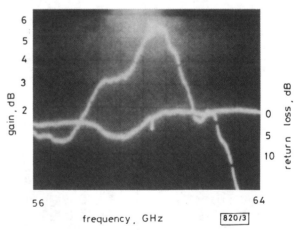

Fig. 3 *HEMT 60 GHz amplifier response curves*

Curve not corrected for losses in test set

Conclusions: The first practical HEMT devices have been fabricated for operation at 60 GHz. The gain achieved from these quarter-micrometre gate-length depletion-mode HEMT exceeds that reported for GaAs MESFETs at this frequency. Further improvement of performance is anticipated with narrower gate length and improved gate parasitics. However, the

ELECTRONICS LETTERS   24th October 1985   Vol. 21   No. 22

potential for useful millimetre-wave amplification with HEMTs has been demonstrated by these results.

Acknowledgments: This work was sponsored by NASA Goddard Space Flight Center under contract NAS5-28587: Mr. Robert Jones, Technical Program Monitor.

J. BERENZ   *5th September 1985*
K. NAKANO
TING-IH HSU
J. GOEL

TRW Electronic Systems Group
One Space Park
Redondo Beach, CA 90278, USA

References

1 WATKINS, E. T., SCHELLENBERG, J. M., HACKETT, L. H., YAMASAKI, H., and FENG, M.: 'A 60 GHz GaAs FET amplifier'. 1983 MTT-S digest, pp. 145–147
2 KIM, B., TSERNG, H. Q., and SHIH, H. D.: 'Millimeter-wave GaAs FETs prepared by MBE', *IEEE Electron Device Lett.*, 1985, **EDL-6,** pp. 1–2
3 BERENZ, J. J., NAKANO, K., and WELLER, K. P.: 'Low noise high electron mobility transistors'. 1984 IEEE microwave and millimeter-wave circuits symposium digest, pp. 83–86
4 BERENZ, J. J.: 'High electron mobility transistors'. 1984 IEEE MTT-S newsletter, Summer 1984, pp. 43–52
5 BERENZ, J., NAKANO, K., SATO, T., and FAWCETT, K.: 'Modulation-doped FET DCML comparator', *Electron. Lett.*, 1985, **21,** pp. 242–243

94 GHz TRANSISTOR AMPLIFICATION USING AN HEMT

Indexing terms: Semiconductor devices and materials, Transistors, Amplifiers

Transistor amplification at 94 GHz has been demonstrated for the first time. A single-stage amplifier employing a high-electron-mobility transistor (HEMT) exhibits a small-signal gain of 3·6 dB and an output power of 3·4 mW with 2 dB gain.

Rapid progress in the development of short-gate-length GaAs field-effect transistors (FETs) and high-electron-mobility transistors (HEMTs) has pushed the operating frequencies of these two devices well into the millimetre-wave regime. 60 GHz amplifiers based on the FET[1-3] and the HEMT[3,4] have been reported. In this letter we present initial results obtained at 94 GHz with single-stage amplifiers utilising HEMTs.

Device description: The HEMTs used in this work take advantage of the enhanced mobility and velocity of electrons in the two-dimensional electron gas formed at a selectively doped GaAs/AlGaAs heterojunction. High performance has been achieved through improvements in the design of the device and material structure as well as through reduction of critical parasitic elements. Devices feature quarter-micrometre-long gates with large cross-sectional area to minimise the gate series resistance.[5] The end-to-end resistance of these gate structures is 80 Ω/mm, the lowest value reported for 0·25 μm gates. Improvements in the ohmic contact metallurgy and alloying conditions have resulted in low contact transfer resistances—typically 0·05 Ω/mm, and as low as 0·03 Ω mm. A maximum extrinsic DC transconductance of 600 mS/mm has been measured for these HEMTs. The results reported in this letter were obtained for a device with a normalised transconductance of 440 mS/mm.

Fig. 1 *0·25 × 50 µm HEMT*

The device geometry consists of a single 50 μm gate stripe fed at the centre. As depicted in Fig. 1, multiple source wire-

bonds have been used to achieve low source inductance. Modelling of 2–20 GHz S-parameter data for devices connected in this fashion indicates an inductance of 10 pH.

Amplifier performance: A 94 GHz microstrip test fixture was developed for device evaluation. The fixture, shown in Fig. 2, utilises optimally shaped E-field probe transitions from waveguide to 50 Ω microstrip transmission lines fabricated on 0·13 mm-thick quartz substrates. With a through-connection, the end-to-end fixture loss (including 9 cm of WR-10 rectangular waveguide and 1 cm of microstrip) is 2·0 dB at 94 GHz, and a return loss of greater than 20 dB is observed

Fig. 2 *94 GHz test fixture*

from 93 to 95 GHz. Although the bandwidth is narrow, the centre frequency can be varied over a wide range by adjusting the micrometers that control the position of the waveguide backshorts.

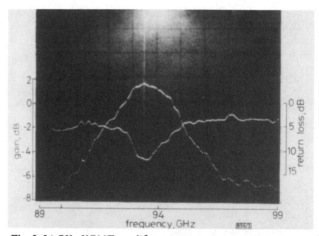

Fig. 3 *94 GHz HEMT amplifier response*

A single-stage amplifier was constructed, where impedance matching of the device input and output was achieved by empirically altering the metal microstrip pattern on the quartz substrates and tuning the backshort positions. The amplifier response shown in Fig. 3, not corrected for test fixture dissipative loss, was obtained with the HEMT biased at $V_{ds} = 4·0$ V and $I_d = 13$ mA. The power saturation characteristic of this amplifier is plotted in Fig. 4. The output power saturates at more than 5 mW, and an output power of 3·4 mW with

2 dB power gain is observed.

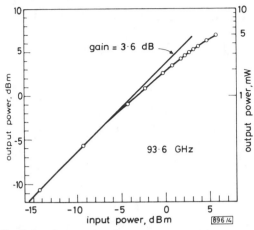

Fig. 4 *Power saturation characteristic*

Conclusion: Transistor amplification at 94 GHz has been demonstrated for the first time. A single-stage HEMT amplifier has exhibited a gain of 3·6 dB and an output power of 3·4 mW with 2 dB gain. Although this performance is modest, this work clearly demonstrates the potential of the HEMT for useful transistor amplification at frequencies up to 100 GHz. With improvements arising from a reduction in gate lenth, further optimisation of the device structure and reduction of parasitic elements, greatly improved 94 GHz performance can be expected in the near future.

Acknowledgment: This work was sponsored by AFWAL Wright–Patterson AFB under contract F33615-84-C-1472.

P. M. SMITH
P. C. CHAO   *5th June 1986*
K. H. G. DUH
L. F. LESTER
B. R. LEE

Electronics Laboratory
General Electric Company
Syracuse, NY 13221, USA

References

1 WATKINS, E. T., SCHELLENBERG, J. J., HACKETT, L. H., YAMASAKI, H., and FENG, M.: 'A 60 GHz FET amplifier'. MTT-S digest, 1983, pp. 145–147
2 KIM, B., TSERNG, H. Q., and SHIH, H. D.: 'Millimeter-wave GaAs FETs prepared by MBE', *IEEE Electron Device Lett.*, 1985, **EDL-6**, pp. 1–2
3 SMITH, P. M., CHAO, P. C., MISHRA, U. K., PALMATEER, S. C., DUH, K. H. G., and HWANG, J. C. M.: 'Millimeter-wave power performance of 0·25 µm HEMTs and GaAs FETs'. Proc. of IEEE/Cornell conf. on advanced concepts in high speed semiconductor devices and circuits, Ithaca, NY, 1985, pp. 189–194
4 BERENZ, J., NAKANO, K., HSU, T.-I., and GOEL, J.: 'HEMT 60 GHz amplifier', *Electron. Lett.*, 1985, **21**, pp. 1028–1029
5 CHAO, P. C., SMITH, P. M., PALMATEER, S. C., and HWANG, J. C. M.: 'Electron-beam fabrication of GaAs low-noise MESFETs using a new tri-layer resist technique', *IEEE Trans.*, 1985, **ED-32**, pp. 1042–1046

FET's and HEMT's at Cryogenic Temperatures—Their Properties and Use in Low-Noise Amplifiers

MARIAN W. POSPIESZALSKI, SENIOR MEMBER, IEEE, SANDER WEINREB, FELLOW, IEEE, ROGER D. NORROD, MEMBER, IEEE, AND RONALD HARRIS

Abstract —This paper reviews the performance of a number of FET's and HEMT's at cryogenic temperatures. Typical dc characteristics and X-band noise parameters are presented and qualitatively correlated wherever possible with other technological or experimental data. While certain general trends can be identified, further work is needed to explain a number of observed phenomena. A design technique for cryogenically cooled amplifiers is briefly discussed, and examples of realizations of L-, C-, X-, and K-band amplifiers are described. The noise temperature of amplifiers with HEMT's in input stages is usually less than half of that for all-FET realizations, setting new records of performance for cryogenically cooled, multistage amplifiers.

I. INTRODUCTION

THE FEASIBILITY of cooling GaAs FET amplifiers has been very well documented [1], [2]. Recently, very low noise temperatures for cryogenically cooled HEMT's have also been reported [3], [4], [34]. It has long been recognized, however, that the cryogenic performance of both HEMT's and FET's may not be inferred from room-temperature performance. In fact, for both HEMT's and FET's several different phenomena were observed which render some devices useless for cryogenic applications. Besides low-noise applications, the emerging importance of cooling of HEMT's and FET's for high-speed applications makes it important at least to list the observed phenomena, even without full explanation. While certain general trends can be identified, their explanation is not possible, even qualitatively, without detailed knowledge of the device structure and processing. For obvious reasons, this information is not always easily available. This paper is, therefore, intended to provide a comprehensive view to what is observed in HEMT's and FET's at cryogenic temperatures. The second section deals with FET's, the third with HEMT's, the fourth describes briefly a computer-aided design technique of cryogenically cooled amplifiers and gives examples of amplifiers built with cryogenically well-behaved FET's and HEMT's for L- through K-band frequency range.

Manuscript received April 24, 1987; revised September 15, 1987.

The authors are with the National Radio Astronomy Observatory, Charlottesville, VA 22903. The observatory is operated by Associated Universities, Inc. under contract with the National Science Foundation.

IEEE Log Number 8718363.

II. FET's

Sample transistors of the following types have been tested at cryogenic temperatures: MGF1412, MGF1405 (Mitsubishi), NE75083, NE04583, NE71083 (NEC), FSC10FA, FSX02FA (Fujitsu), 2SK525 (Sony). While the choice of transistors is hardly balanced, it follows the past history of good cryogenic performance [1], [2], [18]–[20].

Table I presents the comparison of noise performance of FET's that were found most useful for cryogenic application. Table II gives the room-temperature noise parameters of the same FET's. All were measured by the method described in [12] with estimated accuracy of T_{min}: ±9 K and ±1.5 K, R_{opt}: ±1.5 Ω and ±0.7 Ω, X_{opt}: ±4 Ω, g_n: ±1.5 mS and ±0.7 mS, at 297 K and 12.5 K, respectively. A description of the test fixture used both for dc and noise measurement is given in [3], [4], and [13]. Although the data of Tables I and II are self-explanatory, the following comments could be useful.

There appears to be no correlation between room- and cryogenic-temperature values of minimum noise temperature T_{min}. It can be explained [1] by a large difference in the relative contribution of thermal noise in parasitic resistances to the value of T_{min} at room and cryogenic temperatures. However, as subsequently demonstrated, large differences between cryogenic dc characteristics for different FET's which are not understood should be considered another contributing factor.

The spread in minimum noise temperature between transistors of the same type but from different lots is much greater than from within the same lot. In fact, in the latter case the spread may be as low as 2 K. For repeatable cryogenic performance of amplifiers, it is always useful to use transistors from the same lot.

Relatively poor cryogenic noise performance for FET's with otherwise orderly behavior can usually be traced to the poor pinch-off characteristic at the cryogenic temperatures, not necessarily noticeable at the room temperature. As an example, a comparison of room- and cryogenic-temperature characteristics of two NE75083 FET's having very different noise performance at cryogenic temperatures (15 K and 23 K at 8.5 GHz) is shown in Fig. 1. Note the small differences in room-temperature characteristics, as

Reprinted from *IEEE Transactions on Microwave Theory and Techniques*, vol. 36, no. 3, pp. 552–560, March 1988.

TABLE I
NOISE PERFORMANCE COMPARISON OF BEST GaAs FET'S AT 8.5 GHz AND 12.5 K

TABLE I
NOISE PERFORMANCE COMPARISON OF BEST GaAs FET'S AT 8.5 GHz AND 12.5 K

| TYPE | GATE | | BIAS | | NOISE PARAMETERS OF SAMPLE FET | | | | | RANGE OF T_{min} [K] | | ASSOC. |
	L μm	W μm	V_{ds} V	I_{ds} mA	T_{min} K	R_{opt} Ω	X_{opt} Ω	g_n mS		MIN	MAX	GAIN dB
MGF1412	.7	400	4.5	10	20	7.1	38	3.7		18	26	12
FSC10FA	.5	400	3	10	20	3.6	32	6.6		15	24	9
NE75083*	.3	300	3	10	15	4.5	32	4.3		15	23	11.1
FSX02FA		200	2	5	17	5.9	24	4.5		-	-	11.5
NE04583	.3	200	1.5	5	19	3.7	42	6.1		-	-	11.4

*NE75803 production has been discontinued.

TABLE II
ROOM TEMPERATURE NOISE PERFORMANCE OF FET'S OF TABLE I AT 8.5 GHz

Device	V_{ds} V	I_{ds} mA	T_{min} K	R_{gopt} Ω	X_{gopt} Ω	g_n mS	G_{AS} dB
MGF1412	3	10	122	13.4	40	11.5	9.0
FSC10FA	2.5	10	125	10.7	33	12.8	7.3
NE75083	3	10	89	9.4	32	8.4	9.7
FSX02FA	2.5	10	94	9.4	23	10.3	10.3
NE04583	3	10	84	8.2	42	12.1	10.3

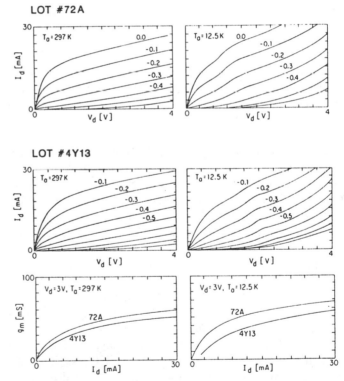

Fig. 1. The dc characteristic at room and cryogenic temperatures of two NE75083 FET's of the same type with different cryogenic noise performance at 8.5 GHz. Lot #72A FET exhibited $T_{min} = 15$ K. Lot #4Y13 FET exhibited $T_{min} = 23$ K.

opposed to large differences at 12.5 K between these two transistors. Worse pinch-off characteristics for the lot 4Y13 in comparison with lot 72A can be qualitatively explained by different crystal quality and/or electrical characteristics of the interface region between the semi-insulating buffer layer and the active layer. The importance of the quality of the interface for low-noise room-temperature operation has been known [22], [23]. It appears from this example that the cryogenic noise performance is much more sensitive to characteristics of the interface region than the room-temperature performance.

For most of the FET's, the dc measured transconductance g_m does not vary appreciably upon cooling. It usually goes up by about 20 percent of the room-temperature value. However, examples can be found to the contrary, as is shown in Fig. 2, where dc measured transconductance g_m is plotted as a function of ambient temperatures for the transistor samples of Tables I and II. A notable exception is the MGF1412 transistor (Fig. 2), where an increase in

159

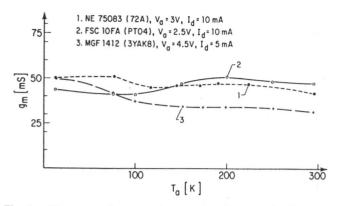

1. NE 75083 (72A), V_a = 3V, I_d = 10 mA
2. FSC 10FA (PT04), V_a = 2.5V, I_d = 10 mA
3. MGF 1412 (3YAK8), V_a = 4.5V, I_d = 5 mA

Fig. 2. The measured transconductance as a function of ambient temperature for some FET samples of Tables I and II.

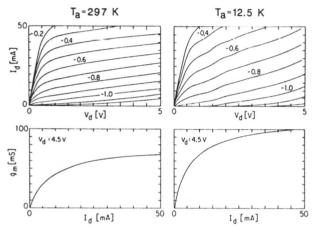

Fig. 3. Typical dc characteristics at room and cryogenic temperatures of MGF1412 FET (3YAK8) exhibiting T_{min} = 20 K at f = 8.5 GHz and T_a = 12.5 K.

transconductance as large as 50 percent of the room-temperature value was observed; this is about the same as for good cryogenic HEMT's (30 to 60 percent increase [3], [4]). An example of the dc characteristic of the MGF1412 FET is shown in Fig. 3. For all FET's, together with an increase in transconductance, a reduction of small-signal shunt drain resistance is observed (Figs. 1 and 3).

Theoretical studies [5]–[7] predict linear dependence of the minimum noise temperature on the gate length, even for submicron gate devices [6], [7]. In this light, comparison of the gate dimensions (published by the manufacturer) with the noise performance of the best FET's (Table I), which reveals no correlation, indicates the importance of quality and/or structure of epitaxial GaAs from the point of view of cryogenic applications. In particular, the superb quality of MGF1412 epitaxial GaAs and/or structure is quite apparent. The cryogenic dc data, notably the $g_m = f(I_{ds})$ characteristic, also support this conclusion. Bearing in mind that the existence of trapping centers, especially at the channel substrate interface, could greatly influence the properties of cryogenic FET's, it is interesting to point out that MGF1402/12 FET's have been found to exhibit the lowest $1/f$ noise spectra at room temperature [24], also attributed to the existence of traps.

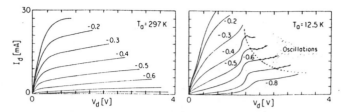

Fig. 4. Example of unusual distortion of I–V characteristics at cryogenic temperatures observed for NE71083 FET's.

Not all transistors present such an orderly behavior when cooled as those in Figs. 1 and 3. Unusual distortion of the I–V characteristic (example of NE71083 is shown in Fig. 4) or oscillations which are most likely caused by the Gunn type instabilities in GaAs (a number of MGF 1405 transistor samples) are sometimes present. Other relevant data can be found in [14].

III. HEMT's

The sample HEMT's of the following type have been tested: H-503-P70 (Gould-Dexel), JS8901-AS (Toshiba), 2SK676 (Sony), and FHR01FH (Fujitsu). Samples of their dc characteristics at room and cryogenic temperatures are shown in Figs. 5 and 6, and a comparison of their noise performance is given in Table III. All these HEMT's had to be illuminated with light to be time-invariant, memoryless devices at cryogenic temperatures. A red-light-emitting diode has been used for illumination. With the notable exception of FHR01FH, all failed to pinch off properly at the cryogenic temperatures, exhibiting a large portion of drain current not controlled by the gate voltage. Judging from the available manufacturer information [8]–[10], this effect seems to be linked to the very high doping (2–3×10^{18} cm^{-3}) of the AlGaAs layer. Devices with moderate doping, as reported in [3] and [34], do not exhibit this effect.

The dc characteristics at 12.5 K have been measured also for a dark Fujitsu HEMT, which was previously illuminated, and are shown in Fig. 6. In this case, it is important to specify the measuring procedure, since in the dark the device possesses memory and is not time-invariant, as indicated earlier. The I–V characteristics of Fig. 6 were taken stepping the drain voltage from 0 to 2.5 V for each of the gate voltages, starting at $V_{gs} = -0.9$ V. It is shown here to demonstrate that the illumination greatly affects the measured characteristics at relatively small drain voltages, while at $V_{ds} = 2.5$ V the effect of illumination becomes insignificant. The "wiggles" of I–V characteristics of an illuminated HEMT ($V_d < 0.75$ V) have been determined to be the result of an insufficient amount of light reaching the intrinsic HEMT (hence, the HEMT is not a memoryless device in this region), as they were taken for packaged HEMT. Care was taken to eliminate parasitic oscillations as a probable cause of this effect. Also, sample characteristics of an FHR01X chip, sufficiently illuminated and dark, were taken and are shown in Fig. 7.

The observed dc behavior of the Fujitsu HEMT at 12.5 K is qualitatively the same as recently reported in [25] and is linked to the existence of DX centers in doped AlGaAs.

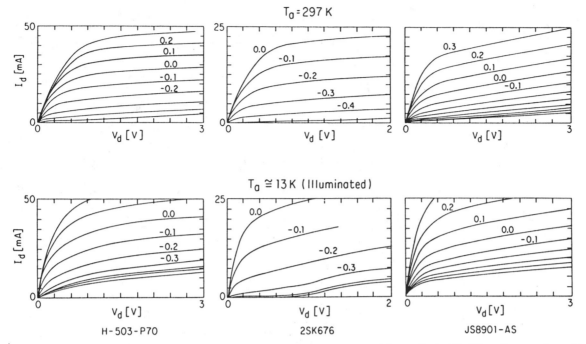

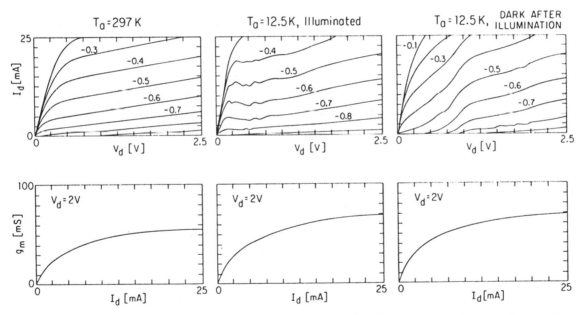

Fig. 5. Examples of room and cryogenic temperature characteristics of commercially available HEMT's. All were taken with light illumination.

Fig. 6. The dc characteristic of FHR01FH HEMT with excellent cryogenic noise performance, demonstrating the effect of light illumination. Note large influence of light at small drain voltages. (Compare Fig. 7.)

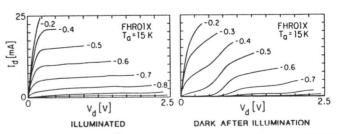

TABLE III
CRYOGENIC NOISE PERFORMANCE OF SAMPLE HEMT's AT
$T_a = 12.5$ K, $f = 8.5$ GHz

DEVICE	H503-P70	JS8901-AS	2SK676	FHR01FH
T_{MIN} [K]	25	27	14	9.4

Fig. 7. The comparison of dc characteristics at 12.5 K for FHR01X chip, for which the amount of light reaching AlGaAs layer was sufficient. (Compare Fig. 6.)

(a)

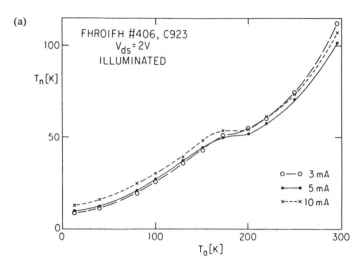

(b)

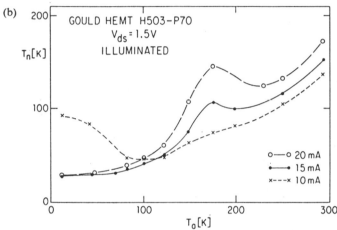

Fig. 8. The noise temperature of a single-stage (a) FHR01FH and (b) H-503-P70 amplifiers versus ambient temperature. Source impedance is optimal for cryogenic temperature.

TABLE IV
NOISE PARAMETERS OF FHR01FH HEMT AT $f = 8.5$ GHz

$T_a = 12.5$ K

I_{ds} mA	V_{ds} V	T_{min} K	R_{gopt} Ω	X_{gopt} Ω	g_n mS	Assoc. Gain dB
3	2	9.4	4.3	18.3	2.7	11.8
5	2	10.3	4.6	17.0	2.6	13.1
10	2	13.0	5.3	16.3	2.7	13.4
15	2	16.3	5.7	15.6	3.2	13.7

$T_a = 297$ K

I_{ds} mA	V_{ds} V	T_{min} K	R_{gopt} Ω	X_{gopt} Ω	g_n mS	Assoc. Gain dB
5	2	81	8.6	18.4	12.5	10.8
10	2	78	10.5	17.5	9.4	11.7
15	2	81	11.0	16.7	10.1	12.5

For the illuminated HEMT's, the dc and noise data taken at a number of different temperatures between 297 K and 12.5 K reveal the same qualitative behavior and sensitivity to light as reported in [3]. This has again been linked to the existence of traps but, as yet, is not fully understood. As an example, the dependences of the noise temperature T_n of the single-stage 8.5-GHz amplifier employing the Gould and Fujitsu HEMT's measured with light illumination as a function of ambient temperature are shown in Fig. 8. Both characteristics exhibit a "hump" around $T_a \simeq 175$ K, which is very small for Fujitsu HEMT with excellent cryogenic performance, and very large for Gould HEMT with poor cryogenic performance.

The noise parameters of the FHR01FH HEMT both at room and cryogenic temperature are given in Table IV. The minimum noise temperature $T_{min} = 9$ K at 8.5 GHz and $T_a = 12$ K is, within the measurement error, equal to the values reported previously for GE HEMT's [3], [34]. However, recent measurements on experimental GE HEMT's of similar structure indicates $T_{min} = 5.5$ K at 8.5 GHz and $T_{min} = 1.4$ K at 1.4 GHz [35]. The noise temperature of FHR01FH was found to be almost independent of drain voltage between 2 V and 3 V, i.e., the region in which

the dependence of dc characteristics on illumination is very small (Fig. 6). However, if cooled down in the dark, it required illumination to achieve low-noise state. A typical difference in noise temperature prior to and after illumination was 3 K.

An interesting comparison of noise parameters of the Fujitsu HEMT and FET with the same gate periphery, measured at the same bias, is shown in Table V. Although the packages of both devices are different (FH and FA styles), the noise parameters T_{min} and $4NT_0$ remain the same for a chip with a source bond wire inductance only, as they are invariant upon lossless transformation at input and output. Both these parameters for a FHR01FH HEMT are approximately half those for a FSX02FA FET. The ratio $4NT_0/T_{min}$ is a measure of correlation between a pair of noise sources representing noise properties of a two-port [17]; it would be equal to one in the case of a perfect correlation. The ratio is usually larger for FET's than for HEMT's (compare Table V), indicating stronger correlation for the latter, which agrees with computer modeling of noise in HEMT's and FET's [6], [26].

IV. AMPLIFIER EXAMPLES

A number of amplifiers were constructed for radio astronomy applications, using both FET's and HEMT's with good cryogenic performance. It was found that in the

TABLE V
COMPARISON OF NOISE PARAMETERS OF FUJITSU HEMT AND FET

									Assoc.
	I_{ds}	V_{ds}	T_{min}	R_{gopt}	X_{gopt}	g_n	$4NT_0$	$\dfrac{4NT_0}{T_{min}}$	Gain
	mA	V	K	Ω	Ω	mS	K		dB
FHR01FH	3	2	9.4	4.3	18.3	2.7	13.5	1.4	11.8
FSX02FA	3	2	17.0	4.9	24.5	5.1	29.0	1.7	10.8

T_a = 12.5 K, f = 8.5 GHz

$T_n = T_{min} + T_0(g_n/R_g)|Z_g - Z_{gopt}|^2$, $N = g_n R_{opt}$, $T_0 = 290$ K

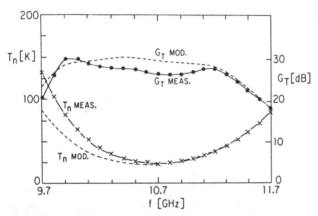

Fig. 9. Comparison of computer-predicted and measured noise temperature and gain of the 10.7 GHz, three-stage amplifier with NE75083 transistors at 12.5 K. All transistors are biased at $V_d = 3$ V, $I_d = 10$ mA. At 10.7 GHZ, $T_n = 26$ K.

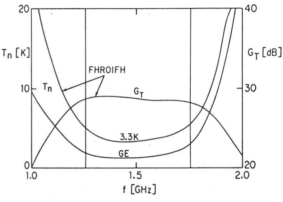

Fig. 10. Typical noise and gain characteristics of cryogenically cooled, three-stage, L-band amplifier with FHR01FH in input stage and MGF1412's in subsequent stages. The noise characteristic of an amplifier with an experimental GE HEMT in input stage is also shown for comparison. Design of similar all-FET amplifier has been described in [2] and [30].

process of computer-aided design of cryogenically cooled amplifiers with well-behaved FET's and/or HEMT's, the room-temperature S-parameter data and cryogenic noise parameter data can be used with accuracy sufficient for practical applications. The change in amplifier gain upon cooling can be accounted for by the changes in transconductance and small-signal drain resistance (compare Figs. 2 and 3).

The example of computer-predicted and measured cryogenic performance of an all-FET (NE75083), three-stage, 10.7-GHz amplifier is shown in Fig. 9 [29]. In this example, the following frequency dependence of the noise parameters of a chip was assumed:

$$T_{min} = Af \qquad R_{opt} = \frac{B}{f} \qquad X_{opt} = \frac{C}{f} \qquad g_n = Df^2 \quad (1)$$

and the constants A, B, C, and D were determined from the measurement at 8.5 GHz (Table I) by de-embedding the elements of the package equivalent circuit using computer routines [15], [16], [29]. The frequency dependence of the minimum noise temperature below 20 GHz given by (1) is confirmed by a number of studies [5]–[7], [26]–[28],

both for HEMT's and for FET's, while that of R_{opt}, X_{opt}, and g_n comes from the analysis by Pucel, *et al.* [5] for intrinsic chip (parasitic resistances excluded) under the approximation $\omega^2 C_{sg}^2 R_i^2 \ll 1$. This frequency dependence of the noise parameters is also the one preserving the fundamental inequality [17], [32]:

$$T_{min} \leqslant 4NT_0 \qquad \text{where } N = R_{opt}g_n \quad (2)$$

at all frequencies. It should be noted that for this model the ratio $4NT_0/T_{min}$ is not only invariant through lossless reciprocal two-ports at input and output, but is also frequency independent.

In Figs. 10 through 14, the typical characteristics of cryogenically cooled L-, C-, X-, and K-band amplifiers built with FET's and those built with Fujitsu HEMT's in input stages are compared. Noise temperature of L-band and X-band amplifiers built with experimental GE HEMT's is also shown for comparison. In general the HEMT amplifiers have noise temperature lower by a factor of two or more than all FET realizations. The minimum noise temperature in the band for C- through X-band

(a)

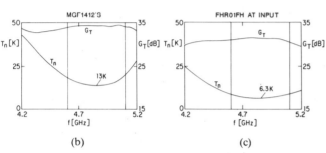

(b)   (c)

Fig. 11. Cryogenically cooled, three-stage, C-band amplifier. (a) Photograph. (b) Typical characteristics of all-FET realization (MGF1412's) [21]. (c) Typical characteristic for FHR01FH in input stage and MGF1412's in the following stages.

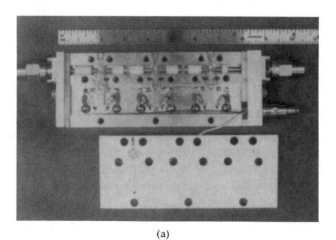

(a)

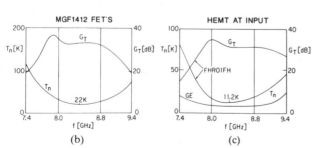

(b)   (c)

Fig. 12. Cryogenically cooled, three-stage, 8.4 GHz amplifier. (a) Typical characteristic for all-FET realization (MGF1412's [13]. (b) Typical characteristic for FHR01FH in input stage and MGF1412's in the following stages. The noise characteristics of an amplifier with an experimental GE HEMT is also shown for comparison.

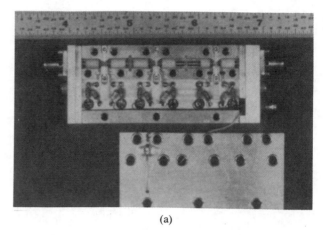

(a)

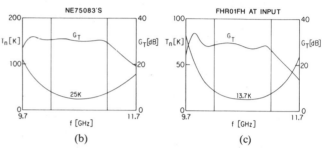

(b)   (c)

Fig. 13. Cryogenically-cooled, three-stage, 10.7 GHz amplifier. (a) Photograph. (b) Typical characteristic for all-FET realization (NE75083's). (c) Typical characteristics for FHR01FH in input stage with NE75083 in the following stages.

(a)

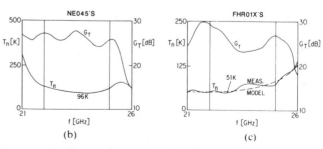

(b)   (c)

Fig. 14. Cryogenically cooled, four-stage, 23 GHz amplifier: (a) Photograph. (b) Typical characteristics for all-FET realization (NE045 chips). (c) Typical characteristics for FHR01X chips in first two stages and NE045 chips in the following stages. The noise and gain data are for the amplifier with cold isolators both at the input and output. The noise predicted by the computer model for two-stage FHR01X amplifier with cold isolator at the input is also shown for comparison.

realizations is approximately proportional to frequency with proportionality factor 1.3 K/GHz and 2.6 K/GHz for Fujitsu HEMT and FET realizations, respectively. This translates into 1.1 K/GHz and 2.2 K/GHz factors for the minimum noise temperature of FHR01FH HEMT and best FET, respectively, directly confirming the validity of expression (1) for T_{min}. The results at L-band slightly deviate from the proportional dependence, most likely due to the influence of $1/f$ noise. On the other hand, K-band results still closely follow it. For example, the noise temperature of a cold, single-stage, FHR01X microstrip amplifier at 22.5 GHz predicted using the noise and signal model based on room-temperature S-parameter measurement, X-band noise measurement, and relations (1) was 33 K, precisely as measured; for the chip with bond wire inductance only, the model predicts $T_{min} = 26$ K at this frequency. Furthermore, computer-predicted noise temperature for the two-stage, K-band FHR01X amplifier with isolator at the input is compared in Fig. 14 with the measured noise temperature of a four-stage experimental amplifier, demonstrating a good agreement.

More detailed information concerning the design, construction, and performance of some of the amplifier examples presented briefly in this paper can be found in a number of NRAO reports[1] and related papers [2], [13]–[16], [21], [29], [30].

IV. CONCLUSIONS

The properties of commercially available FET's and HEMT's at cryogenic temperatures have been described. Several phenomena were observed which await full explanation. In this light, testing of the devices at cryogenic temperatures could provide additional insight into the device physics. This would be of great importance not only for low-noise, but also for high-speed applications.

Noise testing of best FET's and an HEMT showed that the minimum noise temperature of Fujitsu HEMT at 12.5 K is approximately half of the best FET at any given frequency and for the C- to K-band frequency range is given by the approximate relation $T_{min} = 1.1 * f(\text{GHz})$ kelvins. There is, however, room for improvement—recent results with GE HEMT's [35] with structure similar to that published in [3] and [34] exhibited $T_{min} = 5.5$ K at $f = 8.5$ GHz and $T_a = 12.5$ K.

The examples of cryogenically cooled, multistage amplifiers built for radio astronomy applications from L- through K-band represent the current state of the art and at frequencies below X-band their performance is comparable to 4 K masers, with enormous reduction in cost, not only of the microwave parts but also of the cryogenic systems [33].

ACKNOWLEDGMENT

Part of this work was done under cooperative agreement between the National Radio Astronomy Observatory and the Jet Propulsion Laboratory to allow the use of the Very Large Array for the purpose of signal reception from Voyager during the Neptune flyby.

The excellent technical assistance of B. Lakatosh, K. Crady, and R. Simon of NRAO is gratefully acknowledged.

REFERENCES

[1] S. Weinreb, "Low-noise, cooled GASFET amplifiers," *IEEE Trans. Microwave Theory Tech.*, vol. MTT-28, pp. 1041–1054, Oct. 1980.

[2] S. Weinreb, D. Fenstermacher, and R. Harris, "Ultra low-noise, 1.2–1.7 GHz, cooled GASFET amplifiers," *IEEE Trans. Microwave Theory Tech.*, vol. MTT-30, pp. 849–853, June 1982.

[3] M. W. Pospieszalski, S. Weinreb, P. C. Chao, U. K. Mishra, S. C. Palmateer, P. M. Smith, and J. C. M. Hwang, "Noise parameter and light sensitivity of low-noise, high electron mobility transistors at 300 K and 12.5 K," *IEEE Trans. Electron Devices*, vol. ED-33, pp. 218–223, Feb. 1986.

[4] S. Weinreb and M. Pospieszalski, "X-band noise parameters of HEMT devices at 300 K and 12.5 K," in *Proc. 1985 Int. Microwave Symp.* (St. Louis, MO), June 1985, pp. 539–542.

[5] R. A. Pucel, H. A. Haus, and H. Statz, "Signal and noise properties of GaAs microwave FET," *Advances in Electronics & Electron Physics*, vol. 38, L. Morton, Ed. New York: Academic Press, 1975.

[6] B. Carnez, A. Cappy, R. Fauquembergue, E. Constant, and G. Salmer, "Noise modeling in submicrometer-gate FET's," *IEEE Trans. Electron Devices*, vol. ED-28, pp. 784–789, July 1981.

[7] T. M. Brookes, "The noise properties of high electron mobility transistors," *IEEE Trans. Electron Devices*, vol. ED-33, pp. 52–57, Jan. 1986.

[8] A. Swenson, J. Herb, and M. Yung, "First commercial HEMT challenges GaAs FET's," *Microwave & RF*, vol. 24, p. 107, Nov. 1985.

[9] K. Tanaka, H. Takakuwa, F. Nakamura, Y. Mori, and Y. Kato, "Low-noise microwave HIFET fabricated using photolithography and MOCVD," *Electron. Lett.*, vol. 22, pp. 487–488, Apr. 1986.

[10] Y. Kamei, H. Kawasaki, S. Hori, K. Shibata, M. Higashima, M. O. Watanabe, and Y. Ashisawa, "Extremely low-noise 0.25-µm-gate HEMT's," *Inst. Phys. Conf. Ser.*, no. 79, ch. 10, pp. 541–546, 1986.

[11] M. Iwakuni, M. Niori, T. Saito, T. Hamabe, H. Kurihara, K. Jyoshin, and M. Mikuni, "A 20 GHz Peltier-cooled, low-noise HEMT amplifier," in *Proc. IEEE-MTTS 1985 Int. Microwave Symp. Dig.*, (St. Louis, MO), June 1985, pp. 551–553.

[12] M. W. Pospieszalski, "On the measurement of noise parameters of microwave two-ports," *IEEE Trans. Microwave Theory Tech.*, vol. MTT-34, pp. 456–458, Apr. 1986.

[13] M. W. Pospieszalski, "Low-noise, 8.0–8.8 GHz, cooled, GASFET amplifier," NRAO Electronics Division Internal Report No. 254, National Radio Astronomy Observatory, Charlottesville, VA, Dec. 1984.

[14] M. W. Pospieszalski, "X-band noise performance of commercially-available GaAs FET's at room and cryogenic temperatures," NRAO Electronics Division Internal Report No. 260, National Radio Astronomy Observatory, Charlottesville, VA, May 1986.

[15] J. Granlund, "FARANT on the HP9816 Computer," NRAO Electronics Division Internal Report No. 250, National Radio Astronomy Observatory, Charlottesville, VA, Aug. 1984.

[16] D. L. Fenstermacher, "A computer-aided analysis routine including optimization for microwave circuits and their noise," NRAO Electronics Division Internal Report No. 217, National Radio Astronomy Observatory, Charlottesville, VA, July 1981.

[17] M. W. Pospieszalski and W. Wiatr, "Comment on 'Design of Microwave GaAs MESFET's for Broadband, Low-Noise Amplifiers'," *IEEE Trans. Microwave Theory Tech.*, vol. MTT-34, p. 194, Jan. 1986.

[18] G. Tomassetti, S. Weinreb and K. Wellington, "Low-noise, 10.7 GHz, cooled, GASFET amplifier," NRAO Electronics Division Internal Report No. 222, National Radio Astronomy Observatory, Charlottesville, VA, Nov. 1981.

[19] M. Sierra, "15-GHz, cooled GaAsFET amplifier—Design background information," NRAO Electronics Division Internal Report No. 229, National Radio Astronomy Observatory, Charlottesville, VA, June 1982.

[20] S. Weinreb and R. Harris, "Low-noise, 15 GHz, cooled GaAsFET amplifiers," NRAO Electronics Division Internal Report No. 235,

[1]National Radio Astronomy Observatory Electronics Division Internal Reports are distributed to interested persons upon request.

National Radio Astronomy Observatory, Charlottesville, VA, Sept. 1983.

[21] R. D. Norrod and R. J. Simon, "Low-noise, 4.8 GHz, cooled GaAs FET amplifier," NRAO Electronics Division Internal Report No. 259, National Radio Astronomy Observatory, Charlottesville, VA, Feb. 1986.

[22] T. M. Brookes, "Noise in GaAs FET's with a non-uniform channel thickness," *IEEE Trans. Electron Devices*, vol. ED-29, pp. 1632–1634, Oct. 1982.

[23] T. Suzuki, A. Nara, M. Nakatani, T. Ishi, S. Mitsui, and K. Shirahata, "Highly reliable GaAs MESFET's with static mean NF of 0.89 dB and a standard deviation of 0.07 dB at 4 GHz," in *Proc. 1979 MTT-S Int. Microwave Symp.* (Orlando, FL), May 1979, pp. 393–395.

[24] A. N. Riddle and R. J. Trew, "Low frequency noise measurement of GaAs FET's," in *Proc. 1986 Int. Microwave Symp.* (Baltimore, MD), June 1986, pp. 79–82.

[25] A. Kastalsky and R. A. Kiehl, "On the low temperature degradation of (AlGa)As/GaAs modulation doped field-effect transistors," *IEEE Trans. Electron Devices*, vol. ED-33, pp. 414–423, Mar. 1986.

[26] A. Cappy, A. Vanoverschelde, M. Schortgen, C. Versnaeyen, and G. Salmer, "Noise modeling in submicrometer-gate, two-dimensional, electron-gas, field-effect transistors," *IEEE Trans. Electron Devices*, vol. ED-32, pp. 2787–2795, Dec. 1985.

[27] H. Fukui, "Design of microwave GaAs MESFET's for broadband, low-noise amplifiers," *IEEE Trans. Microwave Theory Tech.*, vol. MTT-27, pp. 643–650, July 1979.

[28] H. Fukui, "Optimal noise figure of microwave GaAs MESFET's," *IEEE Trans. Electron Devices*, vol. ED-26, pp. 1032–1037, July 1979.

[29] M. W. Pospieszalski, "Design and performance of cryogenically-cooled, 10.7 GHz amplifiers," Electronics Division Internal Report No. 262, National Radio Astronomy Observatory, Charlottesville, VA, June 1986.

[30] S. Weinreb, D. Fenstermacher, and R. Harris, "Ultra low-noise, 1.2–1.7 GHz, cooled GASFET amplifiers," Electronics Division Internal Report No. 220, National Radio Astronomy Observatory, Charlottesville, VA, Sept. 1981.

[31] S. Weinreb, "Noise parameters of NRAO 1.5 GHz GASFET amplifiers," Electronics Division Internal Report No. 231, National Radio Astronomy Observatory, Charlottesville, VA, June 1983.

[32] W. Wiatr, "A method of estimating noise parameters of linear microwave two-ports," Ph.D. dissertation, Warsaw Technical University, Warsaw, Poland, 1980 (in Polish).

[33] S. Weinreb, R. Norrod, and M. W. Pospieszalski, "Compact cryogenic receivers for 1.3 to 43 GHz range," presented at the Cambridge Symposium No. 129, Smithsonian Astrophysical Observatory, Cambridge, MA, May 11–15, 1987.

[34] P. C. Chao, P. M. Smith, U. K, Mishra, S. C. Palmateer, J. C. M. Hwang, M. W. Pospieszalski, T. Brooks, and S. Weinreb, "Cryogenic noise performance of quarter-micrometer gate-length high-electron-mobility transistors," *IEEE Trans. Electron Devices*, vol. ED-32, p. 2528, Nov. 1985.

[35] K. H. G. Duh, M. W. Pospieszalski, W. F. Kopp, P. Ho, A. Jabra, P. C. Chao, P. M. Smith, L. F. Lester, J. M. Ballingall, and S. Weinreb, "Ultra-low-noise cryogenic high-electron-mobility transistors," *IEEE Trans. Electron Devices*, vol. ED-35, pp. 249–256, Mar. 1988.

EXTREMELY HIGH GAIN, LOW NOISE InAlAs/InGaAs HEMTs GROWN BY MOLECULAR BEAM EPITAXY

P. Ho, P.C. Chao, K.H.G. Duh, A.A. Jabra,
J.M. Ballingall and P.M. Smith

Electronics Laboratory, General Electric Co.,
Syracuse, New York 13221

ABSTRACT

High performance InAlAs/InGaAs planar-doped HEMTs lattice-matched to InP have been fabricated with a 0.25 μm T-gate. A maximum extrinsic transconductance g_m of 900 mS/mm, corresponding to an intrinsic g_m of 1640 mS/mm, was obtained at room temperature. There exist no kinks in the I-V characteristics as have been observed previously. RF measurements at 18 GHz yielded a minimum noise figure of 0.5 dB with an associated gain of 15.2 dB and a maximum stable gain, MSG, of 20.9 dB. This is the best 18 GHz noise performance ever reported. The MSG vlaue is the highest reported for any HEMT. At 58 GHz, the devices exhibited 1.2 dB minimum noise figure with 8.5 dB associated gain. Both our 18 GHz and 58 GHz minimum noise figures correspond well with the Fukui-type frequency dependence exhibited by other HEMT material systems and FETs fabricated in our laboratory. At 63 GHz, a maximum available gain MAG of 15.4 dB was measured from a single-stage amplifier. This value, extrapolated at -6 dB/octave, yielded a maximum frequency of oscillation f_{max} of 380 GHz which is the highest f_{max} ever reported for any transistor. A three-stage HEMT amplifier exhibited an average noise figure of 3.0 dB with a gain of 22.0±0.2 dB from 60-65 GHz. This is the best result ever reported for three-stage HEMT LNA.

INTRODUCTION

Due to the large conduction band discontinuity between $In_{.52}Al_{.48}As$ and $In_{.53}Ga_{.47}As$ and the superb carrier transport properties in the $In_{.53}Ga_{.47}As$ channel, the InAlAs/InGaAs HEMT lattice-matched to InP has drawn considerable attention in recent years. The higher two-dimensional electron gas (2-DEG) density, electron mobility [1-6] and peak velocity, and hence higher device transconductance and gain make the InAlAs/InGaAs HEMT system more attractive than the conventional AlGaAs/GaAs and AlGaAs/InGaAs/GaAs pseudomorphic HEMTs for microwave/millimeter-wave low noise and high gain applications. The improved DC and microwave performance have been demonstrated in the InAlAs/InGaAs HEMT devices [2-6]. For instance, Fathimulla et al. [2] reported an MSG of 14.3 dB for a HEMT compared to 12 dB for an MESFET measured at 26.5 GHz for 0.7 μm devices. Recently, Mishra and his co-workers [6] reported a minimum noise figure of 1.4 dB with an associated gain of 8.5 dB at 63 GHz for 0.2 μm gate-length devices. In this paper, we report new results for 0.25 μm gate-length InAlAs/InGaAs HEMTs with record high gain and low noise microwave/millimeter-wave performance.

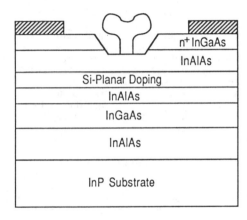

Figure 1.   Cross-section of the InAlAs/InGaAs HEMT.

DEVICE STRUCTURE AND FABRICATION

The devices were fabricated on selectively doped InAlAs/InGaAs heterostructures grown by molecular beam epitaxy (MBE) in a Varian Gen II system on Fe-doped (100)InP substrates. The cross-section of the device channel is shown in Figure 1. The MBE layers consisted of an undoped InAlAs buffer layer, an InGaAs channel, a thin undoped InAlAs spacer layer, a Si planar-doped layer, an undoped InAlAs barrier layer and finally a doped InGaAs contact layer. The planar-doped structure was used to increase the device transconductance, improve the device breakdown voltage and reduce the short channel effect [7]. The growth rates of InAlAs and InGaAs were both about 1 μm/h. The beam equivalent pressure ratio of As_4 to group III was 15 to 1 throughout the entire layer growth. The substrate growth temperature was 510±10°C. Hall measurements of this HEMT structure yielded a 2-DEG density of 2.4×10^{12} cm^{-2} with electron mobilities of 9,000 and 23,000 cm^2/V·s at room temperature and 77 K respectively.

50 μm wide devices with 0.25 μm T-shaped gates were fabricated by direct-write electron-beam lithography. AuGeNi was used for source and drain ohmic contacts by rapid thermal annealing. The gate was wet chemically recessed to achieve the desired full channel current. The gate was formed using Ti/Pt/Au metal. A typical source resistance of 0.5 Ω·mm was achieved for these devices.

Reprinted from *IEEE Int. Electron Devices Meeting, 1988, Tech. Dig.*, pp. 184–186, 1988.

DEVICE PERFORMANCE AND DISCUSSION

The DC source-drain I-V characteristics for the 0.25 μm InAlAs/InGaAs HEMT devices are shown in Figure 2. Extrinsic transconductance, g_m, as high as 900 mS/mm was obtained at room temperature with very good pinchoff characteristics. This corresponds to a maximum intrinsic g_m of 1640 mS/mm. The output conductance, g_o, was measured to be 47 mS/mm. This yields a voltage gain factor g_m/g_o of ~19.

As illustrated in Figure 2, in addition to the relatively low output conductance there exist no kinks in the I-V characteristics as have been observed by other groups [2,5]. Schaff and Eastman [8] have suggested that the absence of kinks is due to the quality of the InAlAs buffer layer, which they believe is enhanced with a low arsenic flux during growth, minimizing deep levels. Fathimulla et al. [2] reported that kinks can be simply due to the real-space electron transfer by thermionic emission of hot electrons from the InGaAs into the lower mobility InAlAs. Further investigation is needed to clarify this issue.

RF measurements at 18 GHz yielded a minimum noise figure (F_{min}) of 0.5 dB at room temperature with an associated gain (G_a) of 15.2 dB. This is the best 18 GHz noise performance ever reported. A maximum stable gain, MSG, of 20.9 dB was also obtained at 18 GHz. This MSG value is to our knowledge the highest value obtained for any HEMT, and is attributed to the small gate-drain feedback capacitance and high transconductance of the device. The current gain h_{21} of the devices was determined from the S-parameters measured over a range of 2-20 GHz. An extrinsic unity current gain cutoff frequency f_T of ~70 GHz was extrapolated from the h_{21} values. The f_T value is limited by the parasitic input capacitance in the device due to the small device size (50 μm gate-width). A higher f_T can be expected by using a larger gate-width device where the parasitic capacitance is relatively small.

Figure 3 shows the noise performance of the device in a single-stage amplifier from 55 to 65 GHz. The data have been corrected for test fixture loss. As can be seen in the figure, the device offered flat gain and noise responses over a wide range of 56-65 GHz. The device exhibited a 1.2 dB minimum noise figure with 8.5 dB associated gain at 58 GHz. These values compare favorably to the values of $F_{min} = 1.4$ dB and $G_a = 8.5$ dB at 63 GHz reported recently by Mishra et al. [6] for their 0.2 μm InAlAs/InGaAs HEMT devices.

At 63 GHz, a maximum available gain MAG of 15.4 dB was also measured from a single-stage amplifier. This value, extraploated at -6 dB/octave, yielded a maximum frequency of oscillation, f_{max}, of 380 GHz which is the highest f_{max} ever reported for any transistor. The extremely high f_{max} is mainly attributed to the very high g_m and small gate-drain feedback capacitance C_{gd} (2.27×10^{-2} pF/mm).

Both our 18 GHz and 58 GHz minimum noise figures correspond well with the Fukui-type frequency dependence exhibited by other HEMT material systems and MESFETs fabricated in our laboratory, as illustrated in Figure 4. The results from this work together with the best reported data for AlGaAs/GaAs HEMTs [9] are summarized in Table 1 for comparison. It can be seen from Figure 4 and Table 1, that the InAlAs/InGaAs HEMTs clearly demonstrate superior high gain and low noise performance compared to conventional AlGaAs/GaAs HEMTs for microwave/millimeter-wave applications.

A three-stage low noise amplifier (LNA) was designed and fabricated using three discrete 0.25 μm x 50 μm devices. The LNA exhibited flat noise and gain performance over the range of 60 to 65 GHz as illustrated in Figure 5. The LNA yielded an average noise figure of 3.0 dB with a gain of 22.0±0.2 dB in the test band. This is the best result ever reported for V-band LNA performance.

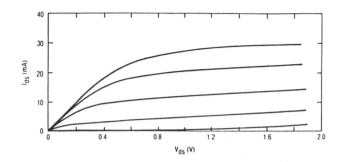

Figure 2.   I-V characteristics of the 0.25 μm InAlAs/InGaAs HEMT.

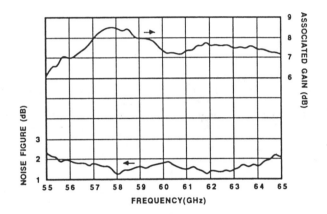

Figure 3.   Noise performance of a single-stage amplifier using the 0.25 μm InAlAs/InGaAs HEMT.

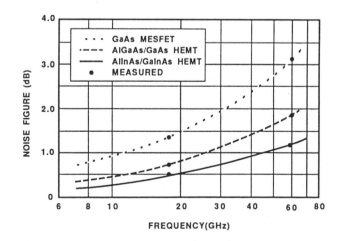

Figure 4.   Minimum noise figure versus frequency for 0.25 μm FETs and HEMTs fabricated at General Electric. All devices exhibit a Fukui-type frequency dependence.

Further improvements in the device performance can be achieved by optimizing the processing steps, carefully chosing the gate-length to gate-to-channel separation (or gate aspect) ratio and further reducing the gate length. A minimum noise figure of less than 1 dB at 94 GHz with an f_{max} of 450 GHz is projected.

CONCLUSIONS

State-of-the-art gain and noise performance have been obtained from 0.25 μm gate-length InAlAs/InGaAs/InP HEMTs. By reducing the gate-length and optimizing the aspect ratio, the device performance can be further improved. This work clearly demonstrates the great potential of the InAlAs/InGaAs HEMTs for high performance microwave/millimeter-wave high gain and low noise applications.

ACKNOWLEDGEMENT

The authors would like to thank Dr. A.W. Swanson, Manager of the Advanced Materials and Devices Laboratory, General Electric Company for his strong support of this work. The authors also wish to acknowledge Drs. L.F. Eastman and W.J. Schaff, and Mr. Y.-C. Pao for useful technical discussions.

REFERENCES

[1] Y. Nakata, S. Sasa, Y. Sugiyama, T. Fujui and S. Hiyamizu, Jap. J. Appl. Phys. 26(1) L59 (1987).

[2] A. Fathimulla, J. Abrahams, T. Loughran and H.Hier, IEEE Elect. Dev. Lett. 9(7) 328 (1988).

[3] A.S. Brown, U.K. Mishra, J.A. Henige and M.J. Delaney, J. Vac. Sci. Technol. B6(2) 678 (1988).

[4] G.I. Ng, D. Pavlidis, M. Quillec, Y.J. Chan, M.D. Jaffe and J. Singh, Appl. Phys. Lett. 52(9) 728 (1988).

[5] U.K. Mishra, A.S. Brown, L.M. Jelloian, L.H. Hackett and M.J. Delaney, IEEE Elect. Dev. Lett. 9(1) 41 (1988).

[6] U.K. Mishra, A.S. Brown, S.E. Rosenbaum, M.J. Delaney, S. Vaughn and K. White, presented at the 46th Device Research Conference, University of Colorado at Boulder, June 20-22, 1988.

[7] P.C. Chao, P.M. Smith, K.H.G. Duh, J.M. Ballingall, L.F. Lester, B.R. Lee, A.A. Jabra and R.C. Tiberio, 1987 IEDM Technical Digest, p. 410.

[8] W.J. Schaff and L.F. Eastman, private communication.

[9] K.H.G. Duh, P.C. Chao, P.M. Smith, L.F. Lester, B.R. Lee, J.M. Ballingall and M.Y. Kao, 1988 IEEE MTT-S Digest, p. 923.

Table 1. DC and Microwave Performance of 0.25 μm AlGaAs/GaAs HEMTs and InAlAs/InGaAs/InP HEMTs.

	AlGaAs/GaAs	InAlAs/InGaAs/InP
Extrinsic g_m (mS/mm)	600	900
Intrinsic g_m (mS/mm)	835	1640
g_m/g_o	29	19
MSG @ 18 GHz (dB)	14.0	20.9
F_{min} @ 18 GHz (dB)	0.7	0.5
G_a @ 18 GHz (dB)	13.8	15.2
F_{min} @ 60 GHz (dB)	1.8	1.2
G_a @ 60 GHz (dB)	6.4	8.5
MAG @ 60 GHz (dB)	11.2	15.4
f_{max} (GHz)	230	380
Reference	[9]	This Work

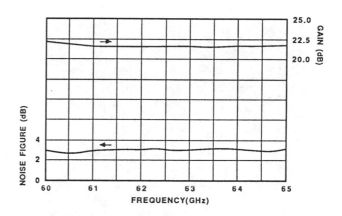

Figure 5. Measured noise and gain characteristics of V-band three-stage HEMT LNA.

Section 1.2: Broad Band Amplifiers

WIDEBAND HEMT BALANCED AMPLIFIER

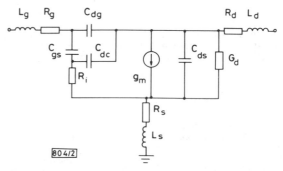

Indexing terms: Semiconductor devices and materials, Transistors, Amplifiers

The high electron mobility transistor (HEMT) has demonstrated great potential for high-gain and low-noise applications, achieving a noise figure and current gain cutoff frequency f_T superior to that of the GaAs MESFET. The letter presents the practical use of an HEMT in a hybrid wideband balanced amplifier covering 8.5 to 16 GHz producing 9.7 ± 0.5 dB gain and 2.6 dB midband noise figure. The performance is also compared to a GaAs MESFET stage.

Device: Devices were fabricated on AlGaAs/GaAs heterostructure layers grown by molecular beam epitaxy (MBE). The layer structure consists of a 300 Å-thick n^+ GaAs cap layer, Si-doped at a carrier concentration of $2 \times 10^{18}/cm^3$, a 500 Å $1 \times 10^{18}/cm^3$ Si-doped AlGaAs layer and a 1 μm undoped GaAs buffer. No spacer was used in order to maximise the device's two-dimensional electron gas sheet charge density (and hence current capability) and transconductance g_m.

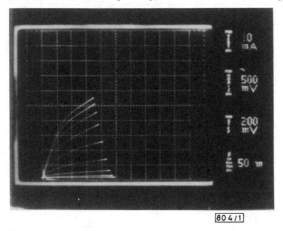

Fig. 1 *DC I/V characteristics of 0.25 μm $\times$ 150 μm HEMT*

$V_{gs} = +0.8$ V to -0.8 V in 0.2 V steps

Devices with low-resistance T-shaped cross-section gates 0.25 μm long $\times$ 150 μm wide were fabricated using direct write electron-beam lithography. The AuGeNi/Ag/Au source and drain ohmic contact metallurgy was alloyed in a rapid thermal annealing system. After patterning the resist for the gate level prior to deposition of the Ti/Pt/Au gates, a channel recess was chemically etched completely through the GaAs cap layer and partially into the AlGaAs layer. A source-drain separation of 2 μm was employed with the gate 0.3 μm from the source metallisation. The device geometry consists of a single centre-fed gate stripe, yielding an effective unit finger length of 75 μm.

The DC I_{ds} against V_{ds} characteristics of a device are shown in Fig. 1. A normalised transconductance g_m of 320 mS/mm is obtained at $V_{ds} = 2.5$ V, $I_{ds} = 25$ mA and $V_{gs} = 0$ V. The transconductance is strongly dependent on gate voltage, falling off rapidly with decreasing drain current (a soft pinch-off). The characteristics are less linear than for a GaAs MESFET.

A device was mounted to a test fixture and de-embedded S-parameters were measured from 2 to 18 GHz. An equivalent circuit model was then fitted to the data, yielding the parameters given in Fig. 2. This model (with adjustments for wire-bond length) was then utilised in the circuit design.

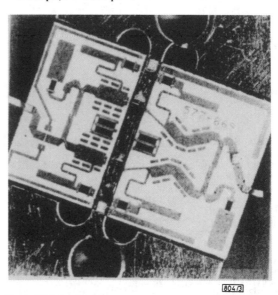

Fig. 2 *Equivalent circuit of 0.25 μm $\times$ 150 μm HEMT at a bias condition of 2.5 V/25 mA*

$L_g = 0.40$ nH, $R_g = 2.29$ Ω, $C_{gs} = 0.25$ pF, $C_{dg} = 0.0248$ pF, $C_{dc} = 0.0241$ pF, $R_i = 1.63$ Ω, $g_m = 48.076$ mS, $G_d = 4.764$ mS, $R_s = 3.17$ Ω, $L_s = 0.026$ nH, $R_d = 4.18$ Ω, $L_d = 0.18$ nH, $C_{ds} = 0.045$ pF, $\tau = 0.711$ ps

Fig. 3 *Wideband HEMT balanced amplifier*

Amplifier: Fig. 3 is a photograph of the balanced hybrid HEMT amplifier. The cascadable cell measures 0.4 $\times$ 0.28 in (10.2 $\times$ 7.1 mm). The matching circuits on both input and output utilise a T-configuration of series transmission lines and shunt stubs. The shunt circuit is designed so that a reactive match is obtained at the high end of the frequency band. To accommodate device gain slope, the resistors terminating both the gate and drain are only partially bypassed at the low end of the band, resulting in a lossy compensating network and improved stability. With a device as 'hot' as an HEMT, the tendency is to oscillate at sub-band frequencies when reactively terminated. With the 50/60 Ω DC-coupled terminations, this condition is avoided. The circuits are on 15 mil (381 μm) $\varepsilon_r = 10.0 \pm 2\%$ alumina with 40 Ω/$\square$ laser-trimmed NiCr thin-film resistors and 1.5×10^{-4} in (3.8 μm) of Au plate. The Lange coupler is a four-finger interdigitated design with nominal 1.8 mil (46 μm) line width and 0.8 mil (20 μm) space. The coupler is terminated in a wideband 3-section AC-coupled termination with better than 20 dB return loss from 6 to 18 GHz.

The devices were biased for maximum gain, i.e. $V_{ds} = 2.5$ V, $V_{gs} = 0$ V and $I_{ds} = 25$ mA/device. The gain curve of Fig. 4 was measured without tuning. The gain is 9.7 ± 0.5 dB from 8.5 to 16 GHz. The noise figure is less than 3.8 dB across the band with a minimum of 2.6 dB at midband. The return loss

Reprinted with permission from *Electronics Letters*, vol. 22, no. 14, pp. 747–749, July 1986.

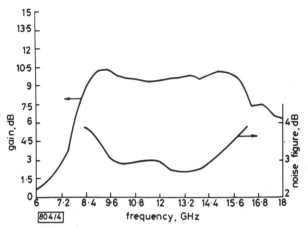

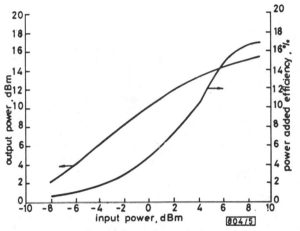

Fig. 4 *Measured gain/noise figure of wideband HEMT balanced amplifier*

$p_{in} = -5.337$dBm

Fig. 5 *Measured output power/power-added efficiency of wideband HEMT balanced amplifier*

$f = 9.5$ GHz

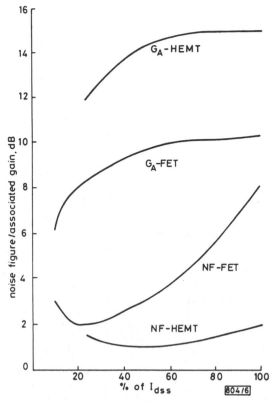

Fig. 6 *Drain current dependence of noise figure and associated gain at 12 GHz*

0.5 μm GaAs FET against 0.25 μm HEMT

By way of comparison, the gain was 6.0 ± 0.5 dB, the noise figure 6.2 dB max/5.5 dB min and the DC dissipation 330 mW.

An interesting comparison between noise figure/gain trade-offs for HEMTs and GaAs MESFETs is given in Fig. 6. Whereas the gain is compromised at a noise figure bias or the noise figure is compromised at a gain bias for the GaAs MESFET, the HEMT shows a much more benign characteristic. As such, in a cascade of stages with multistage contributions to noise figure, the HEMT will show substantially improved performance.

Acknowledgments: Special thanks to P. M. Smith and U. Mishra for providing the HEMT devices used in these experiments.

J. J. KOMIAK   *12th May 1986*
General Electric Company
Electronics Laboratory
Electronics Park, Syracuse, NY 13221, USA

curves are excellent (>20 dB), primarily due to the balanced configuration, the coupler/termination and the similarity of the devices. The amplifier was also power tested, producing a $+13$dBm 1 dB compression point and 17% power-added efficiency, as shown in Fig. 5. Note, however, that device efficiency is much better, as the balanced amplifier not only includes coupler loss, but 37% of the DC dissipation occurs in the 60 Ω drain resistors.

A similar self-biased balanced amplifier was designed around the Avantek AT-10600 0.5 μm $\times$ 250 μm GaAs FET biased at 3 V/30 mA (gain bias). This device has similar C_{gs}, C_{ds} and G_d, and hence matching circuit tuning was minimal.

BROADBAND HEMT AND GaAs FET AMPLIFIERS FOR 18 - 26.5 GHz

Kiyoyasu Shibata, Bunichiro Abe, Hisao Kawasaki,
Shigekazu Hori and Kiyoho Kamei

Microwave Solid-State Department
Komukai Works, Toshiba Corporation
1 Komukai-Toshiba-Cho, Saiwai-Ku
Kawasaki, 210 JAPAN

ABSTRACT

Two types of broadband amplifiers operating over 18 to 26.5 GHz have been developed by using newly developed 0.4-μm gate HEMTs and conventional 0.25-μm gate GaAs FETs. The four-stage HEMT amplifier exhibits a noise figure of $\leq$ 7.2 dB and a gain of 19.3 $\pm$ 1.8 dB and the five-stage GaAs FET amplifier exhibits a noise figure of $\leq$ 12 dB and a gain of 22.7 $\pm$ 2.2 dB over 18 to 26.5 GHz. The minimum noise figures in the measured frequency range are 5.0 dB and 7.5 dB for the HEMT and GaAs FET amplifiers, respectively. No essential difference is found between the amplifiers in input/output VSWR, output power and temperature variation of noise figure and gain.

Introduction

The progress of GaAs FET performance has made it possible to build GaAs FET amplifiers operating up to millimeter-wave region as reported by Rosenberg et al.[1] and Watkins et al.[2]. Considering the fact that the GaAs FET is approaching its theoretically predicted performance with finest geometries feasible at present, almost little can be expected to boost up the amplifier performance using GaAs FETs. Recently, extensive work has been done in many laboratories on HEMT (High Electron Mobility Transistor) that performs much better than the GaAs FET [3],[4].

This paper reports on two types of broadband amplifiers operating from 18 to 26.5 GHz built by using newly developed 0.4-μm gate HEMTs and conventional 0.25-μm gate GaAs FETs. The superiority of HEMT amplifier is shown by this work. The following sections describe the HEMT and GaAs FET employed, the unit amplifier design, and the RF performance of the amplifiers.

Device

Fig. 1 shows the schematic cross-sectional views of the HEMT (a) and the GaAs FET (b) used in the present work. The HEMTs [5] have been fabricated on epitaxial wafers grown by MBE. The wafer is made of a 1-μm thick undoped GaAs layer, a 300-Å thick n-$Al_{0.3}Ga_{0.7}As$ layer and a 500-Å thick n-GaAs layer on semi-insulating GaAs substrate. A recess structure is formed to control the drain current by etching the n-GaAs layer and part of n-$Al_{0.3}Ga_{0.7}As$ layer. The gate electrode with a length of 0.4 μm and a width of 200 μm is formed by Al/Ti to a thickness of 0.6 μm. The gate delineation is done by a direct electron-beam lithography.

The GaAs FETs are selected from conventional recessed-gate GaAs FETs (Toshiba JS8830-AS), whose gate electrode is also delineated by a direct electron-beam lithography and has a length of 0.25 μm and a width of 200 μm. Fig. 2 shows the top view of HEMT chip. Two gate bonding

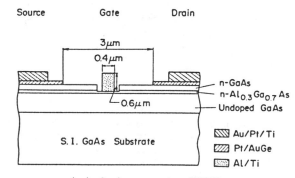

(a) 0.4-μm gate HEMT

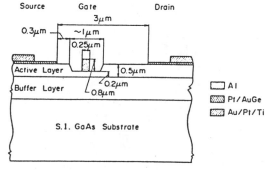

(b) 0.25-μm gate GaAs FET

Fig. 1 Schematic cross-sectional view of HEMT (a) and GaAs FET (b).

Reprinted from *1985 IEEE MTT-S Int. Microwave Symp., Tech. Dig.*, pp. 547–550, 1985.

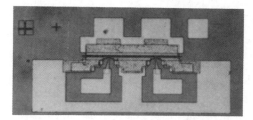

Fig. 2   Top view of HEMT chip.

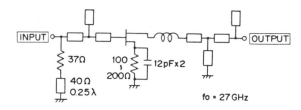

(a) Single-ended

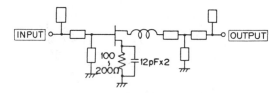

(b) Half of balanced-type

Fig. 4   Equivalent circuit of unit amplifiers.

pads are provided to reduce the gate resistance.

Fig. 3 shows the frequency dependence of microwave noise figure (NF) and associated gain (Ga) for the HEMT and GaAs FET chips measured with a microstrip test fixture. The HEMTs and GaAs FETs have shown typical values of NF = 1.4 dB with Ga = 9.7 dB and NF = 1.8 dB with Ga = 9.0 dB, respectively, at 18 GHz. At higher frequencies, the HEMT has exhibited an NF of 2 dB at 26 GHz, and the FET has an NF of 3.3 dB at 30 GHz. It has been found that the HEMT has 0.5-dB better noise figure and 1-dB higher gain than the GaAs FET at 8, 12 and 18 GHz.

Unit Amplifier Design

In broadband amplifier applications below 20 GHz, a balanced-type unit amplifier consisting of two single-ended amplifiers with two 3-dB hybrids at input and output ports is commonly used. Since the maximum available gain of FET decreases and the 3-dB hybrid loss increases with higher frequency, we have designed and evaluated a single-ended unit amplifier as well as a balanced one for our application. Fig. 4 shows the equivalent circuits of the single-ended unit amplifier (a) and the half of the balanced-type unit amplifier excluding 3-dB hybrids (b).

The single-ended unit amplifier has been designed to give good input and output VSWRs and gain flatness. Since a pure

reactive matching network cannot realize the desired results, a resistor with a quarter-wavelength shorted stub is shunted at the input port to achieve a good input VSWR and gain flatness. The balanced-type unit amplifier has been designed only from a gain flatness point of view. In both unit amplifiers, source resistors and RF-bypass capacitors are incorporated for single power supply operation. Circuit parameters of two types of amplifiers have been optimized by a computer simulation to give a gain flatness of less than 1 dB and VSWRs of less than 2.0. In this optimization, the S-parameters of the HEMT and FET shown in Fig. 5 have been used. The S-parameters have been calculated up to 30 GHz using the equivalent circuit element values determined from 2 to 18 GHz S-parameter measurements.

The predicted gains of single-ended unit amplifier are 7.0 dB and 6.3 dB for the HEMT and FET amplifiers, respectively, with an inband ripple of less than 1 dB over 18 to 26.5 GHz. The predicted gains of balanced-type unit amplifier are 7.5 dB and 6.3 dB for the HEMT and the FET amplifiers, respectively, with an inband ripple of 0.6 dB excluding the loss due to 3-dB hybrids.

The 3-dB hybrid designed for the balanced-type unit amplifier is an interdigitated coupler with four strips [6]. Alumina substrate with 0.38mm thickness has been used considering the radiation loss of microstrip and the etching accuracy of the narrow gaps of the hybrid. The fabricated 3-dB hybrid has shown an insertion loss of 0.4 dB and VSWR of ≤ 2.0 over 18 to 26.5 GHz.

Fig. 6 shows the top view of the single-ended (a) and balanced-type (b) unit amplifiers. The unit amplifiers have been fabricated on a 0.38-mm thick alumina substrate with Ti/Pt/Au microstrip lines and Ta_3N_4 thin film resistors. The substrate has a size of 2.5 x 6 mm and is

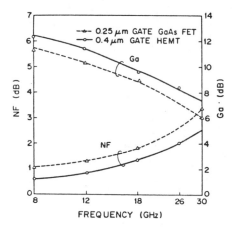

Fig. 3   Measured minimum noise figure (NFmin) and associated gain (Ga) for HEMT and GaAs FET.

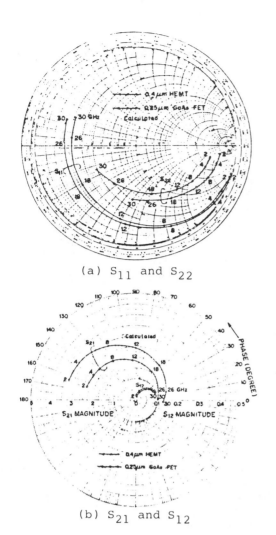

(a) S_{11} and S_{22}

(b) S_{21} and S_{12}

Fig. 5   Calculated S-parameters up to
30 GHz for HEMT and GaAs FET.

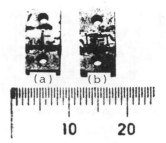

Fig. 6   Top view of single-ended
(a) and balanced-type (b)
unit amplifiers.

Table 1   Performance of single-ended and
balanced-type unit amplifier
measured over 18 - 26.5 GHz.
I_{DS} : drain current.

	0.4μm HEMT		0.25μm FET	
	I_{DS}(mA)	GAIN (dB)	I_{DS}(mA)	GAIN (dB)
SINGLE-ENDED	13	6.2±0.6	22	5.5±0.7
BALANCED-TYPE	26	5.7±0.6	46	5.6±0.6

soldered to a 0.5-mm thick Kovar carrier.
Table 1 summarizes the measured performance
of the single-ended and balanced-type
amplifiers.   The single-ended unit
amplifiers using the HEMT and GaAs FET have
given gains of 6.2 ± 0.6 dB and 5.5 ± 0.7
dB under a DC supply of 13 mA x 3 V and 22
mA x 3 V, respectively, over 18 to 26.5
GHz.   The balanced-type unit amplifiers
employing the HEMT and GaAs FET have shown
gains of 5.7 ± 0.6 dB and 5.6 ± 0.6 dB
under a DC supply of 26 mA x 3 V and 46 mA
x 3 V, respectively, over the same
frequency range.   Since the used HEMT has
lower drain current than that used in the
simulation, the measured gains of the HEMT
unit amplifiers are 1 dB less than the
predicted values.

Multi-Stage Amplifiers
Two types of multi-stage amplifiers
have been built by cascading four HEMT unit
amplifiers and five GaAs FET unit
amplifiers in order to realize a total
amplifier gain of 20 dB. Fig. 7 shows the
configuration of the amplifiers. In both

amplifiers, the balanced-type unit
amplifiers are used for input and output
stages to obtain good VSWR.   In each
amplifier, two single-ended amplifiers are
used for interstage to achieve higher gain.
Fig. 8 shows the inside view of the
HEMT amplifier measuring 22 x 22 x 48 mm.
Two rows of metal blocks on the inner side
of the lid are for suppressing a waveguide
mode within the housing. The spacing
between the rows is 3 mm and the metal
blocks make contact with Kovar carrier of
the unit amplifier through conducting
rubber. A DC voltage regulator is installed
in the backside of the housing, and the
amplifier can operate for an external
supply voltage of 9 - 20 V.
Fig. 9 shows the measured noise figure
and gain of the four-stage HEMT amplifier
at a supply voltage and current of 15 V and
96 mA. A noise figure of ≤ 7.2 dB and a
gain of 19.3 ± 1.8 dB are obtained over the

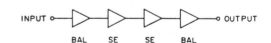

(a) 5-stage GaAs FET amplifier

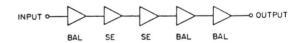

(b) 4-stage HEMT amplifier

Fig. 7   Multi-stage amplifier configuration
using GaAs FETs (a) and HEMTs (b).

174

Fig. 8   Inside view of HEMT amplifier
(lower) and inner of lid (upper).

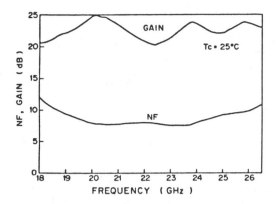

Fig. 10   Measured noise figure (NF) and
gain of 5-stage GaAs FET amplifier.

frequency range of 18 to 26.5 GHz. Fig. 10 shows the measured noise figure and gain of the five-stage GaAs FET amplifier at a supply voltage and current of 15 V and 165 mA. A noise figure of $\leq$ 12 dB and a gain of 22.7 $\pm$ 2.2 dB are obtained over the same frequency range. In the measured frequency band, the minimum noise figure is 5.0 dB for the HEMT amplifier and 7.5 dB for the FET amplifier. It has been found that HEMTs are superior to GaAs FETs in terms of the noise figure and gain. Input and output VSWRs are $\leq$ 2.0 for both types of amplifiers over 18 to 26.5 GHz.

The temperature dependence of noise figure and gain has been measured for both amplifiers. It has been found that the noise figure changes against temperature at a rate of $\simeq$ 0.013 dB/°C for both amplifiers and the gain changes at a rate of -0.02 dB/°C and -0.03 dB/°C for the HEMT and GaAs FET amplifiers, respectively. The output power at 1-dB gain compression is 12 dBm and 13 dBm for the HEMT and GaAs FET amplifiers, respectively.

Conclusion

Broadband amplifiers operating over 18 to 26.5 GHz have been developed by using 0.4-µm gate HEMTs and 0.25-µm gate GaAs FETs. Through this work, it has been

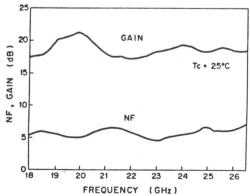

Fig. 9   Measured noise figure (NF) and
gain of 4-stage HEMT amplifier.

demonstrated that the amplifier using the HEMT shows better performance than the GaAs FET amplifier in terms of noise figure and gain. Both types of amplifiers have shown nearly the same characteristics with respect to VSWR, output power and temperature variation. Because of its inherent advantage, the HEMT will be widely used above Ku band in the broadband applications as well as the low-noise applications.

Acknowledgement

The authors would like to thank Dr. Ohtomo, S. Okano and Y. Kimura for their encouragement and helpfull discussions. They are also grateful to Dr. T. Nakanisi, H. Mashita and Y. Ashizawa for supplying the MBE wafers.

References

[1] J. Rosenberg, P. Chye, C. Huang and G. Policky, "A 26.5 - 40.0 GHz GaAs FET Amplifier," in 1982 IEEE MTT-S Int. Microwave Symp. Dig., pp. 166 -168.

[2] E. T. Watkins, J. M. Schellenberg, L. H. Hackett, H. Yamazaki and M. Feng, "A 60 GHz GaAs FET Amplifier," in 1983 IEEE MTT-S Int. Microwave Symp. Dig., pp. 145 - 147

[3] J. J. Bereng, K. Nakano and K. P. Weller, "Low Noise High Electron Mobility Transistors," in 1984 IEEE MTT-S Int. Microwave Symp. Dig., pp. 98 - 101.

[4] K. Kamei, S. Hori, H. Kawasaki, T. Chigira and K. Kawabuchi, "Quarter Micron Low Noise GaAs FET's Operable up to 30 GHz," IEDM Dig. Tech. Papers, pp. 102 - 105, 1980.

[5] K. Kamei, S. Hori, H. Kawasaki, K. Shibata, H. Mashita and Y. Ashizawa, "Low Noise High Electron Mobility Transistor," 11th Int. Symp. on GaAs and Related Compounds, France, Sept. 1984.

[6] Y. Tajima and S. Kamihashi, "Multiconductor Couplers," IEEE Trans. MTT, vol. MTT-26, No. 10, Oct. 1978

GaAs HEMT Distributed Amplifiers

Yasushi ITO[†], *Member*

SUMMARY A 0.1 to 18 GHz hybrid distributed amplifier using GaAs HEMTs with 0.3 micron gate length and 200 micron gate width has been realized. This amplifier exhibits 9.0 ± 0.5 dB of gain and less than 5.4 dB of noise figure. This letter describes the design approach which maximizes the gain-bandwidth product of the hybrid distributed amplifier, adding several FETs.

1. Introduction

GaAs HEMT has superior performances in gain and noise figure to GaAs MESFET. This letter describes the improvement of the performance of the distributed amplifier by replacing the conventional GaAs MESFET with high performance, 0.3 micron gate length GaAs HEMT, and the design approach which maximizes the gain-bandwidth product of the hybrid distributed amplifier, adding several FETs. As the number of FETs employed in the distributed amplifier is increased, gain does not increase monotonically and the upper cutoff frequency is lowered. These are primarily due to the effect of the gate loading. Therefore we first discuss the gate loading and examine the tradeoffs between design variables, such as the FET used, the number of FETs, and the cutoff frequency. And finally, we present there exists an optimum number of FETs employed in the hybrid distributed amplifier when the upper cutoff frequency is determined for a given FET.

2. Circuit Design

A schematic diagram of the hybrid distributed amplifier is shown in Fig. 1. The gate and drain transmission lines can be modeled as a *m*-derived low-pass filter which are shown in Fig. 2[(1)]. Using this equivalent circuit, the gain expression for the distributed amplifier with n FETs can be obtained as

$$G = \frac{g_m^2 Z_g Z_d}{4} \cdot \left(\frac{e^{-nAg} - e^{-nAd}}{A_g - A_d} \right)^2 \quad (1)$$

where Z_g, Z_d and A_g, A_d are the characteristic impedance and the attenuation per T-section of the gate and

drain lines, and g_m is the transconductance of the FET, respectively[(2)]. Now the bandwidth of the distributed amplifier is expressed as a cutoff frequency at which the attenuation of the gate line is equal to 3 dB. The attenuation (a_n) of the distributed amplifier with n FETs can be obtained from Fig. 2 as

$$a_n = nAg = n \ln \left((R^2 + I^2 + K^2) \pm 2K \left(R \cos \frac{\varphi}{2} \right. \right.$$
$$\left. \left. + I \sin \frac{\varphi}{2} \right) \right) \quad (2)$$

$$R = 1 - \frac{\omega^2 L_g C_{gs}(1 - \omega^2 L_i C_{gs})/2}{(1 - \omega^2 L_i C_{gs})^2 + (\omega R_i C_{gs})^2} \quad (3)$$

$$I = \frac{\omega^3 L_g C_{gs}^2 R_i / 2}{(1 - \omega^2 L_i C_{gs})^2 + (\omega R_i C_{gs})^2} \quad (4)$$

$$K = ((R^2 - I^2 - 1)^2 + 4R^2 I^2)^{1/4} \quad (5)$$

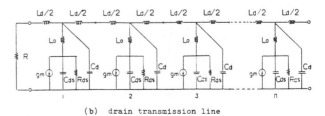

Fig. 1 Schematic diagram of the hybrid distributed amplifier.

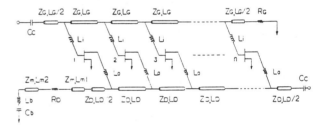

(a) gate transmission line

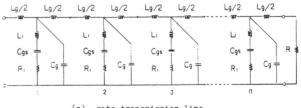

(b) drain transmission line

Fig. 2 Equivalent circuit of the gate and drain transmission line.

Manuscript received August 13, 1987.
Manuscript revised September 25, 1987.
† The author is with Tokyo Keiki Co., Ltd., Tokyo, 144 Japan.

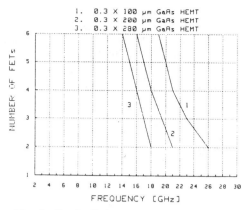

Fig. 3 Tradeoffs between design variables.

Table 1 Intrinsic parameters of the GaAs HEMT.

GaAs HEMT	Lg X Wg	Ri	Cgs
NE20100	0.3 X 100 micron	7.0 ohm	0.12 pF
NE20200	0.3 X 200 micron	4.0 ohm	0.20 pF
NE20300	0.3 X 280 micron	3.0 ohm	0.28 pF

$$\varphi = \tan^{-1} \frac{2RI}{R^2 - I^2 - 1} \qquad (6)$$

where L_g is the microstrip inductance, L_i is the bond-wire inductance, and R_i, C_{gs} are gate-to-source resistance and capacitance of the FET, respectively[3]. In the gate line, R_i plays a predominant role on the attenuation and high-frequency rolloff characteristic of the distributed amplifier.

The objective of our design is to maximize the gain of the distributed amplifier while the upper cutoff frequency (bandwidth) is fixed to 18 GHz, increasing n.

In this approach, the tradeoffs between design variables, such as the FET used (R_i, C_{gs}) and the number of FETs (n), and the cutoff frequency (f_c) must be examined. Figure 3 and Table 1 display these tradeoffs, which are derived from Eqs. (2)-(6). In this figure, it is assumed that $L_g = 0.4\ nH$ and $L_i = 0.1\ nH$ for a high cutoff frequency. When the upper cutoff frequency is determined to 18 GHz, the optimum number is $n=6$ for 0.3×100 micron GaAs HEMT and $n=3$ for 0.3×200 micron GaAs HEMT, where the gain-bandwidth product of the distributed amplifier is maximized, respectively. In the case of 0.3×280 micron GaAs HEMT, $n<2$ means it is impossible to realize the distributed amplifier. The optimum number can be obtained from the point of gain maximum by using Eq. (1). Now it is assumed that A_d is about 0.1 and g_m is 45 mS for 0.3×200 micron GaAs HEMT. From Eqs. (1)-(6), gain is obtained to be 7.6 dB, 8.5 dB, 7.0 dB at 18 GHz for $n=2, 3, 4$, respectively. Therefore the optimum number is $n=3$ for 0.3×200 micron GaAs HEMT. The distributed amplifier with three 0.3×200 micron GaAs HEMTs has a large margin in the cutoff frequency as compared to that with six 0.3×100 micron GaAs HEMTs, while both amplifiers have

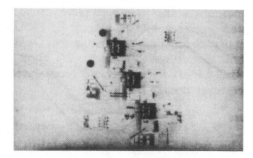

Fig. 4 Photograph of the distributed amplifier.

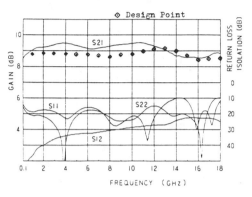

Fig. 5 Predicted, measured gain, measured return loss and isolation.

the same gate periphery of 600 micron. Thus we design here the hybrid distributed amplifier, using three 0.3×200 micron GaAs HEMTs.

3. Circuit Fabrication and Performance

A photograph of the amplifier appears in Fig. 4. Thin film circuit is fabricated on 0.4 mm thickness alumina substrate with TaN_x/Cr/Au microstrip lines. This circuit has a size of 4.6×6.0 mm. Plated through holes are used to achieve high integrity and to allow both RF/DC grounds for the FET bias and RF bypass capacitors to be surface-mounted. The diameter of these plated through holes is approximately 0.2 mm.

Measured gain, return loss, and isolation performance of the distributed amplifier are shown in Fig. 5. Predicted gain performance is also plotted in Fig. 5. Over the 0.1 to 18 GHz band, this amplifier exhibits 9.0 ± 0.5 dB of gain, better than 10 dB of return loss, and greater than 22 dB of isolation. Measured and predicted gain are in good agreement. It can be observed in Fig. 5 that gain does not roll off steeply below 2 GHz. This is due to high g_m(45 mS) of the HEMT and high value (> 40 pF) of the source capacitor for the FET bias. Over the 0.1 to 18 GHz band, less than 5.4 dB of noise figure can be achieved. All data presented was measured at VDS = 2V and IDS ≃ 10 mA.

4. Conclusion

In this letter, we have described the design approach

which maximizes the gain-bandwidth product of the distributed amplifier, adding several FETs. Using this design approach and high performance of GaAs HEMT, the hybrid distributed amplifier with three 0.3×200 micron GaAs HEMTs has demonstrated excellent performances over decade bandwidth.

References

(1)   Y. Ito and A. Takeda : "A 0.8 to 18 GHz distributed amplifier using hybrid circuits", IEICE Technical Report, **MW86**-134 (Feb. 1987).

(2)   Y. Ayasli, et al. : "A monolithic GaAs 1-13-GHz traveling-wave amplifier", IEEE Trans. Microwave Theory & Tech., **MTT-30**, 7 (July 1982).

(3)   Y. Ito : "0.8 to 18 GHz hybrid distributed amplifiers using GaAs HEMTs", IEICE Technical Report, **MW87**-17 (May 1987).

A 2–20 GHz High-Gain Monolithic HEMT Distributed Amplifier

STEVE G. BANDY, CLIFFORD K. NISHIMOTO, CINDY YUEN, ROSS A. LARUE,
MARY DAY, JIM ECKSTEIN, ZOILO C. H. TAN, CHRISTOPHER WEBB,
AND GEORGE A. ZDASIUK, MEMBER, IEEE

Abstract — A low-noise 2–21 GHz monolithic distributed amplifier utilizing 0.35-micrometer-gate-length HEMT devices has achieved 12 ± 0.5 dB of gain. This represents the highest gain reported for a distributed amplifier using single FET gain cells. A record low noise figure of 3 dB was achieved midband (4–12 GHz). The circuit design utilizes five HEMT transistors of varying width with gates fabricated by E-beam lithography. The same amplifier fabricated with 0.35-μm-gate-length MESFET's in place of the HEMT devices resulted in 9.5 ± 0.5 dB of gain across the 2–20 GHz band. This record performance level for a MESFET distributed amplifier is used to determine that the use of HEMT devices rather than the small gate lengths is primarily responsible for the HEMT amplifier performance.

I. INTRODUCTION

THIS PAPER is the first to report results of a working monolithic distributed amplifier utilizing 0.35-μm-gate-length HEMT's. A number of authors have previously reported reactively tuned, narrow-band amplifiers using discrete high electron mobility transistors (HEMT's) [1], [2]. Also, many workers have reported monolithic distributed amplifiers utilizing 0.50-μm-gate-length MESFET's that have achieved 7–8 dB gain and a noise figure of approximately 5 dB in the 2–20 GHz band [3]. By replacing the conventional MESFET with high-performance, 0.35-μm-gate-length HEMT's, significant improvements in gain and noise figure are achievable. Fig. 1 shows the HEMT device structure used in this work. The AlGaAs epitaxial layer is doped higher ($\sim 2 \times 10^{18}$/cm^3) than the active layer in a MESFET; hence HEMT's usually possess higher intrinsic transconductance (g_m), as high as 500 mS/mm but with a concomitant higher gate capacitance (C_{gs}). Additionally, the gate-to-drain breakdown voltage for the HEMT may be reduced because of the higher doping, impacting the power performance.

II. CIRCUIT DESIGN

Fig. 2 shows the equivalent circuit model for discrete 0.3-$\times$150-μm-gate-length HEMT's fabricated in our lab. The model is for maximum gain bias conditions. The devices were first characterized by *s*-parameter measurements in the common-gate, common-source, and

Manuscript received May 8, 1987; revised July 31, 1987. This work was supported in part by the Naval Air Development Center under NRL Contract N00014-86-C-2048.

The authors are with the Device Laboratory, Varian Research Center, Palo Alto, CA 94303.

IEEE Log Number 8717251.

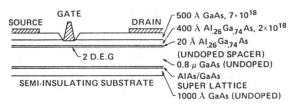

CROSS-SECTION OF 1/4 μ HEMT FET

Fig. 1. Cross section of high electron mobility transistor fabricated in the lab. A noise figure of 1.35 dB and an associated gain of 12 dB were measured at 18 GHz.

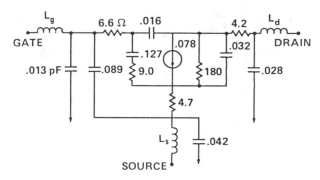

Fig. 2. Equivalent circuit model of HEMT device biased for maximum gain. The intrinsic transconductance is 520 mS/mm.

common-drain configurations from 2 to 18 GHz. A single equivalent circuit model is then constrained to fit the three sets of data. Details of the procedure are described elsewhere [4], but it should be noted that this particular device had a g_m of 520 mS/mm. Because the doping is high, C_{gs} would normally be high, but is kept to a minimum by reducing the gate length.

The same procedure was used to obtain the device model for the same HEMT biased for minimum noise, resulting in an intrinsic g_m of 56 mS (373 mS/mm), a C_{gs} of 0.085 pF, and an r_{ds} of 238 Ω. The remaining parameters were close to the Fig. 2 values.

Both the high-gain and low-noise models were used in the design of five HEMT distributed amplifiers in the 2–20 GHz band. Maximum gain, gain flatness, and minimum return loss were traded off against each other in the amplifier design optimizations. Finally, two amplifier designs for each model were obtained, one for maximum gain and one for maximum return loss.

Reprinted from *IEEE Transactions on Electron Devices*, vol. ED-34, no. 12, pp. 2603–2609, December 1987.

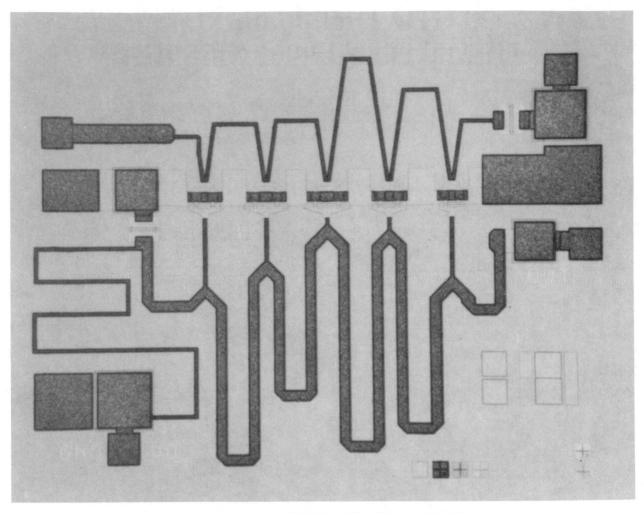

Fig. 3. Photomicrograph of HEMT amplifier. Chip size: 2.3 × 1.7 mm.

CAD was used to optimize 24 circuit design parameters including five HEMT gate widths and 12 gate and drain line lengths. By allowing the HEMT's to vary in width and the gate and drain line sections to vary in length, the gain is enhanced by 1–2 dB and an improved return loss is achieved when compared to the conventional constant-K design. Additional details of this variable device width design approach are found elsewhere [4], but generally the wider HEMT's are located in the center of the amplifier.

Using π-gate-configured FET's, the total device periphery is 620 μm for the amplifier using the maximum gain bias model and 698 μm for the amplifier using the minimum noise bias model. A photomicrograph of the complete chip for the amplifier based on the maximum gain bias model is shown in Fig. 3. The transmission line widths are 12.7 μm for the gate line and 45.2 μm for the drain line. To be noted are the variable lengths of the gate and drain line sections. The longest section in the gate line is 1117 μm (35.4° at 10 GHz) and the longest section in the drain line is 1914 μm (60.6° at 10 GHz). The MMIC has on-chip gate and drain line terminations and biasing networks.

If the gate and drain line losses per unit length were zero or equal, filter theory would suggest equal-width devices except for tapering on the ends to maximize return loss, as shown in Fig. 4(a). Fig. 4(b) shows a simple gain expression for a two-FET amplifier which is optimized for relative device widths and generalized to successive device pairs in a multiple-FET distributed amplifier. If W_n and W_{n+1} are the successive device widths of the nth and $(n+1)$th FET's and if r_{gl} and r_{dl} are the gate and drain line shunt resistances per unit length, then

$$\frac{W_n}{W_{n+1}} \cong \frac{r_{dl}}{r_{gl}} \tag{1}$$

for optimum gain. R_0 is the transmission line characteristic impedance, and g_{ml} the transconductance per unit width W. The MESFET distributed amplifiers fabricated in our labs ($L_g \leqslant 0.6$ μm) have been dominated by drain-line loss ($r_{dl} < r_{gl}$), and have had the ideal profile shown in Fig. 4(a) skewed towards increasing device widths so that the last width is nearly equal to the middle device widths, as shown in Fig. 4(b) and predicted by (1). On the other hand, the superior output conductance of the HEMT device (as will be discussed in Section IV) enables the ratio in (1) to be nearly unity at the upper band edge, thereby returning the width distribution back to the ideal profile of Fig. 4(a). This is indeed the case, as shown for the opti-

Filter Theory

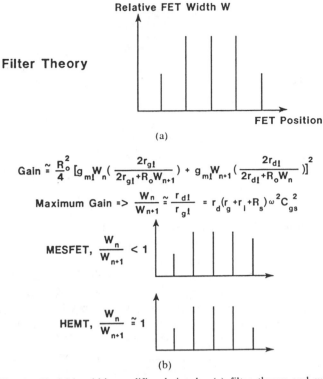

(a)

$$\text{Gain} \cong \frac{R_o^2}{4}\left[g_{ml}W_n\left(\frac{2r_{gl}}{2r_{gl}+R_oW_{n+1}}\right)+g_{ml}W_{n+1}\left(\frac{2r_{dl}}{2r_{dl}+R_oW_n}\right)\right]^2$$

$$\text{Maximum Gain} \Rightarrow \frac{W_n}{W_{n+1}}\cong\frac{r_{dl}}{r_{gl}}=r_d(r_g+r_i+R_s)\omega^2C_{gs}^2$$

MESFET, $\dfrac{W_n}{W_{n+1}}<1$

HEMT, $\dfrac{W_n}{W_{n+1}}\cong 1$

(b)

Fig. 4. Variable-width amplifier design by (a) filter theory and with (b) transmission line loss.

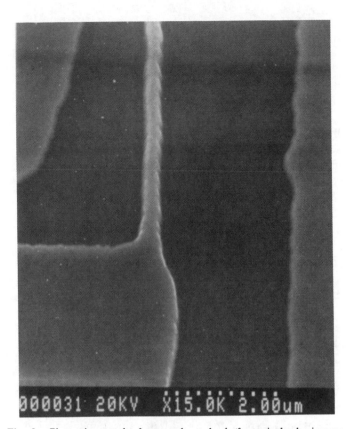

Fig. 5. Photomicrograph of gate and gate lead of a typical submicrometer HEMT device.

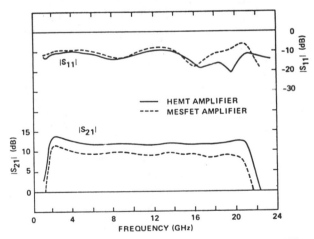

Fig. 6. Gain and return loss of working HEMT and MESFET amplifiers biased for maximum gain.

mized design of Fig. 3, and verifies the superior output conductance of HEMT's over that for MESFET's.

III. FABRICATION

The HEMT layers shown in Fig. 1 were grown by MBE at 565°C. The GaAs is unintentionally doped n-type at about $1\times10^{14}/\text{cm}^3$. A five-period AlAs/GaAs superlattice is grown midway into the buffer layer in an attempt to reduce dislocations and improve surface morphology.

Standard processing techniques were used for most of the HEMT distributed amplifier fabrication. Isolation was achieved with a 2500-Å mesa etch. Fig. 5 shows a photomicrograph of a typical submicron gate fabricated in the lab using E-beam lithography. Gate exposures in PMMA were achieved with a commercial E-beam exposure system, after which a shallow gate recess etch to the first layer of AlGaAs was performed. In order to achieve good grounding of the source, 50×50-μm backside vias were fabricated using reactive ion etching. All other process steps used conventional metallization, liftoff, and pulse plating techniques.

IV. MEASURED GAIN PERFORMANCE

As shown in Fig. 3, the layout of the circuit is compatible with RF wafer probing techniques. Fig. 6 shows the gain and return loss obtained for the amplifier design using the maximum gain bias model, using the Cascade Microtech RF wafer prober. 12 ± 0.5 dB of gain over the 2–21-GHz bandwidth was achieved. The drain and gate bias were adjusted for maximum gain ($V_d=3.5$ V, $V_g=-0.15$ V, $I_{ds}=106$ mA). The gain, bandwidth, and return

loss agree well with the simulated result, indicating that capacitive and inductive coupling between the lines, which was not modeled, is not significant (a run with deep RIE-etched grooves between the lines showed no discernible difference in performance).

Another design version using the same maximum gain bias model gave a better return loss of -14 to -15 dB for the same 12 dB of gain, but the gain experienced a sudden step drop to 10.5 dB at 18 GHz. When biased for maximum gain, the design version using the low-noise bias model showed only ~0.5 dB less gain than shown in Fig. 6, with the return loss being essentially the same.

The HEMT device is particularly suited for achieving high gain in distributed amplifiers because of its higher transconductance and lower output conductance (per unit g_m) compared with the MESFET. The higher transconductance is a result of the increased saturated velocity of the two-dimensional electron gas (2DEG) layer [5], [6] and the fact that AlGaAs can be doped higher than GaAs without compromising the gate breakdown voltage (due to the larger bandgap). The lower output conductance is a result of the two-dimensional nature of the conduction electrons.

The high transconductance means smaller device widths, which translates to smaller MMIC chip sizes, especially for mm-wave applications.

The distributed amplifier gain is usually 1–2 dB below the maximum available gain (MAG) at the upper band-edge frequency. At this frequency, the power to the gate line termination falls to nearly zero, and the power to the drain line termination is mostly reflected due to the mismatch caused by the increasing drain line impedance as the line cutoff frequency is approached.

Ignoring feedback, MAG is given by

$$\text{MAG} \cong \frac{g_m^2}{4\omega^2 C_{gs}^2 r_i g_{ds}} \qquad (2)$$

so one would expect the amplifier gain to be 1 or 2 dB below this expression when evaluated at the upper band edge. In this expression, r_i is the input series resistance and g_{ds} is the output conductance. The amplifier gain, then, is roughly proportional to

$$G \propto \left(\frac{g_m}{C_{gs}}\right)^2 (g_m r_{ds}) \qquad (3)$$

assuming $r_i \propto g_m^{-1}$. Fig. 7 shows the $g_m r_{ds}$ factor comparison between 0.35-μm-gate-length HEMT's and MESFET's fabricated in our lab when biased at zero gate bias. HEMT's appear to enjoy an advantage of about 1.5 for this factor. It also appears that r_{ds} for HEMT devices is inversely proportional to g_m. Even though r_{ds} is quite low in Fig. 2, the $g_m r_{ds}$ product is still superior to that for MESFET's. Because the epitaxial layer thickness a is so much smaller for HEMT devices, the larger L_g/a ratios result in higher values of r_{ds} [8, fig. 13].

If the gate side wall capacitance is accounted for in the manner of [13], the other factor in (3) is given by

$$\frac{g_m}{C_{gs}} \cong \frac{v_s}{L_g + \pi a} \qquad (4)$$

where a is the epitaxial thickness under the gate. For $L_g = 0.35$ μm and $a = 600$ Å for MESFET's and 300 Å for HEMT's, for example, (4) gives a factor of 1.2 improvement for HEMT's for the same saturated velocity. This factor, along with the $g_m r_{ds}$ factor, gives a 3.4-dB improvement in gain for HEMT's by (3). The superior performance of the HEMT may thus be due more to these factors than to an increased saturated velocity, and it may be that a pulse-doped MESFET may perform nearly as well.

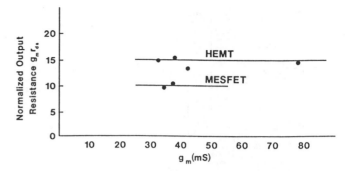

Fig. 7.   Output resistance comparison between HEMT's and MESFET's ($L_g \cong 0.35$ μm).

While it is true that the small 0.35-μm gate lengths are partially responsible for the improved performance of Fig. 6, simulations show that due to the increase of g_{ds} with decreasing gate length, only marginal gain improvement using such small gate lengths is possible with MESFET's (see Section VI). This necessitates going to a scheme such as the cascode configuration [4] to circumvent this problem. The lower output conductance of the HEMT device enables the advantages of a shorter gate length to be realized in the distributed amplifier configuration without resorting to a more complicated and potentially unstable gain cell. Simulations using the HEMT device in the cascode configuration indicate that gains of 15 dB over the 2–20 GHz bandwidth can be achieved.

V.   Amplifier Noise Performance

This section will discuss HEMT low-noise performance, HEMT suitability for broad-band low-noise applications, and the measured noise performance of the distributed amplifier.

The narrow-band noise performance of discrete HEMT devices fabricated in our facility is typically a minimum noise figure of 1.25–1.5 dB at 18 GHz, with an associated gain of 11–12 dB for gate lengths in the 0.3–0.35 μm range.

The noise figure of a FET device can be expressed by [8]

$$F = 1 + \frac{r_n}{R_s} + \frac{g_n}{R_s}\left[\left(R_s + \sqrt{R_{s,\text{opt}}^2 - \frac{r_n}{g_n}}\right)^2 + (X_s - X_{s,\text{opt}})^2\right]$$

where r_n and g_n are the noise resistance and conductance, respectively, $Z_s = R_s + jX_s$ is the source impedance, and $R_{s,\text{opt}} + jX_{s,\text{opt}}$ is the optimum source impedance for minimum noise. The equation shows that, under conditions of noise mismatch such as encountered in a distributed amplifier, the largest noise bandwidths occur when g_n and the ratio $X_{s,\text{opt}}/R_{s,\text{opt}}$ are the smallest. Empirical data have been taken [9] which demonstrate that HEMT devices have lower values of both g_n and the ratio $X_{s,\text{opt}}/R_{s,\text{opt}}$ than MESFET's. Accordingly, HEMT devices should be better suited for broad-band low-noise applications.

Fig. 8 compares the gate bias dependence of the minimum noise figure and associated gain at 12.4 GHz for both the HEMT and MESFET devices in a 50-Ω system

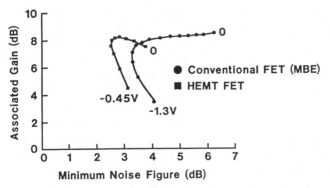

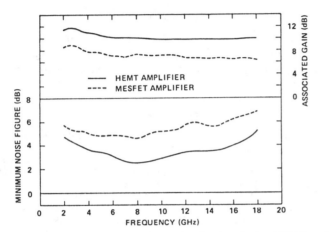

Fig. 8. 12-GHz FET performance in a 50-Ω system with varying gate bias.

Fig. 9. Noise performance and associated gain of the HEMT and MESFET amplifiers.

(no input or output matching), which gives some indication of what would be experienced in the environment of a distributed amplifier. Both devices have a 0.4-μm gate length and the HEMT has a 50-Å AlGaAs spacer layer. The HEMT has an advantage of approximately 1 dB in both the minimum noise figure and the associated gain under these conditions.

Fig. 9 shows the minimum noise performance for the amplifier. The low value of approximately 3 dB over the 4–12 GHz midband region exceeds previously reported performance for distributed amplifiers by several dB for the same bandwidth [3], [10], [11]. The associated gain of around 10 dB is only 2 dB below the maximum gain. Fig. 9 shows the usual noise skirts due to the gate termination resistor on the low-frequency end and the transmission line cutoff frequency on the high end of the band.

The maximum gain bias noise performance is only slightly higher, being 0.2 dB and 0.5 dB higher at the low and high ends of the band, respectively.

The amplifier was designed for maximum gain rather than minimum noise figure for purposes of achieving a multistage gain of 20 dB with as few stages as possible. The minimum noise figure for the amplifier using the minimum noise bias model was measured to be ~ 0.5 dB higher than that for the amplifier using the maximum gain bias model. This result emphasizes the need for an accurate noise parameter model for the HEMT device in

order to achieve optimum low-noise performance. Optimization using the considerations in [12] in conjunction with a suitable computer simulation program for noise sources should give a design which further reduces the noise figure.

VI. MESFET Amplifier Performance

To ascertain what part of the gain improvement of the HEMT amplifier over previously published results [3], [10], [11] is due to the use of 0.35-μm gate lengths and what part is due to the use of HEMT devices, an amplifier run was made with MESFET devices (MBE growth with $5 \times 10^{17}/cm^3$ active layer doping) having 0.35-μm gate lengths.

An optimized CAD simulation for a distributed amplifier using the device model for the 0.35-μm-gate-length MESFET results in 9.5-dB gain from 2 to 20 GHz. Fig. 6 shows the measured gain response for the MESFET version of the amplifier using the same HEMT amplifier mask set. This 9.5 ± 0.5-dB measured performance is the highest reported for a MESFET amplifier, and is, of course, primarily the result of the 0.35-μm gate lengths employed. The fact that this gain is virtually the same as that obtained for the optimized CAD simulation indicates that very little improvement would be achieved with a MESFET optimized layout. The 9.5-dB response indicates an improvement of only 1.5 dB over the 8-dB gain obtained for the distributed amplifier employing 0.5-μm-gate-length FET's previously fabricated in our lab, and far short of the 12-dB performance of the HEMT version.

Even more notable is the difference in noise performance between the use of MESFET's and HEMT's in distributed amplifier applications. Fig. 9 shows a difference of ~ 2 dB lower noise figure for the HEMT amplifier version. As discussed previously, since the HEMT version was not optimized for noise, it is probably safe to say that this large difference would probably not be significantly reduced by a design optimized for MESFET gain in the same way that the Fig. 3 design was optimized for HEMT gain.

Also shown in Fig. 9 is that the associated gain for the MESFET amplifier was measured to be 7 dB, which is 3 dB lower than the associated gain for the HEMT amplifier.

It seems clear, then, that the bulk of the performance improvement measured for the HEMT distributed amplifier is due to the use of HEMT devices in place of MESFET devices.

VII. Measured Output Power Performance

The 1-dB compression power and the third-order intermodulation intercept point were measured for the HEMT and MESFET amplifiers at 10 GHz, as shown in Fig. 10. When biased at maximum gain, the 1-dB compression power for the HEMT amplifier was measured to be 17 dBm at 10 GHz. The third-order intermodulation intercept point was measured to be 27 dBm under the same conditions. The 10-dB difference between these two measurements is close to the theoretical difference of 10.6 dB [7].

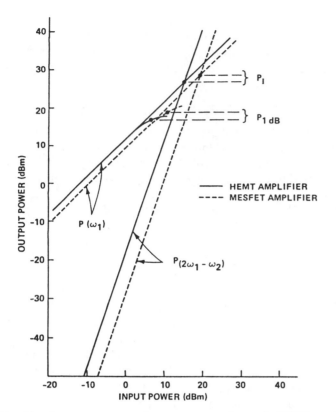

Fig. 10. Output power performance comparison of HEMT and MESFET amplifiers.

Fig. 10 shows that for close to the same bias current and voltage, MESFET distributed amplifiers have around 2 dB higher compressed power and third-order intercept levels when biased for maximum gain. Presumably, the lower values for the HEMT version are due to a much higher nonlinearity in the g_m versus gate bias characteristic. Occasionally, the HEMT amplifier power levels fall even further (sometimes by 5 dB) below those of the MESFET amplifier, perhaps because of increased gate recess etch nonuniformity over the width of a single device. HEMT devices, because of their much higher doping and resulting thinner layers, show an increased sensitivity to etch depth variations over the FET width than do MESFET's.

VIII. CONCLUSIONS

The results presented in this paper indicate that by using high-performance, low-noise HEMT's with gates defined by E-beam lithography, record improvements in gain and noise figure can be achieved for broad-band distributed amplifiers. A gain of 12 dB $\pm$ 0.5 dB from 2 to 21 GHz and a midband noise figure of 3 dB have been achieved on a working HEMT amplifier. This represents the highest gain and lowest noise figure reported for a distributed amplifier using single FET's for the gain cell.

The use of MESFETs with E-beam-defined gates in the same amplifier configuration resulted in 9.5 $\pm$ 0.5 dB of gain across the 2–20 GHz band. This represents a record performance level for a MESFET distributed amplifier. Comparison with the HEMT amplifier performance indicates that the bulk of the performance improvement measured for the HEMT amplifier over previously published MESFET performance levels is due to the use of HEMT devices in place of MESFET devices rather than the use of 0.35-μm gate lengths. The HEMT device is able to outperform its MESFET counterpart with the same gate length in distributed amplifier applications by virtue of its higher transconductance and lower output conductance.

ACKNOWLEDGMENT

The authors would like to acknowledge the technical support given by B. Knapp, C. Shih, and H. Luong. They also wish to thank M. Glenn for his measurement expertise.

REFERENCES

[1] M. Sholley and A. Nichols, "60 and 70 GHz (HEMT) amplifiers," in *1986 IEEE MTT-S Dig.*, p. 463–465.

[2] J. J. Berenz, W. Nakano, and K. P. Weller, "Low noise high electron mobility transistors," in *Proc. 1984 IEEE Microwave and Millimeter-Wave Monolithic Circuits Symp.*, p. 83–86.

[3] T. McKay, J. Eisenberg, and R. E. Williams, "A high-performance 2–18.5-GHz distributed amplifier—Theory and experiment," *IEEE Trans. Microwave Theory Tech.*, vol. MTT-34, pp. 1559–1568, Dec. 1986.

[4] R. A. LaRue *et al.*, "A 12-dB high-gain monolithic distributed amplifier," *IEEE Trans. Microwave Theory Tech.*, vol. MTT-34, pp. 1542–1547, Dec. 1986.

[5] T. J. Drummond *et al.*, "Enhancement of electron velocity in modulation-doped (Al,Ga)As/GaAs FETs at Cryogenic Temperatures," *Electron. Lett.*, vol. 18, p. 1057–1058, 1982.

[6] T. Hanoguchi and O. Miyatsuji, "Negative differential mobility and velocity overshoot," presented at Eng. Foundation Second SDHT Conf., Hawaii, Dec. 1986.

[7] G. L. Heiter, "Characterization of nonlinearities in microwave devices and systems," *IEEE Trans. Microwave Theory Tech.*, vol. MTT-21, p. 797–805, 1973.

[8] R. A. Pucel, H. A. Haus, and H. Statz, "Signal and noise properties of gallium arsenide microwave field effect transistors," in *Advances in Electronics and Electron Physics*, vol. 38, L. Marton, Ed. New York: Academic Press, 1975, p. 195–265.

[9] S. Weinreb and M. Pospiezalski, "X-band noise parameters of HEMT devices at 300K and 12.5K," in *1985 IEEE MTT-S Dig.*, p. 539–542.

[10] W. Kennan, T. Andrade, and C. C. Huang, "A 2–18 GHz monolithic distributed amplifier using dual-gate GaAs FET's," *IEEE Trans. Electron. Devices*, vol. ED-31, p. 1926–1930, 1984.

[11] Y. Ayasli *et al.*, "2–20 GHz GaAs traveling-wave amplifier," *IEEE Trans. Microwave Theory Tech.*, vol. MTT-32, p. 71–77, 1984.

[12] K. B. Niclas and B. A. Tucker, "On noise in distributed amplifiers at microwave frequencies," *IEEE Trans. Microwave Theory Tech.*, vol. MTT-31, p. 661–668, 1983.

[13] A. M. Goodman, "Metal-semiconductor barrier height measurement by the differential capacitance method-one carrier system," *J. Appl. Phys.*, vol. 34, p. 329–338, 1963.

A MONOLITHIC 3 to 40 GHz HEMT DISTRIBUTED AMPLIFIER

C. Yuen, C. Nishimoto, M. Glenn, Y. C. Pao, S. Bandy and G. Zdasiuk

Varian Research Center, Device Laboratory
Palo Alto, CA   94303

ABSTRACT

A monolithic 3 to 40 GHz distributed amplifier has been developed using 0.25 µm HEMTs with mushroom gate profile as active devices. This amplifier consists of five 0.25 µm HEMTs with variable gate widths and on chip biasing circuit. The chip size is 2.3 x 0.8 mm. A measured gain of 8-dB from 3 to 33 GHz and a measured midband noise figure of 3-dB from 10 to 30 GHz were achieved. These are the best reported results for an amplifier over this bandwidth.

INTRODUCTION

HEMTs have demonstrated superior gain and noise figure performance over conventional MESFETs [1]. State-of-the-art gain and noise figure performance has been achieved from monolithic HEMT amplifiers up to Q-band [2-3].

The distributed amplifier is known for its broadband flat gain and low input/output VSWR performance. Monolithic distributed amplifiers have been developed up to 45 GHz using MESFETs as active devices [4].

This paper describes the results of a 3 to 40 GHz monolithic distributed amplifier using 0.25 µm HEMTs having a mushroom gate profile as the active devices. A gain of 8-dB from 3 to 33 GHz and a midband noise figure of 3-dB from 10 to 30 GHz were measured, which are the best results reported to date.

DEVICE CONSIDERATIONS

Figure 1 shows the HEMT epitaxial growth structure used in this amplifier. These HEMT layers were grown by MBE at 580°C on GaAs.

The I-V characteristics of a discrete 0.25 x 150 µm HEMT is shown in Figure 2. The gate is π–configured with a triangular gate profile.

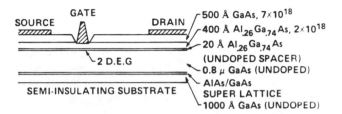

Fig. 1   Cross section of HEMT device.

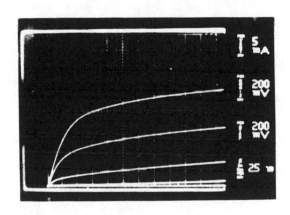

Fig. 2. I-V Characteristics of 0.25 x 150 µm HEMT device.

This device has a typical I_{dss} of 170 mA/mm, a g_m of 520 mS/mm at maximum gain bias and 380 mS/mm at minimum noise bias, a pinch-off voltage of 0.8V and a breakdown voltage of 5V.

Figure 3 shows the equivalent circuit model for this HEMT biased for minimum noise figure. The device was first characterized by the S-parameter measurements in the common-gate, common-source and common-drain configurations from 2-18 GHz. An equivalent circuit model is then constrained to fit the three sets of data.

This 0.25 x 150 µm HEMT has a measured minimum noise figure of 1.35-dB and an associated gain of 12-dB at 18 GHz.

Reprinted from *Rec. of the IEEE GaAs Integrated Circuits Symp.*, pp. 105–108, 1988.

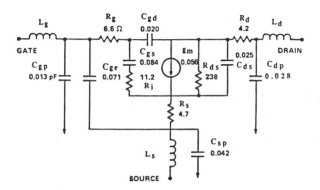

Fig. 3 Equivalent circuit model of 0.25 x 150 μm HEMT device.

CIRCUIT DESIGN

A 3-40 GHz monolithic distributed amplifier was designed using 0.25 μm HEMTs with a triangular gate profile. Figure 4 shows the schematic layout of this amplifier which consists of five 0.25 μm HEMTs with variable gate widths and on-chip biasing circuits. The electrical lengths of the transmission lines are given for 30 GHz. The HEMTs are T-configured and the total gate width is 475 μm. Twenty-four circuit parameters, including five HEMT gate widths and twelve gate and drain line lengths, were optimized for flat gain performance from 2 to 40 GHz. A simulated gain of 6.5-dB and an input/output return loss of better than 15-dB from 3 to 40 GHz were obtained

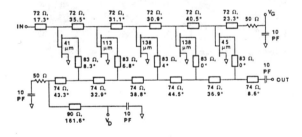

Fig. 4 Schematic circuit diagram for monolithic 3-40 GHz distributed amplifier.

By allowing the HEMTs to vary in width and the gate and drain line sections to vary in length, the gain is enhanced by 1-2-dB and an improved return loss is achieved when compared to the conventional constant-K design [2]. Generally the wider HEMTs are located in the center of the amplifier.

AMPLIFIER FABRICATION

Standard processing techniques were used for most of the HEMT distributed amplifier fabrication. Isolation was achieved with a 2500 Å mesa etch. The 0.25 μm gate was written with a Cambridge EBMF 10.5 E-beam machine using PMMA resist. The gate recess etch was a self-aligned wet etch process. The gate cross section was triangular with a gate resistance of 470 Ω/mm. A sputtered SiO_2 layer of 2000 Å was used as the capacitor dielectric material. In order to achieve good RF grounding, 60 x 60 μm backside vias were incorporated using reactive ion etching. All other process steps used conventional metallization, liftoff and pulse-plating techniques.

Figure 5 shows a photograph of the 3 to 40 GHz distributed amplifier. The chip size is 2.3x 0.8 mm, and the layout of the circuit is compatible with Cascade Microtech RF wafer-probing.

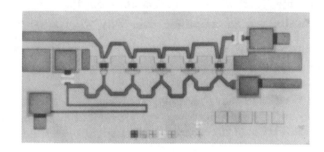

Fig. 5 Photograph of monolithic 3-40 GHz distributed amplifier.

MEASURED GAIN PERFORMANCE

A gain of approximately 6-dB from 3 to 33 GHz and an input return loss of better than 10-dB from 2-40 GHz (biased at V_d = 3V, V_g = -0.4V and I_{ds} = 45 mA) were measured using the Cascade Microtech RF wafer prober, as shown by the solid curves in Fig. 6. Measurement was accomplished using the coax-based (2.4 mm) HP-8510B network analyzer with full error corrections from 45 MHz to 40 GHz. The measured gain agrees well with the simulated result, except at the high end of the band where the measured gain starts to fall at 33 GHz. This gain degradation could be due to any combination of higher values of gate resistance, source inductance [5], and transmission line loss than the values used in the simulation.

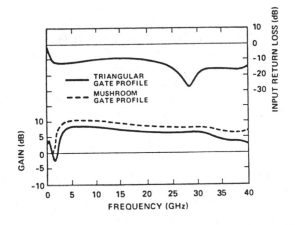

Fig. 6 Measured gain and input return loss performance of the monolithic 3-40 GHz distributed amplifiers.

The HEMT device is particularly suited for achieving high gain because of its higher transconductance and lower output conductance (per unit g_m) compared with the MESFET. The higher transconductance is a result of the increased saturated velocity of the two-dimensional electron gas layer and the fact that AlGaAs can be doped higher than GaAs without compromising the gate breakdown voltage (due to the larger bandgap). The lower output conductance is a result of the two-dimensional nature of the conduction electrons and the thinner epitaxial layer thickness.

MEASURED NOISE FIGURE PERFORMANCE

This amplifier has the characteristic noise performance of a distributed amplifier [6] as shown by the solid curve in Fig. 7. It has a high noise figure at low frequencies due to the gate

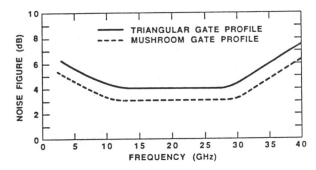

Fig. 7 Measured noise figure performance of the monolithic 3-40 GHz distributed amplifiers.

termination resistor. The noise figure near the cutoff frequency rises because of the increasing impedance of the transmission line as cutoff is approached. In the midband, the amplifier has a low noise figure.

The solid curve in Fig. 7 shows approximately 4-dB measured noise figure for the distributed amplifier from 10 to 30 GHz when biased at minimum noise bias condition.

The amplifier noise figure was measured using the HP noise measurement system and the Cascade prober. There is a measurement discontinuity at 26.5 GHz, where the measurement apparatus was changed. Measurement below 26.5 GHz was accomplished using the HP346C noise source and a coax-based setup. Above 26.5 GHz, the measurement was performed using the waveguide-based test setup and the HP-R347B noise source.

The HEMT device has demonstrated superior low-noise performance and suitability for broadband low-noise applications [7]. The lower noise figure performance for HEMTs compared to MESFETs is due to the higher cutoff frequency (i.e., higher g_m/C_{gs}) and the higher correlation coefficient between the drain and gate noise current sources, subtracting partially the gate noise from the drain noise and reducing the intrinsic noise figure in the HEMT [7]. In addition, the HEMTs have a lower noise conductance, which results in reduced sensitivity of the noise figure to changes in source impedance and therefore permits low-noise performance over a wider bandwidth.

PERFORMANCE OF AMPLIFIER WITH MUSHROOM GATE PROFILE

The 0.25 μm gate with a mushroom gate profile has been developed recently using the tri-level resist and E-beam lithography. Fig. 8 shows a SEM photograph of a typical mushroom gate profile. The DC gate resistance of a 0.25 μm gate with a mushroom gate profile (70 Ω/mm) is reduced to almost 1/7 of that for a 0.25 μm gate with a triangular gate profile (470 Ω/mm).

The 0.25 x 150 μm π-configured HEMT with a mushroom gate profile has a measured noise figure of 0.95 ± 0.2-dB and an associated gain of 13-dB at 18 GHz, comparable to the performance reported by others [1] for a 0.25x 150 μm T-configured HEMT.

A 3-40 GHz distributed amplifier with a mushroom gate profile was fabricated on HEMT

Fig. 8 SEM photograph of a typical mushroom gate profile.

material. This amplifier gave 8-dB of gain from 3 to 33 GHz and a 3-dB noise figure over the 10-30 GHz band (shown in Figs. 6 and 7 as dashed curves). These are the best reported results for a MMIC amplifier over this bandwidth. Compared to the amplifier with a triangular gate profile, the amplifier with the mushroom gate profile has 2-dB higher gain and 1-dB lower noise figure in the 3-40 GHz bandwidth, due to the lower R_g of the device.

MEASURED OUTPUT POWER PERFORMANCE

The 1-dB compression power was measured for the 3-40 GHz HEMT distributed amplifier at 18 GHz. When biased at maximum power, the i-dB compression power for the HEMT amplifier was measured to be 13-dBm at 18 GHz.

DISCUSSION AND CONCLUSIONS

A monolithic low-noise distributed amplifier using HEMT technology for the active device has been developed from 3 to 40 GHz with about 6-dB of gain from 3-33 GHz and approximately 4-dB for the noise figure from 10-30 GHz. By replacing the triangular gate profile with a mushroom gate profile, the amplifier achieved 8-dB gain from 3 to 33 GHz with a 3-dB midband noise figure from 10 to 30 GHz. These are the best reported results for a MMIC amplifier over this bandwidth.

The 3-40 GHz HEMT distributed amplifier has a measured 1-dB compression power of 13-dBm at 18 GHz.

Devices with shorter gate length (i.e., 0.1 µm) has been developed using E-beam lithography, which will improve the gain and noise per-formance of the amplifier further.

With improvement in the noise figure of the device (for example, a 0.1 µm HEMT with a mushroom gate profile), along with the improved circuit topology designed for minimum noise, a midband noise figure around 2-dB for a broadband 2-40 GHz MMIC distributed amplifier should be achievable.

ACKNOWLEDGMENT

The authors would like to acknowledge B. Knapp and C. Shih for their technical support, and M. Day, I. Zubeck and Z. Tan for their assistance in the E-beam lithography.

REFERENCES

1. P. M. Smith, P. C. Chao, K. H. G. Duh, L. F. Lester, B. R. Lee and J. M. Ballingall, "Advances in HEMT Technology and Applications", 1987 IEEE MTT-S International Microwave Symposium Digest, p. 749.

2. S. Bandy, C. Nishimoto, C. Yuen, R. LaRue, M. Day, J. Eckstein, Z. Tan, C. Webb and G. Zdasiuk, "A 2-20 GHz High-Gain Monolithic HEMT Distributed Amplifier", IEEE Trans. Microwave Theory and Tech., MTT-35(12), 1494 (1987).

3. C. Yuen, M. Riaziat, S. Bandy and G. Zdasiuk, "Application of HEMT Devices to MMICs", Microwave Journal, Aug. 1988.

4. M. J. Schindler, J. P. Wendler, A. M. Morris and P. A. Lamarre, "A 15 to 45 GHz Distributed Amplifier Using 3 FETs of Varying Periphery", GaAs IC Symposium 1986 Technical Digest, p. 67.

5. C. Yuen, S. Bandy, S. Salimian, C. B. Cooper III, M. Day and G. Zdasiuk, "Via Hole Studies on a Monolithic 2-20 GHz Distri-buted Amplifier", IEEE Trans. Microwave Theory and Tech., MTT-36(7), 1191 (1988).

6. C. S. Aitchison, "The Intrinsic Noise Figure of the MESFET Distributed Amplifier", IEEE Trans. Microwave Theory and Tech., 33(6) (1985).

7. A. Cappy, "Noise Modeling and Measurement Techniques", IEEE Trans. Microwave Theory and Tech., 36(1), 1 (1988).

2 TO 42 GHz FLAT GAIN MONOLITHIC HEMT DISTRIBUTED AMPLIFIERS

Patrice GAMAND, Alain DESWARTE, Michel WOLNY,
Jean-Christophe MEUNIER, Patrick CHAMBERY

LEP : Laboratoires d'Electronique et de Physique Appliquée
A member of the Philips Research Organization
3, avenue Descartes, 94451 LIMEIL-BREVANNES, France,
Phone (1)45.69.96.10

Abstract

Ultra wide band monolithic HEMT amplifiers have been designed and fabricated using MOVPE heterostructures. A cascode configuration has been used to reduce the coupling effects and therefore to improve the frequency bandwidth of the amplifier. A 6 ± 1 dB gain have been obtained from 2 to 42 GHz. In particular, a gain ripple better than ± 0.5 dB up to 26.5 GHz has been currently measured directly on wafer. The I/O VSWR's are less than 2.5 over the all bandwidth. The average noise figure of the amplifier is 4.2 at 12 GHz rising up 5.2 at 18 GHz. The chip size is 2.3 x 0.9 mm2.

INTRODUCTION

Wide frequency bandwidth in distributed monolithic amplifiers has been achieved using quater micron MESFET's (1-2). The first HEMT monolithic distributed amplifier ever reported (3) covers the 2 to 20 GHz frequency band and uses 0.3 µm devices on MBE material. We report here the first monolithic 2 to 42 GHz distributed amplifiers using 0,5 µm MOVPE HEMT's and cascode stages.

The advantages of the Cascode configuration have also been widely described in the literature. However at frequencies above 30 GHz, characteristic responses of the amplifier are very sensitive to the parasitic elements such as discontinuities and pads capacitances for instance but also the the decoupling circuit of the gate of the second transistor. This particular aspect has been studied and is discussed further in this paper.

0,5 µm MOVPE HEMT'S

Figure 1 shows the HEMT epitaxial heterostructure used in the amplifier. The gate is T-configurated and the drain to source spacing is 3 µm. The device was characterized by on wafer S parameter measurements and a MAG of 5.4 dB has been extrapolated at 40 GHz for a 0.5 x 100 µm transistor. The cut off frequency Ft is typically 32 GHz at Idss/3 for this device.

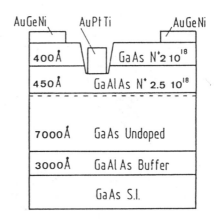

Figure 1 : Structure of the HEMT

Due to the high transconductance of HEMT's, the coupling between gate and drain lines in distributed amplifier has to be reduced to avoid unstability problems.

The use of cascode pairs combines several advantages : high gain block, lower feedback coupling and therefore wider bandwidth.

However accurate RF decoupling of the common gate transistor is utterly required. At frequencies below 30 GHz, the effect of the decoupling network on the gain of the amplifier is very small but becomes critical at high frequency.

Figure 2 presents simulated gain on an amplifier as a function of the length of the transmission line used to connect the gate at the DC blocking capacitor. It can be seen that a small change of the length or a shift in the reference plane between the capacitor and the line leads to a large variation of the gain response. Therefore, to optimize the gain response, the MIM capacitor has been modelled as a distributed element (4) and the line length has been adjusted to provide a flat gain.

Reprinted from *Rec. of the IEEE GaAs Integrated Circuits Symp.*, pp. 109–111, 1988.

The amplifier has thus been optimized by taking the distributed nature of the MIM capacitors into account and the topology of the transistor. In addition, the length of the lines has been varied to compensate for the losses along the transmission lines. The use of integrated resistor to bias the amplifier has been possible because of the low current level of the device used. The lower end of the frequency bandwidth has thus been extended. The topology of the amplifier has been optimized in order to reduce the sensitivity of the gain response to the coupling effects between components.

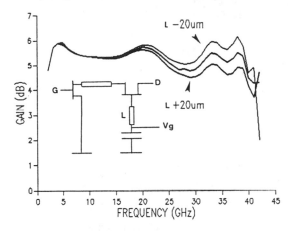

Figure 2 : Effect of the decoupling circuit on the gain response

CIRCUIT TECHNOLOGY

A MMIC technological process using MOVPE HEMT's has been developed for a variety of microwave and mm-wave applications. A total gate width of 850 μm is accomodated in this circuit. Figure 3 presents a microphotograph of the circuit. The chip size is 0.9 x 2.3 mm^2.
An excellent 3dB/mm^2 real estate efficiency has thus been obtained for a 2 to 40 GHz amplifier as can be seen in figure 4. A further reduction of

the chip size associated with an increase of the cut off frequency of the amplifier (i.e. of the transistor) allow to predict monolithic wide band amplifiers with a real estate efficiency X bandwidth product up to 150 dB x GHz/mm^2.

CIRCUIT PERFORMANCE

The measurements have been carried out successively over two frequency ranges : using an on wafer measuring system up to 26.5 GHz and a scalar network analyser up to 42 GHz. No additional matching element external to the chip has been used.

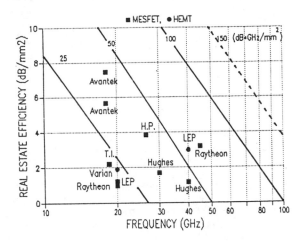

Figure 4 : Real estate efficiency of monolithic distributed amplifiers

Figure 5 presents the typical gain and I/O VSWR's from 2 to 40 GHz. 6 ± 1 dB gain is obtained with VSWR's less than 2.5 over the all bandwidth.
5 dB gain have also been measured at 42 GHz. A very good agreement between measured and calculated gain responses has been observed as shown in figure 5.

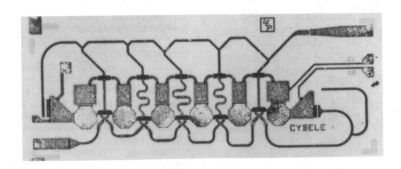

Figure 3 : Microphotograph of the amplifier
Chip size : 2 mm^2

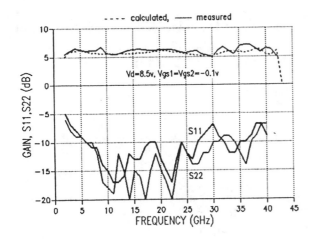

Figure 5 : Gain and return losses
of the amplifiers

Figure 6 presents the noise figure of the amplifier up to 18 GHz. 4 dB noise figure have been measured at low frequency rising up to 5.2 dB at 18 GHz. As previously reported (3), HEMT's monolithic wideband amplifiers exhibit lower noise figure than MESFET's amplifiers.

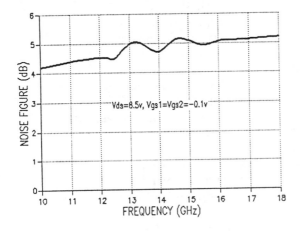

Figure 6 : Noise figure
as a function of frequency

CONCLUSION

Flat gain monolithic distributed amplifiers, using MOVPE heterostructures, have been fabricated. Typical 6 ± 1 dB gain have been measured from 2 to 40 GHz band, with I/O VSWR's less than 2.5 and 5 dB gain have also been measured at 42 GHz. These results show that MMIC's using HEMT's are very promising for mm-wave applications.

REFERENCES

1 R. Pauley, P. Asher, J. Schellenberg,
   H. Yamasaki
   "2 to 40 GHz monolithic distributed
   amplifier"
   GaAs IC Symposium, nov. 1985

2 M.J. Schindler, J.P. Wendler, A.M. Morris,
   R.A. Lamarre
   "A 15 to 45 GHz distributed amplifier using 3
   FET's of varying periphery"
   GaAs IC Symposium, oct. 1986

3 C. Nishimotom, R. Larue, S. Bandy, M. Day,
   J. Eckstein, C. Webb, C. Yuen, G. Zdasiuk
   "A 2-20 GHz, high gain, monolithic HEMT
   distributed amplifier"
   IEEE Microwave and Millimeter Wave Monolithic
   Circuits Symposium, June 1987

4 J.P. Mondal
   "An experimental verification of a single
   distributed model of MIN capacitors for MMIC
   applications"
   IEEE Trans. on Microwave Theory and
   Techniques, vol. MTT-35, n° 4, April 1987

An Investigation of the Power Characteristics and Saturation Mechanisms in HEMT's and MESFET's

MATTHIAS RAINER WEISS, MEMBER, IEEE, AND DIMITRIS PAVLIDIS, SENIOR MEMBER, IEEE

Abstract—A new method for large-signal transistor analysis is presented, which is based on the harmonic-balance approach, but is able to use: 1) input data from measured S-parameters instead of dc or pulsed dc characteristics and 2) a large-signal equivalent circuit with "harmonic elements." The topology of this circuit is nearly identical to commonly used small-signal equivalent circuits; its application enables a detailed interpretation of the computed results, which are very precise due to the use of small-signal S-parameters. The large-signal model is applied to HEMT's and MESFET's; the saturation mechanisms of both types of transistors are investigated with the model, and the operation difference are discussed. The importance of including higher harmonic signal components in the large-signal analysis is also shown.

NOMENCLATURE

C_{dg}	Gate–drain capacitance.
C_{gs}	Gate–source capacitance.
d_i	Spacer-layer thickness.
D_I	Implantation dose.
d_0	AlGaAs-layer thickness.
E	Implantation energy.
F_c	Critical electric field.
G_{dg}	Gate–drain conductivity.
G_{ds}	Gate–source conductivity.
G_{gf}	Gate forward conductivity.
G_{opt}	Optimum gain.
$I_{dg}, I_{dg}(t)$	Gate–drain current.
I_{dgi}	Fourier components of I_{dg}.
$I_{ds}, I_{ds}(t)$	Drain–source current.
I_{dsi}	Fourier components of I_{ds}.
$I_{gs}, I_{gs}(t)$	Gate current.
I_{gsi}	Fourier components of I_{gs}.
l_{Cdg}	Current through C_{dg}.
I_{Cgs}	Current through C_{gs}.
I_{Gd}	Current through C_{ds}.
I_{Gdg}	Current through G_{dg}.
I_{Ggf}	Current through G_{gf}.
I_{gm}	Current through g_m.
L_g	Gate length.
N, N_d	Doping concentration.
N^+	Doping of the "+" layer.
PAE	Power-added efficiency.
P_{in}	Input power.
P_{out}	Output power.
P_{02}	Second harmonic output power.
P_{03}	Third harmonic output power.
Q	Transistor charge.
R_d	Drain resistance.
R_g	Gate resistance.
R_i	Intrinsic gate resistance.
R_p	Projected range (LSS theory).
R_s	Source resistance.
$u_{\gamma i}$	($\gamma = g, d, dg$) Fourier components of V_g, V_d, V_{dg}.
V_d	Drain voltage.
V_{dg}	Gate–drain voltage.
V_{ds}	Drain bias.
V_g	Gate voltage.
V_{ge}	Extrinsic gate voltage.
V_{gs}	Gate bias.
v_s	Electron saturation velocity.
V_T	Threshold voltage.
W	Gate width.
W_{tot}	Total gate width.
x	Al mole fraction.
Z_{opt}	Optimum load impedance.
ΔR_p	Standard deviation of R_p.
γ	g, d, dg (index).
τ	Transit time.

Manuscript received August 17, 1987; revised December 3, 1987. This work was supported in part by the U.S. Army Research Office under Grant DAAL03-87-K-007.

The authors are with the Center for High Frequency Microelectronics, Solid State Electronics Laboratory, The University of Michigan, Ann Arbor, MI 48109-2122.

IEEE Log Number 8821840.

I. INTRODUCTION

GaAs MESFET's for a long time have been the traditional three-terminal devices for microwave power applications. Another potential candidate, initially recognized for its excellent low-noise characteristics only, is the high-electron-mobility transistor (HEMT). GaAs/AlGaAs multiple-heterojunction HEMT's have demonstrated very good characteristics for power applications, especially at high frequencies. Saunier and Lee [1] have reported at 32.5 GHz a power of 0.6 W/mm, a gain of 5.4 dB, and a power-added efficiency of 30 percent. The same device at 60 GHz had a power of 0.4 W/mm, a gain of 3.6 dB, and a power-added efficiency of 14 percent. HEMT's are also very well suited for integrated-circuit applications [2], [3]. It is important, therefore, to study

Reprinted from *IEEE Transactions on Electron Devices*, vol. 35, no. 8, pp. 1197–1206, August 1988.

their large-signal characteristics and saturation mechanisms in order to better understand their operation and achieve optimum power performance.

Furthermore, a simultaneous examination of similar effects in MESFET's allows a better understanding of the relative merits of each technology. These aspects have been examined both theoretically and experimentally and are presented in this paper. The influence of physical parameters on device performance has been treated by the authors earlier [4], [5].

The method of large-signal analysis is described in Section II. A discussion of measured and modeled results can be found in Section III. Section IV presents an analysis of HEMT and MESFET saturation mechanisms. The bias dependence of large-signal elements is discussed in Section V. Section VI shows finally the inclusion of high harmonic signals in the analysis and their effect on the computed results.

II. POWER PREDICTION FROM SMALL-SIGNAL MEASUREMENTS

To investigate the large-signal properties of FET's, the small-signal S-parameters of devices realized in our laboratories were measured in the 2–18-GHz range for various gate–source and drain–source bias conditions V_{gs} and V_{ds}.

At each bias point, the elements of the equivalent circuit shown in Fig. 1 were then evaluated to fit the S-parameters. The parasitic resistances R_s, R_d, and R_g were determined by dc measurements. Gate forward conduction and gate–drain breakdown were simulated by conductances G_{gf} and G_{dg} connected in parallel to C_{gs} and C_{dg} capacitances [6] (see Fig. 1). Their values are derived from measured dc diode characteristic during the nonlinear modeling process described in this section.

The transistors were multifinger structures with 0.5- to 0.6-μm gate length and $W = 75$ μm finger width for the HEMT's ($W_{\rm tot} = 150$ μm) and 1-μm gate length and 150-μm finger width for the MESFET ($W_{\rm tot} = 600$ μm).

Interpolation functions fitting the voltage-dependent characteristics of equivalent circuit parameters were then determined. This technique was applied to g_m, G_{ds}, C_{gs}, and C_{dg}, τ, and R_i. The functions have been chosen in a way that the bias dependence of the intrinsic "high-frequency" drain current (I_{ds}) and transistor charge (Q) could be evaluated analytically by integrating the following equations (see the Appendix).

$$g_m = \frac{\partial I_{ds}}{\partial V_g}$$

$$G_{ds} = \frac{\partial I_{ds}}{\partial V_d} = G_d \tag{1a}$$

$$C_{gs} = \frac{\partial Q}{\partial V_g}$$

$$G_{gd} = \frac{\partial Q}{\partial V_{dg}} \tag{1b}$$

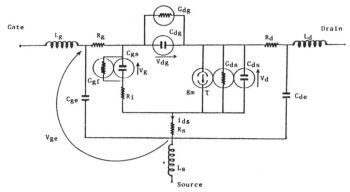

Fig. 1. Large-signal equivalent circuit used for the study of power HEMT's and MESFET's.

where V_g, V_d, and V_{dg} are the intrinsic gate–source, drain–source, and drain–gate voltages, respectively (see Fig. 1).

Instead of the high-frequency methodology used here, the fitting parameters could have been determined by approximating (A1a) and (A2b) to measured dc or low-frequency characteristics. Such an approach, without consideration of high-frequency harmonics, has been presented in [6]. In spite of the fact that this procedure also provides the necessary input data and economizes measurement effort, it introduces considerable error due to the impact of traps and interface states with long time constants. The approach proposed in this paper avoids these problems by characterizing the devices directly at the high frequencies where they are used.

The power-dependent "large-signal" parameters were evaluated as average instantaneous values over a signal period [4], [7]. The flow-chart of Fig. 2 describes the calculations. The following steps were necessary:

Step 1: Small-signal parameters are evaluated at the given bias point: To begin the calculations, the intrinsic bias voltages V_g, V_d are determined from the given dc $I_{ds}(V_{gs}, V_{ds})$ characteristics and R_s, R_d values. The small-signal elements g_m, G_{ds}, C_{gs}, and C_{dg} can be calculated using (1). The complex voltage amplitudes u_{g1}, u_{d1}, and u_{dg1} needed for calculating time-dependent voltages are provided by small-signal network analysis. The instantaneous voltages are calculated using the following equations:

$$V_\gamma(t) = {\rm Re}\left\{ V_{\gamma 0} + u_{\gamma 1}e^{j\omega t} + u_{\gamma 2}e^{j2\omega t} + \cdots \right\}$$

$$\text{with } \gamma = g, d, \text{ or } dg. \tag{2}$$

In this first step, the higher harmonic amplitudes $u_{\gamma 2}$, $u_{\gamma 3}$, $\cdots$ are set to zero.

Step 2: The complex current amplitudes I_{dsi}, I_{gsi}, and I_{dgi} (the index "i" denotes the ith harmonic component) are determined (step (2a)) at each considered harmonic frequency from the time-domain currents $I_{ds}(t)$, $I_{gs}(t)$, and $I_{dg}(t)$ by Fourier transforming (step (2b))

$$I_{ds}(t) = I_{ds}(V_g(t), V_d(t)) \tag{3a}$$

$$I_{gs}(t) = I_{Cgs}(V_g(t), V_d(t)) + I_{Ggf}(V_g(t)) \tag{3b}$$

$$I_{dg}(t) = I_{Cdg}(V_g(t), V_d(t)) + I_{Gdg}(V_{dg}(t)). \tag{3c}$$

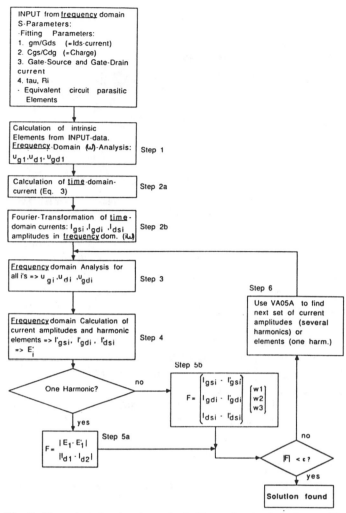

Fig. 2. Flow-chart showing the method of large-signal calculation used in the computer program.

$I_{ds}(V_g, V_d)$ of (3) is directly given by analytical integration of (1) (see the Appendix). I_{Cgs} and I_{Cdg} are the currents flowing through C_{gs} and C_{dg} and can be derived from (1) by using the following relations:

$$I_{Cgs} = \frac{\partial Q}{\partial V_g} \cdot \frac{dV_g}{dt} \qquad (4a)$$

$$I_{Cdg} = \frac{\partial Q}{\partial V_{dg}} \cdot \frac{dV_{dg}}{dt}. \qquad (4b)$$

The currents I_{Gdg} and I_{Ggf} through G_{dg} and G_{gf} are finally determined from measured dc diode characteristics.

Step 3: A network analysis is performed at each considered harmonic frequency, providing the complex voltage amplitudes u_{gi}, u_{di}, and u_{dgi} (the index "i" denotes the ith harmonic component).

Step 4: Using these new amplitudes in (2), one obtains through (3) modified current characteristics and new harmonic complex current components I_{di}, I_{gi}, and I_{dgi}. The values of "harmonic elements" are calculated from

$$g'_{mi} = \left| \frac{I_{gmi}}{u_{gi}} \right| \qquad (5a)$$

$$\tau'_i = \frac{1}{\omega} \arctan \left(\frac{\text{Im} \left(\frac{I_{gmi}}{u_{gi}} \right)}{\text{Re} \left(\frac{I_{gmi}}{u_{gi}} \right)} \right) \qquad (5b)$$

$$G'_{dsi} = \text{Re} \left\{ \frac{I_{Gdi}}{u_{di}} \right\} \qquad (5c)$$

$$C'_{dsi} = \frac{1}{i\omega} \text{Im} \left\{ \frac{I_{Gdi}}{u_{di}} \right\} + C_{ds(ss)} \quad (ss \text{ is small signal}) \quad (5d)$$

$$G'_{gfi} = \text{Re} \left\{ \frac{I_{Gsi}}{u_{gi}} \right\}$$

$$C_{gsi} = \frac{1}{i\omega} \text{Im} \left\{ \frac{I_{gsi}}{u_{gi}} \right\} \qquad (5e)$$

$$C'_{dgi} = \text{Re} \left\{ \frac{I_{dgi}}{u_{dgi}} \right\}$$

$$C_{dgi} = \text{Im} \left\{ \frac{I_{dgi}}{u_{dgi}} \right\} \cdot \frac{1}{i\omega}. \qquad (5f)$$

In the above equations the complex current amplitudes I_{gmi} and I_{Gdi} through g_m and G_{ds} are determined by Fourier transformation of the following equations:

$$I_{gm}(t) = \int_t g_m(t) \frac{dV_g}{dt} \, dt \qquad (6a)$$

$$I_{Gd}(t) = \int_t G_d(t) \frac{dV_d}{dt} \, dt. \qquad (6b)$$

The remaining currents of (5) can be determined using (3).

Step 5: Initial current amplitudes I or harmonic elements E are compared to those of step 4. If the difference is small enough, the calculation is stopped and a solution is found. Otherwise, a next set of current amplitudes and harmonic elements is determined using the Harwell libraries optimization program VA05A (step 6).

III. APPLICATION OF THE LARGE-SIGNAL MODEL TO HEMT's AND MESFET's

Fig. 3(a), (b) compares the measured and calculated power characteristics of a HEMT and an ion-implanted MESFET at the basic and higher harmonic frequencies in order to demonstrate the accuracy of the model; different load values are used for terminating the HEMT's and the device parameters are listed in Tables I and II. A single-heterojunction HEMT was used for this study. Devices of this type show good power performance, particularly at high frequencies. Recent pseudomorphic material developments [8] show, for example, a capability of 0.43 W/mm and 28-percent efficiency at 60 GHz.

Calculations neglecting higher harmonics underestimate the saturation power. The agreement can be improved by taking into account two or more harmonics.

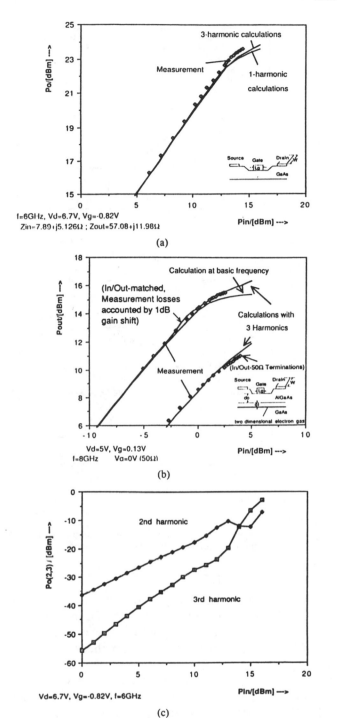

(a)

(b)

(c)

Fig. 3. Comparison of measured output power (squares) with predicted powers (continuous lines) versus input power (a), (b). Calculations take into account either three harmonics or only the basic frequency and dc-signal components: (a) 1 μm $\times$ 600 μm MESFET (see Table I). (b) 0.5 μm $\times$ 150 μm HEMT (see Table II). (c) Simulated second and third harmonic powers versus input-power P_{in} of a 1 μm $\times$ 600 μm ion-implanted MESFET.

TABLE I
NOMINAL PARAMETERS OF INVESTIGATED MESFET's

Parameter	1μ-FET
L_g/ [μm]	1.0
W/ [μm]	600
N^+: E/ [keV]	60
N^+: D_I/ [10^6/m^2]	2.0
N^+: R_p/ [μm]	0.051
N^+: ΔR_p/ [μm]	0.029
N: E/ [keV]	250
N: D_I/ [10^6/m^2]	0.6
N: R_p/ [μm]	0.222
N: ΔR_p/ [μm]	0.079
r/ [Å]	2000<r<2700
V_T/ [V]	-2.5

(L_g is the gate length, W is the gate width, N is the channel doping, N^+ is the N^+ doping, E is the implantation energy, D_I is the implantation dose, R_p is the projected range (LSS-theory), ΔR_p is the standard deviation of R_p, r is the recess, and V_T is the threshold voltage).

TABLE II
PHYSICAL HEMT PARAMETERS

Parameter	0.6μm HEMT
W [μm]	150
L_g [μm]	0.6 (0.5-0.6)
d_o [Å]	516 (900)
d_i [Å]	22
N_d [10^{23}/m^3]	6 (7)
x	0.22
v_s [m/sec]	1.66x10^5
F_c [V/m]	35 000
μ [m^2/(Vsec)]	0.6
R_s [Ω]	6.64
R_d [Ω]	8.62
C_{ds} [fF]	34
V_T [V]	-0.75

(W is the gate width, L_g is the gate length, d_0 is the AlGaAs-layer-thickness, d_i is the spacer-layer thickness, N_d is the doping concentration, x is the Al mole fraction in AlGaAs, v_s is the electron saturation velocity, F_c is the saturation field, μ is the electron mobility, and V_T is the threshold voltage. "( )" indicates expected technology specifications.)

Fig. 3(c) shows the calculated second- and third-harmonic output power (P_{02}, P_{03}) of the MESFET generated by transistor nonlinearities. At very small input powers, P_{02} versus P_{in} and P_{03} versus P_{in} characteristics show as expected a 2:1 and 3:1 slope, respectively. At higher input levels, the harmonic power components deviate from this slope. This happens when the time-domain currents and voltages shift with respect to the time reference changing from symmetric to asymmetric waveforms. Sine or cosine terms are then suppressed in the Fourier transformation resulting in nonregular $P_0(2, 3)$ versus P_{in} characteristics. Similar harmonic characteristics were also reported previously [9].

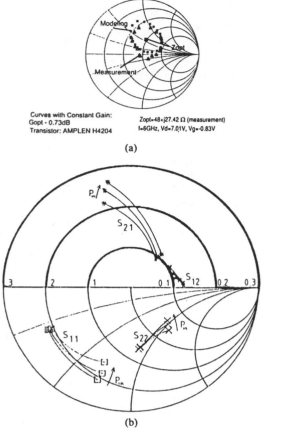

Curves with Constant Gain:
Gopt - 0.73dB
Transistor: AMPLEN H4204

Zopt=48+j27.42 Ω (measurement)
f=6GHz, Vd=7.01V, Vg=-0.83V

(a)

(b)

Fig. 4. (a) Comparison of measured (triangles) and simulated (squares) load-pull contours of a MESFET (1 μm × 600 μm device, Table I). (b) S-parameters calculated from the large-signal equivalent circuit. (f = 4–12 GHz, P_{in_1} = 0 dBm, P_{in_2} = 13 dBm, P_{in_3} = 16 dBm, V_d = 9 V, V_g = −0.82 V.)

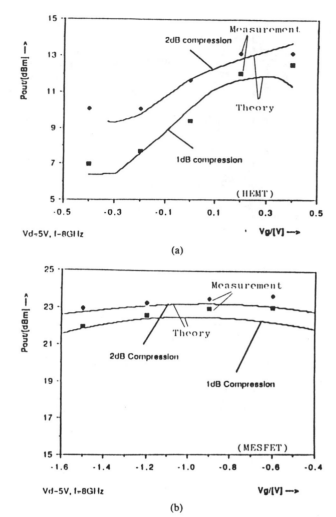

(a)

(b)

Fig. 5. Output power at 1- and 2-dB gain compression as function of dc bias for (a) HEMT and (b) MESFET. (Solid line: model; discrete points: measurement.)

Fig. 4(a) shows measured and calculated load-pull curves of the same MESFET. It demonstrates the validity of this model also for the prediction of load contours. After the transistor has been characterized by dc and S-parameters the model can be used for the calculation of any load-pull condition and therefore avoids long and complicated characterizations. Another feature of large-signal modeling using small-signal S-parameters is that high-precision de-embedding techniques can be used [10]. In our experience, this is much more accurate than de-embedding devices mounted in power packages, where the extraction of impedances at the device level is more difficult. The theoretical results were obtained without taking into consideration higher harmonics. The gain corresponding to the load contours is 0.75 dB smaller than the gain G_{opt} with optimum power at 1-dB compression. The latter occurs with an optimum load Z_{opt}. The agreement between calculations and measurements is excellent, suggesting that the inclusion of higher harmonics was not necessary for these devices. As the harmonic content of the component increases (e.g., the HEMT results of Fig. 3(b)), higher order harmonic calculations become necessary for load-pull predictions.

The model is capable of predicting all four S-parameters as a function of input power. Fig. 4(b) shows the S-

parameters of a MESFET (Table I) as a function of input power. These were obtained from the large-signal equivalent circuits of the device and are given between 4 and 12 GHz. All S-parameters show for this study a dependence on power level.

To further confirm our model also for different bias conditions, Fig. 5 shows the gate-bias dependence of HEMT and MESFET output power at 1- and 2-dB gain compression. To achieve a better measurement reproducibility and eliminate inaccuracies due to imprecisely determined loads, devices were terminated by 50 Ω at both input and output. Good agreement between simulation and measurement can be observed. The output power increases with V_{gs} until V_{gs} = −0.6 V for the MESFET and V_{gs} = 0.1 to 0.4 V for the HEMT. The increase of power with V_{gs} follows the observed gain enhancement. This results from the transconductance increase, which does not vary significantly with input power in this region of bias. Larger positive gate bias leads to output power reduction due to an input power limitation by the partial forward operation of the gate–source diode. This is discussed later in Section V and illustrated by Fig. 8(b) where the har-

196

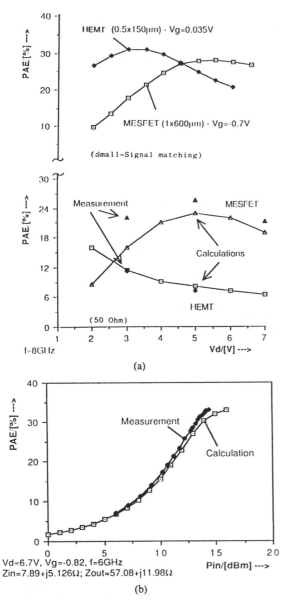

Fig. 6. Calculated and measured power-added efficiency (PAE) of HEMT and MESFET (devices are small-signal matched or terminated with 50 Ω). (a) As function of V'_d; (b) as function of input power.

monic G_{gf} of the HEMT is plotted as a function of V_{gs} at 1- and 2-dB gain compression.

The output power is found to increase with drain voltage [11] and saturate later for MESFET's (at about V_{ds} = 5 V). This is explained by the change of G_{ds} nonlinearity with voltage and will be discussed in Section IV.

Fig. 6(a) depicts the simulated and measured power-added efficiencies (PAE) of a MESFET and a HEMT as a function of drain bias for arbitrary loads. The matching impedances correspond to the complex conjugate S_{11} or S_{22} parameters. Complete matching taking into consideration the nonunilateral characteristics of the device was impossible for stability reasons. Gate bias was chosen to be V_{gs} = 0.035 V for the HEMT and −0.7 V for the MESFET. A comparison between theoretical and experimental data is shown for the case of 50-Ω termination.

Under matched conditions, the HEMT has a maximum

PAE of 31.2 percent at V_{ds} = 3.0 V. For higher drain voltages, the output power saturates with V_{ds} while the dc power is further increasing, leading to decreased PAE. The MESFET reaches its maximum PAE of 28.2 percent at V_{ds} = 5.5 V, which is typical of their slower output power saturation with V_{ds} in comparison to HEMT's. Although both devices are seen to have similar PAE at the investigated frequency of 8 GHz, HEMT's are found to be superior at small operating voltages. The theoretical data shown in Fig. 6(a) with matched loads are supported by the experimental data shown in Fig. 6(b). Here, the MESFET PAE is plotted versus input power for a selected V_{ds}, V_{gs} bias.

Although the HEMT and MESFET devices investigated here are different in gate width and pinchoff voltage, the conclusions of this section remain valid. This was tested by applying the physical device models as described in [4] and [5].

IV. SATURATION MECHANISMS

The device simulations with our model show that there are basically four different mechanisms that may cause power saturation of transistors. These can be classified according to bias conditions and are illustrated for HEMT's and MESFET's by regions I to IV of Fig. 7. They result from the physical limitations of the V_g and V_d voltage swings at the borders of the I–V characteristics. A large region in this figure shows that the indicated saturation mechanism is predominant over a wide range of bias conditions.

Region I: Here, the V_{ds} bias is close to the linear I_{ds}–V_{ds} region. Any V_{ds} swings under large-signal conditions will easily lead to partial operation in the linear I_{ds}–V_{ds} region. This limits the output power and causes FET saturation. In terms of device parameter characteristics, this operation is indicated by large power-dependent (increasing) G_{ds} values.

Region II: Here, a near-zero or even positive gate bias V_{gs} is chosen in order to obtain large I_{ds} currents. A large input power swing will cause partial forward operation of the gate–source diode and will induce output power saturation. This effect is associated with large values of gate–source forward conductance.

Region III: A large negative V_{gs} bias is employed in this near-to-pinchoff region. Large input powers will cause even closer operation to pinchoff. Since g_m is here very small, the transistor cannot provide high gains. The output power saturates due to large-signal g_m values decreasing with input level.

Region IV: The drain–source bias V_{ds} is here relatively large and close to the breakdown voltage. Large output voltage swings will be limited by gate–drain avalanche and will lead to saturation of the output power. This saturation mechanism is associated to gate–drain conductance increase (reverse-biased diode).

In practice, all four saturation mechanisms described above are normally present, but generally only one is predominant, depending on the bias condition. Capacitive re-

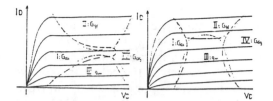

Fig. 7. HEMT and MESFET operating regions according to saturation mechanism by (I) G_{ds}, (II) G_{gf}, (III) g_m, and (IV) G_{dg}.

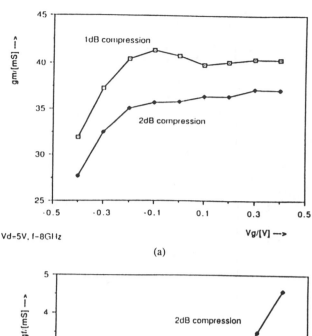

Vd=5V, f=8GHz

(a)

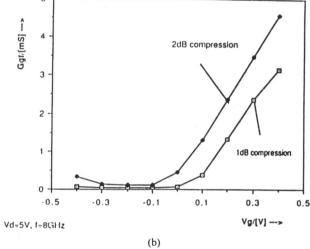

Vd=5V, f=8GHz

(b)

Fig. 8. First-harmonic (a) g_m and (b) G_{gf} of HEMT at 1- and 2-dB gain compression.

actances do not contribute significantly to saturation at the considered power levels, since the limitations for the V_g and V_d voltage swings are mainly due to current rather than charge voltage dependence. They may, however, influence the higher harmonic power. In the following we will describe the HEMT and MESFET large-signal features using Fig. 7(a) and (b): Region I is very large for MESFET's, extending to high V_{ds} values and including $I_{dss}/2$ operation. For V_{gs} larger than the gate voltage corresponding to $I_{dss}/2$, saturation is dominated by partial forward operation of the gate–source diode (region II).

HEMT's have much smaller regions I and II because of two reasons: 1) g_m compression for near-zero or even positive V_{gs} (caused by partial modulation of electrons in the low-mobility AlGaAs donor layer) results in a g_m decrease at high powers. This is much more important in HEMT's than in MESFET's and starts earlier than the G_{ds} increase. Therefore, the HEMT region III is very large, confining regions I and II to smaller areas. 2) The Schottky-barrier voltage is approximately 0.25 V larger in HEMT's than in MESFET's (for the Al mole fraction $x = 0.22$ of our devices). Slightly higher input voltage swings therefore can be tolerated without causing partial forward operation of the gate–source diode and result in a smaller region II.

Region IV of HEMT's is finally larger than for MESFET's because their avalanche breakdown voltage is usually smaller. This is due to the high doped AlGaAs region, which has usually about 10 times higher doping than MESFET channels.

V. Bias Dependence of Large-Signal Elements

Fig. 8(a), (b) shows HEMT ''basic frequency'' g_m and G_{gf} as a function of V_{gs} bias for two different gain compressions. These figures illustrate also quantitatively the saturation mechanism results of Fig. 7(b). Starting from pinchoff and moving toward 0 V (Fig. 8(a)), one sees initially a small difference between $g_m(-1$ dB$)$ and $g_m(-2$ dB$)$. This is explained by the fact that, at this region, the mechanism of saturation is due to both G_{gf} (see Fig. 8(b)) and g_m rather than g_m only (boarder of regions II and III). In practice, this region of operation is associated with a phenomenon called gain expansion where transistor gain increases with input power. At higher V_{gs}'s, the difference between the two g_m curves becomes larger, indicating that, in this bias region, g_m influences more the output rather than input saturation characteristics of the transistor (region III of Fig. 7). For even larger V_{gs}'s, the g_m differences decrease again. Here the

G_{gf} increase is the predominant reason for the HEMT's gain compression and power saturation (region II of Fig. 7). The similarity of g_m versus V_{gs} characteristics for different drain voltages suggests that the predominant saturation mechanisms are due to g_m and G_{gf} rather than G_{ds} and at low V_{dsi} G_{dg}.

In contrast to HEMT's, the compressed g_m of MESFET's does not change significantly with input drive and never exceeds 10 percent of the small-signal g_m value. This is particularly true for bias between $V_{gs} = -1.5$ V and $V_{gs} = -0.5$ V. In this range, MESFET g_m is not the predominant reason for saturation. For V_{gs} still smaller (close to pinchoff), g_m is very small. Power increases will lead here to larger g_m and therefore gain expansion.

Fig. 8(b) illustrates the saturation mechanism by partial forward operation of the gate–source diode. For V_{gs} near pinchoff, G_{gf} is zero. When V_{gs} becomes larger than about 0 V for the HEMT and -1 V for the MESFET, the G_{gf} corresponding to 1-dB gain compression starts increasing. This occurs at lower gate voltages when the transistor is driven stronger by the input signal (operation at 2-dB gain compression).

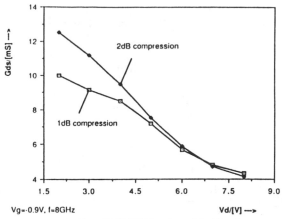

Fig. 9. First-harmonic G_{ds} of MESFET at 1- and 2-dB gain compression.

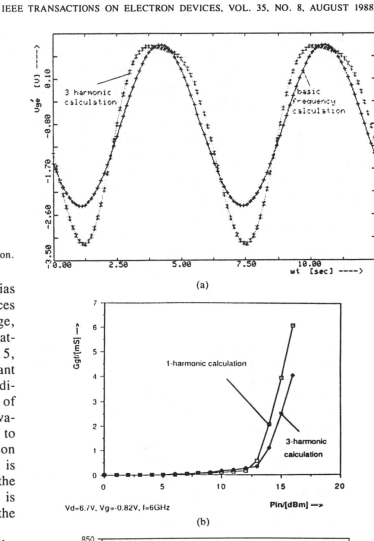

(a)

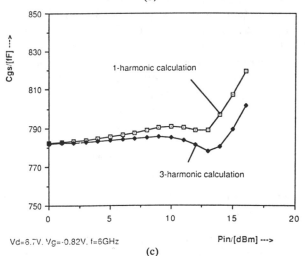

(b)

(c)

Fig. 10. (a) External gate–source voltage V_{ge} of MESFET at basic frequency ($+$) and with consideration of three harmonics ($*$). (b) Gate–source diode conductance G_{gf} of 1 μm × 600 μm MESFET versus input power. (c) Gate–source capacitance C_{gs} of 1 μm × 600 μm MESFET versus input power.

Fig. 9 shows MESFET G_{ds} as a function of drain bias at 1- and 2-dB gain compression. The large differences observed for $V_{ds} < 5$ V indicate that, in this bias range, the MESFET G_{ds} is at least partially responsible for saturation. However, for V_{ds} increasing beyond $V_{ds} = 5$, $\cdots$, 6 V (where the power remains more or less constant with V_{ds}), the difference between the two G_{ds} curves diminishes. In this bias region, partial forward operation of the gate–source diode (G_{gf}) and later gate–drain avalanche (G_{dg}) become increasingly important and lead to transistor saturation. Considering the first saturation mechanism of output power (G_{gf}) and noting that it is independent of the output voltage swing, one expects the output power to show the same independence. If V_{ds} is increased further, the G_{dg} nonlinearity intervenes in the saturation resulting in an output power decrease.

The transition from G_{ds} to G_{gf} saturation occurs at the same drain bias ($V_{ds} = 5$ V) where the PAE has its maximum. Since the output power is no longer limited by G_{ds} increase (related to higher V_{ds}'s) and the input power cannot be further increased due to G_{gf} saturation, PAE saturates or even decreases due to the additional dc power level (V_{ds} increase).

The output power of HEMT's reaches its maximum for drain voltages relatively close to the knee voltage, especially at very high frequencies. This is, however, not the case in MESFET's, where larger V_{ds} values are necessary to obtain optimum output power. If the operating bias is chosen for maximum gain, then both HEMT's and MESFET's show power saturation due to G_{gf} (region II of Fig. 7). With bias conditions suitable for maximum output power, this saturation is seen to occur due to either G_{gf}, G_{ds} for MESFET's (regions II and I) or G_{gf}, g_m for HEMT's (regions II and III).

VI. INFLUENCE OF HARMONIC SIGNAL COMPONENTS

Fig. 10(a) shows time-domain characteristics of the extrinsic gate–source voltage V_{ge} of a MESFET (see Fig. 1), calculated by neglecting higher harmonics ($+$) and by accounting for the first and second harmonic components ($*$). The inclusion of higher harmonics results in flattened peak characteristics for positive voltages, indicating that the partial forward operation of the gate–source diode generates harmonic signal components. Moreover, the negative peak values of V_{ge} are found to be larger, suggesting earlier reaching of pinchoff conditions. The amplitude corresponding to the flattened positive cycle is smaller than the basic frequency amplitude. The resulting gate–source diode current I_{gs} determined using exponential-law diode characteristics therefore is smaller than expected by considering the basic frequency only. Similarly, the gate–diode forward conductance G_{gf} has a smaller value when higher harmonics are considered as shown in Fig. 10(b). This results in output power saturation at larger input levels. The G_{gf} full-harmonic characteristics are in fact responsible for the better experimental and theoretical agreement in the results of Fig. 3(c). The model can also predict the power dependence of the reactive circuit elements. Fig. 10(c) shows C_{gs} as a function of input power. C_{gs} increases with power, but not as pronounced as G_{gf}. Calculations neglecting higher harmonic components overestimate the C_{gs}. A comparison of Figs. 10(b) and (c) shows that G_{gf} rather than C_{gs} is the predominant saturation mechanism.

VII. Conclusions

A large-signal analysis has been presented and applied to HEMT's and MESFET's. By comparing the obtained results to experimental power saturation characteristics and load-pull contours, good agreement was demonstrated between theory and experiment. The saturation mechanisms of MESFET's and HEMT's have been analyzed: While MESFET saturation is predominantly due to a G_{ds} and G_{gf} increase with the input power, HEMT's saturate primarily because of a g_m decrease. Based on our results, the power added efficiency (PAE) of HEMT's and MESFET's can be explained as follows. The HEMT's reach large maximum PAE values at low drain bias V_{ds}, where dc powers are small, while the MESFET gives maximum PAE at larger V_{ds}. Here, dc powers are larger and reduce the PAE. Second and third harmonics have been included in the large-signal analysis. The neglection of higher harmonics was shown to lead to wrong predictions of saturated output power. This is associated with an overestimation of the gate–source current.

Appendix
Interpolation Functions for HEMT and MESFET Characteristics

A. Interpolation Functions for MESFET $I_{ds} - g_m - G_{ds}$ ($I_{d0} = 100$ mA)

$$I_{ds}(V_d, V_g) = I_{d0}\left\{ f \tanh(P_1 V_d) + P_2 V_d \right\},$$
$$f = c_1 V_{gs}^2 + c_2 V_{gs}^3$$
$$V_{gs} = V_g + V_T$$
$$P_1 = (a_1 + a_2 V_g^2)$$
$$P_2 = a_3 + a_5 V_g$$
$$V_T = a_4 + a_9 V_d$$
$$c_1 = a_6 + a_8 V_d$$
$$c_2 = a_7 + a_{10} V_d \tag{A1a}$$

$$g_m(V_d, V_g) = \frac{\partial I_{ds}}{\partial V_g} = I_{d0}\left\{ \frac{\partial f}{\partial V_g} \tanh(P_1 V_d) \right.$$
$$\left. + f \frac{P_1' V_d}{\cosh^2(P_1' V_d)} + P_2' V_d \right\},$$
$$\frac{\partial f}{\partial V_g} = 2c_1 V_{gs} + 3c_2 V_{gs}^2$$
$$P_1' = 2a_2(a_1 + a_2 V_g)$$
$$P_2' = a_5. \tag{A1b}$$

$$G_{ds}(V_d, V_g) = \frac{\partial I_{ds}}{\partial V_d} = I_{d0}\left\{ \frac{\partial f}{\partial V_d} \tanh(P_1 V_d) \right.$$
$$\left. + f \frac{P_1}{\cosh^2(P_1 V_d)} + P_2 \right\},$$
$$\frac{\partial f}{\partial V_d} = c_1' V_{gs}^2 + c_2' V_{gs}^3$$
$$+ (2c_1 V_{gs} + 3c_3 V_{gs}^2)V_T'$$
$$c_1' = a_8$$
$$c_2' = a_{10}$$
$$V_T' = a_9. \tag{A1c}$$

B. Interpolation Functions for MESFET $Q - C_{gs} - C_{dg}$ ($Q_0 = 100$ fC)

$$Q(V_d, V_g) = Q_0\left\{ f*(x_1 + x_1) + g \right\}.$$
$$x_1 = V_g, \quad x_2 = V_d - V_g$$
$$f = a_3 + a_4 x_1 + a_8 x_1^2$$
$$g = \sqrt{Q_1} * \sqrt{Q_2}$$
$$Q_1 = a_1 + a_2 x_1 + a_5 x_1^2$$
$$Q_2 = a_6 + a_7 x_2 \tag{A2a}$$

$$C_{gs}(V_d, V_g) = -\frac{\partial Q}{\partial V_g} = -Q_0\left\{ \frac{\partial f}{\partial x_1}(x_1 + x_2) \right.$$
$$\left. + f + \frac{\partial g}{\partial x_1} \right\},$$
$$\frac{\partial f}{\partial x_1} = a_4 + 2a_8 x_1,$$
$$\frac{\partial g}{\partial x_1} = \frac{Q_1' \sqrt{Q_2}}{2\sqrt{Q_1}},$$
$$Q_1' = a_2 + 2a_5 x_1 \tag{A2b}$$

$$C_{dg}(V_d, V_g) = \frac{\partial Q}{\partial V_{gd}} = Q_0\left\{ f + \frac{\partial g}{\partial x_2} \right\},$$
$$\frac{\partial g}{\partial x_2} = \frac{Q_2' \sqrt{Q_1}}{2\sqrt{Q_2}},$$
$$Q_2' = a_7 \tag{A2c}$$

C. Interpolation Functions for MESFET's: R_i, τ

$$R_i(V_d, V_g) = A \tanh(BV_d),$$

$$A = a_1 + a_2 V_g + a_3 V_g^2$$

$$B = a_4 + a_5 V_g + a_6 V_g^2$$

$$\cdot (V_d \geq 0.5 \text{ V}) \qquad \text{(A3a)}$$

$$\tau(V_d, V_g) = A \tanh(D[V_d - B] + CV_d),$$

$$A = a_1 + a_2 V_g$$

$$B = a_4 + a_5 V_g + a_8 V_g^2,$$

$$C = a_6$$

$$D = a_7 + a_3 V_g. \qquad \text{(A3b)}$$

D. HEMT functions: I_{ds}, g_m, G_{ds} ($I_{d0} = 100$ mA)

The respective functions are given in [5].

E. Interpolation functions for HEMT's: Q, C_{ss}, C_{dg} ($Q_0 = 100$ fC)

The respective functions are given in [5].

ACKNOWLEDGMENT

The authors would like the thank P. Chaumas and P. Resneau for help in characterization and V. Pavlidis for her understanding. The support of the University of Michigan computing staff is greately acknowledged.

REFERENCES

[1] P. Saunier and J. W. Lee, "High-efficiency millimeter-wave GaAs/GaAlAs power HEMT's," *IEEE Electron Device Lett.*, vol. EDL-7, no. 9, pp. 503–505, Sept. 1986.

[2] C. A. Liechti, "Heterostructure transistor technology—A new frontier in microwave electronics," in *Proc. EMC*, (Paris 1985), pp. 21–29.

[3] T. J. Drummond, W. T. Masselink, and H. Morkoç, "Modulation-doped GaAs (Al, Ga) As heterojunction field-effect transistors: MODFET's," *Proc. IEEE*, vol. 74, no. 6, pp. 773–822, June 1986.

[4] M. Weiss and D. Pavlidis, "Power optimization of GaAs implanted FET's," *IEEE Trans. Microwave Theory Tech.*, vol. MTT-35, no. 2, pp. 175–188, Feb. 1987.

[5] ——, "The influence of device physical parameters on HEMT large-signal characteristics," *IEEE Trans. Microwave Theory Tech.*, vol. 36, pp. 239–249, Feb. 1988.

[6] Y. Tajima and P. D. Millter, "Design of broad-band power GaAs FET amplifiers," *IEEE Trans. Microwave Theory Tech.*, vol. MTT-32, pp. 261–267, Mar. 1984.

[7] M. Weiss and D. Pavlidis, "A comparative study of HEMT and MESFET large-signal characteristics and saturation mechanisms," in *Proc. IEEE MTT-S Int. Microwave Symp.* (Las Vegas, NV, June 1987), pp. 553–556.

[8] P. M. Smith, P. C. Chao, K. H. G. Duh, L. F. Lester, B. R. Lee, and J. M. Ballingall, "Advances in HEMT technology and appliances," in *Proc. IEEE MTT-S Int. Microwave Symp.* (Las Vegas, NV, June 1987), vol. II, pp. 748–752.

[9] H. A. Willing and C. Rauscher, "A technique for predicting large-signal performance of a GaAs MESFET," *IEEE Trans. Microwave Theory Tech.*, vol. MTT-26, no. 12, pp. 1017–1023, Dec. 1987.

[10] M. Parisot, Y. Archambault, D. Pavlidis, and J. Magashack, "Highly accurate design of spiral inductors for MMIC's with small size and high cutoff frequency characteristics," in *Proc. Microwave Millimeter-Wave Monolithic Circuits Symp.* (San Francisco, 1984), pp. 91–95.

[11] H. M. Macksey *et al.*, "Dependence of GaAs power MESFET microwave performance on device and material parameters," *IEEE Trans. Electron Devices*, vol. ED-24, no. 2, pp. 113–122, Feb. 1977.

Power Performance of Microwave High-Electron Mobility Transistors

P. M. SMITH, MEMBER, IEEE, U. K. MISHRA, MEMBER, IEEE, P. C. CHAO, S. C. PALMATEER, AND J. C. M. HWANG, SENIOR MEMBER, IEEE

Abstract—The large-signal performance of 0.25-μm gate-length high-electron mobility transistors (HEMT's) operating at 15 GHz is reported. At a drain voltage of 5 V, an output power of 135 mW (0.34 W/mm) has been obtained with 8-dB associated gain and 37-percent power-added efficiency. Furthermore, in class C operation a power-added efficiency of 57 percent has been measured. These results are attributed to high transconductance (300 mS/mm), high gain (12-dB maximum stable gain at 18 GHz), low knee voltage (1.25 V), relatively high full channel current (280 mA/mm at V_{ds} = 2 V), and good RF drain–source breakdown voltage (9 V).

THE HIGH-electron mobility transistor (HEMT) has recently demonstrated great potential as a microwave device for high frequency and low-noise applications [1]–[4], achieving noise figure and current gain cutoff frequency f_t superior to that of the GaAs MESFET. However, the large-signal properties, (e.g., power handling capability, efficiency) of the HEMT have not yet been reported. In this letter, we present measured power performance of 0.25-μm gate-length HEMT's at 15 GHz for a variety of bias and tuning conditions.

Devices were fabricated on AlGaAs/GaAs heterostructure layers grown by molecular beam epitaxy (MBE). The layer structure consists of a 300-Å-thick n$^+$ GaAs cap layer, Si doped at a carrier concentration of 2×10^{18} cm^{-3}, a 500 Å, 1×10^{18} cm^{-3} Si-doped AlGaAs layer, and a 1-μm undoped GaAs buffer. The heterostructure has a sheet charge density N_s, in excess of 10^{12} cm^{-2} and a mobility of 20 000 cm$^2 \cdot$V$^{-1} \cdot$s^{-1} as determined by a Hall measurement at 77 K. While the use of a spacer layer results in a higher value of low-field mobility [4], we have used no spacer in order to maximize the two-dimensional electron gas sheet charge density (and hence current-carrying capability) and transconductance g_m of the device.

Devices with low resistance T-shaped cross-sectional-gates [5] 0.25 μm long and 400 μm wide were fabricated using direct write electron beam lithography. The AuGeNi/Ag/Au source and drain ohmic contact metallurgy was alloyed in a rapid thermal annealing system. After patterning the resist for the gate level and prior to deposition of the Ti/Pt/Au gates, a channel recess was chemically etched completely through the GaAs cap layer and partially into the AlGaAs layer. A source–drain electrode separation of 2 μm is employed, and the gate is located 0.3 μm from the source

Manuscript received November 5, 1984; revised December 6, 1984.
The authors are with the General Electric Company, Electronics Laboratory, Syracuse, NY 13221.

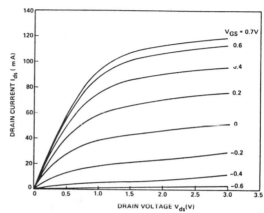

Fig. 1. DC drain I–V characteristics of a 0.25 × 400-μm HEMT.

metallization. The device geometry consists of a single gate stripe fed at two points, yielding an effective unit finger width of 100 μm.

The dc drain characteristics of one device are shown in Fig. 1. A maximum normalized transconductance g_m of 300 mS/mm is obtained at V_{ds} = 2 V, V_{gs} = 0 V, and g_m is seen to be strongly dependent on gate voltage, falling off rapidly with decreasing drain current, resulting in a "soft" pinchoff. The characteristics are less linear (i.e., more variation in transconductance as a function of drain current) than those of a GaAs MESFET of comparable design [6]. Compared with the GaAs FET, the transconductance g_m of the HEMT is reduced at high drain current due to parallel conduction in the AlGaAs layer, and decreases more rapidly at low drain current. As expected for short gate devices, output conductance is relatively high (18 mS/mm), and as a result one must carefully define the saturated current I_{dss}, full channel current I_f, and knee voltage V_k. At V_{ds} = 2 V, an I_f and I_{dss} of 280 and 115 mA/mm, respectively, are noted. This value of full channel current is comparable to the 250–450 mA/mm normally obtained for GaAs power FET's. The full channel knee voltage is approximately 1.25 ± 0.25 V.

15-GHz power measurements were performed using a calibrated semiautomatic power test station. The output power and power-added efficiency are plotted in Fig. 2 as functions of drain voltage for a constant gate bias. The devices were tuned for maximum output power at a gain of approximately 7 dB. A maximum efficiency of 46 percent is observed at V_{ds} = 2.5 V, and this decreases to 27 percent at V_{ds} = 6 V. Output power rises monotonically with drain voltage, reaching a saturated value of 0.35 W/mm at a drain

Reprinted from *IEEE Electron Device Letters*, vol. EDL-6, no. 2, pp. 86–87, February 1985.

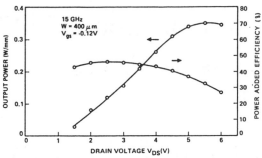

Fig. 2. Drain voltage dependence of 15-GHz output power and power-added efficiency. Device is tuned for maximum output power at 7-dB gain.

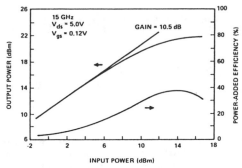

Fig. 3. Output power and efficiency as functions of input power at $V_{ds} = 5.0$ V, $V_{gs} = -0.12$ V.

bias of 5.5 V. The output power versus drain voltage characteristic indicates a knee voltage of 1.25 V and an RF drain–source breakdown voltage of roughly 9 V [7]. (We differentiate here between RF and dc breakdown voltage since time-dependent phenomena, (e.g., carrier trapping, Gunn domain formation, and impact ionization) can cause the two to differ significantly [8].)

This breakdown voltage may seem low compared to that of longer gate GaAs FET's, but is higher than one might expect, considering the short gate length (0.25 μm) and small gate–drain separation in the present structure (as a result of the self-aligned, shallow (~ 500-Å) channel recess, the spacing between the gate and the drain n$^+$ GaAs ledge is less than 0.1 μm). Note that calculations performed for GaAs FET's indicate that the gate–drain avalanche breakdown voltage should be roughly proportional to gate length [9].

The output power and efficiency versus input power at $V_{ds} = 5$ V are presented in Fig. 3. At low input power levels (i.e., small signal) the gain is 10.5 dB. At a gain of 8 dB, the output power is 21.3 dBm (135 mW, or 0.34 W/mm) with a power-added efficiency of 37 percent. The device exhibits a relatively soft power saturation behavior, and the output power at 1-dB gain compression is significantly lower (1.7 dB) than that at saturation. Device nonlinearity is due in part to the nonlinear dc characteristics, but may also be related to the unpassivated surface.

With a fixed drain bias of 2.5 V, the device was then tuned for maximum power-added efficiency as gate voltage was varied from forward conduction ($+0.6$ V) to beyond pinchoff (-1.25 V). Drain current and power-added efficiency are plotted in Fig. 4 as functions of gate voltage. Note that a maximum power-added efficiency of 57 percent (with an associated gain of 5.4 dB, drain efficiency of 80 percent, and output power of 34 mW) is obtained at a gate voltage of -0.75 V, where the device is slightly beyond pinchoff. This is the highest power-added efficiency reported for a transistor at this frequency, and the first reported operation of a microwave transistor in a class B or class C mode at such a high frequency. These results are believed to be due primarily to the high gain of the device.

In summary, the 0.25-μm HEMT has demonstrated extremely high efficiency (57-percent maximum power-added in class C operation) and good power-handling capability (0.34 W/mm with 8-dB associated gain) at 15

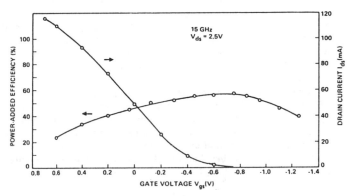

Fig. 4. Drain current and power-added efficiency versus gate voltage, with $V_{ds} = 2.5$ V. Device is tuned for maximum efficiency.

GHz. This performance is attributed to high device gain, low knee voltage, reasonable current-carrying capability, and fairly high drain–source breakdown voltage. These initial results indicate that the HEMT holds great promise for power applications at high frequencies.

REFERENCES

[1] K. Joshin, T. Mimura, Y. Yamashita, K. Kosemura, and J. Saito, "Noise performance of microwave HEMT," in *IEEE MTT-S Int. Symp. Dig.*, June 1983, pp. 563–565.

[2] J. J. Berenz, K. Nakano, and K. P. Weller, "Low noise high electron mobility transistors," in *IEEE MTT-S Int. Symp. Dig.*, June 1984, pp. 98–101.

[3] P. C. Chao, T. Yu, P. M. Smith, J. C. M. Hwang, S. Wanuga, W. H. Perkins, H. Lee, L. F. Eastman, and E. D. Wolf, "Quarter-micron gate length microwave high-electron mobility transistor," *Electron. Lett.*, vol. 19, pp. 894–895, 1983.

[4] P. Delescluse, M. Laviron, J. Chaplart, D. Delagebeaudeuf, and N. T. Linh, "Transport properties in GaAs — Al$_x$Ga$_{1-x}$As heterostructures and MESFET application," *Electron. Lett.*, vol. 17, pp. 342–344, 1981.

[5] P. C. Chao, P. M. Smith, S. Wanuga, W. H. Perkins, R. Tiberio, and E. D. Wolf, "Electron-beam fabrication of quarter-micron T-shaped gate FET's using a new tri-layer resist system," in *IEEE IEDM Tech. Dig.*, Dec. 1983, p. 613.

[6] P. C. Chao, unpublished data.

[7] For further explanation, see J. V. DiLorenzo and W. R. Wisseman, "GaAs power MESFET's: design, fabrication, and performance," *IEEE Trans. Microwave Theory Tech.*, vol. MTT-27, pp. 367–378, 1979.

[8] R. S. Pengelley, *Microwave Field-Effect Transistors—Theory, Design and Applications.* New York: Wiley, 1982, p. 91.

[9] W. R. Curtice, "Investigation of voltage breakdown in GaAs MESFET's with and without gate recess and including surface depletion effects," in *Proc. Eighth Biennial Cornell Elect. Eng. Conf.*, 1981, pp. 209–222.

Microwave Power Double-Heterojunction HEMT's

KOHKI HIKOSAKA, ASSOCIATE MEMBER, IEEE, YASUTAKE HIRACHI, MEMBER, IEEE,
AND MASAYUKI ABE, MEMBER, IEEE

Abstract—The RF and dc characteristics of microwave power double-heterojunction HEMt's (DH-HEMT's) with low doping density have been studied. Small-signal RF measurements indicated that the cutoff frequency and the maximum frequency of oscillation in DH-HEMT's with 0.8–1 μm gate length and 1.2 mm gate periphery are typically 11–16 GHz and 36–41 GHz, respectively. However, the cutoff frequency in DH–HEMT's degrades strongly with increasing drain bias voltage. This may be caused by both effects of increasing effective transit length of electrons and decreasing average electron velocity, due to Gunn domain formation. In large-signal microwave measurement, the DH-HEMT (2.4 mm gate periphery) delivered a maximum output power of 1.05 W with 2.8 dB gain and 0.58 W with 1.6 dB gain at 20 and 30 GHz, respectively. These are the highest output powers yet reported for HEMT devices. For the dc characteristics, the onset of two-terminal gate breakdown voltage is found to correlate with the drain current I_{dss} and recessed length, and three-terminal source–drain breakdown characteristics near pinchoff are limited by the gate–drain breakdown. A simple model on gate breakdown voltage in HEMT is also presented.

I. INTRODUCTION

HIGH ELECTRON mobility transistors (HEMT's) based on modulation-doped AlGaAs/GaAs have considerable potential in the fields of digital and analog devices. Because of the higher average electron velocity ($\bar{V}$; 1.5–2 times over GaAs MESFET's) in two-dimensional electron gas (2-DEG), HEMT's can provide higher frequency operation and higher power-gain performance ($G_a \propto \bar{V}^2$) than GaAs MESFET's. Hence, they offer possibilities for microwave power devices. In order to realize a power device using HEMT's, both high current density and high gate breakdown voltage are required. These requirements are, however, rather difficult to meet with normal single-heterojunction structures ($N_s < 1 \times 10^{12}$ cm^{-2}) based on high doping density (more than 1×10^{18} cm^{-3}). One of the approaches to solve the problems is to use a double-heterojunction structure [1]–[3] to increase current density and a low doping concentration to increase gate breakdown voltage. Consequently, we have developed a selectively low-doped double-heterojunction structure [4] for high-power HEMT's.

The microwave power performance of various HEMT devices has already been demonstrated by the authors [1], [4] and several laboratories [5], [6], mostly using double-heterojunction or multiple-channel structures [6], [7]. Ac-

tually, these structures have offered 2-DEG concentration as high as 1–3×10^{12} cm^{-2} for improving current density and reducing source resistance. Moreover, another advantage of such structures is electron confinement due to effective barrier potential of AlGaAs to hot-electron injection into the substrate, and thereby it results in small drain conductance [3] for short-gate-length HEMT's.

In this paper, we study the RF and dc characteristics of microwave power double-heterojunction HEMT's with low doping density. For dc characteristics, the breakdown behavior is mainly discussed.

II. DEVICE FABRICATION

Selectively doped epi-layers for double-heterojunction structures were grown on semi-insulating GaAs substrates by MBE at a growth temperature of 680°C. The thicknesses of layers such as the GaAs channel layer and the n-AlGaAs layers were initially optimized in order to obtain high 2-D electron concentration ($N_s > 1 \times 10^{12}$ cm^{-2}). The doping density was also chosen to achieve high gate breakdown voltage, which was partly supported by our experiment [4]. The double-heterojunction structure with optimized design [4] consists of a 10–15-nm Si-doped AlGaAs layer (5×10^{17} cm^{-2}), a 10-nm undoped AlGaAs spacer layer, a 10-nm undoped GaAs channel layer, a 15-nm Si-doped AlGaAs layer (5×10^{17} cm^{-3}) and a 30-nm graded Si-doped AlGaAs layer (2×10^{17} cm^{-3}). A GaAs top layer (2×10^{17} cm^{-3}) with a thickness of 200 nm also provided to facilitate the formation of ohmic contacts and to reduce source resistance. The Hall effect measurement in DH-HEMT yielded a sheet electron concentration as high as 1.2–1.5×10^{12} cm^{-2} at both 300 and 77 K. The electron mobility was typically 6800 cm^2/V · s at 300 K and 48 000 cm^2/V · s at 77 K. Power DH-HEMT's with recessed gates were fabricated using the above epi-structure. The recessed depth was about 0.14 μm and the gate electrodes were formed on the recessed n-GaAs layer. The basic process of fabrication of power devices was similar to the previously reported one [4]. Fig. 1 shows the typical device pattern of the power device ($W_g = 1.2$ mm). The devices fabricated had Al gates with 0.8–1 μm gate length and 1.2–2.4 mm total gate periphery. The unit gate width was 60–100 μm. The source–drain and source–gate spacings are 5 and 2 μm, respectively. The FET chips were then mounted on chip carriers, and gate and drain pads were connected to 20-μm Au bonding wire in order to measure the RF characteristics discussed below.

Manuscript received October 7, 1985; revised January 11, 1986. This work was supported by the Ministry of International Trade and Industry (MITI) of Japan.

The authors are with Fujitsu Laboratories Ltd., Fujitsu Ltd., 10-1 Morinosato-Wakamiya, Atsugi 243-01, Japan.

IEEE Log Number 8607884.

Reprinted from *IEEE Transactions on Electron Devices*, vol. ED-33, no. 5, pp. 583–589, May 1986.

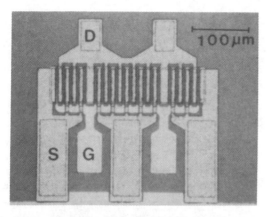

Fig. 1. A device pattern of a power double-heterojunction HEMT (DH-HEMT) with 1-μm gate length and 1.2-mm gate periphery.

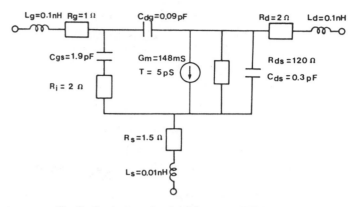

Fig. 3. Equivalent circuit of the power DH-HEMT.

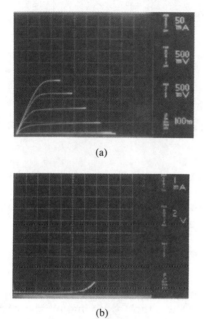

Fig. 2. (a) Drain I–V characteristics of power DH-HEMT with 1-μm gate length and 1.2-mm gate periphery. The scales are 50 mA/div, 0.5 V/div, and −0.5 V/step, for drain current, drain voltage, and gate voltage, respectively. (b) Reverse gate–drain charactistics of power DH-HEMT. The scales are 1 mA/div for gate–drain current and 2 V/div for gate–drain voltage.

III. dc CHARACTERISTICS

Typical drain I–V characteristics of a power DH-HEMT with 1-μm gate length and 1.2-mm gate width are shown in Fig. 2(a). Saturation current I_{dss} is 230 mA and saturation voltage is about 1.3 V. Also, maximum drain current was about 270 mA. The current is 1.5–2 times as high as that in a conventional HEMT, because of much higher sheet electron concentration in DH-HEMT's. Normalized transconductance g_m is typically 100–120 mS/mm, which is rather low due to low doping density in active layers, while output conductance g_d is typically 3–7 mS/mm. Our results show that this output conductance is 1/2–2/3 lower than that in conventional GaAs MESFET (8–10 mS/mm) with the same device dimension. The improvement of output conductance is probably due to the confinement of channel electrons by the heterojunction

barrier of the bottom AlGaAs, which will reduce the short-channel effect. The gate–drain reverse characteristics are shown in Fig. 2(b). The breakdown voltage is about 14 V at a leakage current of 1 mA. This breakdown voltage is significantly higher than that of conventional HEMT's with highly doped layers (1–2 $\times$ 10^{18} cm^{-3}), which is typically 4–8 V. The gate breakdown characteristics in DH-HEMT are also discussed in Section V.

IV. RF CHARACTERISTICS

A. Small-Signal Measurement

Small-signal S-parameter measurements for microwave power DH-HEMT's, covering a range of 2–18 GHz, were made using an automatic network analyzer. These S-parameters were then applied to the equivalent circuit elements, current gain, and maximum unilateral gain. Fig. 3 shows the lumped element of an equivalent circuit in power DH-HEMT ($L_g = 1$ μm and $W_g = 1.2$ mm) for operation in common-source configuration. The bias condition is $V_{ds} = 5$ V and current level is 120 mA. At this bias condition, the gate capacitance was 1.9 pF and the source resistance was about 1.5 Ω. The feedback capacitance C_{dg} was about 0.1 pF. The ratio of C_{dg}/C_{gs} is nearly 5 percent in this structure. This ratio is considerably lower than that in conventional HEMT's [8], [9], which typically shows 15–30 percent. The superiority is caused by the result of the low doping density (2 $\times$ 10^{17} cm^{-3}) in surface layers and relatively large spacing between the gate and drain (2 μm). The current gain and unilateral gain performance under the high drain voltage is shown in Fig. 4. The bias condition was $V_{ds} = 9$ V and $I_{dss} = 100$ mA. An extrapolation of the curves at 6 dB/octave yielded a 11 GHz cutoff frequency f_t and 37-GHz maximum frequency of oscillation f_{max}. The cutoff frequency obtained is a little lower than the expected one. This degradation is in part caused by the large parasitic capacitance of the gate pad area in the power device. The parasitic capacitance, except for fringing capacitance of gate electrode, reached about 15 to 20 percent of total gate capacitance. This parasitic component may be due to relatively shallow oxygen isolation (approximately 0.5–0.6 μm deep) and due to the fact that the implanted depth profile does not entirely cover the buffered undoped GaAs

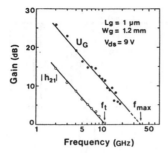

Fig. 4. Current gain and unilateral gain as a function of frequency.

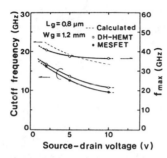

Fig. 5. Cutoff frequency f_t as a function of drain bias voltage in DH-HEMT and GaAs MESFET power devices ($L_g = 0.8$ μm and $W_g = 1.2$ mm). The dashed line shows the estimation of f_t.

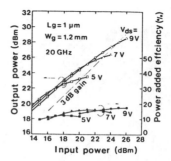

Fig. 6. Output power and power added efficiency of the DH-HEMT as a function of input power for several drain–source bias voltages.

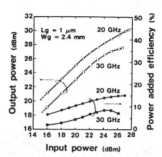

Fig. 7. Output power and power added efficiency as a function of input power at 20 and 30 GHz for a 2.4-mm gate–periphery DH-HEMT.

epi-layer. Therefore, if we can minimize the parasitic capacitance to the order of 5 percent, the f_t should be as high as 15 GHz.

The most interesting result on the *S*-parameter measurement is that the cutoff frequency f_t in power DH-HEMT degrades strongly with the increasing drain bias voltage. Fig. 5 shows an example of such f_t characteristics in a power device with 0.8μm gate length and 1.2-mm gate periphery (the figure also includes f_{max}'s in DH-HEMT's). The results indicate that f_t decreases dramatically from 16.5 to 11 GHz when V_{ds} is increased from 2 to 10 V. This effect indicates possibility of reduced electron velocity, due to a real space transfer of hot-electron injection into the bottom AlGaAs layer. However, this might be explained by both effects [10], [11] of an increase in the transit length (or channel length) of electrons (L_e) and reduction of electron velocity due to Gunn domain formation. The figure also includes the f_t characteristics in conventional GaAs power MESFET with the same device dimensions as DH-HEMT, and the simple estimation of f_t value ($= L_e/2\pi\overline{V}$) in DH-HEMT's. The f_t's are derived from assumptions that the effective transit length of electrons ($L_e > L_g$) is determined by an extra spreading of the depletion region at the drain end of the gate, and the average electron velocity $\overline{V}$ is deduced from the measured dc transconductance ($g_m = \epsilon\overline{V}/d$). The decrease in the calculated f_t's due to each effect was nearly same percentage. From a comparison of these results, it seems that the degradation of f_t is caused by the increase of effective transit length of electrons and the reduction of average electron velocity associated with the stationary Gunn domain formation, but is probably not due to the reduced electron velocity with respect to a real space

transfer effect, from the comparison to GaAs MESFET's. The qualitative discrepancy between measured and calculated f_t's in Fig. 5 might be due to the parasitic capacitance described earlier and parasitic series resistance in the power device.

B. Large-Signal Measurement

Next, to evaluate the microwave power performance of DH-HEMT's the chips were mounted on chip-carriers with input and output transmission lines and open stabs fabricated on Al_2O_3 substrates. The microwave power measurements were done in the same systems as reported in [12]. The power capabilities of the DH-HEMT with 1-μm gate length and 1.2-mm gate width are plotted in Fig. 6, as a function of input power for various drain bias voltages. It is noteworthy that the DH-HEMT with 1-μm gate length delivered an output power of 630 mW (0.53 W/mm) with 3-dB gain at 20 GHz. At a fixed gain of 3 dB, the output power is also shown to increase almost linearly with increasing drain bias voltage. The device exhibits a relatively soft power saturation behavior, which is caused by soft breakdown characteristics of drain current and nonlinear dc characteristics. This soft power saturation behavior, which may degrade intermodulation distortion IM3, does not seem an essential feature of HEMT devices since it depends on epi-structure or chip to chip. So, further optimization of epi-structure will be needed. For maximum output power performance, the DH-HEMT produced 1.05 W with 2.8-dB gain and 0.58 W with 1.6-dB gain at 20 and 30 GHz, respectively, where the device has a 1-μm gate length and a 2.4-mm gate periphery (see Fig. 7).

V. Gate–Drain Breakdown Consideration

Next, we consider the gate–drain breakdown characteristics in the DH-HEMT. The maximum power capability in a three-terminal FET device is proportional to the product of maximum channel current $I_{f\max}$ and drain breakdown voltage V_b. It is generally accepted that in a GaAs MESFET the breakdown voltage V_b is limited by the gate–drain avalanche breakdown, where Wemple *et al.* [13] have shown that gate–drain avalanche voltage can be achieved by factors of 2–3 larger than those predicted by bulk avalanche. However, such a mechanism of drain-source breakdown characteristics has not been studied for HEMT devices. In this section, we show some aspects of drain–source breakdown characteristics and report that near the pinchoff region, source–drain breakdown voltage V_b is limited by the gate–drain breakdown characteristics in HEMT devices, whose mechanism is very similar to GaAs MESFET's. Also, a simple model of gate–drain breakdown voltage in recessed structure is presented in the appendix. The model is modified from the lateral spreading model presented by Wemple *et al.* [13].

Fig. 8(a) shows simple dc measurement of drain *I–V* characteristics of a 1.2-mm power DH-HEMT near the pinchoff region as a function of gate bias voltage, Fig. 8(b) shows characteristics of the same device under more negative gate bias conditions. The device has a drain current I_{dss} of 220 mA, and pinchoff voltage of about 2.5 V. The striking feature of these characteristics is that the source-drain breakdown voltage increases with an increase in the negative gate bias voltage up to about -3 V, but it shifts to a lower value, with a step slightly larger than those corresponding to a negative gate bias step, as the gate voltage exceeded about -3.5 V. The observed maximum drain voltage V_b near pinchoff is around 17 V, and at this point the gate bias voltage is about -3 V. Two-terminal gate–drain reverse characteristics and three-terminal gate leakage current characteristics are also shown in Fig. 8(c). It can be seen from the comparison of Fig. 8(b) and (c) that excess drain current, which is much greater than the gate leakage current, flows at a gate bias voltage of -2.5 V. However, the excess drain current does correspond well to the gate leakage current under more negative gate bias conditions, for instance, the above two currents are almost coincident at a gate bias of -3.5 V. It is also confirmed that the three-terminal source–drain current and two-terminal gate leakage current coincide well with one another, although there is a slight discrepancy (about 0.5–0.6 V) of gate–drain voltage between them, especially under more shallow gate bias conditions. We do not know what causes the discrepancy, but we speculate that there is a slightly different depletion-region distribution between two-terminal and three-terminal measurement. These results indicate that the source–drain breakdown characteristics is limited by the onset of gate–drain breakdown, although the unknown excess drain current at more shallow gate bias near pinchoff is observed (which may be responsible for chan-

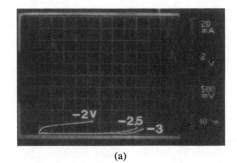

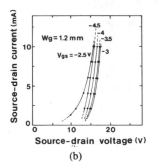 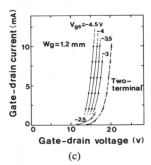

Fig. 8. (a)Drain *I–V* characteristics near pinchoff region. The scales are 20 mA/div for source–drain current and 2 V/div for source–drain voltage. The gate bias voltages are applied by -2, -2.5, and -3 V. (b) Drain *I–V* characteristics under more negative gate bias conditions. (c) Two-terminal gate–drain reverse characteristics (dashed line) and three-terminal gate–drain leakage current characteristics (solid line) in DH-HEMT.

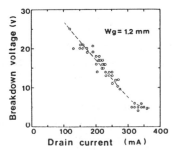

Fig. 9. Two-terminal gate–drain breakdown voltages as a function of drain current.

nel breakdown), but it still does not limit the maximum drain voltage. For the gate–drain breakdown mechanism in the DH-HEMT, it will be dominated by the avalanche breakdown because of low doping configuration, but there is a possibility of a mixed avalanche-tunneling mechanism.

The two-terminal gate-drain breakdown voltages in a power DH-HEMT with simple recessed structure, which is defined at a fixed gate current of 1 mA for 1.2 mm gate periphery, are plotted in Fig. 9 as a function of drain current I_{dss}. The key experimental results are the general trend toward rapidly increasing gate breakdown voltages with decreasing drain current I_{dss}, which coincides with the results from GaAs MESFET's. However, the breakdown voltages do not decrease any further when the I_{dss} exceeds about 300 mA. According to the two-dimensional lateral spreading model [13], this means that much-improved breakdown voltages are achievable if the drain current is sufficiently small, but that for higher drain current region,

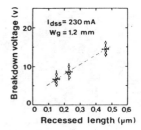

Fig. 10. Two-terminal gate–drain breakdown voltages as a function of recessed length.

one-dimensional bulk avalanche can be expected. In the experiment, this critical point of current level corresponds to about 280 mA/mm, which is smaller than the value of 350 mA/mm in GaAs MESFET's, but this is partly responsible for the effect of recessed length in the gate region. Fig. 10 also shows the two-terminal gate–drain breakdown voltages as a function of recessed length in gate region under a fixed drain current of 230 mA. The data indicate that the gate breakdown voltages increase from about 7 to 14 V with the increase of the recessed length from about 0.15 to 0.45 μm, respectively. However, such dependence could not be observed under high-current conditions around about 350 mA, although such a high current level includes a parallel condition in surface layers. Therefore, this result suggests that, in the low-current region affecting gate breakdown, the gate breakdown voltage is not limited only by the amount of the drain current, but it also correlates with the recessed length. This can be understood in terms of the lateral spreading of gate depletion region recessed structure, shown in our model in the appendix.

VI. Conclusion

The RF and dc characteristics of microwave power double-heterojunction HEMT's with low doping density have been studied. The results obtained are as follows:

1) The cut-off frequency f_t and maximum frequency of oscillation f_{max} in DH-HEMT with 0.8–1-μm gate length and 1.2-mm gate periphery are 11–16 GHz and 36–41 GHz, respectively. These show good performance in 0.8–1-μm gate-length HEMT's.

2) The cutoff frequency is observed to degrade strongly with increasing source–drain voltage. This can be explained by increasing the effective transit length of electrons and reduction of the average electron velocity, due to Gunn domain formation, but probably is not due to a real space transfer effect.

3) The microwave power capability of the DH-HEMT is remarkable. The 1.2-mm gate-periphery device produced an output power of 630 mW (0.53 W/mm) with a 3-dB gain at 20 GHz. The measured maximum output power was 1.05 W at 20 GHz and 0.58 W at 30 GHz for a 2.4-mm gate-periphery DH-HEMT. These are the highest output power yet reported for HEMT devices.

4) The source–drain breakdown characteristics in the DH-HEMT is limited by the onset of gate–drain breakdown, although an excess drain current near pinchoff was

observed. It was also confirmed that two-terminal gate–drain breakdown voltages correlate remarkably well with drain current I_{dss} and recessed length. A simple model of gate breakdown voltage, which can be explained by the above feature, was presented.

Appendix

Here we present a simple model of gate–drain breakdown voltages in HEMT devices with a recessed-gate structure. The model is basically similar to the lateral spreading model proposed by Wemple *et al.* [13], except for consideration of the recessed-gate structure, and including the contribution of an electric field perpendicular to the surface undr the gate. The latter is introduced because conventional HEMT's have much higher doping densities than those of GaAs MESFET's, and such an electrical field component cannot be ignored. For HEMT's as well as GaAs MESFET's, the depletion region will also widen from the gate edge toward the drain when an extra gate–drain bias voltage was applied beyond pinchoff voltage. This is because the extra gate–drain potential must exist in the gate–drain spacing region when the channel was cut off. In practice, the extra potential drop is supported by the lateral spreading of the depletion region, since the channel current under the cutoff condition must be zero, when the channel current is given by $J_{ch} = eN_s \mu_n(dF/dx)$, where F is quasi-Fermi potential in the channel. That is, if the region with $dF/dx \neq 0$ exists between the gate–drain spacing, the N_s must be equal to zero, and also if the region with $N_s \neq 0$ exists, dF/dx must be zero.

The basic assumptions of our model are 1) the electric field $E_\parallel$ (transverse electric field) associated with lateral spreading of depletion region terminates under the gate; 2) the surface layer between gate and drain is depleted a fixed depth by midgap pinning of Fermi level (0.8 eV), and its depth is independent of gate–drain bias voltage. In addition, 3) the electric field $E_\perp$ (longitudinal electric field) perpendicular to the surface, which is due to the depletion region beneath the gate and the electric field $E_\parallel$ (by assumption 1)), are summed up for simplicity, and the total amount of both electric field components induces the critical electric field, which in turn induces gate breakdown voltage. Fig. 11 shows the schematic diagram of our breakdown model in HEMT's. According to Wemple's assumption and our consideration, electric field $E_\parallel$ and $E_\perp$ terminate on the effective length L_{eff} ($L_{eff} < L_g$) under the gate, where L_{eff} is some unknown distance, and is an adjustable parameter of the model. However, we can roughly estimate the length L_{eff} using avalanche electric field and also using experimental results reported by Wemple *et al.* By applying Gauss' law to the total ionized charge in the lateral depletion region, the transverse electric field $E_\parallel$ along the surface is given by

$$\epsilon E_\parallel L_{eff} = eN[tD + t_1(x_B - D)] \qquad (A1)$$

where e is the electric charge, ϵ is the material dielectric constant, N is the doping density, D is the recessed length,

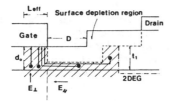

Fig. 11. Schematic diagram of a breakdown model in conventional HEMT.

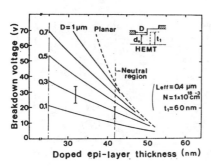

Fig. 12. Calculated gate breakdown voltage characteristics as a function of doped epi-layer thickness for various recessed lengths.

t is depleted epi-layer thickness in the recessed region, x_b is the length of the lateral depletion region, and t_1 is the depleted epi-layer thickness in the nonrecessed region. On the other hand, the longitudinal electric field per unit length ($E_\perp$) beneath the gate is given by

$$\epsilon E_\perp = eN d_0 \tag{A2}$$

where d_0 is the doped layer thickness under the gate. From assumption 3), we can express the critical electric field E_a as

$$E_a = E_\parallel + E_\perp. \tag{A3}$$

Combining (A1), (A2), and (A3), and then applying Poisson's equation, one obtains the following expression for the gate breakdown voltage.

$$V_B = \frac{\epsilon L_{\text{eff}} E_a^2}{2eN t_1}\left(1 - \frac{eN d_0}{\epsilon E_a}\right)^2$$

$$+ \left(1 - \frac{t}{t_1}\right)\left\{ D E_a\left(1 - \frac{eN d_0}{\epsilon E_a}\right) - \frac{eNtD^2}{2\epsilon L_{\text{eff}}}\right\}. \tag{A4}$$

In (4), the first term corresponds to a planar structure when t_1 is replaced by d_0, and the second term is related to a recessed structure. Moreover, the first term, neglecting the term in parentheses, is coincident with Wemple's model. An example of calculations in our model is shown in Fig. 12, where the breakdown voltages are given as a function of doped epi-layer thickness for various recessed lengths. Here, we assumed that L_{eff} was about 0.4 μm long

from the results by Wemple *et al.*, and the error bar corresponds to the case where L_{eff} changes from 0.2 to 0.6 μm. It is apparent from the results that the breakdown voltage decreases rapidly with increasing epi-layer thickness, which in turn implies an increasing current density, whereas the breakdown voltage is improved by adopting a large recessed length.

ACKNOWLEDGMENT

The authors wish to thank T. Misugi, M. Kobayashi, A. Shibatomi, and T. Mimura for their encouragement. They are also very grateful to K. Joshin, M. Iwakuni, and N. Hidaka for their support in microwave measurement, and T. Yamada for his technical assistance in device fabrication.

REFERENCES

[1] K. Hikosaka, J. Saito, T. Mimura, and M. Abe, "High frequency and high power heterojunction FET's," in *Proc. Nat. Conv.(Dept. Semiconductor Mater.) Inst. Electron. Commun. Eng. Japan*, no. 268, Sept. 1983.

[2] K. Inoue and H. Sakaki, "A new highly conductive (AlGa)As/GaAs/ (AlGa)As selectively doped double heterojunction field effect transistor (SD-DH-FET)," *Japan. J. Appl. Phys.*, vol. 23, p. L61, 1984.

[3] N. H. Sheng, C. P. Lee, R. T. Chen, and D. L. Miller, "GaAs/ AlGaAs double heterostructure high electron mobility transistor," in *Proc. IEDM Tech. Dig.*, pp. 352–354, Dec. 1984.

[4] K. Hikosaka, Y. Hirachi, T. Mimura, and M. Abe, "A microwave power double-heterojunction high electron mobility transistor," *IEEE Electron Device Lett.*, vol. EDL-6, pp. 341–343, July 1985.

[5] P. M. Smith, U. K. Mishra, P. C. Chao, S. C. Palmateer, and J. C. M. Hwang, "Power performance of microwave high electron mobility transistors," *IEEE Electron Device Lett.*, vol. EDL-6, pp. 86–87, Feb. 1985.

[6] A. K. Gupta, R. T. Chen, E. A. Sovero, and J. A. Higgins, "Power saturation characteristics of GaAs/AlGaAs high electron mobility transistors," in *IEEE Microwave and Millimeter-Wave Monolithic Circuits Symp. Dig.*, pp. 50–53, June 1985.

[7] N. H. Sheng, C. P. Lee, R. T. Chen, D. L. Miller, and S. J. Lee, "Multiple-channel GaAs/AlGaAs high electron mobility transistors," *IEEE Electron Device Lett.*, vol. EDL-6, pp. 307–310, June 1985.

[8] K. Joshin, T. Mimura, Y. Yamashita, K. Kosemura, and J. Saito, "Noise performance of microwave HEMT," in *IEEE MTT-S Int. Symp. Dig.*, pp. 563–565, June 1983.

[9] L. H. Camnitz, P. J. Tasker, H. Lee, D. van der Merwe, and L. F. Eastman, "Microwave characterization of very high transconductance MODFET," in *IEDM Tech. Dig.*, pp. 360–363, Dec. 1984.

[10] R. W. H. Engelmann and C. A. Liechti, "Gunn domain formation in the saturated current region of GaAs MESFET's," in *IEDM Tech. Dig.*, pp. 351–354, Dec. 1976.

[11] ——, "Bias dependence of GaAs and InP MESFET parameters," *IEEE Trans. Electron Devices*, vol. ED-24, pp. 1288–1296, Nov. 1977.

[12] Y. Hirachi, Y. Takeuchi, M. Igarashi, K. Kosemura, and S. Yamamoto, "A packaged 20 GHz 1-W GaAs MESFET with a novel via-hole plated heat sink structure," *IEEE Trans. Microwave Theory Tech.*, vol. MTT-32, no. 3, pp. 309–316, Mar. 1984.

[13] S. H. Wemple, W. C. Niehausm, H. M. Cox, J. M. Dilorenzo, and W. O. Schlosser, "Control of gate–drain avalanche in GaAs MESFET's," *IEEE Trans. Electron Devices*, vol. ED-27, pp. 1013–1018, June 1980.

HIGH-EFFICIENCY POWER 2DEGFETS BASED ON A SURFACE UNDOPED LAYER n-AlGaAs/GaAs SELECTIVELY DOPED STRUCTURE FOR Ka-BAND

Indexing terms: Semiconductor devices and materials, FETs, Microwave devices and components

We have evaluated the power performance of surface undoped structure n-AlGaAs/GaAs 2DEGFETs at Ka-band. This unique configuration 2DEGFET with a 0·5 μm gate length showed a 143 W/mm output power at 28·5 GHz. Furthermore, 7 dB linear gain and 21% power-added efficiency were attained, which are the best data in the Ka-band frequency range reported to date.

Introduction: There has recently been greatly increased demand for high-frequency operating devices, especially for those used in satellite communication systems. Two-dimensional electron gas field-effect transistors (2DEGFETs) highlight their superior high-frequency low-noise and high-power performances to GaAs MESFETs because of a large saturation velocity.[1-7] However, there was a serious problem with regard to low gate and drain breakdown voltage, ascribed to a highly doped layer just under the gate, from the aspect of practical use.

In our previous work,[2] to improve the breakdown voltage and simultaneously achieve higher-frequency operation, we proposed the surface undoped layer 2DEGFETs based on an n-AlGaAs/GaAs selectively doped structure with a low input capacitance. Then, the fabricated 0·5 μm-gate FETs demonstrated such noticeable performances as $\sim$11 V reverse gate breakdown voltage, $\sim$16 V drain breakdown voltage, 45 GHz current gain cutoff frequency and 0·95 dB noise figure at 12 GHz.

This letter reports the notable power performance of the previously proposed 2DEGFETs in the Ka-band frequency range.

Device structure and fabrication process: As shown in Fig. 1, the FET structure is composed of undoped GaAs (7500 Å),

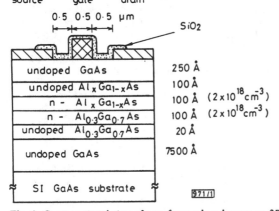

Fig. 1 *Cross-sectional view of a surface undoped structure 2DEGFET*

undoped $Al_{0.3}Ga_{0.7}As$ (20 Å), Si-doped $Al_{0.3}Ga_{0.7}As$ (2×10^{18} cm^{-3}, 100 Å), Si-doped $Al_xGa_{1-x}As$ with graded AlAs mole fraction (2×10^{18} cm^{-3}, 100 Å), undoped $Al_yGa_{1-y}As$ (100 Å) and undoped GaAs (250 Å) on a semi-insulating GaAs substrate grown by molecular beam epitaxy (MBE). The Si-doped $Al_{0.3}Ga_{0.7}As$ layer was minimised in thickness, which is still sufficient to supply the maximum 2DEG density, and the surface undoped GaAs layer was employed to reduce the influence of DX-centres as much as possible.[1] The fabrication process, which we call a closely spaced electrode structure FET process, was briefly as follows.

A 0·5 μm-long gate and 0·4 μm-thick gate electrode were made with Al by a side-etching technique. Then, Ni/AuGe ohmic metals were evaporated through the same photoresist mask as employed in the side-etching process, and alloyed with those semiconductor layers after the lift-off process. Finally, the device surface was passivated with a 1500 Å-thick SiO_2 film.

DC and power performances: In Fig. 2 typical drain current/voltage characteristics of the 2DEGFET at room temperature

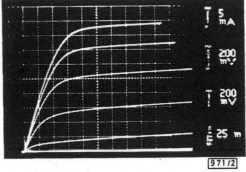

Fig. 2 *Drain current/voltage characteristics of present FET*

$L_g = 0.5$ μm, $W_g = 280$ μm

are shown. The gate length is 0·5 μm and the gate width is 280 μm. The maximum transconductance was as sufficiently high as 240 mS/mm, despite the thick layer between the gate and the heterointerface. In addition, the input capacitance was greatly diminished. Therefore, a noted high-frequency performance is expected.

The power performance of the FET was tested at Ka-band. Fig. 3 describes the output power and the power-added efficiency as functions of the input power at 28·5 GHz. The device was biased at 2·9 V between the source and the drain. Note that the linear gain was as high as 7 dB, and a maximum power-added efficiency of 21% was obtained. These are the best data ever reported in this frequency range. Furthermore, the saturation output power was 16dBm, i.e. 143 mW/mm. Such an FET with high linear gain and high power efficiency plays a great important role in high output power systems, especially from the viewpoint of low power consumption. The uniquely designed 2DEGFET, emphasising

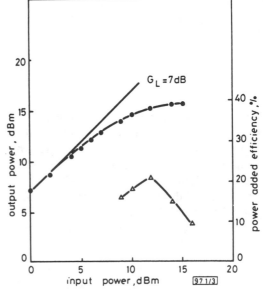

Fig. 3 *Output power-added efficiency against input power at 28·5 GHz*

$L_g = 0.5$ μm, $W_g = 280$ μm, $V_{DS} = 2.9$ V

the lowering of the input capacitance without significant transconductance degradation, demonstrated the highest cutoff frequency f_T in the 0·5 μm-gate FETs, where the saturation velocity deduced from f_T was as large as $1·4 \times 10^7$ cm/s, as previously reported.[2] As a result, the pronounced power performance could be realised, since a maximum power gain has basically a square relationship with a cutoff frequency and hence saturation velocity under the short channel approximation.[7]

The frequency dependence of the output power from 27·3 GHz to 29·5 GHz is shown in Fig. 4 with the input power as a parameter. In this frequency range, an almost flat output power was realised in terms of precisely matched external circuits, which will guarantee stable and reliable operation in satellite communication equipment.

In summary, the power performance of the surface undoped structure 2DEGFET has been reported for the first time. The unique configuration FET showed the marked large-signal performance at Ka-band, especially in the power gain and the

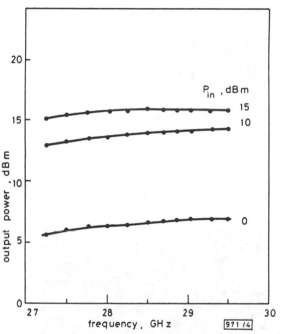

Fig. 4 *Output power as a function of frequency at Ka-band with input power as a parameter*

$L_g = 0·5\ \mu$m, $W_g = 280\ \mu$m, $V_{DS} = 2·9$ V

power-added efficiency, which indicates that the FET will be very promising for high-frequency power device applications. In future it will be of great importance to raise the channel electron density, for instance by utilising an AlInAs/InGaAs material system[8] and/or a highly doped channel.[9] Moreover, the device structure should be more precisely optimised to achieve small distortion, i.e. good linearity in the input/output performance.

Acknowledgment: The authors are grateful to Dr. T. Itoh and H. Miyamoto for their helpful discussion. They also express their great appreciation to Dr. Y. Takayama, Dr. I. Haga and Dr. S. Aihara for their support and encouragement during the course of this work.

H. HIDA
Y. AKIBA*
Y. SUZUKI
H. TOYOSHIMA
K. OHATA

23rd June 1986

Microelectronics Research Laboratories
NEC Corporation
4-1-1, Miyazaki, Miyamae-ku
Kawasaki, Kanagawa 213, Japan

*Microwave & Satellite Communications Division
NEC Corporation
4035, Ikebe-cho, Midori-ku
Yokohama, Kanagawa 226, Japan

References

1 HIDA, H., MIYAMOTO, H., OHATA, K., ITOH, T., BABA, S., and OGAWA, M.: 'Planar AlGaAs/GaAs selectively doped structure with high performances and high stabilities'. Proc. 11th int. symp. on GaAs and related compounds, Sept. 1984, pp. 551–556
2 HIDA, H., K. SUSUKI, Y., and TOYOSHIMA, H.: 'A new low noise AlGaAs 2DEGFET with a surface undoped layer', *IEEE Trans.*, 1986, **ED-33**, pp. 601–607
3 KAMEL, K., KAWASAKI, H., HORI, S., SHIBATA, K., HIGASHIURA, M., WATANABE, O. M., and ASHIZAWA, Y.: 'Extremely low-noise 0·25 μm HEMTs'. Proc. 12th int. symp. on GaAs and related compounds, Sept., 1985, pp. 541–546
4 TANAKA, K., TAKAKUWA, H., NAKAMURA, F., MORI, Y., and KATO, Y.: 'Low-noise microwave HIFET fabricated using photolithography and MOCVD', *Electron. Lett.*, 1986, **22**, pp. 487–488
5 MISHRA, U. K., PALMATEER, S. C., CHAO, P. C., SMITH, P. M., and HWANG, J. C. M.: 'Microwave performance of 0·25-micron gate length high electron mobility transistors', *IEEE Electron Dev. Lett.*, 1985, **EDL-6**, pp. 142–145
6 SMITH, P. M., MISHRA, U. K., CHAO, P. C., PALMATEER, S. C., and HWANG, J. C. M.: 'Power performance of microwave high-electron mobility transistors', *ibid.*, 1985, **EDL-6**, pp. 86–87
7 HIKOSAKA, K., HIRACHI, Y., and ABE, M.: 'Microwave power double-heterojunction HEMT's', *IEEE Trans.*, 1986, **ED-33**, pp. 583–589
8 HIROSE, K., OHATA, K., MIZUTANI, T., ITOH, T., and OGAWA, M.: '700 mS/mm 2DEGFET fabricated from high mobility MBE-grown *n*-AllnAs/GaInAs heterostructures'. Proc. 12th int. symp. on GaAs and related compounds, Sept. 1985, pp. 529–534
9 HIDA, H., OKAMOTO, A., TOYOSHIMA, H., TAHARA, S., and OHATA, K.: 'New high current drivability MIS-like FETs utilizing highly doped thin GaAs channel'. 44th annual device research conference digest, IVA-6

GaAs/AlGaAs Heterojunction MISFET's Having 1-W/mm Power Density at 18.5 GHz

BUMMAN KIM, MEMBER, IEEE, HUA QUEN TSERNG, SENIOR MEMBER, IEEE, AND J. W. LEE

Abstract—The previously reported GaAs/AlGaAs heterojunction MISFET with an undoped AlGaAs layer as an insulator has been further optimized for power operation at upper *Ku* band. A 300-μm gate-width device generated 320 mW of output power with 33-percent efficiency at 18.5 GHz. The corresponding power density exceeds 1 W/mm. When optimized for efficiency, the device has achieved a power added efficiency of 43 percent at 19 GHz.

IT IS well known that to improve the gate–drain breakdown voltage of an FET either an insulating or a semi-insulating layer can be used between the Schottky metal and the active channel. We reported the first power operation of GaAs/ AlGaAs MISFET at *X* band [1]. In that device, an undoped AlGaAs layer grown by MBE was used as the gate insulator. A power density of 0.84 W/mm was obtained at 10 GHz from an unoptimized device. In this paper, results of further device structure optimization are presented.

Fig. 1 shows the channel structure of the MISFET and a photograph of a 300-μm gate-width device used for this work. A gate length of 0.5 μm is shown. The device has four epitaxial-layers: 1-μm-thick undoped GaAs buffer layer, 0.06-μm-thick n-GaAs (doped to $6 \times 10^{17}/cm^3$), 0.03-μm-thick undoped AlGaAs layer (as gate insulator), and 0.05-μm-thick n$^+$ contact layer. The epitaxial layers were sequentially grown on a Cr-doped substrate using a commercial Riber MBE-2300 system. The active layer was heavily doped to increase the channel current and the transconductance without having short-channel effects. The new channel structure shown in Fig. 1 has a higher gate length to channel thickness ratio, compared to the one in [1] resulting in better high-frequency performance. The FET's were fabricated using the standard MESFET process. Mesa isolations were followed by source and drain ohmic contacts (with a source–drain spacing of 4 μm). Patterns with 2-μm-wide recess and 0.5-μm-long gates were defined by the electron beam. The gate metals were evaporated Ti/Pt/Au having a total thickness of 0.5 μm. The source and drain ohmic contacts were alloyed through the undoped AlGaAs layer, which results in a relatively high input resistance (~ 10 Ω for a 300-μm-wide FET). It is possible to further reduce the contact resistances by using n$^+$ implant at the contact areas.

The dc *I–V* characteristics of the MISFET are shown in Fig. 2. The transconductance is about 145 mmho/mm (as

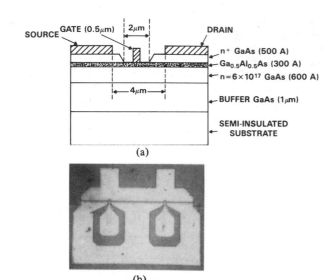

(a)

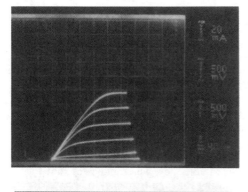

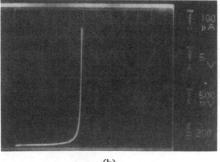

(b)

Fig. 1. GaAs/AlGaAs MISFET. (a) GaAs/GaAlAs MISFET channel structure. (b) Photograph of a 300-μm gate-width MISFET.

Fig. 2. Current–voltage characteristics of a 300-μm MISFET. (a) Drain current–voltage curves with stepped gate voltages. (b) Gate-to-drain breakdown current versus voltage curve.

Manuscript received July 22, 1986; revised September 15, 1986.
The authors are with the Central Research Laboratories, Texas Instruments Incorporated, Dallas, TX 75265.
IEEE Log Number 8611349.

Reprinted from *IEEE Electron Device Letters,* vol. EDL-7, no. 11, pp. 638–639, November 1986.

compared with 87 mmho/mm for the device of [1]). The maximum channel current was 150 mA for the 300-μm gate-width FET. This current is about 15 percent higher than that of the previous device when normalized to the same gate width. The estimated f_t is 21 GHz. This f_t is about 15 percent higher than that of a MESFET having similar device parameters. Although the increase in f_t is lower than the 40-percent increase predicted by Hill [2] and the 22 percent increase reported by Mimura *et al.* [3], it is still quite significant. The breakdown voltage remains high (20–23 V) even though a very thin AlGaAs layer (300 Å) was used.

The microwave performance of the heterojunction MISFET described above was measured at 18.5 GHz. Conventional microstrip matching circuits fabricated on 10-mil-thick quartz substrate were used. Simple quarter-wave impedance transformers were used for evaluating the device. When the device was optimized for small-signal operation, a gain of 9 dB was obtained. The small-signal gain is comparable to that of a conventional MESFET having a similar gate geometry. Improved power performance was obtained for the MISFET with a drain bias in excess of 10 V. Fig. 3 shows the gain compression curve of the MISFET. The 300-μm gate-width device generated 320 mW of output power with 5.5-dB gain. The power added efficiency was 33 percent. The power density exceeds 1 W/mm, which is the highest power density ever reported for a GaAs FET at this frequency. The operation biases are 13 V for the drain and -1.5 V for the gate. Devices from other processed slice, when optimized for high-efficiency operation, have achieved power added efficiencies as high as 43 percent with 0.4-W/mm power density at 19 GHz. The input impedance of the MISFET will be higher than that of a MESFET, since the gate-to-source capacitance is lower. Due to the higher breakdown voltage, the operating voltage will be higher, resulting in a high output impedance. These higher input/output impedance levels will be advantageous for a large gate-width FET, since the device/circuit combining efficiency will be improved at higher impedance levels.

In conclusion, high-efficiency, high-power density opera-

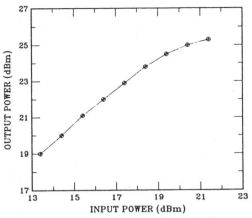

Fig. 3. Gain compression curve of a 300-μm MISFET at 18.5 GHz.

tion of AlGaAs/GaAs heterojunction MISFET's has been demonstrated at upper *Ku*-band frequencies. Power densities in excess of 1 W/mm and efficiencies as high as 43 percent were obtained. Further reduction of the gate length and optimization of the channel parameters will extend the operation to millimeter-wave frequencies. Due to the higher operating drain voltage, this type of device is expected to generate higher power than conventional MESFET's at millimeter-wave frequencies.

ACKNOWLEDGMENT

The authors wish to thank K. Bradshaw for electron-beam lithography, S. F. Goodman, J. M. Ramzel, L. J. Schoellman, and P. Tackett for technical assistance, and J. Fuller for editing of this manuscript.

REFERENCES

[1] B. Kim, H. Q. Tserng, and H. D. Shih, "Microwave power GaAs MISFET's with undoped AlGaAs as an insulator," *IEEE Electron Device Lett.*, vol. EDL-5, no. 11, p. 494, 1984.
[2] P. M. Hill, "A comprehensive analytical model for III-V compound MISFET's," *IEEE Trans. Electron Devices*, vol. ED-32, no. 11, p. 2249, 1985.
[3] T. Mimura *et al.*, "GaAs microwave MOSFET's," *IEEE Trans. Electron Devices*, vol. ED-25, no. 6, p. 573, 1978.

A 30-GHz 1-W Power HEMT

KOHKI HIKOSAKA, MEMBER, IEEE, NORIO HIDAKA, YASUTAKE HIRACHI, MEMBER, IEEE, AND
MASAYUKI ABE, MEMBER, IEEE

Abstract—Millimeter-wave power high electron mobility transistors (HEMT's) employing a multiple-channel structure have been fabricated and evaluated in the *R*-band frequency range. An output power of 1.0 W (a saturated output power of 1.2 W) with 3.1-dB gain and 15.6-percent efficiency was achieved at 30 GHz with a 0.5-μm gate-length and 2.4-mm gate-periphery device. At 35 GHz, a 2.4-mm device delivered 0.8 W with 2.0-dB gain and 10.7-percent efficiency. These are the highest output power figures reported to date for single-chip power FET's in the 30-GHz frequency range.

I. INTRODUCTION

POWER capabilities of high electron mobility transistors (HEMT's) at microwave and millimeter-wave frequencies have been steadily reported by the authors and several laboratories [1]–[7]. The advantages of HEMT's over GaAs MESFETs for power applications are high power gain and high efficiency. Those advantages are due to a high-current-gain cutoff frequency resulting from higher electron velocity, which further lead to greater output power. Our previous data on a 1-μm gate-length HEMT [5], a double-heterojunction type, generated 1.05-W output power at 20 GHz and 0.58 W at 30 GHz. High-gain and high-efficiency performance has also been reported by Saunier and Lee [6] from 10 to 60 GHz and by Sovero *et al.* [7] at 35 GHz.

In this paper, we demonstrate a single-chip power HEMT having a capability of 1-W output power for *R*-band operation. A multiple-channel structure is employed in order to have a larger current drive capability.

II. DEVICE FABRICATION

Selectively doped epilayers for a multiple-channel structure were grown by MBE at 600°C. The layer structure is shown in Fig. 1. It consists of a threefold double-heterojunction (or six-channel) structure. The AlGaAs layers are Si doped to a concentration of 1×10^{18} cm^{-3} and the AlAs mole fraction is 0.25. A 3-nm-thick GaAs top layer is provided for ohmic contacts and to reduce parasitic resistance. The as-grown multiple-channel structure has a sheet electron concentration of 5.2×10^{12} cm^{-2} and an electron mobility of 59 500 cm^2/V·s at 77 K. Power HEMT's with recessed gates were fabricated. The fabrication process was similar to previously reported ones [4], [5]. The 0.5-μm gates were defined by electron-beam lithography. The recessed depth was 75 nm and Al gates were formed on the recessed i-GaAs channel layers.

Manuscript received July 7, 1987. This work was supported by the Ministry of International Trade and Industry (MITI) of Japan.

The authors are with Fujitsu Laboratories Ltd., 10-1 Morinosato-Waka-miya, Atsugi 243-01, Japan.

IEEE Log Number 8717501.

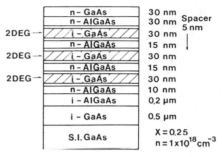

Fig. 1. Schematic cross section of as-grown multiple-channel structure.

The resulting sheet electron concentration in the channel was 2.8×10^{12} cm^{-2}. The devices fabricated had 60-μm unit gate widths and 1.2–2.4-mm total gate peripheries. A photograph of a power HEMT with a 2.4-mm gate width is shown in Fig. 2.

III. DC CHARACTERISTICS

The dc characteristics for a 2.4-mm device are shown in Fig. 3. The saturated drain current is 800 mA and the maximum drain current is 1.27 A (530 mA/mm). Double peaks in transconductance are observed for full current swing. These result from two-channel conduction of i-GaAs layers, which sandwich an n-AlGaAs layer with parasitic conduction. The channel conduction due to the top i-GaAs layer is eliminated by the surface fermi-level pinning. The transconductance under a reverse gate bias was 120–130 mS/mm, and under a forward gate bias it was 170–180 mS/mm. The series resistance in the input terminal was 0.8 Ω ·mm. The gate-to-drain breakdown voltage was 7–8 V at a leakage current of 1 mA.

IV. RF CHARACTERISTICS

We studied the cutoff frequency characteristics of a multiple-channel HEMT with a gate dimension of $0.8 \times 100 \ \mu$m^2. Fig. 4 shows cutoff frequency versus drain current characteristics. Two peaks in the cutoff frequencies were observed. The f_t's are 21.5 and 20 GHz. These peaks in the cutoff frequencies result from the two-channel conduction, and the location is correspondent to that of each peak transconductance. The average cutoff frequency for 0.1–0.9 $I_{f \max}$ was about 19 GHz. This may be higher than that of a single-heterojunction HEMT because of less degradation of f_t's in the forward gate-bias region.

To measure power performance, the devices were mounted on chip carriers. The performance of a 1.2-mm device at 30 GHz was an output power of 0.7 W with 3.4-dB gain and 23.1-

Reprinted from *IEEE Electron Device Letters,* vol. EDL-8, no. 11, pp. 521–523, November 1987.

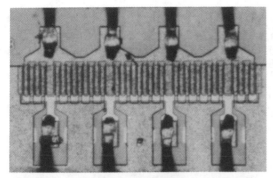

Fig. 2.   Photograph of a power HEMT with 0.5-μm gate length and 2.4-mm gate periphery (the magnification is $\times 160$).

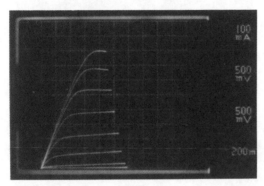

Fig. 3.   Drain I–V characteristics of a 2.4-mm gate-periphery power HEMT. The scales are 100 mA/div for drain current, 0.5 V/div for drain voltage, and -0.5 V/step for gate voltage.

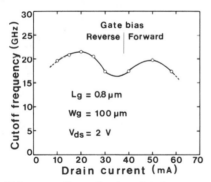

Fig. 4.   Cutoff frequency f_t as a function of drain current at a fixed drain voltage of 2 V. The device has a 0.8-μm gate length and a 100-μm gate width.

percent efficiency. The output power at 1-dB gain compression point was 560 mW and the linear gain was 5.2 dB. The saturated output power was 740 mW (620 mW/mm) with 2.7-dB gain and 20.4-percent efficiency. The output power density is the highest yet reported for an HEMT with a useful large gate width.

Devices with a 2.4-mm gate periphery were also tested. The results are shown in Fig. 5. At 30 GHz, output power was 1 W with 3.1-dB gain and 15.6-percent efficiency at a drain voltage of 8 V. The output power at 1-dB gain compression point was 950 mW and linear gain was 4.2 dB. The saturated output power was 1.2 W with 2-dB gain and 12-percent efficiency. No significant gain expansion occurred in the multiple-channel HEMT as indicated in Fig. 5. Therefore, lower intermodulation distortion is expected. Different chips were evaluated at 35 GHz. Output power was 0.8 W with 2-dB gain and 10.7-

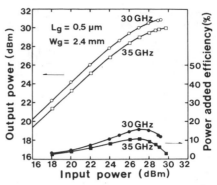

Fig. 5.   Output power and power-added efficiency as a function of input power at 30 and 35 GHz for a 2.4-mm gate-periphery HEMT.

percent efficiency at a drain voltage of 7.3 V. The output power at 1-dB gain compression point was 690 mW and the linear gain was 3.4 dB. The saturated output power was 950 mW with 1.0-dB gain and 7.7-percent efficiency. It is significant to point out that the power performance in the HEMT's is the best yet reported for single-chip power FET's in the frequency range.

In comparison with our previous data [5], the output power obtained was more than doubled at 30 GHz. The improved output power is attributed to high current drive capability and high power gain due to reduced parasitic resistance (owing to high sheet electron concentration) as well as reduced gate length. In addition, we believe that a lower drain bias voltage resulting from high current density in a multiple-channel structure also yields higher power gain performance, because the cutoff frequency in an HEMT degrades strongly with increasing drain voltage [5] (this phenomenon is very similar to that in a GaAs MESFET). Hence, if the product of maximum drain current and drain breakdown voltage is equivalent, an FET with higher drain current and lower drain breakdown voltage gives much higher power gain and then produces higher output power than an FET with lower drain current and higher drain breakdown voltage.

V. Conclusion

High output power operation of multiple-channel HEMT's has been demonstrated in the R-band frequency range. Output power in excess of 1 W obtained at 30 GHz using a 2.4-mm gate-periphery device. This output power performance is the best yet reported for single-chip power FET's.

Acknowledgment

The authors thank T. Misugi, M. Kobayashi, H. Ishikawa, T. Mimua, and K. Kondo for their encouragement and support. They are also grateful to H. Ishiwari, K. Kosemura, and Y. Yamashita for the electron-beam lithography, K. Ogasawara, J. Saito, and K. Nanbu for supplying MBE material, and T. Yamada for his technical assistance in device fabrication.

References

[1] K. Hikosaka, J. Saito, T. Mimura, and M. Abe, "High frequency and high power heterojunction FETs," in *Proc. Nat. Conv* (Dept. Semiconductor Mater.) *Inst. Electron. Commun. Eng. Japan,* no. 268, Sept. 1983.
[2] P. M. Smith, U. K. Mishra, P. C. Chao, S. C. Palmateer, and J. C. M.

Hwang, "Power performance of microwave high electron mobility transistors," *IEEE Electron Device Lett.*, vol. EDL-6, pp. 86–87, Feb. 1985.

[3] A. K. Gupta, R. T. Chen, E. A. Sovero, and J. A. Higgins, "Power saturation characteristics of GaAs/AlGaAs high electron mobility transistors," in *IEEE Microwave and Millimeter-Wave Monolithic Circuits Symp. Dig.*, June 1985, pp. 50–53.

[4] K. Hikosaka, Y. Hirachi, T. Mimura, and M. Abe, "A microwave power double-heterojunction high electron mobility transistor," *IEEE Electron Device Lett.*, vol. EDL-6, pp. 341–343, July 1985.

[5] K. Hikosaka, Y. Hirachi, and M. Abe, "Microwave power double heterojunction HEMT's," *IEEE Trans. Electron Devices*, vol. ED-33, pp. 583–588, May 1986.

[6] P. Saunier and J. W. Lee, "High-efficiency millimeter-wave GaAs/GaAlAs power HEMT's," *IEEE Electron Device Lett.*, vol. EDL-7, pp. 503–505, Sept. 1986.

[7] E. Sovero, A. K. Gupta, J. A. Higgins, and W. A Hill, "35-GHz performance of single and quadruple power heterojunction HEMT's," *IEEE Trans. Electron Devices*, vol. ED-33, pp. 1434–1438, Oct. 1986.

35-GHz Performance of Single and Quadruple Power Heterojunction HEMT's

E. SOVERO, MEMBER, IEEE, ADITYA K. GUPTA, ASSOCIATE MEMBER, IEEE,
J. A. HIGGINS, SENIOR MEMBER, IEEE, AND W. A. HILL

Abstract—The potential of single hetrojunction (SHJ) and quadruple heterojunction (QHJ) HEMT devices to provide power amplification at the *Ka*-band frequencies has been measured. The power level observed, from QHJ devices that have gate lengths of 0.5 μm and gate widths of 200 μm, has been over +20 dBm when gain is compressed below the small signal level by 2 dB. The small-signal gain was 5.2 dB at 35 GHz. The power level demonstrated by the SHJ devices is lower than that of the QHJ devices due to the lower "two-dimensional electron gas" sheet carrier density. Our measurements have shown a saturated power level of +15.3 dBm devices of the same geometry as the above-mentioned QHJ devices. The power performance in both cases (QHJ and SHJ) has been obtained with high efficiencies of 38 and 21 percent, respectively. These performance data represent the highest levels of gain and power reported at *Ka*-band frequencies from transistors that employ a 0.5-μm geometry.

I. Introduction

SINCE the introduction of the concept of high electron mobility transistors (HEMT), these devices have been the best prospect for obtaining low-noise amplification at millimeter-wave frequencies. The potential for low-noise devices has been amply demonstrated in experimental results [1], [2]. The realization that this technology could provide power amplification at moderate to high power levels and at high frequencies such as the *Ka*-band is a more recent consideration [3], [4]. In this paper we present measured data on the power gain and available power levels from HEMT devices at 35 GHz. It is the conclusion from this work that the quadruple heterojunction type of HEMT device provides a superior performance because it not only has a higher sheet carrier concentration in the "two-dimensional gas" of high mobility electrons, but also because it enhances two other important device parameters, i.e., it provides a lower level of device output conductance and it also exhibits a higher drain breakdown voltage.

The power transistors discussed in this work are of the dimensions:

gate length	0.5 μm
gate width	200 μm
gate drain spacing	2.0 μm.

These devices, when implemented on an epitaxial layer

Manuscript received January 27, 1986; revised May 16, 1986.
The authors are with the Rockwell International Corporation Microelectronic Research and Development Center, Thousand Oaks, CA 91360.
IEEE Log Number 8609955.

system that provides four layers of high-mobility electrons, have produced an output power level that is over 100 mW. The small-signal gain of these devices is 5.2 dB at 35 GHz, and when operating at an output level of over 100 mW (+20 dBm) the devices still have 3.2 dB of gain. The power-added efficiency of the transistors at the high-power level is over 38 percent. Single heterojunction power HEMT's have shown a maximum output power level, at 35 GHz and under similar conditions of gain compression, of over 35 mW (+15.3 dBm). For the SHJ device the power added efficiency was 17.2 percent. Extrapolation of these results leads to an expectation that power levels of greater than 0.5 W per millimeter of gate periphery should be expected from the multiple heterojunction type of device.

II. Device Fabrication

Molecular-beam epitaxial growth was used to produce the active layers for all the devices reported on in this paper. The layer structures of both the single HJ and the quadruple HJ devices are shown in Fig. 1. It is seen that the QHJ has three AlGaAs layers, called the doping layers as they provide all the free carriers in the neighboring undoped GaAs layer. The AlGaAs layers are doped to a level of over 10^{18} cm^{-3} by Si. In contrast with the SHJ layer system, the undoped high-mobility regions of the QHJ layer system are very narrow, and this is possibly an advantage both in terms of speed and breakdown voltage. In both types of layer system there are, on the top surface, a thin and very highly doped layer of GaAs that assists in the formation of low-resistance ohmic contacts. In both the SHJ and QHJ cases the mole fraction of Al in the AlGaAs in 24 percent.

The SHJ layer is seen to have a sheet carrier concentration of just about 1×10^{12} cm^{-2}, whereas the QHJ layer has over 3.2×10^{12} cm^{-2} carriers, a number that is above the level available from the MESFET technology. The mobility of the carriers in the SHJ system is, however, slightly higher than those in the QHJ system, a fact that is thought to be due, in some degree, to the generally greater difficulty in growing good interfaces on an AlGaAs layer [5].

Fig. 2(a) reveals the layout of the devices and the fact that the gate is accessed at two points in its 200 μm of width. This means that the individual gate finger segment

Reprinted from *IEEE Transactions on Electron Devices*, vol. ED-33, no. 10, pp. 1434–1438, October 1986.

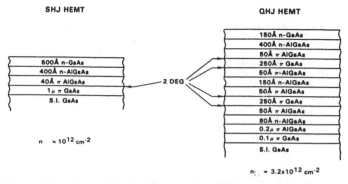

Fig. 1. Epitaxial structure of single and quadruple heterojunction HEMT's.

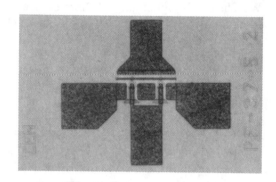

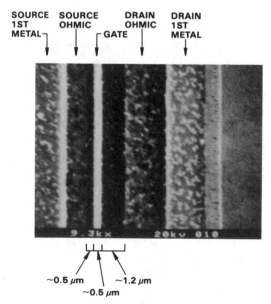

Fig. 2. Photo showing the device structure. Total gate width is 200 μm. Source electrodes are grounded through the use of via holes.

width is 50 μm. As shown in Fig. 2(b), the gate is defined to be 0.5 μm in length by the use of electron-beam lithography, and it is placed much closer to the source electrode than to the drain electrode in order to reduce the drain to gate capacitance. This off-center placement is particularly important in power transistors. The gate is recessed ~500 Å into the active layer by a wet chemical etch.

The ohmic contacts are formed using a very straightforward Au-Ge/Ni alloy process. Passivation and metal crossovers for access to the source islands, which are on the inside of the device, are provided through the use of polyimide coatings. The most significant aspect of these devices for millimeter waves is that the chip thickness has been reduced to ~ 125 μm by lapping and the backside metallized to provide the ground plane of a microstrip system. The source connections are through the chip directly to the ground plane using via hole technology thus providing a nearly ideal low inductance in the source feedback path. The vias that are not obvious from the figure are placed in the wing pads on both sides of the device. The gate and drain connections are designed to provide microstrip format that is of a 50-Ω impedance.

III. DC MEASUREMENTS

DC *I–V* characteristics of representative devices are shown in Fig. 3, and the variation of extrinsic transconductance (G_m) and drain current (I_d) with gate voltage for these devices is shown in Fig. 4.

The maximum drain current (I_f) for the SHJ HEMT is 80 mA (400 mA/mm) and for the QHJ HEMT is 104 mA (520 mA/mm). QHJ HEMT's with even higher currents can be fabricated by reducing the amount of gate recess. However, this also reduces the drain breakdown voltage, and no significant increase in output power is obtained. Although the peak G_m in the SHJ HEMT, 63 mS (315 mS/mm), is considerably higher than that in the QHJ HEMT, 42 mS (210 mS/mm), the large signal variation in G_m is much lower in the QHJ HEMT. For a 0.1 I_f to 0.9 I_f swing in I_{ds}, the G_m variation in the SHJ HEMT is from 18 to 63 mS, while it is only 26 to 42 mS for the QHJ HEMT. Thus, lower intermodulation distortion is expected in the QHJ HEMT and has been confirmed by data presented earlier [4].

Another advantage for the QHJ HEMT, obvious from Fig. 3, is the improvement in device output conductance. At 0.5 I_{dss}, the dc output conductance of the QHJ HEMT is 3.9 mS/mm versus 12.8 mS/mm for the SHJ HEMT. Carrier confinement by the n-AlGaAs layer between the 250-A undoped GaAs layer (Fig. 1) is credited for this improvement. However, measurements of $S22$ at 1 GHz indicate that output conductance increases to 12.2 mS/mm for the QHJ HEMT and 16.5 mS/mm for the SHJ HEMT at microwave frequencies. A better understanding of the device is needed to explain these variations.

Drain breakdown voltage near pinchoff was measured for several devices of each kind. All devices had soft breakdown characteristics, with the SHJ HEMT's limited to 6–8 V and the QHJ HEMT's limited to 10–12 V. The breakdown voltage of low-current QHJ HEMT's was consistently higher than that of higher current devices, indicating that further optimization in the epitaxial layer and device structure is required to achieve simultaneously the goals of higher current and higher breakdown voltage.

IV. MILLIMETER-WAVE MEASUREMENTS

For 35-GHz measurements, the devices were mounted on a specially designed carrier that was inserted between

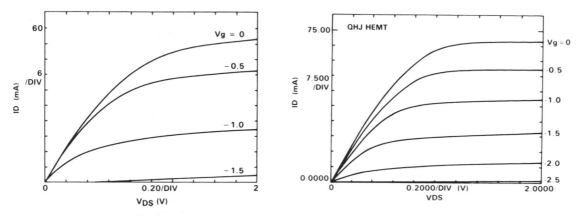

Fig. 3. DC saturation characteristics of SHJ and QHJ HEMT's.

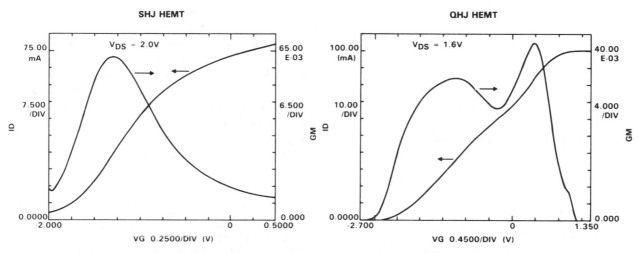

Fig. 4. Transconductance and current characteristics of SHJ and QHJ HEMT's.

two stepped waveguide transitions. The transitions match the WR28 waveguide (= 300-Ω impedance) to a microstrip line (50 Ω). This is accomplished with three transmission line sections of progressively reduced waveguide heights plus a section of ridge waveguide, all one quarter-wave long at 35 GHz. The completed design is shown in Fig. 5. In the photograph the three different regions can be identified; the waveguide steps, a ridge waveguide section, and the chip carrier. The initial region, made up of three quarter-wave long steps, is designed to reduce the height and impedance of the waveguide (WR 28) from 0.140 to 0.034 in and from 300 to 72 Ω, respectively. The second region consists of a quarter-wave ridge waveguide section. The ridge itself is formed by a removable anodized aluminum plunger that doubles as a dc bias contact and also acts as the final transition to the microstrip line on the chip carrier. The chip carrier consists of two 0.090-in-long sections of 50-Ω line on 0.010-in-thick alumina microstrip. The device is placed in an 0.020-in gap between the 50-Ω lines. Single bond wires are used to connect the gate and drain to the input and output microstrips. Via holes connect the source directly to the ground plane of the carrier. A symmetric transition couples the output microstrip back to the waveguide.

Fig. 5. The 35-GHz waveguide test fixture used to couple power into and out of the device.

The device under test is placed on the chip carrier and in the test set described above. Input and output slide screw tuners serve as matching elements. The tuners and device holder were independently tested and their insertion loss was measured at 35 GHz with a vector network analyzer. The tuners and test holder, with a through-line

219

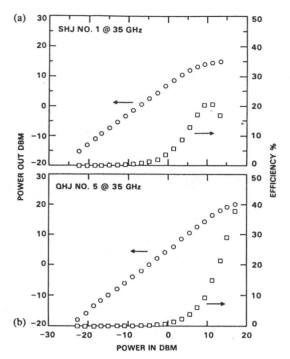

Fig. 6. (a) Power output versus power input for a SHJ HEMT device. (b) Power output versus power input for a QHJ power HEMT device.

in place of the device under test, showed less than 1-dB total insertion loss. From symmetry considerations the losses are split equally between the input and output network. For the measurements the tuners were set for maximum output power.

Plots of 35-GHz power performance are provided in Fig. 6(a) and b for the SHJ and QHJ devices, respectively. Included in these data are the power added efficiencies at each measurement level. The maximum power obtained from a 200-μm QHJ HEMT was over +20 dBm (525 mW/mm) with 3 dB of gain and a power-added efficiency of 38.3 percent. For the SHJ HEMT corresponding performance levels were +15.3 dBm (170 mW/mm) with 1.6 dB gain and a power-added efficiency of 21 percent. These measurements were performed at a drain bias

of +6.00 V. This voltage, which is substantially lower than the drain breakdown for the QHJ HEMT, was chosen as it is the voltage above which only a marginal increase in power level is available, but at a significant cost in efficiency and gain.

It is notable that the optimum point of operation can be well below the drain breakdown voltage. This is particularly true for the QHJ devices, opening the possibility of using these devices in high efficiency class *B* mode.

V. CONCLUSIONS

Power amplification for moderate power levels can be obtained from HEMT technology at *Ka*-band frequencies. High gain, high power level, and high power-added efficiency have been obtained at 35 GHz from HEMT devices employing a gate length of no less than 0.5 μm. The use of quadruple heterojunction high-mobility layer systems leads to improved performance through higher breakdown voltage and lower output conductance levels evident both at dc and RF. It is expected, based on these data, that further development of the HEMT power technology will lead to transistors with an output power density of approximately 1 W/mm at *Ka*- and possibly *Q*-band frequencies in the near future.

REFERENCES

[1] P. C. Chao, S. C. Palmateer, P. M. Smith, U. K. Mishra, K. H. G. Duh, and J. C. M. Hwang, "Millimeter-wave low-noise high electron mobility transistors," *IEEE Electron Device Lett.*, vol. EDL-6, no. 10, pp. 531–533, Oct. 1985.
[2] A. K. Gupta, E. A. Sovero, R. L. Pierson, R. D. Stein, R. T. Chen, D. L. Miller, and J. A. Higgins, "Low noise high electron mobility transistors for monolithic microwave integrated circuits," *IEEE Electron Device Lett.*, vol. EDL-6, no. 2, pp. 81–82, Feb. 1985.
[3] K. Hikosaka, Y. Hirachi, T. Mimura, and M. Abe, "A microwave power double-heterojunction high electron mobility transistor," *IEEE Electron Device Lett.*, vol. EDL-7, no. 7, pp. 341–343, July 1985.
[4] A. K. Gupta, R. T. Chen, E. A. Sovero, and J. A. Higgins, "Power saturation characteristics of GaAs/AlGaAs high electron mobility transistors," to be published.
[5] T. J. Drummond, J. Klem, D. Arnold, R. Fisher, R. E. Thorpe, W. G. Lyons, and H. Morkoç, "Use of a superlattice to enhance the interface properties between two bulk heterolayers," *Appl. Phys. Lett.*, vol. 42, p. 615, 1983.

High-Efficiency Millimeter-Wave GaAs/GaAlAs Power HEMT's

PAUL SAUNIER, MEMBER, IEEE, AND J. W. LEE

Abstract—The power, gain, and efficiency of 0.5-μm gate-length, 75- and 50-μm gate-width multiple heterojunction high electron mobility transistors (HEMT's) have been evaluated from 10 to 60 GHz. At 10 GHz, with a source-to-drain voltage as low as 2.4 V, the device delivers a power density of 0.37 W/mm with 13.4-dB gain and 60.8-percent efficiency. At 60 GHz, a 50-μm device gave 0.4 W/mm with 3.6-dB gain and 14-percent efficiency. The power density and efficiency of these 0.5-μm gate-length HEMT's above 40 GHz are the best reported for a three-terminal device. Fundamental frequency oscillations up to 104 GHz were observed when a device was bonded as a free-running oscillator.

I. INTRODUCTION

EXCELLENT low-noise performances have been demonstrated with the high electron mobility transistor (HEMT) up to 60 GHz [1], [2]. Its usefulness for power generation has only been investigated recently [3]–[5]. The work presented here shows that 0.5-μm gate-length HEMT's can produce useful power levels with good gain and excellent efficiency up to 60 GHz. The devices described here have a gate width of 75 and 50 μm; the gate length is 0.5 μm and the source-to-drain spacing is 2 μm.

II. DEVICE FABRICATION

The MBE grown material structure is shown in Fig. 1. It provides six layers of high mobility electrons. The AlGaAs layers are doped to over 1×10^{18} by Si and the Al concentration is 25 percent. On top of these layers a 300-Å-thick layer of GaAs doped in the high 1×10^{18} with Si is used to facilitate ohmic contacts and reduce parasitic source and drain resistances. The resulting sheet carrier concentration is over 3×10^{12} cm^{-2}. The mask set used includes 75- and 50-μm gate-width devices of the Pi gate configuration. The 75-μm devices have, respectively, one, two, and three gate feeds. The 50-μm devices have one gate feed. Source–drain spacing is 3 μm. Fig. 2 is a photograph of a three gate-feed 75-μm device.

Device fabrication is conventional. Isolation is performed by boron implant. Ohmic contacts are formed using Au–Ge/Ni/Au metallization alloyed at 450 C for 3 min. The 0.5-μm gates are defined and recessed before Ti/Pt/Au evaporation. Fig. 3 shows a cross-sectional view of the HEMT. Passivation is done with silicon nitride. Metal overlay and air–bridge plating complete the front side processing. The slice is lapped to 3 mils before backside metallization and dicing.

Manuscript received May 9, 1986; revised June 16, 1986.
The authors are with Texas Instruments Incorporated, Dallas, TX 75265.
IEEE Log Number 8610210.

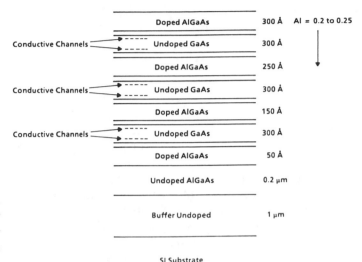

Fig. 1. Multiple HEMT structure.

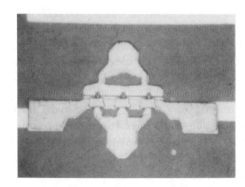

Fig. 2. The 75-μm device with three gate-feeds.

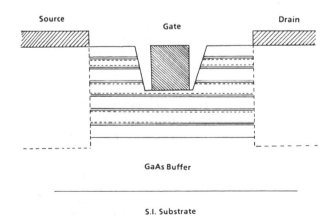

Fig. 3. Multiple HEMT cross sections.

Reprinted from *IEEE Electron Device Letters,* vol. EDL-7, no. 9, pp. 503–505, September 1986.

III. DC MEASUREMENTS

The dc characteristics for a 75-μm device are shown in Fig. 4. The saturation current is 450 mA/mm and the maximum current with +0.4 V on the gate before conduction in the AlGaAs layers occurs is 600 mA/mm. The maximum transconductance is 453 mS/mm. The source resistance is about 0.28 $\Omega \cdot$mm. Breakdown voltage is about 5–6 V. The observed variation in transconductance is much lower than in a single heterojunction HEMT. Thus lower intermodulation distortion is expected as reported in [5].

IV. MICROWAVE AND MILLIMETER-WAVE MEASUREMENTS

All the devices are soldered on a horseshoe and source grounding is done by multiple bond wires. For the measurements at 10, 15, and 20.5 GHz, a fixture with 50-Ω input and output lines on quartz and 3.5-mm connectors is used. Tuning is achieved by using the proper bond wire length to connect the gate and drain pad to the 50-Ω microstrip lines.

At 10 GHz, a maximum efficiency of 60.8 percent is obtained with a drain voltage of 2.4 V; the corresponding gain and power are 13.4 dB and 0.37 W/mm. Maximum power of 0.6 W/mm with 13.6-dB gain and 44.3-percent efficiency is achieved with 3.5 V on the drain.

At 15 GHz a maximum efficiency of 49 percent with 9-dB gain and 0.51 W/mm is achieved with 3-V drain bias. The corresponding small signal gain is 11 dB.

At 20.5 GHz a maximum efficiency of 42 percent is achieved with 2.5-V drain bias. The corresponding power is 0.38 W/mm with 6-dB gain and the small signal gain is 10 dB. When the drain voltage is increased to 3.5 V, a maximum efficiency of 35 percent is achieved, and the corresponding power is 0.5 W/mm with 7.3-dB gain. The small signal gain is 11 dB.

For the measurements at 32.5, 40.5, 50, and 60 GHz, test fixtures with fin line transitions from waveguide to microstrip line are used. The 50-Ω transmission lines on quartz substrate are used with gate and drain bond wire length optimized for the different frequencies. Chip tuning is used for final performance optimization. At 32.5 GHz, with 5 V on the drain, a power of 0.6 W/mm with 5.4-dB gain and 30-percent efficiency is achieved. The corresponding gain compression curve is shown in Fig. 5. The small signal gain is 5.1 dB.

At 40.5 GHz, a power of 0.56 W/mm with 4.2-dB gain and 25.8-percent efficiency is obtained with a 5-V drain voltage. The small signal gain is 5.6 dB and 0.6 W/mm is obtained with 3.5-dB gain and 24-percent efficiency.

At 50 GHz, a power of 0.39 W/mm with 4.2-dB gain and 16.3-percent efficiency is achieved with a 5-V drain bias. The small-signal gain is 5.1 dB and 0.47 W/mm is obtained with 3 dB.

At 60 GHz a 50-μm device gave a 0.4 W/mm power with 3.6-dB gain and 14-percent efficiency.

A 75-μm device was bonded as a free-running oscillator. Fundamental oscillations as high as 104 GHz were observed. Although the maximum output power was only 0.03 mW, oscillations at this frequency started with a drain voltage as low as 0.5 V. No major differences were observed in the

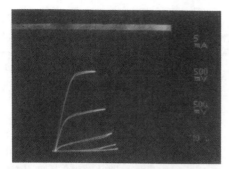

Fig. 4. HEMT dc characteristics.

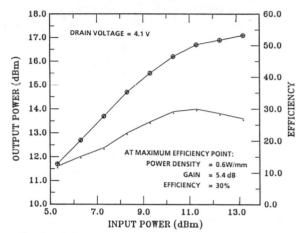

Fig. 5. Gain compression of a 75-μm HEMT at 32.5 GHz.

TABLE I
RF PERFORMANCE SUMMARY

FREQUENCY (GHz)	POWER DENSITY (W/mm)	GAIN (dB)	EFFICIENCY (%)
10	0.37	13.4	60.8
15	0.48	9.8	49
20.5	0.38	6.0	42
32.5	0.6	5.4	30
40.5	0.56	4.2	25.8
50	0.39	4.2	16.3
60	0.4	3.6	14

performances of the different 75-μm device or the 50-μm device.

Table I summarizes the results (power density, gain, and efficiency) obtained from 10 to 60 GHz.

V. CONCLUSIONS

High gain, high efficiency, and respectable power density have been demonstrated with 0.5-μm gate-length multiple structure HEMT's from 10 to 60 GHz. Very high transconductance, high current capability, and low source resistance permit these devices to operate with high efficiency at low drain voltages.

ACKNOWLEDGMENT

The authors wish to thank G. Oliver for technical assistance and D. N. McQuiddy for supporting this work.

REFERENCES

[1] P. C. Chao, P. M. Smith, U. K. Mishra, S. C. Palmateer, K. H. G. Duh, and J. C. M. Hwang, "Quarter-micron low-noise high electron mobility transistors," in *Proc. 1985 IEEE/Cornell Conf. Advanced Concepts in High Speed Semiconductor Devices and Circuits.*

[2] J. Berentz, "Millimiter wave HEMT technology," presented at the NSF Workshop on Electromagnetics, Univ. of Texas at Arlington, Jan. 29, 1986.

[3] P. Saunier and H. D. Shih, "A novel double heterostructure HEMT for microwave application," presented at WOCSEMMAD 1985.

[4] K. Hikosaka, Y. Hirachi, T. Mimura, and M. Abe, "A microwave power double-heterojunction mobility transistor," *IEEE Electron Device Lett.*, vol. EDL-7, no. 7, pp. 341–342, July 1985.

[5] A. K. Gupta, R. T. Chen, E. A. Sovero, and J. A. Higgins, "Power saturation characteristics of GaAs/AlGaAs high electron mobility transistors," presented at the IEEE 1985 Microwave and Millimiter-Wave Monolithic Circuits Symp.

223

Section 1.4: Mixers

Large-Signal Time-Domain Simulation of HEMT Mixers

GUAN-WU WANG, IKUROH ICHITSUBO, MEMBER, IEEE,
WALTER H. KU, YOUNG-KAI CHEN, AND
LESTER F. EASTMAN, FELLOW, IEEE

Abstract—A large-signal HEMT model and a time-domain nonlinear circuit analysis program have been developed. In this work a systematic method to simulate HEMT mixers and design them for maximum conversion gain is presented. The transconductance-compression effect reduces the mixer's conversion gain at high frequencies. Simulation results from mixers designed to operate at 10, 20, and 40 GHz show that a reduction in parasitic conduction in the AlGaAs layer significantly increases the conversion gain.

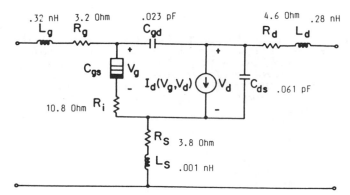

Fig. 1.   Large-signal model of the HEMT with values of the linear elements.

I. INTRODUCTION

The high electron mobility transistor (HEMT) is superior to the GaAs MESFET when used in low-noise microwave amplifiers. Recently the power performance of HEMT's with a single heterojunction [1] and double heterojunctions [2] have been investigated. Maas reported a HEMT mixer that operates at 45 GHz [3]. These promising results indicate that the HEMT will no longer be confined to small-signal applications.

For both the small-signal and the large-signal application of GaAs FET's, computer-aided design (CAD) is very valuable in the design process, especially in the design of monolithic integrated circuits, where tuning the circuits to optimize performance is impractical. The accuracy and efficiency of computer aided design and simulation rely on a well-developed device model. In this paper we present a large-signal HEMT model and a systematic CAD method in the time domain for the design and simulation of HEMT mixers. The effect on mixer performance by transconductance compression, which is a distinct feature of the HEMT [4], has also been studied.

II. A LARGE-SIGNAL MODEL OF THE HEMT

Under large-signal operation, the element values of the HEMT equivalent circuit vary with time and become dependent on the terminal voltages. A large-signal model can be derived by considering the main nonlinear elements of the equivalent circuit. As shown in Fig. 1, the elements of the large-signal HEMT model assumed to be nonlinear in this work are the gate-to-source capacitance, C_{gs}, and the drain current source, I_d. Other circuit elements are assumed to be linear.

The drain current source is represented by the current–voltage equations derived in [5]. From the experimental results, C_{gs} of a HEMT behaves quite similarly to that of a MESFET [6], [7], because considerable charges in addition to the two-dimensional electrons are modulated by the terminal voltages. Hence a valid approach is to use a Schottky diode equation and

$$C_{gs}(V_g) = \frac{C_{gs0}}{\left(1 - \dfrac{V_g}{V_{bi}}\right)^m} \qquad (1)$$

Manuscript received July 10, 1987; revised December 3, 1987.
G.-W. Wang, Y.-K. Chen, and L. F. Eastman are with the School of Electrical Engineering, Cornell University, Ithaca, NY 14853.
I. Ichitsubo was with the School of Electrical Engineering, Cornell University, Ithaca, NY. He is now with Toshiba Corporation, Tokyo, Japan.
W. H. Ku was with the School of Electrical Engineering, Cornell University Ithaca, NY. He is now with the Department of Electrical Engineering, University of California at San Diego, La Jolla, CA.
IEEE Log Number 8719438.

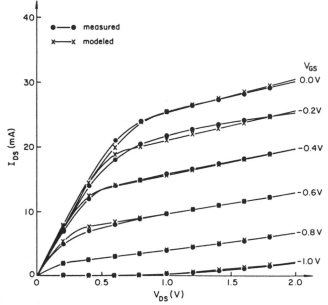

Fig. 2.   Measured and modeled DC I–V characteristics of the HEMT.

where C_{gs0} and m are model parameters adjusted to fit the measured values. V_{bi} is the built-in voltage for the Schottky gate and V_g is the internal gate voltage.

Compared to the large-signal MESFET models given in [8] and [9], two diodes representing forward gate conductive current and gate–drain breakdown current are neglected in our HEMT model; hence this model is valid only when these two currents are absent in device operation, e.g., in a mixer. The nonlinearity of gate-to-drain capacitance, C_{dg}, is also neglected [7].

A GE HEMT with 0.25 μm gate length and 150 μm gate width was chosen [1]. Bias-dependent S parameters and dc characteristics were used to construct the large-signal model. Small-signal equivalent circuits were extracted from the measured S parameters at two bias points: $V_{GS} = -0.2$ V, $V_{DS} = 2.0$ V and $V_{GS} = -0.5$ V, $V_{DS} = 2.0$ V. C_{gs0} and m in (1) were determined by a simple optimization program to fit (1) to the values of C_{gs} in the bias-dependent small-signal equivalent circuits. For $C_{gs0} = 0.178$ pF and $m = 0.67$, the model gives a close fit; therefore

$$C_{gs}(V_g) = \frac{0.178}{\left(1 - \dfrac{V_g}{0.8}\right)^{0.67}} \qquad (\text{pF}). \qquad (2)$$

Use was made of dc current–voltage characteristics to determine the parameters in the expression for $I_d(V_g, V_d)$. The modeled and measured I–V characteristics are shown in Fig. 2.

Reprinted from *IEEE Transactions on Microwave Theory and Techniques*, vol. 36, no. 4, pp. 756–759, April 1988.

224

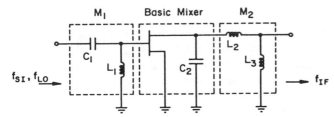

Fig. 3. Circuit diagram of the designed mixer.

TABLE I
ELEMENT VALUES OF THE MIXER MATCHING NETWORKS.

Freq.	C_1(pF)	C_2(pF)	L_1(nH)	L_2(nH)	L_3(nH)
10	0.08	2.2	1.23	1.89	0.91
20	0.08	2.2	0.33	1.84	0.96
40	0.1	2.2	0.1	1.99	0.87

The modeling procedure and model parameters have been discussed in [5].

The values of the linear elements in the large-signal HEMT model are shown in Fig. 1. The S parameters calculated from the large-signal model are compared with the measured S parameters. The r.m.s. errors of the modeled and measured S parameters are 6.8 percent and 7.2 percent for the bias at $V_{GS} = -0.2$ V and $V_{GS} = -0.5$ V, respectively. These results are comparable to those obtained from using small-signal equivalent circuits to model the S parameters.

III. DESIGN AND SIMULATION

Owing to the nonlinear nature of the problem, to design a mixer with maximum conversion gain, a systematic computer-aided design method proves to be very helpful. A time-domain nonlinear circuit analysis program, CADNON [10], suitable for microwave FET circuit analysis, has been developed. This program utilizes Aprille and Trick's shooting algorithm [11] to reduce the convergence time for a nonlinear circuit to reach steady state. With this program, we can simulate the input and output impedances of the device at the desired radio frequency (RF) and intermediate frequency (IF) under large-signal local oscillator (LO) condition. The maximum conversion gain can then be obtained by simultaneous adjustment of the matching networks for both impedance matching and optimum load conditions. The RF power level is set at -30 dBm and the bias is chosen at $V_{GS} = -0.8$ V and $V_{DS} = 2.0$ V. Fig. 3 shows the circuit diagram of the mixers, and the element values of the matching networks are summarized in Table I for each operating frequency. The IF is 2 GHz in all cases.

For the mixer designed to operate at $f_{RF} = 10$ GHz and $f_{LO} = 8$ GHz, the optimal LO power is 2 dBm for a maximum conversion gain of 9 dB (Fig. 4). This result compares favorably with the maximum conversion gain on the order of 6 dB for MESFET mixers at X-band [12]–[14]. The input VSWR at 10 GHz and the output VSWR at 2 GHz are 1.1 at maximum conversion gain. These low VSWR's ensure that the mixer is optimized for conversion gain.

The conversion gain versus LO power of the mixer operating at $f_{RF} = 20$ GHz and $f_{LO} = 18$ GHz is also shown in Fig. 4. A

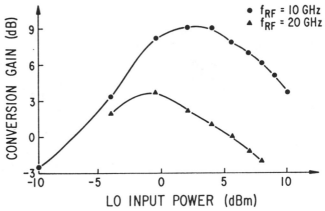

Fig. 4. Conversion gain versus LO power of the 10 GHz mixer and 20 GHz mixer.

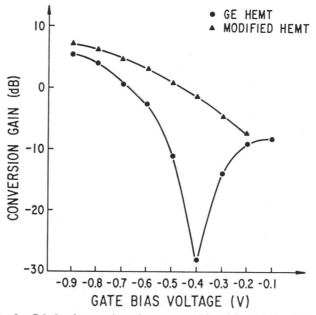

Fig. 5. Calculated conversion gain versus gate bias of the 20 GHz mixer with the HEMT and the modified HEMT.

maximum conversion gain of 3.6 dB with an input VSWR of 1.05 and an output VSWR of 1.1 is achieved at an LO power of -0.46 dBm.

To investigate the feasibility of millimeter-wave mixers using HEMT's, a HEMT mixer is designed and simulated at 40 GHz. Instead of conversion gain, the mixer shows a minimum conversion loss of 2 dB at LO power equal to 2 dBm, with input and output VSWR's of 1.1 and 1.15. The low VSWR's imply that we can hardly obtain more conversion gain from the HEMT mixer.

Conversion gain versus gate bias voltage of the 20 GHz mixer is shown in Fig. 5. The conversion gain shows a minimum at $V_{GS} = -0.4$ V, which is close to the bias point for maximum dc transconductance. This anomalous phenomenon has never been observed in MESFET mixers and is caused by transconductance compression.

IV. EFFECTS OF TRANSCONDUCTANCE COMPRESSION ON MIXER PERFORMANCE

A distinct feature of the HEMT as compared to the GaAs MESFET is the strong transconductance compression frequently

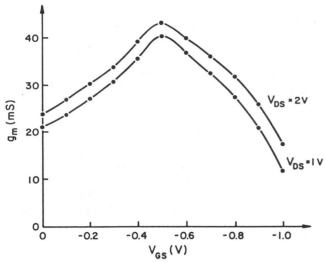

Fig. 6. Calculated transconductance versus gate bias of the HEMT.

V. Conclusions

HEMT mixers have been designed and analyzed effectively in the time domain with a CAD technique. The HEMT shows potential for application as a millimeter-wave mixer which offers conversion gain. Transconductance compression is found to degrade the performance of the HEMT mixer. In addition, to avoid the reduction of conversion gain, the gate bias has to be properly chosen for HEMT's that exhibit transconductance compression.

Acknowledgment

The authors would like to thank Dr. P. C. Chao of General Electric Co. for helpful discussions and measurement data.

References

[1] P. M. Smith, U. K. Mishra, P. C. Chao, S. C. Palmateer, and J. C. M. Hwang, "Power performance of microwave high electron mobility transistors," *IEEE Electron Device Lett.*, vol. EDL-6, pp. 86–87, Feb. 1985.

[2] K. Hikosaka, Y. Hirachi, T. Mimura, and M. Abe, "A microwave power double-heterojunction high electron mobility transistor," *IEEE Electron Device Lett.*, vol. EDL-6, pp. 341–343, July 1985.

[3] S. A. Maas, "45 GHz active HEMT mixer," *Electron. Lett.*, vol. 21, no. 2, pp. 86–87, Feb. 1985.

[4] K. Lee, M. S. Shur, T. J. Drummond, and H. Morkoc, "Parasitic MESFET in (Al, Ga)As/GaAs modulation doped FET's and MODFET characteristics," *IEEE Trans. Electron Devices*, vol. ED-31, pp. 29–35, Jan. 1984.

[5] G. W. Wang and W. H. Ku, "An analytical and computer-aided model of the AlGaAs/GaAs high electron mobility transistor," *IEEE Trans. Electron Devices*, vol ED-33, pp. 657–663, May 1986.

[6] L. H. Camnitz, P. J. Tasker, P. A. Maki, H. Lee, J. Huang, and L. F. Eastman, "The role of charge control on drift mobility in AlGaAs/GaAs MODFET's," presented at Tenth Biennial Cornell Conf., Cornell University, July 1985.

[7] Y. K. Chen, D. C. Radulescu, G. W. Wang, A. N. Lepore, P. J. Tasker, and L. F. Eastman, "Bias-dependent microwave characteristics of an atomic planar doped AlGaAs/InGaAs/GaAs double heterojunction MODFETs," presented at 1987 IEEE MTT-S Int. Microwave Symp., Las Vegas, June 1987.

[8] L. O. Chua, and Y. W. Sing, "Nonlinear lumped circuit model of GaAs MESFET," *IEEE Trans. Electron Devices*, vol. ED-30, pp. 825–833, July 1983.

[9] A. Materka and T. Kapcprzak, "Computer calculation of large-signal GaAs FET amplifier characteristics," *IEEE Trans. Microwave Theory Tech.*, vol. MTT-33, pp. 129–135, Feb. 1985.

[10] I. Ichitsubo, G. W. Wang, W. H. Ku, and Z. Li *CADNON User's Manual*.

[11] J. A. Aprille, Jr., and T. N. Trick "Steady-state analysis of nonlinear circuits with periodic inputs," *Proc. IEEE*, vol. 60, pp. 108–114, Jan. 1972.

[12] R. A. Pucel, D. Masse, and R. Bera, "Performance of GaAs MESFET mixers at X-band," *IEEE Trans, Microwave Theory Tech.*, vol. MTT-24, pp. 351–360, June 1976.

[13] O. Kurita and K. Morita, "Microwave MESFET mixers," *IEEE Trans. Microwave Theory Tech.*, vol. MTT-24, pp. 361–366, June 1976.

[14] S. A. Maas, "Theory and analysis of GaAs MESFET mixers," *IEEE Trans. Microwave Theory Tech.*, vol. MTT-32, pp. 1402–1406, Oct. 1984.

observed in the normal bias range, because of the parasitic conduction in the AlGaAs layer [4].

Fig. 6 shows the calculated g_m versus V_{GS} of the HEMT. Transconductance compression is observed when the gate biasing voltage exceeds -0.5 V. Since g_m is nearly symmetrical with V_{GS}, when the gate is biased at -0.4 V, the odd components of the Fourier series expansion of the $g_m(t)$ will be reduced. The conversion gain of the HEMT mixer decreases accordingly.

To study the transconductance compression effects, the same HEMT is used in the 20 GHz mixer circuit. But the parasitic conduction is removed from the $I_d(V_g, V_d)$ model by setting the threshold voltage for parasitic conduction much larger than the maximum gate voltage during mixer operation. The conversion gain as a function of gate bias is shown in Fig. 5 for this modified HEMT and all the other operating conditions unchanged. The dip in the original conversion gain characteristics shown in Fig. 5 disappears, and the maximum conversion gain increases by 2.5 dB. With the modified HEMT in the 40 GHz mixer, an increase of the maximum conversion gain from -2 dB to 0.75 dB is observed. This suggests that a mixer for millimeter-wave application, which provides conversion gain instead of conversion loss, is possible by using properly designed HEMT's. This result also agrees with the experimental result by Maas, who obtained 1.5 dB conversion gain in a 45 GHz HEMT mixer [3].

Design and Performance of a 45-GHz HEMT Mixer

STEPHEN A. MAAS, MEMBER, IEEE

Abstract — A 45-GHz single-ended HEMT mixer has been developed with unity gain and 7–8-dB SSB noise figure over a 2-GHz bandwidth, including an IF amplifier, and 2-dBm output intermodulation intercept. This paper describes its design, structure, and performance. High performance has been achieved by careful attention to the design of the input and output embedding networks. This is the first reported HEMT mixer and the first active mixer above 30 GHz.

I. INTRODUCTION

GaAs MESFET mixers have conversion gain and have achieved noise figures and intermodulation levels superior to those of diode mixers at microwave frequencies [1], [2]. They are particularly valuable for use in small, lightweight, integrated receivers and in GaAs integrated circuits. The use of FET mixers above 30 GHz has not been explored, however, and the usefulness of high-electron mobility transistors (HEMT's) as mixers has also not been addressed. This paper describes the design and performance of the first reported HEMT mixer and the first active mixer at 45 GHz. It shows that such mixers can achieve conversion gain, and have noise and intermodulation performance that compares favorably with that of diode mixers.

HEMT's have several demonstrable and potential advantages over GaAs MESFET's for use in millimeter-wave mixers. The major advantage is that the high mobilities achieved in the two-dimensional electron gas result in substantially higher transconductance (in this case over 300 mS/mm), hence higher conversion gain and lower noise. The large increase in transconductance with only moderate cooling may also result in very low-noise cooled mixers. The one disadvantage is that the peaked transconductance versus gate voltage characteristic of currently available devices results in intermodulation levels that are higher than those of MESFET's, but still comparable to those of diode mixers.

II. HEMT DEVICE

The HEMT device used for this mixer was designed for low-noise amplifier applications above 30 GHz. It is described in detail by Berenz [3]. The active layer is grown at TRW by molecular-beam epitaxy, and the source and drain contacts are ion-implanted for low resistance. The dimensions of the recessed gate are 0.25 μm by 60 μm, and it is rectangular in cross section. The gate is defined by electron-beam lithography, and it is offset slightly toward the source to minimize drain/gate capacitance. The device's transconductance, as a function of gate voltage, is shown in Fig. 1; its peak value is 28 mS. Similar devices from the same manufacturing lot have achieved noise figures below 3 dB with 6-dB associated gain for a single-stage amplifier at frequencies near 40 GHz [4].

III. MIXER DESIGN

The mixer is designed according to the principles outlined in [2]. In designing the mixer's input and output matching circuits, it is important to present the optimum terminations to the HEMT gate and drain not only at the RF, IF, and LO frequencies, but at all significant LO harmonics and mixing frequencies. In particular, for a conventional downconverter, it is important to short-circuit the LO frequency and its harmonics at the drain and the IF frequency at the gate. The number of LO harmonics which are significant is sometimes problematical; however, for a 45-GHz mixer it is safe to assume that only two or at most three harmonics are significant. The input is matched at the RF frequency; no improvement in noise figure has been obtained experimentally by mismatching the input as is done in FET and HEMT amplifiers.

The IF output impedance of a gate-driven HEMT mixer is usually very high, because it is dominated by the HEMT's drain/source resistance. This resistance varies over the LO cycle from a low of several hundred ohms to an open circuit when the device is turned off. Measurements and numerical simulations using the techniques in [2] indicate that the output impedance of a well-designed, strongly-pumped FET or HEMT mixer is on the order of 1000–3000 Ω, and is in practice nearly impossible to match. In the rare instances when the IF can be matched (e.g., at IF frequencies below 100 MHz with bandwidths of a few MHz), the high load impedance may cause instability. For the broad-band microwave IF's invariably required by modern communications receivers, it is necessary to employ a different matching rationale. The IF should be designed to present a load impedance to the HEMT of 50–150-Ω resistive, depending upon the desired gain and circuit realizability limitations. The drain is biased to the same voltage that would be used in amplifier operation; the gate bias point is near the gate turn-on voltage.

Manuscript received November 11, 1985; revised February 11, 1986. This work was supported in part by TRW, Inc., with independent research and development funds.

The author was with TRW, Electronic Systems Group, Redondo Beach, CA. He is now with Aerospace Corporation, Los Angeles, CA 90009.

IEEE Log Number 8608327.

Reprinted from *IEEE Transactions on Microwave Theory and Techniques*, vol. MTT-34, no. 7, pp. 799–803, July 1986.

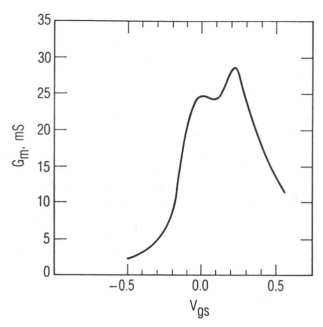

Fig. 1.   DC HEMT transconductance.

The purpose of the broad-band short at the drain is twofold: first, it guarantees large-signal stability of the device even under hard LO input drive by minimizing feedback at the LO frequency and its harmonics. Second, it ensures that the drain voltage remains constant at the bias value, so the HEMT remains in its saturation region throughout the LO cycle. The drain/source resistance therefore remains high and relatively constant, and the variation of the drain/gate and gate/source capacitances, although not zero, is minimized. The result is that the dominant parameter in the frequency conversion process is the fundamental frequency component of the transconductance, and its magnitude is maximized. The gate termination at the IF frequency does not appreciably affect conversion gain if the mixer is otherwise well designed; however, it does affect noise figure significantly. Shorting the gate reduces the gain through the device at the IF frequency, and therefore prevents the amplification of noise from the input termination or the bias circuit at the IF frequency.

It is possible to estimate the conversion gain and input impedance of a MESFET or HEMT mixer very easily and with good accuracy if its input and output matching networks are well designed and its IF frequency is low compared to the RF. The RF input impedance can be estimated from device parameters as follows:

$$Z_{in} = R_{in} + \frac{1}{j\omega C_j}. \tag{1}$$

This input impedance is a good starting value for input circuit design. The conversion gain G_c is given by (8) in [1] as follows:

$$G_c = \frac{g_1^2}{4\omega^2 C_j^2} \frac{\overline{R}_d}{R_{in}}. \tag{2}$$

In this expression $\overline{R}_d$ is defined in [1] as the time-averaged drain resistance, although, since R_d is an open circuit over part of the LO cycle, it is more precisely the inverse of the time-averaged drain admittance. g_1 is the magnitude of the fundamental component of the transconductance waveform. Since (2) assumes a matched output, only half the current from the drain current source circulates in R_L. For this mixer, $R_L \ll \overline{R}_d$ so the full output current circulates in R_L and

$$G_c = \frac{g_1^2}{\omega^2 C_j^2} \frac{R_L}{R_{in}}. \tag{3}$$

Assuming that the gate is biased near the device's turn-on voltage, the transconductance waveform is a half sinusoid, and $g_1 \cong G_{m\,max}/4$ so

$$G_c = \frac{G_{m\,max}^2}{16\omega^2 C_j^2} \frac{R_L}{R_{in}} \tag{4}$$

where $G_{m\,max}$ is the maximum transconductance, R_L is the IF load impedance at the drain of the device, R_{in} is the total resistance in the input loop of the device, the sum of gate, source, and intrinsic resistances, C_j is the gate/source capacitance at the turn-on voltage, and ω is the RF radian frequency. Equation (4) is valid only for the optimum short-circuit terminations discussed above. Equation (4) also assumes that the voltage drop across the source resistance R_S is negligible (satisfied if $g_1 R_S \ll 1$) and the gate/drain capacitance is negligible. The latter is satisfied if the drain terminations are short circuits.

Equation (4) shows that FET mixers, unlike FET amplifiers, achieve higher gain as the gate width is increased, because $G_{m\,max}$ increases in proportion to width and the device can be designed so that $\omega C_j R_{in}$ stays approximately constant. The limit to gain improvement by increasing gate width is that input impedance drops with increasing gate width, and matching becomes progressively more difficult in practice to achieve. Gain also increases with R_L. The limit to increasing R_L, other than circuit realizability limitations, is that the mixer may become unstable. For the HEMT mixer, $R_{in} = 7.8$ Ω, $C_j = 0.074$ pF, $R_L = 110$ Ω, and $G_{m\,max} = 0.028$ S, (4) predicts approximately 2-dB conversion gain. This is very close to the 1.5-dB measured gain, which necessarily includes circuit losses. More accurate conversion gain and input/output impedance calculations, or calculations for other types of mixers (e.g., upconverters) can be obtained using the methods in [2].

One should recognize that (4) predicts conversion gain achievable in practice, not the maximum available gain (MAG), which is usually unachievable, and is defined only if the circuit is unconditionally stable. Numerical simulations which include all the device parasitics, especially the gate/drain capacitance, indicate that FET mixers are often only conditionally stable. A load resistance below 150 Ω, however, is almost invariably adequate to insure stability.

It is clear from (4) that a wider device would be preferable to that used here. Also, it appears likely that conversion gain can be achieved with similar devices at even

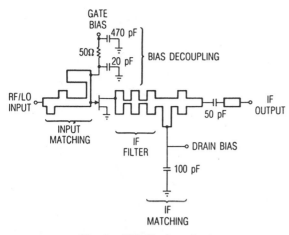

Fig. 2. HEMT mixer circuit.

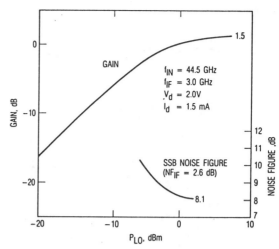

Fig. 4. Conversion gain and noise figure for the narrow-band mixer. Gate bias is varied with LO level to maintain constant drain current.

Fig. 3. 45-GHz HEMT mixer. The RF and LO are applied through a common WR-22 waveguide input.

higher frequencies. Reducing gate resistance by developing a "T" gate would also improve gain significantly, and might also improve noise figure. Matching difficulties in a wide device with a highly reactive input could be overcome through the use of on-chip matching, as is sometimes done with power FET's.

The HEMT mixer circuit is shown in Fig. 2 and a photograph of the mixer is shown in Fig. 3. The input circuit is realized in microstrip on a 0.010-in-thick fused silica substrate, with an integral waveguide/microstrip transition. The substrate metalization consists of a chrome adhesion layer with a 75-μin gold layer. The input matching circuit is a simple open-circuit stub, designed using (1) and appropriate estimates for the bond-wire reactances. DC gate bias is applied through a decoupling circuit which

also supplies the IF frequency short; since the input circuit has no dc connection to ground, no input dc block is needed. The output matching circuit is realized on a 0.010-in finely-polished alumina substrate to minimize the size of the IF matching elements. It consists of a filter, which supplies the required LO harmonic shorts to the drain, and a simple stub matching arrangement for the IF frequency. The filter was designed as a conventional stepped-impedance low-pass structure, and computer-optimized to achieve the desired terminations. Its IF-frequency S-parameters were then used to design the stub matching circuit to achieve the 110-Ω resistive load. DC drain bias is applied through the stub. Bond-wire reactances at the gate and drain, and especially the source, are minimized by using very short multiple connections.

The mixer is single-ended. For test purposes, the LO and RF signals must be combined by a filter diplexer or directional coupler. In a systems application, a pair of these mixers would be used in a balanced configuration, with the LO and RF combined via an input hybrid.

IV. PERFORMANCE

Fig. 4 shows the conversion gain and noise figure of the mixer as a function of LO level, with the mixer tuned for narrowband (approximately 200 MHz) operation near 44.5 GHz. The IF frequency is 3.0 GHz and the LO is 41.5 GHz. Gate bias in Fig. 4 has been adjusted to maintain a constant drain current at all LO levels, but no retuning has been performed. Varying the gate bias this way, which can be done easily in practice, minimizes the variation of gain with LO level. The noise figure in Fig. 4 includes the effects of a 2.5-dB IF amplifier.

Fig. 5 shows the conversion gain and noise figure of the mixer tuned for 2-GHz bandwidth. In this figure, the IF frequency is fixed at 3.0 GHz and has a 200-MHz bandwidth. The LO and RF frequencies are varied, but the tuning, bias, and LO level are held constant. Fig. 6 shows the input return loss under the same conditions.

Fig. 7 shows the two-tone intermodulation characteristics. Under the same bias and tuning conditions as those

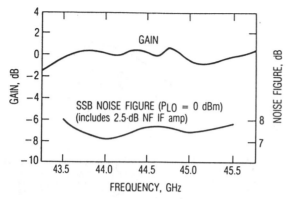

Fig. 5. Noise figure and conversion gain over a 2-GHz bandwidth. The IF is fixed at 3.0 GHz and the LO frequency is varied.

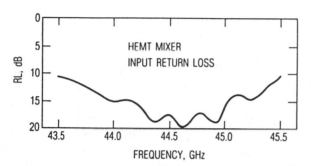

Fig. 6. HEMT mixer input return loss. IF frequency is fixed.

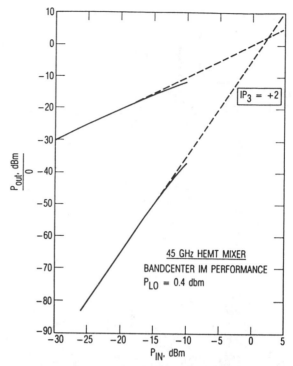

Fig. 7. Intermodulation curves for the HEMT mixer under optimum gain bias conditions. LO power is 0.4 dBm.

for Fig. 5, but at 0-dBm LO power, the output third-order intermodulation intercept is 2 dBm. This level is comparable to that of single-ended diode mixers, which are typically about 0–5 dBm, but the HEMT requires less LO power. It is difficult to compare this result to measured intermodulation levels in microwave MESFET mixers, because of differences in device size and frequency, and the paucity of published data. The transconductance/gate voltage nonlinearity of the HEMT is clearly stronger than that of most MESFET's, however, so it is to be expected that nonlinear distortion would be worse. The main cause of the mildly disappointing intermodulation level is the sharp peak in the transconductance curve, which increases the harmonic content of the transconductance waveform. It is possible to improve the intercept point to 8 dBm by biasing the mixer so that the most nonlinear part of the G_m/V_{gs} curve is avoided, but the reduced transconductance variation reduces the conversion gain by 5–6 dB. A device with a less strongly peaked transconductance characteristic would show the same gain with better linearity. It might also have better linearity in amplifier applications.

The noise figure and intermodulation performance of this mixer compares favorably to that of diode mixers at the same frequency. A very good waveguide diode mixer with whisker-contacted diodes typically would have a 4-dB conversion loss and a 4.5-dB SSB noise figure, but a 5-dB loss and a 5.5-dB noise figure is more expectable. With a 2.5-dB IF noise figure, the receiver noise figure would be 7–8 dB. Conversion loss of integrated mixers is significantly higher, in the range 6–10 dB, with resulting receiver noise figures above 10 dB [5], [6]. The noise figure of this

mixer with the same IF is 7–8 dB, but it is achieved in a small, reliable, low-cost microstrip component with conversion gain, lower LO power, and better manufacturing repeatability. Furthermore, performance may improve as more suitable devices for mixer application become available. Being a FET device, it is also amenable to monolithic integration.

It is true that the noise figure of the HEMT mixer is higher than that of an amplifier using the same device. For this reason one might conclude that it would be better to use the HEMT solely in a multistage amplifier ahead of a prosaic diode mixer. This reasoning is valid if noise figure is the sole performance criterion; however in many modern space communications systems, intermodulation and spurious response susceptibility are as important as noise, and these are exacerbated by the use of high-gain, low-noise preamplifiers. The levels of third-order intermodulation products at the mixer IF, for example, increase 3 dB for each decibel of preamplifier gain. Therefore, even if a preamp must be used, a low-noise HEMT mixer affords the same noise figure with less preamplifier gain, hence better receiver dynamic range.

V. CONCLUSIONS

This paper has described the design and performance of the first active HEMT millimeter-wave mixer. Its performance compares favorably with that of diode mixers with less LO power in an inexpensive, reliable, microstrip circuit. Performance could be improved through the use of a wider device with a "T" gate, and a less strongly peaked transconductance characteristic. Since HEMT amplifiers show dramatic gain and noise figure improvements on even moderate cooling, it is likely that cooled HEMT mixers also might achieve excellent noise performance.

ACKNOWLEDGMENT

The author would like to thank K. Nakano and J. Berenz for supplying the HEMT devices, M. Sholley and R. Sawires for the use of their waveguide/microstrip transition design, and B. Allen for the use of his device characterization data.

REFERENCES

[1] R. A. Pucel, D. Masse, and R. Bera, "Performance of GaAs MESFET mixers at *X* band," *IEEE Trans. Microwave Theory Tech.*, vol. MTT-24, pp. 351–360, June 1976.
[2] S. Maas, "Theory and analysis of GaAs MESFET mixers," *IEEE Trans. Microwave Theory Tech.*, vol. MTT-32, pp. 1402–1407, Oct. 1984.
[3] J. J. Berenz, K. Nakano, and K. P. Weller, "Low-noise high-electron mobility transistors," in *1984 IEEE Int. Microwave Symp. Dig.*, pp. 98–101.
[4] M. Sholley *et al.*, "36–40-GHz HEMT low-noise amplifier," in *1985 IEEE Int. Microwave Symp. Dig.*, pp. 555–558.
[5] L. T. Yuan and P. G. Asher, "A *W*-band monolithic balanced mixer," in *Proc. 1985 GaAs Monolithic Circuits Symp.*, pp. 71–73.
[6] K. Chang *et al.*, "*V*-band low-noise integrated circuit receiver," *IEEE Trans. Microwave Theory Tech.*, vol. MTT-31, pp. 146–154, Feb. 1983.

A WIDE BAND DISTRIBUTED DUAL GATE HEMT MIXER

M. LaCon   K. Nakano   G. S. Dow

TRW Inc.
One Space Park, Redondo Beach, CA. 90278

Abstract

This paper presents a systematic design, fabrication and evaluation of a distributed wide-band monolithic mixer using dual gate HEMTs. The measurement results show the mixer operates from 5 to 18 GHz with a conversion loss from 3 to 5 dB. This is the first reported wide band distributed mixer using dual gate HEMT devices. The mixer chips developed are key components for EW applications. With our systematic design approach and HEMT technology, development of ultra wide-band mixers up to 40 GHz can be realized.

1 Introduction

The introduction of the traveling wave concepts has provided a new technique for wide-band downconverters at microwave frequencies. With the advent of Al-GaAs/GaAs HEMTs, interest in a new class of wide-band mixer circuits has increased. The distributed mixer described here provides a method of achieving performance that is better than more conventional diode based circuits, and in a manner that is amenable to a MMIC implementation. In this paper, the design of the distributed mixer, fabrication and evaluation are presented. There is good agreement between the simulation and measurements.

2 Design

The first step in the design is to derive an accurate dual gate HEMT model. It begins with the construction of the bi-directional IV curves[1,2]. These curves, as shown in Figure 1, are used to determine the operating current and voltage of the two transistors in the dual gate HEMT. A linear model of the dual gate HEMT is derived from the single gate HEMT S-parameters measured at this operating current and voltage. The linear model is used to synthesize the RF, LO, and IF matching networks. In this design, the input and output capacitances of the transistors are combined with high impedance microstrip lines to form an artificial transmission line terminated in 50 ohms at each port of the mixer[3]. A schematic of the mixer is shown in Figure 2.

To simulate and optimize the mixer design, a non-linear model for the dual gate HEMT is also derived. The characteristic of the non-linear elements in the linear model is determined based on .25 X 80 uM Al-GaAs/GaAs HEMT single gate device measurement data at various bias conditions. Measured results of some key elements, ie., transconductance (Gm), saturation resistance (Rsat), and gate capacitance (Cgs) are shown in Figure 3, Figure 4, and Figure 5 respectively. The non-linear simulation was performed by using time domain analysis.

Reprinted from *Rec. of the IEEE GaAs Integrated Circuits Symp.*, pp. 173–176, 1988.

3 Fabrication

The circuit was fabricated on an Al-GaAs/GaAs heterostructure HEMT wafer grown with MBE. The device structure Figure 6 utilizes a TRW base line planner HEMT process.

The process begins with oxygen ion implantation to obtain device isolation. This implantation process is critical for uniform EBL gate processing. Ohmic contacts are deposited using gold-germanium metalization. A contact resistance of less than .08 ohm-mm is achieved by using rapid thermal alloying. Thin film resistors (nichrome) are deposited by a standard lift-off process for the bias networks and distributed network terminations. A thin layer of metal (Ti-Au) is deposited to form the first level metal, this layer is used to form the matching networks and bottom plate of the capacitors. Electron beam lithography is used to define the .2 to .25 uM gate length resist patterns. Gate recess etching is performed followed by gate metal deposition. A thin dielectric film (silicon-dioxide) is deposited to form the MIM capacitors. The capacitors are used in DC blocking and RF by-passing functions. The top metal is defined using an airbridge process. This form of interconnection reduces cross-over parasitics and improves circuit yield. Lift-off techniques are used in both dielectric and top metal process steps. The substrate thickness is reduced by lapping the wafer to .1 millimeter. The substrate vias and backside metalization complete the process. The distributed dual gate mixer chip is illustrated in Figure 7. The chip measures .75 X 1.38 MM and contains four .25 X 80 uM dual gate HEMT.

4 Results

A comparison of non-linear simulation and measured results shows good agreement. A plot of the measured and simulated results are shown in Figure 8. The mixer has a measured conversion loss of 3 to 5 dB when operating at an RF frequency of 5 to 18 GHz and an IF frequency of 1 GHz. Also shown in the Figure is the simulated conversion loss based on the non-linear model. The close match not only verifies our model but also ensures that the design approach developed can be extended to millimeter frequencies.

5 Conclusion

A compact wide-band mixer has been described. Based on the size of the chip and the performance, this circuit will be a very attractive candidate for many wide-band downconverter applications.

6 Acknowledgements

The authors would like to express their gratitude to the people who helped make the results of this paper possible: Louis Liu, Po-Hsin Liu, Anchin Han, Susan Hertweck, Linda Klamecki, Rosie Dia, Laurie DeLuca, and Frank Higa.

References

[1] C. Tsironis and R. Meierer, "Dual Gate MESFET Mixers," IEEE Trans. Microwave Theory Tech., vol. MTT-32 No. 3, March 1984.

[2] J. Scott and R. Minasian, "A Simplified Microwave Model of the GaAs Dual Gate MESFET," IEEE Trans. Microwave Theory Tech., vol. MTT-32 No. 3, March 1984.

[3] O. A. Tang and C. Aitchison, "A Very Wide-band Microwave MESFET Mixer Using the Distributed Mixing Principle," IEEE Trans. Microwave Theory Tech., vol. MTT-33 No. 12, December 1985.

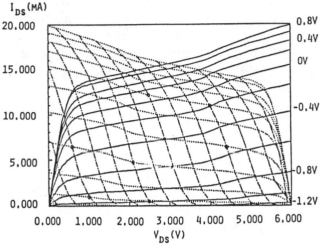

Figure 1: Bi-directional IV Curves Using Single Gate HEMT.

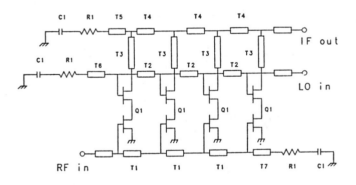

Figure 2: Schematic of Distributed Dual Gate Mixer.

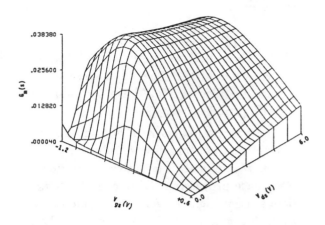

Figure 3: Plot of Transconductance (gm) of .25 X 80 uM Al-GaAs/GaAs HEMT.

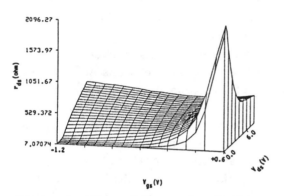

Figure 4: Plot of Saturation Resistance (Rsat) of .25 X 80 uM Al-GaAs/GaAs HEMT.

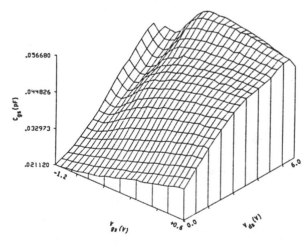

Figure 5: Plot of Gate Capacitance (Cgs) of .25 X 80 uM Al-GaAs/GaAs HEMT.

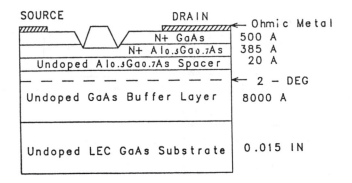

Figure 6: Al-GaAs/GaAs HEMT Device Structure.

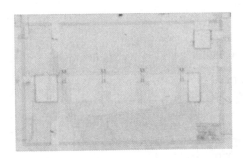

Figure 7: Distributed Dual Gate Mixer Chip.

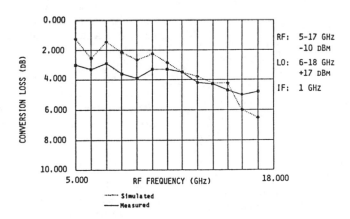

Figure 8: Conversion Loss of Mixer Chip.

SIS Mixer to HEMT Amplifier Optimum Coupling Network

SANDER WEINREB, FELLOW, IEEE

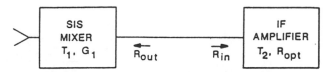

Fig. 1. General configuration discussed in this note. The mixer is described by noise temperature T_1, exchangeable gain G_1, and output resistance R_{out}. The amplifier is described by noise temperature T_2, which is a function of R_{out}; an input resistance R_{in}, which has no effect upon G_1 or noise performance; and an optimum generator resistance R_{opt}. Changing the impedance transformation of the amplifier input network changes both R_{in} and R_{opt}, but by use of lossless feedback, R_{in} can be changed independent of R_{opt}.

Abstract — The coupling network between a superconductor–insulator–superconductor (SIS) mixer and a high-electron-mobility-transistor (HEMT) amplifier is investigated from the point of view of minimizing the overall noise temperature and also increasing the saturation level of the mixer. The effect of a negative output impedance of the mixer upon the amplifier noise is considered and an optimum negative source resistance is found. The amplifier noise at this optimum negative source resistance is shown to be related to the noise wave coming out of the amplifier input terminals.

Key words: SIS, HEMT, low-noise, negative resistance.

I. INTRODUCTION

Superconductor–insulator–superconductor (SIS) mixers have become the device of choice for cryogenic (4 K) low-noise applications in the frequency range of 70 to 250 GHz [1]. The mixer is usually followed by a cryogenic low-noise FET or HEMT amplifier, and receiver single-sideband noise temperatures of under 100 K at 115 GHz have been achieved [2], [3]. Theoretical minimum noise temperatures are an order of magnitude lower [1] and experimental laboratory results with narrow-band, low-frequency IF amplifiers have confirmed this [4].

The general configuration discussed in this note is shown in Fig. 1. The design of an IF amplifier driven by an SIS mixer requires special consideration for the following reasons:

1) The output resistance of an SIS mixer may be negative and the effect of this negative source impedance upon the IF amplifier noise should be understood.

2) The SIS mixer has a very limited output voltage and saturates at a very low input power level (of the order of nW's), which can be increased if the IF amplifier input resistance is made very low.

3) The internal noise in the SIS mixer is very low and the receiver noise is critically dependent upon realizing a low-noise contribution from the IF amplifier.

4) For stability reasons, an SIS mixer with a negative output resistance may require an IF amplifier input impedance which is low or high dependent upon the reactance versus frequency slope in the circuit.

An important point that is often not realized by developers of SIS mixers is that the source impedance Z_{opt} which minimizes the IF amplifier noise can be chosen independently of the IF amplifier input impedance Z_{in}. It has been common practice to use IF amplifiers which have been designed to have $Z_{opt} = Z_{in} = 50\ \Omega$. This is not required; Z_{in} can be made low for reason 2) and Z_{opt} can be chosen to give best noise performance. An example of a feedback amplifier designed for $Z_{opt} = Z_{in} = 50$ is given in [14]; the feedback and input network could be changed to give, for example, $Z_{opt} = 200$ and $Z_{in} = 20$.

In this paper we will assume that a lossless coupling network exists between the FET or HEMT device and the mixer; the impedance transformation of this network will then be optimized so that the resulting Z_{opt} at the mixer plane minimizes the IF noise contribution. If the mixer has an output impedance, Z_{out},

with positive real-part, we simply equate $Z_{opt} = Z_{out}$ and are finished. Lossless feedback would then be utilized to make the amplifier input impedance low; this feedback would have small effects upon Z_{opt}, but the input coupling network could be adjusted to compensate and preserve the $Z_{opt} = Z_{out}$ condition. We will find Z_{opt} for the negative source resistance case and compute the resulting noise performance. Finally, the effects of transmission lines and isolators will be considered.

SIS mixers can be realized with positive output resistance, input match, and a small amount of gain (a few dB, SSB) [5]. The output resistance may be high, and an IF amplifier with high R_{opt} and low R_{in} appears to be desirable. If an isolator is inserted between the mixer and the IF amplifier, then $R_{opt} = R_{in}$ at the mixer–isolator interface [6]. Good results may be obtained, but the optimization which may result from $R_{opt} \neq R_{in}$ is not realized.

It is important to consider the effect of a transmission line of characteristic impedance Z_0 between the mixer and IF amplifier in the case $R_{opt} \neq R_{in}$. If $Z_0 = R_{opt} \neq R_{in}$, the noise performance will be independent of IF frequency, but the gain will depend upon the frequency-dependent transformation of R_{in} caused by the transmission line. On the other hand, if $Z_0 = R_{in} \neq R_{opt}$, the gain will be independent of IF frequency, but the noise will not. The required solution is either to make the transmission line short (for wide bandwidth) and integrate the IF amplifier with the mixer, or to make the transmission line one-half wavelength long at the IF center frequency (with of the order of 30 percent bandwidth).

There may be no advantage to operating the SIS mixer with negative output resistance but should this occur, there are large effects which can increase or decrease the overall noise dependent upon the noise properties of IF amplifier active device; this topic is discussed in the next section.

II. NEGATIVE RESISTANCE CASCADING THEORY

The conventional noise figure cascading formula of Friis [7] uses concepts of available power and available gain. In the case of a negative output resistance, both of these quantities and the second-stage noise figure become infinite and the first-stage noise figure becomes indeterminate. Fortunately, this situation was recognized by Haus and Adler 30 years ago, and they made modifications to the theory [8], [9]. The modifications are somewhat strange and involve the concepts of exchangeable power and exchangeable gain, which are understandable mathematically but are somewhat obscure from a physical point of view. The important physical concept is that the conventional minimum noise temperature T_{min} of an amplifier is the minimum with respect to variation of the source impedance *within the positive resistance plane*. When the source impedance is allowed to vary

Manuscript received March 16, 1987; revised July 21, 1987.
The author is with the National Radio Astronomy Observatory, Charlottesville, VA 22903. The observatory is operated by Associated Universities, Inc., under contract with the National Science Foundation.
IEEE Log Number 8716920.

Reprinted from *IEEE Transactions on Microwave Theory and Techniques,* vol. MTT-35, no. 11, pp. 1067–1069, November 1987.

into the negative resistance plane, a new noise minimum noise temperature, which we will call T_{neg}, is revealed. In the formalism of Haus and Adler, T_{neg} is also negative but is divided by a negative exchangeable gain to provide a positive contribution of the second stage to the overall noise temperature. For minimum noise, $|T_{neg}|$ should be minimized, and this is an entirely different case from the minimization of T_{min}. In terms of the correlated noise voltage- and current-source model of a noisy amplifier [10], the negative resistance allows complete cancellation of the correlated noise components from the two sources. In terms of the noise wave model of a noisy amplifier [11], the source reflection coefficient >1 produced by a negative resistance allows the noise wave coming out of the amplifier input to more completely cancel the correlated portion of the ingoing noise wave. The minimum noise temperature T_{neg} does not occur when the cancellation is complete because the magnitude of the uncorrelated noise must also be considered.

A formal application of the Haus/Adler noise theory to the case of an SIS mixer is as follows. The exchangeable power P_e of a current source I_s with internal shunt resistance R_s is given by

$$P_e = |I_s|^2 \cdot R_s / 4 \tag{1}$$

and the exchangeable gain G_1 of the mixer is the ratio of exchangeable power at the IF output terminals to exchangeable power of the RF source. If R_s is negative, the exchangeable power is negative; if R_s is positive, the exchangeable power is positive and is equal to the available power of the source. For an SIS mixer with negative IF output resistance driven from a positive RF source resistance, the exchangeable gain is negative. In terms of the gain G_0 into normalizing impedance Z_0 (typically 50 Ω), $G_1 = G_0/(1 - |\Gamma_{out}|^2)$, where $\Gamma_{out} = (Z_{out} - Z_0)(Z_{out} + Z_0)$ is the output reflection coefficient of the mixer.

The noise temperature of the mixer, T_1, without IF amplifier is given by

$$T_1 = N_e / (G_1 \cdot k \cdot \Delta f) \tag{2}$$

where N_e is the exchangeable noise power at the mixer IF output terminals due to noise sources within the mixer.

The noise temperature of the cascade of mixer and IF amplifier is then given by

$$T_{12} = T_1 + T_2 / G_1 \tag{3}$$

where T_2 and G_1 are both negative if the mixer has negative output resistance. Both T_1 and G_1 are properties of the mixer and are independent of the IF load impedance, but the IF noise temperature T_2 is dependent upon the mixer output impedance Z_{out}. For a given Z_{out}, we wish to find the lossless coupling network which minimizes T_2. For this purpose the IF noise temperature T_2 is most conveniently expressed in the form

$$T_2 = T_{min} + N T_0 |Z_{out} - Z_{opt}|^2 / (R_{out} R_{opt}) \tag{4}$$

where T_{min}, N, and Z_{opt} are four noise parameters describing the IF amplifier and $T_0 = 290$ K. The noise parameter N is equal to $g_n R_{opt}$, where g_n is the noise conductance. This form is used because T_{min} and N are invariant to variation of a lossless coupling network which can be used to transform Z_{opt} to whatever value minimizes T_2. By differentiating T_2 with respect to R_{opt}, we find the optimum positive values of R_{opt} and the resulting minima of T_2 as follows:

$$R_{opt} = R_{out} \qquad T_2 = T_{min} \qquad \text{for } R_{out} > 0 \tag{5}$$

$$R_{opt} = -R_{out} \qquad T_2 = T_{neg} = T_{min} - 4NT_0 \qquad \text{for } R_{out} < 0. \tag{6}$$

For all cases $X_{opt} = X_{out}$. It can be shown as a property of noise

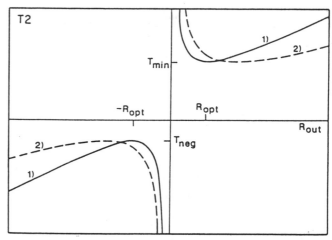

Fig. 2. The general form of the noise temperature of an amplifier as a function of generator resistance R_{out} is shown above. Adjustment of the impedance transformation of the amplifier input network can change the location of R_{opt}, as from curve 1) to 2), but not the value of T_{min}. For negative generator resistance, the noise temperature is negative and has a maximum value of T_{neg} at a generator resistance of $-R_{opt}$. The temperature, T_{neg}, is related to the noise wave coming out of the *input* port of the amplifier and may have absolute value higher or lower than T_{min} dependent upon noise properties of nonreciprocal or active elements in the amplifier.

parameters [6] that $T_{min} < 4NT_0$ and thus that the optimum T_2 for negative R_{out}, which we will call T_{neg}, is always negative. Furthermore, since T_{neg} is a maximum, T_2 is always negative for negative R_{out}. It is also interesting to note that the optimum value of R_{opt} is a positive value equal to the absolute value of R_{out}, and that T_{neg} is invariant to a lossless coupling network since both T_{min} and N are invariant. Thus, T_{neg} is a property of the active device or nonreciprocal elements in the amplifier. These results are illustrated in Fig. 2.

A further physical interpretation can be given to T_{neg} by examination of the noise wave model of a noisy amplifier [11]. This model represents the noise in the amplifier by two correlated noise waves having temperatures, T_a and T_b, coupled in and out of the amplifier input terminals, respectively; close inspection reveals that $T_b = -T_{neg}$ if $Z_{opt} = Z_0$. (Note that in [11] $T_d = 4NT_0)/(1 - |\Gamma_0|^2)$.) The total noise coming out of the amplifier input port is not just T_b, but also contains a contribution of T_a reflected from the amplifier input terminals. However, in the case of an amplifier with an ideal input isolator with termination temperature T_i, then $T_b = -T_{neg} = T_i$. In this case we see that T_{neg} is independent of T_{min}. A very noisy amplifier with a cold input isolator could produce very little noise when driven by a negative resistance with magnitude close to the isolator characteristic impedance. This case can be understood as a negative-resistance amplifier which precedes the amplifier. Another viewpoint is to consider the isolator as part of the mixer with impedances chosen for high gain. In practice, the only problem with this case is that the isolator forces the input resistance R_{in} to be equal to R_{opt}, and as R_{out} becomes close to $-R_{opt}$ for low noise, it will also become close to $-R_{in}$ and produce high and unstable gain. A better approach may be to design an amplifier with low R_{in} for stable gain and low $T_b = -T_{neg}$ and $R_{opt} = -R_{out}$ (instead of low T_{min} and $R_{in} = R_{opt}$, as is normal practice).

Some examples of measured values of noise parameters at cryogenic temperatures are given in Table I. The MGF1412 values are for a complete three-stage amplifier with an input coupling network, while the FHR01FH values refer to the transistor alone. The low value of $-T_{neg}$ at 8.4 GHz for the HEMT is encouraging.

TABLE I

Transistor	Mitsubishi MGF1412 GASFET	Fujitsu FHR01FH HEMT
Reference	[12]	[13]
Temperature	15 K	12.5 K
Frequency	1.6 GHz	8.4 GHz
T_{min}	7.4 K	10.3 K
$4NT_0$	16.0 K	13.9 K
$-T_{neg}$	8.6 K	3.6 K
R_{opt}	50	4.6
X_{opt}	0	17.0

REFERENCES

[1] J. R. Tucker and M. J. Feldman, "Quantum detection at millimeter wavelengths," *Rev. Mod. Phys.*, vol. 57, no. 4, Oct. 1985.

[2] L. R. D'Addario, "An SIS mixer for 90–120 GHz with gain and wide bandwidth," *Int. J. Infrared and Millimeter Waves*, vol. 5, pp. 1419–1422, 1984.

[3] S.-K. Pan, M. J. Feldman, A. R. Kerr, and P. Timbie, "Low-noise, 115 GHz receiver using superconducting tunnel junctions," *Appl. Phys. Lett.*, vol. 43, no. 8, pp. 786–788, Oct. 15, 1983.

[4] D. W. Face, D. E. Prober, W. R. McGrath, and P. L. Richards, "High-quality tantalum superconducting tunnel junctions for microwave

[5] mixing in the quantum limit," *Appl. Phys. Lett.*, vol. 48, no. 16, pp. 1098–1100, Apr. 21, 1986.

[5] A. R. Kerr, private communication, Feb. 17, 1987.

[6] M. W. Pospieszalski, "On the noise parameters of isolator and receiver with isolator at the input," *IEEE Trans. Microwave Theory Tech.*, vol. MTT-34, pp. 451–453, Apr. 1986.

[7] H. T. Friis, "Noise figures of radio receivers," *Proc. IRE*, vol. 32, pp. 419–422, July 1944.

[8] H. A. Haus, and R. B. Adler, "An extension of the noise figure definition," *Proc. IRE*, vol. 45, pp. 690–691, May 1957.

[9] H. A. Haus and R. B. Adler, "Optimum noise performance of linear amplifiers," *Proc. IRE*, vol. 46, pp. 1517–1533, Aug. 1958.

[10] H. Rothe and W. Dahlke, "Theory of noisy four-poles," *Proc. IRE*, vol. 44, pp. 811–818, June 1956.

[11] R. P. Meys, "A wave approach to the noise properties of linear microwave devices," *IEEE Trans. Microwave Theory Tech.*, vol. MTT-26, pp. 34–37, Jan. 1978.

[12] S. Weinreb, "Noise parameters of NRAO 1.5 GHz GASFET amplifiers," Electronics Division Internal Report No. 231, National Radio Astronomy Observatory, Charlottesville, VA, Dec. 1982.

[13] M. Pospieszalski and S. Weinreb, "FET's and HEMT's at cryogenic temperatures—Their properties and use in low-noise amplifiers," in *1987 IEEE MTT-S Int. Microwave Symp. Dig.*, vol. II (Las Vegas, NV), June 1987, pp. 955–958.

[14] S. Weinreb, D. L. Fenstermacher, and R. W. Harris, "Ultra-low-noise 1.2- to 1.7-GHz cooled GaAs FET amplifiers," *IEEE Trans. Microwave Theory Tech.*, vol. MTT-30, pp. 849–853, June 1982.

MONOLITHIC INTEGRATION OF HEMTS AND SCHOTTKY DIODES FOR MILLIMETER WAVE CIRCUITS

W-J. Ho, E.A. Sovero D.S. Deakin, R.D. Stein, G.J. Sullivan, and J.A. Higgins

Rockwell International Science Center
1049 Camino Dos Rios, Thousand Oaks, CA 91360

T.N. Trinh and R.R. August

Rockwell International Corporation, Autonetics Sensor & Aircraft System Division,
3370 Miraloma Avenue, Anaheim, CA 92803

ABSTRACT

This paper reports the first monolithic integration of high quality Schottky diodes ($F_c > 1250$ GHz) and high performance, low noise HEMT devices ($F_{max} > 120$ GHz, $F_{min} < 2$ dB at 18 GHz) and amplifiers (18 dB gain at 20 GHz) on the same wafer. MBE-grown AlGaAs/GaAs HEMTs show dc transconductance of up to 456 mS/mm for an 0.35 μm gate length with excellent uniformity. Noise figures of 2 dB, with associated gain of 13 dB at 18 GHz, are uniform over a wide range of device currents at room temperature and decline monotonically to 0.4 dB at 28K. At room temperature, fabricated two-stage amplifiers show 18 dB gain at 20 GHz. Hybrid microstrip branch line mixers fabricated with the high frequency diodes demonstrate 7.5 dB loss at 87-96 GHz.

INTRODUCTION

The need for low-noise microwave and millimeter-wave receivers has prompted extensive research in the field of high electron mobility transistors (HEMTs) (1-4). After they were first successfully fabricated in 1980 (5), HEMTs have demonstrated their superior gain and noise figure performance over conventional MESFETs because of their higher average velocity and mobility (6). GaAs/AlGaAs HEMTs using higher doping concentrations in the AlGaAs layer than used in MESFET have resulted in increased transconductance. Thinner GaAs drift region layers reduce the output conductance. HEMTs may be optimized to provide much better low temperature noise performance (7). To implement mixers at mm-wave frequencies, Schottky diodes are presently preferred over MESFETs or HEMTs, because the Schottky diodes yield better mixer performance (8-10). The layer structure required for high performance Schottky diodes, however, is significantly different from that used in HEMTs which, to date, has precluded the use of both device types on the same chip. This paper reports the first monolithic integration of high quality Schottky diodes ($F_c > 1250$ GHz) and high performance, low noise HEMT devices ($F_{max} > 120$ GHz, $F_{min} < 2$ dB at 18 GHz) and amplifiers (18 dB gain at 20 GHz) on the same wafer. Monolithic integration is important because it minimizes parasitic losses and cuts down the cost since all the high frequency circuitry is on one chip. Monolithic combination of the diodes and HEMTs is significant because it allows for the design of millimeter-wave receiver front-ends which may include low noise amplifiers, diode mixers, low noise IF amplifiers and varactor controlled HEMT VCOs all on the same chip.

DEVICE STRUCTURE

The device layer structure grown by MBE shown in Figure 1 consists of diode layers superimposed on top of HEMT layers. A GaAs/AlAs superlattice was grown to stop outdiffusion of impurities and dislocations from the substrate and reduce the output conductance. A conventional AlGaAs/GaAs layer structure was subsequently grown for the HEMT application. A 100Å GaAlAs layer is used to separate the diode from the HEMT layers; it provides an etch stop for reactive-ion-etching. The diode layers consist of a thick N^+ layer with a 5×10^{18} cm^{-3} donor concentration to reduce the series resistance, a 750Å Schottky barrier layer, and a surface n^+ GaAs contact layer. The Schottky barrier layer is moderately doped to avoid high electric fields that might cause unwanted electron heating. But, the dopant density is low enough to prevent electron tunneling through the barrier which might be another source of noise. The device structure with monolithic integration of HEMTs and diodes is shown in Figure 2. The Schottky diodes were formed on the top mesa layers and HEMTs were fabricated on the underlying layers. A 2000Å thick layer of low dielectric constant sputtered silicon dioxide was deposited on top of the mesa. The Schottky gate junction and the ohmic contacts were formed separately by etching holes through the insulator layer. The first metal was deposited on top of the insulator and the second metal was plated on using an air-bridge technology or using a polyimide as the intermetal dielectric. The uppermost unwanted diode layers were removed by selective reactive-ion-etching in freon 12 and helium plasma, which stopped at the AlGaAs layer (11). Contrast-enhanced optical lithography was used for the most of the process with the exception of the HEMT gates. The 0.35 μm long gate was formed by two-layer electron-beam-lithography, and the gate was recessed by selective reactive-ion-etching to control the uniformity of HEMT threshold voltage. AuGe/Ni/Au ohmic contact metal was alloyed at 450°C. Ti/Pt/Au was evaporated as Schottky gate metal for the diodes and HEMTs. Devices were passivated by a layer of polyimide, which also served as a crossover insulator for the gold plating step, wherein all gate feeds were brought out to a common bonding pad. The capacitors were formed by plasma-enhanced CVD silicon nitride. After the front side process was completed, the wafer was mounted on glass disc and thinned down to 75 μm. Through substrate, backside via-

Reprinted from *Rec. of the IEEE GaAs Integrated Circuits Symp.*, pp. 301–304, 1988.

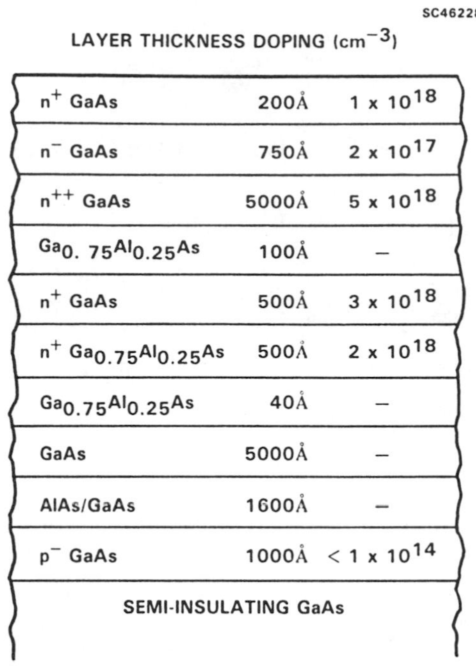

SC46228

LAYER THICKNESS DOPING (cm^{-3})

n^+ GaAs	200Å	1×10^{18}
n^- GaAs	750Å	2×10^{17}
n^{++} GaAs	5000Å	5×10^{18}
Ga$_{0.75}$Al$_{0.25}$As	100Å	—
n^+ GaAs	500Å	3×10^{18}
n^+ Ga$_{0.75}$Al$_{0.25}$As	500Å	2×10^{18}
Ga$_{0.75}$Al$_{0.25}$As	40Å	—
GaAs	5000Å	—
AlAs/GaAs	1600Å	—
p^- GaAs	1000Å	$< 1 \times 10^{14}$
SEMI-INSULATING GaAs		

Fig. 1 Device layer structure grown by MBE on semi-insulating GaAs with diode layers superposed on top of HEMT layers.

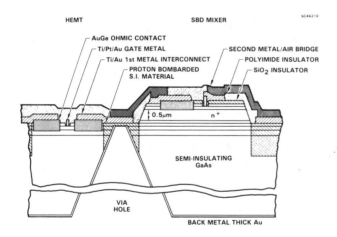

Fig. 2 Cross sectional view of a monolithic HEMT and Schottky barrier diodes.

holes were fabricated by wet chemical etching, and then plated with gold for source grounding.

HEMT DEVICES PERFORMANCE

One of the significant achievements of this integration technology is that the HEMT layer system and device performance is preserved during the removal of the diode layers. A factor contributing to this is the 100Å AlGaAs etch-stop layer which stops radiation damage from reactive-ion-etching from penetrating into the HEMT

layers. In this report, a conventional AlGaAs/ GaAs HEMT device structure was implemented for device applications. Data was measured on the devices with 0.35 μm gate length and 75 μm width. Uniformity of device parameters (derived from equivalent circuit fitting of S-parameters measurements) is shown in Table 1. The variation of extrinsic transconductance (G_m) and drain current (I_d) with gate voltage is shown in Figure 3. Peak transconductance of 456 mS/mm was measured. Figure 4 shows the variation of device noise figure and associated gain with drain current at 18 GHz. The noise figure was measured with an HP-346C noise source and an HP-8970A noise meter using a HP transistor test fixture. The minimum noise figure achieved in this device is 2 dB with associated gain of 13 dB, which are uniform over a range of device currents. Figure 5 shows unilateral gain (U), maximum stable gain (MSG), and the current gain (H_{21}) as a function of frequency for a 150 μm HEMT. The maximum frequency (F_{max}) of 120 GHz was extrapolated from unilateral gain at 6 dB per octave slope and the cut-off frequency (F_t) of 35 GHz was extrapolated from the current gain (H_{21}). Excellent low temperature performance was demonstrated in Figure 6 as the noise figures of 2 dB at 300K decrease monolithically to 0.4 dB at 28K. The associated gains increase with temperature. The two-stage amplifiers fabricated through this integration technology show an 18 dB minimal gain at 20 GHz at room temperature (Figure 7) and a 2.5 dB amplifier noise figure.

Table 1
Statistical Results of 5 Equivalent
Circuits for 75 μm Devices

Parametric	G_m	C_{gs}	C_{gd}	R_{in}	R_{ds}
Unit	ms	fF	fF	ohm	ohm
Average	34.8	136.05	18.4	10.2	569
Std. Dev.	2.8	18.80	2.1	3.3	49

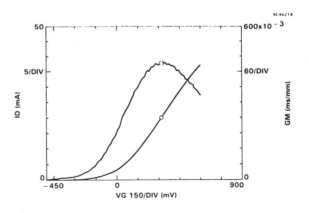

Fig. 3 Drain current and transconductance vs gate voltage for an 0.35 μm HEMT fabricated with the integration technology of HEMT and diodes.

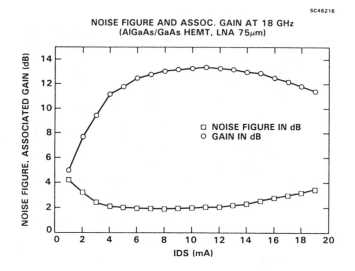

Fig. 4   Noise figure and associated gain (both in dB) vs drain current at 18 GHz for an AlGaAs HEMT with an 0.35 μm gate length.

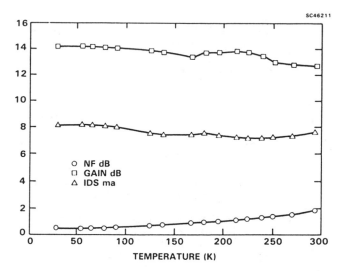

Fig. 6   Measurements of noise figure, associated gain and source-drain current vs temprature at 21 GHz.

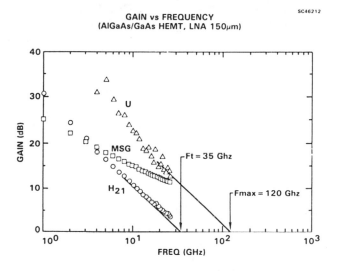

Fig. 5   Unilateral gain (U), current gain (H_{21}) and maximum stable gain (MSG) as a function of frequency for a 150 μm low-noise amplifier.

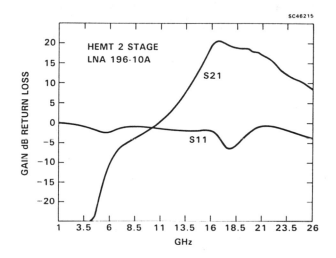

Fig. 7   Two-state amplifier produced 17.94 dB gain at 20 GHz.

DIODE PERFORMANCE

Diode dc characteristics show excellent uniformity of ideality factor of 1.13 with a barrier height of 0.73 V. For high frequency characterization, the Schottky barrier diode was probed using the Cascade prober in conjunction with the HP-8510 Network Analyzer from 1 to 26 GHz. The obtained S22 data was fed into Touchstone optimization program to fit the equivalent circuit. The series resistance (R_s) was thus measured to be 5 Ω at 10 mA of forward current. The total capacitance, which includes parastics parallel capacitance (C_p) and junction capacitance (C_{jo}) was measured to be 25 fF(C_p = 20 fF and

C_{jo} = 5 fF). The "cut-off" frequency (Fc)," which is defined as:

$$F_C = 1/[6.28*R_s*(C_p + C_{jo})] \qquad (1)$$

was found to be higher than 1250 GHz. A hybrid micro-strip branch line mixer fabricated with the high frequency diodes in a hybrid fixture demonstrated 7.5 dB conversion loss at 87-96 GHz (Figure 8).

CONCLUSIONS

The concept and the technology have been successfully demonstrated to produce monolithic integration of diodes for mixers and HEMTs for the low noise amplifiers on the

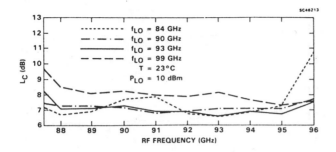

Fig. 8   Conversion loss performance of the nonbiased mixer circuit at room temperature.

same wafers. Excellent high frequency performance of diodes ($F_c > 1250$ GHz) and low noise characteristics ($F_{min} < 2$ dB at 18 GHz) of HEMTs at low temperatures have been achieved without sacrifice of their individual performance.

ACKNOWLEDGEMENTS

The authors would like to express their appreciation to S. Pittman, R. Pierson, E. Peterson, R. Bernescut, R. Anderson and S. Skylstad for their assistance in wafer processing. They would also like to thank Dr. P. Asbeck for his valuable technical discussion.

REFERENCES

(1)   C. Yuen, C. Nishimoto, M. Glenn, Y.C. Pao, S. Bandy and G. Zdasiuk, IEEE MTT-S Digest, 247 (1988).

(2)   P.M. Smith, P.C. Chao, K.H.G. Duh, L.F. Lester, B.R. Lee and J.M. Ballingall, 1987 IEEE MTT-s International Microwave Symposium Digest, p. 749.

(3)   J.M. Schellenberg, M.V. Maher, S.K. Wang, K.G. Wang and K.K. Yu, 1987 IEEE HEMT Digest, p. 44.

(4)   E.A. Sovero, A.K. Gupta and J.A. Higgins, IEEE Electron Dev. Lett. EDL-7(3), 179 (1986).

(5)   T. Mimura, S. Hiyamizu, T. Fujii and K. Nanbu, Jpn. J. App;l. Physi. 19, 2255 (1980).

(6)   Y. Ito, 1987 International European Microwave Symposium, p. 832.

(7)   K.H. Duh, M.W. Pospierszalski, W.F. Kopp, P. Ho, A.A. Jabra, P.C. Chao, P.M. Smith, L.F. Lester, J.M. Ballingall and S. Weibreb, IEEE Trans. on Electron Dev. 35(3), 249 (1988).

(8)   M. Sholley, S. Mass, B. Allen, R. Sawires, A. Niochols and J. Abell, Microwave J., p. 121, Aug. 1985.

(9)   M. Mccoll, IEEE Trans. on Microwave Theory and Tech., p. 54, Jan. 1977.

(10)   H.R. Fetterman, B.J. Clifton, P.E. Tannewald, and C.D. Parker, Appl. Phys. Lett. 24, 70-72, Jan. 1974.

(11)   T. Mimura, K. Joshin, S. Hiyamizu, K. Hikosaka and M. Abe, Jpn. J. Appl. Phys. 20, L598 (1981).

HEMT MILLIMETER WAVE AMPLIFIERS, MIXERS AND OSCILLATORS

MICHAEL SHOLLEY*, BARRY ALLEN*, STEVE MAAS°, ARTHUR NICHOLS*

ABSTRACT:

This paper describes the design, and development of state-of-the-art millimeter wave low noise components, each of which is based upon High Electron Mobility Transistor (HEMT) Technology and are well suited for integrated millimeter-wave receivers. These components include 40, 60, and 70 GHz amplifiers, an active 45 GHz mixer and a dielectric resonator stabilized oscillator. The 36 to 40 GHz three stage amplifier achieved a gain of 15 to 17 dB with an associated noise figure of 4.0 to 4.6 dB. The 56 to 62 GHz amplifier had a gain of 4.5 to 6.5 dB with a noise figure of 6.0 dB measured at 57.5 GHz and the 70 GHz amplifier had 4 to 5 dB gain with a noise figure of 7.8 dB measured at 71.0 GHz. The oscillator was stable to 17 ppm°C over 20°C to 70°C and had an output power of 0 dBm. The mixer achieved unity gain, 7-8 dB SSB noise figure that includes the IF amplifier and 2 dBm output intermodulation intercept. For future design, methods have been developed to characterize a HEMT device for gain and noise figure through 44 GHz. These results represent entirely new approaches to millimeter-wave component design and advance the state-of-the-art in performance capabilities.

INTRODUCTION

The increased use of EHF millimeter-wave communications systems for use in both military and commercial communications satellites, has developed a need for receiver components with extremely low noise figures, wider bandwidths and higher gains than previously available. During the past two years TRW has developed High Electron Mobility Transistor (HEMT) amplifiers, mixers and oscillator components (2). The use of these HEMT devices has allowed the design of low noise amplifiers at frequencies greater than 70 GHz. The use of HEMT amplifiers as receiver preamplifiers allows an all MIC type receiver construction, which eliminates the need for waveguide components, and also improves the sensitivity of the receiver by lowering the noise figure of the system. These receivers have achieved improved noise performance over what is available with the more commonly used diode mixer. TRW has developed several receivers based on this principle. To further improve the performance of these components, extensive device modeling techniques have been developed which involve S-parameter measurements and the fitting of models. This paper describes several of the HEMT based components and the device modeling techniques used in the characterization of the HEMT device.

40 GHz Amplifier

The three stage 40 GHz low noise amplifier was designed utilizing TRW's HEMT device and conventional microstrip fabrication techniques as shown in Figure I. The amplifier consists of three single ended amplifier stages cascaded onto an aluminum supercarrier. Gain and noise figure for the amplifier are presented in Figure II. To optimize the noise figure and gain the use of isolators or hybrid coupled stages were avoided.

* TRW Electronics Systems Group, Redondo Beach, CA 90278

° Currently with Aerospace Corp., Los Angeles, CA 90009
  S. Maas was employed at TRW during the time this work was performed.

Reprinted with permission from *Conf. Proc. Military Microwaves 1986*, pp. 517–522, June 1986.

60 and 70 GHz Amplifiers

The 60 and 70 GHz amplifiers were designed and constructed utilizing the expertise and fabrication techniques that were developed during the construction of the 40 GHz amplifier. The HEMT devices used in these amplifiers had ≤ 0.25 μM gate lengths, with gate widths of 80 μM. The method of direct write E-beam lithography was used to produce the gates on epitaxial layers that were grown by molecular beam lithography as shown in Figure III. Amplifiers have been designed at frequencies of 30 GHz and 40 GHz with noise figures of 2.2 dB at 30 GHz and 3.0 dB at 40 GHz respectively. As shown in Figure IV the amplifiers were constructed in two parts using invar, with the lower section of the housing containing the bottom waveguide portion, input and output substrates, amplifier circuitry and biasing components, while the upper section contains the amplifier cavity, waveguide components and lid to complete the enclosure. The input and output matching circuits are realized in microstrip on 0.010" thick quartz substrate with chrome gold metallization.

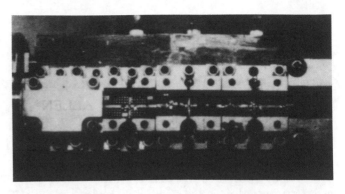

FIGURE 1.   PHOTOGRAPH OF THREE STAGE AMPLIFIER

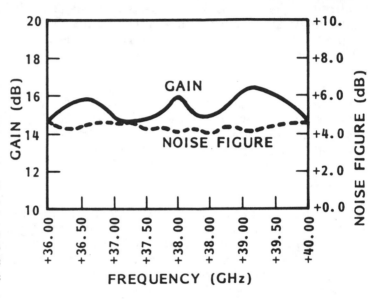

FIGURE 2.   GAIN AND NOISE FIGURE OF THREE STAGE AMPLIFIER

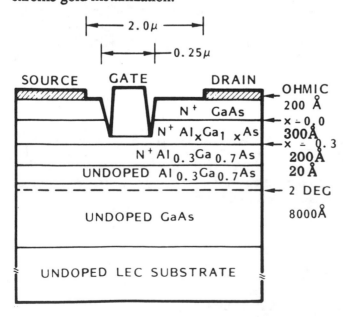

FIGURE 3.   HEMT DEVICE CROSS SECTION

Gate and drain bias were applied through decoupling circuits that consist of $\lambda/4$ high impedance lines and $\lambda/4$ open stubs for RF rejection, optimized to insure out of band stability. A broadband 60 GHz amplifier was designed to cover the 56 to 62 GHz band. As shown in Figure V the gain was 4.5 to 6.5 dB. The noise figure measured at 57.5 GHz was 6.0 dB and the 1 dB compression point was 8.5 dBm. For purposes of assembly and testing of the amplifiers, waveguide to microstrip transitions were developed to cover the 56 to 65 GHz band and the 69 to 75 GHz band [3]. The 70 GHz amplifiers were developed as higher frequency variations of the 60 GHz design. Gain for the 70 GHz single stage amplifier is shown in Figure VI. The associated noise

244

figure for this amplifier was 7.8 dB measured at 70.0 GHz, while the 1 dB compression point was 7.5 dBm.

HEMT TEST FIXTURES

To provide accurate linear circuit modeling for HEMTs, a series of special test fixtures were developed for S-parameter and noise figure measurements at frequencies through 44 GHz. A photograph of the 44 GHz fixture is shown in Figure VII. The split design of the fixture allows accurate loss measurement to be made on the fixture so that true device nosie figure can be calculated. The fixtures include a bias network and tuning pads for noise matching the device. Versions of the noise figure test fixture have been tested at both 8 and 18 GHz. This particular fixture uses E-plane probe microstrip to waveguide transitions with over 4 GHz bandwidth. For the "thru" fixture connection, the total loss from flange to flange was <1.2 dB at 44 GHz, including the bias network. Using these fixtures the noise figure measured at 44 GHz is shown in Figures VIII.

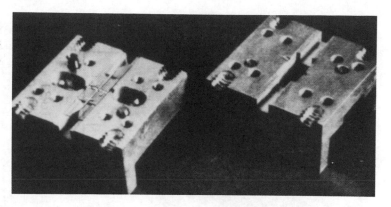

FIGURE 4. 60 GHz AMPLIFIER CONSTRUCTION

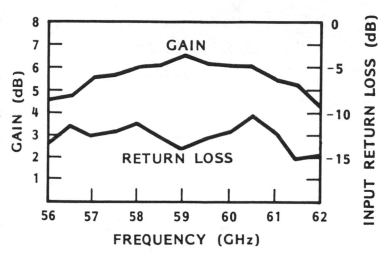

FIGURE 5. GAIN OF 60 GHz AMPLIFIER

As an example of the data obtained from these fixtures Figure IX gives a plot of the measured S-parameters and the fit for a linear circuit model. The de-embedded model and model parameters are given in Figure X. The noise figure data at 18 GHz was taken at the bias level that produced the lowest 44 GHZ noise figure. By reducing the drain voltage, the minimum noise figure at 18 GHz was reduced to 1.1 dB with 10.2 dB associated gain. Much of the noise from this device is thermal noise due to the series gate resistance. Improve HEMT processing is under development to reduce the gate resistance and decrease the millimeter wave noise figure.

HEMT OSCILLATOR

The HEMT oscillator was designed to operate at a frequency of 44 GHz, utilizing a dielectric resonator to supply 8ppm/°C stability at room temperature. A photograph of the oscillator is shown in Figure XI. The oscillator was fabricated on highly polished 0.010" alumina substrate with chrome gold metallization. Tuning of the oscillator was accomplished by varying the thickness of the quartz

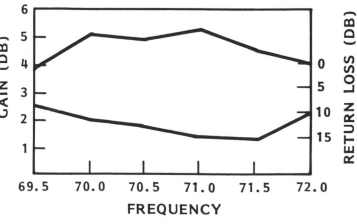

FIGURE 6. GAIN OF 70 GHz AMPLIFIER

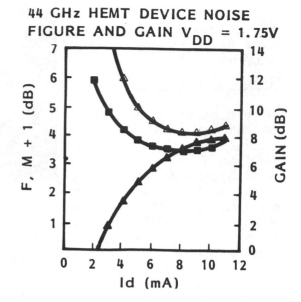

44 GHz HEMT DEVICE NOISE
FIGURE AND GAIN V_{DD} = 1.75V

FIGURE 7.   44 GHz S-PARAMETER TEST
FIXTURE

FIGURE 8.   44 GHz NOISE FIGURE

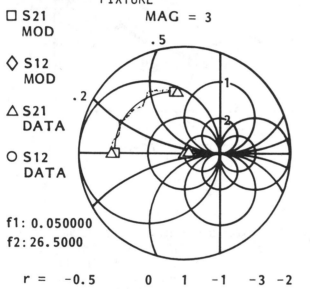

□ S21 MOD

◇ S12 MOD

△ S21 DATA

○ S12 DATA

MAG = 3

f1: 0.050000
f2: 26.5000

r =   -0.5   0   1   -1   -3 -2

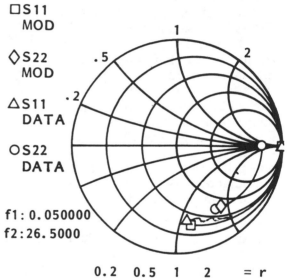

□ S11 MOD

◇ S22 MOD

△ S11 DATA

○ S22 DATA

f1: 0.050000
f2: 26.5000

0.2   0.5   1   2   = r

FIGURE 9.   PLOT OF S-PARAMETERS

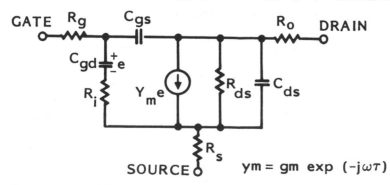

FIGURE 10.
DE-EMBEDDED MODEL AND
MODEL PARAMETERS

y_m = g_m exp (-$j\omega\tau$)

R_g	= 5.9 Ω	R_{ds}	= 500 Ω
C_{gd}	= 0.070 pF	R_s	= 0.7 Ω
R_i	= 5.3 Ω	C_{ds}	= 0.038
C_{gs}	= 0.0045 pF	R_o	= 7.2 Ω
g_m	= 22.5 ms	τ	= 0.6 psec

spacer under the resonator and the position of the resonator between the drain and gate lines. Data for the oscillator is shown in Figure XII and XIII. The DC efficiency and power consuption are 5% and 1 mW respectively.

HEMT Mixer

As a potential replacement for diode mixers a HEMT mixer was developed as shown in Figure XIV. Due to the non-availability of acceptable millimeter-wave FET devices and the difficulty in their design and poor performance, MESFET mixers above 30 GHz have not been developed. But with the availability of TRW's HEMT devices and the superior gain and noise performance achieved at frequencies greater than 40 GHz shows the possibility of designing HEMT mixers at these same frequencies. The mixer was a narrowband (200 MHz) IF near 3 GHz, and a low side tunable LO [4]. The mixer gain and noise figure vs LO power is shown in Figure XV. The mixer achieved a 0 dB ± 1 dB conversion loss, with a noise figure of 7.1 to 8.0 dB as shown in Figure XVI.

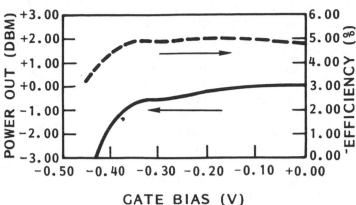

FIGURE 12.   OSCILLATOR OUTPUT POWER VS GATE VOLTAGE AND FREQUENCY

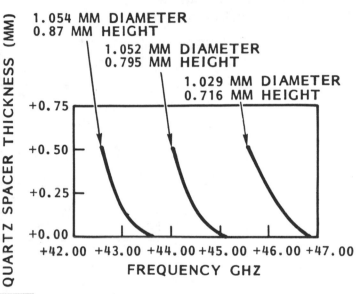

FIGURE 13.   PLOT OF FREQUENCY VS SPACE THICKNESS

FIGURE 11.   PHOTOGRAPH OF HEMT OSCILLATOR

FIGURE 14.   PHOTOGRAPH OF HEMT MIXER

247

Conclusions

The authors have constructed state-of-the-art millimeter wave components utilizing HEMT technology that are suitable for use in EHF communications receivers. These components include amplifiers, mixers and oscillators. The use of HEMT devices and device modeling techniques combined with advanced RF microwave technology makes it possible to construct these receivers at even higher frequencies of operation. These receivers will have superior RF performance over diode mixer based technology and also possess the benefits of reduced size, weight, and power.

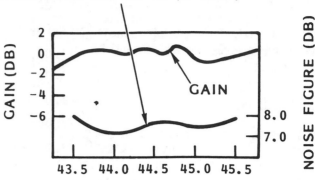

SSB NOISE FIGURE (P_{LO} = 0 DBM)
(INCLUDES 2.5 DB NF IF AMP)

FIGURE 15.   GAIN AND NOISE FIGURE VS LO POWER

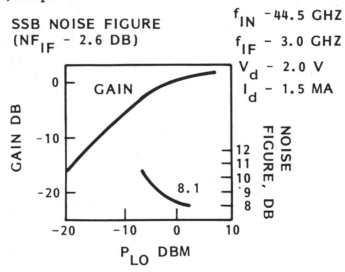

FIGURE 16.   GAIN AND NOISE FIGURE VS FRFQUENCY

Acknowledgement

This work was performed at TRW in the Communication Systems Development Department under several independent research and development program.

REFERENCES:

1)   Sholley, M., et al., "36.0 - 40.0 GHz HEMT Low Noise Amplifier" IEEE MTT Symposium, June 1985.
2)   Sholley, M., Maas, S., Allen, B., Nichols, A., Sawires, R., Abell, J., "HEMT mm - Wave Amplifiers, Mixers and Oscillators," Microwave Journal, Vol. 28, No. 8, August 85. MTT Symposium, June 1985.
3)   Van Heuven, J.H.C., "A New Integrated Waveguide Microstrip Transition," IEEE Transactions on Microwave Theory and Techniques, March 1976.
4)   Maas, S.A., "Theory and Analysis of GaAs MESFET mixers," IEEE Trans. MTT, Vol. MTT-32, No. 10, p1402 (Oct., 1984).

COMPARATIVE STUDY OF PHASE NOISE IN HEMT AND MESFET MICROWAVE OSCILLATORS

M. POUYSEGUR[*] - J. GRAFFEUIL[**] - J.F. SAUTEREAU[**] - J.P. FORTEA[***]

* Laboratoire d'Automatique et d'Analyse des Systèmes du CNRS
7, Avenue du Colonel Roche - 31077 TOULOUSE CEDEX

** Université Paul Sabatier
118, Route de Narbonne - 31062 TOULOUSE CEDEX
*** Centre National d'Etudes Spatiales
18, Avenue Edouard Belin - 31400 TOULOUSE

Summary

Several identical X band cavity stabilized MESFET and HEMT oscillators are presented. Their phase noise and some other noise data are reported. Under exactly the same oscillating conditions, the MESFET oscillators exhibit the best phase noise performance not only because of their lower low frequency noise but also because of a better linearity which provides a smaller LF noise conversion in the microwave frequency range.

Introduction

Solid state oscillators are widely used in most microwave equipment and their spectral purity specifications are among the hardest to satisfy. The reason is that microwave solid state devices, because of their reduced sizes, inherently exhibit a large low frequency (L.F) excess noise which is up-converted by the device non-linearities in the microwave frequency range.

This drawback is particularly noticeable in field effect devices. On the contrary, it is less critical in bipolar devices which exhibit a lower excess noise. Unfortunately silicon bipolar transistors are unable to provide good R.F performances above the X-band and bipolar heterojunction transistors are not commercially available yet.

Therefore only two competitors subsist, i.e., the GaAs MESFET and the GaAs - GaAlAs HEMT. A significant comparison of their phase noise performances has not been made yet since it necessitates very stringent requirements so as to be significant. Indeed, the phase noise performance of an oscillating two-port device depends at least on two main factors :

(i) the low frequency excess noise of the device (LF noise)

(ii) the topology and electrical performance of the circuit in which the device is embedded.

Therefore a significant comparison of the phase noise between two different devices requires firstly that a full characterization of their low frequency noise is completed before they are put into oscillation and secondly that they are embedded in strictly identical oscillating circuits.

This paper reports data obtained on several MESFET and HEMT oscillators complying with the aforementioned requirements. The origins of the observed difference in the oscillator's phase noise are finally discussed in terms of device properties.

Device description

Most of the oscillators were made with MESFET's or HEMT's commercially available at some major manufacturers. They are low noise figure devices featuring a gate length of about 0.5 μm and a gate width of about 300 μm. The HEMTs are usually processed using MBE. For comparison purposes a laboratory device, MOCVD processed by L.E.P. (France), is also included. We also expect the results obtained with some other HEMTs to be available by the time of the conference. Among all such devices a particular attention will be paid to the comparison between the GOULD and THOMSON MESFET and HEMT since these two devices feature an identical layout and similar electrical characteristics.

For the eight different devices available at the time of printing the most significant electrical data are given in table I.

	A MESFET S 8818 THOSHIBA	B MESFET 1405 MITSHUBISHI	C MESFET GOULD OXL0503	D HEMT GOULD MPD H503	E HEMT FUJITSU FHO1FH	F HEMT LEP EP 742	G MESFET THOMSON TC 1403	H TEGFET THOMSON LA313 A
Z (μm)	300(?)	300(?)	300	300	300(?)	200	300	300
L (μm)	0.5(?)	0.5(?)	0.3	0.5	0.5	0.55	0.55	0.55
V_T(V)	-1	-0,47	-0,8	-0,75	-0,91	-0,65	-1.67	-1.72
IDSS(mA)	29	20	27	23	33	17	55	73
Gass at NF db	10	9	9.5	10.5	10	10.5	9.3	9.6
NF(12 GHz) db	1.5	1.3	1.9	1.2	1.1	1.2	2	1.7

Table I

Reprinted from *1987 IEEE MTT-S Int. Microwave Symp., Tech. Dig.*, vol. 2, pp. 557–560, 1987.

All the devices are used into a 70 mil ceramic package.

Oscillator structure and measurements

For comparison purposes it is essential for the oscillating devices to be embedded in exactly the same circuits : the best way to fulfil such a condition is in fact to use only one circuit, whatever the device. Such a circuit, operating at any X-band frequency, is a cavity stabilized feedback structure realized with discrete elements, as shown in Figure 1.

The devices are used in a grounded source configuration. The packaged transistors are inserted into a classical coaxial test fixture. This technique ensures easy substitution of one device for another.

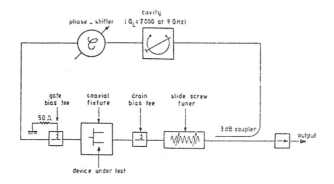

Figure 1 : Block diagram of the cavity stabilized HEMT/MESFET oscillator. The cavity loaded Q is about 2000 at 9 GHz.

For different devices to oscillate at a constant given frequency (9 GHz) when successively embedded in a given circuit, it is requested that they exhibit very similar RF characteristics. This condition was particularly well fulfilled with the GOULD MESFET and its HEMT counterpart where correct oscillating conditions were met without any further tuning of the tuner and/or the phase shifter (see Figure 1) when substituting one device for the other. The other devices necessitated a slight adjustment of the phase shifter to get the oscillation.

The gate bias voltage is fixed at zero volt and the drain bias voltage is adjusted between 3 and 5 V so as to obtain a given power of about 6 mW whatever the device on test.

The S.S.B. phase noise L(f) is firstly measured with an improved frequency discriminator technique at baseband frequencies f between 1 kHz and 100 kHz. The

corresponding noise spectra for the eight devices of table I are given in Figure 2.

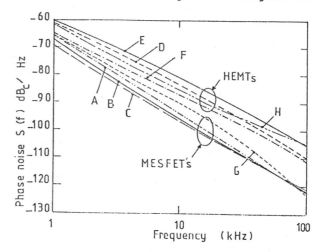

Figure 2 : Phase noise spectra of four different HEMTs compared with four different MESFETs with similar sizes and electrical performance (see Table I). (9 GHz) Vgs=0 V, 3 V < Vds <5 V. A, B, C, G are data from MESFETs of table I and D, E, F, H are data from HEMTs of Table I.

Secondly a white low frequency noise of known spectral density $Sv(f)$ is injected through the gate bias tee (see Figure 1) into the HEMT or the MESFET and the related excess phase noise $Le(f)$ is carefully checked. Next the ratio of the excess frequency fluctuation $f_e(f)_{RMS}$ in $Hz_{RMS}/\sqrt{Hz}$ by the LF noise voltage in $V_{RMS}/\sqrt{Hz}$, called the upconversion coefficient $K(f)$ at the baseband frequency f, is given by :

$$K(f) = \Delta f_e(f) / (\sqrt{Sv(f)}) \qquad (1)$$

where $\Delta fe(f)$ is obtained from :

$$Le(f)_{dBc/Hz} = 20 \log \left[\Delta fe(f) / (\sqrt{2} f) \right] \quad (2)$$

This coefficient $K(f)$ will be particularly useful for any further discussion on the phase noise since we believe that it is highly significant of the non-linear effects responsible for the noise upconversion in the device.

Low frequency noise

Apart from any microwave measurement, the devices are also inserted between two 50 ohms RF loads to prevent any oscillation and their input low frequency noise voltage spectral density $Se(f)$ is measured between 10 Hz and 100 kHz at the same DC operating points as those previously selected for phase noise investigations.

The eight different LF noise spectra are given in Figure 3. They indicate that in the frequency range of interest (1 kHz-100 kHz) :

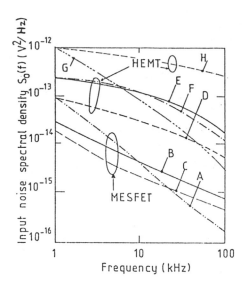

Figure 3 : Input voltage low frequency noise spectral density of the eight different devices involved in Table I and Figure 2. Bias conditions are identical to those used for phase noise measurements (Vgs=0 V, Vds 3 V).

(i) HEMTs exhibit 4 to 20 dB more noise at 10 kHz than the MESFETs
(ii) the noise spectral density varies as $1/f^\alpha$ with $0.5 \leqslant \alpha \leqslant 1.5$ for MESFET's and $0 < \alpha \leqslant 1$ for HEMT's. The existence of a noise bulge in the 1-10 kHz frequency range for most of the HEMTs results from some generation-recombination noise on DX centers at the GaAs/GaAlAs interface [1, 2].

Discussion and conclusion

The phase noise spectra displayed in Figure 4 makes it possible to state, first, that the phase noise of most of the MESFET's is about -95 dBc at 10 kHz off the carrier, which is very near from the state of the art performance obtained on low noise figure devices [3] although a further noise reduction should be possible when using power or medium power devices [4, 5] or larger Q resonators [3].
The most stringent result obtained from Figure 2 is that the phase noise in HEMT's is in average 10 dB larger than in MESFET's at 10 kHz off the carrier. The evolution of this difference when optimizing the gate bias of each device for a minimum phase noise is currently under investigation. Moreover this difference dramatically increases at larger baseband frequencies since an average -30 dBc/decade slope is observed for MESFETs and must be compared with about -20 dBc/decade slope only for HEMTs. This is the consequence of the larger $\alpha (\alpha \simeq 1)$ of MESFETs. The Lorentzian LF noise spectrum of DX centers in HEMTs results in more phase noise at larger baseband frequencies (10 kHz - 100 kHz). To get better insight into the way in

which this LF noise is upconverted, the average upconversion coefficient (which is almost a constant between 1 kHz and 100 kHz) is given in table 2.

Device	A	B	C	D	E	F	G	H
K(MHz/V)	1.25	0.82	0.9	1.5	1.3	1	0.55	0.65

Table 2

Table 2 shows that in average the upconversion is 3 dB larger in HEMTs than in MESFETs. For example, the GOULD devices exhibit 3dB difference in upconversion and a 10dB difference in LF noise which correctly accounts for the 12dB difference observed in the phase noise at 10 kHz off the carrier. Also worth mentioning is the fact that HEMTs from LEP and THOMSON (F,H) exhibits the smallest K which accounts for the smallest phase noise of these devices in comparison with other HEMTs.
It is therefore possible to conclude that HEMTs are not suitable yet for low noise oscillator applications because :
(i) they exhibit a larger low frequency noise partly due to DX centers in the GaAlAs layer, responsible for GR noise in the 10 kHz frequency range and therefore for the extra phase noise at the corresponding baseband frequencies
(ii) they feature a larger upconversion of this noise in the microwave range which probably denotes a non-linearity stronger than in MESFETs.
However future HEMTs will probably overcome these drawbacks provided that :
(i) a more matured technology results in less LF noise
(ii) a precise analysis of the non-linear behavior results in a better design of the device and of the embedding circuits which may differ from those known to ensure a minimum phase noise in MESFET's oscillators.

Aknowledgements : The authors are indebted to Mr. WOLNY (LEP - France) and Mr. BRIERE (THOMSON DAG - France) for supplying some of the devices used in this work.

References

(1) S.M.J. Liu, M.B. Das, C.K. Penk, J. Klem, T.S. Henperson, W.F. Kopp and H. Morkoç, "Low noise behaviour of InGaAs Quantum-Well-Structured Modulation-Doped FET's from 10^{-2} to 10^{8} Hz", IEEE Trans electron devices vol ED-33, pp 576-581, 1986

(2) J.M. Dieudonne, M. Pouysegur, J. Graffeuil and J.L. Cazaux, "Correlation between low-frequency noise and low-temperature performance of two dimensional electron gas FET's", IEEE Trans electron devices, vol ED-33, pp 572-575, 1986

(3) G. Lan, D. Kalokitis, E. Mykietyn, E. Hoffman, F. Sechi, "Highly stabilized ultra low noise FET oscillator with dielectric resonator", MTT-S Digest, pp 83-86, 1986

(4) J. Graffeuil, K. Tantrarongroj and J.F. Sautereau, "Low frequency noise physical analysis for the improvment of the spectral purity of GaAs FET's oscillators", Solid State Elect., vol 25, pp 367-373, 1982

(5) J. Graffeuil, A. Bert, M. Camiade, A. Amana and J.F. Sautereau, "Ultra low noise GaAs MESFET microwave oscillators", Proc. 7th international conference on Noise in physical systems and 1/f noise, Elsevier Science Publishers, pp 329-332, 1983

A STUDY OF THE RELATION BETWEEN DEVICE LOW-FREQUENCY NOISE AND OSCILLATOR PHASE NOISE FOR GaAs MESFETs

Hans Rohdin, Chung-Yi Su and Charles Stolte

Hewlett-Packard Laboratories
Palo Alto, California

ABSTRACT

An analytical model for oscillator noise resulting from active device LF noise is presented. We apply it to a number of GaAs MESFET oscillators finding good quantitative agreement, and demonstrating several ways of reducing the phase noise. We show evidence that after having reduced the effect of the normally dominant device LF noise source, a residual LF noise source starts to dominate the phase noise. The best phase noise result for the 5GHz oscillators is $S_\phi(1kHz) = -75dB/Hz$.

INTRODUCTION

The work reported here is a continuation of a recent study of the origin of low-frequency (LF) noise in GaAs MESFETs [1]. There it was shown that the major source is trap generation-recombination in a depletion region. In this paper we address the problem of how the GaAs MESFET LF noise is related to the phase noise of oscillators using these devices, and if the reduction in LF noise translates into a reduction in phase noise. For this purpose we have developed an analytical model and applied it to a variety of GaAs MESFET oscillators with a wide spread in phase noise.

THEORY

We take the Kurokawa approach [2] and extend it to include LF noise. It is assumed that the device LF noise can be represented by a small fluctuation, ε, in some device parameter, E. It is also assumed that E has the effect of modulating the device impedance and thus its reflection coefficient. This appears to be a reasonable assumption to make, since ε is a low-frequency fluctuation and is the driving force for the amplitude (A) and phase (ϕ) fluctuations which, in the original Kurokawa approach, are assumed to have a modulating effect. We can solve for the spectral densities of the amplitude ($a=\delta A/A_0$) and phase fluctuations, as well as for their cross-spectral density:

$$S_a(\omega) = S_\varepsilon(\omega)\frac{1}{1+(\omega/\omega_1)^2}(k\frac{\partial P_0}{\partial E_0})^2 \qquad (1)$$

$$S_\phi(\omega) = S_\varepsilon(\omega)\left[\frac{1}{\omega^2}(\frac{\partial\omega_0}{\partial E_0})^2 + \right.$$
$$\left. + \frac{1}{1+(\omega/\omega_1)^2}\left[2\frac{\xi}{\omega_1}(\frac{\partial\omega_0}{\partial E_0})(k\frac{\partial P_0}{\partial E_0}) + \xi^2(k\frac{\partial P_0}{\partial E_0})^2\right]\right] \qquad (2)$$

$$2\mathrm{Im}\{S_{a\phi}(\omega)\} = S_\varepsilon(\omega)\frac{2}{\omega(1+(\omega/\omega_1)^2)}(k\frac{\partial P_0}{\partial E_0})(\frac{\partial\omega_0}{\partial E_0}) \qquad (3)$$

where $k=\ln(10)/20$; P_0 is the output power of the oscillator in dBm; E_0 is the stationary value of E; S_ε is the noise spectral density of E; ω is the modulation frequency; and ω_0 is frequency of oscillation. The four parameters entering in these equations are given below in terms of various derivatives of the device and resonator reflection coefficients Γ_d and Γ_r, respectively.

$$\frac{\partial\omega_0}{\partial E_0} = \left[(A\frac{\partial\Gamma_d^{-1}}{\partial A})\times(\frac{\partial\Gamma_d^{-1}}{\partial E})\right]\div\left[(A\frac{\partial\Gamma_d^{-1}}{\partial A})\times(\frac{d\Gamma_r}{d\omega} - \frac{\partial\Gamma_d^{-1}}{\partial\omega})\right] \qquad (4)$$

$$k\frac{\partial P_0}{\partial E_0} = -\left[(\frac{\partial\Gamma_d^{-1}}{\partial E})\times(\frac{d\Gamma_r}{d\omega} - \frac{\partial\Gamma_d^{-1}}{\partial\omega})\right]\div\left[(A\frac{\partial\Gamma_d^{-1}}{\partial A})\times(\frac{d\Gamma_r}{d\omega} - \frac{\partial\Gamma_d^{-1}}{\partial\omega})\right] \qquad (5)$$

$$\omega_1 = -\left[(A\frac{\partial\Gamma_d^{-1}}{\partial A})\times(\frac{d\Gamma_r}{d\omega} - \frac{\partial\Gamma_d^{-1}}{\partial\omega})\right]\div\left[|\frac{d\Gamma_r}{d\omega} - \frac{\partial\Gamma_d^{-1}}{\partial\omega}|^2\right] \qquad (6)$$

$$\xi = \left[(A\frac{\partial\Gamma_d^{-1}}{\partial A})\cdot(\frac{d\Gamma_r}{d\omega} - \frac{\partial\Gamma_d^{-1}}{\partial\omega})\right]\div\left[(A\frac{\partial\Gamma_d^{-1}}{\partial A})\times(\frac{d\Gamma_r}{d\omega} - \frac{\partial\Gamma_d^{-1}}{\partial\omega})\right] \qquad (7)$$

The derivatives are evaluated at the operating point, and × and · stand for vector and scalar product in the complex Γ-plane, respectively.

For GaAs MESFETs we assume $E=V_{gs}$. This makes it easy to test the model since (1) the measured LF noise is represented as an equivalent input (gate) noise voltage $e_g(\omega)$ [1]; (2) V_{gs} is easy to modulate; and (3) the sensitivity $\partial P_0/\partial V_{gs}$ of the output power with respect to the gate bias is negligible for typical GaAs MESFET oscillators. Thus the amplitude noise is negligible, and we can focus on the phase noise, which, to a good approximation, will be given by

$$S_\phi(\omega) = \left[\frac{e_g(\omega)}{\omega}\frac{\partial\omega_0}{\partial V_{gs}}\right]^2 \qquad (8)$$

The phase noise can thus be predicted from the LF noise spectral density measured in [1], and the easily measured sensitivity, $\partial\omega_0/\partial V_{gs}$, of the carrier frequency with respect to the gate bias. The analytical expression for the latter is given in equation (4). There are four ways to reduce the phase noise: (1) reduce the device LF noise; (2) use an improved device structure to minimize $\partial\Gamma_d^{-1}/\partial E$; (3) make $d\Gamma_r/d\omega$ as large as possible, i.e. use a resonator with large Q; and (4) use a large signal design and optimize the intersection of the device and resonator trajectories at the

Reprinted from *1984 IEEE MTT-S Int. Microwave Symp., Tech. Dig.*, pp. 267-269, 1984.

operating point. In our experiments we exploited the first three.

EXPERIMENTAL

In Fig. 1 the phase noise, S_ϕ, of three simple microstrip oscillators ($f_o \approx 5\text{GHz}$) using devices with varying magnitude and slope of the LF noise is shown. The sloped lines are the predicted phase noise using a straight line fit to the measured LF noise, and the easily measured carrier frequency sensitivity with respect to the gate bias. Equation (8) predicts the phase noise, and the differences in magnitude and slope of the LF noise are reflected in the phase noise. This gives us confidence in the model, and supports the conclusion in [1] concerning the origin of LF noise, indicating that the charge fluctuation occurs in the gate depletion region.

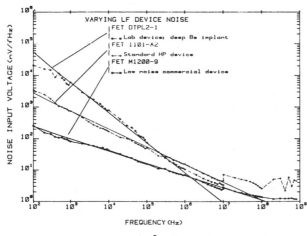

a

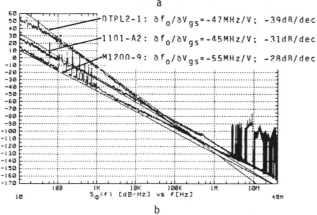

b

Fig. 1. (a) Equivalent input noise voltage of three very different devices; and (b) phase noise of microstrip oscillators using these devices. The slope lines are related through equation (8).

Fig. 2 illustrates how a reduction in device LF noise translates into, in this case, a 10 dB improvement in phase noise relative to a device with typical level of LF noise. The reduction is predicted well by the model.

Fig. 3 illustrates how improving the device structure can lead to a 15dB reduction in phase

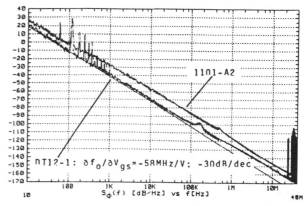

Fig. 2. The effect on the phase noise of reducing the device LF noise (DT12-1). The slope line is the predicted phase noise.

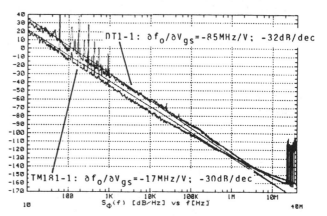

Fig. 3. The effect on the phase noise of using a device (TM181-1) with improved structure, versus using a standard device (DT1-1).

noise compared to a typical device. This is accomplished, in spite of a fairly large device LF noise, by a substantial reduction in $\partial\omega_o/\partial V_{gs}$.

The third way to lower the phase noise is to use a resonator with higher Q than that of simple microstrip matching. We chose to use a 5GHz dielectric resonator for this purpose, and positioned this close to the output microstrip line of the oscillator at a position where oscillation occurs, and where the sensitivity factor in equation (8) is minimized, typically to ~1MHz/V. The phase noise of three such oscillators; using a standard device, a low noise device and a device with improved structure, is shown in Fig. 4. There is a significant reduction in phase noise (20-30dB) to a limit that appears to be typical for GaAs FET dielectric resonator oscillators. There is very little difference in phase noise between the oscillators. Furthermore, using equation (8), the predictions are considerably lower than the experimental values. After evaluating possible reasons for this descrepancy, including possible violations of the assumptions made, we hypothesize that a residual LF noise source must start to dominate the phase noise once the normally dominant charge fluctuation in the gate depletion has been neutralized by sufficiently reducing $\partial\omega_o/\partial V_{gs}$.

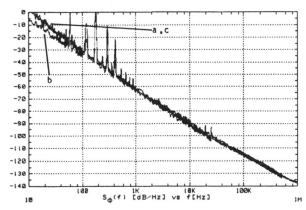

Fig. 4. The phase noise of a 5GHz dielectric resonator oscillator using a (a) standard device; (b) low-noise device; and (c) device with improved structure.

Since the effect of a residual noise source may differ with oscillator configuration we studied a modified version of the 5GHz dielectric resonator oscillator. In Fig. 5 the phase noise of such oscillators, using three different devices, is shown. The phase noise is lower than the limit reached earlier, and the predictions are much better, except for device (c) where there is significant phase noise in spite of a zero sensitivity factor. The phase noise of the oscillator using device (b) and (c) $(S_\phi(1kHz) = -75dB/Hz)$ is as low as the best result [3] we have seen in the literature for GaAs MESFET oscillators.

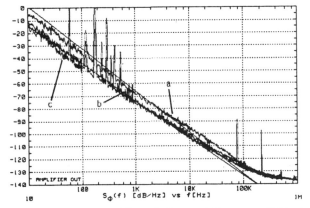

Fig. 5. Same as Fig. 4 but with a modified oscillator configuration.

The case of device (c) in Fig. 5 illustrates that we still reach a limit below which we cannot lower the phase noise by merely decreasing $\partial\omega_0/\partial V_{gs}$. This phenomenon is illustrated more effectively in Fig. 6, where the FM noise Δf_{rms} (directly related to the phase noise S_ϕ by $\Delta f_{rms}=2\Pi\omega/S_\phi$) of an oscillator using a device with very interesting structure in the LF noise is shown. In Fig. 6a the device is biased for non-zero sensitivity factor. The predictions (the dots) are very good, and the structure in the device LF noise is reproduced in the phase noise. Compare this to Fig 6b where we have biased the device for zero sensitivity factor. The phase noise is reduced (but not zero!) and the structure

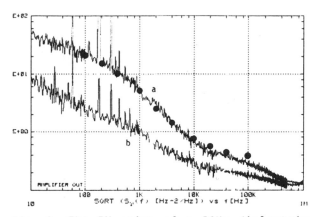

Fig. 6. The FM noise of a 5GHz dielectric resonator oscillator using a device with pronounced structure in the LF noise, biased for (a) non-zero and (b) zero $\partial f_0/\partial V_{gs}$. The dots are the predicted values.

has all but disappeared, indicating that a new LF noise source with, in this case, a different spectrum takes over.

SUMMARY AND CONCLUSIONS

We have presented an analytical model for the oscillator noise resulting from LF device noise. Applied to GaAs MESFETs, it gives clear guidelines for low-noise design. The model has been experimentally verified for a number of GaAs MESFET oscillators. We have demonstrated reduced phase noise by reducing the device LF noise, and by reducing the sensitivity factor $\partial\omega_0/\partial V_{gs}$. The latter was accomplished by improving device structure and by using high-Q dielectric resonators. The results are consistent with the device LF noise being mainly due to trap generation-recombination in the gate depletion region. However, we have presented evidence that once the effect of this noise source has been neutralized by proper design, a secondary LF noise source determines the phase noise. Using devices with reduced LF noise and improved structure, we have demonstrated phase noise as low as $S_\phi(1kHz) = -75dB/Hz$, corresponding to £(1kHz) = -78dBc/Hz.

ACKNOWLEDGEMENTS

We greatfully acknowledge the contribution of many people within Hewlett-Packard, including: Rick Powell, Bob Temple, Dieter Scherer, and Steve Kofol for technical assistance and advice; Brian Hughes, Dean Nicholson, Jerry Gladstone, John Boyles and Bob Archer for stimulating discussions; T.S. Tan and Marina Bujatti for supplying some very interesting devices.

REFERENCES

1. Chung-Yi Su, Hans Rohdin and Charles Stolte, 1983 IEDM Tech. Dig., 601.
2. K. Kurokawa, Bell Syst. Tech. J. 48, 1937 (1969).
3. J. Graffeuil, A. Bert, M. Camiade, A. Amana and J.F. Sautereau, Noise in Physical Systems and 1/f Noise, Elsevier, 1983, p. 329.

Part 2
Digital MODFET Circuits

Modulation-Doped GaAs/AlGaAs Heterojunction Field-Effect Transistors (MODFET's), Ultrahigh-Speed Device for Supercomputers

PAUL M. SOLOMON AND HADIS MORKOÇ, SENIOR MEMBER, IEEE

Abstract—In the past few years, a new transistor has appeared on the scene, made of GaAs and AlGaAs, which now holds the record as the fastest logic switching device, switching at speeds of close to ten trillionths of a second (10 ps). The device evolved from the work on GaAs-AlGaAs superlattices (thin alternating layers of differing materials sharing the same crystalline lattice) pioneered by L. Esaki and R. Tsu at IBM in the late 1960's. They realized that high mobilities in GaAs could be achieved if electrons were transferred from the doped and wider band-gap AlGaAs to an adjacent undoped GaAs layer, a process now called modulation doping. R. Dingle, H. L. Stormer, A. C. Gossard, and W. Wiegmann of AT&T Bell Labs, working independently, were the first to demonstrate high mobilities obtained by modulation doping in 1978, in a GaAs-AlGaAs superlattice. Realizing that such a structure could form the basis for a high-performance field-effect transistor (Bell Labs Patent 4163237, filed on April 24, 1978), researchers at various labs in the United States (Bell Labs, University of Illinois, and Rockwell), Japan (Fujitsu), and France (Thomson CSF) began working on this device. In 1980, the first such device with a reasonable microwave performance was fabricated by the University of Illinois and Rockwell, which they called a modulation-doped FET or MODFET. The same year Fujitsu reported the results obtained in a device with a 400-μm gate which they called the "high electron mobility transistor" or HEMT, in the open literature. Thomson CSF published shortly thereafter calling their realization a "two-dimensional electron gas FET" or TEGFET, and Bell Labs followed, using the name "selectively doped heterojection transistor" or SDHT. These names are all descriptive of various aspects of the device operation as we will discuss in the text. For the sake of internal consistency will call it MODFET, hereafter.

In this paper we review the principals of MODFET operation, factors affecting its performance, optimization of the device, and comparison with other high-performance compound and elemental semiconductor devices. Finally, the remaining problems and future challenges are pointed out.

I. INTRODUCTION

ELECTRON devices with ever-increasing speed are used either as switches or amplifiers. As advanced semiconductor preparation and processing tools become available and are combined with ingenious device synthesis, the frequency of operation and switching speeds are constantly being challenged. In switching devices, such as those used in digital circuits, the current flowing through the devices is used to drive the subsequent stages. The speed with which such a switching operation takes place is primarily determined by how fast the capacitances associated with the device and interconnects can be charged and discharged. It is clear that the interconnect capacitance plays an important role and must be minimized by proper circuit design. This presentation will not deal with the interconnect capacitances per se, but will instead concentrate on the device itself. It must, however, be pointed out that in a large-scale integrated circuit, devices and interconnects must be considered simultaneously.

The switching speed of the device is primarily determined by how fast an input pulse can be transmitted to the output. The transit time through the device, "intrinsic propagation delay," and input and output capacitance charging times are added to give the switching time of the device. This implies that for a fast switching time, the capacitances and the transit time through the devices must be made smaller. The issue of capacitance should be treated in the context of the available current since larger amounts of current, if available, can charge and discharge capacitances faster. The transit time can be made smaller by either reducing the current path length by making the terminals closer together or by increasing the speed at which the carriers travel.

The speed of the carriers, for low electric fields, is given by the product of the mobility and electric field; however, in short-channel field-effect transistors (FET's) electric fields are quite large and the carrier velocity reaches some limiting value. As the device dimensions are pushed to submicrometer range, electron velocities greater than equilibrium values, or "over-shoot," effects can be obtained. Since the current is proportional to the carrier velocity as well as the carrier density, carrier density must be increased if one wants a larger current to charge and discharge the capacitances faster. Since the transport properties of electrons are better than those of holes, we will concentrate on the use of electrons as carriers or n-channel devices.

In conventional metal–semiconductor FET's (MESFET's), the electrons are obtained by incorporating donor impurities which share the same space with electrons and interact with them (Fig. 1). Increased electron concentration, necessary for the high currents required for high speed, also means increased donor concentration which leads to more electron–donor interaction, called ionized impurity scattering. A conclusion that can be drawn from this is that one must pay a price for

Manuscript received February 25, 1984. The work on MODFET's at the University of Illinois was funded by the Air Force Office of Scientific Research.

P. M. Solomon is with the IBM Thomas J. Watson Research Center, Yorktown Heights, NY 10598.

H. Morkoç is with the Department of Electrical Engineering and Coordinated Science Laboratory, University of Illinois, Urbana, IL 61801.

Reprinted from *IEEE Transactions on Electron Devices*, vol. ED-31, no. 8, pp. 1015–1027, August 1984.

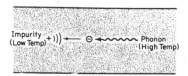

Fig. 1. In bulk semiconductors, the electrons (n-type material) and positively charged donor impurities share the same space. As a result, interaction of electrons and positive ions is inevitable. In small-geometry devices, the thickness of the conducting channel is reduced and the dopant concentration is increased, leading to increased scattering which becomes more dominant at low lattice temperatures. At high lattice temperatures electrons are scattered by the vibrating atoms (phonon scattering) and either gain or lose energy.

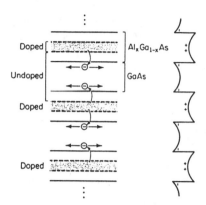

Fig. 2. Multiple-interface AlGaAs GaAs modulation-doped structures where only the center region (shaded) of AlGaAs layers is doped with Si donors. Since the bottom of the conduction band energy in GaAs is smaller than donor energy level in AlGaAs (right-hand side), the electrons diffuse into GaAs layers where they are confined because of the energy barrier. Positive signs indicate the ionized donors and the negative ones represent the transferred electrons. If the parameters are chosen correctly, all of the free electrons will be located in GaAs layers where they show enhanced transport parallel to the heterointerfaces.

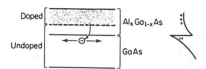

Fig. 3. Single-interface heterostructure used for MODFET's. The structure with AlGaAs grown on top of GaAs, "normal modulation-doped structure," is the one that is used commonly for FET's. The diagram on the right-hand side shows the conduction band edge with respect to distance.

large electron concentrations since they are associated with large donor concentrations with their deleterious effects. In fact, the peak velocity of electrons in GaAs goes down from 2.1×10^7 cm/s for pure GaAs down to 1.7×10^7 cm/s for GaAs with 10^{17}-cm^{-3} donors.

In general, as FET's become smaller, thinner channel layers and higher electron concentrations are required. The requirement for large electron concentration without the deleterious effects of donors can be met by novel heterojunctions. A heterojunction composed of AlGaAs and GaAs layers can be structured so that the donors are introduced only into the larger bandgap (AlGaAs) material [1], [2]. The heterojunction lineup is such that the energy of the electrons donated to the AlGaAs layer is higher in the AlGaAs than in the adjacent GaAs (Fig. 2). The electrons originally introduced into the AlGaAs layer then diffuse to the lower energy GaAs layer where they are confined due to the energy barrier at the heterointerface as shown in Fig. 2. This technique of "modulation doping" is a perfect means of introducing electrons into the GaAs layer without the adverse effects of donors.

Having the electrons confined at the heterointerface in a "two-dimensional electron gas" very close to the gate and a perfect interface leads to very high mobilities and large electron velocities at very small values of drain voltage [3]. This in turn leads to extremely fast charging times of capacitors with

small power consumption. These advantages are enhanced by almost a factor of two when cooled to 77 K, which is conceivable for larger supercomputer systems. Minimum switching speeds of <10 ps per gate should be possible in a few years, corresponding to typical switching delays of <30 ps in a large computer, compared to switching speeds of >500 ps found in the fastest computers today [4]. The principles of operation of MODFET's are similar to that of Si MOSFET and the models developed for MODFET's benefited greatly from Si MOSFET models [5].

With 1-μm gate lengths and using conventional MESFET technology, propagation delay times as low as 12.2 ps at 300 K as measured by ring oscillators (logic inverters connected in a recirculating loop) with a power–delay product of 13.7 fJ have been obtained [6]. Frequency dividers have also been demonstrated at 77 K with operation frequencies of up to 8 GHz [7].

How Modulation Doping Works

These structures are prepared by molecular-beam epitaxy, which is an ultrahigh vacuum semiconductor deposition technique with control on the atomic scale of both the dopants and constituents forming the semiconductor itself [8]. This is achieved by blocking or not blocking the beam flux with a mechanical shutter controlled with computer to allow the formation of alternating heterolayers as thin as about 10 Å each, about 3–4 atomic layers. In the case of modulation-doped structures intended for FET's, single-interface structures with larger layer thicknesses are used (Fig. 3).

The region of the $Al_x Ga_{1-x} As$ depleted of electrons forms a positive space charge region which is balanced by the electrons confined at the heterointerface. The resulting electric field perpendicular to the interface reaches values over 10^5 V/cm and causes a severe band bending, particularly in GaAs because the electrons are confined to a space of about 80 Å thick (Fig. 4). The electron energies are increased by their quantum-mechanical confinement, and discrete quantum-electric subbands are formed, each subband corresponding to a discrete state of the electron's perpendicular momentum (or to a discrete number of standing waves in the electron wave function, see the inset of Fig. 4) [9]. Even though the electrons and donors are separated spatially, their close proximity allows an electrostatic interaction called Coulomb scattering. By setting the donors away from the interface, Coulombic scattering by donors can be reduced (Fig. 5). This was demonstrated by the University of Illinois group with a resultant understanding that increased set back leads to enhanced electron mobilities and concomitantly reduced electron transfer [10]. The amount of electron

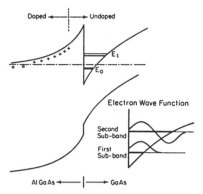

Fig. 4. Energy band diagram of a single-interface modulation-doped structure. Donor impurities are in the AlGaAs layer and set back (between 20 and 500 Å depending on the application) from the interface to reduce Coulombic interaction between the electrons that diffuse into the GaAs and remaining ionized donors in the AlGaAs. The depleted region in the AlGaAs layer is positively charged and is balanced by the electrons in GaAs in equilibrium. The large electric field present in GaAs severely bends the conduction band and forms a quasi-triangular potential leading to quantum electric subband. In MODFET's, the first subband at energy E_0 is filled completely (solid dark line) whereas the second subband at energy E_1 is partially filled (dark and light lines). Inset shows the electron wave functions associated with the first and second subbands.

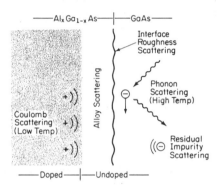

Fig. 5. Scattering mechanisms present in modulation-doped structures. The ionized donor impurities in AlGaAs scatter electrons that diffused to GaAs from AlGaAs through the force of attraction (Coulomb interaction) which is noticeable at cryogenic temperatures. Interaction of electrons with vibrating host atoms in GaAs leads to energy and momentum gain or loss (phonon scattering), which is important when the lattice temperature is high. In inverted modulation-doped structures, the heterointerface is not of high quality, e.g., not atomically smooth and perhaps contaminated with impurities, which causes scattering of the electrons. The overlap of electron wave function with AlGaAs allows interaction with potential perturbation caused by the random distribution of Al Atoms (alloy scattering). This process is very small and may affect the electron mobility only at low temperatures. Finally, unintentionally introduced impurities in GaAs, acceptor type in GaAs grown by MBE, scatter the electrons transferred from AlGaAs, but their effect is reduced substantially because high-mobility electrons can surround the residual acceptors and screen them. This is why 4-K mobilities in modulation-doped structures with about 5×10^{14} cm^{-3} residual ions are more than an order of magnitude larger than those obtained in high-purity bulk GaAs with only about mid 10^{13} cm^{-3} ionized impurities.

transfer is determined by the donor density in AlGaAs, conduction band edge discontinuity and the amount of set back.

Even though electrons and donors are separated at room temperature, electrons interact with the lattice vibrations, "phonons" in GaAs. The phonon scattering limited mobility in GaAs is 8500–9000 cm^2/V · s which sets an upper limit to the

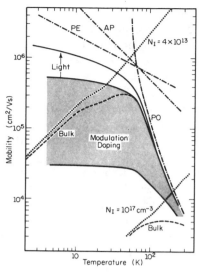

Fig. 6. Electron mobility versus lattice temperature of modulation-doped structures having equivalent electron concentrations larger than 10^{18} cm^{-3} in bulk GaAs. For comparison GaAs with ion concentration of 4×10^{13} cm^{-3} (lowest obtained) is also shown. The mobility of modulation-doped structures is limited by piezoelectric and acoustic phonon scattering at low temperatures and polar optical scattering at high temperatures. Because of large electron gas concentrations needed for MODFET's, both subbands are filled and the undoped layer thickness is small. Low-temperature electron mobilities obtained are about 30 000–60 000 cm^2/V · s. When the undoped layer thickness is increased, the Coulomb scattering and the electron gas concentration drop, leading to mobilities in excess of 10^6 cm^2/V · s at 4 K. With light, the mobility increases even more because of increased Fermi energy and possible neutralization of scattering centers in AlGaAs within about 100 Å of the interface. The mobilities obtained in single-interface structures are generally larger than those in modulation-doped superlattices.

mobility obtainable in modulation-doped structures at 300 K [11]. At low lattice temperature, however, the phonon scattering in both bulk GaAs and modulation-doped structures is reduced. In bulk GaAs, ionized impurity scattering becomes dominant to the point of reducing the 50-K mobility from 3×10^5 cm^2/V · s with $N_I = 4 \times 10^{13}$ cm^{-3} [12] to 5000 cm^2/V · s with $N_I = 10^{17}$ cm^{-3} (Fig. 6).

Background impurities ($\sim 10^{14}$ cm^{-2}) are still present in the bulk GaAs, even in the modulation-doped structures, and these would limit mobilities to about 10^5 cm^2/V · s if it were not for the effect of electrostatic screening of the impurities by the electrons in the 2-D gas layer. In fact mobilities of about 2×10^6 cm^2/V · s have been obtained at low temperatures [13]. Fig. 6 shows the influence of the various electron scattering mechanisms on mobility, and how modulation doping effectively eliminates the previously dominant impurity scattering component. Though of a secondary nature, mechanisms such as alloy scattering and interface roughness scattering do play a role in determining the mobility. Such high mobilities, while indicative of extremely good interfaces are not really essential for devices since the electron velocity is the dominant factor [14]. High mobilities, however, have recently made it possible to observe quantum Hall effects at high magnetic fields. It should be pointed out that basic physics behind the electron gas and its behavior under electric and magnetic fields are to some extent similar to those of Si/SiO$_2$ MOS structures [15].

At the time of this writing many industrial and university

laboratories around the globe, primarily in the United States, Japan, and France, have programs addressing various aspects of MODFET work, ranging from basic understanding through high-speed circuit development. Among the universities, the University of Illinois, University of Minnesota, and Cornell University in the United States, and the University of Tokyo in Japan are currently active in this study. Industrial participants are greater in number and include Bell Laboratories, Honeywell, Hughes, Rockwell, TRW, Hewlett-Packard, Texas Instruments, IBM, General Electric, and Westinghouse of the United States, Fujitsu and Nippon Telephone and Telegraph of Japan, and Thomson CSF of France. Some of the governmental laboratories, e.g., Naval Research Laboratories and Air Force Avionics Laboratories, have some in-house projects underway. It is quite possible that the list is even greater than what has been made available to the authors.

FABRICATION OF MODFET's

The heterojunction structures needed for MODFET's are grown by molecular-beam epitaxy on semi-insulating substrates. First, a nominally 1-μm-thick undoped GaAs layer is grown at a substrate temperature of about 580°C. Gallium flux, which determines the growth rate, is adjusted to yield a growth rate of about 1 μm/h. This rate can be increased if desired to about 5 μm/h by increasing the source temperature. This is followed by the growth of the AlGaAs layer, about 20-60 Å of which is not doped near the heterointerface. The doped AlGaAs layer, about 600 Å thick, may be capped with a doped GaAs layer (200-300 Å thick) or the mole fraction may be graded down to GaAs towards the surface.

Device isolation is in most cases done by chemically etching mesas down to the undoped GaAs layer or to the semi-insulating substrate, or by an isolating implant. The source and drain areas are then defined in positive photoresist and typically AuGe/Ni/Au metallization is evaporated. Following the lift-off, the source-drain metallization is alloyed at or above 400°C for a short time (~1 min) to obtain ohmic contacts. During this process, Ge diffuses down past the heterointerface, thus making contact to the sheet of electrons. In some instances a surface passivation layer of SiO_2 has been used between the terminals.

The gate is then defined and a very small amount of recessing is done by either chemical etching, reactive ion etching, or ion milling. The extent of the recess is dependent upon whether depletion or enhancement mode devices are desired. In depletion mode devices, the remaining doped layer should be just the thickness to be depleted by the gate Schottky barrier. In enhancement mode devices, the remaining doped AlGaAs is much thinner and thus the Schottky barrier depletes the electron gas as well. In test circuits composed of ring oscillators, the switches are of enhancement mode which conduct current when a positive voltage is applied to the gate and the loads are of depletion type. Fig. 7 shows an artistic view of the cross section of a MODFET.

PRINCIPLES OF MODFET OPERATION

A. General Background

The MODFET operation is to some extent analogous to that of the Si/SiO_2 MOSFET. While the basic principles of operation

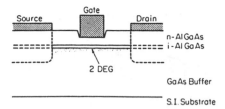

Fig. 7. Cross-sectional view of a MODFET.

are similar, material systems and the details of device physics are different. The most striking difference, however, is the lack of interface states in MODFET structures. In MODFET's the gate metal and the channel are separated by only about 400 Å. This, coupled with the large dielectric constant of $Al_xGa_{1-x}As$ as compared to SiO_2 gives rise to extremely large transconductances. In addition, large electron densities, about 10^{12} cm^{-2}, can be achieved at the interface which leads to high current levels. The effective mass of electrons in GaAs is much smaller than in Si and therefore electron concentrations under consideration raise the Fermi level well up into the conduction band, which is not the case for Si MOSFET's. It is therefore necessary to develop a new model for the MODFET as has been attempted by the Thomson CSF group [16] and by the team at the Universities of Minnesota and Illinois [17]. In order to calculate the current-voltage characteristics of MODFET's, we must first determine the two-dimensional electron gas concentration.

B. Electron Gas Concentration

As indicated earlier, the electrons diffuse from the doped $Al_xGa_{1-x}As$ to the GaAs where they are confined by the energy barrier, Fig. 4, and form a two-dimensional electron gas. This was verified by observing the Shubnikov-de Haas oscillations and their dependence on the angle between the magnetic field and the normal of the sample [18]. The wave vector for such a system is quantized in the direction perpendicular to but not parallel to the interface.

The electric field set up by the charge separation causes a severe band bending in the GaAs layer with a resultant triangular potential barrier where the allowed states are no longer continuous in energy, but discrete. As a result, quantized subbands are formed and a new two-dimensional model is needed to calculate the electron concentration. In most cases the ground subband is filled while the first subband is partially empty. Since the spread in the electron concentration perpendicular to the heterointerface is very small and the density varies, we will refer to the areal density of the electrons from now on.

To determine the electron concentration we must first relate it to the subband energies. The rigorous approach is to solve for the subband energies self-consistently with the solution for the potential derived from the electric charge distribution. This has been done by Stern and co-workers for the silicon-silicon dioxide system in the sixties, and more recently by Ando for the GaAs-AlGaAs system. A workable approximation is to assume that the potential well is perfectly triangular, and that only the ground and first subbands need be considered. Using the experimentally obtained subband populations, adjustments in the parameters can be made to account for the nonconstant electric field and nonparabolic conduction band. Solving Pois-

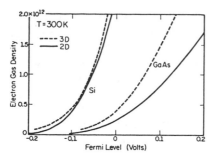

Fig. 8. Variation of the electron gas density with Fermi level as measured from the bottom of the conduction band in GaAs. Since the conduction band density of states in Si is very large, the Fermi level even for the largest sheet carrier concentration, 2×10^{12} cm^{-2}, is still below the conduction band and predictions are reasonably accurate when the problem is treated in a three-dimensional manner and the quantization is neglected. For GaAs, however, the density of states is smaller (or the effective mass is smaller) and the quantization of the electron population at the heterointerface can not be neglected. Models encompassing the two-dimensional (solid lines) nature of the electron population must be utilized.

son's equation in AlGaAs and GaAs layers and using Gauss' law, one can obtain another expression for the sheet electron concentration in terms of structural parameters, e.g., the doping level in AlGaAs, doped and undoped AlGaAs layer thicknesses, and the magnitude of conduction band energy discontinuity or the AlAs mole fraction in AlGaAs.

Analysis of the Fermi level shows that it is a linear function of the sheet carrier concentration n_{so} for $n_{so} \geqslant 5 \times 10^{11}$ cm^{-2}. Taking this into account one can eliminate the iteration process because analytical expressions become available [14], [19]. Another feature that must be considered in the model is the necessity of using the Fermi-Dirac as opposed to the commonly used Maxwell-Boltzmann statistics. This term is particularly important at room temperature because of larger thermal energy and thus larger uncertainty in the position of electrons at the boundary of the depletion region. In the case of Si/SiO$_2$ MOSFET's, three-dimensional analyses work quite well because the Fermi level is not as high; but they fail for MODFET's, as illustrated in Fig. 8 where broken lines take into account the two-dimensionality and the solid lines do not [5].

C. Charge Control and I–V Characteristics

So far we have related the interface charge, which is to carry the current parallel to the heterointerface, to the structural parameters of the heterojunction system. To control and modulate this charge, and therefore the current, a Schottky barrier is placed on the doped AlGaAs layer. The doped AlGaAs is depleted at the heterointerface by electron diffusion into GaAs, but this is limited to about 100 Å for an AlGaAs doping level of about 10^{18} cm^{-3}. It is also depleted from the surface by the Schottky barrier [16], [17]. To avoid conduction through AlGaAs which has inferior transport properties and screening of the channel by the carriers in the AlGaAs, parameters must be chosen such that the two depletion regions just overlap.

In normally on devices the depletion by the gate built-in voltage should be just enough to have the surface depletion extended to the interface depletion. Devices designed for $\sim 10^{12}$ cm^{-2} in the channel and an AlGaAs thickness of ~ 600

Å will be turned off at a gate bias of –1 V. This is the structure used for discrete high-speed analog applications, e.g., microwave low-noise amplifiers, since the power consumption is too high for large-scale integration.

In normally off devices, the thickness of the doped AlGaAs under the gate is smaller and the gate built-in voltage depletes the doped AlGaAs, overcomes the built-in potential at the heterointerface, and depletes the electron gas. No current flows through the device unless a positive gate voltage is applied to the gate. This type of device is used as a switch in high-speed integrated digital circuits because of the associated low power dissipation. The loads may be normally on transistors with the gate shorted to the source, or an ungated "saturated resistor," which has a saturating current characteristic due to the velocity saturation of the carriers.

Away from the cutoff regime, it is quite reasonable to assume that the capacitance under the gate is constant and thus the charge at the interface is linearly proportional to the gate voltage minus the threshold voltage. As threshold voltage is approached, the triangular potential well widens, and the Fermi energy of the electrons is lowered. This change in surface potential with electron concentration subtracts from the change in the applied gate bias, so that a lesser change in potential acts across the AlGaAs layer, reducing the transconductance of the device, and causing the curvature of the gate characteristic near threshold [14], as will be discussed later. This curvature is more pronounced at room temperature, due to the thermal distribution of the electrons; however, some curvature will persist down to the lowest temperatures, due to quantum-mechanical confinement energies. This has profound implications for device operation since it precludes high-speed operation at voltages less than a few tenths of a volt. This means that ultralow power–delay products, similar to those of Josephson junction devices, which operate at a few millivolts only, would not be realized.

Away from cutoff, the charge can be assumed to be linearly proportional to the gate voltage, and in the velocity saturated regime the current will then be linearly proportional to gate voltage and the transconductance will approach a constant (except if the AlGaAs starts conducting). These arguments apply to the velocity saturated MOSFET as well. For the MESFET, in contrast, the transconductance increases with increasing gate biases, since the depletion layer width narrows and modulation of the channel charge increases.

In order to calculate the current-voltage characteristic, one must know the electron velocity as a function of electric field. Since the device dimensions (gate) used are about 1 μm or less, high field effects such as velocity saturation must be considered.

Even though the electrons in MODFET's are located in GaAs and the electron transport in GaAs is well known, there was some confusion in the early days as to what one should expect from MODFET's. There were, in fact, reports, such as the ones from Fujitsu, that this heterojunction structure held promise because of the high mobilities obtained. One should, however, keep in mind that mobilities are measured at extremely small voltages (electric field $\simeq 5$ V/cm). In short-channel MODFET's, the electric field can reach tens of kilovolts per centimeter, making it necessary to understand the high field transport.

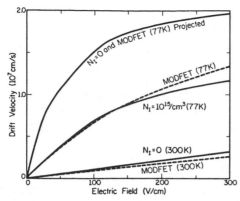

Fig. 9. Velocity versus electric field measured in a single-interface modulation-doped structure at 300 and 77 K. The electron gas concentration at the heterointerface is about 7×10^{11} cm^{-3} and the unintentional background acceptor concentration in GaAs of the MODFET structure is about 10^{14} cm^{-3}. For comparison, calculated velocity field characteristics of bulk GaAs with zero ionized impurity density ($N_I = 0$) at 300 and 77 K and with $N_I = 10^{15}$ cm^{-3} at 77 K are also shown. It is clearly seen that at 300 K, transport properties of the modulation-doped structure with as many electrons as needed for FET's are comparable to the pure GaAs. At 77 K, it is almost comparable to pure GaAs at fields below 300 V/cm and quite comparable at 2 kV/cm and above (estimated from MODFET performance).

Using 400-μm-long conventional Hall bar structures the team at the University of Illinois measured the velocity field characteristics. A dc technique below 300 V/cm and a pulsed technique up to 2 kV/cm were used to measure the current versus field characteristics [20]. Knowing the electron concentration from the same sample by Hall measurements, the electron velocity versus electric field characteristics were calculated on many modulation-doped structures. Above 2 kV/cm the results were not reliable because the electric field was nonuniform as indicated by the voltage between equally spaced voltage probes along the sample.

The velocity versus field characteristics below 300 V/cm for a typical modulation-doped structure is shown in Fig. 9. Also shown are the Monte Carlo calculations performed for lightly doped and ion free bulk GaAs layers [21]. The agreement between the modulation-doped structures and undoped GaAs ($N_I \leqslant 10^{15}$ cm^{-3}) is striking. The agreement at low temperatures is even better at high fields as determined from the MODFET performance. It is clear that having electrons but not the donors in concentrations of about 10^{12} cm^{-2} in modulation-doped structures does not degrade the velocity. The most important aspects of these results are:

1) A quasisaturation of electron velocities is obtained at fields of about 200 V/cm. This implies that the extremely high electron mobilities obtained at very low electric fields have only a secondary effect on device performance.

2) The higher mobilities at low fields help give the device a low saturation voltage and small on-resistance.

3) Since the properties of the pure GaAs are maintained, electron peak velocities over 2×10^7 and 3×10^7 cm/s at 300 and 77 K, respectively, can be obtained. These values have already been deduced using drain current versus gate voltage characteristics in MODFET's.

It can simply be concluded that modulation-doped structures provide current transport which is needed to charge and discharge capacitances, without degrading the properties of pure GaAs. To get electrons in conventional structures, the donors have to be incorporated, which degrades the velocity. From the velocity considerations only, MODFET's offer about 20 percent improvement at 300 K and about 60 percent at 77 K. However, other factors, e.g., large current, large transconductance, and low source resistance improve the performance of MODFET's in a real circuit far beyond the aforementioned figures.

OPTIMIZATION

In a normally off MODFET, the type used for the switches in a circuit, a positive gate voltage is applied to turn the device on. The maximum gate voltage is limited to the value above which the doped AlGaAs layer begins to conduct. If exceeded, a conduction path through the AlGaAs layer which has much inferior properties is created leading to reduced performance. This parasitic MESFET for typical parameters becomes noticeable above a gate voltage of about +0.6 V which determines the gate logic swing. Using alternate methods to improve this shortcoming should be very useful.

Since the ultimate speed of a switching device is determined by the transconductance divided by the sum of the gate and interconnect capacitances, the larger the transconductance, the better the speed is [5]. MOSFET's already exhibit larger transconductances because of higher electron velocity and, in addition, since the electron gas is located only about 400 Å away from the gate metal, a large concentration of charge can be modulated by small gate voltages. The latter comes at the expense of slightly larger gate capacitance. Considering the interconnect capacitances, any increase in transconductance, even with increased gate capacitance, improves this component of the speed.

The transconductance in MODFET's can be optimized by reducing the AlGaAs layer thickness. This must accompany increased doping in AlGaAs, which in turn is limited to about 10^{18} cm^{-3} by the requirement for a nonleaky Schottky barrier. By decreasing the undoped setback layer thickness, one can not only increase the transconductance, but also the current level (through the increased electron gas concentration). There is, of course, a limit to this process as well because thinner setback layers increase the Coulombic scattering. All things considered, a setback layer thickness of about 20–30 Å appears to be the best at the present time, as shown in Fig. 10. Setback layers less than 20 Å led to much inferior performance. Transconductances of about 225 (275 being the best) and 400 mS/mm gate width have been demonstrated at 300 and 77 K, respectively [3]. The theoretical and experimental current levels of MODFET's also depend strongly on the setback layer thickness as shown in Fig. 11 and on the doping level in AlGaAs [5].

For good switching and amplifier devices, a good saturation low-differential conductance in the current saturation region, and a low saturation voltage are needed. These are attained quite well in MODFET's, particularly at 77 K, as shown in Fig. 12. The increased current level at 77 K is attributed to the enhancement of electron velocity. The rise in current would have been more if it were not for the shift in the threshold

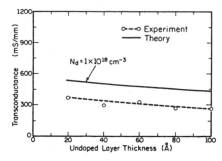

Fig. 10. Since the transconductance is inversely proportional to the gate to electron gas separation, the undoped AlGaAs layer at the heterointerface can influence the transconductance substantially. Considering that the gate to electron gas distance is about 300 A, an undoped layer thickness of greater than 100 A can have a dramatic influence on the transconductance. For best results an undoped layer thickness of about 20–40 A must be used. This imposes stringent requirements on the epitaxial growth process and only molecular-beam epitaxy has so far been able to produce such structures. The circles are the experimental data points while the solid line shows the theory. Below 20 A, the performance degrades.

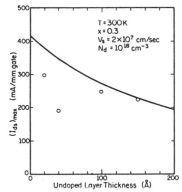

Fig. 11. Maximum drain current is also very sensitive to the undoped AlGaAs layer thickness. For desired large current levels a smaller electron–donor separation is needed to yield a large electron gas concentration. The available data (circles) obtained in n-on MODFET's (University of Illinois) while showing the general trends, should be augmented with more experiments. Maximum current levels of about 300 mA (per millimeter of gate width) at 300 K in n-off MODFET's with a 1-μm gate length is possible. Large current levels obtainable at low voltages lead to fast switching speeds with low power dissipation. Solid line shows the calculated values.

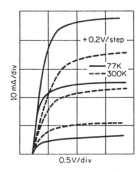

Fig. 12. Drain I–V characteristic of a MODFET with a 300-μm gate width at 300 and 77 K. As indicated, the extrinsic transconductance increases from about 225 (best 275 mS/mm) to 400 mS/mm as the device is cooled to 77 K. The improvement in the drain current observed at 77 K could be much larger if it were not for the positive shift in the threshold voltage. This shift is attributed to electronic defects in AlGaAs and is a subject of current research. (Data are from the University of Illinois).

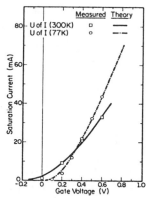

Fig. 13. Drain saturation current with respect to gate voltage of a n-off MODFET at 300 and 77 K. The threshold voltage shifts about +0.1 V as the device is cooled to 77 K which is attributed to freeze out of electrons in the AlGaAs and also to traps. As discussed in the text, the drain current, particularly at 77 K, is proportional to the gate voltage away from cutoff. Near cutoff, the quasitriangular potential well widens, no longer confining the electrons, which in turn results in the observed nonlinear behavior.

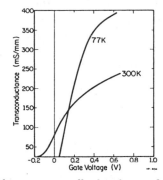

Fig. 14. Transconductance normalized to 1 mm of gate width obtained at 300 and 77 K. The decrease in the rate of increase above a gate bias of +0.7 V at 77 K is observable. This effect is attributed to filling of the second subband and undepleting the doped AlGaAs under the gate, which creates a new conduction path through the inferior quality AlGaAs.

voltage, from about 0 V at 300 K to about $\geqslant 0.1$ V at 77 K, which will further be discussed.

Experimental drain saturation current and transconductance as a function of gate voltage are shown in Figs. 13 and 14. Away from threshold and below Schottky turn-on, the drain saturation current is proportional to the gate voltage and the proportionality constant gets larger at 77 K, again because of enhanced electron velocity. The shift in threshold voltage is apparent both in g_m versus V_g and I_{DS} versus V_g characteristics as the device is cooled to 77 K. Using the model developed and the I_{DS} versus V_g characteristics, average electron velocities as high as 2 and 3×10^7 cm/s were deduced at 300 and 77 K, respectively.

To summarize optimizing the transistor for use in normally off logic, the main parameters to be determined are the Al concentration in the AlGaAs, and the thickness and doping of the AlGaAs layer.

1) Increasing the Al concentration in the AlGaAs increases both the Schottky-barrier height of the gate, and the heterojunction interface barrier. These permit higher forward gate voltages on the device, reduced hot-carrier injection from the GaAs into the AlGaAs, and permit higher electron concentra-

tions in the channel without conduction in the AlGaAs. The concentration of Al in the AlGaAs should therefore be as high as possible consistent with obtaining low ionization energies for the donors, good ohmic contacts, and minimum traps. In present practice it varies from 25 to 30 percent.

2) Maximum voltages on the gate, limited by Schottky diode leakage or by conduction in the AlGaAs, are about 0.8 V at room temperature and about 1 V at liquid nitrogen tempera- ture. Threshold voltages should be about 0.1 V for good noise margins and tolerances.

3) To maximize transconductance (and dc current, since volt- age swings are given) the AlGaAs should be as thin as possible. Thinner AlGaAs implies higher doping to achieve the desired threshold voltage. Doping levels cannot be larger than about 1×10^{18} cm^{-3} because of gate leakage currents.

4) The setback layer should be as narrow as possible with- out compromising transport properties (20–40 Å) since this gives the minimum total AlGaAs thickness and maximum trans- conductance consistent with the above limits. Typical param- eters for a normally off MODFET to satisfy these criteria would be: Al concentration of 30 percent, AlGaAs thickness of 350 Å, setback thickness of 40 Å, and doping of 1×10^{18} cm^{-3}.

PERFORMANCE AND APPLICATIONS

Interest in the MODFET device was aroused almost immedi- ately after the first working circuits were built by Fujitsu in 1980, by the (then) record-breaking delays of 17 ps attained by ring oscillators operating at liquid nitrogen temperature [22]. These results can be explained on the basis of the higher velocities and transconductance, and lower saturation voltages of the device as evidenced from the experimental characteristics of Fig. 11. These results have been improved since then, both at liquid nitrogen and room temperatures.

In the logic application area, using 1-μm gate technology and ring oscillators (about 25 stages), Fujitsu in 1982 reported a τ_D = 12.8 ps switching time at 77 K (power consumption not given) [22] and Thomson CSF reported 18.4 ps with a power dissipation of P_D = 0.9 mW/stage at 300 K [23]. In late 1982, Bell Labs reported $\tau_D \sim$ 23 ps and $P_D \sim$ 4 mW/stage with 1-μm gate technology [24]. Very recently Rockwell reported a switching speed of 12.2 ps at 300 K with 13.6-fJ/stage power– delay product [6]. Rockwell also reported a switching speed of 27.3 ps with 3.9-fJ/stage power–speed product. The much- improved results of Rockwell can be attributed to the low source resistance, $\sim$0.5 Ω · mm, obtained.

MODFET's have recently progressed from no-function cir- cuits, e.g., ring oscillators, to frequency dividers. Bell Labora- tories reported on a type D flip-flop divide-by-two circuit with 1-μm gate technology operating at 3.7 GHz (with 2.4-mW/gate power dissipation and 38-ps/gate propagation delay) at 300 K and 5.9 GHz (with 5.1-mW/gate power dissipation and 18-ps/ gate propagation delay) at 77 K [25]. Fujitsu has also recently reported results on their master–slave direct-coupled flip-flop divide-by-two circuit [7]. At 300 K and with a dc bias of 1.3 V, input signals with frequencies up to 5.5 GHz were divided by two. At 77 K, the frequency of the input signal could be increased to 8.9 GHz before the divide-by-two function was no longer possible. The dissipation per gate was 3 mW and the dc bias voltage was 0.96 V.

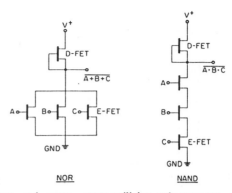

Fig. 15. NOR and NAND gates utilizing enhancement mode drivers (E-FET) and depletion mode loads (D-FET) with gate shorted to source.

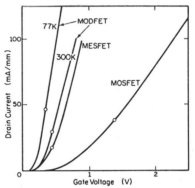

Fig. 16. Saturated drain current versus gate–source voltage for a Si MOSFET, a GaAs MESFET, and a GaAs–AlGaAs MODFET. The MODFET characteristics were measured at temperatures of 300 and 77 K. The MODFET data originated from the University of Illinois, and the MOSFET and MESFET data from the IBM Research Center. The curves were shifted with respect to the voltage axis to simulate operation in a logic inverter with adequate noise margins. The points represent the design load current for that inverter.

All of the preceding circuits have used the simple direct- coupled logic circuit family using enhancement mode drivers and depletion mode loads, or saturated resistor loads. Circuit diagrams of NOR and NAND gates, implemented using direct- coupled logic, are shown in Fig. 15. A simplified delay equa- tion for such a stage is given by

$$\text{delay} = (C_D + C_L) V_L / 2 I_L$$

where I_L is the load current, V_L the logic voltage, C_D the device capacitance, and C_L the load capacitance which includes the wiring capacitance. To achieve high speed we need to develop a high current-to-voltage ratio. Referring to Figs. 13 and 16, this is more than just demanding a high transconductance, which is simply the slope of the drain current versus gate volt- age characteristic. The characteristic should also have a sharp knee so that little of the valuable voltage swing is lost traversing the low transconductance knee region. The sharp turn-on of the MODFET maximizes the load current of the NOR gate for a given noise margin and therefore maximizes speed. The max- imum transconductance is mainly a function of the saturated carrier velocity, but the sharpness of the knee depends strongly on the lower field part of the velocity versus field characteristic (as well as on the charge control characteristics as mentioned previously) and it is in both of these areas that the MODFET excels. While the MODFET possesses good high-speed characteristics at room temperature, these are enhanced at liquid nitrogren temperature.

Low voltages are the key to low-power operation, since the switching energy of the circuit is proportional to CV^2; however, operation at low power supply voltages would require a very tight control over turn-on characteristics of the device. Good uniformity of threshold voltage has been achieved over distances of a few centimeters, the best number being about a 10-mV standard deviation, achieved by Fujitsu [7], and 14 mV over a $2\frac{1}{2}$-in wafer achieved by Honeywell [26]. This control would be sufficient for enhance–deplete logic, if they could be obtained reproducibly. Shifts in threshold voltage due to trapping in the AlGaAs is a yet unsolved problem with regard to liquid nitrogen operation. Recent results, however, look promising.

The higher mobility and hence low on-resistance of the MODFET make it ideal for circuits where logic is performed by serial connection of devices, as illustrated for a NAND gate in Fig. 15. Another example would be in a static memory, where serial devices are used to couple the cells to the sense amplifier and to decode the sense amplifier. Serial switched logic (pass-transistor logic) has the property of higher speed, higher densities, and lower power than conventional NAND or NOR gates.

The combination of a large transconductance per unit width, with low on-resistance is ideal for off-chip drivers. For instance, a MODFET of only 75-μm width at 77 K, or 150-μm width at 300 K would suffice for an off-chip driver into 50 Ω, and with the addition of the good driving characteristics of the predriver stage, would give good overall performance.

The MODFET at room temperature offers a higher speed than the GaAs MESFET (see next section) but more materials complexity. Certainly the Rockwell results of 12.2 ps are very encouraging, although ring oscillator results can be very misleading since they are usually not designed with adequate noise margins for performing general-purpose logic (about a factor of 2 in performance can be gained by operating the ring oscillator in a "small-signal" mode.) Nevertheless, the first applications for MODFET logic will undoubtedly be at room temperature, and at relatively low levels of integration, where its higher speed compared to a GaAs MESFET will give it an edge in small-signal and logic "front end" applications. Its progress in integration level will depend on improvement of the quality of the epitaxial layers.

While the room temperature applications should nurture the MODFET technology initially, the real leverage for the MODFET, which was early appreciated by Fujitsu, is in large digital systems (e.g., supercomputer) at liquid nitrogen temperature. The improvement in MODFET speeds of greater than 50 percent and power-delay products of greater than a factor of two (if lower voltages are used) are not the only driving forces. As important is the improvement in wire resistance ($\sim$10 times) and reduction of electromigration (a wire failure mechanism) with temperature which greatly improve the allowable wire density on-chip and improve system performance. An additional bonus is the improvement in the thermal conductivity of GaAs ($\sim$10 times) at lower temperatures, allowing for larger power dissipations. The problems of cooling the system

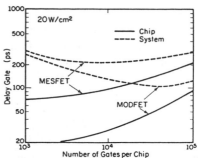

Fig. 17. Fujitsu study showing the projected chip delay and also the average stage delay in a large system, including off-chip delays for both MODFET and MESFET technologies. Power consumption per chip is limited to 20 W/cm². For a given total number of gates the delay associated with the interchip connections decreases or the number of gates per chip is increased. On the other hand, the delay per gate increases as the number of gates per chip is increased because of smaller power dissipation available per gate. The balance of these two competing factors gives rise to an optimum number of gates per chip which is strongly dependent on the individual device performance.

to liquid nitrogen temperature do not seem much more formidable than those already faced by designers in cooling high-performance silicon modules. The advantages will be systems of the future operating at Josephson-like speeds, which will be unattainable by future silicon technology.

The Fujitsu study [22] (see Fig. 17) illustrates the advantage of high speed at low power of the MODFET in a large system. The system delay is comprised of the chip and package delays. As the integration level is increased the on-chip wiring capacitance increases and the power per logic gate is reduced (to stay within chip power constraints) so that the circuit delay increases. The package delay is reduced since the package is used less and less as more of the system fits on a chip, so that the result is a minimum system delay at an optimum level of integration which depends on the speed–power characteristics of the circuit, the power dissipation capability per chip, and the propagation time in the package. The MODFET offers high speed at high integration levels due to two favorable characteristics, its small speed power product and its large current per unit area, which means that smaller area (compared to MOSFET or MESFET) devices can be used to drive on-chip capacitances and to drive off-chip. The problems in achieving high levels of integration will be formidable; yet this is where the large pay-off will be.

While a great majority of the MODFET related research has so far been directed toward logic applications because of distinct advantages over conventional GaAs MESFET's, recently promising results in the area of low-noise amplifiers have become available. Even though this device is being considered for power applications as well, its power handling capabilities are limited by the relatively low breakdown voltage of the gate Schottky barrier. Approaches such as camel gate, which utilizes a p⁺-n⁺ structure on n-AlGaAs for an increased breakdown voltage will have to be advanced before this device could be a good contender in the power FET area.

In the microwave low-noise FET area, using a 0.55-μm gate technology, researchers at Thomson CSF obtained noise figures of 1.26, 1.7, and 2.25 dB at 10, 12, and 17.5 GHz with associated gains of 12, 10.3, and 6.6 dB, respectively [27]. At cryogenic temperatures, the noise performance is enhanced substantially, well below 0.5 dB. Assigning a hard figure, however, is hampered by the inaccuracy of measurements in that range. Programs are currently being initiated to carefully characterize the noise performance at cryogenic temperatures. If the state-of-the-art source resistance were obtained, almost a two-fold improvement over GaAs MESFET's could be expected.

Three-stage amplifiers for satellite communications operating at 20 GHz were constructed by the Fujitsu group with a 300 K (L_g = 0.5 μm) overall noise figure of 3.9 dB and gain of 30 dB [28]. It must be pointed out that these results by no means represent the ultimate from MODFET's. With further improvements in the source resistance, much lower noise figures can be expected.

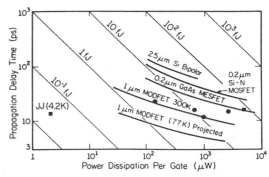

Fig. 18. Delay versus power curve for various technologies obtained in ring oscillators (except for the Josephson junction, where a gate chain was used).

COMPARISON WITH OTHER DEVICES

The MODFET is a field-effect transistor and can conveniently be compared with the GaAs MESFET, and the silicon MOSFET (NMOS and CMOS) all of which are FET technologies. The key advantage of the MODFET, as we have discussed, is its ability to operate at high currents and at low voltages. A comparison of the MODFET, in this regard, to other devices is shown in Fig. 16, where in all cases the best devices (at dimensions of about 1 μm) reported in the open literature were selected. The maximum transconductances for these devices were 80 mS/mm (MOSFET), 230 mS/mm (MESFET), 270 mS/mm (MODFET, 300 K), and 400 mS/mm (MODFET, 77 K). The curves were all displaced horizontally, from the original data, to match the noise margin requirements for logic gates having a logic swing of 0.5 V (MODFET, 77 K), 0.8 V (MESFET and 300-K MODFET), and 2.5 V (MOSFET). The circuit switching energies vary as the square of the voltage, and will be in the ratio of 1:2.6:25 for the technologies considered. The MOSFET could operate at smaller voltages with a better delay–power product, but at the lower voltages its transconductance, hence speed, would be drastically reduced. Comparing the devices for a driver application, for the same 50-Ω driver design, the widths of the different devices would be 75 μm (MODFET, 77 K), 150 μm (MODFET, 300 K), 200 μm (MESFET, 300 K), and 400 μm for the MOSFET.

The other factor in determining device performance is capacitance. Parasitic capacitances are less in the GaAs technology due to the semi-insulating substrate. Silicon-on-sapphire circuits share this advantage. Studies have shown, however, that this advantage is much diminished at higher levels of integration. MOSFET's have larger gate–drain overlap capacitance, but this is no longer an important factor in the comparison due to improved device structures. The intrinsic speed capability is given by transconductance capacitance ratio, and this is related to the average drift velocity of the carriers in the channel, for a given device dimension. As deduced from device characteristics, velocities of 0.8-1 X 10^7 cm/s are characteristic of MOSFET's, 1.2-1.5 X 10^7 cm/s of MESFET's, and 2 X 10^7 cm/s (300 K)-3 X 10^7 cm/s (77 K) of MODFET's.

The scaling potential of the MODFET, MESFET, and MOSFET should be comparable, with minimum dimensions in the 0.1-0.2-μm range, before short-channel effects become severe. Vertical dimensions are reduced in coordination with horizontal dimensions, i.e., the AlGaAs thickness for the MODFET,

the channel depth for the MESFET, and the SiO$_2$ thickness for the MOSFET. Transconductances improve both as a result of the reduced thicknesses and as a result of increased carrier velocities. On the former, the MOSFET is perhaps more extendible than the MODFET, while the AlGaAs thickness of 350 Å is close to the limit set by breakdown, present thicknesses of SiO$_2$ in the MOSFET of 250 Å can be reduced to 50 Å before encountering a tunneling breakdown limit. On the latter, increases in velocity due to velocity overshoot effects are anticipated in both GaAs MESFET and MODFET's as device dimensions are reduced, but to a much lesser extent in Si MOSFET's.

Performance characteristics of various technologies on the basis of ring oscillator results (except for Josephson, where a gate chain was used) are shown in Fig. 18. Josephson junction circuits have been demonstrated with 13-ps switching times and 0.03-fJ power-delay products [29] (see *IEEE Spectrum*, May 1979). The best 300-K figure of silicon NMOSFET's with a 0.3-μm source–drain spacing is 28-ps delay time with a 40-fJ power-delay product [30]. The fastest GaAs self-aligned gate MESFET ring oscillators exhibit a delay time of 15 ps with a power-delay product of 84 fJ [31]. Silicon bipolar nonthreshold logic (0.5 X 5 μm^2) (NTL) circuits have achieved delays of 42 ps at a delay-power product of 20 fJ per gate [32], while the more useful ECL circuits have achieved delays of 96 ps with 96-fJ power-delay product [33], [34]. The GaAs heterojunction bipolar transistor technology, although potentially very fast, has been demonstrated only in I^2L circuits where delays of 200 ps at 2 mW have been obtained [35]. As mentioned previously, these results are rather misleading because the ring oscillators are not usually designed to adequate noise margins, and loading effects are not taken into account. Fig. 19 shows delay estimations for more conservative designs, and in a large-scale integration environment. Excluding the Josephson junction technology, which is very difficult in practice, and requires liquid helium (~4.2 K) which is expensive to maintain and difficult to interface (a much more demanding requirement than liquid nitrogen), only the MODFET combines the advantages of both high speed and low power.

The high-speed market that the MODFET would have to compete in is at present dominated by the Si bipolar transistor at the high-speed low-integration level end, and is increasingly being encroached on by NMOS and especially CMOS at lower

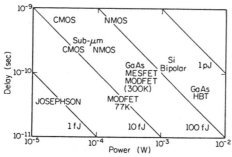

Fig. 19. Delay versus power curve for logic circuits in a large-scale integrated chip taking into account the effect of loading and conservative design for tolerances.

speeds yet much higher levels of integration. The bipolar technology has seen significant improvements, and projections for loaded logic delays are 100 and 50 ps for 1- and 0.5-μm lithographies, respectively. Spectacular improvements in heat removal capability from silicon have recently been demonstrated [36], which should enable the high-power high-speed silicon bipolar circuits to attain higher levels of integration. Bipolar technology, however, is much more complex than FET, including MODFET, technology.

Complimentary MOS has very low power–delay products, since the circuits only dissipate power while switching. The low-voltage advantage of the MODFET at 77 K would more or less balance the switching factor advantage of CMOS, giving these two technologies roughly equal power–delay products; however, the MODFET would still be much faster by about a factor of 10. Because of its technological maturity, the silicon technology may be able to take advantage of smaller channel lengths than the MODFET. Short-channel MOSFET's made (0.15-μm channel lengths by Bell Labs) [37] have a lower transconductance (200 mS/mm) than a 1-μm MODFET. The commercial MOSFET technology is still practiced at dimensions of >1 μm, but the MODFET would have to achieve a rapid learning to compete effectively with silicon. The liquid nitrogen option can be exercised by silicon MOSFET's as well, as shown by IBM researchers where mobilities in MOSFET's increase to 3000 cm^2/V $\cdot$ s. MOSFET peak velocities, even at 77 K are only about 1.2 $\times$ 10^7 cm/s compared with MODFET peak velocities of >3 $\times$ 10^7 cm/s so that the MODFET would retain a speed advantage.

The GaAs MESFET technology has made rapid strides recently, with circuits containing >10 000 devices having been demonstrated. While the MODFET benefits from advances in the GaAs based technology (just as GaAs benefits from advances made in silicon), it still is in a much more primitive state. The applications area of the GaAs MESFET and the MODFET, especially at room temperature, are very similar, placing these technologies in direct competition. The MODFET at room temperature has advantages over the standard GaAs MESFET circuits because of its higher mobility (8000 versus 4000 cm^2/ V $\cdot$ s) which gives it a sharper turn-on, higher peak velocities (by about 20 percent) in undoped versus doped GaAs, and also due to its higher turn-on voltage due to the larger barrier height of the (Ga, Al) As compared with GaAs. An area where the MODFET may not be able to compete with the MESFET is in

radiation hardness, in analog with a MOSFET, due to charge generation in the AlGaAs. Short-channel MESFET's will be able to compete at room temperature with MODFET's in terms of speed, but not at liquid-nitrogen temperature.

Other devices are being investigated by various laboratories for high-speed applications. Vertical-type FET's replace control of horizontal dimensions of the channel length for the much easier control of vertical dimension. Examples are the permeable base transistor [38] (a solid-state analog of a vacuum tube), and the vertical FET's [39] which have a heterojunction "ballistic launcher" source, to increase electron velocities. The devices should be faster eventually than the MODFET but at the cost of considerable fabrication complexity. The vertical FET's swap the problems of vertical with horizontal dimensional control and vice-versa; in particular the problem of threshold voltage control is exacerbated since it now depends on the horizontal dimension. As mentioned earlier, another competitor to the MODFET is the heterojunction bipolar transistor. Again, transit times are controlled by vertical dimension (transit times as small as 1.5 ps have been predicted by a Texas Instruments group) [40]. The heterojunction bipolar transistor should achieve performance in the same range as the MODFET, with the inherent good control of turn-on characteristics of bipolar transistors. Like its silicon counterpart, the heterojunction bipolar transistor would suffer from charge storage effects in saturation. This leaves the circuit designer with the choice of ultrahigh speed, but high-power nonsaturating circuits like ECL ($\sim$30 ps) compared with low-power less area-consuming, but slower, I^2L circuits (>100 ps). A compromise choice is Schottky transistor logic, which is faster than I^2L yet has very low power. The fabrication complexity of the heterojunction bipolar transistor versus that of the MODFET bears the same relationship as silicon bipolar versus FET. None of these alternatives has yet demonstrated high-speed operation in logic circuits.

A comparison summary of the various high-speed technologies is shown in Table I. Compared to the other devices, the MODFET has the advantage of high speed and low power, combined with relatively simple processing and a relaxed lithography requirement. Solutions to the difficult materials problems represent the greatest challenge.

Remaining Problems and Projections

Since MODFET's are large-current and small-voltage devices, the saturation voltage and transconductance are very sensitive to the contact resistance. In fact, the higher the transconductance the more severe the effect of the source resistance becomes. In order to fully take advantage of the device potential, it is essential that extremely low contact resistances be obtained. Not only the contact resistance but parasitic resistances such as the source and drain semiconductor access resistance must be minimized. This could be done using the gate as an ion implantation mask to increase the conductance on each side of the gate but has not yet been done, partially because of degradation of the quality of the channel during annealing to activate donors. Currently many laboratories, both university and industrial, are looking into the degradation mechanism occurring during the annealing step. Although preliminary,

TABLE I
DEVICE COMPARISON GRADED 1 (BEST) THROUGH 5

	GaAs				Si	
	MODFET	MESFET	Vertical FET	HBT	MOSFET [CMOS]	BJT
Speed	1	3	1	2	5	4
Power delay product	1	2	2	4	1	4
Lithographic requirement	2	3	5	1	4	1
Doping control	4	4	4	1	2	1
Processing complexity	2	1	4	5	3	5
Materials problems	4	3	5	4	1	2

the transient annealing technique looks very promising in this regard.

Modulation-doped structures also suffer from the persistent photoconductivity (PPC) effect below 100 K [41]. This is believed to be the result of donor-induced defects in AlGaAs which once ionized exhibit a repulsion towards capture. Recent results, however, appear to indicate that defect-related processes in GaAs as well can play an important role. As a result, increased carrier concentrations, which persist unless the sample is warmed up, are obtained. The electron mobility too, increases with illumination in samples with low areal carrier density. This is tentatively attributed to perhaps neutralizing some defect centers in the depleted AlGaAs near the hetero-interface which then do not cause as much scattering. This PPC effect has been shown to decrease when the AlGaAs layer is grown at high substrate temperatures. It must be kept in mind that in a few instances almost PPC-free MODFET structures have been obtained.

Although for the most part the heterointerface in MODFET's is almost perfect so that interface states encountered in Si MOSFET's do not occur, in a few instances the AlGaAs layer quality is not good enough because of larger concentration of traps. The drain $I-V$ characteristics of a few MODFET's at 77 K collapsed when the device was stressed with a large drain bias of 3 V. Exciting the device with light or raising the temperature cause the device to return to its normal properties. This is attributed to thermionic emission and trap assisted tunneling of electrons into AlGaAs through the barrier and/or PPC related defects in AlGaAs.

It is obvious that the problems associated with the AlGaAs must be minimized so that their influence on device performance is not noticeable. Realizing the importance of the issue, many researchers are looking into sources and causes of the traps and electronic defects in AlGaAs. MODFET's, in contrast to injection lasers, are the first devices utilizing AlGaAs where charge and defect concentrations of about 10^{11} cm^{-2} give rise to unacceptably large adverse effects on the device performance. There are also efforts to explore device structures that are not very sensitive to at least some of the obstacles discussed earlier.

The question of yield and reliability may, however, take a little longer to resolve. For yield, the processing philosophy with regard to GaAs must change. Instrumentation, care, and environment similar to that used for Si IC's must be implemented.

There is also the question of epi defects either introduced by the epi process or present on substrates. Some of those are morphological defects which not only degrade the semiconductor but also cause processing defects. The present state of the art of molecular-beam epitaxy when used with average GaAs substrates is such that only MSI circuits with some success in terms of yield are possible. Efforts are under way to reduce the morphological defect count on the epi. There are already encouraging results which tend to suggest that by the latter part of the decade the substrate quality, the epi morphological quality, the processing that introduces few defects and thus functional circuits with active elements in the mid to upper thousands may be possible.

Like that of MESFET's, the threshold voltage of MODFET's is very sensitive to the epi properties. For a normally off MODFET, a thickness control to about 2 monolayers ($\sim$5 Å) and doping control and AlAs mole fraction control of about 1 percent are needed to control the threshold voltage within about 10 mV. Controls like this have already, though occasionally, been obtained on wafers with slightly less than 3-in diameter. The repeatability of this technology is one of the questions that is also being addressed.

In summary, it is clear that the MODFET has many of the attributes required for high-speed devices, particularly those of the integrated circuits. Present results with moderate numbers of devices are very encouraging and with more effort even better results are expected. In fact, the heterojunction FET with only 1-μm gate length and 3-μm source–drain spacing has surpassed the performance of other techniques, e.g., conventional GaAs with sub 0.5-μm dimensions as shown in Fig. 18. It should be kept in mind that the delay times shown in Fig. 18 are bound to increase substantially in a real circuit with loaded gates. Nevertheless, the MODFET should be capable of providing functional operations in a large system by about a factor of 10 faster than the current state of the art. With more advanced fabrication technologies, even better performance can be expected.

ACKNOWLEDGMENT

H. Morkoç would like to express his sincere thanks to his graduate students, particularly to T. J. Drummond, for carrying out the research reported here. Contributions by and with Prof. M. S. Shur of the University of Minnesota and his student, K. Lee, were invaluable. The authors would also like to thank Dr. C. P. Lee of Rockwell, Dr. N. T. Linh of Thomson CSF, and Dr. M. Helix of Honeywell for providing the results of their work prior to publication. Discussions with Prof. L. F. Eastman of Cornell University were very fruitful.

REFERENCES

[1] R. Dingle, H. Stormer, A. C. Gossard, and W. Wiegmann, "Electron mobilities in modulation doped semiconductor heterojunction superlattices, *Appl. Phys. Lett.*, vol. 31, pp. 665–667, 1978.

[2] L. Esaki and R. Tsu, "Superlattice and negative conductivity in semiconductors," *IBM Internal Res. Rep.*, RC2418, Mar. 26, 1969.

[3] M. Morkoç and P. M. Solomon, "The HEMT: A superfast transistor," *IEEE Spectrum*, vol. 21, pp. 28–35, Feb. 1984.

[4] P. M. Solomon, "A comparison of semiconductor devices for high speed logic," *Proc. IEEE*, vol. 70, no. 5, pp. 489–509, 1982.

[5] K. Lee, M. S. Shur, T. J. Drummond, S. L. Su, W. G. Lyons, R. Fischer, and H. Morkoç, "Design and analysis of modulation doped (Al, Ga)As/GaAs FET's (MODFET's)," *J. Vacuum Sci. Technol.*, vol. JVST B1, pp. 186–189, 1982.

[6] C. P. Lee, D. L. Miller, D. Hou, and R. J. Anderson, "Ultra high speed integrated circuits using GaAs/AlGaAs high electron mobility transistors," *IEEE Trans. Electron Devices*, vol. ED-30, p. 1569, 1983.

[7] K. Nishiuchi, T. Mimura, S. Kuroda, S. Hiyamizu, H. Nishi, and M. Abe, "Device characteristics of short channel high electron mobility transistors (HEMT)," *IEEE Trans. Electron Devices*, vol. ED-30, p. 1569, 1983.

[8] A. Y. Cho and J. R. Arthur, "Molecular beam epitaxy," in *Progress in Solid State Chem.*, vol. 10, 1975, pp. 157–191.

[9] T. Ando, "Self-consistent results for a GaAs/Al$_x$GaAs$_{1-x}$ heterojunction I. Subband structure and light scattering spectra," *J. Phys. Soc. Japan.*, vol. 51, pp. 3872–3899, 1982.

[10] L. C. Witkowski, T. J. Drummond, C. M. Stanchak, and H. Morkoç, "High electron mobilities in modulation doped AlGaAs/GaAs heterojunctions prepared by MBE," *Appl. Phys. Lett.*, vol. 37, pp. 1033–1035, 1980.

[11] D. L. Rode, "Electron mobility in direct gap semiconductors," *Phys. Rev. B*, vol. 2, pp. 1012–1024, 1970.

[12] G. E. Stillman, C. M. Wolfe, and J. O. Dimmock, "Hall coefficient factor for polar mode scattering in n-type GaAs," *J. Phys. Chem. Solids*, vol. 31, pp. 1199–1204, 1970.

[13] H. Heiblum, private communication.

[14] T. J. Drummond, H. Morkoç, K. Lee, and M. S. Shur, "Model for modulation doped Al$_x$GaAs$_{1-x}$/GaAs field effect transistors," *IEEE Electron Device Lett.*, vol. EDL-3, pp. 338–341, 1982.

[15] F. Stern and E. Howard, "Properties of semiconductor surface inversion layers in electric quantum subband limit," *Phys. Rev.*, vol. 163, pp. 816–835, 1967.

[16] D. Delagebeaudeuf and N. T. Linh, "Metal-(n) AlGaAs–GaAs two dimensional gas FET," *IEEE Trans. Electron Devices*, vol. ED-29, pp. 955–960, 1982.

[17] K. Lee, M. S. Shur, T. J. Drummond, and H. Morkoç, "Current voltage and capacitance voltage characteristics of modulation doped field effect transistors," *IEEE Trans. Electron Devices*, vol. ED-30, pp. 207–212, 1983.

[18] H. Stormer, R. Dingle, A. C. Gossard, and W. Wiegmann, "Two-dimensional electron gas at a semiconductor–semiconductor interface," *Solid-State Commun.*, vol. 29, pp. 705–709, 1979.

[19] K. Lee, M. S. Shur, T. J. Drummond, and H. Morkoç, "Two-dimensional electron gas in modulation doped layers," *J. Appl. Phys.*, vol. 54, pp. 2093–2096, 1983.

[20] H. Morkoç, "Current transport in modulation doped (Al, Ga) As/GaAs heterostructure: Applications to high-speed FET's," *IEEE Electron Device Lett.*, vol. EDL-2, pp. 260–261, 1981.

[21] T. J. Drummond, W. Kopp, H. Morkoç, and M. Keever, "Transport in modulation doped structures (Al$_x$Ga$_{1-x}$As/GaAs): Correlations with Monte Carlo calculations (GaAs)," *Appl. Phys. Lett.*, vol. 41, pp. 277–279, 1982.

[22] M. Abe, T. Mimura, N. Yokoyama, and H. Ishikawa, "New technology towards GaAs LSI/VLSI for computer applications," *IEEE Trans. Electron Devices*, vol. ED-29, pp. 1088–1093, 1982.

[23] N. T. Linh, P. N. Tung, D. Delagebeaudeuf, P. Delescluse, and M. Laviron, "High speed-low power GaAs/AlGaAs TEGFET integrated circuits," *IEDM Tech. Dig.*, pp. 582–585, Dec. 1982.

[24] J. V. Dilorenzo, R. Dingle, M. Feuer, A. C. Gossard, R. Hendal, J. C. M. Hwang, A. Katalsky, V. G. Keramidas, R. A. Kiehl, and P. O'Connor, "Material and device considerations for selectively doped heterojunction transistors," *IEDM Tech. Dig.*, pp. 578–581, Dec. 1982.

[25] R. A. Kiehl, M. D. Feuer, R. H. Handle, J. C. M. Hwang, V. G. Keramidas, C. L. Allyn, and R. Dingle, "Selectivity doped heterostructure frequency dividers," *IEEE Electron Device Lett.*, vol. EDL-4, pp. 377–379, 1983.

[26] J. Abrokwah, N. C. Cirillo, M. Helix, and M. Longerbone, "Modulation doped FET: Threshold voltage uniformity of a high throughput 3-inch MBE system," presented at 5th Annual MBE Workshop (Georgia Techn.), Oct. 6–7, 1983; and *J. Vacuum Sci. Technol.*, to be published.

[27] N. T. Linh, M. Laviron, P. Delescluse, P. N. Tung, D. Delagebeaudeuf, F. Diamond, and J. Chevrier, "Low-noise performance of two-dimensional electron gas FET's," in *Proc. 9th IEEE Cornell Biennial Conf.*, to be published.

[28] M. Niori, T. Sito, S. Joshin, and T. Mimura, "A 20 GHz HEMT amplifier for satellite communications," presented at IEEE Int. Solid State Circuit Conf., (New York), Feb. 23–25, (see the digests).

[29] W. Anacker, "Computing at 4 degrees Kelvin," *IEEE Spectrum*, pp. 26–37, May 1979.

[30] G. E. Smith, "Fine line MOS technology for high speed integrated circuits," *IEEE Trans. Electron Devices*, vol. ED-39, p. 1564, 1983.

[31] R. Sadler and L. F. Eastman, "High-speed logic and 300 K with self-aligned submicrometer gate GaAs MESFET's," *IEEE Electron Device Lett.*, vol. EDL-24, pp. 215–217, 1983.

[32] T. Sakai, S. Konaka, Y. Kobayashi, M. Suzuki, and Y. Kawai, "Gigabit logic bipolar technology: Advanced super self aligned process technology," *Electron. Letts.*, vol. 19, pp. 283–284, 1983.

[33] T. H. Ning, R. D. Isaac, P. M. Solomon, D. D. Tang, H. N. Yu, G. C. Feth, and S. K. Wiedmann, "Self-aligned bipolar transistors for high-performance low-power VLSI," *IEEE Trans. Electron Devices*, vol. ED-28, pp. 1010–1012, 1981.

[34] C. P. Snapp, "Advanced silicon bipolar technology yields usable monolithic microwave and high speed digital IC's," *Microwave J.*, pp. 93–103, Aug. 1983.

[35] H. T. Yuan, Texas Instruments, private communication.

[36] D. B. Tuckerman and R. F. W. Pease, "High-performance heat sinking for VLSI," *IEEE Electron Device Lett.*, vol. EDL-2, pp. 126–129, 1981.

[37] W. Fichtner, R. K. Watts, D. B. Fraser, R. L. Johnston, and S. M. Sze, "0.15-µm channel length MOSFET'S fabricated using E-beam lithography," *IEEE Electron Device Lett.*, vol. EDL-3, pp. 412–414, 1982.

[38] C. O. Bozler and G. D. Alley, "Fabrication and numerical simulations of the permeable base transistors," *IEEE Trans. Electron Devices*, vol. ED-27, pp. 1128–1141, 1980.

[39] E. Kohn, N. Mishra, and L. F. Eastman, "Short channel effects in 0.5 µm source–drain spaced vertical GaAs FET's: A first experimental investigation," *IEEE Electron Device Lett.*, vol. EDL-4, pp. 125–127, 1983.

[40] W. R. Frensley, private communication.

[41] T. J. Drummond, W. Kopp, R. Fischer, H. Morkoç, R. E. Thorne, and A. Y. Cho, "Photoconductivity effects in extremely high mobility modulation doped (Al, Ga)As/GaAs heterostructures," *J. Appl. Phys.*, vol. 53, pp. 1238–1240, 1982.

Performance of Heterostructure FET's in LSI

SANDIP TIWARI, MEMBER, IEEE

Abstract—This paper reports a comparative evaluation of circuits based on heterostructure field-effect transistors (HFET's) for delay, noise-margin and power dissipation in unloaded and loaded configurations. n-channel enhancement/depletion (E/D) circuits operating at 300 and 77 K and complementary circuits operating at 77 K are compared with respect to each other. The paper also shows that a modified short-channel MOSFET model gives good agreement with experimental behavior of the devices and is adequate for evaluation. Fan-in (FI) sensitivities of delay are much smaller than fan-out (FO) sensitivities of delay for E/D circuits because of capacitive effects. E/D circuit delays are more fan-out sensitive at 300 K than at 77 K because of lower current capability. The fan-in sensitivity of the delay of complementary circuits is larger and is comparable to that circuit's fan-out sensitivity. Under loaded conditions (FI is 3, FO is 3, capacitance is 0.1 pF) at 77 K, 0.5-μm gate length E/D structures show gate delays near 50 ps and 1.0-μm gate length show gate delays near 75 ps. The circuits at 300 K exhibit a doubling of the gate delay. The complementary circuits offer, at 77 K, a performance of 70 ps at 0.5-μm gate length and 140 ps at 1.0-μm gate length. The significant performance improvements of complementary circuits with reduction of gate lengths to submicrometer dimensions occurs primarily due to reduction in the device capacitances and secondarily due to improvement of current characteristics. They demonstrate noise margins that are more than 50 percent better than their E/D counterpart along with lower power dissipations. The larger noise margin may be a significant advantage because the small logic swings require stringent parasitic resistance and threshold voltage control.

I. INTRODUCTION

THE HETEROSTRUCTURE field-effect transistor (HFET) has made significant progress in recent years and its performance potential in the n-channel modulation-doped field-effect transistor (MODFET) enhancement/depletion transistor circuit (E/D) configuration has been clearly demonstrated [1], [2]. However, its weakness—small logic swings, poor noise margins, and their interactions with threshold and parasitic variations (particularly the resistances) are not well understood. Some of these weaknesses can be reduced in their significance by changes in the configuration of devices and circuits. With the recent independent demonstration of p-channel HFET's by Tiwari and Wang [3], [4] and Stormer *et al.* [5], complementary circuits have also become possible. These circuits are expected to exhibit high noise margins and less sensitivity to threshold voltage and parasitic re-

Manuscript received October 1, 1985; revised December 24, 1985. This paper is based in part on a presentation entitled ''Circuit Performance of HFET's,'' presented at the 10th Biennial IEEE/Cornell Conference on Advanced Concepts in High Speed Devices and Circuits, Ithaca, NY, July 29-31, 1985.

The author is with the IBM Thomas J. Watson Research Center, Yorktown Heights, NY 10598.

IEEE Log Number 8607718.

sistance variations. One possible configuration using n-channel MESFET and p-channel MODFET, although not the fastest possible one, has already been demonstrated [6]. In addition, the HFET configuration employing GaAs as a semiconductor-gate electrode [7]–[9] allows for an increase in logic swings in both the E/D and the complementary configuration because of reduced forward gate conduction.

The interrelationships of the circuit performance and its comparison in large-scale circuit configuration, with practical parasitics, is best understood by gate-delay and noise-margin simulations that include the idiosyncrasies of the devices and the effects of various loadings and parasitics on circuits. This paper uses modifications on a short-channel hot-electron MOSFET model [10], [11] for transistor simulation, and compares the E/D circuits at 77 and 300 K and complementary circuits at 77 K under increased fan-in (FI), fan-out (FO), and capacitive load (C_L) conditions to obtain a realistic comparison. An alternative approach to this model is the charge control model [12], [13]. It has been shown to be similarly effective in modeling the major elements of the device. The present approach is believed to be more suited for circuit simulations because it is physically more intuitive and computationally more tractable. The paper does not include vertical scaling, although advantages may be derived from it. The technological limits of thicknesses and usable dopings, device considerations of hot-electron, tunneling, deep trap effects, etc. will determine the extent of this scaling.

II. TRANSISTOR MODELS

Circuit simulations require unified models in order to be computationally tractable and physically meaningful and interpretable. This work uses the short-channel hot-electron transistor model of Hoefflinger [10], [11] as the basis for modifications to evaluate the charge transport. Charge storage in the devices below the channel pinchoff condition are derived analytically from the charge control behavior, and are extrapolated beyond pinchoff according to the observed device behavior of GaAs MESFET's [14], [15] and GaAlAs/GaAs MODFET's [16]. The device material parameters are derived according to reference [17] for a chosen threshold voltage. The parameters used in this model can generally be derived by performing measurements on the actual device structures [18] and result in a good fit to actual device characteristics observed.

The low-field and the high-field behavior of the carrier velocity is approximated by the hyperbolic relationship

Reprinted from *IEEE Transactions on Electron Devices*, vol. ED-33, no. 5, pp. 554–563, May 1986.

$$v = \frac{\mu_0 E_y}{1 + E_y/E_C} \qquad (1)$$

where μ_0 is the low field velocity, E_y the longitudinal electric field, and E_C a critical electric field which characterizes the crossover from low-field- to high-field-dominated behavior. The saturation velocity v_S for the carriers is given by the product $\mu_0 E_C$. For p-channel devices, this relationship is in excellent agreement [3]. For n-channel devices, the fit is good at 300 K for v_S of 1.5×10^7 cm $\cdot$ s^{-1} and the measured low-field mobility. For 77 K operation of the n-channel devices, with the high low-field mobilities, a fit is obtained in the limited electric field that these devices operate in by using a saturated velocity of 1.8×10^7 cm $\cdot$ s^{-1}. The measured mobilities in short-channel structures are smaller than in the long-channel structures [18] so a low-field mobility of 30 000 cm^2 $\cdot$ V^{-1} $\cdot$ s^{-1} is used for simulations. The measured velocity field characteristics of electron gas at 77 K, quite variable from sample to sample, are about 15-percent accurate to this approximation at velocities around 1×10^7 cm $\cdot$ s^{-1}. For purposes of a comparative evaluation, this is not considered a limitation. The fit is excellent for holes and for electrons at 300 K. No attempt has been made to incorporate the negative differential mobility due to charge transfer into upper valleys because their effect on current has not been directly observed in our work or in published literature. However, the decrease in C_{dg}, the feed-back capacitance, [14]–[16], ascribed to the negative differential mobility in GaAs MESFET's [14], is modeled by decreasing C_{dg} exponentially beyond pinchoff.

For an intrinsic device structure, using the velocity field behavior, the drain current I_D for the drain–source voltage (V_{DS}) less than the pinchoff voltage (V_{DSS}) can be shown to be

$$I_D = \frac{\beta_0}{1 + V_{DS}/E_C L} (V_G - V_{DS}/2) V_{DS} \qquad (2a)$$

where

$$\beta_0 = \frac{\epsilon_{ins} \mu_0 W}{t L}. \qquad (2b)$$

V_G is the excess gate–source voltage above the threshold voltage ($V_{GS} - V_T$), L is the gate length, β_0 is the transconductance constant with ϵ_{ins} and t as the permittivity and thickness of the insulator GaAlAs, and W is the gate width. V_{DSS}, at any excess gate bias V_G, is obtained from the expression

$$V_{DSS} = \frac{(g + \frac{1}{2})}{(g + 1)} E_C L \left(\sqrt{1 + \frac{2g(g + 1)}{(g + \frac{1}{2})^2} \frac{V_G}{E_C L}} - 1 \right). \qquad (3)$$

This expression is derived [11] from the observation that pinchoff in a given structure is the result of transfer of channel control from the gate electrode to the drain electrode. Physically, this occurs when the longitudinal elec-

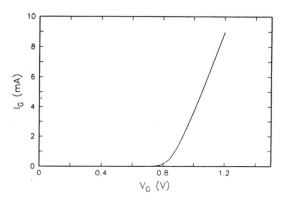

Fig. 1. Modeled gate current (I_G) as a function of gate to source voltage (V_{GS}) at 300 K for an n-channel device with gate length L of 1.0 μm and gate width normalized to 1.0 mm.

tric field (drain control) becomes much larger than the transverse electric field (gate control), and mathematically, can be related by the expression

$$\frac{qN_B}{f\epsilon_S} + \frac{q\epsilon_{ins} V_{DSS}^2}{2 f \epsilon_S t E_C L} \frac{g + 1}{g} = g \frac{E_C}{L}. \qquad (4)$$

The factor g is dependent on the permittivities, the gate length L, the background doping N_B, the electron charge q, the critical electric field E_C, and the ratio f of the field changes in the longitudinal and transverse direction at pinchoff. It can be independently extracted in practical structures from observed characteristics. Beyond pinchoff, the output conductance is obtained from the pinchoff region modulation by the drain voltage, which results in approximately linear characteristics beyond pinchoff with the slope derived from the above.

The parasitic MESFET action in MODFET structures is incorporated by the calculation of the threshold gate voltage at which it begins to occur, and then modeling it as a channel due to a sheet charge spaced at a mean depth in GaAlAs with a low-field mobility of 1000 cm^2 $\cdot$ V^{-1} $\cdot$ s^{-1} and saturated velocity of 0.65×10^6 cm $\cdot$ s^{-1}. In the modeled characteristics of n-channel enhancement HFET with parameters described later, this parasitic action causes transconductance degradation at gate voltages above 0.7 V. High-barrier Schottky diodes are included to model the injection into GaAlAs from the gate electrode. This models forward gate conduction into the parasitic MESFET channel. Fig. 1 shows modeled diode characteristics at 300 K for 1-μm gate length device with device width normalized to 1 mm. This modeled characteristic was based on experimental measurements on devices using 0.3 aluminum molefraction. The barrier height in measured structures is close to 1 eV. This conduction limits the logic swing in the simulated characteristics of the circuits to near 0.9 V. Parasitics are introduced into this intrinsic model to generate practical structures. An example of its accuracy can be seen in Fig. 2(a) and (b) where published [19] output characteristics of transistors are compared to modeled characteristics. Source resistance for the MODFET's, which was not known from an *in situ* measurement, was used as a fitting parameter. The

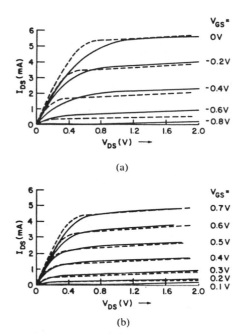

(a)

(b)

Fig. 2. (a) Measured (solid line, [19]) and modeled (dashed line) output characteristics of a depletion-mode device with $\mu_0 = 5700$ cm$^2 \cdot$ V$^{-1} \cdot$ s^{-1}, $L = 1.1$ μm, $W = 50$ μm. (b) Measured (solid line, [19]) and modeled (dashed line) output characteristics of an enhancement-mode device with $\mu_0 = 5700$ cm$^2 \cdot$ V$^{-1} \cdot$ s^{-1}, $L = 1.1$ μm, $W = 50$ μm.

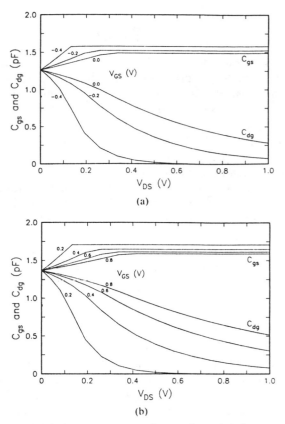

(a)

(b)

Fig. 3. (a) Modeled gate–source capacitance (C_{gs}) and drain–gate capacitance (C_{dg}) at 300 K as a function of drain–source voltage (V_{DS}) and gate–source voltage (V_{GS}) for the standard n-channel depletion-mode device with V_T of -0.6 V and width normalized to 1.0 mm. (b) Modeled gate–source capacitance (C_{gs}) and drain–gate capacitance (C_{dg}) at 300 K as a function of drain–source voltage (V_{DS}) and gate–source voltage (V_{GS}) for the standard n-channel enhancement-mode device with V_T of 0.0 V and width normalized to 1.0 mm.

model deviates most at pinchoff and near subthreshold. This is to be expected because both punchthrough and weak inversion are not modeled and material characteristics are not known.

The transistor characteristics generated in this way are neither square-law nor linear-law in gate control behavior at 300 K and closer to linear law at 77 K for n-channel devices. This is also observed in practice, and results from pinchoff at 77 K occuring closer to saturated velocities for carriers. The p-channel devices show closer to square-law behavior [3] because of their lower mobilities.

The intrinsic gate–source capacitance (C_{gs}) and gate–drain capacitance (C_{dg}) are derived below pinchoff from the integrated channel charge. These are given by

$$C_{gs} = \frac{2}{3} C_{ins} \frac{3V_G^2 - 2V_G V_{DS}}{(2V_G - V_{DS})^2} \tag{5}$$

and

$$C_{dg} = \frac{2}{3} C_{ins} \frac{3V_G^2 + V_{DS}^2 - 4V_G V_{DS}}{(2V_G - V_{DS})^2} \tag{6}$$

before pinchoff. Here C_{ins} is the capacitance associated with the insulator GaAlAs and given by the expression

$$C_{ins} = \frac{\epsilon_{ins} WL}{t}. \tag{7}$$

Beyond pinchoff, C_{gs} is assumed constant and C_{dg} is allowed to decrease. Fig. 3(a) and (b) shows these capacitances (C_{gs} and C_{dg}) for n-channel depletion mode ($V_T = -0.6$ V) and enhancement-mode ($V_T = 0.0$ V) devices of 1-μm gate length and 1-mm normalized gate width at 300

K. Other characteristics of these devices are discussed later in Section III, Circuit Simulation. The gate–source capacitance (C_{gs}) is fairly constant as a function of drain–source bias [14] and [15] and is comparable to Arnold's [16] microwave measurements on MODFET's of slightly different design (1.28 pF/mm for a depletion-mode device at $V_{GS} = -1.0$ V, $V_{DS} = 5$ V, and $L = 1.0$ μm, 1.62 pF/mm for an enhancement-mode device at $V_{GS} = +0.3$ V, $V_{DS} = 5$ V, and $L = 1.6$ μm; V_T of devices is not known). Both types of devices show C_{dg} rapidly becoming less than one-tenth of C_{gs} at $V_{DS} > V_{DSS}$—the channel pinchoff condition. This is in agreement with experimental measurements [14]–[16]. The parasitic capacitances due to depletion and fringing capacitances from gate–source C_{gsp}, gate–drain C_{dgp} and drain–source C_{ds} are incorporated through approximation of depletion regions as cylindrical extensions in GaAlAs, and fringing through air and dielectric [20]. Fig. 4 shows the general model elements used for simulation in ASTAP [21].

For GaAs gate HFET structures, the parasitic MESFET effect is ignored, and the forward gate conduction suppressed because of the presence of insulating GaAlAs. Comments with respect to the GaAs gate HFET are qualitative because the optimal device structure limited by

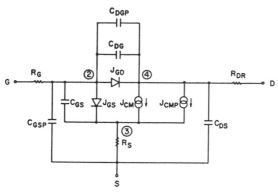

Fig. 4. Device model used for simulation showing the intrinsic and parasitic channel elements, the conducting diodes for MODFET's and other elements.

TABLE I
SUMMARY OF DEVICE PARAMETERS

		n-chan. enhanc.	n-chan. deplet.	p-chan. enhanc.
Low Field Mobility μ_0,cm^{2}V^{-1}s^{-1}	77 K	30,000	30,000	6,500
	300 K	9,000	9,000	-----
Saturation Velocity v_s,cms^{-1}	77 K	1.8x10^7	1.8x10^7	1.0x10^7
	300 K	1.5X10^7	1.5x10^7	-----
Threshold Voltage,V_T,V		0.0	-0.6	0.0
L_{gs} μm	SA	0.0	0.0	0.0
	NSA	1.0	1.0	1.0
L_{dg} μm	SA	0.0	0.0	0.0
	NSA	1.0	1.0	1.0
Ext. Sheet Res.,Ω	77 K	150	150	1500
	300 K	1000	1000	----
Contact Res.,Ω.mm	SA	0.0	0.0	0.0
	NSA	0.2	0.2	0.2

GaAlAs: $\mu_0 = 1000$ cm^{2}V^{-1}s^{-1}, $v_s = 0.65 \times 10^7$ cms^{-1}

            doping = 1 x 10^{18} cm^{-3}

GaAs: background doping = 5 x 10^{14} cm^{-3}, p-type

practical considerations of semiconductor thicknesses, parasitic resistances, etc. is not yet known.

A transistor model generated with the above elements is believed to be accurate within 15 percent for evaluating circuit elements. This limits the accuracy of the performance evaluation to a similar value. Predictions of better accuracy will have to await technological maturity leading to optimal and reproducible material and electronic parameters. The object of this paper is to emphasise comparative evaluation and performance determining elements. This evaluation is very meaningful under the limitations posed by the accuracy of this model.

III. CIRCUIT SIMULATIONS

Table I summarizes the parameters of transistor structures used for evaluating circuit delays. Nominal parameters are employed—a constant doping of 1×10^{18} cm^{-3} for GaAlAs of 0.3 mole fraction and a background p-type doping of 5×10^{14} cm^{-3} for GaAs. The p-channel devices are assumed to be fabricated using GaAlAs of 0.5 mole fraction. The threshold of the enhancement-mode device is assumed to be 0.0 V, and that for the n-channel depletion HFET load is assumed to be −0.6 V. The choice of enhancement-mode device is derived from the necessity of maximizing the logic swing. The thickness of GaAlAs and electron charge in GaAlAs are derived as in [17]. The Schottky-barrier height on GaAlAs is assumed to be 1.0 eV following measurements on experimental structures. The mobility of n-channel device is based on experimental observations of short-gate-length structures made on material that otherwise shows a factor of 2 or larger mobility [18] in larger gate-length configurations.

The devices and circuits do not function appreciably faster when the mobility is raised above 20 000 cm^2 · V^{-1} · s^{-1}, provided the saturated velocity is kept constant. This is to be expected because at pinchoff the carriers achieve near-saturated velocity at mobilities greater than 20 000 cm^2 · V^{-1} · s^{-1}, and neither the current capability nor the transconductance of the device changes appreciably thereafter. Circuits are also compared for their intrinsic capability derived by eliminating parasitic resis-

tances (we will call this self-aligned structure), and in the presence of a parasitic resistance of 0.2 Ω · mm in all devices (the non-self-aligned structure). In the latter, a gate–source and gate–drain spacing of 1.0 μm with sheet resistance of 1000 Ω at 3000 K and 100 Ω at 77 K for electron gas, and 1500 Ω for hole gas is also assumed. Table II summarizes some of the resulting transconductances ($V_{GS} = 0.6$ V) for enhancement-mode devices at 77 and 300 K for the n-channel case, and at 77 K for the p-channel case. The transconductances for the depletion-mode devices are slightly worse than those of the enhancement-mode devices shown in the Table II because of thicker GaAlAs used for adjusting the threshold voltage. It is seen from the table that the transconductances for p-channel are around half as much as the n-channel case. This is mostly due to the differences in saturation velocity and mobility of the different carriers.

NOR circuits schematically shown in Fig. 5(a) and (b) are used for simulations of both circuit families. The implementation of NAND circuits in E/D circuit configuration leads to a slower performance because of additional resistances of the enhancement-mode device; hence, the choice of NOR E/D circuit. The resistances associated with p-channel devices are higher than the n-channel devices. The NOR implementation, with series combination of p-channel devices, is therefore expected to be slower than the NAND implementation for complementary circuits. An added effect of the slightly larger resistive drops in the p-channel device is the change in logic high voltage depending on number of inputs used in the NOR gate. However, this resistive effect is not as severe as in silicon because the hole-gas mobilities in GaAs are much larger than the holes mobilities in silicon. Supply voltages as low as 0.8 V are still adequate for switching to occur in these gates although with poorer noise margins. For the low power supply voltages required in circuits based on HFET's using doped GaAlAs (due to Schottky-barrier

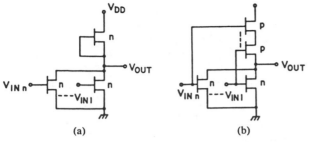

Fig. 5. (a) NOR implementation of E/D circuits simulated in this work. (b) NOR implementation of complementary circuits simulated in this work.

TABLE II
DEVICE TRANSCONDUCTANCES
$(V_T = 0.0$ V, $V_{GS} = 0.6$ V.)

	L=1.0 μm		L=0.5 μm	
	SA	NSA	SA	NSA
n-channel 77 K	420	360	460	400
n-channel 300 K	310	260	350	280
p-channel 77 K	195	145	230	170

TABLE III
COMPARISON OF PERFORMANCE CHARACTERISTICS

	E/D, 77 K				E/D, 300 K				Compl., 77 K			
	L=1.0 μm		L=0.5 μm		L=1.0 μm		L=0.5 μm		L=1.0 μm		L=0.5 μm	
	SA	NSA	SA	NSA	SA	NSA	SA	NSA	SA	NSA	SA	NSA
Delay, pS at 0.6 V	12.6	18.1	7.30	9.10	21.0	31.1	9.80	15.4	21.5	32.7	9.90	15.5
Delay, pS at 0.8 V	13.4	18.8	7.70	9.50	21.7	35.6	10.2	18.0	22.3	31.2	10.6	15.4
Delay/FO Ps/FO	12.0	13.5	5.50	6.45	15.9	22.4	7.40	11.0	19.0	25.4	8.90	12.0
Delay/FI pS/FI	3.70	3.80	1.15	1.85	5.15	7.30	2.05	3.45	13.6	20.1	5.70	8.75
Delay/C_L pS/0.1pF	20.8	21.4	16.8	19.7	26.8	50.8	23.0	34.1	12.2	15.6	10.4	13.7
Loaded Delay, pS	67.2	73.2	37.8	45.8	90.5	145.7	52.1	81.0	99.6	137.7	50.2	70.5
Power/Gate mW, at .8 V	1.50	1.28	1.52	1.48	0.66	0.54	0.75	0.69	----	----	----	----

conduction), NAND implementation should be preferable to NOR implementation. The simulated complementary performance are therefore slightly more pessimistic. In addition, unlike E/D circuits, complementary circuits do not allow use of large supply voltages because of conduction in the gate junction of the p-channel device. Many of these effects lose their severity when GaAs gate HFET's are used because GaAs gate devices allow larger gate voltages without excessive gate conduction.

The enhancement-mode n-channel device is chosen to be of a constant width (20.0 μm) for simulations, and the load n-channel and p-channel devices optimized in width for minimum delay and large noise-margin at 0.8 V bias for a variety of parasitic variations. The latter result in optimal width changes in a range of 15 percent. A constant load width of 12.0 μm for E/D circuits at 300 and 77 K, and 30.0 μm for complementary circuits at 77 K is used because it is close to the center of minimum delay distribution. For simulating loaded performance, a standard condition of FI of 3, FO of 3, C_L = 0.1 pF was assumed. This condition approximates a generalized loading condition of a 10 000-gate array chip. A summary of the salient results is provided in Table III and a critical discussion follows.

IV. E/D CIRCUITS AT 77 K

For unloaded self-aligned ring oscillators, delays of 7.7 ps are predicted at 0.5-μm gate lengths in E/D circuits at 77 K. At 1.0-μm gate lengths this number approximately doubles to 13.4 ps due to an increase in the capacitance and a slight decrease in the transconductance. The power dissipation corresponding to minimum delays are slightly above 1 mW. The simulated delays are generally seen to be 1.0 to 3.0 ps lower for all E/D circuits when saturated resistor loads are used instead of depletion HFET loads. This occurs because capacitance changes and bias changes in the depletion load device result in some transfer of charge. Hendel et al. [22] have observed near 10-ps delays for saturated resistor loads at gate lengths of 0.7 μm in quite good agreement with simulated performance. The recent results of Cirillo et al. [23] of 8.4 ps at 3.3-mW power at sub-1.0-μm gate lengths also lie within the accuracy of simulated results, but are difficult to interpret because of the reportedly large threshold voltage (0.4 V) of the enhancement-mode device.

Parasitic resistances increase the gate delay in practical devices by 50 percent at 1.0-μm gate length and 30 percent at 0.5-μm gate length. Fig. 6 shows intrinsic gate delay as a function of supply voltage for a 0.5-μm gate-length circuit with varying fan-out condition. It decreases in the beginning and then increases as the voltage is raised further for the FO = 1 curves. The decrease is an artifact of the load width (and hence current) optimization for 0.8 V, the increase comes about from the charging up of the load device capacitance to a higher voltage. From Table III, the fan-out sensitivity of intrinsic gate delay is 5.5 ps/

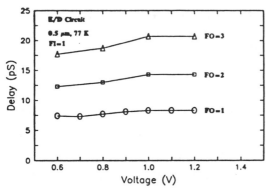

Fig. 6. Delay versus supply voltage characteristics for a 0.5-μm self-aligned E/D circuit at 77 K showing the effect of fan-out.

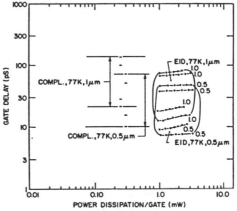

Fig. 7. Power-delay curve for circuits (E/D and complementary) operating at 77 K. The various curves and lines correspond to the simulated un loaded and loaded circuit delays with self-aligned and non-self-aligned devices. The two gate lengths, 0.5 and 1.0 μm, define two regions of the expected performance.

FO for 0.5 μm and 12.0 ps/FO for 1.0-μm gate length, which are slightly smaller than the unloaded ring-oscillator gate delay. It is slightly smaller because only the enhancement gate of the fan-out device is charged and discharged. This fan-out sensitivity increases with applied bias voltage for the same reasons as the gate delays. The fan-in sensitivity of the intrinsic structures is 1.15 and 3.7 ps/FI at 0.5 and 1.0-μm gate length, respectively. This sensitivity is small because both drain–gate capacitances and drain–source capacitances of the fan-in gate's enhancement device are significantly smaller than the gate–source capacitance encountered in the fan-out condition for the enhancement-mode device, and because the effect of the load device is also small. The resistive effects are also fairly small. The intrinsic gate-delay sensitivity to capacitive load is 16.8 ps/0.1 pF at 0.5 μm and 20.8 ps/0.1 pF for 1.0-μm circuits. The 25-percent increase in the sensitivity with gate length occurs predominantly due to the decrease in transconductance and current capability of the device at higher gate lengths.

The resistive effect on this sensitivity is also fairly small. Under a standardized loaded environment of FI = 3, FO = 3, C_L = 0.1 pF, the loaded intrinsic gate delays are 37.8 ps for 0.5 μm and 67.2 ps for 1.0 μm. For non-self-aligned device structures these delays are 45.8 ps for 0.5-μm and 73.2 ps for 1.0-μm devices. An estimate of the power-delay range for E/D circuits can be obtained from Fig. 7 where a summary of simulated delay results are plotted as a function of power dissipated per gate for 0.5 and 1.0-μm gate length. Both the self-aligned and non-self-aligned cases under unloaded and fully loaded conditions are included. This allows definition of the range over which these circuits would be expected to operate for integrations up to 10 000 gates per chip and with varying technologies for implementation of the device structure.

V. E/D CIRCUITS AT 300 K

With the raising of temperature, the mobility and saturation velocity of electron gas reduce without an appreciable change in capacitances. This leads to increased gate delays for otherwise similar structures. The decrease in

transconductance of the device and the increase in resistance in the nonpinched region of transistor characteristics also causes smaller logic swings and increased gate-delay sensitivities. Unloaded intrinsic gate delays of 10.2 ps are predicted at 0.5-μm gate lengths, which nearly double to 21.7 ps at 1.0-μm gate lengths. This is a result of an increase in capacitances and a decrease in transconductances in these devices. The delays at 77 K are 1.0–3.0 ps lower for all E/D circuits when saturated resistor loads are used. Lee et al. [24] have observed near-12-ps delays for saturated resistor loads at gate lengths below 1.0 μm in quite good agreement with modeled predictions. The role of parasitic resistances is seen to be significant for the practical devices with nearly a 50-percent worsening of delays. The increase in the sheet resistance of electron gas from 150 to 1000 Ω increases the parasitic resistances by 50 percent more than at 77 K. For non-self-aligned structures, gate delays in unloaded structures lie in between 18.0 and 35.6 ps for gate-lengths in between 0.5 and 1.0 μm, respectively. The fan-out sensitivity of the self-aligned gate-delay is 7.4 and 15.85 ps/FO for 0.5- and 1.0-μm gate lengths, respectively. Similar to 77 K, this is slightly less than the intrinsic gate delay of an unloaded structure for the same reasons. For non-self-aligned structures, this increases to 11.0 and 22.35 ps/FO at 0.5- and 1.0-μm gate lengths, a nearly 50-percent increase due to resistances. The power dissipation at 300 K is smaller by about 40 percent because with the rise in temperature, the device current capability is reduced resulting from reduced carrier velocities and higher resistances.

The fan-in sensitivity, as at 77 K, is much smaller than fan-out sensitivity because the feedback capacitance is much smaller. It is 2.05 and 5.15 ps/FI for intrinsic structures at 0.5 and 1.0-μm gate lengths, respectively, and increases to 2.45 and 7.3 ps/FI when parasitic resistances are incorporated. Gate-delay sensitivity to capacitive loading is 23.0 and 26.5 ps/0.1 pF at 0.5- and 1.0-μm gate lengths, respectively, for self-aligned delays, in-

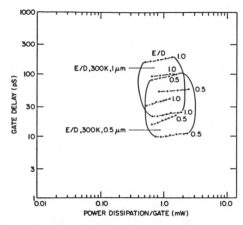

Fig. 8. Power-delay curve for E/D circuits operating at 300 K. The various curves correspond to the simulated unloaded and loaded circuit delays with self-aligned and non-self-aligned devices. The two gate lengths, 0.5 and 1.0 μm, define two regions of the expected performance.

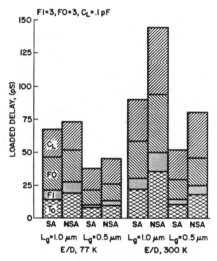

Fig. 9. Bar chart showing comparative contribution to the loaded delay from the intrinsic gate-delay (T_0), fan-in (FI), fan-out (FO), and load capacitance (C_L) for E/D circuits operating at 77 and 300 K. SA denotes self-aligned structures and NSA denotes non-self-aligned structures.

creasing to 34.1 and 50.8 ps/0.1 pF with the inclusion of parasitic resistances. The intrinsic sensitivity of delay to capacitance is 25 to 40 percent worse than at 77 K owing to the decrease in transconductance and the higher resistances in the pre-pinchoff region of the transistor. The loaded gate delays at 0.5- and 1.0-μm gate lengths are 52.1 and 90.5 pS, respectively, for intrinsic delay and 81.0 and 145.7 ps, respectively, when parasitic resistances are included. Fig. 8 shows the power-delay range of E/D circuits at 300 K. This figure, like Fig. 7, shows performance under a unloaded and loaded conditions for 0.5- and 1.0-μm gate lengths. When compared with Fig. 7, it clearly shows the power trade-off of temperature due to smaller drive capability and the delay increase resulting from the decrease in power dissipation at 300 K.

Fig. 9 is a bar chart showing the loaded circuit delay of E/D circuits operating at 77 and 300 K identifying the individual gate contribution (T_0), the contribution due to increased fan-in (FI), the contribution due to increased

fan-out (FO), and the contribution due to load capacitance (C_L). The contribution from load capacitance and fan-out are the dominating components in loaded delays irrespective of the temperature of operation. At 0.5-μm gate length, the fan-out effect is slightly less important than the load capacitance effect. However, it does not scale because transistor parasitic capacitances are significant.

VI. COMPLEMENTARY CIRCUITS AT 77 K

The required width of the p-channel enhancement device is about 50-percent larger than the n-channel enhancement device that results in a larger capacitance. In the NOR circuit implementation, the resistance of p-channel device degrades the performance of the gate with increased fan-in unlike the E/D case. Thus, the complementary circuits will always perform slower than the E/D circuits based on electron gas with similar drive capability. However, they show significantly better noise margins, have smaller dissipations and have smaller sensitivity to capacitivite loading because of a larger designed current capability, making their performance approach that of E/D circuits in large-scale integrated circuits. In addition, the effect of p-channel resistance with increased fan-in is smaller than the Si MOSFET case because of better hole transport properties.

The self-aligned gate delays are 10.6 and 22.3 ps for 0.5- and 1.0-μm gate-length circuits, respectively, with the doubling of delay with gate length due to increased capacitance and smaller transconductances. The 0.5-μm device has a delay that approaches that of the 0.5-μm E/D circuit at 77 K because at such short gate lengths the capacitances increasingly are dominated by fringing capacitances and do not scale any smaller. At 1.0-μm gate length, however, the effect of C_{gs} and C_{dg} from channel charge is significant enough to make the complementary circuit slightly less than half as fast as its E/D counterpart. The resistances affect the performance significantly (nearly 50 percent) because of the poorer hole transport properties (sheet resistance of 1500 Ω) that lead to a larger effect on the logic operation than the E/D counterpart case. The fan-out delay sensitivity in the structures is 8.9 and 19.0 ps/FO for 0.5- and 1.0-μm gate lengths, respectively, increasing to 11.2 and 25.4 ps/FO under the influence of parasitic resistances. These sensitivities are significantly larger than the E/D counterpart because of the larger capacitance loads that such a fan-out condition entails.

The delay fan-in sensitivity for 0.5- and 1.0-μm gate length is 5.7 and 13.6 ps/FI, respectively, for the intrinsic structure and 8.8 and 20.1 ps/FI when resistances are taken into account. These are significantly larger than the E/D case because the p-channel device of the fan-in gate forms a much larger capacitive load than does the n-channel depletion HFET with gate shorted to source. Also, the resistance affects the sensitivity similar to the other situations described. The sensitivity to a load capacitance of this logic is less than that of the E/D case because while the loading is the same, the complementary gates de-

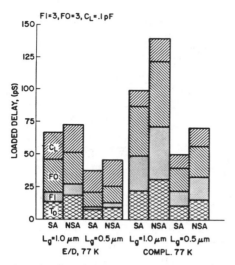

Fig. 10. Bar chart showing comparative contribution to the loaded delay from the intrinsic gate-delay (T_0), fan-in (FI), fan-out (FO), and load capacitance (C_L) for E/D circuits and complementary circuits operating at 77 K. SA denotes self-aligned structures and NSA denotes non-self-aligned structures.

signed for proper circuit performance allow for a higher current to pass during switching transient than their E/D counterpart. This leads to a delay sensitivity to capacitance load at 0.5- and 1.0-μm gate-lengths of 10.4 and 12.2 ps/0.1 pF, respectively, increasing to 13.7 and 15.6 ps/0.1 pF when parasitic resistances are included. The loaded gate delays for 0.5- and 1.0-μm gate lengths is 50.2 and 99.6 ps, respectively, for intrinsic structures and 70.5 and 137.7 ps when resistances are included. Fig. 7 shows a summary of the range of these characteristics and compares it to the corresponding E/D circuit case. The power dissipation of complementary circuits varies widely with circuit applications, but will in general be less than the E/D counterpart. Complementary circuits show a comparable performance to E/D circuits at 0.5-μm gate length. This is obtained because of the better sensitivity of the complementary technology to large capacitive loads and larger improvement in device characteristics with gate-length scaling (larger currents and smaller capacitances).

Fig. 10 shows a bar chart showing the loaded circuit delay of E/D circuits and complementary circuits operating at 77 K identifying the individual gate contribution (T_0), the contribution due to increased fan-in (FI), the contribution due to increased fan-out (FO), and the contribution due to load capacitance (C_L). The contribution from load capacitance is less dominant in the complementary circuits because the current drive during switching transients is larger for same n-channel device widths. However, this is compensated by the increased fan-out and fan-in effect resulting from larger widths of the p-channel device and the circuit configuration. Since this involves transistor capacitances, the 1-μm gate-length complementary circuits have significantly larger delays than the 1-μm E/D circuits. But, because transistor capacitances reduce at 0.5 μm and current drive capability improves, the fan-out and fan-in effect reduce signifi-

cantly at this dimension allowing complementary circuits to operate closer to the 0.5-μm E/D circuit gate delays.

VII. GaAs GATE STRUCTURES

The complementary structure is technologically most easily achieved [25] in the GaAs gate HFET structure because it involves only a change of doping of the semiconductor layer for gate control, because such a structure has less diffusion resulting from the lack of doping in GaAlAs, and because it has a natural threshold control. In addition, regrowth techniques can also be easily applied toward fabrication of the material structure. From circuit considerations, the major advantage of GaAs gate structure is in larger logic swings and hence higher noise margins due to lack of injection from gate electrode and parasitic conduction in GaAlAs. The logic swing in MODFET circuits is limited by the latter phenomena to about 0.9–1.0 V. Since the lower limit of usable GaAlAs thickness is not known as a function of mole fraction and temperature, and since delays increase with logic swing, no attempt is made to optimize the performance. In simulations, it is seen that the lack of parasitic conduction also leads to a less rapid degradation of delays with higher supply voltages, which is a particularly serious problem for modulation-doping based structures. This allows higher noise margins to be achieved (at 77 K) without a large increase in delay. Since the thickness of GaAlAs can be reduced below that required for modulation-doped structures, this delay could be reduced further in the optimal case and a higher noise margin achieved with it. For similarly designed GaAlAs structures, the delays obtained for less than 1.0-V supply voltages are comparable in the two cases.

The principal problem of GaAs gate HFET structures compared to the doped HFET structures appears to be parasitic resistance. Unlike silicon MOSFET's, the poorer insulating and barrier properties of GaAlAs require an undoped region between the gate edge and the doped source and drain regions. The parasitic resistance of this becomes a crucial determinant in the current control of the structure and hence in the logic delay.

VIII. NOISE MARGINS

For the purposes of understanding the role of noise margins and parasitic resistances toward determining proper logic operation, transfer characteristics of gates are compared for intrinsic operation and with the introduction of parasitic resistances up to 0.8 $\Omega \cdot$ mm. The enhancement devices were assumed to have a threshold of 0 V, and no subthreshold behavior was modeled. Both these are quite important in understanding of the margins. In practice, enhancement mode devices may be driven off harder with a designed threshold voltage. The process would be somewhat compensated by the subthreshold behavior of these devices. The approximations made allow us to make only rough comparisons for the circuits.

Table IV summarizes the noise-margins obtained for a 0.8-V bias voltage design. Introduction of parasitic resis-

TABLE IV
NOISE MARGINS AND LOGIC VOLTAGES

			V_{H}-V^* V	V^*-V_L V	N.M. V	N.M. Percent
E/D, 77 K	L=1 μm	SA	.284	.391	.338	42.2
		NSA	.072	.288	.180	22.5
	L=.5 μm	SA	.279	.392	.336	41.9
		NSA	.068	.270	.169	21.1
E/D, 300 K	L=1 μm	SA	.264	.360	.312	39.0
		NSA	.124	.315	.220	27.7
	L=.5 μm	SA	.286	.383	.335	41.9
		NSA	.122	.320	.221	27.6
Compl., 77 K	L=1 μm	SA	.266	.439	.353	44.1
		NSA	.278	.405	.342	42.7
	L=.5 μm	SA	.288	.437	.363	45.3
		NSA	.304	.405	.355	44.3

V^* is defined as the voltage at which transfer characteristics show a voltage gain of -1. V_H is the logic high voltage and V_L is the logic low voltage.

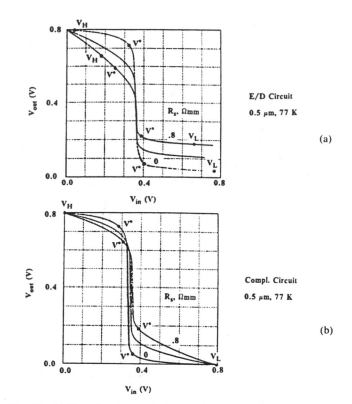

Fig. 11. (a) Transfer characteristics of a 0.5-μm self-aligned E/D circuit at 77 K showing the effect of parasitic resistance on logic voltages. (b) Transfer characteristics of a 0.5-μm self-aligned complementary circuit at 77 K showing the effect of parasitic resistance on logic voltages.

tances degrades the noise margins of E/D circuits to nearly 25 percent and thus imposes a very stringent requirement on threshold voltage variation. For a 0.8-V supply design the requirement that noise margin be at least 10 times larger than the threshold voltage fluctuations, requires that the σV_T be kept below 20 mV. On the other hand, because one of the devices is off in complementary circuits during logic high or low case and parasitic resistance effects are low, the noise margins remain larger than 40 percent allowing for a σV_T of the 32 mV. Fig. 11 (a) and (b) shows the effect more drastically, comparing the situation of 0.5-μm E/D and complementary circuits. While the logic high output part in the E/D case is dominated by the parasitic resistance of the load device and active device current for the 0.8-$\Omega \cdot$ mm case, it is not for the complementary case. Under conditions of high fan-out and fan-in such a low output high voltage would reduce further and may be insufficient. As discussed in the GaAs gate structures section, the GaAs gate HFET's allow larger noise margins in both E/D and complementary circuit implementations because of the larger supply voltages possible. That may be their single most important advantage in addition to the material and electronic sensitivity advantages.

IX. CONCLUSIONS

This paper showed that a modified short channel MOSFET model gives good agreement with experimental behavior of the HFET. It reported a comparative evaluation of gate delays, loading effects, noise margins, and power dissipation for HFET E/D circuits operating at 77 and 300 K and complementary circuits operating at 77 K. It was shown that E/D circuits are more fan-out sensitive at 300 than at 77 K with fan-in sensitivities much smaller in both cases. Under loaded conditions at 77 K, 0.5-μm gate-length E/D structures showed gate delays near 50 ps and

1.0-μm gate length showed gate delays near 75 ps. The circuits at 300 K exhibit a doubling of the gate delay. The complementary circuits showed, at 77 K, loaded gate-delays near 70 pS at 0.5-μm gate length and 140 ps at 1.0-μm gate length. Thus, 0.5-μm complementary structures show a slightly better performance than 1-μm E/D structures at the same 77 K temperature. However, at larger gate lengths the complementary circuit performance degrades rapidly and is significantly worse than the E/D circuit performance of similar or slightly larger gate lengths. They demonstrate noise margins that are more than 50 percent better than their E/D counterpart with power dissipations that are lower. The larger noise margin may be a significant advantage because the small logic swings that exist in HFET circuits require stringent parasitic resistance and threshold voltage control.

REFERENCES

[1] S. Kuroda, T. Mimura, M. Suzuki, M. Kobayashi, K. Nishiuchi, A. Shibatomi, and M. Abe, "New device structure for 4Kb HEMT SRAM," in *Tech. Dig. GaAs IC Symp.* (Boston), p. 125, 1984.
[2] M. Abe, T. Mimura, K. Nishiuchi, A. Shibatomi, and M. Kobayashi, "HEMT LSI technology for high speed computers," in *Tech. Dig. GaAs IC Symp.* (Phoenix), p. 158, 1983.
[3] W. I. Wang and S. Tiwari, "p-channel Ga$_{0.5}$Al$_{0.5}$As/GaAs MODFET's," in *Proc. Device Research Conf.* (Santa Barbara, CA), June 1984.
[4] S. Tiwari and W. I. Wang, "p-channel MODFET's using GaAlAs/GaAs hole gas," *IEEE Electron Device Lett.*, vol. EDL-5, no. 8, p. 333, Aug. 1984.
[5] H. L. Stormer, K. Baldwin, A. C. Gossard, and W. Wiegmann, "Modulation-doped field-effect transistor based on two-dimensional hole gas," *Appl. Phys. Lett.*, vol. 44, no. 11, p. 1062, June 1, 1984.

[6] R. A. Kiehl and A. C. Gossard, "Complementary p-MODFET and n-HBMESFET (Al, Ga)As transistor," *IEEE Electron Device Lett.*, vol. EDL-5, no. 12, p. 521, Dec. 1985.

[7] J. J. Rosenberg, 1980, patent applied for.

[8] K. Matsumoto, M. Ogura, M. Wada, T. Hashizume, T. Yao, and Y. Hayashi, "n$^+$-GaAs/undoped GaAlAs/undoped GaAs field-effect transistor," *Electron Lett.*, vol. 20, no. 11, p. 462, May 24, 1984.

[9] P. M. Solomon, C. M. Knoedler, and S. L. Wright, "A GaAs gate heterojunction FET," *IEEE Electron Device Lett.*, vol. EDL-5, p. 379, Sept. 1984.

[10] B. Hoefflinger, H. Sibbert, and G. Zimmer, "Model and performance of hot-electron MOS transistors for VLSI," *IEEE Trans. Electron Devices*, vol. ED-26, no. 4, p. 513, Apr. 1979.

[11] B. Hoefflinger, "Output characteristics of short-channel field-effect transistors," *IEEE Trans. Electron Devices*, vol. ED-28, no. 8, p. 971, Aug. 1981.

[12] M. S. Shur, T-H. Chen, C. H. Hyun, and P. N. Jenkins, "Designs and simulation of self-aligned modulation doped AlGaAs/GaAs ICs," in *Tech. Dig. ISSCC* (New York), p. 264, 1985.

[13] C. H. Hyun, M. S. Shur, J. H. Baek, and N. C. Cirillo, presented at 1985 IEEE/Cornell Conference on High Speed Devices and Circuits.

[14] R. W. H. Engelmann and C. A. Liechti, "Gunn domain formation in the saturated current region of GaAs MESFET's," in *IEDM Tech. Dig.* (Washington), p. 351, 1976.

[15] R. W. H. Engelmann and C. A. Liechti, "Bias dependence of GaAs and InP MESFET parameters," *IEEE Trans. Electron Devices*, vol. ED-24, no. 11, p. 1288, Nov. 1977.

[16] D. Arnold, W. Kopp, R. Fischer, J. Klem, and H. Morkoç, "Bias dependence of capacitances in modulation-doped FET's at 4 GHz," *IEEE Electron Device Lett.*, vol. EDL-5, no. 4, p. 123, Apr. 1984.

[17] S. Tiwari, "Threshold and sheet concentration sensitivity of high electron mobility transistors," *IEEE Trans. Electron Device*, vol. ED-31, no. 7, p. 879, July 1984.

[18] S. Tiwari, P. Solomon, S. L. Wright, unpublished.

[19] C. P. Lee and W. I. Wang, "High-performance modulation-doped GaAs integrated circuits with planar structures," *Electron Lett.*, vol. 19, no. 5, p. 155, Mar. 3, 1983.

[20] R. A. Pucel, H. A. Haus, and H. Statz, "Signal and noise properties of gallium arsenide microwave field-effect transistors," *Adv. Electron. Electron Phy.*, vol. 38, p. 195, 1975.

[21] Advanced Statistical Analysis Program (ASTAP) Program Reference Manual, Pub. No. SH20-1118-0, IBM Corp., White Plains, NY.

[22] R. H. Hendel, S. S. Pei, C. W. Tu, B. J. Roman, J. J. Shah, and R. Dingle, "Realization of sub-10 picosecond switching times in selectively doped (Al, Ga)As/GaAs heterostructure transistors," in *IEDM Tech. Dig.* (San Francisco), p. 857, 1984.

[23] N. C. Cirillo and J. K. Abrokwah, "8.5-picosecond ring oscillator gate delay with self-aligned gate modulation-doped n$^+$-(Al, Ga)As/GaAs FET's, in *Proc. Device Research Conf.* (Boulder), June 1985.

[24] C. P. Lee, D. Hou, S. J. Lee, D. L. Miller, and R. J. Anderson, "Ultra high speed digital integrated circuits using GaAs/GaAlAs high electron mobility transistors," in *Tech. Dig. GaAs IC Symposium* (Phoenix), p. 162, 1983.

[25] J. Y. F. Tang, and S. Tiwari, IBM Tech. Discl. Y0884-0281, 1984.

Invited

HEMT LSI Circuits

M. Abe, T. Mimura, K. Nishiuchi, A. Shibatomi and M. Kobayashi

Fujitsu Laboratories Ltd., Fujitsu Limited

1677, Ono, Atsugi 243-01, Japan

Self-aligned High Electron Mobility Transistor (HEMT) technology based on selectively doped GaAs/AlGaAs heterojunction structure for LSI circuits is described. Internal logic delay of 22 ps per gate at 77K was achieved at a fan-out of about 2, roughly three times faster than that of GaAs MESFET technology. HEMT 1Kb and 4Kb sRAM circuits have been successfully developed. HEMT 1Kb sRAM demonstrated address access time of 0.87 ns with power dissipation of 360 mW at 77K. Projected performance of HEMT 4Kb sRAM is sub-ns address access time using 1 µm gate device and 2 µm line process technology.

I. Introduction

The information processing in 1990 will need ultra-high-speed computers, requiring high speed LSI circuits with logic delays of sub-100 ps.[1,2]

High Electron Mobility Transistor (HEMT) technology has opened the door on new possibilities for ultra-high-speed LSI/VLSI applications.[3-6] Due to the super-mobility GaAs/AlGaAs heterojurction structure, HEMT is especially attractive for low temperature operations at liquid nitrogen temperature. In 1981, a HEMT ring oscillator with the gate length of 1.7 µm demonstrated 17.1 ps switching delay with 0.96 mW power dissipation per gate at 77K, indicating that switching delay below 10 ps will be achievable with 1 µm gate devices.[4] Switching delay of 12.2 ps with 1.1 mW power dissipation per gate has already been obtained with 1 µm-gate HEMT at room temperature.[7]

This paper presents the recent advances in HEMT technology for high performance LSI circuits, i.e., self-alignment fabrication technology, HEMT performances, logic and memory LSI circuits.

II. Self-Alignment Fabrication Technology

Figure 1 is a cross-sectional view of a typical self-aligned structure of enhancement-mode (E) and depletion-mode (D) HEMTs forming an inverter for DCFL circuit configuration. The basic epilayer structure consists of a 600 nm undoped GaAs layer, a 30 nm $Al_{0.3}Ga_{0.7}As$ layer doped to 2×10^{18} cm^{-3}

with Si, and a 70 nm GaAs top layer successively grown on a semi-insulating substrate by MBE. The low field electron mobility was found from Hall measurements to be 7200 cm^2/V.s at 300K and 38000 cm^2/V.s at 77K. The concentration of two Dimensional Electron Gas (2DEG) was 1.0×10^{12} cm^{-2} at 300K and 8.4×10^{11} cm^{-2} at 77K. A thin $Al_{0.3}Ga_{0.7}As$ layer to act as a stopper against selective dry etching is embedded in the top GaAs layer to fabricate E-and D-HEMTs in the same wafer. By adopting this new device structure, we can apply the selective dry etching of GaAs to AlGaAs to achieve precise control of the gate recessing process for E- and D-HEMTs.

The fabrication of E/D-HEMT circuits starts with etching shallow mesa islands down to the undoped GaAs layer to localize the active region. Next the source and drain for E- and D-HEMTs are metallized with AuGe/Au eutectic alloy to

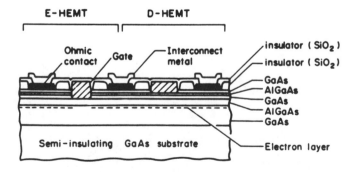

Fig. 1 Cross-sectional view of the self-aligned structure of E/D-HEMTs forming an inverter for DCFL circuit configuration.

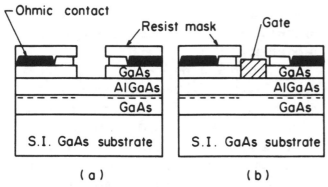

Fig. 2 Self-alignment fabrication: (9) dry recessing of n-GaAs layer, and (b) gate metallization.

form ohmic contact with the 2DEG. Then selective dry etching using an etching gas composed of CCl_2F_2 and He[8] is carried out to remove the GaAs top layer, exposing the top surface of the thin $Al_{0.3}Ga_{0.7}As$ stopper. A significantly high selectivity ratio of more than 260 was achieved at a power density of 0.18 W/cm^2, where the etching rate of GaAs was about 520 nm/min and that of AlGaAs was as low as 2 nm/min.

Figure 2 shows a step in self-aligned gate HEMT fabrication. Both the recessed structure to control the threshold voltage of the devices and the Schottky contact for the gate can be fabricated by using the same resist pattern, as shown in this figure. As a result, the Schottky gate contact and n-GaAs top layer for ohmic contact are self-aligned to achieve high speed performance.

III. HEMT Device Performances

Figure 3 shows the gate length dependence of device characteristics. The values for V_T and K were obtained by fitting the measured drain-current gate-voltage characteristics of the square root of drain-current versus gate-voltage relationship at V_{ds} of 1 V. Dependence of K-factor and transconductance g_m of E-HEMTs on gate length is measured at both 77K and 300K, and K is plotted in Fig. 3. Dashed line indicates L_g^{-1}-dependence of K-factor, estimated from the gradual channel approximation. Below 1 μm gate length at 300K, K-factor drops off from L_g^{-1}-dependence. Velocity saturation effect and parasitic source resistances probably play a significant role in these results. The 0.5 μm-gate E-HEMT at 77K exhibits g_m of

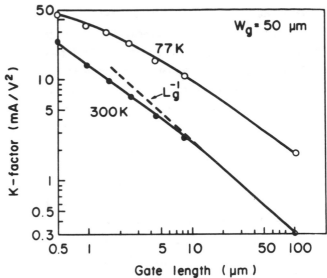

Fig. 3 Dependence of K-factor on gate length Lg at 77K and 300K, respectively.

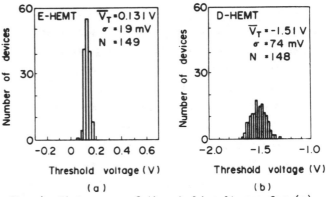

Fig. 4 Histograms of threshold voltages for (a) E-HEMTs and (b) D-HEMTs.

500 mS/mm, which is the highest value ever reported for any FET type device. No significant variation in threshold voltages with gate length was observed in the range from L_g = 10 μm to L_g = 0.5 μm. This indicates that reducing the geometry of HEMTs is an easily acceptable way to increase performance without causing problems regarding short channel effects.

Histograms of threshold uniformities for E- and D-HEMTs are shown in Fig. 4. The standard deviations in threshold voltages, measured for 149 E-HEMTs and 148 D-HEMTs distributed over an area of 15 mm x 30 mm, are 19 mV and 74 mV, respectively. The threshold sensitivity defined by the differential threshold voltage to the thickness of AlGaAs layer can be calculated based on the simple model, to be 70 mV/nm at V_T = 0.13 V. As shown in Fig. 4(a), the deviation in the maximum to minimum threshold voltages for the E-HEMT is 140 mV. This corresponds to a thickness deviation of only 2 nm over the wafer, indicating excellent controllability of MBE growth and the device fabrication process. The ratio of standard deviation of threshold voltage to the logic voltage swing (0.5 V for DCFL) is 3.8%. This strongly supports the viability of these technologies for realizing ICs with LSI/VLSI level complexities.

IV. Fundamental Logic and LSI Circuits

To evaluate the high speed capability of HEMTs in complex logic circuits, a single-clocked divide-by-two circuit based on the master-slave flip-flop consisting of eight DCFL NOR-gates, one inverter and four output buffers was fabricated. The circuit has a fan-out of up to 3 and 0.5 mm-long interconnects, giving a more meaningful indication of the overall performance of HEMT ICs than that obtained with a simple ring oscillator. The basic gate consists of 0.5 μm x 20 μm-gate E-HEMT and saturated resistors as loads. Divide-by-two operation is demonstrated at up to 8.9 GHz at 77K and up to 5.5 GHz at 300K.[2] The values of 8.9 GHz and 5.5 GHz respectively correspond to internal logic delays per gate of 22 ps with power dissipation of 2.8 mW at 77K, and 36 ps with power dissipation of 2.9 mW at 300K, with an average fan-out of about 2. The speed-power performances of ring oscillators and frequency divider circuits are summarized in Tables I and II. Figure 5 compares switching delay and power dissipation of a variety of

frequency dividers.[2,7,10] The switching speed of HEMT is roughly three times as fast as that of GaAs MESFET.

The HEMT 1K x 1b fully decoded static RAM has been successfully developed with E/D type DCFL circuit configuration.[9] The memory cell is a 6-transistor cross-coupled flip-flop circuit with switching devices having gate lengths of 2.0 μm. For peripheral circuits, a 1.5 μm gate switching device was chosen. In order to obtain high-speed operation, sufficiently high operating current was assigned to peripheral circuits. As a result, peripheral circuits, with 15% of the total device count, dissipate 85% of the chip power. To obtain a good noise margin and guarantee stable logic operation in the HEMT DCFL circuit, which has a small logic swing, special attention was paid to the power supply. The RAM has a total device count of 7244. The design rule for these lines is 3 μm line width and spacing at minimum. Minimum size of the contact hole is 2 μm x 2 μm. The memory cell is 55 μm x 39 μm ($2145 \, \mu m^2$), and the chip size is 3.0 mm x 2.9 mm. Normal read-write operation was confirmed both at 300K and 77K. At 300K, the row address access time was 3.4 ns and chip dissipation power P_{chip} of 290 mW was obtained with a supply voltage V_{dd} of 1.30 V. Sub-nanosecond access operation of access time of 0.87 ns with P_{chip} of 360 mW was achieved with V_{dd} of 1.60 V at 77K. With device parameters designed, HEMT 1Kb static RAMs will achieve sub-500 ps access operation with power dissipation of 1 W.

Table I Ring oscillators speed-power performance for HEMT device approaches.

Source (ref)		Approach	Gate length & width (μm x μm)	Switching delay (ps)	Speed-power product (fJ)	Fan-in / Fan-out
Fujitsu	(4)	HEMT (77K)	1.7 x 13	17	16	1/1
	(2)	HEMT (300K)	0.5 x 20	15	18	1/1
		HEMT (300K)	0.5 x 20	25	4	1/1
Thomson CSF	(5)	TEGFET (300K)	0.7 x 20	18	17	1/1
AT & T Bell Lab.	(6)	SDHT (77K)	1 x 125	18	141	1/1
		SDHT (300K)	1 x 125	30	135	1/1
Rockwell	(7)	HEMT (300K)	1 x 20	12	14	1/1

Table II HEMT frequency divider performance.

Source (ref)		Device approach		Circuit approach	Max.freq. (GHz)	Td (ps)	Pd Td (fJ)
Fujitsu	(2)	0.5μm HEMT	(77K)	MS-FF, 1/2 (NOR)	8.9	22	62
			(300K)		5.5	36	104
AT & T Bell Lab.	(10)	1 μm SDHT	(77K)	D-FF, 1/2 (NOR)	5.9	34	170
			(300K)		3.7	54	173
Rockwell	(7)	1 μm HEMT	(77K)	D-FF, 1/4 (NOR)	5.2	38	30
			(300K)		3.6	56	26

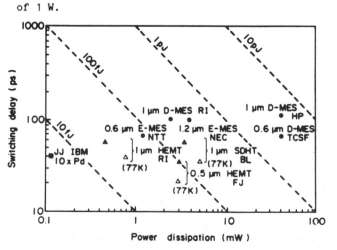

Fig. 5 Switching delay and power dissipation of a variety of frequency dividers. The symbol ● denotes GaAs MESFET at 300K, ▲ and Δ denote HEMTs at 300K and 77K, respectively, and ■ denotes Josephson junction devices.

284

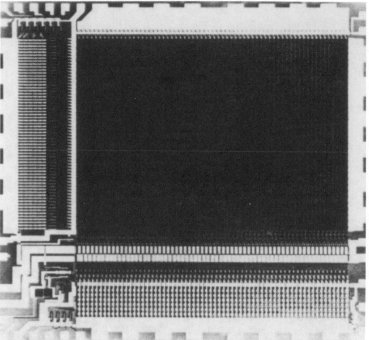

Fig. 6 Microphotograph of HEMT 4K x 1b static RAM, which measures 4.76 mm x 4.35 mm and contains 26864 E/D-HEMTs.

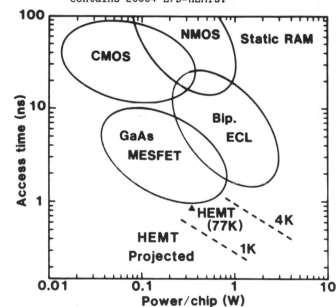

Fig. 7 Address access time and power dissipation of RAM chip.

The HEMT 4K x 1b fully decoded static RAM has also been successfully fabricated using the technology described above, and tested. Figure 6 shows a microphotograph of the 4K x 1b sRAM. The memory cell is 55 μm x 39 μm, the chip is 4.76mm x 4.35 mm, and 26864 HEMTs are integrated in a 4K bit static RAM. Normal read-write operation was confirmed both at 300K and 77K. Figure 7 shows the address access time and power dissipation of sRAM. By using 1-μm gate devices and 2-μm design rule technology, subnanosecond address access time can be projected for the 4Kb sRAM.

V. Summary

Recent advances in HEMT technology for high performance LSI circuits are presented with focus on self alignment fabrication technology using new device structure, controllability in device parameters, logic and memory LSI circuits. A HEMT 1Kb sRAM has been developed to achieve address access time of 0.87 ns with power dissipation of 360 mW at 77K, using 1.5-μm gate devices and 3-μm line process. Under the same technology, HEMT 4Kb sRAM has also been successfully fabricated and normal read-write operation been confirmed. With device technology of 1-μm gate devices and 2-μm line process, HEMT 4Kb sRAM should achieve subnanosecond access operations.

ACKNOWLEDGMENT

The authors wish to thank Dr. T. Misugi, Dr. O. Ryuzan, Dr. M. Fukuta, and T. Kotani for encouragement and support. This work is supported by the Agency of Industrial Science and Technology, MITI of Japan in the frame of National Research and Development Project "Scientific Computing System".

REFERENCES

1) M. Abe et al.: IEEE Trans. Electron Devices ED-29 (1982) 1088.

2) M. Abe et al.: IEEE GaAs IC Symposium Tech. Digest (1983) 158.

3) T. Mimura et al.: Japan. J. Appl. Phys., 19 (1980) L225.

4) T. Mimura et al.: Japan J. Appl. Phys., 20 (1981) L598.

5) P. N. Tung et al.: Electronics Lett., 18 (1982) 517.

6) J. V. DiLorenzo et al.: Int. Electron Devices Meeting Tech. Digest (1982) 578.

7) C.P. Lee et al.: IEEE GaAs IC Symposium Tech. Digest (1983) 162.

8) K. Hikosaka et al.: Japan. J. Appl. Phys., 20 (1981) L847.

9) K. Nishiuchi et al.: ISSCC Digest Tech. Papers (1984) 48,314.

10) R. A. Kiehl et al.: IEEE Electron Device Lett., EDL-4 (1983) 377.

SPEED POWER IN PLANAR TWO-DIMENSIONAL ELECTRON GAS FET DCFL CIRCUIT: A THEORETICAL APPROACH

Indexing terms: Logic and logic design, Field-effect transistors

A comparative study is made on the speed power of the two-dimensional electron gas FET (TEGFET) and the conventional GaAs FET. It is shown that the TEGFET is three times faster than the GaAs FET. Propagation delay times of 10 and 6 ps are predicted at 300 and 77 K, respectively, for a gate length of about 1 μm. Effects of electron velocity and mobility are given.

Planar enhancement-mode (E-mode) two-dimensional electron gas FETs (TEGFETs) have been shown to exhibit high performance.[1] One contribution to this performance is the low access resistance resulting from the large difference between the Schottky barrier height (1 eV) and the surface potential (0·33 eV) of the AlGaAs uppermost layer.[2] The capability of fabricating planar E-mode TEGFET opens a very interesting approach in the field of integrated circuits. Recently, planar E-mode transistors have been used in a LPFL exhibiting 19·1 ps of propagation delay time at 300 K.[3] This result compares fairly well with the 53·4 ps obtained on a more complicated recessed gate structure[4] having, nevertheless, a gate length 2·4 times larger. The LPFL circuit mentioned above has ungated transistors as loads, which present a small contribution to the node capacitance.

These experimental results have demonstrated the superiority of the TEGFET over the conventional GaAs FET. But it is not clearly understood which parameter has the highest contribution to the performances of the TEGFET: electron mobility, electron velocity, access resistance etc. This letter:

(i) gives the calculated speed-power of DCFL circuits made of TEGFETs

(ii) compares this performance to that of GaAs FET DCFL circuit

(iii) demonstrates that the charge control law of the two-dimensional electron gas by the Schottky gate plays an important role in the performance of the TEGFET.

The calculations of the propagation delay time t_{pd} and the speed-power product in GaAs FET DCFL circuit is classical,[5] so they will not be described in detail. For the TEGFET circuit we used the analytical model we have recently developed[6] and which can be summarised as follows:

(a) The gate capacitance C_G is constant since the two-dimensional electron gas is located at the interface of the heterojunction, whatever the gate voltage.

(b) The saturated drain current I_{DS} obeys a linear relationship against V_G. Under nonsaturation conditions, the drain current will be approximated by the linear relationship with the drain voltage.

(c) The source resistance is calculated by taking into account the contact resistance R_c and the sheet resistance $R_\square$. The calculation of $R_\square$ has been developed elsewhere;[2] the conduction in the two-dimensional electron gas and in the AlGaAs layer are both taken into account. A planar E-mode transistor is considered.

The calculations of speed and power are basically classical. The inverter consists of a driver TEGFET loaded by a con-

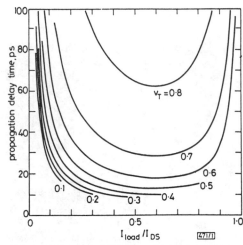

Fig. 1 *Propagation delay time of TEGFET DCFL circuit at 300 K for various threshold voltages*

stant current source I_L (DCFL circuit). The high logic voltage V_H is assumed as clamped at the value ϕ_{S2} (1 V) of the directly biased Schottky gate of the next stage driver. The low-level logic voltage is considered as $\leq V_T$.

The rise and fall delay times correspond to a voltage swing of one half the total logic voltage swing. Using the piecewise approximation for the TEGFET characteristics at $V_G = V_H$ and accounting for various possibilities concerning the hierarchy of the three voltages V_{DS} (VH), ϕ_{s2}, V [$V = 1/2(V_L + V_H)$], the propagation delay time is easily obtained.

The following parameters are maintained constant in all our calculations: Al concentration in AlGaAs = 0·3, doping concentration of AlGaAs = 10^{23} m^{-3}, source to drain spacing = $2·5 \times 10^{-6}$ m, gate width = 2×10^{-5} m, contact resistance $R_c = 10^{-10}$ Ωm^2, supply voltage $V_{DD} = 2$ V.

Fig. 1 shows the dependence at 300 K of t_{pd} on the ratio I_{load}/I_{DS} for various values of V_T. The parameters used for the calculations are mobility of the two-dimensional electron gas = 0·8 m^{2}V^{-1}s^{-1}, $L_G = 10^{-6}$ m, $v = 1·5 \times 10^5$ m s^{-1}. In fact, the actual value of the saturation electron velocity in the two-dimensional electron gas is not known. From the DC characteristics of TEGFETs and calculations reported in Reference 6 we often deduce a saturation velocity of 1·1 to $1·2 \times 10^5$ m s^{-1}, which is comparable to the electron velocity in the channel of a GaAs FET. We think that with an improved material quality, the saturation velocity can reach $1·5 \times 10^5$ m s^{-1}, since electron velocity increases as impurity concentration decreases.

It can be noted in Fig. 1 that the minimum t_{pd} is reached for $I_{load}/I_{DS} \sim 0·5$ and $V_T \sim 0·3$ V as in the GaAs FET.[5] The value of this t_{pd} is about 8·7 ps, and will be commented on further. The corresponding power dissipation is 8 fJ. As our calculations do not take into account parasitic capacitances, these values cannot be taken as absolute values. To give a more realistic approach of our study, a comparison between the TEGFET and GaAs FET has been made, the basic calculations of t_{pd} and power-delay product being the same in both cases. Moreover, to have a more valuable comparison, we have chosen for the TEGFET and GaAs FET the same values of mobility (0·5 m^{2}V^{-1}s^{-1}), velocity (1·1 $\times$ 10^5 m s^{-1}) and access resistance. The access resistance of the GaAs FET is determined by assuming a recessed gate.

Fig. 2 compares the performance of TEGFET to GaAs FET for $L_G = 1$ μm. An optimum t_{pd} has been obtained with $V_T = 0·3$ V and 0·2 V for the TEGFET and GaAs FET, respec-

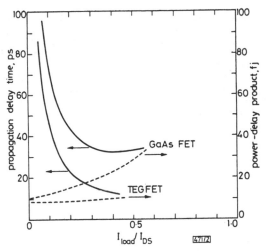

Fig. 2 *Comparison between GaAs FET and TEGFET*

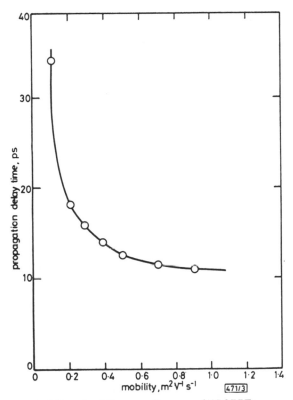

Fig. 3 *Effect of mobility on performance of TEGFETs*

The calculated power-delay product corresponding to 13 ps is 10 fJ for a supply voltage of 2 V. Improvement of the power dissipation can be obtained by reducing the supply voltage. This aspect of the problem will be treated elsewhere.

The superiority of the TEGFET over GaAs FET can be understood through the charging and discharging CR process: the higher the current, the lower the delay time. Because of its $I_{DS} - V_{DS}$ linear relationship and the large value of the Schottky barrier height which allows a large positive bias, the TEGFET delivers more current than the GaAs FET for the same threshold voltage V_T. The capacitance of the TEGFET which is constant is roughly equivalent to the average capacitance of the GaAs FET.

Since the I/V linear relationship is an important factor, one would imagine improving the speed of GaAs FETs by linearising their characteristics through the choice of a right electron concentration profile. The ideal profile for a GaAs FET which would give similar I/V characteristics to the TEGFET is an $n^-/n^+/SI$ structure with $n^- \sim 10^{22}$ m^{-3}, 50 nm thick and $n^+ \sim 3 \times 10^{24}$ m^{-3}, 5 nm thick. Such a doping level would give a mobility of 0·2 m^{2}V^{-1}s^{-1} which deteriorates the FET performance (see Fig. 3). Moreover, the Schottky barrier height is still low on GaAs, so that the delay time of such an ideal profile in GaAs would be higher than 25 ps, to be compared to 10 ps in TEGFET.

Cooling the TEGFET to 77 K will improve mobility and electron velocity. If low field Hall mobilities as high as 10 m^{2}V^{-1}s^{-1} has been measured at 77 K, one has to be careful about the drift mobility which is reduced by a factor of 3 at field of 200 Vcm^{-1}.[7] A higher field exists between gate and source, and therefore the drift mobility in a TEGFET would be drastically reduced compared to the Hall mobility. Lacking more precise data, we have taken a value of mobility of 3 m^{2}V^{-1}s^{-1} at 77 K. By assuming an electron velocity of $2·3 \times 10^5$ ms^{-1} at 77 K, which can be deduced from TEGFET DC characteristics, we have found a delay time of 6 ps for a gate length of 1 1 μm.

In conclusion, in addition to the possibility of fabricating planar E-mode transistors and low node capacitance loads, the superiority of TEGFET over GaAs FET results from different points:

(i) The large charging and discharging current due to the specific charge control of the two-dimensional electron gas by the Schottky gate reduces the delay times by a factor of 3 compared to the GaAs FET.
(ii) The high electron mobility contributes to reduce parasitic resistances and therefore improves the delay time by a factor of 30%. The predicted propagation delay times corresponding to a gate length of about 1 μm are 10 and 6 ps at 300 and 77 K, respectively.

Acknowledgments: This work is partially supported by the DRET. The authors would like to thank P. N. Tung for stimulating discussion.

D. DELAGEBEAUDEUF   *29th April 1982*
NUYEN T. LINH

Thomson-CSF Central Research Laboratories
Domaine de Corbeville, 91401 Orsay, France

References

1 LINH, N. T., DELAGEBEAUDEUF, D., LAVIRON, M., DELESCLUSE, P., and CHAPLART, J.: 'Low noise two-dimensional electron gas MESFETs'. GaAs and related compounds symp., Oiso, Japan, Sept. 1981

2 DELAGEBEAUDEUF, D., LAVIRON, M., DELESCLUSE, P., and TUNG, P. N.: 'Planar enhancement-mode two-dimensional electron gas FET as-

tively. It can be noted that:

(i) t_{pd} and power-delay products of TEGFET are three times less than those of GaAs FETs.
(ii) The value of 33 ps calculated for GaAs FET is close to the experimental value of 40 ps (gate length of 1·2 μm).[7] So, we believe that experimental data on TEGFET will be close to our calculated results (13 ps). Note that the low value of mobility used in Fig. 2 corresponds to a pessimistic situation for TEGFET. The effect of mobility on delay time is represented in Fig. 3. It can be noted that mobilities lower than 0·5 m^{2}V^{-1}s^{-1} strongly affect t_{pd}, but the improvement of t_{pd} between 0·5 and 0·9 m^{2}V^{-1}s^{-1} is only 30%. Taking into account the data of Figs. 1 and 3 we estimate that 10 ps is a realistic value of t_{pd} in a 1 μm gate length TEGFET DCFL inverter.

sociated with a low AlGaAs surface potential', *Electron. Lett.*, 1982, **18**, pp. 103–105

3 TUNG, P. N., DELAGEBEAUDEUF, D., LAVIRON, M., DELESCLUSE, P., CHAPLART, J., and LINH, N. T.: 'High speed two-dimensional electron gas FET logic', *ibid.*, 1982, **18**, pp. 109–110

4 MIMURA, T., JOSHIN, K., HIYAMIZU, S., KIKOSAKA, K., and ABE, M.: 'High electron mobility transistor logic', *Jpn. J. Appl. Phys.*, 1981, **20**, pp. L598–L600

5 INO, M., HIRAYAMA, M., and OHMORI, M.: 'Analysis for optimum threshold voltage and load current of E-D type GaAs DCFL circuits', *Electron. Lett.*, 1981, **17**, pp. 522–523

6 DELAGEBEAUDEUF, D., and LINH, N. T.: 'Metal-(N)AlGaAs-GaAs two-dimensional electron gas FET', *IEEE Trans.*, 1982, **ED-29**, to be published

7 DRUMMOND, T. J., KEEVER, M., KOPP, W., MORKOÇ, H., HESS, K., and STREETMAN, B. G.: 'Field dependence of mobility in $Al_{0.2}Ga_{0.8}As$/GaAs heterojunction at very low fields', *Electron. Lett.*, 1981, **17**, pp. 545–546

HIGH-PERFORMANCE MODULATION-DOPED GaAs INTEGRATED CIRCUITS WITH PLANAR STRUCTURES

Indexing terms: Integrated circuits, Semiconductor devices and materials

A planar process of using proton isolation for modulation-doped GaAs/GaAlAs integrated circuits is presented. Devices with very low contact resistance ($\sim 5 \times 10^{-7}$ $\Omega\,cm^2$) and very high transconductance (~ 210 ms/mm) have been achieved. DCFL ring oscillators with 10 μm switching FETs have been operated at $\tau = 36.6$ ps at room temperature and 27.3 ps at 77 K. The lowest speed-power product is 2.43 fJ per gate.

Modulation-speed GaAs/Ga$_{1-x}$Al$_x$As heterostructure devices have been demonstrated having excellent device characteristics at both room temperature and liquid nitrogen temperature.[1,2] The advantages in mobility enhancement and the structure itself have resulted in devices with very high transconductances and very high switching speeds.[3,4] However, because the devices are fabricated on epitaxial layers some process techniques still need to be refined before they can be used for reliable circuit fabrication. Two of the most immediate problems one has to face in fabricating a circuit with the modulation-doped structures are device isolation and ohmic contacts. All the work on modulation-doped devices reported so far uses mesa isolation,[2,3] which is normally done by chemical etching. However, this type of isolation results in a nonplanar structure, which is not desirable for large circuit fabrication. Ohmic contact to a modulation-doped structure is another area which has drawn significant attention because the contact has to be done on an easily oxidised wide-bandgap GaAlAs layer and a high contact resistance can limit the performance of a device. In this letter we present a planar process which uses proton bombardment for device isolation and an ohmic contact scheme which gives very low contact resistance.

The modulation-doped structures were grown by molecular beam epitaxy on one inch square (100)-oriented LEC semi-insulating GaAs substrates. The heterostructure consisted of 0.8 μm undoped GaAs, 70 Å undoped Ga$_{0.7}$Al$_{0.3}$As spacer layer, 500 Å N-Ga$_{0.7}$Al$_{0.3}$As layer with a doping concentration of 1×10^{18} cm^{-3}, and a very thin (~ 50 Å) undoped GaAs layer for surface protection. The Hall mobilities of this structure were about 5700 cm^2/Vs and 80000 cm^2/Vs at 300 K and 77 K, respectively. The uniformity of the growth as determined by sheet resistance measurements was 1260 $\Omega/\square$ ± 4.2% over the entire wafer. The fabrication procedure of the device starts with ohmic contacts for the source and the drain areas. After the photolithography step the contact areas were cleaned and slightly etched and immediately loaded in the vacuum chamber for AuGe/Ni evaporation. The sample was alloyed in an H$_2$ atmosphere at 490°C for 3 min. Following ohmic contact, proton bombardment was used to isolate the devices. Photoresist was used as the proton mask and the protons were implanted with a dose of 4×10^{15} cm^{-2} at 70 keV. After a gate recessing step the Schottky metal, Ti/Pt/Au, was evaporated and lifted off to form the gates for the enhancement-mode FETs. Following that, another Schottky metallisation step was done to form the gates of the depletion mode FETs and the first level interconnections. Then an SiO$_2$ isolation layer was deposited and the second level metal was defined and ion-milled.

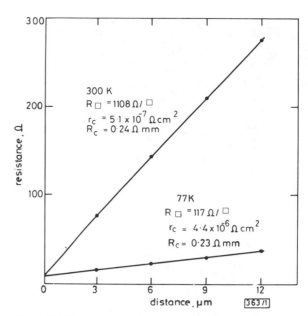

Fig. 1 *TLM contact resistance measurement of contacts to a modulation-doped structure at 300 K and 77 K*

Excellent ohmic contact has been obtained using the above procedure. Fig. 1 shows the result from the transmission-line model (TLM) measurement, in which the resistances between 50 μm-wide ohmic pads with different gaps were measured. At 300 K, the epilayer has a sheet resistance of 1108 $\Omega/\square$. Specific contact resistance of 5.1×10^{-7} $\Omega\,cm^2$ and contact resistance of 0.24 $\Omega\,mm$ were obtained. These contact resistance values are the best reported on modulation-doped structures to date. At 77 K, due to mobility enhancement, the sheet resistance of the epilayers reduced to 117 $\Omega/\square$, which is about ten times less than the sheet resistance at room temperature. The contact resistance and the specific contact resistance were 4.4×10^{-6} $\Omega\,cm^2$ and 0.23 $\Omega\,mm$, respectively. Although the contact resistance is about the same as that in room temperature, the percentage of the contribution of contact resistance to the total device resistance is much larger because of the reduction in the channel resistance at 77 K.

Both enhancement-mode FETs (E-FETs) and depletion-mode FETs (D-FETs) have been fabricated on the same modulation-doped epilayer structure but with different amounts of gate recessing. The gates of the FETs were 1.1 μm long, and the source-to-drain spacing was 4 μm. Fig. 2 shows the I/V characteristics of an E-FET and a D-FET fabricated from the same wafer. Room-temperature transconductances of 210 mS/mm and 175 mS/mm were obtained for the E-FET and the D-FET, respectively. The transfer characteristics, I_{DS} against V_{GS}, of these two FETs are shown in Fig. 3. Straight-line behaviour was obtained, meaning the transconductance is high over a wide range of gate voltages. The 210 mS/mm is the highest transconductance value ever reported for enhancement-mode GaAs FETs with similar dimensions. Recent reports by Su *et al.* have shown the same transconductance value for a modulation-doped E-FET,[5] but their device has a 3 μm source-to-drain spacing, which is 1 μm less than in our devices. If we reduce this dimension in our device to 3 μm and take into account the reduction of the source resistance (based on a measured sheet resistance value of 1200 $\Omega/\square$), the transconductance of our device would be 246 mS/mm.

Reprinted with permission from *Electronics Letters*, vol. 19, no. 5, pp. 155–157, March 1983.

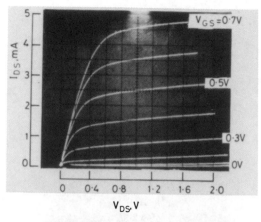

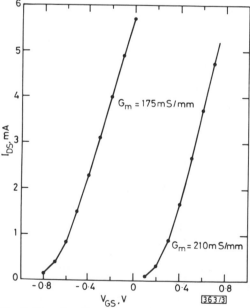

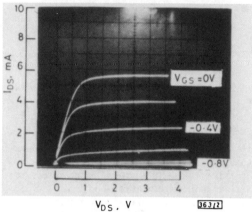

Fig. 2 *I/V characteristics of a modulation-doped E-FET (upper curve) and a modulation-doped D-FET (lower curve) at room temperature*

Fig. 3 *Transfer characteristics, I_{DS} against V_{GS}, of modulation-doped devices shown in Fig. 2*

D-FET was measured at $V_{DS} = 2.5$ V; E-FET was measured at $V_{DS} = 1.6$ V

DCFL ring oscillators with 9 stages have also been fabricated. The enhancement-mode switching FETs were 10 μm wide, and the depletion mode loads were 3 μm wide. The source-to-drain spacing and the gate length were the same as described above. The speed-power performance of a typical ring oscillator operated at room temperature is shown in Fig.

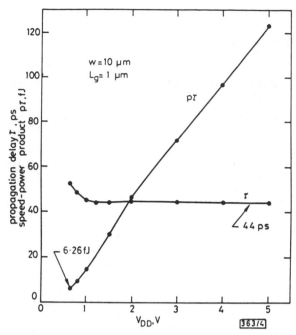

Fig. 4 *Speed-power performance of a ring oscillator with a modulation-doped structure*

4. The propagation delay and the speed-power product were measured as functions of the supply voltage, V_{DD}. As the Figure shows, the propagation delay stays constant at about 44 ps for V_{DD} above 1 V and the speed-power product increases linearly with V_{DD}. This stability in performance, a characteristic of modulation-doped devices, indicates that modulation-doped devices have better saturation characteristics than those of conventional MESFETs. The best speed performance observed for ring oscillators operated at room temperature corresponds to $\tau = 36.6$ ps at $V_{DD} = 2$ V with power dissipation of 1.2 mW and $p\tau = 44.8$ fJ. The lowest speed-power product observed was 2.43 fJ at $V_{DD} = 0.5$ V with a propagation delay of 63 ps. At 77 K the lowest propagation delay was 27.3 ps with a speed-power product of 3.95 fJ. These performance results are the best ever reported for ring oscillators with 10 μm switching FETs. Thomson CSF and Fujitsu have reported faster modulation-doped ring oscillators, but they used much wider devices (20 μm and 33 μm, respectively) and a tighter source-to-drain spacing.[3,5]

In conclusion, high-performance GaAs ICs with modulation-doped structures have been presented. Proton implantation was used for device isolation to achieve a planar structure. Excellent ohmic contacts with specific contact resistance in the 10^{-7} Ω cm² range and very high transconductance of about 210 mS/mm have been achieved. Ring oscillators with 10 μm switching FETs have been shown to operate at $\tau = 36.6$ ps at 300 K and 27.3 ps at 77 K. The lowest speed-power product was 2.43 fJ. With an improved FET design much better circuit performance is expected.

Acknowledgments: The authors would like to thank D. Hou, Y. D. Shen, R. Anderson and D. L. Miller for many helpful discussions.

C. P. LEE   *28th January 1983*
W. I. WANG*

Microelectronics Research & Development Center
Rockwell International
Thousand Oaks, CA 91360, USA

* Present address: IBM Thomas J. Watson Center, Yorktown Heights, NY 10598, USA

References

1 MIMURA, T., HIYAMIZU, S., JOSHIN, K., and HIKOSAKA, K.: 'Enhancement-mode high electron mobility transistors for logic applications', *Jpn. J. Appl. Phys.*, 1981, **20**, pp. L317–L319

2 LAVIRON, M., DELAGEBEAUDEUF, D., DELESCLUSE, P., ETIENNE, P., CHAPLART, J., and LINH, N. T.: 'Low noise normally on and normally off two-dimensional electron gas FETs', *Appl. Phys. Lett.*, 1982, **40**, pp. 530–532

3 ABE, M., MIMURA, T., YOKOYAMA, N., and ISHIKAWA, H.: 'New technology towards GaAs LSI/VLSI for computer applications', *IEEE Trans.*, 1982, **ED-29**, pp. 1088–1094

4 TUNG, P. N., DELESCLUSE, P., DELAGEBEAUDEUF, D., LAVIRON, M., CHAPLART, J., and LINH, N. T.: 'High-speed low-power DCFL using planar two-dimensional electron gas FET technology', *Electron. Lett.*, 1982, **18**, pp. 517–519

5 SU, S. L., FISCHER, R., DRUMMOND, T. J., LYONS, W. G., THORNE, R. E., KOPP, W., and MORKOÇ, H.: 'Modulation-doped AlGaAs/GaAs FETs with high transconductance and electron velocity', *ibid.*, 1982, **18**, pp. 794–796

OPTIMIZATION OF HEMTS IN ULTRA HIGH SPEED GaAs INTEGRATED CIRCUITS

S.J. Lee[*][†], C.R. Crowell[†] and C.P. Lee[*]

[*]Rockwell International, Microelectronics Research and Development Center,
Thousand Oaks, CA 91360

[†]University of Southern California, Department of Materials Science and
Electrical Engineering, Los Angeles, CA 90007

ABSTRACT

At high gate voltages, a high electron
mobility transistor may have two current conduction channels. One is at the GaAs/ GaAlAs interface and the other is in the n-type GaAlAs
layer. Conduction in the latter channel causes
degradation of the device transconductance. It is
shown in this paper that the threshold voltage for
turning on the GaAlAs channel can be controlled by
the layer parameters of the HEMT structures.
Optimization of a HEMT in an integrated circuit
requires control over both the 2-DEG channel and
the GaAlAs channel to fit each particular circuit
configuration.

1. INTRODUCTION

In modulation doped GaAlAs/GaAs heterostructures, high electron mobility and high carrier
density can coexist because of the spatial separation between the charge carriers in the undoped
GaAs region and the ionized impurities in the n-
type GaAlAs layer.[1] Field effect transistors,
called high electron mobility transistors or
HEMTs, fabricated with these structures show better performance than conventional GaAs MESFETs.
Ring oscillators with propagation delay as low as
12.2 ps and divide-by-two circuits with 5.5 GHz
clock rate have been demonstrated.[2,3] Although
much work has been reported on the physics of the
modulation doped structure and the performance of
experimental circuits and devices, very little
work has been reported on the design of HEMT
structures for integrated circuits. It is the
purpose of this paper to report on the optimization of HEMTs design from both circuit and fabrication points of view.

2. CONDUCTION IN 2-DEG CHANNEL VS CONDUCTION IN GaAlAs "CHANNEL"

A high electron mobility transistor structure
(see Fig. 1) typically consists of a undoped GaAs
layer, a undoped GaAlAs spacer layer, and a n-type
GaAlAs layer sequentially grown on LEC-undoped
GaAs substrate by molecular beam epitaxy. In a
HEMT, under normal operating conditions, the
GaAlAs layer between the Schottky gate and the
undoped GaAs layer is totally depleted. The

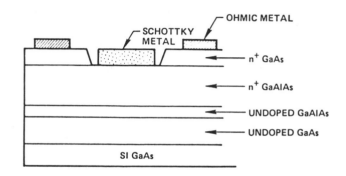

Figure 1 The cross section of a HEMT.

device current is only dependent on the flow of a
two dimensional electron gas (2-DEG) at the GaAs/
GaAlAs interface, and because of the high mobility
of the 2-DEG the device transconductance is
high. As the forward gate voltage is increased to
induce higher carrier density in the 2-DEG at the
interface, the device may reach a point where the
GaAlAs layer is not totally depleted. When the
GaAlAs layer is not totally depleted, a conducting
channel will occur in the n-type GaAlAs layer and
the sheet density of the 2-DEG at the interface
will reach its maximum. The overall device transconductance will degrade under this operating
condition because free carriers with low mobility
in the n-type GaAlAs will be modulated by the gate
voltage but the 2-DEG will not be modulated
further. Therefore, in order to fully exploit the
high mobility in the 2-DEG, one needs to choose an
appropriate layer structure for HEMTs so that the
conduction channel in the GaAlAs layer will not be
formed and the device remains only in the 2-DEG
conduction regime. To illustrate the phenomenon
described above, the band diagram of a HEMT is
shown in Fig. 2. The 2-DEG at the GaAs/GaAlAs
interface and the free carriers in n-type GaAlAs
layer can be considered as two FET channels in
parallel controlled by the same gate. This model
gives rise to two threshold voltages for a HEMT
device: one is the gate voltage required to turn
on the 2-DEG channel and the other is the gate

Reprinted from *IEEE Int. Electron Devices Meeting, 1983, Tech. Dig.*, pp. 103–106, 1983.

voltage required to turn on the n-type GaAlAs channel. The gate voltage swing for HEMT high gain operation is therefore the difference between the threshold voltages of the two FETs.

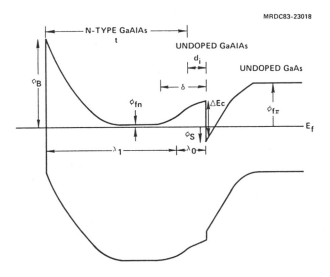

MRDC83-23018

Figure 2 The band diagram of a HEMT.

3. ANALYSIS OF DEVICE CHARACTERISTICS

In the following, the relationship between the gate voltage swing and the layer parameters of the HEMT structure is derived. We first analyze the basic properties of the two dimensional electron gas and follow by determining the threshold voltage for current in the GaAlAs channel.

a. Saturation 2-DEG Carrier Concentration

Assuming a triangular potential well at the GaAs/GaAlAs interface, the relationship between the density of the two dimensional electron gas and the interface potential, ϕ_s, at the potential well notch (Fig. 2) is given by,[4]

$$n_s = D.K.T. \log$$

$$(1 + \exp(-\frac{\gamma_0 n_s^{2/3} + \phi_s}{k_T})) \cdot (1 + \exp(-\frac{\gamma_1 n_s^{2/3} + \phi_s}{k_T}))$$

(1)

where $\gamma_0 = 1.16 \times 10^{-19}$, $\gamma_1 = 1.49 \times 10^{-9}$, D = 3.24×10^{13} cm^2 V^{-1} and the Boltzman constant K = 8.616×10^{-5} eV/K with n_s being expressed in number of electrons per square centimeter. Consider first an equilibrium GaAlAs heterostructure

with infinitely thick n-type GaAs layer and an undoped GaAlAs spacer layer with a thickness of d_i. The interface potential at the potential well notch, ϕ_s, is given by

$$\phi_s = \phi_{fn} + \frac{qN_d}{2\epsilon} (\delta^2 - d_i^2) - \Delta Ec$$

(2)

where ϕ_{fn} is the Fermi level, N_d is the concentration of the ionized impurities in GaAlAs, ΔEc is the discontinuity in the conduction band edge and δ is the depletion region thickness in the GaAlAs layer (Fig. 2). In such an equilibrium heterojunction, charge neutrality reuires that the sheet concentration of the two dimensional electron gas be equal to the number of the depleted charges in the GaAlAs layer. The sheet concentration obtained for an infinitely thick GaAlAs layer is the maximum that can be obtained in the HEMT structure. Therefore,

$$n_{s\ max} = N_d (\delta - d_i)$$

(3)

after substituting Eq. (3) into Eq. (2), we can express the maximum sheet concentration of the 2-DEG as a function of λ_0, the effective distance of the source of 2-DEG from the 2-DEG interface:

$$\lambda_0 = (\frac{\delta + d_i}{2})$$

(4)

and

$$n_{s\ max} \cdot \lambda_0 = \frac{\epsilon}{q} (\phi_s + \Delta E_c - \phi_{fn})$$

(5)

Since ϕ_s can be expressed as a slowly varying function of n_s by Eq. (1), Eq. (5) states that the maximum sheet concentration of the 2-DEG is roughly inversely proportional to the distance between the source of the 2-DEG and the interface channel, thus the closer the source of 2-DEG the higher the density for the 2-DEG in the channel. This relationship is valid for any layer parameters in a HEMT structure once ΔE_c and ϕ_{fn} are determined. The quantitative result for $n_{s\ max}$ vs λ_0 is shown in Fig. 3.

b. Gate Voltage Swing for HEMT Operation

In the above section it was assumed that GaAlAs layer was infinitely thick. This defined a maximum 2-DEG channel carrier concentration. However, in a typical HEMT structure without gate bias, the sheet density of the 2-DEG is less than this maximum carrier concentration. As the gate bias is increased to enhance the 2-DEG, the density of 2-DEG may reach its maximum and the thickness of GaAlAs depletion layer under Schottky gate is smaller than $(t + d_i - \delta)$, where t is the n-type GaAlAs thickness. The GaAlAs conduction channel is therfore turned on. The threshold voltage for this condition is given by

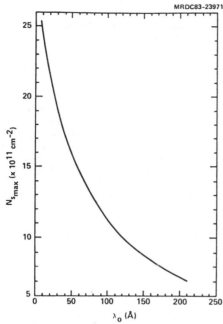

Figure 3   Maximum 2-DEG density vs the effective distance between the source of 2-DEG and the interface.

$$V_{T_1} = \phi_B - \frac{qN_d}{2\varepsilon} (t + d_i - \delta)^2 - \phi_{fn} \qquad (6)$$

In order to calculate V_{T_1}, one has to determine the depletion thickness, δ. The depletion thickness, δ, can be obtained by substituting Eq. (2) and Eq. (3) into Eq. (1) to reduce Eq. (1) to an equation linking three parameters, N_d, d_i and δ. Given N_d and d_i, we can solve it for δ or equivalently $\delta - d_i$. Figure 4 shows the calculated results for two different cases, $N_d = 5 \times 10^{17}$ cm^{-3} and $N_d = 1 \times 10^{18}$ cm^{-3}. The depletion thickness, $\delta - d_i$, of the n-type GaAlAs and the maximum concentration of 2-DEG, $n_{s\ max}$, decrease as the thickness of the spacer layer, d_i, increases, correspondingly, the threshold voltage to turn on the GaAlAs channel also decreases.

The threshold for turning on the 2-DEG channel is dependent on the thickness of the n-type GaAlAs layer, t, and is given by,[4]

$$V_{T_0} = \phi_B - \Delta E_c - \frac{qN_d}{2\varepsilon} t^2 - \phi_{fn} \qquad (7)$$

The gate voltage swing for HEMT operation can then be obtained by subtracting V_{T_o} from V_{T_1},

$$\Delta V_G = \Delta E_c + \frac{q}{\varepsilon} \eta_{s\ max} \cdot \lambda_1 \qquad (8)$$

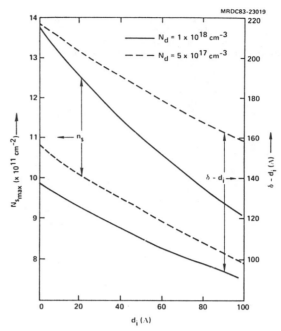

Figure 4   The maximum 2-DEG density and depletion region thickness vs the thickness of spacer layer.

where λ_1 is effective the separation of the gate electrode from the source of the carriers,

$$\lambda_1 = t - \frac{\delta - d_i}{2} \qquad (9)$$

Provided that the threshold voltage, V_{T_0}, can be conveniently adjusted, the charge control Eq. (8) states that the gate voltage swing for a given maximum channel conductance is proportional to the separation of the gate electrode from the source of the carriers. Figure 5 shows the gate voltage swing vs the thickness of the spacer layer for different values of V_{T_0}, 0.05 V and 0.2 V and two different doping concentrations N_d, 5×10^{17} cm^{-3} and 1×10^{18} cm^{-3}. It shows that the gate voltage swing is larger for smaller V_{T_0} and lower N_d due to larger λ_1 in these device structures.

4.   OPTIMIZATION OF DEVICE STRUCTURE IN INTEGRATED CIRCUITS

Optimization of HEMT structure in integrated circuits requires consideration of the optimum logic levels for logic gates. The output "low" voltage of a direct coupled inverter or logic gate ranges from 0.05 V to 0.2 V depending on the characteristics of the pull-up FET and the switching FET and the parasitic resistance of the switching FET. The output "high" voltage can be equal to the supply voltage, or it can be clamped by the Schottky gate of the next stage, depending on the value of the supply voltage. When design

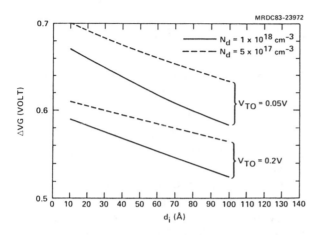

Figure 5 The relationship between the gate voltage swing and the thickness of spacer layer for different V_{TO}.

ing a HEMT device for a digital integrated circuit, the threshold voltage for turning on the 2-DEG channel should be higher than the voltage level of the low state and the gate voltage swing for HEMT operation should be close to the output voltage swing of a logic gate.

In the following we present a procedure to optimize a HEMT structure in integrated circuits. We start by choosing a saturated channel conductance in the device, i.e., $n_{s\ max}$ for the 2-DEG. By solving Eq. (5), we find λ_o, which is the first geometrical parameter related to the separation between the source of the 2-DEG and the interface. The choice of $n_{s\ max}$ will also determine the minimum possible parasitic resistance in the device. Following the choice of $n_{s\ max}$, we select a suitable gate voltage swing, ΔV_G. Combining ΔV_G and $n_{s\ max}$ and solving Eq. (8) we obtain another geometric parameter, λ_1, which is the separation between the gate electrode and the source of 2-DEG. The choice of voltage swing should reflect the obtainable performance and characteristics of an integrated circuit operation. Notice that the sum of λ_0 and λ_1, is the total thickness of GaAlAs layer. Finally, Eq. (7) can be formulated in the following form

$$\lambda_2 = \frac{t^2}{\delta - d} = (\phi_B - \Delta Ec - V_{T_o} - \phi_{fn})$$

$$\cdot \frac{2\varepsilon}{q\eta_{s\ max}} \qquad (10)$$

The choice of the threshold voltage V_{T_0} and $\eta_{s\ max}$ can give us the last geometrical parameter

λ_2, and also determine a noise margin for integrated circuit operation. Knowing three geometrical parameters of the structure, λ_0, λ_1, and λ_2, we can solve for t, d_i and δ; then the required doping concentration, N_d, can be obtained from Eq. (3).

To summarize the above procedure, we start with choices of three circuit parameters: the saturated channel conductance, the gate voltage swing and the threshold voltage of 2-DEG channel. Those choices lead to three geometrical parameters for the device. By solving these geometrical parameters, the layer parameters for the device, t, d_i and N_d can be obtained. However, in choosing the circuit parameters, one should not be unrealistic otherwise the layer parameters will become negative.

5. CONCLUSION

The layer parameters of a HEMT structure should be carefully chosen to fit the characteristics of a particular circuit; this will require simultaneous control of the 2-DEG channel and the GaAlAs channel. A HEMT structure with desired, values of $n_{s\ max}$, ΔV_G and V_{TO} can be fabricated by controlling the doping concentration, the thickness of the n-type GaAlAs and the thickness of undoped GaAlAs spacer layer.

6. REFERENCES

1. H.L. Stormer, R. Dingle, A.C. Gossard, W. Wiemann and M.d. Starge, "Two Dimensional Electron Gas at a Semiconductor-Semiconductor Interface," Solid State Commun. 29, pp. 705-709, 1979.
2. C.P. Lee, D.L. Miller, D. Hou and R.J. Anderson, "Ultra High Speed Integrated Circuit Using GaAs/GaAs High Electron Mobility Transistors," Device Research Conf. June 1983.
3. K. Nishiuchi, T. Mimura, S. Koroda, S. Hiyamizu, H. Nishi and M. Abe, "Device Characteristics of Short Channel High Electron Mobility Transistor (HEMT)," Device Research Conf. June 1983.
4. D. Delagebeaudeuf and N.T. Linh, "Metal-(n)AlGaAs-GaAs Two-dimensional Electron Gas FET," IEEE Trans. Electron Devices, ED-29, pp. 955-961, June 1982.
5. K. Lee, M. Shur, T. Drummond and H. Morkoc, "Electron Density of the Two-dimensional Electron Gas in Modulation Doped Layers," J. Appl. Phys. 54, pp. 2093-2096, April 1983.

Gate-Length Dependence of the Speed of SSI Circuits Using Submicrometer Selectively Doped Heterostructure Transistor Technology

NITIN J. SHAH, SHIN-SHEM PEI, MEMBER, IEEE, CHARLES W. TU, AND RICHARD C. TIBERIO

Abstract—Frequency dividers and ring oscillators have been fabricated with submicrometer gates on selectively doped AlGaAs/GaAs heterostructure wafers. A divide-by-two frequency divider operated up to 9.15 GHz at room temperature, dissipating 25 mW for the whole circuit at a bias voltage of 1.6 V, with gate length ~ 0.35 μm. A record propagation delay of 5.8 ps/gate was measured for a 0.35-μm gate 19-stage ring oscillator at 77 K, with a power of 1.76 mW/gate, and a bias voltage of 0.88 V. The maximum switching speed at room temperature was 10.2 ps/gate at 1.03 mW/gate and 0.8 V bias, for a ring oscillator with the same gate length. With a range of gate lengths on the same wafer fabricated by electron-beam lithography, a clear demonstration of gate-length dependence on the propagation delay was observed for both dividers and ring oscillators.

I. INTRODUCTION

SELECTIVELY doped heterostructure transistors (SDHT's) have remarkable device properties, especially with submicrometer gates and operating at low temperatures. Peak transconductances of 450 mS/mm at 300 K and 580 mS/mm at 77 K have been measured [1], and SDHT's have favorable low-noise microwave device properties [2]. Small-scale SDHT digital circuits have demonstrated ultrahigh-speed operation, e.g., ring oscillators operating at 77 K with 8.5 ps/gate [3] and frequency dividers functioning at 6.3 GHz at 300 K and 13.0 GHz at 77 K [4]. Recently, a 4 × 4 bit multiplier circuit of 150-driver-gate complexity was reported with a multiplication time of 1.6 ns at room temperature [5]. The aim of this work was to fabricate small-scale SDHT circuits to investigate the effect of gate length on speed of operation. This was done in an unique way, by fabricating devices with different gate lengths adjacent to each other on the same wafer, using direct-write electron-beam lithography. Circuits with varying gate lengths were then tested to provide a clear demonstration of the influence of gate length on device and circuit performance.

Manuscript received October 17, 1985; revised January 29, 1986. This work was supported in part by the Air Force Wright Aeronautical Laboratories, Avionics Laboratory, Wright-Patterson AFB, OH, under Contract F 33614-83-C1067, and in part by the U.S. National Science Foundation under Grant ECS 8200312 to the National Research and Resource Facility for Submicron Structures.

N. J. Shah, S.-S. Pei, and C. W. Tu are with AT&T Bell Laboratories, Murray Hill, NJ 07974.

R. C. Tiberio is with the National Research and Resource Facility for Submicron Structures, Ithaca, NY 14853.

IEEE Log Number 8607897.

II. PROCESSING

Processing of the SDHT wafers was done with a conventional mesa-isolation recess-gate fabrication technology. The wafers were semi-insulating GaAs substrates with molecular-beam expitaxial layers of undoped GaAs (buffer layer), undoped AlGaAs (spacer layer), n^+ AlGaAs (donor layer), and n^+ GaAs (cap layer). The typical sheet carrier density and mobility at 77 K was 8 to 10 × 10^{11} cm^{-2} and 70 000 to 90 000 cm^2/V · s, for these structures. All the metallization levels were patterned by liftoff, and wet chemical etching was used to adjust the threshold voltage of driver transistors and the current in the saturated resistor loads.

The definition of the driver gates was accomplished by direct-write electron-beam lithography. Adjacent dies on the wafer were exposed under different electron-beam conditions, so there were different gate lengths on the completed wafer. The fabrication of the loads was done with conventional optical lithography, so that the load characteristics were identical for both the long- and short-gate drivers. This scheme ensured that a legitimate comparison of gate lengths could be performed without influence of nonuniformity in processing or material across a wafer, or from wafer to wafer.

III. CIRCUITS

The circuits were implemented in direct-coupled FET logic (DCFL), with either 50- or 25-μm drivers and 25- or 12-μm saturated resistor loads. Discrete transistors had 3.5-μm source–drain spacing, and the gates were aligned in the center of this separation. The dc properties of transistors were measured over the complete wafer, to assess the effect of varying gate length. Nineteen-stage ring oscillators were fabricated with an inverter as the output buffer. The ring oscillators were tested both at constant bias, and then optimized for speed by varying the bias voltage. The output of the ring oscillators was monitored with a shielded coaxial probe, feeding into a spectrum analyzer.

The complementary clock frequency divider was a master-slave flip-flop design with four directly-coupled AND-NOR gates [6]. The transistors for the AND-NOR function were dual-gate devices with 6-μm source–drain spacing and gate to gate separation of 1 μm. The set-reset flip-flop

Reprinted from *IEEE Transactions on Electron Devices*, vol. ED-33, no. 5, pp. 543–547, May 1986.

296

was made with two cross-coupled AND-NOR gates, and the output of the first (master) flip-flop was fed into the second (slave) flip-flop. The inverted output of the slave was fed into the master, thereby giving the divide-by-two function. The output buffers were two inverters with the same device width as the drivers within the circuit. The average fan-out for the circuit was 2.5.

All the circuit and device testing was done on-wafer. The high-frequency input probes were terminated with a 50-Ω resistor at the probe tip, and the complementary clock was generated externally with a power splitter and adjustable delay line. An offset voltage V_{off} was superimposed onto the input clocks of the divider, and a bias voltage V_{dd} was put on the supply rail. The output of the divider circuits was probed with a shielded co-axial probe, feeding into a spectrum analyzer. The dc offset voltage, bias voltage, and the amplitude of the input clock were adjusted for optimal divider performance.

IV. DIVIDER CIRCUITS

The master-slave divider circuits (Fig. 1) were first tested for functionality under constant bias conditions, and then optimized for maximum frequency. Under the constant conditions, dependence of frequency on gate length was observed.

The maximum input frequency for a 0.35-μm-gate divider was 9.15 GHz, at room temperature, with offset voltage of 2.47 V, bias voltage of 1.60 V, and total power of 25 mW. This divider was functional from less than 5 GHz up to its maximum frequency without the need to change either of the offset or bias voltages. A number of dividers were functional up to over 8.5 GHz, and operated over an input frequency range of 2 GHz to over 8 GHz.

Previous results from our work with the same recess-gate process using optical lithography yielded a maximum dividing frequency of 5.5 GHz [6] for 1.0-μm gates and 6.3 GHz [4] for 0.7-μm gates, in comparison to 9.15 GHz for this work, with 0.35-μm gates fabricated by electron-beam lithography. Before the present work, the fastest room-temperature dividers in III-V compound semiconductor technology was a divide-by-four with heterojunction bipolar devices, operating at 8.6 GHz and 210 mW [7], and the fastest divider in any semiconductor technology was the 9.1 GHz at 554 mW for a divide-by-eight in super-self-aligned silicon bipolar devices [8]. Therefore, the submicrometer-gate SDHT result is comparable in speed to the fastest room-temperature dividers reported, but with much less power. No low-temperature data were available for these circuits.

V. RING OSCILLATORS

The propagation delay of 19-stage ring oscillators with driver:load width ratio of 2:1 was measured on an automated test system at a constant bias $V_{dd} = 2$ V at room temperature. A histogram of the results is presented in Fig. 2. There is a clear bimodal distribution of delay times, with medians at 16.7 ps/gate for the faster circuits

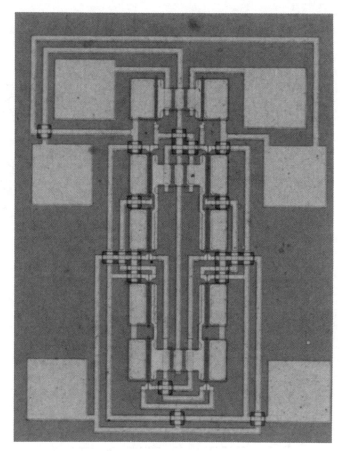

Fig. 1. Microphotograph of the master-slave frequency divider (the pads are 75 μm square).

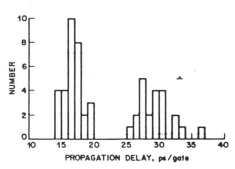

Fig. 2. Histogram of ring oscillator delay times at constant bias of 2 V measured at 300 K.

and 29.1 ps/gate for the slower circuits. Identification of the individual circuits led to a direct correspondence of short gates with short propagation delay and long gates with slower circuits. The range of gate lengths of the faster circuits was 0.35–0.4 μm, as measured by SEM, whereas the gate length of the slower circuits lay in the range 0.8–0.9 μm. The dependence of propagation delay and power dissipation of V_{dd} has been plotted in Fig. 3(a), (b) for a 0.8- and 0.35-μm gate ring oscillator. The power (~2.4 mW/gate at $V_{dd} = 2$ V) is almost the same for both devices, whereas the propagation delay is dominated by the effect of the shorter gate length.

The bimodal speed distribution due to gate length was also evident at 77 K. The fastest switching speed at 77 K

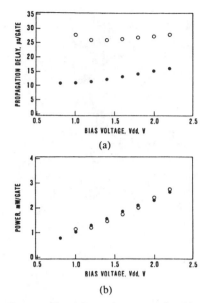

Fig. 3. Graph of propagation delay and power against bias voltage for two ring oscillators adjacent to each other on the wafer: open circles show 0.8-μm gate length, closed circles show 0.35-μm gate length.

TABLE I
MEAN PROPAGATION DELAY AS A FUNCTION OF GATE LENGTH AND TEMPERATURE

Mean propagation delay ps/gate	Gate length μm	Temperature Kelvin
24.4	0.8 to 0.9	300
12.6	0.35 to 0.4	300
13.9	0.8 to 0.9	77
7.0	0.35 to 0.4	77

was 5.8 ps/gate, at bias voltage 0.88 V and power of 1.76 mW/gate with a gate length of 0.35 μm. This is the fastest known switching time for any semiconductor device. The corresponding speed for a ring oscillator with the same gate length operating at 300 K was 10.2 ps/gate, at 0.8 V bias and 1.03 mW/gate. Ten ring oscillators with gate length below 0.4 μm were measured at 77 K, and of those, all had a minimum propagation delay below 8 ps/gate, and five were below 7 ps/gate. The mean value of the best switching time of ring oscillators tested at 300 and 77 K are given in Table I, for the two gate length populations from two SDHT wafers. It should be noted that by reducing the gate length roughly by one-half, the propagation delay at either of the measuring temperatures also fell by a half.

It is important to realize the limitations of ring oscillators with fan-in and fan-out of 1 as an evaluation of the performance of high-speed digital circuits. The matching of the driver threshold voltage and the load characteristics is critical in obtaining the best performance, as evidenced by the lower switching speed of the 11-ps/gate measurements from the initial results from this work [9]. However, the clear influence of gate length on the speed of our present circuits is valid for two reasons. Firstly, the different gate length circuits were fabricated adjacent to each

other on the same wafer for direct comparison. Secondly, fully functional divider circuits demonstrating gate length dependence on speed were also measured on the wafer, indicating that there was sufficient noise margin for the correct operation of DCFL digital circuits.

VI. DISCRETE DEVICES

The dc characteristics of discrete single-gate transistors with 3.5-μm source–drain spacing were measured for both long- and short-gate devices. The typical values of threshold voltage ($\sim$ 50 mV), peak transconductance ($\sim$ 190 mS/mm), and source resistance ($\sim$ 2.4 $\Omega \cdot$ mm) were similar for transistors adjacent to the long- and short-gate-length circuits on the wafer. The only significant difference was that the output resistance of the enhancement mode devices was $\sim$ 250 $\Omega \cdot$ mm for the 1-μm gates and fell to $\sim$ 150 $\Omega \cdot$ mm for the sub-half-micrometer devices. This decrease was expected, due to the "short-channel" effect for short-gate-length field-effect transistors. In comparison, the output resistance of GaAs MESFET's with 1-μm gates fabricated by a similar recess-gate technology is typically $\sim$ 120 $\Omega \cdot$ mm. Therefore, the degradation of the output resistance of short-channel SDHT devices is not critical for these circuits.

SEM micrographs of 0.35- and 0.8-μm gates are shown in Fig. 4(a) and (b), respectively. These devices were part of the ring oscillator circuits that demonstrated the gate-length dependence of speed. The recess gate structure for both devices had negligible lateral etching, indicating good control over the resist profile with the electron-beam lithography, as well as good resist adhesion during wet chemical etching.

VII. DISCUSSION

In contrast to our devices, Cirillo *et al.* [10] reported a strong dependence on gate length of dc properties of transistors made with a self-aligned refractory gate ion-implant MODFET technology. The peak transconductance and output conductance rose significantly over the gate length range of 2–0.5 μm, and the threshold voltage dropped by 200 mV for a change in gate length of 1–0.5 μm. We did not observe these effects, except for the rise in output conductance, over gate lengths of 1–0.35 μm, with the recessed gate technology. This difference in the behavior of the two types of devices is explained by the geometry and precise distribution of active donors in the transistors. Specifically, the lower source resistance in a self-aligned structure increases the extrinsic transconductance, which is desirable for higher speeds, but also makes the discrete device properties more sensitive to gate length. Also, estimation of the true gate length of the device is difficult, due to lateral scattering of implanted ions. The large source resistance of our devices may mask any intrinsic improvement in transconductance from making submicrometer gates.

The only major difference between short and long gate lengths in our recessed gate process is the capacitance associated with the gate. The delay associated with a gate

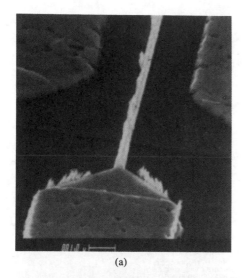

(a)

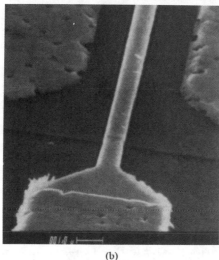

(b)

Fig. 4. SEM micrographs of gates from ring oscillator circuits, with 0.35- and 0.8-μm gate lengths.

9.15 GHz at room temperature with 25-mW power dissipation at 1.6 V, and gate length of 0.35 μm. This speed is comparable to the fastest reported divider circuits at room temperature, but at a considerably lower power. A clear gate-length dependence was observed from both dividers and ring oscillators. A minimum propagation delay of 5.8 ps/gate (1.76 mW/gate, 0.88 V), at 77 K and 10.2 ps/gate (1.03 mW/gate, 0.80 V) at 300 K for 0.35-μm gates was recorded.

These results are a demonstration of the ultrahigh-speed, low-voltage, and low-power devices that can be achieved with a submicrometer-gate selectively doped heterostructure transistor technology.

ACKNOWLEDGMENT

The electron-beam lithography was done at the National Research and Resource Facility for Submicron Structures, Ithaca, NY, with the assistance of B. Whitehead and D. Costello. We are grateful to V. J. Scarpelli and L. D. Urbanek for assistance in testing, R. F. Kopf for material growth, and P. F. Sciortino for processing. We wish to thank A. S. Jordan, L. J. Varnerin, and E. D. Wolf for their encouragement and advice.

REFERENCES

[1] L. H. Camnitz, P. J. Tasker, H. Lee, D. Van der Merwe, and L. F. Eastman, "Microwave characterization of very high transconductance MODFET," in *IEDM Tech. Dig.*, pp. 360–363, Dec. 1984.

[2] H. Goronkin and V. Nair, "Comparison of GaAs MESFET noise figures," *IEEE Electron Device Lett.*, vol. EDL-6, pp. 47–49, 1985.

[3] N. C. Cirillo and J. K. Abrokwah, "Ultra-high-speed ring oscillators based on self-aligned-gate modulation-doped n$^+$ (Al,Ga)As/GaAs FETs," *Electron. Lett.*, vol. 21, pp. 772–773, 1985.

[4] R. H. Hendel, S. S. Pei, C. W. Tu, B. J. Roman, N. J. Shah, and R. Dingle, "Realization of sub-10 picosecond switching times in selectively doped (Al,Ga)As/GaAs heterostructure transistors," in *IEDM Tech. Dig.*, pp. 857–858, Dec. 1984.

[5] A. R. Schlier, S. S. Pei, N. J. Shah, C. W. Tu, and G. E. Mahoney, "A high-speed 4 × 4 bit parallel multiplier using selectively doped heterostructure transistors," presented at the GaAs I.C. Symp., Nov. 1985.

[6] R. H. Hendel, S. S. Pei, R. A. Kiehl, C. W. Tu, M. D. Feuer, and R. Dingle, "A 10 GHz frequency divider using selectively doped heterostructure transistors," *IEEE Electron Device Lett.*, vol. EDL-5, pp. 406–408, 1984.

[7] P. M. Asbeck, D. L. Miller, R. J. Anderson, R. N. Deming, R. T. Chen, C. A. Liechti, and F. H. Eisen, "Application of heterojunction bipolar transistors to high speed, small-scale digital integrated circuits, in *Proc. GaAs I.C. Symp.*, pp. 133–136, Oct. 1984.

[8] M. Suzuki, K. Hagimoto, T. Ichino, and S. Konaka, "A 9-GHz frequency divider using Si-bipolar super self-aligned process technology," *IEEE Electron Device Lett.*, vol. EDL-6, pp. 181–183, 1985.

[9] N. J. Shah, S. S. Pei, C. W. Tu, R. H. Hendel, and R. C. Tiberio, "11 ps ring oscillators with submicron selectively doped heterostructure transistors," *Electron. Lett.*, vol. 21, pp. 151–152, 1985.

[10] N. C. Cirillo, J. K. Abrokwah, and S. A. Jamison, "A self-aligned gate modulation-doped (Al,Ga)As/GaAs FET IC process," in *Proc. GaAs I.C. Symp.*, pp. 167–170, Oct. 1984.

is proportional to (transconductance/gate capacitance): (g_m/C_g). For a fixed g_m, reduction is gate length reduces the intrinsic gate capacitance and hence the gate delay. Therefore, we attribute the exceptional speed of the submicrometer-gate devices to the reduction in the gate capacitance of the SDHT. It should be noted that for larger circuits (e.g., [5]), the circuit speed is limited both by gate capacitance and interconnect capacitance. In this case, the high transconductance of SDHT's [1] also contributes to the speed performance.

VIII. CONCLUSION

We have demonstrated submicrometer-gate static frequency dividers with a maximum operating frequency of

Close Drain–Source Self-Aligned High Electron Mobility Transistors

N. H. SHENG, M. F. CHANG, C. P. LEE, D. L. MILLER, AND R. T. CHEN

Abstract—A new self-aligned process for high-electron mobility transistors has been developed. This process uses a photoresist dummy gate to self-align the ohmic contacts. The resulting structure has very closely spaced source and drain. For 1-μm gate transistors, device transconductances as high as 320 ms/mm have been achieved at room temperature.

THE USE of self alignment techniques in the gate fabrication of GaAs MESFET's has attracted a lot of attention recently. It not only improves the device performance, but also reduces the alignment error during the gate process. Although there are several self-alignment techniques developed [1]-[3], they are mainly developed for ion-implanted MESFET's and cannot be easily applied to the process of high-electron mobility transistors (HEMT's). Because the current transport of HEMT's is controlled by a high-mobility two-dimensional electron gas (2-DEG) at the interface of the GaAs/GaAlAs epilayers, any disturbance to the properties of the epilayers by process can degrade the device performance. The most commonly used self-alignment techniques in MESFET fabrication involve ion implantation of n-type impurities in regions next to the gate. The devices have to be annealed at high temperatures to activate the implanted species. The ordinary thermal annealing procedure is detrimental to the 2-DEG at GaAs/GaAlAs interface in HEMT's. Several attempts have been carried out to fabricate self-aligned HEMT's by ion implantation [4], [5]. Flash annealing techniques had to be used to avoid the disturbance to the GaAs/GaAlAs interfaces. Although this self-aligned structure does have low source resistance (at room temperature), but because of the implanted donors in the channel regions are just beside the gate, the high-mobility two-dimensional electrons in these regions are destroyed by the implantation. At low temperatures these added three-dimensional carriers may not help the device performance because the reduction of source resistance due to mobility enhancement in a two-dimensional electron system will not happen in this structure. In this paper a different self-aligned HEMT structure is presented. No ion implantation is required in this structure. Regular Schottky metal can be used for the gates and gate recess can be easily incorporated in the process.

The structure and the fabrication procedure for the new self-aligned HEMT is shown in Fig. 1. After the epi-layer growth by MBE, a dual-level photoresist, consisting of PMMA and

Manuscript received August 19, 1985; revised October 2, 1985.

The authors are with Rockwell International Corporation, Microelectronics Research and Development Center, Thousand Oaks, CA 91360.

IEEE Log Number 8406790.

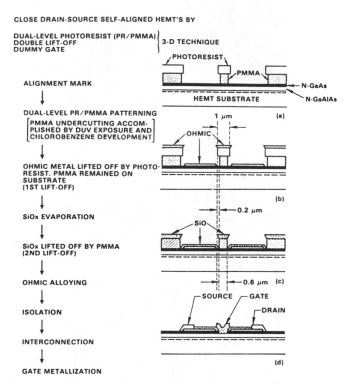

CLOSE DRAIN-SOURCE SELF-ALIGNED HEMT'S BY

DUAL-LEVEL PHOTORESIST (PR/PMMA) DOUBLE LIFT-OFF DUMMY GATE — 3-D TECHNIQUE

ALIGNMENT MARK

DUAL-LEVEL PR/PMMA PATTERNING [PMMA UNDERCUTTING ACCOMPLISHED BY DUV EXPOSURE AND CHLOROBENZENE DEVELOPMENT]

OHMIC METAL LIFTED OFF BY PHOTORESIST. PMMA REMAINED ON SUBSTRATE (1ST LIFT-OFF)

SiOx EVAPORATION

SiOx LIFTED OFF BY PMMA (2ND LIFT-OFF)

OHMIC ALLOYING

ISOLATION

INTERCONNECTION

GATE METALLIZATION

Fig. 1. Fabrication procedure for close drain–source self-aligned HEMT's.

regular photoresist, is first spun on and patterned. A T-bar resist structure is formed in the gate area by using deep UV exposure and chlorobenze development. Using this T-bar dummy gate as a mask, the ohmic metals for the source and the drain are deposited. After that, the PMMA dummy gate is removed, the gate region is chemically wet recessed, and the real gate metal is deposited. In the finished device, the separation between the source and the gate is very small and is controlled to be the amount of PMMA undercut. Experimentally, we have been able to achieve a 0.2-μm separation controllably. The minimum gate length which can be achieved using this technique is 0.6 μm. The device characteristics presented in this paper were taken from a 1-μm gate length device because it has higher yield than the submicrometer devices. In this process, since the photoresist dummy gate is used as the mask for self-aligned ohmic contacts and the real metal gate is put down after the ohmic metal process, tha gate recess can be easily done after the dummy gate is removed. The photographs of the cross section of a device after photoresist T-bar gate formation, after ohmic metal lift-off, and after SiOx deposition are shown in Fig. 2.

Excellent device characteristics have been achieved using

Reprinted from *IEEE Electron Device Letters,* vol. EDL-7, no. 1, pp. 11–12, January 1986.

STEP 1. T-BAR FORMATION

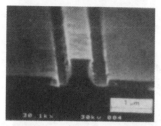

STEP 2. OHMIC METAL LIFT-OFF BY TOP PHOTORESIST

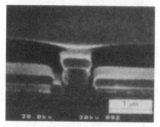

STEP 3. AFTER SiO$_x$ DEPOSITION

Fig. 2. SEM microphotographs of the cross sections of a self-aligned HEMT during fabrication.

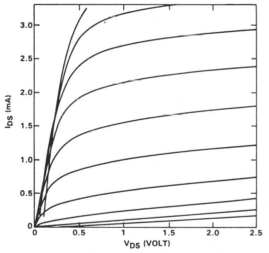

Fig. 3. I–V characteristics of a close drain–source self-aligned HEMT. The gate width is 50 μm. The gate-to-source voltage (V_{GS}) corresponding to the bottom curve is 0 V and increased by 0.1 V for each step.

this self-aligned HEMT structure. The I–V characteristics of a self-aligned HEMT with 50 Å of undoped AlGaAs spacer layer are shown in Fig. 3. The peak transconductance of the device is 320 ms/mm. The sub-threshold current conduction behavior of the self-aligned HEMT's has also been studied. Fig. 4 shows the semi-log plot of the drain current versus gate voltage of the same self-aligned HEMT device. An exponen-

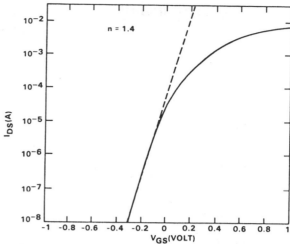

Fig. 4. Semi-log plot of the drain current versus gate voltage of a self-aligned HEMT with $V_{DS} = 1.5$ V. The exponential behavior at low drain current shows the subthreshold conduction.

tial behavior is observed in the subthreshold region. If we use the equation $I = I_0 \exp{(qVg/nkT)}$ to describe such behavior, the n factor obtained is 1.4 for this 1-μm gatelength device. This value is about the same as that of a regular nonself-aligned HEMT and is much lower than that of an ion-implanted MESFET and an ion-implanted self-aligned HEMT [6] ($n = 2$). This low n value, meaning sharp turn-off characteristics, indicates that there is no lateral diffusion of n-type impurities in the channel under the gate. In ion-implanted self-aligned process, however, lateral diffusion often causes problems resulting in short channel effect and poor cut-off characteristics.

In conclusion, the close drain–source self-aligned process offers a new process for self-aligned HEMT's. It does not require ion-implantation and high-temperature annealing. Ordinary Schottky metal can be used for the gates and gate recess can be easily achieved. Because ohmic contact is used for self-alignment instead of n-type implantation, short-channel effect due to lateral diffusion of n-type impurities is greatly reduced.

REFERENCES

[1] N. Yokoyama, T. Onishi, H. Onodera, T. Shinoki, and A. Shibatomi, "A GaAs 1K static RAM using Tungsten-Silicide gate self-alignment technology," in ISSCC Dig. Tech. Pap., Feb. 1983, pp. 44–45.
[2] K. Yamasaki, K. Asai, and K. Kurumada, "GaAs directed MESFET's with self-aligned implantation for n-layer technology (SAINT)," IEEE Trans. Electron Devices, vol. ED-29, pp. 1772–1777, 1982.
[3] H. M. Levy and R. E. Lee, "Self-aligned submicron gate digital GaAs integrated circuits," IEEE Electron Device Lett., vol. EDL-4, pp. 102–104, 1983.
[4] N. C. Cirillo, J. K. Abrokwah, and S. A. Jamason, "Self-aligned gate modulation-doped AlGaAs/GaAs FET IC process," in Tech. Dig., GaAs Ic Symp., 1984, pp. 167–170.
[5] C. Kocot and A. Watica, "Self-aligned modulation-doped FET's," presented at the 21st Ann. Workshop on Compound Semiconductor Microwave Material and Devices, Fort Lauderdale, FL, Feb. 1985.
[6] M. F. Chang, N. H. Sheng, C. P. Lee, and R. T. Chen, "Self-aligned substitutional gate process for HEMT's," presented at the 21st Annu. Workshop Compound Semiconductor Microwave Material and Devices, Fort Lauderdale, FL, Feb. 1985.

Threshold Voltage Behavior for WSi/Al$_x$Ga$_{1-x}$As/GaAs MIS-Like Heterostructure FET

Kunihiro ARAI, Takashi MIZUTANI and Fumihiko YANAGAWA

NTT Atsugi Electrical Communication Laboratories,
3–1, Morinosato Wakamiya, Atsugi-shi, Kanagawa 243–01

(Received June 11, 1985; accepted for publication July 20, 1985)

WSi/Al$_x$Ga$_{1-x}$As/GaAs MIS-like heterostructure FET (MIS-HFET) is compared with HEMT fabricated using the same MBE apparatus. Threshold voltage (V_T) variation through a wafer is much less for MIS-HFET (± 0.05 V) than for the present HEMT (± 0.15 V $\sim \pm 0.4$ V). In addition, threshold voltage shift as lowering the temperature from 300 K to 85 K is also much less for MIS-HFET (0.05 V) than for HEMT (0.22 V). As it is simpler for MIS-HFET to obtain high V_T uniformity and small V_T dependence on temperature than for HEMT, MIS-HFET is believed to be attractive for LSI application.

§1. Introduction

In the past few years, FETs utilizing high mobility 2-dimensional electron gas induced in a potential well at the Al$_x$Ga$_{1-x}$As–GaAs interface[1] have been intensively studied, because of its potentiality for very high speed and low power dissipation logic circuits. Such an FET realized first (HEMT) was essentially composed of a selectively doped Al$_x$Ga$_{1-x}$As and an undoped GaAs layer.[2] Although it showed inherent high performance,[3-4] it has some difficulties for LSI application. First, it requires stringent control of both Al$_x$Ga$_{1-x}$As impurity concentration and layer thickness to obtain a desired threshold voltage, V_T.[5] Second, V_T depends strongly on temperature.[6] Third, V_T is strongly influenced by light at low temperature such as 77 K.[7]

Recently, new MIS (metal-insulator-semiconductor)-like heterostructure FETs (here, we call them MIS-HFETs), which are expected to get rid of the above difficulties, were proposed (BIFET,[8] SAHFET,[9] SISFET,[10] GaAs gate heterojunction FET[11]). The essential difference of these FETs from HEMT is that the Al$_x$Ga$_{1-x}$As layers are undoped. Actually, MIS-HFET was experimentally shown to be free from the threshold voltage instability under illumination.[11] Furthermore, in a small sample area (5×10 mm^2), V_T standard deviation for MIS-HEFT is reported to be small (0.013 V).[10] However, thereshold voltage variation through a wafer, which is important for LSI fabrication, is not clarified, yet. Temperature dependence of V_T, which is important for chilled IC application, is also left unclear. In addition, reported performances for these FETs are not good compared with HEMT.

This letter makes experimental comparison between MIS-HFET and HEMT, which are fabricated using the same MBE apparatus. It shows MIS-HFET has better V_T uniformity through a wafer and smaller temperature dependence of V_T than HEMT. It also reports that good performance (270 mS/mm at room temperature) can be obtained for MIS-HFET. Thus, MIS-HFET appears to be attractive for LSI application.

§2. Device Structure and Fabrication

A schematic cross section of the present MIS-HFET is shown in Fig. 1. GaAs/Al$_x$Ga$_{1-x}$As/GaAs structure was grown by MBE on a 2 inch (100)-oriented semi-insulating GaAs substrate. The structure consists of a 300 nm undoped GaAs buffer layer, a 60 nm undoped Al$_x$Ga$_{1-x}$As ($x=0.43$) layer, and a 10 nm undoped GaAs cap layer. The epitaxial growth was performed rotating the GaAs substrate around its center axis.

A 200 nm WSi layer was deposited by the DC sputtering method and etched by RIE (SF$_6$) to form a gate electrode. Sourced/drain n$^+$ regions self-aligned to the gate were then implanted with Si at 100 keV to a dose of 3×10^{13} cm^{-2}. Subsequently, the implanted Si atoms were activated by halogen lamp annealing at 980°C without any duration time. Finally, source and drain ohmic contacts were formed by evaporating AuGe(200 nm)/Ni(30 nm) sequentially and annealing at 450°C for 1 min. The sheet resistance of the implanted region was 440 Ω/□, and the contact resistance was 0.27 Ωmm. The gate width of the device was 20 μm. The gate length was 1 μm or 1.5 μm.

As a reference, HEMT was also fabricated with a 30 nm Al$_x$Ga$_{1-x}$As ($x=0.3$) and 5 nm GaAs cap layer. These layers were doped with Si to $1 \sim 2 \times 10^{18}$ cm^{-3}.

§3. Results and Discussion

The threshold voltage variation for MIS-HFET on a 2 inch wafer is compared with that for HEMT in Fig. 2. The threshold voltage for MIS-HFET is independent of

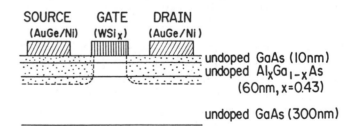

Fig. 1. Schematic cross section of MIS-HFET.

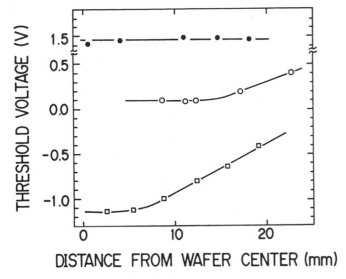

Fig. 2. Threshold voltage variation with the distance from the wafer center. The solid circles represent results for MIS-HFET. The open circles and the open squares represent result for HEMTs with different average threshold voltage. The gate length is 1.5 μm.

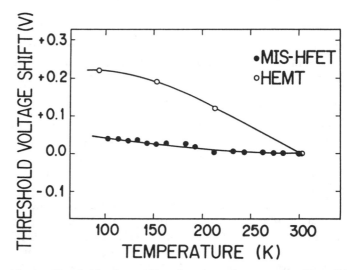

Fig. 3. Threshold voltage shift as functions of temperature. The solid circles and the open circles represent results for MIS-HFET and HEMT, respectively. The gate length is 1.5 μm.

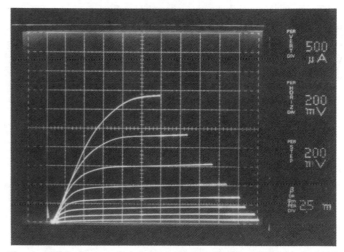

Fig. 4. FET characteristics for MIS-HFET. V_{Gmax} is 3.2 volts with steps of 0.2 volts. The gate length and width are 1 μm and 20 μm, respectively.

the position on the wafer and V_T variation is as small as ±0.05 V arround its average value. On the contrary, the threshold voltage for HEMT is observed to increase with the distance from the wafer center to the edge. The larger the threshold voltage, the smaller the V_T variation: from 0.1 to 0.4 volts for one sample in contrast with from −1.15 to −0.35 volts for the other. From the HEMT threshold voltage variation data, one can estimate the variation of the product $N \cdot d^2$ to be 25%, where N is the impurity concentration and d is the $Al_xGa_{1-x}As$ layer thickness. This corresponds to N variation within 25% and to d variation within 13%. It is important to note that these variations do not affect the V_T for MIS-HFET as shown in Fig. 2. This shows the expected V_T independence from the MBE film parameters.

There are several papers reporting high uniformity for the MBE film thickness d and doping N: within 1.0% over the central 60 mm,[12] or within 2.0% over the central 2.5 inch[5] of 3 inch substrates. These data may result in the V_T variation less than ±0.02 ~ ±0.04 V though a 2 inch wafer for HEMT with average V_T of 0.1 V, if process caused V_T scatter is neglected. However, such d and N uniformity is obtained by using elaborate MBE apparatus with improved Knudsen cell structure and the source-substrate distance twice as large as conventional one,[12] or with two Ga sources placed diametrically opposite to one another.[5] Thus, it is simpler for MIS-HFET to obtain V_T uniformity than for HEMT. Further V_T uniformity improvement for MIS-HFET is believed to be achieved by improving the fabrication process such as gate length control. The V_T insensitivity on the MBE film parameters d and N also suggests good V_T reproducibility for MIS-HFET.

Threshold voltage dependences on temperature are compared for MIS-HFET and HEMT in Fig. 3. For both devices, the threshold voltage increases gradually with lowering the temperature from 300 K to 85 K. However, the amount of the threshold voltage shift is much less for MIS-HFET than for HEMT: 0.05 V for MIS-HFET and

0.22 V for HEMT. The present V_T shift for HFET is nearly equal to the value (0.25 V) reported by A. J. Valois et al.[6]

The cause of the threshold voltage increase for HEMT is attributed to temperature dependent occupation of deep donor traps in $Al_xGa_{1-x}As$ layer.[6] These traps are well known to be related to Si atoms doped in the $Al_xGa_{1-x}As$ layer.[13] The small temperature dependence for MIS-HFET is believed to be due to Si atom absence in the $Al_xGa_{1-x}As$ layer.

The threshold voltage for the present MIS-HFET is about 1.5 V as shown in Fig. 2. On the other hand, it is estimated to be about 0.8 V from the device structure. The cause of the difference between the observed and the estimated V_T might be crystal defects caused by the lamp annealing, interfacial states, or others. Further study is necessary to clarify it.

Typical transconductance g_m is 145 mS/mm at room temperature and 210 mS/mm at 85 K. The best data for g_m at room temperature is 270 mS/mm as shown in Fig. 4. This is the largest value for MIS-HFET ever reported. The large g_m, as well as the better V_T uniformity and the smaller V_T temperature sensitivity for MIS-HFET than for the present HEMT, makes MIS-HFET attractive for LSI application.

§4. Conclusion

WSi/Al$_x$Ga$_{1-x}$As/GaAs MIS-HFET was fabricated and the threshold voltage was experimentally compared with that of HEMT fabricated using the same MBE apparatus. Threshold voltage variation through a wafer for MIS-HFET (± 0.05 V) was mush less than that for the present HEMT (± 0.15 V $\sim \pm 0.4$ V). Threshold voltage shift with lowering temperature from 300 K to 85 K was also much less for MIS-HFET (0.05 V) than for HEMT (0.22 V). From these results and the good performance obtained for the present MIS-HFET (270 mS/mm), MIS-HFET appears to be attractive for LSI application.

Acknowledgments

The authors wish to express their thanks to K. Oe for his valuable discussions and MBE film growth. They also thank T. Honda for ion implantation, M. Suzuki for advising WSi depositon and RIE, and K. Maezawa for lamp annealing.

References

1) R. Dingle, H. L. Stormer, A. C. Gossard and W. Wiegmann: Appl. Phys. Lett. **33** (1978) 665.
2) T. Mimura, S. Hiyamizu, T. Fujii and K. Nanbu: Jpn. J. Phys. **19** (1980) L225.
3) T. Mimura, K. Joshin, H. Hiyamizu, K. Hikosaka and M. Abe: Jpn. J. Appl. Phys. **20** (1981) L598.
4) C. P. Lee, D. L. Miller, D. Hou and R. J. Anderson: 41st Device Research Conference, Paper IIA-7, Burlington, VT, June 1983.
5) J. K. Abrokwah, N. C. Cirillo, Jr., M. J. Helix and M. Longerbone: J. Vac. Sci. Technol. **B2** (1984) 252.
6) A. J. Valois, G. Y. Robinson, K. Lee and S. Shur: J. Vac. Sci. Technol. **B1** (1984) 190.
7) T. J. Drummond, R. J. Fischer, W. F. Kopp, H. Moroc, K. Lee and M. S. Shur: IEEE Trans. Electron Devices **ED-30** (1983) 1806.
8) T. J. Drummond, W. Kopp, D. Arnold, R. Fisher and H. Morkoc: Electron. Lett. **19** (1983) 986.
9) Y. Katayama, M. Morioka, Y. Sawada, K. Ueyanagi, T. Mishima, Y. Ono, T. Usagawa and Y. Shiraki: Jpn. J. Appl. Phys. **23** (1984) L150.
10) K. Matsumoto, M. Ogawa, T. Wada, N. Hashizume, T. Yao and Y. Hayashi: Electron. Lett. **20** (1984) 462.
11) P. M. Solomon, C. M. Knoedler and L. Wright: IEEE **EDL-5** (1984) 379.
12) A. Shibatomi, J. Saitou, M. Abe, T. Mimura, K. Nishiuchi and M. Kobayashi: IEDM Tech. Dig. (1984) p. 340.
13) M. O. Watanabe, K. Morizuka, M. Mashita, Y. Ashizawa and Y. Zohta: Jpn. J. Appl. Phys. **23** (1984) L103.

A High Threshold Voltage Uniformity MIS-Like Heterostructure FET Using n$^+$-Ge as a Gate Electrode

KUNIHIRO ARAI, TAKASHI MIZUTANI, MEMBER, IEEE, AND FUMIHIKO YANAGAWA, MEMBER, IEEE

Abstract—A new n$^+$-Ge/ undoped-Al$_x$Ga$_{1-x}$As/ undoped-GaAs MIS-like heterostructure FET (n$^+$-Ge-HFET), using n$^+$-Ge layer as a gate electrode material, is shown to have a high threshold voltage uniformity (σV_{TH} = 11 mV) over a large sample area of a 2-in wafer quadrant. This is thought to come from the FET structure, for which the threshold voltage is principally determined by the difference in the electron affinities of Ge and GaAs. The high V_{TH} uniformity, as well as the positive FET characteristics (g_m = 170 mS/mm, V_{TH} = 0.25 V), makes n$^+$-Ge-HFET very attractive for LSI application.

I. INTRODUCTION

IN THE PAST few years, FET's utilizing two-dimensional electron gas at Al$_x$Ga$_{1-x}$As/undoped-GaAs interface [1] have been intensively studied, because of their great potential in the field of high-speed and low-power devices. A remarkably high switching speed of 8.5 ps is attained [2] by one such FET using an Al$_x$Ga$_{1-x}$As layer doped with n-dopant (HEMT). [3]. However, stringent control of both the Al$_x$Ga$_{1-x}$As thickness and the impurity concentration is indispensable in order to ensure V_{TH} uniformity and reproducibility. This is because the threshold voltage for HEMT depends strongly on these MBE film parameters [4].

Within the past year, MIS-like heterostructure FET's, which are characterized by the use of an undoped Al$_x$Ga$_{1-x}$As layer as opposed to the n$^+$-doped Al$_x$Ga$_{1-x}$As layer of HEMT's, have become the object of intensive research [5]–[8]. The advantages of these FET's are that high V_{TH} uniformity over a large sample area and V_{TH} reproducibility can be obtained. This is because the threshold voltage is principally determined by a gate electrode material and independent of the Al$_x$Ga$_{1-x}$As layer thickness.

Several materials have been proposed for the gate electrode of MIS-like heterostructure FET: Ti/Pt/Au [5], n$^+$-GaAs [6], [7], and WSi$_x$ [8]. However, the threshold voltages of these FET's are too high (~ 0.8 V) [8] or too low (0.0 V) [6], [7] for fabricating LSI. Recently, the authors proposed a new n$^+$-Ge gate MIS-like heterostructure FET (n$^+$-Ge-HFET) [9], using n$^+$-Ge as a gate electrode material. This FET is superior, owing to a threshold voltage of 0.25 V, making it well suited for LSI application.

There are a few reports discussing the V_{TH} uniformity for MIS-like heterostructure FET. Small V_{TH} standard deviation

of 0.013 V is reported [6] in a restricted sample area of 5 × 10 mm^2. V_{TH} variation from the 2-in wafer center to the edge is reported to be ±0.05 V, a figure considerably lower than for HEMT's (±0.15 ~ ±0.4 V) fabricated using the same MBE apparatus [8]. However, high V_{TH} uniformity over a large sample area, which is one of the most attractive features for MIS-like heterostructure FET, has not been proved yet.

This letter reports that the new n$^+$-Ge gate MIS-like heterostructure FET is proved to have a high uniformity (σV_{TH} = 11 mV) over a large sample area of a 2-in wafer quadrant. This result, as well as good FET characteristics (g_m = 170 mS/mm, V_{TH} = 0.25 V), makes n$^+$-Ge-HFET quite utilizable for LSI application.

II. STRUCTURE OF N$^+$-GE-HFET

A schematic cross section of n$^+$-Ge-HFET is shown in Fig. 1(a). This FET is fabricated using a MBE grown structure composed of n$^+$-Ge(200 nm)/undoped GaAs(5 nm)/undoped Al$_x$Ga$_{1-x}$As(30 nm, x = 0.5)/undoped GaAs(300 nm). Energy-band diagrams for n$^+$-Ge HFET are shown in Fig. 1(b). When the gate voltage is 0 V, the conduction band edge of the GaAs layer is situated at $\delta\chi = \chi_{Ge} - \chi_{GaAs}$ above fermi energy, where χ_{Ge} and χ_{GaAs} are the electron affinities for Ge and GaAs, respectively. By applying positive gate voltage V_G larger than the threshold voltage V_{TH}, one can induce electron gas in the potential well at the Al$_x$Ga$_{1-x}$As/GaAs interface. The threshold voltage V_{TH} is approximately equal to $\delta\chi$ and independent of the Al$_x$Ga$_{1-x}$As layer thickness. There are several reports on $\delta\chi$: 0.05 eV [10] and 0.16 eV [11]. Thus n$^+$-Ge-HFET is expected to exhibit a threshold voltage of about 0.1 V in a uniform and reproducible manner.

III. FABRICATION

The MBE films were grown on a 2-in (100)-oriented semi-insulating GaAs substrate. During the growth period, the substrate was rotated for uniformity improvement. The dopant of the n$^+$-Ge was As and the carrier concentration measured by the Hall method was 5 × 10^{19} cm^{-3}.

The fabrication process of n$^+$-Ge-HFET is quite simple. First, a 200-nm WSi$_x$ layer is deposited by the sputter method. Then, WSi$_x$ and Ge layers are etched by RIE (CF$_4$) to form a gate. Next, source and drain n$^+$ regions, self-aligned to the gate, are formed by the Si ion implantation (100 kV and 1 × 10^{14} cm^{-2}) and lamp annealing method. Finally, source and drain ohmic contacts are formed by evaporating AuGe(200

Manuscript received November 1, 1985; revised December 30, 1985.
The authors are with NTT Electrical Communications Laboratories, 3-1, Morinosato Wakamiya, Atsugi-shi, Kanagawa Pref., 243-01 Japan.
IEEE Log Number 8607677.

Reprinted from *IEEE Electron Device Letters*, vol. EDL-7, no. 3, pp. 158–160, March 1986.

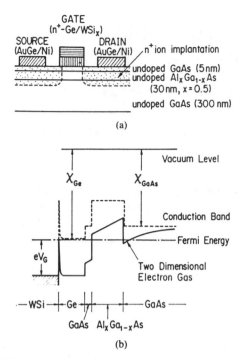

(a)

(b)

Fig. 1 (a) Schematic cross section for n⁺-Ge-HFET, and (b) n⁺-Ge-HFET energy-band diagrams.

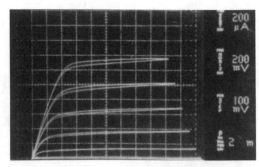

Fig. 2. Characteristics for n⁺-Ge-HFET at room temperature. $V_{G\,max}$ is 0.8 V with steps of 0.1 V.

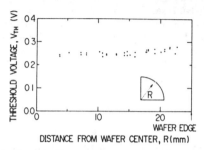

Fig. 3. Threshold voltage as a function of the distance R from the 2-in. wafer center. Sample shape and the meaning of R are shown schematically.

nm)/Ni(30 nm) and sintering. The endpoint detection of the RIE process is easily accomplished because of the large etching rate ratio of Ge to GaAs. The lamp annealing was performed at 700°C for 16 s. The FET's were fabricated on a sample of quadrant of a 2-in wafer. The gate length and width are 1.5 and 20 μm, respectively.

IV. RESULTS AND DISCUSSION

Fabricated FET's exhibit good characteristics at room temperature, as shown in Fig. 2. Typical transconductance is 170 mS/mm. Contact resistance, the sheet resistance of the n⁺ implanted layer, and source resistance are 0.13 Ω·mm, 650 Ω/□, and 1.4 Ω·mm, respectively. The sheet resistance of the gate is 9 Ω/□. Gate forward current is sufficiently small (20 μA at V_{GS} = 0.8 V, 200 μA at V_{GS} = 1.3 V). Gate-to-source reverse breakdown voltage is about 6 V. The threshold voltage is 0.25 V and larger than the expected value of 0.1 V. Interdiffusion at the Ge/GaAs interface may be responsible for this threshold voltage shift. Further investigation is necessary to clarify this hypothesis. At 77 K, transconductance of the FET is enhanced to 260 mS/mm, about 1.5 times larger than that at 300 K [9]. Threshold voltage shift between room temperature and 77 K is less than 0.03 V [9], much smaller than the reported value for HEMT (0.25 V) [12].

Threshold voltage uniformity over a quadrant of a 2-in wafer measured at room temperature is shown in Fig. 3. The standard deviation of the threshold voltage is 11 mV over the sample area. One of the reasons for such a high V_{TH} uniformity is thought to be the FET structure, which produces a V_{TH} insensitivity in regard to the thickness and the impurity concentration of the $Al_xGa_{1-x}As$ layer. The other reason is though to be the simplicity of the fabrication process itself, which is believed to suppress process-induced V_{TH} scattering.

V. SUMMARY

N⁺-Ge gate heterostructure FET (n⁺-Ge-HFET) is proved to have high V_{TH} uniformity (σV_{TH} = 11 mV) over a large sample area (quadrant of a 2-in wafer). This uniformity is thought to come from the following factors. First, the FET structure, which produces a V_{TH} insensitivity in regard to the thickness and the impurity concentration of the $Al_xGa_{1-x}As$ layer. Second, the simplicity of the fabrication process itself. The high V_{TH} uniformity, as well as the good FET characteristics (g_m = 170 mS/mm, V_{TH} = 0.25 V), makes n⁺-Ge gate MIS-like heterostructure FET very attractive for LSI application.

ACKNOWLEDGMENT

The authors wish to express their gratitude to K. Oe and N. Kondo for their valuable discussions and MBE film growth. In addition, grateful acknowledgment is made to S. Fujita for the FET pattern design, as well as his fruitful comments. The authors also thank Y. Yamane for his advice about Ge film growth, H. Yamazaki regarding ion implantation, H. Sugawara for advising WSi$_x$ deposition, and finally, K. Maezawa for his help with lamp annealing.

REFERENCES

[1] R. Dingle, H. L. Stormer, A. C. Gossard, and W. Wiegmann, "Electron mobilities in modulation-doped semiconductor heterojunction superlattices," *Appl. Phys. Lett.*, vol. 33, p. 665, 1978.
[2] N. C. Cirillo, Jr. and J. K. Abrokwah, "8.5-picosecond ring oscillator gate delay with self-aligned gate modulation-doped n⁺-(Al, Ga)As/GaAs FET," presented at the 43rd Device Res. Conf., Paper IIA-7.
[3] T. Mimura, K. Joshin, S. Hiyamizu, K. Hikosaka, and M. Abe, "High-electron mobility transistor logic," *Japan J. Appl. Phys.*, vol. 20, p. L598, 1981.
[4] J. K. Abrokwah, N. C. Cirillo, Jr., M. J. Helix, and M. Longerbone, "Modulation-doped FET threshold voltage uniformity of a high-

throughput 3-in. MBE system," *J. Vac. Sci. Technol.*, vol. B2, p. 252, 1984.

[5] Y. Katayama, M. Morioka, Y. Sawada, K. Ueyanagi, T. Mishima, Y. Ono, T. Usagawa, and Y. Shiraki, "A new two-dimensional electron gas field-effect transistor fabricated on undoped-AlGaAs-GaAs heterostructure," *Japan J. Appl. Phys.*, vol. 23, p. L150, 1984.

[6] K. Matsumoto, M. Ogura, T. Wada, N. Hashizume, T. Yao, and Y. Hayashi, "n$^+$-GaAs/undoped GaAlAs/undoped GaAs field-effect transistor," *Electron. Lett.*, vol. 20, p. 462, 1984.

[7] P. M. Solomon, C. M. Knoedler, and S. L. Wright, "A GaAs Gate heterojunction FET," *IEEE Electron Device Lett.*, vol. EDL-5, p. 379, 1984.

[8] K. Arai, T. Mizutani, and F. Yanagawa, "Threshold voltage behavior for WSi/Al$_x$Ga$_{1-x}$As/GaAs MIS-like heterostructure FET," *Japan J.*

Appl. Phys., vol. 24, p. L623, 1985.

[9] K. Arai, T. Mizutani, and F. Yanagawa, "An n$^+$-Ge Gate MIS-like Heterostructure FET," presented at the 12th Int. Symp. GaAs and Related Compounds, Karuizawa Japan, Sept. 23–26, 1985 (to be published in *Inst. Phys. Conf. Ser.*).

[10] J. M. Ballingall, C. E. C. Wood and L. F. Eastman, "Electrical measurements of the conduction band discontinuity of the abrupt Ge-GaAs⟨100⟩ heterojunction," *J. Vac. Sci. Technol.*, vol. B1, p. 675, 1983.

[11] R. S. Bauer, presented at the Conf. Phys. Chem. Semiconductor Interfaces, Monterey, CA, 1982.

[12] A. J. Valois, G. Y. Robinson, K. Lee, and M. S. Shur, "Temperature dependence of the I–V characteristics of modulation-doped FET's," *J. Vac. Sci. Technol.*, vol. B1, p. 190, 1983.

Threshold-Voltage Stability of *P*-Channel AlGaAs/GaAs MIS-like Heterostructure FETs

Makoto Hirano and Naoto Kondo

NTT Electrical Communications Laboratories, 3-1, Morinosato,
Wakamiya, Atsugi-shi, Kanagawa Pref., 243-01

(Received May 6, 1986; accepted for publication June 21, 1986)

Threshold-voltages V_{th} of *p*-channel AlGaAs/GaAs MIS-like heterostructure FETs (*p*-MIS HFETs) were found to be constant -0.5 (± 0.05) V at 77 K, independent of the AlGaAs layer thickness (8 to 30 nm), when the impurity concentration of the buffer layer was a smaller value than 5×10^{15} cm^{-3}. The standard deviation of V_{th} was measured to be 24 mV at 77 K with the 3 μm gate FET devices in a quartering 2-inch-diameter wafer. Turn-on voltages of *p*-channel diodes were measured with 1 MHz to be in good agreement with the threshold voltage of the FET devices.

§1. Introduction

A basic study of complementary logic circuits composed of *n*-channel and *p*-channel AlGaAs/GaAs heterostructure FETs (*n*-HFETs and *p*-HFETs)[1-8] has been recently started.[9] The circuits are promising in their high speed operation and low power dissipation, because *n*-HFETs and *p*-HFETs, both employing high-mobility two-dimensional-carrier-gas, can be expected to have high performance, particularly at low temperatures.

In order to implement integrated circuits, device simplicity is required. We proposed a *p*-channel MIS-like HFET (a *p*-MIS HFET),[5] analogous to an *n*-channel MIS-like HFET (an *n*-MIS HFET),[10,11] which has a very simple structure fabricated on an undoped AlGaAs/ GaAs heterostructure. The ring oscillations of the complementary circuits composed of *n*-MIS HFETs and *p*-MIS HFETs were observed.[9]

Stability of the threshold-voltage V_{th}, particularly the V_{th} uniformity over a wafer, is also essential to integrate the logic circuits. The *p*-MIS HFET devices, like the *n*-MIS HFETs,[11] are expected to be much more stable in threshold-voltage than the modulation-doped devices: Since the entire crystal layers are undoped, the V_{th} of MIS-like HFETs should be defined as the Schottky barrier height of the gate metal, a constant peculiar to the material. Nevertheless, the intrinsic stability of *p*-MIS HFET V_{th} has not been experimentally confirmed yet.

This paper deals with the threshold-voltage stability of *p*-MIS HFETs. The independence of the threshold voltage on the AlGaAs layer thickness is presented.

§2. Experiments

The sample crystals were grown by molecular beam epitaxy (MBE) on (100) oriented semi-insulating GaAs substrates. The epitaxial layers consisted of a 1 μm undoped (the unintentionally-doped impurity concentration was a smaller value than 5×10^{15} cm^{-3}) GaAs buffer layer, an Al$_{0.5}$Ga$_{0.5}$As layer and a 5 nm undoped GaAs cap layer. Samples with various AlGaAs layer thicknesses (8 to 30 nm) were grown with the intension of investigating the layer thickness dependence of the *p*-MIS HFET's V_{th}.

Wafers with a Si-doped buffer layer, and a various thickness (8 to 45 nm) of the Al$_{0.5}$Ga$_{0.5}$As layer, were also grown to observe the V_{th} change depending on the impurity concentration in the buffer layer, as will be mentioned in section 3. The doping concentration was about 5×10^{16} cm^{-3}. The only 50 nm thick buffer layer near the heterointerface was left undoped as the channel layer.

A schematic cross section of *p*-channel MIS-like heterostructure FETs (*p*-MIS HFETs) is shown in Fig. 1. They were fabricated with a self-aligned-gate technique. To form a Schottky barrier, a 0.2 μm WSi$_x$ layer was deposited by sputtering, and the layer was patterned by etching with SF$_6$ plasma. To form source and drain areas, 50 keV Be ions were implanted at a dose of 6×10^{13} cm^{-2} using the gate metal as a mask. The samples were then annealed using a halogen lamp system at 800°C. The peak temperature holding time of the system was 4 seconds. After that, to form an ohmic contact, AuZnNi/Ti/Au were deposited and alloyed at 520°C for 30 sec.

p-Channel MIS-like heterostructure diodes (*p*-MIS HDs) were also fabricated in a planar structure on the wafers with an undoped buffer layer and various AlGaAs layer thicknesses, on which the FETs were fabricated. The diodes were structured by a similar self-aligned technique under the same processing conditions as for the *p*-MIS HFET devices.

§3. Results and Discussion

A schematic view of the energy band of *p*-MIS HFETs is shown in Fig. 2. The voltage accumulating 2DHG is defined as the Schottky barrier height ϕ_{Mp}, a constant

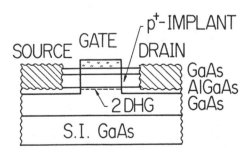

Fig. 1. Schematic cross section of the *p*-MIS HFET.

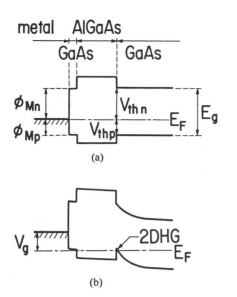

Fig. 2. Energy band diagram of the device. (a) with $V_g = 0$ V, (b) with V_g.

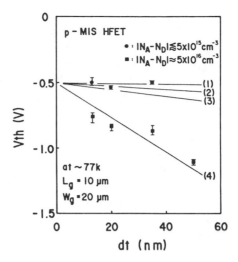

Fig. 3. Threshold-voltage V_{th} vs layer thickness d_t between gate and channel. Circle points: normal p-MIS HFETs (the unintentionally-doped impurity concentration in the buffer layer is a smaller value than 5×10^{15} cm^{-3}). Square points: the devices with the Si-doped buffer layer (the doping concentration is about 5×10^{16} cm^{-3}). Solid lines indicate the calculated values. (1) for $|N_A - N_D| = 1 \times 10^{15}$ cm^{-3}. (2) for $|N_A - N_D| = 5 \times 10^{15}$ cm^{-3}. (3) for $|N_A - N_D| = 1 \times 10^{16}$ cm^{-3}. (4) for $|N_A - N_D| = 5 \times 10^{16}$ cm^{-3}.

peculiar to the material: The AlGaAs layer beneath the Schottky metal is undoped and plays a role as an insulator. The band potential profile in the AlGaAs layer is approximately linear for the layer thickness, and the potential-drop in that layer is negligible if the layer is sufficiently thin, 10 to 50 nm, in comparison with buffer layer thickness (1 μm).

In order to characterize the threshold-voltage V_{th} of the p-MIS HFETs, the devices were fabricated on wafers with various layer thicknesses of AlGaAs. V_{th} was determined by extrapolating the $I^{1/2} - V$ plot to $I = 0$ with 10 μm gate devices. In order to eliminate the effect of the gate-source leakage current to determine V_{th} precisely, the measurement was performed at 77 K.

The determined V_{th} of the devices with various layer thicknesses are summarized in Fig. 3, shown as circle points. V_{th} of the samples were almost contstant, -0.5 (± 0.05) V, a value independent of the AlGaAs layer thickness. This value of V_{th} is in good agreement with the usual metal-Shottky-barrier-height ϕ_{Mp} for p-GaAs,[12] which is almost equivalent to the difference of GaAs energy band gap (1.4 V) and the metal-Schottky-barrier-height ϕ_{Mn} (around 0.9 V) for n-GaAs.[12] Since V_{th} did not change with the layer thickness, it can be said that V_{th} of p-MIS HFET devices was defined only as ϕ_{Mp}, as a material constant.

If the buffer layer is doped to be n-type with the intension to suppress the short-channel-effect, V_{th} can be slightly changed, as in Si-MOS FET devices. This V_{th} change was also investigated with the wafers of a Si-doped buffer layer. The V_{th} change depending on the impurity concentration in the buffer layer is approximately described as follows, according to the V_{th} formula for Si-MOS FETs:[13]

$$V_{th} = \phi_{Mp} + (4\varepsilon_1 q(N_A - N_D)\phi_{Mp})^{1/2}/C_i.$$

Where

$$C_i = \varepsilon_1/d_1 + \varepsilon_2/d_2 \approx \varepsilon_2/d_t$$

Here, ϕ_{Mp} is metal-Schottky-barrier-height for p-GaAs

(at the valence band), q is electron charge, N_D is donor concentration in the buffer layer, N_A is acceptor concentration in the buffer layer, C_i is the capacitance between the gate and the channel, ε_1 is permittivity of GaAs (the cap layer), ε_2 is permittivity of AlGaAs, d_1 is the thickness of the cap layer, d_2 is the thickness of the AlGaAs layer, and d_t is total thickness between the gate and the channel.

For normal p-MISFETs, since $|N_A - N_D|$ is a smaller value than 5×10^{15} cm^{-3}, the second term of the V_{th} formula should be negligible when compared to the first term.

For the devices with a Si-doped buffer layer, if $|N_A - N_D|$ reaches the level around 5×10^{16} cm^{-3} by the doping, V_{th} can slightly change from the ϕ_{Mp}, the intrinsic threshold voltage of p-MIS HFETs. According to the above fomula, such V_{th} changes will depend on the layer thickness d_t between gate and channel.

The measured V_{th} of the p-MIS HFET devices, which were fabricated on the wafers with a Si-doped buffer layer, are summarized in Fig. 3. The tendency of V_{th} change against the layer thickness d_t is almost in agreement with the calculation using the above-mentioned formula.

In order to confirm the voltage accumulating the hole carriers at the heterointerface, the heterostructure diodes (p-HDs) were fabricated and their C-V characteristics were measured at 77 K. Figure 4 shows the summarized C-V characteristics of p-MIS HDs fabricated on the same undoped wafers on which the p-MIS HFETs were fabricated. Since the characteristics were measured at 77 K, the influence of the leakage current was minimized in this case also. In addition, since the measuring frequency was 1 MHz, the influence of the surface level or the interface level would be neglected even if they exist. The turn-on voltage V_T of the p-MIS HDs were in good agreement with the threshold voltage V_{th} of the FETs fabricated on the same wafers. Thus, V_T of p-MIS HDs were also independent of the AlGaAs layer thickness. The measured capacitance C_i of the insulator layer are shown in Fig. 4. They were almost equal to the values estimated from the nominal values of the permittivity of AlGaAs and GaAs and their layer thickness.

The distribution of V_{th} in a quartering 2-inch-diameter wafer was measured at 77 K with 3 μm gate p-MIS HFETs. As shown in Fig. 5, the average of V_{th} is -0.45 V and its standard deviation is 24 mV. Since the V_{th} deviation of n-channel MIS-like devices was confirmed to be as small a value as 11 mV,[11] the V_{th} deviation of p-channel devices could be reduced to such a value by improving the process technology.

It is noted that the above-mentioned uniformity of p-MIS HFETs can be obtained without any particular growth control such as thickness uniformity control or doping uniformity control. Thus, p-MIS-like devices have potential usefulness for implementing integrated complementary logic circuits with their superior

reproductivity and uniformity of threshold voltage, as n-MIS-like devices.[11]

§4. Conclusion

P-channel AlGaAs/GaAs MIS-like heterostructure FETs (p-MIS HFETs) with various AlGaAs layer thicknesses (8 to 30 nm) were fabricated in order to confirm the threshold-voltage stability of these devices. The threshold voltages V_{th} of the devices were -0.5 (± 0.05) volts at 77 K, independent of layer thickness, when the impurity concentration in the buffer layer was a smaller value than 5×10^{15} cm^{-3}. P-channel MIS-like heterostructure diodes (p-MIS HDs) were also fabricated on the same wafers with various layer thicknesses and their C–V characteristics were measured with 1 MHz at 77 K. Their turn-on voltages V_T were in good agreement with V_{th} of the FETs.

Concerning threshold-voltage uniformity, the distribution of the V_{th} of 3 μm gate p-MIS HFETs was measured at 77 K in a quartering 2-inch-diameter wafer. The standard deviation of V_{th} was 24 mV.

The V_{th} change from its intrinsic value was also investigated with the devices of an n-type doped buffer layer. It was confirmed that the V_{th} change can be estimated in a similar manner to Si-MOS FET devices.

The stability of threshold voltage of p-MIS HFETs described in this paper indicates the potential usefulness of p-HFETs for integrated complementary logic circuits combining them with n-channel MIS-like devices.

Acknowledgements

The authors are indebted to K. Oe for his valuable advice on MBE growth, F. Yanagawa, K. Arai and T. Mizutani for their helpful discussions on MIS HFET's threshold-vloltage. They also thank T. Uchida of Shizuoka University for his assistance in FET characteristics measurements.

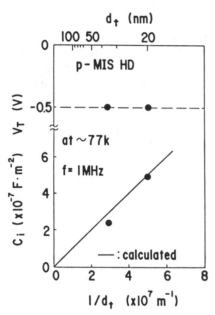

Fig. 4. C–V characteristics of p-MIS heterostructure diodes. Measurement frequency is 1 MHz. V_T: turn-on voltage. C_i: capacitance of the insulator (cap GaAs and AlGaAs) layer. d_t: thickness of the insulator layer (a sum of the cap GaAs layer thickness and the AlGaAs layer thickness).

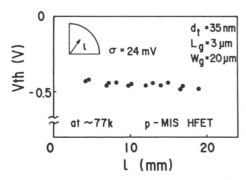

Fig. 5. Threshold-voltage deviation in a quartering 2-inch-diameter wafer. V_{th}: threshold-voltage. l: distance from the wafer center.

References

1) H. L. Störmer, K. Baldwin, A. C. Gossard and W. Wiegmann: Appl. Phys. Lett. **44** (1984) 1062.
2) S. Tiwari and W. I. Wang: IEEE Trans. Electron Device Lett. **EDL-5** (1984) 333.
3) R. A. Kiehl and A. C. Gossard: IEEE Trans. Electron Device Lett. **EDL-5** (1984) 521.
4) M. Hirano, K. Oe and F. Yanagawa: Jpn. J. Appl. Phys. **23** (1984) L868.
5) K. Oe, M. Hirano, K. Arai and F. Yanagawa: Jpn. J. Appl. Phys. **24** (1985) L335.
6) K. Oe, M. Hirano, F. Yanagawa and K. Tsubaki: to be published in Surf. Sci.
7) N. C. Cirillo. Jr., M. S. Shur, P. J. Vold, J. K. Abrokwah, R. R. Daniels and O. N. Tufte: IEEE, IEDM 85, 12-2 (1985) 317.
8) M. Hirano, K. Oe and F. Yanagawa: IEEE **ED-33** (1986) 620.
9) T. Mizutani, S. Fujita and F. Yanagawa: Electron. Lett. **21** (1985) 1116.
10) Y. Katayama, M. Morioka, Y. Sawada, K. Ueyanagi, T. Mishima, Y. Ono, T. Usagawa and Y. Shiraki: Jpn. J. Appl. Phys. **23** (1985) 1116.
11) K. Arai, T. Mizutani and F. Yanagawa: IEEE Trans. Electron Device Lett. **EDL-7** (1986) 158.
12) C. A. Mead and W. G. Spitzer: Phys. Rev. **A134** (1964) 713.
13) S. M. Sze: Physics of Semiconductor Devices (John Wiley and Sons, Inc., New York, 1969).

High Performance (AlAs/n-GaAs Superlattice)/GaAs 2DEGFETs with Stabilized Threshold Voltage

Toshio BABA, Takashi MIZUTANI, Masaki OGAWA and Keiichi OHATA[†]

Fundamental Research Laboratories, NEC Corporation
1-1, Miyazaki 4-chome, Miyamae-ku, Kawasaki, Kanagawa 213
[†]Microelectronics Research Laboratories, NEC Corporation
1-1, Miyazaki 4-chome, Miyamae-ku, Kawasaki, Kanagawa 213

(Received July 18, 1984; accepted for publication July 21, 1984)

Novel two dimensional electron gas FETs, which have a short period AlAs/n-GaAs superlattice for an electron supplying layer, were successfully fabricated for the first time using a molecular beam epitaxy and a conventional recessed gate process. The threshold voltage changes with changes in temperature (77–300 K) and resulting from light illumination are effectively suppressed ($\Delta V_T \approx 0.1$ V). This stability is ascribed to low deep electron trap concentration in the superlattice. The high intrinsic transconductances of 325 mS/mm (77 K) and 146 mS/mm (300 K) are obtained for 3 μm gate FETs.

Two dimensional electron gas field effect transistors (2DEGFETs) with n-type $Al_xGa_{1-x}As$ ($x \approx 0.3$) and undoped GaAs modulation doped heterojunction have exhibited high transconductances,[1] high speed switching[2] and low noise figures.[3] However, the threshold voltage instabilities of these 2DEGFETs, due to temperature change or light illumination, have been pointed out.[4,5] The threshold voltage changes toward the positive direction with the decrease in temperature and moves toward the negative direction according to a light illumination at a low temperature (less than 200 K). These shifts in threshold voltage are mainly caused by deep electron traps in n-$Al_xGa_{1-x}As$, which are generally called DX-centers.[6] At a low temperature, some electrons are captured by the deep traps, causing the decrease in 2DEG concentration and threshold voltage shift. Therefore, it is desired to eliminate DX-centers in n-$Al_xGa_{1-x}As$ for the threshold voltage stabilization. The authors postulated that DX-centers are related to the lattice distortion around the donor impurity due to the random mixture of Al and Ga in $Al_xGa_{1-x}As$. Based on this hypothesis, a novel short period AlAs/n-GaAs superlattice with Si doped only in the GaAs layer has been proposed instead of conventional alloyed n-$Al_xGa_{1-x}As$. The elimination of the deep traps, together with high donor activation coefficient and low thermal activation energy (< 10 meV), has been realized.[7]

This letter proposes novel 2DEGFETs with high quality AlAs/n-GaAs superlattice as an electron supplying layer, and describes the outstanding characteristics of novel 2DEGFETs with threshold voltages insensitive to the temperature change and light illumination.

The layer structures were grown on (100) oriented semi-insulating GaAs substrates by MBE. The GaAs and AlAs growth rate were 0.5 μm/h. A 0.5 μm undoped GaAs buffer layer, 400 Å multiple layers and 500 Å n$^+$-GaAs top layer were successively grown on a substrate at 520°C. The multiple layers consisted of one period undoped AlAs/GaAs for a spacer layer (40 Å), three period AlAs/n-GaAs superlattice (3 × 40 Å) and graded superlattice which had a constant n-GaAs layer thickness (23 Å) and gradually reduced AlAs layer thicknesses (12–3 Å) toward the surface. The doping concentration was about 2×10^{18}

cm^{-3} in the superlattices and 6×10^{18} cm^{-3} in the n$^+$-GaAs top layer. The electron mobility and sheet carrier concentration at 77 K were estimated by Hall measurement for a wafer which had no n$^+$-GaAs top layer. Electron mobility was 89000 cm^2/V·s and electron concentration was 8×10^{11} cm^{-2}. The 2DEG existence at the superlattice and undoped GaAs heterointerface was also confirmed by Shubnikov-de Haas measurements at 4.2 K.

Figure 1 illustrates a cross-sectional view of a novel 2DEGFET stucture fabricated by conventional recessed gate process with 4.5 μm source-gate spacing. Non-alloyed contacts with AuGe/Ni/Au were used for source and drain electrodes, because the alloying condition has not been optimized yet. For this reason, the graded superlattice was formed to reduce the energy band discontinuity between the superlattice and n$^+$-GaAs top layer. The linear relation between drain current and voltage at a low electric field indicated that the present non-alloyed contacts were ohmic.

Figure 2 shows typical drain current-voltage characteristics for normally-on type novel 2DEGFETs with 3 μm gate length and 200 μm gate width at 300 K in the dark (a), 77 K in the dark (b), and under the flashlight illumination (c).

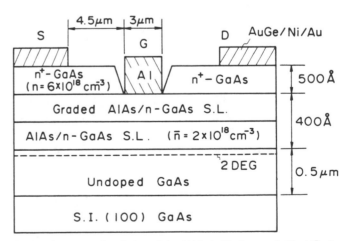

Fig. 1. A cross-sectional view of the (AlAs/n-GaAs superlattice)/GaAs 2DEGFET. One period undoped AlAs/GaAs is inserted between the superlattice and undoped GaAs buffer layer for a spacer layer.

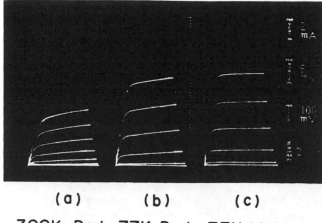

(a)　　(b)　　(c)

300K, Dark 77K, Dark 77K, Light

(Lg = 3 μm , W = 200 μm)

Fig. 2. Drain current-voltage characteristics of the novel 2DEGFET with 3 μm gate length and 200 μm gate width. Scales are 2 mA/div., 0.5 V/div. and −0.1 V/step for the drain current, drain voltage and gate voltage, respectively. The gate voltage of the uppermost trace is 0 V.

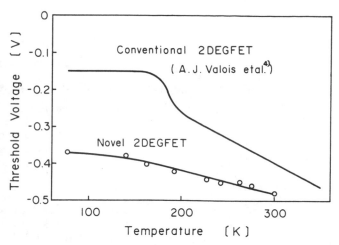

Fig. 3. Threshold voltage dependence on temperature for novel 2DEGFET and the conventional unit.

Transconductance enhancement at 77 K is clearly observed. This is due to the enhancement of low field mobility and high field saturation velocity. Threshold voltage differences among (a), (b) and (c) are very small.

The device parameters for 3 μm gate FET are listed in Table I. Contact resistance R_c and source-to-gate spacing resistance R_{sg} were estimated by gateless structures with different ohmic electrode spacings. Maximum transconductance g_{mmax} was obtained from the drain current-voltage chracteristics. Using these three parameters, intrinsic transconductance g_{m0} was calculated. High intrinsic transconductances, 325 and 146 mS/mm, were obtained at 77 K and 300 K, respectively. The authors also fabricated 1 μm and 0.5 μm gate FETs. However, due to the high contact resistances ($\sim 2 \, \Omega \cdot$ mm) for non-alloyed contacts, marked characteristics were not obtained.

Figure 3 shows the temperature dependence of the threshold voltage between 77 and 300 K. Threshold voltages were defined from the linear relationship between the square root of saturated drain current and gate voltage. Data for conventional 2DEGFETs, which were reported by A. J. Valois et al,[4] are also plotted in the figure. In conventional 2DEGFETs, a large threshold voltage shift was observed. The threshold voltage rapidly moved in a positive direction from 300 K to about 160 K, and the voltage shift was about 0.3 V. This shift resulted from the decrease in 2DEG concentration, which was caused by the

electron capture in deep traps in the $Al_xGa_{1-x}As$ layer at a low temperature. On the contrary, the novel 2DEGFET threshold voltage changed more gently, and an overall change as small as 0.1 V was achieved. In addition, the shift resulting from light illumination at 77 K was nearly suppressed (less than 0.03 V), as shown in Fig. 2(b) and (c). In conventional 2DEGFETs, the threshold voltage changed by 0.4 V toward a negative direction under light illumination, due to electron emission from the deep traps in $Al_xGa_{1-x}As$.

Inverted structure 2DEGFETs using the superlattice were also fabricated. Almost the same superior results were obtained. These results lead to the conclusion that low deep electron trap concentration in the AlAs/n-GaAs superlattice made the threshold voltage stable withstanding on temperature change and light illumination.

A small threshold voltage shift (0.1 V) was still observed in the present 2DEGFETs. This shift is considered to be due to the few deep electron traps, which still remained in the superlattice. Optimizing the sperlattice structure and MBE growth condition, further reduction in deep trap concentration and smaller threshold voltage shift will be realized.

Beyond the deep electron trap elimination effect discussed above, the superlattice has high doping capability. The reported highest electron concentration value was $1-2 \times 10^{18}$ cm^{-3} for Si doped $Al_xGa_{1-x}As$.[8] On the other hand, very high carrier concentration, 6×10^{18} cm^{-3}, was obtained for the AlAs/n-GaAs superlattice in the preliminary experiments. Taking advantage of the higher doping capability in the present superlattice, it is possible to reduce the superlattice thickness for 2DEGFETs, which is the dominant factor for determining high transconductance. Consequently, extremely high transconductance, greater than 700 mS/mm, would be realized for submicrometer gate devices by reducing the contact resistance to less than 0.2 $\Omega \cdot$ mm.

This paper has proposed novel two dimensional electron gas field effect transistors using the AlAs/n-GaAs superlattice. Very small threshold voltage shift, 0.1 V, was realized versus temperature change (77–300 K) and under light illumination for normally-on type 3 μm gate

Table I. Device parameters for (AlAs/n-GaAs superlattice)/GaAs 2DEGFET with 3 μm gate length and 200 μm gate width.

Temperature T [K]	Resistance		Transconductance	
	Contact R_c [$\Omega \cdot$ mm]	Source-to-gate R_{sg} [$\Omega \cdot$ mm]	Maximum g_{mmax} [mS/mm]	Intrinsic g_{m0} [mS/mm]
77	2.18	0.24	182	325
300	1.90	0.64	106	146

2DEGFETs. This stability is ascribed to low deep electron trap concentration in the superlattice. High transconductance (g_{m0} = 325 mS/mm [77 K], 146 mS/mm [300 K]) was also obtained. Since the novel 2DEGFETs have the capability to realized high transconductance, owing to the higher doping capability of the superlattice, these devices are very appropriate for use in high speed and low noise discrete devices and integrated circuits.

The authors wish to acknowledge the support of M. Miyamoto and H. Hida in fabricating the devices. They also thank N. Kawamura and T. Ito for useful discussions.

References

1) T. Mimura, K. Nishiuchi, M. Abe, A. Shibatomi and M. Kobayashi: *Int Electron Devices Meeting, 1983*, Tech. Digest (IEEE, New York, 1983) p. 99.

2) C. P. Lee, D. L. Miller, D. Hou and R. J. Anderson: *Device Research Conf., Vermont, 1983*, IIA-7.

3) K. Ohata, H. Hida, M. Miyamoto, M. Ogawa, T. Baba and T. Mizutani: *1984* IEEE Int. MTT-S Microwave Symp. Digest, Tech. Paper, p. 434.

4) A. J. Valois, G. Y. Robinson, K. Lee and M. S. Shur: J. Vac. Sci. & Technol. **B1** (1983) 190.

5) R. Fischer, T. J. Drummond, W. Kopp and H. Morkoç: Electron. Lett. **19** (1983) 789.

6) D. V. Lang, R. A. Logan and M. Jaros: Phys. Rev. **B19** (1979) 1015.

7) T. Baba, T. Mizutani and M. Ogawa: Jpn. J. Appl. Phys. **22** (1983) L627.

8) R. Fischer, T. J. Drummond, R. E. Thorne, W. G. Lyons and H. Morkoç: Thin Solid Films **99** (1983) 391.

Radiation effects on modulation-doped GaAs-Al$_x$Ga$_{1-x}$As heterostructures

D. C. Tsui

Department of Electrical Engineering and Computer Science, Princeton University, Princeton, New Jersey 08544

A. C. Gossard and G. J. Dolan

Bell Laboratories, Murray Hill, New Jersey 07974

(Received 23 August 1982; accepted for publication 27 October 1982)

The effects of 35-keV electron beam and ^{60}Co gamma radiation on modulation-doped GaAs-Al$_x$Ga$_{1-x}$As heterostructures were studied by measuring the transport and the quantum transport of, and the field effect on, the two-dimensional (2D) electrons in GaAs at the heterojunction interface. While the γ radiation in doses up to 1.3×10^6 rad causes no appreciable changes in the 2D transport properties, the electron beam irradiation reduces the electron mobility. This reduction in electron mobility is $\sim 50\%$ for electron doses of 10^{-10} C/μm^2.

PACS numbers: 61.80.Fe, 61.80.Ed

There has been in recent years a great deal of interest in modulation-doped GaAs-Al$_x$Ga$_{1-x}$As heterostructures[1-6] with which a quasi-two-dimensional electron gas (2DEG), with extremely high mobility, can be made to exist in GaAs at the GaAs-Al$_x$Ga$_{1-x}$As interface. Such structures are grown by molecular beam epitaxy (MBE) using the "modulation doping" technique[1] which selectively places the donor impurities inside the Al$_x$Ga$_{1-x}$As. The 2DEG results from ionization of the donors inside the Al$_x$Ga$_{1-x}$As and is con-

fined by a potential well in GaAs at the GaAs-Al$_x$Ga$_{1-x}$As heterojunction interface. This achievement of spatially separating the electrons from their ionized donor impurities, together with that of high quality lattice-matched single-crystal interface, has made it possible to attain extremely high electron mobilities ($\mu \gtrsim 10$ m^2/Vs) at densities corresponding to several times 10^{23}/m^3 in bulk GaAs. These structures, in addition to providing a unique system for studying new physical phenomena arising from 2D elec-

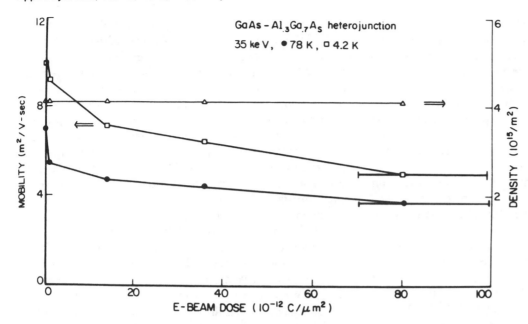

FIG. 1. Effect of 35-keV electron beam irradiation on the mobility and density of the two-dimensional electron gas at the interface of a modulation-doped GaAs-Al$_x$Ga$_{1-x}$As heterostructure.

trons,[7] promise devices for very high-speed applications. In particular, field-effect transistors (FET's) fabricated with such structures, using Al$_x$Ga$_{1-x}$As as the gate insulator, have already been demonstrated to operate at speeds previously achieved only with Josephson junctions.[8]

In this letter we wish to report an investigation of the effect of 35-keV electron and ^{60}Co gamma radiation on these new semiconductor heterostructures. We measured the dc transport and the quantum transport in high magnetic fields of the 2DEG using standard Hall bridges and FET's as a function of the radiation dosage. While the γ radiation in doses up to 1.3×10^6 rad produces no appreciable change in the measured properties of the devices, the electron beam causes a reduction in electron mobility, which is $\sim 50\%$ for electron doses of 10^{-10} C/μm^2. At the dose level usually used in electron beam lithography ($\lesssim 1 \times 10^{-12}$ C/μm^2), the structure suffered $< 10\%$ reduction in mobility. These results, when compared with those known for silicon metal-oxide-semiconductor (MOS) structures,[9] confirm the high quality of the GaAs-Al$_x$Ga$_{1-x}$As heterojunction interface and suggest its suitability for use in devices operating under radiation environment. They also provide for the first time the information needed to assess the employment of electron beam and x-ray technologies in the processing of devices from these structures.

The GaAs-Al$_x$Ga$_{1-x}$As heterostructure consists of a 1-μm-thick undoped GaAs (p-type, with $p \sim 10^{20}$/m^3) and a 800-Å-thick Al$_x$Ga$_{1-x}$As ($x = 0.3$), selectively doped with Si donors. In addition, a 200-Å cap layer of doped GaAs was grown on top of the structure to facilitate ohmic contact through the Al$_x$Ga$_{1-x}$As to the 2DEG at the GaAs-AlGa$_{1-x}$As interface in the undoped GaAs. The Si dopant in the Al$_x$Ga$_{1-x}$As approximately 1.5×10^{24}/m^3, is placed at 100 Å away from the interface in the Hall bridge sample and 50 Å in the FET sample. The structure was grown by MBE on Cr-doped semi-insulating GaAs substrates. The FET was fabricated using alloyed In as source-drain contacts and Al as the gate electrode.

Figure 1 shows the effect of 35-keV electron irradiation on the heterostructure, as observed in the dc transport of the 2DEG. The density and the mobility are obtained from standard Hall effect measurements. However, quantum transport measurements were also carried out before and after each irradiation, and the density obtained from the Shubnikov-de Haas (SdH) effect[2] is the same as that from the low field Hall measurements. It is clear from Fig. 1 that while the electron mobility shows appreciable degradation, the density shows no change with irradiation. This constancy of the density was also evident in the quantum transport data, where the period of the SdH oscillations was not changed by the irradiation. The amplitude of the SdH oscillations also showed decreases with successive irradiation after the second irradiation, consistent with the decrease in mobility seen

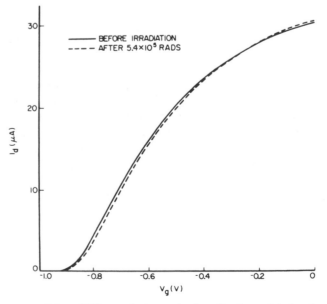

FIG. 2. Effect of ^{60}Co γ radiation on the I_d vs V_g of a modulation-doped GaAs-Al$_x$Ga$_{1-x}$As heterojunction FET.

in Fig. 1. However, an increase in amplitude was observed after the first irradiation, although a marked decrease was seen in the Hall mobility. Moreover, the sample shows a well defined quantitized Hall resistance plateau of 12.9 Ω (i.e., the Hall resistance $\rho_{xy} = h/2e^2$) centered at ~ 85 kG.[10] No appreciable change in the plateau width was observed in this experiment, contrary to previous experiments on silicon metal-oxide-semiconductor field-effect transistors (MOS-FET's) where an increase in plateau width with decreasing mobility[11] was observed.

Irradiation of the heterostructure with ^{60}Co γ radiation produced no measurable change in its 2D transport properties. However, a small but noticeable effect has been observed in the FET's fabricated on the GaAs-Al$_x$Ga$_{1-x}$As heterostructure, using Al$_x$Ga$_{1-x}$As as the gate insulator.[12] Figure 2 shows the drain current (I_d) versus gate voltage (V_g) characteristics of such a device. Its conduction threshold voltage (V_t) was -0.93 V before irradiation, -0.92 V after the first irradiation with 10^4 rad, and -0.91 V after the second irradiation with 2×10^4 rad. Both V_t and I_d vs V_g characteristics remain unchanged with still larger radiation doses up to 1.3×10^6 rad. The initial shift in V_t corresponds to a change in charge at the interface of $\lesssim 2 \times 10^{14}/\text{m}^2$, much smaller than that seen in Si-MOS structures.[9] It is clear from this radiation hardness of the device that the GaAs-Al$_x$Ga$_{1-x}$As heterostructures should be advantageous for high-speed applications in a radiation environment.

The dominant effect of electron and γ radiation on semiconductor devices arises from their creation of electron-hole pairs. In the case of Si-MOS, hole trapping and the generation of interface states are the main causes for conduction threshold shift and mobility deterioration in the devices. Although the detailed mechanisms for these degradation phenomena are still unclear, their origin is in the amorphous nature of the Si-SiO$_2$ interface. In GaAs-Al$_x$Ga$_{1-x}$As heterostructures, the interface is formed by epitaxially grown lattice-matched single crystals. It lacks the defect centers, such as dangling bonds, to act as trapping centers. Since this high perfection of the interface is already evident in its high electron mobility and in the lack of evidence for surface roughness scattering, it is not surprising in retrospect to find these samples so resistant to ionizing radiation.

The decrease in electron mobility after exposure to electron beam irradiation is most noticeable after the first two exposures at low doses. The mobility reduction appears to saturate at $\sim 50\%$ of its value before irradiation for subsequent exposures at higher doses (Fig. 1). It should be pointed out that even after the highest dose used in this experiment, the electron mobility is still much higher than that obtainable in bulk GaAs. For example, after irradiation with a dose

of 8×10^{-12} C/μm^2, the mobility is ~ 5 m^2/Vs at 4.2 K and ~ 4 m^2/Vs at 78 K which is still an order of magnitude higher than that in bulk GaAs with comparable density of several times $10^{23}/\text{m}^3$. The fact that this mobility reduction is not accompanied by any change in density shows conclusively that it is not caused by increase of charged scattering centers close to the interface. On the other hand, the energy of the electron beam in this experiment is sufficiently high to create lattice displacement at the interface, for example, which can give rise to the observed mobility reduction. In any case, the origin of this mobility reduction and the possible existence of experimental procedures to anneal out this effect are currently under investigation.

In summary, we have investigated the effect of 35-keV electron and ^{60}Co γ radiation on modulation-doped GaAs-Al$_x$Ga$_{1-x}$As structures, which have a layer of high mobility two-dimensional electron gas at the GaAs-Al$_x$Ga$_{1-x}$As interface. We found that while the γ radiation causes no appreciable change in the transport properties of the electron gas, the electron beam irradiation causes some mobility reduction without any change in electron density. The reduced electron mobility after irradiation is still an order of magnitude higher than that in bulk GaAs. We conclude by pointing out that, unlike Si-MOS, the modulation-doped GaAs-Al$_x$Ga$_{1-x}$As heterostructures are highly resistant to radiation damage and thus compatible to electron beam lithograph and x-ray lithography techniques for device fabrication and should be advantageous for high-speed device applications in radiation environment.

We thank Dr. A. Kamgar and Dr. H. L. Störmer for discussions. Work at Princeton University was supported by the Office of Naval Research.

[1] R. Dingle, H. L. Störmer, A. C. Gossard, and W. Wiegmann, Appl. Phys. Lett. **33**, 65 (1978).

[2] D. C. Tsui and R. A. Logan, Appl. Phys. Lett. **35**, 99 (1979).

[3] S. Hiyamizu, T. Mimura, T. Fujii, and K. Nanbu, Appl. Phys. Lett. **37**, 805 (1980).

[4] L. C. Witkowski, T. J. Drummond, C. M. Stanchak, and H. Morkoc, Appl. Phys. Lett. **37**, 1033 (1980).

[5] D. Delagebeaudeuf, P. Delescluse, P. Etienne, M. Laviron, J. Chaplart, and N. T. Linh, Electron. Lett. **16**, 667 (1980).

[6] S. Jadapawira, W. I. Wang, P. C. Chao, C. E. Wood, D. W. Woodard, and L. F. Eastman, IEEE Electron. Dev. Lett. **2**, 14 (1981).

[7] See, for example, D. C. Tsui, H. L. Störmer, and A. C. Gossard, Phys. Rev. Lett. **48**, 1559 (1982).

[8] See, for example, T. Mimura, Surf. Sci. **113**, 561 (1982).

[9] E. H. Snow, A. S. Grove, and D. J. Fitzgerald, Proc. IEEE **55**, 1168 (1967).

[10] D. C. Tsui and A. C. Gossard, Appl. Phys. Lett. **37**, 550 (1981).

[11] H. L. Stormer, D. C. Tsui, and A. C. Gossard, Surf. Sci. **113**, 32 (1982).

[12] D. C. Tsui, A. C. Gossard, G. Kaminsky, and W. Wiegmann, Appl. Phys. Lett. **39**, 712 (1981).

The Potential of Complementary Heterostructure FET IC's

RICHARD A. KIEHL, MEMBER, IEEE, MARY A. SCONTRAS, DAVID J. WIDIGER, MEMBER, IEEE, AND
W. MICHAEL KWAPIEN, MEMBER, IEEE

Abstract—The performance capability of AlGaAs/GaAs complementary heterostructure FET (C-HFET) integrated circuits has been evaluated in computer simulations. The study is focused on C-HFET designs in which static currents and gate-leakage currents are sufficiently low to take full advantage of the speed, power dissipation, and logic function capabilities of CMOS-like circuitry. ASTAP computer simulations for loaded NAND and NOR circuits are examined over a wide supply-voltage range at both 300 and 77 K in order to determine the potential of various MODFET, MISFET, and SISFET approaches as well as the prospects of future designs.

While performance is limited by FET threshold and gate leakage in present C-HFET approaches, the speed of properly designed 0.7-μm C-HFETs at 300 K is projected to be 3× faster than comparable 300 K Si-CMOS circuits. C-HFET circuits at 77 K are projected to be more than 4× faster than 77 K Si-CMOS circuits. It is also found that properly designed C-HFET's could operate at speeds close to those of DCFL n-channel HFET circuits while dissipating only 1/10 of the power.

I. INTRODUCTION

NMOS once dominated the world of high-performance silicon circuits, with CMOS playing a minor supporting role in low-power applications. In the 1980's, however, CMOS has emerged as the prevailing technology for not only low-power applications, but many high-performance applications as well. Thus, the performance limit of CMOS scaled to submicrometer gate lengths [1] and operated at low temperatures [2], [3] has become an important issue, and this raises questions as to the potential of complementary circuits based on materials other than silicon.

NMOS-like circuits, such as direct-coupled FET logic (DCFL) MESFET's, presently dominate in the development of high-performance GaAs integrated circuits. N-channel circuits have been more heavily favored in GaAs than in silicon due to the greater disparity in electron and hole transport properties in GaAs. Whereas the electron mobility in GaAs is nearly 10 times higher than the electron mobility in the channel of a silicon MOSFET, the hole mobility in p-channel GaAs FET's is only about twice the value for Si MOSFET's. Nonetheless, since complementary circuit performance is related to some av-

erage of the electron and hole transport properties, substantial performance advantages may still be possible in GaAs or other III-V complementary circuits.

Complementary GaAs circuits were first demonstrated with n- and p-channel JFET's [4]. The low power dissipation and radiation tolerance obtained by this approach make it attractive for static RAM's in certain applications where performance is not critical. However, the low transconductance of the p-channel JFET makes this approach less attractive for more general high-speed applications. The same limitation applies to complementary circuits based on GaAs MESFET's, where the p-channel transconductance is limited by the bulk hole mobility, as in the case of the JFET.

The achievement of enhanced transconductance in p-channel AlGaAs/GaAs MODFET's [5], [6], together with the demonstration of high-speed switching in p-MODFET ring oscillator circuits [7], has focused attention on various complementary heterostructure FET (C-HFET) approaches. The first demonstrated C-HFET structure was based on the integration of n-MESFET's and p-MODFET's [8]. Later, C-HFET's based on n-SISFET/p-MISFET [9], n-MISFET/p-MISFET [10], and n-SISFET/p-SISFET [11] combinations were also reported. A common feature of the various C-HFET approaches is that they make use of the enhanced mobility for holes confined in an undoped region at the interface of an AlGaAs/GaAs heterostructure to achieve improved performance, particularly at low temperatures.

The reported C-HFET approaches illustrate the flexibility in structure design that can be attained by exploiting the wide range of possibilities in heterostructure bandgap engineering and device architecture. Computer simulations for complementary circuits based on MODFET's [12] indicate that high speed is possible with such C-HFET's at 77 K. However, both the experimental approaches and computer analyses reported to date are limited in their applicability to room-temperature operation and in the extent to which they meet the design constraints of a truly competitive complementary circuit approach. In this paper, we discuss these design constraints, examine the suitability of the reported C-HFET structures on the basis of these constraints, and evaluate the potential performance of C-HFET circuits that incorporate device structures meeting these constraints.

Manuscript received March 18, 1987; revised July 27, 1987.
R. A. Kiehl is with the IBM Thomas J. Watson Research Center, Yorktown Heights, NY 10598.
M. A. Scontras, D. J. Widiger, and W. M. Kwapien are with the IBM General Technology Division, East Fishkill Facility, Hopewell Junction, NY 12533.
IEEE Log Number 8717017.

Reprinted from *IEEE Transactions on Electron Devices*, vol. ED-34, no. 12, pp. 2412–2421, December 1987.

II. Design Constraints of a Competitive C-HFET Technology

Properly designed complementary circuits offer a number of advantages over n-channel circuits and these advantages have been highly exploited in the design of Si CMOS circuits. Hence, if a C-HFET approach is to be competitive with other circuit technologies, it should not merely provide complementary switching action, but should be designed to offer a range of advantages similar to those of Si CMOS.

Low power dissipation is a basic advantage of complementary circuits. In an ideal complementary circuit, where the static power is zero, power is dissipated only during the dynamic switching process. The dynamic power P is given by

$$P = \tfrac{1}{2} CV^2(p/N\tau)$$

where C, V, p, N, and τ are the load capacitance, supply voltage, switching probability, number of delays per clock cycle, and delay per stage, respectively. The dynamic power dissipation is low as a result of the small switching frequency p/N encountered in typical circuit designs. The switching frequency is small for two reasons. First, the clock frequency f_c, which is equal to $1/(N\tau)$, is limited by the number of stages per logic chain, as well as time reserved for on-chip/off-chip communication. Typically, N is 10 to 20. Second, the probability p that a gate switches when clocked is considerably less than unity since few combinations of input states dictate a change in the output state. A value for p of about 25 percent is typical. As a result of these factors, the average gate switches only once in 40 to 80 propagation delays even when the circuit as a whole is operated at maximum speed, resulting in a dynamic power that is typically two orders of magnitude below the static power dissipation of n-channel circuits under similar operating conditions.

In order to realize low total power dissipation, the static power must also be small. One source of static power is subthreshold current. The existence of subthreshold current means that the n- and p-channel FET thresholds V_{tN} and V_{tP} must be offset from the low logic level by some amount. Since larger offsets result in reduced speed (as will be discussed later), a minimum offset is desired. In Si CMOS circuits, where the supply voltage is in the 3–5 V range, this offset is typically 20 percent of the supply voltage. In C-HFET circuits a somewhat larger percentage offset may be necessary since the supply voltages of interest are lower than those for Si CMOS.

A second source of static power in C-HFET circuits is gate-leakage current. While the high oxide barrier insures low gate leakage in Si MOSFET's, gate leakage is a potential in problem C-HFET circuits due to the relatively small barrier heights (less than 0.3 V for electrons in AlGaAs/GaAs heterojunctions) achievable in heterostructure designs. The Schottky-barrier gates of typical MODFET's present an additional problem, since such gates become highly conductive at a forward voltage of only about 0.8 V. Gate leakage must be small enough to allow the gate to swing to the full supply voltage without producing significant static power dissipation. Leakage must also be small enough to satisfy stability conditions in such circuits as static RAM's. As a rule, gate leakage should be maintained at a level at least three orders of magnitude below the maximum drain current.

High noise immunity and high process tolerance are other advantages of complementary circuits. Noise margins are high due to the high gate voltage swing (from ground to the supply voltage) and the push–pull action of the n- and p-FET's, which serve to provide a large logic swing and sharp transfer characteristic even in gates designed for maximum speed. However, the flow of gate current acts to reduce the logic swing and softens the transfer characteristics and, hence, should be avoided from this point of view (as well as from the view of static power, as mentioned earlier). A high noise margin means that high process tolerance can be obtained in complementary circuits. However, a high process tolerance also requires that the FET threshold offsets are sufficient to accommodate threshold variations related to materials or processes. Thus, the required threshold offset is larger than that given by subthreshold current considerations alone.

Enhanced logic function is an additional advantage of complementary circuits. Logic function, which relates to the ability to achieve high performance with simple and efficient circuit designs, is high in complementary circuits for several reasons. First, the push–pull switching action means that circuit performance is much less sensitive to the width ratio between load and driver FET's than in the case of n-channel FET logic. This provides increased flexibility in complementary circuit design. Second, the existence of two types of switches with complementary switching voltages (n- and p-channel FET's) allows the development of improved circuit designs. An example of this is the lossless pass transistor, which is realized by wiring n- and p-channel FET's in parallel and allows pass transistors to be connected in series, greatly simplifying many circuit designs. A third form of enhanced logic function relates to the simplicity with which dynamic circuits can be realized in complementary designs. The ability to make use of pre-charging techniques in memory design and to incorporate dynamic logic blocks in a circuit design is highly useful for improving overall circuit performance. These considerations place additional constraints on FET threshold and gate leakage in C-HFET circuits. The most severe constraint is that related to dynamic circuits, where FET thresholds and gate leakage must be such that static currents are several orders of magnitude below drive currents for proper circuit operation.

Thus, design constraints on static currents and, hence, on FET threshold and gate leakage must be met in order to realize a competitive C-HFET technology, i.e., one that offers the full potential of such a circuit approach.

III. Various C-HFET Approaches

In this section we briefly describe the various C-HFET structures demonstrated to date and examine these approaches in light of the design constraints outlined above.

The n-MESFET/p-MODFET approach [8] is based on a three-layer n-GaAs/p-AlGaAs/i-GaAs heterostructure in which the n-GaAs layer is selectively removed in regions where p-MODFET's are fabricated. n-MESFET's are fabricated on the entire heterostructure with the p-AlGaAs layer serving as an electron confinement layer. This form of device architecture has the advantage of a one-step growth process and a simple fabrication procedure. The FET thresholds can be adjusted in the process to desired values. A drawback of this approach, however, is that gate leakage is high for gate voltages beyond the Schottky-barrier turn-on voltage. Although turn-on voltages as high as 1.5 V have been obtained in p-MODFET's under some conditions, the reproducibility of turn-on voltages above 0.8 V (the Schottky-barrier height) is in question. Furthermore, the turn-on voltage of the n-MESFET in this structure is limited to about 0.7 V. Cooled operation produces little increase in the turn-on voltage. Hence, under our design constraints this approach is restricted to operation at supply voltages less than about 0.7 V, both at 300 and 77 K.

The n-SISFET/p-SISFET [11] approach is based on regrowth of the n-GaAs/i-AlGaAs/i-GaAs n-SISFET layers in recesses etched through p-GaAs/i-AlGaAs/i-GaAs p-SISFET layers formed in an earlier growth step. While the need for two separate growth steps is undesirable, this structure has the advantage of lower gate leakage at 77 K than MESFET or MODFET structures. Based on experimental data on n-SISFET gate leakage [13], operation to supply voltages as high as 1.5 V should be possible at 77 K with acceptable gate leakage. Unfortunately, gate leakage is high at room temperature and this limits the supply for 300 K operation to about 0.6 V. While the fixed threshold of the SISFET is advantageous for achieving high uniformity and control, the threshold value (0 V) results in high static currents and represents a drawback of this approach.

The n-SISFET/p-MISFET approach [9], [14] is based on a three-layer n-Ge/i-AlGaAs/i-GaAs heterostructure. The device architecture used in this approach is similar to that of the n-MESFET/p-MODFET approach, with the n-Ge layer selectively removed in regions, allowing both devices to be fabricated with a single growth. Like the n-SISFET/p-SISFET structure, this structure has low gate leakage at 77 K. In addition, the use of Ge rather than GaAs for the gate of the n-SISFET results in a V_{tN} of 0.15 V, while the use of metal for the p-channel gate produces a V_{tP} of -0.6 V. Thus, this approach has the further advantage of low static leakage current. The n-channel threshold is near to the desired value for low static current and high speed. However, a drawback of this approach is that the p-channel threshold is higher than desired. This results in lower p-channel current drive for a given supply

voltage and, hence, reduced speed (as will be discussed in detail later).

The n-MISFET/p-MISFET [10], [15] has the advantage of having the simplest structure. It is based on a two-layer i-AlGaAs/i-GaAs heterostructure that serves to define the channel in both the n- and p-HFET's, in accordance with the dopant of the contact implant. As in the previous two structures, the FET thresholds are fixed by the energy-band lineups between the metal and semiconductor layers. In the present structure this results in V_{tN} = 0.8 V and $V_{tP} = -0.6$ V. Although this insures low static power, such large thresholds again increase delay making it necessary to operate at higher supply voltages for high speed. The extent to which the supply voltage can be increased is limited, however, by leakage current. (It should be noted that gate leakage in MISFET's is not improved over SISFET's, since the drain current and gate current characteristics with respect to gate voltage in the MISFET are both shifted by the threshold voltage.) Thus, the high threshold voltages represent a drawback of this approach.

It is seen that modifications in the various C-HFET structures are needed if they are to meet the design constraints of the previous section. With regard to threshold voltage, the simplest modification is the introduction of doping in the layers to shift the thresholds to the desired values. Gate leakage in the SISFET and MISFET structures can be reduced to some extent by the choice of materials for the layers. In addition, improvement in the basic heterostructure design is possible. In what follows, we investigate the potential performance of C-HFET approaches which satisfy the constraints on threshold voltage and gate leakage.

IV. C-HFET Model

The C-HFET's modeled in the computer simulations are assumed to be ion-implanted self-aligned-gate n- and p-HFET's based on AlGaAs/GaAs heterostructures. The device models for the n- and p-HFET's are modifications of a model previously developed for the simulation of ion-implanted self-aligned-gate GaAs MESFET circuits. Under the restrictions discussed below and in later sections, the model is applicable to MODFET, MISFET, SISFET, and other HFET device structures.

The HFET model topology is shown in Fig. 1. The relation for the current source is the usual result obtained by applying the Shockley gradual channel approximation for the case of a conducting layer that is thin compared to the gate-to-charge spacing. This approach, which differs from that of typical MESFET models, is used to represent the channel formed by a confined electron (or hole) gas with no parallel AlGaAs conduction. A two-piece velocity-field relation, characterized by a low-field mobility and a high-field saturation velocity, is assumed.

We implement the gradual channel approximation as outlined by Grebene *et al.* [16] without considering channel-length shrinkage effects, giving the result

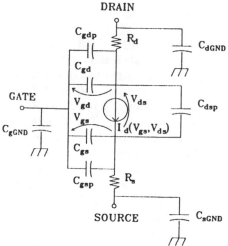

Fig. 1. Topology of HFET model.

TABLE I
HFET MODEL PARAMETERS

HFET	T	μ	v_{sat}	R_c	R_{sh}	g_{me}
	°C	$\dfrac{cm^2}{Vs}$	$\dfrac{cm}{s}$	Ω	$\Omega/\square$	$\dfrac{mS}{mm}$
n	300	8000	1.2×10^7	0.15	150	280
	77	20,000	2.1×10^7	0.15	150	435
p	300	400	2.4×10^6	0.30	300	55
	77	4000	7.6×10^6	0.30	300	165

$$I_d = K\left\{(V_{gs} - V_t)^2 \\ - \max\left[(V_{gs} - V_t - V_{ds})^2, (I_d/G_{m0})^2\right]\right\} \quad (1)$$

where I_d is the current source value, V_t is the threshold potential, V_{gs} and V_{ds} are the gate-to-source and drain-to-source potentials, respectively, K and G_{m0} are the low-current square-law factor and the high-current velocity-saturation transconductance, and max is a function that returns the larger of its two arguments. In implementing the max function, a smoothing algorithm is used near transitions, thereby providing device characteristics free of sharp knees. The network analysis program ASTAP [17] is used in the simulation, allowing (1) to be implemented directly into the model.

The quantities K and G_{m0} for the charge control model have the relations

$$K = \frac{\mu \epsilon w_g}{2hl_g} \quad (2)$$

and

$$G_{m0} = \frac{v_{sat} \epsilon w_g}{h} \quad (3)$$

where μ is the low-field mobility, v_{sat} is the effective saturation velocity, ϵ is the dielectric constant, w_g is the gate width, h is the gate-to-channel distance, and l_g is the gate length. For large μ, the saturated drain current is dominated by velocity saturation, with the result

$$I_d = V_{gs}G_{m0} \quad (4)$$

which is the same result obtained by Delagebeaudeuf *et al.* [18]. For small μ, as in the case of the p-channel device, the effects of operation below velocity saturation are more important.

The parameters used in the velocity-field characteristics are shown in Table I. The mobilities μ for electrons and holes are taken from reported Hall mobilities for typical n- and p-MODFET's. The saturation velocities v_{sat} are

effective values representing channel velocities at typical channel fields. In the case of holes, where little published data is available, v_{sat} has been estimated from the velocity-field dependence of holes in bulk GaAs at 300 K and the mobility of holes in p-MODFET's at 77 K. Although this effective hole saturation velocity is somewhat in doubt, the p-HFET characteristics should not be critically sensitive to the saturation velocity due to the role played by pinchoff in the FET current saturation.

The gate-to-charge distance is taken to be fixed at 475 Å for both the n- and p-HFET's. The 475-Å gate-to-charge distance is chosen to provide a reasonable tradeoff between high transconductance (thin layers) and low gate leakage (thick layers), although no quantitative optimization of this tradeoff has been carried out. Taking into account the fact that the carrier gas is separated from the heterointerface by about 75 Å, the gate-to-charge distance corresponds to a MODFET or MISFET with an AlGaAs layer thickness of approximately 400 Å. In the case of a SISFET, where there is additional charge separation due to depletion in the semiconductor gate, this corresponds to a device with an AlGaAs thickness of about 350 Å. These AlGaAs layer thicknesses are typical of reported MISFET and SISFET structures and somewhat larger than those in typical MODFET structures.

In this study we are concerned only with performance under the constraint of negligible gate leakage. Hence, gate current is assumed to be zero in the n- and p-HFET models. Because the gate voltage swings from zero to the supply voltage, the assumption of negligible gate current is valid only for supply voltages below a certain limit, which depends on the particular device structure as well as on the operating temperature. Thus, in interpreting the simulation results for a specific HFET approach in later sections, we will restrict our attention to a supply voltage range in which leakage is negligible and our model is valid.

The gate-to-contact spacing for the 0.7-μm gate-length FET's is taken to be 1.0 μm. The contact resistance values and sheet resistance values of the n^+ and p^+ regions given in Table I are consistent with what has already been obtained for n-HFET's and what is expected [19] to be achievable for p-HFET's in ion-implanted self-aligned-gate structures.

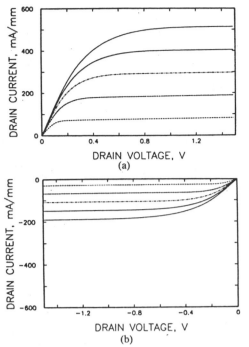

Fig. 2. Current–voltage characteristics for (a) n-HFET and (b) p-HFET at 77 K. Gate voltage range is 0 to 1.5 V in 0.25-V steps.

The 77 K current–voltage characteristics resulting from the device are shown in Fig. 2. The figure shows the external characteristics, i.e., those including the source resistances of 0.30 and 0.60 Ω · mm for the n- and the p-HFET, respectively. Values for the external transconductance G_{me} of the devices are shown in Table I for both 300 and 77 K.

The voltage-dependent capacitances C_{gs} and C_{gd} are determined by considering the effect of gate-to-source and gate-to-drain potentials on the channel charge. From the gradual channel approximation one obtains for the channel charge

$$Q = \frac{\mu \epsilon^2 w_g^2}{3 I_d} \left\{ (V_{gs} - V_t)^3 \right. $$
$$\left. - \max \left[(V_{gs} - V_t - V_{ds})^3, (I_d / G_{m0})^3 \right] \right\}. \quad (5)$$

The capacitances are then determined simply as

$$C_{gs} = \left. \frac{\partial Q}{\partial V_{gs}} \right|_{V_{gd}} \quad (6)$$

and

$$C_{gd} = \left. \frac{\partial Q}{\partial V_{gd}} \right|_{V_{gs}} \quad (7)$$

where $V_{gd} = V_{gs} - V_{ds}$ is the gate-to-drain potential. As in the case of (1), the max function in (5) eliminates discontinuities in capacitances during transitions from the linear region to the saturation region. Note that each capacitance is a function of both V_{gs} and V_{gd} and that, within our model, the capacitance vanishes at the current saturation point regardless of whether saturation is caused by

channel pinchoff or velocity saturation, as desired on physical grounds.

Values for parasitic capacitance elements in the model are estimated from experimentally determined values for ion-implanted self-aligned-gate MESFET's. The parasitic gate-to-source capacitance C_{gsp} and gate-to-drain capacitance C_{gdp} are both equal to 0.25 fF/μm (capacitance per unit gate width), representing values typical of sidewall-assisted gate processes. The parasitic drain-to-source capacitance C_{dsp} is 0.03 fF/μm. The parasitic gate-to-ground capacitance C_{gGND} is 0.05 fF/μm and the parasitic contact-to-ground capacitances C_{sGND} and C_{dGND} are both equal to 0.13 fF/μm. These values are estimates for the case of multilevel wiring.

Further details on the device model will be given elsewhere [20].

V. CIRCUIT SIMULATION ASSUMPTIONS

The ASTAP simulations have been carried out for inverter, NAND, and NOR circuits for various fan-in/fan-out (FI/FO) and capacitive loading conditions. An interconnect capacitance C_l equal to 20 fF is used in evaluating ring oscillator delays and a value of 100 fF is assumed in the case of loaded NAND and NOR circuits.

The case of a loaded inverter (C_l = 100 fF and FI = FO = 1) was used as the basis for choosing the basic gate widths used in the simulations. The gate widths for this case were chosen to give approximately equal pull-up and pull-down times and an input capacitance C_{in} equal to approximately $0.5 C_l$. While the condition of equal pull-up and pull-down times does not result in the minimum average delay, it is an important design constraint in most practical circuits. $C_{in} = 0.5 C_l$ represents a common loading condition near the point where increased gate width results in increased power-delay product.

The widths chosen for the loaded inverters were also used in the unloaded ring oscillator simulations and are shown in Table II. For loaded NAND and NOR circuits, the loaded inverter widths are maintained in parallel-connected FET's, while the widths are scaled with FI in series-connected FET's in order to maintain equal pull-up and pull-down times. Choosing the widths in this way provides a consistent design rule centered around the commonly used design point of $C_{in} = 0.5 C_l$. It should be noted, however, that, due to differences in C_{in} and C_l for the various cases, further optimization of power and delay is possible. This power-delay tradeoff is examined by scaling the widths of the n- and p-FET's for a fixed width ratio and load capacitance.

Simulation results are presented for two different cases of HFET threshold. While lower thresholds result in shorter delays, the FET thresholds must be offset from the low logic voltage as a result of subthreshold current and process tolerance considerations discussed previously. For most of the results presented in the following sections, the threshold voltage of the n- and p-HFET's is set equal to +0.2 and −0.2 V, respectively. Since the subthreshold

TABLE II
C-HFET CIRCUIT PARAMETERS

T	CKT	W_n	W_p	FI = FO	C_l
300	RO	4	27	1	20
	NAND	12	27	3	100
	NOR	4	81	3	100
77	RO	8	21	1	20
	NAND	24	21	3	100
	NOR	8	63	3	100

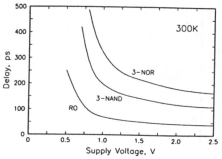

Fig. 3. Propagation delay as a function of supply voltage at 300 K for unloaded (C_l = 20 fF) ring oscillator and loaded (C_l = 100 fF) 3-NAND and 3-NOR circuits. $V_{tN} = -V_{tP} = 0.2$ V.

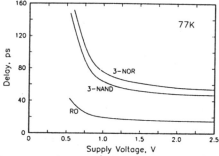

Fig. 4. Propagation delay as a function of supply voltage at 77 K for unloaded (C_l = 20 fF) ring oscillator and loaded (C_l = 100 fF) 3-NAND and 3-NOR circuits. $V_{tN} = -V_{tP} = 0.2$ V.

current in AlGaAs/GaAs HFET's drops at a rate of approximately 80 mV/decade at 300 K (20 mV/decade at 77 K), these thresholds would result in a static current level more than three orders of magnitude below the FET drive current level over a process window of 100 mV. Results are also presented for the case of $V_t = 0.2V_{dd}$. While the reduced offsets for this case at supply voltages less than 1 V are probably not sufficient for 300 K operation, they may be reasonable for 77 K operation with a tightly controlled fabrication process.

In general, the delay of a given stage depends on the characteristics of its driving circuit and output circuits. We calculate delays from ring oscillator simulations in which all gates (including those representing fan-in and fan-out) are identical. Under the common CMOS design rule that FI and FO are limited to three, this procedure gives the worst-case delay for 3-NAND and 3-NOR (FI = FO = 3) circuits.

VI. SIMULATION RESULTS

Fig. 3 shows calculated delays at 300 K as a function of supply voltage. Results are shown for an unloaded ring oscillator (RO) and loaded 3-NAND and 3-NOR circuits. It is seen that the delay is long at low supply voltages, but drops rapidly as the supply is increased, eventually reaching a low and relatively constant value. While the delay rolls off at a voltage of about 0.8 V for the ring oscillator, a supply of about 1.0 V is needed to reach rolloff in the case of the 3-NAND circuit. The rolloff point is highest and least sharp for the 3-NOR circuit, where the low transconductance of the p-HFET plays a more dominant role in determining switching speed. The calculated delays for the RO, 3-NAND, and 3-NOR circuits at a 1.0-V bias are 73, 209, and 326 ps, respectively. At a supply of 1.5 V, where the delays are more fully saturated, these values drop to 52, 143, and 220 ps, respectively.

Results for 77 K operation are shown in Fig. 4. It is seen that the supply voltage dependence of the delay is similar to the case at room temperature, but that the delay rolloff point occurs at a slightly lower supply voltage and saturates more fully at 77 K. The main difference between the two cases is the value of delays, which are about three times smaller at 77 K. At a 1.0-V bias the RO, 3-NAND,

and 3-NOR delays are 20, 64, and 77 ps, respectively. At a supply of 1.5 V, the delays drop to 17, 52, and 62 ps, respectively.

While the experimental C-HFET results reported to date are for quite different device and circuit parameters than those of this study, particularly with regard to gate length, FET transconductances, threshold voltages, and width ratios, it is of some value to compare the minimum ring oscillator delays observed experimentally with those of the simulations. At 300 K, delays of 94 and 50 ps have been obtained for 1.5- [14] and 1.0-μm [15] gates, respectively. At 77 K, a delay of 64 ps has been reported for 1.5-μm gates. Scaling these results to the 0.7-μm gate lengths used in the present simulations gives values in reasonably good agreement with the simulated results.

The supply voltage dependence of the delay is critical in determining the performance of a specific C-HFET approach, since the supply voltage in an actual circuit will be limited by gate leakage. Since the transconductance of the HFET's increases rapidly for gate voltages above the threshold voltage (0.2 V), it might be expected that the delays would drop at lower supply voltages than seen in Figs. 3 and 4—especially at 77 K, where the transconductances of the n- and p-HFET are both relatively high. The reason that this is not the case is that gate voltage swings far beyond threshold are needed to develop high current drive.

The increase in current drive with increased gate swing is illustrated in Fig. 5(a), where a plot of drain current versus gate voltage is given (for an arbitrary FET thresh-

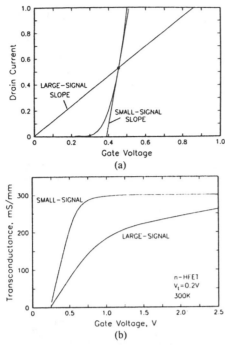

Fig. 5. (a) Drain current versus gate voltage characteristic for an arbitrary threshold voltage. The slopes of the two lines in the figure give the small-signal and large-signal transconductances. (b) Small- and large-signal transconductance in n-HFET.

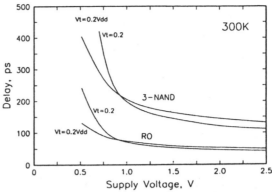

Fig. 6. Comparison of 300 K propagation delays for $V_{tN} = -V_{tP} = 0.2$ V and $V_{tN} = -V_{tP} = 0.2 V_{dd}$. ($C_l = 100$ fF.)

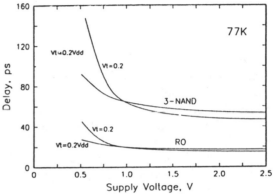

Fig. 7. Comparison of 77 K propagation delays for $V_{tN} = -V_{tP} = 0.2$ V and $V_{tN} = -V_{tP} = 0.2 V_{dd}$. ($C_l = 100$ fF.)

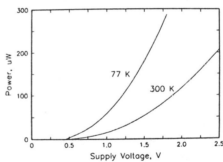

Fig. 8. Dynamic power dissipation versus supply voltage for the 3-NAND circuits with $C_l = 100$ fF. A switching factor p/N equal to 2.5 percent is assumed.

old). At gate voltages above threshold, the drain current increases rapidly and, hence, the small-signal transconductance G_{mS} is high. However, the switching speed is related to the current drive and the amount of charge that needs to be supplied to the loaded output node. Hence, the large-signal transconductance G_{mL}, defined as the ratio of drain current to gate voltage (see Fig. 5(a)), is a more relevant parameter. As shown in Fig. 5(b), G_{mL} increases much less rapidly with gate voltage than G_{mS}. The increase in G_{mS} at low temperatures produces an increase in G_{mL}. However, since the gate-voltage dependence of G_{mL} is tied to the FET threshold, little difference is seen in the delay rolloff point for 300 and 77 K.

The gate voltage dependence of G_{mL} depends strongly on the FET threshold, with higher thresholds producing a slower increase in G_{mL}. Thus, the optimum threshold is the minimum value satisfying process window and subthreshold current constraints. In the above, we have taken $V_{tn} = -V_{tp} = 0.2$ V. However, the effect of reduced FET threshold on the delay versus supply-voltage characteristic can be seen in Figs. 6 and 7, which compare delays for $V_{tN} = -V_{tP} = 0.2V_{dd}$ with those of our usual assumption $V_{tN} = -V_{tP} = 0.2$ V. It is seen that the delays in the case of $V_{tN} = -V_{tP} = 0.1$ V are about 1/2 to 2/3 those for $V_{tN} = -V_{tP} = 0.2$ at a supply voltage of 0.5 V.

The power dissipation of 3-NAND circuits at 300 and 77 K is shown in Fig. 8 for $C_l = 100$ fF. In calculating the dynamic power, it is assumed that the switching probability $p = 25$ percent and the number of delays per clock cycle $N = 10$, which are values appropriate for 3-NAND circuits clocked at a clock frequency f_c equal to $1/(10\tau)$. Consistent with the assumptions of $V_{tN} = -V_{tP} = 0.2$ V

and negligible gate leakage, the static power is taken to be zero.

The influence of gate-width and load capacitance on delay and power dissipation can be seen in Fig. 9. The curves in this figure are generated by scaling the p- and n-HFET gate widths while keeping the width ratio w_p/w_n fixed at 2.25 for 300 K and 0.88 for 77 K. Symbols in the figure indicate factor-of-two scaling, beginning with $w_n = 3$ μm at the left of the figure. The results for the nominal design widths used elsewhere in this paper (see Table II) are indicated by solid symbols. This figure allows the power-delay tradeoff and loading sensitivity of the C-HFET circuits to be examined. For example, it can be seen that doubling the gate widths from the nominal val-

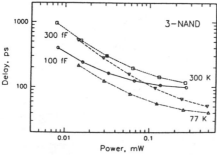

Fig. 9. Delay versus power dissipation for $C_l = 100$ fF and $C_l = 300$ fF at a supply voltage of 1.25 V. Symbols indicate factor-of-two scaling of gate widths. A switching factor p/N equal to 2.5 percent is assumed.

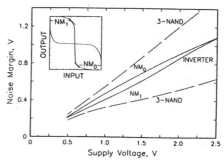

Fig. 10. Static noise margins NM_1 and NM_0 for FI = FO = 1 and FI = FO = 3 at 300 K. Results at 77 K are similar.

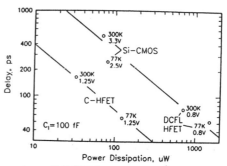

Fig. 11. Comparison of C-HFET performance with that of other circuit technologies. FI = FO = 3, $C_l = 100$ fF, and $p/N = 2.5$ percent. Straight lines show constant power delay.

ues results in a 47-percent increase in speed at 77 K for $C_l = 300$ fF. At 300 K and a load of $C_l = 100$ fF, the speed for this width doubling increases by only 31 percent due to the lower drive current and lower load capacitance in this case.

The noise margins of the 3-NAND circuits are shown as a function of supply voltage in Fig. 10 for 300 K. (The margins at 77 K are approximately the same.) The margins shown in the figure are the usual static noise margins, NM_0 and NM_1, defined from the static transfer characteristics for the 0 and 1 logic states, respectively. The margin in the case of the 3-NAND circuit is dependent on the number of inputs that are switched. In the event that all three inputs are switched together, the margin is split roughly equally between the two logic states, as shown by the solid curves in Fig. 10. The noise margin in this case is about the same as in the case of a simple inverter. In the event that only one input of the 3-NAND circuit is switched, the margin is split unequally with NM_0 favored over NM_1. The dashed curves in Fig. 10 show margins in the case of the bottommost transistor switching, which represents a worst case situation in which NM_1 is its lowest. As can be seen from the figure, these conditions give a range in noise margin from 0.32–0.55 V at supply of 1.0 V and from 0.41–0.88 at a supply of 1.5 V. These values of static noise margin are about 2 or 3 times higher than typical values in n-channel MESFET or HFET circuits. Comparison of the threshold voltage windows achievable with the various technologies involves consideration of cross-coupling noise and supply voltage variations, as well as a specification of the minimum acceptable delay, and is beyond the scope of the present analysis.

VII. COMPARISON OF C-HFET PERFORMANCE WITH THAT OF OTHER TECHNOLOGIES

The projected performance of C-HFET circuits is compared to results for silicon CMOS circuits and n-channel HFET circuits in Fig. 11. We examine the case of FI = FO = 3, a load capacitance C_l equal to 100 fF, and a switching factor p/N of 2.5 percent. The C-HFET power-delay points are those for the nominal gate widths given in Table II. The power-delay points for silicon CMOS are results from computer simulations [21] of optimized CMOS [2] at a gate length of 0.7 μm (0.5-μm channel

length) and a gate width of 10 μm. The n-HFET points in the figure are from simulations [12] of n-channel MODFET DCFL circuits scaled to 0.7-μm gate lengths with an E-mode FET width of 20 μm.

Fig. 11 shows that C-HFET circuits offer power-delay products substantially lower than either Si-CMOS or DCFL n-HFET's. In particular, the C-HFET circuits operate at 1/5 to 1/8 the power-delay of the Si-CMOS circuits. This is due, in part, to the lower supply voltage of the C-HFET circuits. It is seen that C-HFET's offers 3× faster operation with lower power dissipation at 300 K and more than 4× faster operation with comparable dissipation at 77 K.

The power-delay of the C-HFET circuits is also seen to be substantially lower than that of DCFL n-HFET circuits. The power-delay advantage of C-HFET's over n-HFET's is primarily the result of the lower power dissipation of complementary circuits operated at low switching factors. It is seen that C-HFET circuits at our nominal gate widths can operate at speeds close to those of n-HFET circuits at 77 K, while dissipating only about 1/10 the power. At 300 K, the speed of the C-HFET circuits is about half that of DCFL n-HFET's. In this case, it would be necessary to trade increased power for decreased delay in scaled C-HFET designs (see Fig. 9) in order to obtain speeds comparable to those of the n-HFET circuits.

VIII. CONCLUSIONS

Design constraints on FET threshold voltages and gate leakage must be met if C-HFET's are to reach their full potential as a high-performance circuit technology. These constraints arise from the need to maintain low power dissipation and high logic function at supply voltages high enough to provide short switching delays. Improvements in present C-HFET structures are needed to meet these constraints. The greatest limitation of present structures is excessively high leakage current at room temperature.

The projected performance of properly designed AlGaAs/GaAs C-HFET circuits shows significant power-delay advantage over both Si-CMOS and DCFL n-HFETs. Results from computer simulations indicate that C-HFET circuits could operate at speeds $3\times$ faster than comparable Si-CMOS circuits at 300 K and more than $4\times$ faster than such circuits at 77 K. The results also show that C-HFET circuits could operate at speeds close to those of DCFL n-HFET circuits while dissipating only $1/10$ the power.

C-HFET's are ideally suited for applications requiring high speed with a power dissipation of about 0.1 mW/circuit or less. For example, at an integration level of 30 000 circuits and a dissipation of 4.5 W/chip (0.15 mW/circuit), 0.7-μm C-HFET circuits operated at a supply voltage of 1.25 V potentially offer loaded delays as low as 140 ps at 300 K and 88 ps at 77 K (FI = FO = 3, C_l = 300 fF).

These promising results suggest that integrated circuits in GaAs and related compounds may eventually favor complementary designs for many high-performance applications.

ACKNOWLEDGMENT

The authors wish to thank C. J. Kircher, S. Tiwari, P. M. Solomon, C. J. Anderson, R. H. Dennard, T. I. Chappell, and M. G. Smith of the IBM Thomas J. Watson Research Center for useful discussions regarding this work.

REFERENCES

[1] R. H. Dennard and M. R. Wordeman, "MOSFET miniaturization—From one micron to the limits," *Physica*, vol. 129B, pp. 3–15, 1985.

[2] J. Y.-C. Sun, Y. Taur, R. H. Dennard, S. P. Klepner, and L. K. Wang, "0.5 μm-channel CMOS technology optimized for liquid-nitrogen-temperature operation," in *IEDM Tech. Dig.*, pp. 236–239, Dec. 1986.

[3] R. C. Jaeger and F. H. Gaensslen, "MOS devices and switching behavior," in *Low Temperature Electronics*, R. K. Kirschman, Ed. New York: IEEE Press, 1986.

[4] R. Zuleeg, J. K. Notthoff, and G. L. Troeger, "Double-implanted GaAs complementary JFET's," *IEEE Electron Device Lett.*, vol. EDL-5, pp. 21–23, Jan. 1984.

[5] H. L. Störmer, K. Baldwin, A. C. Gossard, and W. Wiegmann, "Modulation-doped field-effect transistor based on two-dimensional hole gas," *Appl. Phys. Lett.*, vol. 44, pp. 1062–1064, June 1984.

[6] S. Tiwari and W. I. Wang, "p-Channel MODFET's using AlGaAs/GaAs hole gas," *IEEE Electron Device Lett.*, vol. EDL-5, pp. 333–335, Aug. 1984.

[7] R. A. Kiehl and A. C. Gossard, "p-channel (Al,Ga)As/GaAs modulation-doped logic gates," *IEEE Electron Device Lett.*, vol. EDL-5, pp. 420–422, Oct. 1984.

[8] ——, "Complementary p-MODFET and n-HB MESFET (Al,Ga)As transistors," *IEEE Electron Device Lett.*, vol. EDL-5, pp. 512–523, Dec. 1984.

[9] T. Mizutani, S. Fujita, and F. Yanagawa, "Complementary circuit with AlGaAs/GaAs heterostructure MISFETs using high mobility two dimensional electron and hole gases," in *Proc. Int. Symp. GaAs and Related Compounds* (Karuizawa), *Int. Phys. Conf. Series No. 79*, pp. 733–734, Sep. 1985.

[10] N. C. Cirillo, Jr., M. S. Shur, P. J. Vold, J. K. Abrokwah, R. R. Daniels, and O. N. Tufte, "Complementary heterostructure insulated gate field effect transistors (HIGFETs)," in *IEDM Tech. Dig.*, pp. 317–320, Dec. 1985.

[11] K. Matusmoto, M. Ogura, T. Wada, T. Yao, Y. Hayashi, N. Hashizume, M. Kato, N. Fukuhara, H. Hirashima, and T. Miyashita, "Complementary GaAs SIS FET inverter using selective crystal regrowth technique by MBE," *IEEE Electron Device Lett.*, vol. EDL-7, pp. 182–184, Mar. 1986.

[12] S. Tiwari, "Performance of heterostructure FETs in LSI," *IEEE Trans. Electron Devices*, vol. ED-33, May 1986.

[13] P. M. Solomon, S. L. Wright, and C. Lanza, "Perpendicular transport across (Al,Ga)As and the Γ to L transition," *J. Superlattices Microdevices*, 1987.

[14] T. Mizutani, S. Fujita, M. Hirano, and N. Kondo, "Circuit performance of complementary heterostructure MISFET inverter using high mobility 2DEG and 2DHG," in *Dig. GaAs IC Symp.*, pp. 107–110, Oct. 1986.

[15] R. R. Daniels, R. Mactaggart, J. K. Abrokwah, O. N. Tufte, M. Shur, J. Baek, and P. Jenkins, "Complementary heterostructure insulated gate FET circuits for high-speed, low power VLSI," in *IEDM Tech. Dig.*, pp. 448–451, Dec. 1986.

[16] A. B. Grebene and S. K. Ghandhi, "General theory for pinched operation of the junction-gate FET," *Solid-State Electron.*, vol. 12, pp. 573–589, 1969.

[17] W. T. Weeks, A. J. Jimenez, G. W. Mahoney, D. Mehta, H. Qassemzadeh, T. R. Scott, "Algorithms for ASTAP—A network analysis program," *IEEE Trans. Circuit Theory*, vol. CT-20, pp. 628–634, Nov. 1973.

[18] D. Delagebeaudeuf and N. T. Linh, "Charge control of the heterojunction two-dimensional electron gas for MESFET application," *IEEE Trans. Electron Devices*, vol. ED-28, pp. 790–795, 1981.

[19] K. Maezawa and K. Oe, "Electrical characteristics of Be-implanted GaAs activated by rapid thermal annealing," *IEEE Electron Device Lett.*, vol. EDL-7, pp. 13–15, Jan. 1986.

[20] D. J. Widiger, J. D. Feder, W. M. Kwapien, M. A. Scontras, and C. J. Anderson, "An accurate circuit model for the GaAs MESFET," to be published.

[21] R. H. Dennard, private communication.

HIGH-SPEED LOW-POWER DCFL USING PLANAR TWO-DIMENSIONAL ELECTRON GAS FET TECHNOLOGY

Indexing terms: Logic and logic design, Field-effect transistors

Planar two-dimensional electron gas FETs (TEGFETs) have been shown to have ultra-high speed and low power in DCFL circuits operating at room temperature: 18·4 ps at 900 μW and 32·5 ps at 62 μW. The latter result and the simplicity of the process involved are compatible with VLSI requirements.

Introduction: The first GaAs integrated circuit was performed with BFL technology.[1] Since this work, various logic types such as SDFL, E-JFET logic, LPFL and DCFL have been investigated in view of the reduction of the power dissipation and the propagation delay time for LSI applications. Table 1 summarises the principal results obtained in GaAs FET technology.

Table 1 STATE-OF-THE-ART OF GaAs FET LOGIC CIRCUITS

Logic circuit	t_{pd}	Power	$t_{pd} X P$	Reference
	ps	mW	fJ	
BFL	34	41	1394	2
	80	3·1	248	3
	108	1·5	162	3
SDFL	156	0·17	26·5	4
	75	2·26	169·5	4
LPFL	95	2·5	235	5
DCFL	30	1·9	57	6
	78	0·05	4·1	6
	39·5	4·05	160	7

Recently, a SDFL 8 × 8-bit binary divider multiplier with 1008 gates per chip was fabricated.[8] It represents the highest circuit density ever made with GaAs FETs. For a higher density of integration, the power dissipation per gate must be below 100 μW and the corresponding propagation delay time less than 100 ps to beat silicon devices. Only DCFL circuits can satisfy these criteria. The GaAs FET DCFL technology has encountered many difficulties. The most severe is due to the fact that the Schottky height (0·8 eV) is close to the free surface potential (0·6 eV), which would lead to high access resistances if processes such as recessed gate or self-aligned implanted n^+ layers are not used. But the complexity of these processes would not be compatible with VLSI or would give a yield too low for fabrication of LSI circuits. Recently, we have shown that the two-dimensional electron gas FET (TEGFET) can be used in a planar structure in LPFL circuits with a delay time of 19·1 ps per gate and a power dissipation of 6·6 mW.[9] The simplicity of the process involved in this technology and its high performance open an interesting approach in high-speed logic circuits.

This letter reports high-speed and lower-power results obtained on planar TEGFET DCFL circuits.

TEGFET technology: Si-doped AlGaAs-undoped GaAs heterostructures have been grown by molecular beam epitaxy.[10] While high-mobility material (8500, 105 000, 187 000 cm^2 V^{-1} s^{-1} at 300, 77 and 20 K, respectively) has been grown, attention has mostly been focused on the growth of layers having the right composition and doping level necessary to control the threshold voltage V_T at about 0 to +0·2 V.

TEGFETs present a large Schottky barrier height (1 eV) and a low surface potential (0·33 eV),[11] so that access resistance in a planar TEGFET is quite low and planar DCFL circuits can be achieved with a good yield. The simplicity of the processing which does not need any recessed gate or self-aligned ion implantation is compatible with VLSI application.

Because of the large difference between the Schottky barrier height and the AlGaAs surface potential, ungated transistors can be used as saturable loads. This approach, which gives a low node capacitance, is also favourable for large-scale integration.

In the following Section, the dimensions of the transistors are: source-drain spacing 2 μm, gate length 0·7–0·8 μm and gate width 20 or 30 μm. The loads have a source-drain spacing of 2 μm and a width adapted to that of the driver.

Results: Two samples have been evaluated. Table 2 gives the characteristics of the layer.

11-stage ring oscillators have been fabricated on sample A.

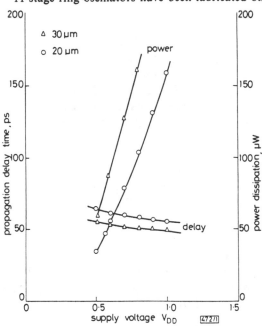

Fig. 1 *Dependence on supply voltage of power and delay of sample A circuits*

Two types of circuit having, respectively, 20 and 30 μm of driver width have been tried. Fig. 1 represents the dependence on the supply voltage V_{DD} of the propagation delay time and the power dissipation. The highest figure of merit is 64·7 ps, 35·8 μW, 2·3 fJ for the 20 μm circuit and 55 ps, 59·5 μW, 3·3 fJ for the 30 μm one. According to these results, circuits with 5 μm transistor width would give a figure of merit less than 1 fJ.

On sample B, 23-stage ring oscillators having 20 μm-wide

Table 2 CHARACTERISTICS OF STUDIED WAFERS

	Sample A	Sample B
Al concentration	0·23	0·3
AlGaAs electron conc. (cm^{-3})	10^{18}	$1·8 \times 10^{17}$
AlGaAs thickness (Å)	300	550
Spacer layer (Å)	140	75
μ (300 K) (cm^2 V^{-1} s^{-1})	6200	5900
μ (77 K) (cm^2 V^{-1} s^{-1})	32 000	39 000
n_s (300 K) (cm^{-2})	7×10^{11}	5×10^{11}
Threshold voltage (V)	+0·2	+0·1

Reprinted with permission from *Electronics Letters*, vol. 18, no. 12, pp. 517–519, June 1982.

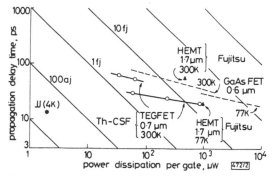

Fig. 2 *Comparison between DCFL circuits*

drivers have been achieved. From over 50 measured circuits, it was found that the propagation delay times are between 18·4 and 36 ps and 50% of them are between 22 and 27 ps. The best figures of merit correspond to 16·6 fJ (18·4 ps, 900 μW, 1·8 V), 5·6 fJ (23 ps, 243 μW, 1 V) and 2 fJ (32·5 ps, 62 μW, 0·65 V). These figures compare favourably with those reported in Table 1 concerning GaAs FET.

Some experiments have been done at 77 K. It was noted that the circuit performances are light-sensitive. This phenomenon is related to the persistent photoconductivity observed in the AlGaAs-GaAs heterojunction itself. In dark, circuits operate 30% faster at 77 K than at 300 K. This improvement does not correspond to that we can expect from the increase of electron velocity at low temperature. These circuits designed for room temperature operation are not well adapted to low temperature working condition.

Conclusion: The preceding results show that planar TEGFET technology using ungated loads is highly appropriate for producing very-low-power and high-speed DCFL integrated circuits satisfying the VLSI condition, i.e. power dissipation below 100 μW and gate delay below 100 ps. The simplicity of the involved process gives a high yield which was illustrated by the great number of circuits obtained though the technology is not optimised.

Comparison with other published results[12] presented in Fig. 2 shows the process performed by the TEGFET in the race towards high-speed and low-power integrated circuits. These experimental data do not present a limit since our calculations predict that 10 ps can be reached at 300 K with 1 μm gate length TEGFETs.[13]

Acknowledgments: This work is partly supported by the DRET. The valuable technical assistance of A. Rannou, J. Frentzel and M. Colombier are appreciated.

PHAM N. TUNG
P. DELESCLUSE
D. DELAGEBEAUDEUF
M. LAVIRON
J. CHAPLART
NUYEN T. LINH

29th April 1982

Thomson-CSF Central Research Laboratory
Domaine de Corbeville, 91401 Orsay, France

References

1 VAN TUYL, R., and LIECHTI, C.: 'High-speed integrated logic with GaAs MESFETs', *IEEE J. Solid-State Circuits*, 1974, **SC-9**, pp. 269–276
2 GREILING, P. T., LUNDGREEN, R. E., KRUMM, C. F., and LOHR, R. F.: 'Why design logic with GaAs and how?', *Microwave System News*, 1980, Jan., pp. 48–60
3 TUNG, PHAM N., GLOANEC, M., and NUZILLAT, G.: 'High density submicron gate GaAs MESFET technology'. 7th European Workshop on active microwave semicond. dev., Greece, Oct. 1981
4 EDEN, R. C., WELCH, B. M., ZUCCA, R., and LONG, S. I.: 'The prospect for ultra high speed VLSI GaAs digital logic', *IEEE J. Solid-State Circuits*, 1979, **SC-14**, pp. 299–317
5 NUZILLAT, G., BERT, G., NGU, T. P., and GLOANEC, M.: 'Quasi-normally-off MESFET logic for high performance GaAs ICs', *IEEE Trans.*, 1980, **ED-27**, pp. 1102–1109
6 MIZUTANI, T., KATO, N., ISHIDA, S., OSAFUME, K., and OHMORI, M.: 'GaAs gigabit logic using normally-off MESFETs', *Electron. Lett.*, 1980, **16**, pp. 315–316
7 YAMASAKI, K., ASAI, K., MISUTANI, T., and KURUMADA, K.: 'Self-align implantation for n^+-layer technology (SAINT) for high-speed GaAs ICs', *ibid.*, 1982, **18**, pp. 119–121
8 LEE, F. S., EDEN, R. C., LONG, S. I., WELCH, B. M., and ZUCCA, R.: 'High speed LSI GaAs integrated circuits'. Proc. IEEE Int. Conf. circuits and computers, 1980, pp. 697–700
9 TUNG, PHAM N., DELAGEBEAUDEUF, D., LAVIRON, M., DELESCLUSE, P., CHAPLART, J., and LINH, NUYEN T.: 'High speed two-dimensional electron gas FET logic', *Electron. Lett.*, 1982, **18**, pp. 109–110
10 DELESCLUSE, P., LAVIRON, M., CHAPLART, J., DELAGEBEAUDEUF, D., and LINH, NUYEN T.: 'Transport properties in GaAs-Al$_x$Ga$_{1-x}$As heterostructures and MESFET applications', *ibid.*, 1981, **17**, pp. 342–344
11 DELAGEBEAUDEUF, D., LAVIRON, M., DELESCLUSE, P., and LINH, NUYEN T.: 'Planar enhancement mode two-dimensional electron gas FET associated with a low AlGaAs surface potential', *ibid.*, 1982, **18**, pp. 103–105
12 HIYAMIZU, S., MIMURA, T., and ISHIKAWA, T.: 'MBE-grown GaAs/N-AlGaAs heterostructures and their application to high mobility transistors', *Jpn. J. Appl. Phys.*, 1982, **21**, Suppl. 21-1, pp. 161–168
13 DELAGEBEAUDEUF, D., and LINH, NUYEN T.: 'Speed power in planar two-dimensional electron gas FET DCFL circuit: a theoretical approach', see pp. 510–512

ULTRA HIGH SPEED DIGITAL INTEGRATED CIRCUITS USING GaAs/GaAlAs HIGH ELECTRON MOBILITY TRANSISTORS

C.P. Lee, D. Hou, S.J. Lee, D.L. Miller and R.J. Anderson

Rockwell International Science Center
Microelectronics Research and Development Center
Thousand Oaks, CA 91360

ABSTRACT

Ultra high speed integrated circuits using planar processed high electron mobility transistors have been fabricated. The lowest propagation delay measured from 25-stage ring oscillators at room temperature is 12.2 ps. Divide-by-four circuits with two T-connected D-flip flops are capable of operating at 3.48 GHz for 0.47 mW per gate at room temperature. The speed-power products achieved are the lowest ever reported.

INTRODUCTION

Modulation doped GaAs/GaAlAs high electron mobility transistors (HEMTs) have attracted much attention recently for their potential applications in ultra high speed integrated circuits. The high electron mobility and the MOSFET-like structure have resulted in HEMT devices with transconductances two to three times higher than those of conventional GaAs MESFETs (1,2). The speeds of HEMT ring oscillators demonstrated by several laboratories have already exceeded the highest speeds achieved by all other semiconductor technologies (3-6). In this paper the high speed performance of ring oscillators and frequency dividers fabricated in this laboratory is presented. At room temperature, a 12.2 ps propagation delay has been achieved. This is the best result ever reported for any digital circuit at any temperature.

DEVICE PROCESSING

The modulation doped structures used for device fabrication were grown by MBE on one inch square LEC semi-insulating GaAs substrates. The layers were, starting from the substrate, undoped GaAs (~ 1 μm), undoped $Ga_{0.7}Al_{0.3}As$ spacer (~90Å), N type $Ga_{0.7}Al_{0.3}As$ (~500Å, 1×10^{18} cm^{-3}) and undoped GaAs (~40Å). The low field electron mobility of these structures was typically around 5000 cm^2 V^{-1} s^{-1} at 300K and 100,000 cm^2 V^{-1} s^{-1} at 77K. The fabrication procedure for the devices consisted of: (1) ohmic contact formation, (2) proton implantation for device isolation, (3) delineation of Schottky gates and first-level interconnect, and (4) isolation dielectric deposition and second-level interconnect formation. The ohmic contact was AuGe/Ni, which typically had a contact resistance of 0.5 Ω-mm. The Schottky metal and the first-level interconnect were

Ti/Pt/Au, and the second level metal was Ti/Au. All the lithographic steps were done with a censor 10X wafer stepper. A cut away view of a processed inverter gate is shown in Fig. 1. Because of the use of proton implantation for device isolation, the process results in a planar structure which is desirable for high density, large circuit fabrication.

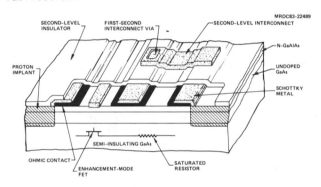

Fig. 1 Cutaway view of a HEMT logic gate.

Both enhancement mode (E-mode) and depletion mode (D-mode) HEMTs have been fabricated. For 1μm gate devices the transconductances were typically 170 ms/mm. The best value obrained was 210 ms/mm. Figure 2 shows the I-V characteristics of a typical 50 μm wide E-mode HEMT. The supplied gate voltage was from 0 to 0.6 V. Usually no gate leakage was observed when V_{gs} < +1 V. The high gate turn-on voltage is a characteristic of the Schottky gate on GaAlAs.

RING OSCILLATOR PERFORMANCE

DCFL (direct-coupled FET logic) ring oscillators have been fabricated for the evaluation of the switching speed. They have 25 inverter stages with fan-in and fan-out equal to one. Two types of loads have been used in the ring oscillator design: saturated resistors and depletion mode FETs. A photograph of a ring oscillator (with saturated resistor loads) is shown in Fig. 3. The switching transistors are 20μm wide with 0.9μm long gates and the saturated resistor loads are 7μm wide. In this design, the oscillation wave form is measured through an inverter stage and a 60μm wide buffer FET, which is connected as a source follower with a 50Ω load. Faster speeds

Reprinted from *Rec. of the IEEE GaAs Integrated Circuits Symp.*, pp. 162–165, 1983.

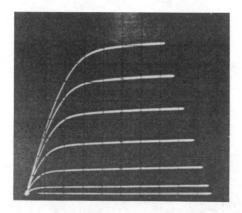

Fig. 2   I-V characteristics of a typical 50 μm
wide enhancement mode HEMT. V_{DS} = 0.2
V/DIV, I_{DS} = 0.5 mA/DIV, V_{gs} = 0.1
V/STEP.

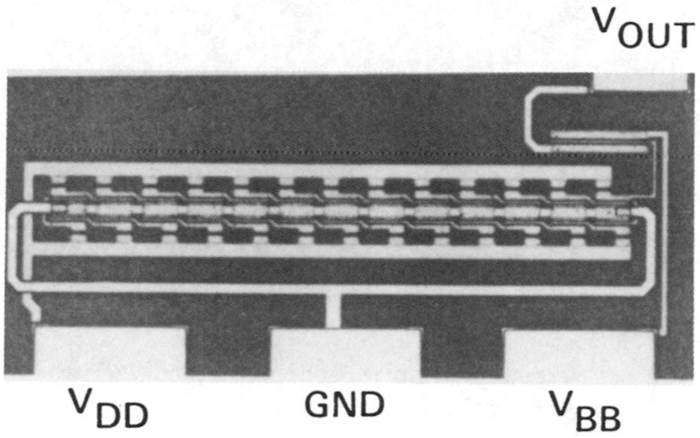

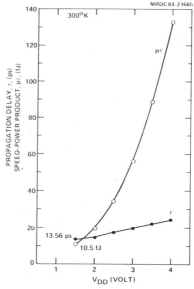

Fig. 4   Speed-power performance of a ring
oscillator.

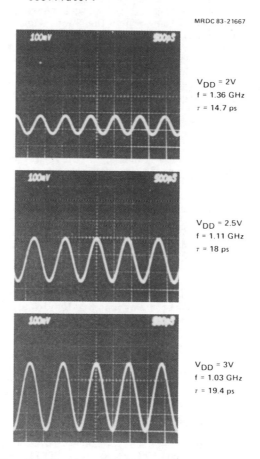

V_{OUT}

V_{DD} GND V_{BB}

Fig. 3   Photograph of a 25 stage ring oscillator.

were normally observed for ring oscillators with
saturated resistor loads than those with
depletion-mode FET loads. This is largely due to
the fact that the saturated resistors do not have
gate capacitance as the FETs do.

Very high ring oscillator speeds have been
measured. Figure 4 shows the propagation delay
and the speed-power product of a ring oscillator
measured at room temperature. At V_{DD} = 1.5 V the
propagation delay was only 13.6 ps. In the entire
measurement range, from V_{DD} = 1.5 V to 4 V, the
propagation delay was between 13.6 ps and 24.2 ps.
The oscillation waveforms measured at V_{DD} = 2 V,
2.5 V and 3 V are shown in Fig. 5. The frequencies were 1.36 GHz, 1.11 GHz and 1.03 GHz, corresponding to propagation delays of 14.7 ps, 18 ps
and 19.4 ps, respectively. The highest speed observed at room temperature was 12.2 ps at a speed-
power product of 13.7 fJ. This speed is the highest ever measured for any digital circuit at any
temperature, exceeding the highest values reported
for HEMT ring oscillators of 16.8 ps at 300K and
12.8 ps at 77K (6). It is comparable to the top
speed of Josephson junction devices measured at
4K.

V_{DD} = 2V
f = 1.36 GHz
τ = 14.7 ps

V_{DD} = 2.5V
f = 1.11 GHz
τ = 18 ps

V_{DD} = 3V
f = 1.03 GHz
τ = 19.4 ps

Fig. 5   Oscillation waveforms of a ring
oscillator measured at V_{DD} = 2 V, 2.5 V
and 3 V.

FREQUENCY DIVIDER PERFORMANCE

Ring oscillators are useful test vehicles for evaluation of the switching speed of a device technology under ideal loading conditions, with fan-in and fan-out equal to one. However, they can not provide sufficient information for more complex circuits in which a larger number of fan-ins and fan-outs are required. In this section we present the results of HEMT divide-by-four circuits, which use D-flip flops as building blocks and have gates with fan-ins and fan-outs equal to three.

The divide-by-four circuits use two T-connected D-flip flops as binary counters to perform the frequency division. Direct-coupled-FET-logic (DCFL) was used for circuit design. A basic NOR gate consists of several enhancement-mode switching transistors and a saturated resistor load. The circuit diagram and the photograph of a fabricated circuit are shown in Fig. 6. The output signal goes through two buffer inverter states and a 300 μm wide buffer FET, and is detected with the buffer FET connected as a source follower with a 50Ω load. Besides the buffer stages there are totally 12 gates to perform the logic functions. The total signal delay time in this circuit configuration is 5 τ_D, where τ_D is the propagation delay per gate. The average fan-in and fan-out are 2.2 and 2 respectively. There are other configurations of flip-flops, such as master-slave flip-flops, which theoretically can operate faster than the configuration used here. However, they usually require complementary clock inputs and more complex design. Therefore, the simpler D-type flip-flop was chosen for this divider circuit, with the understanding that overall circuit speed was being traded for design and test simplicity.

The sizes of the logic gates vary with the loading conditions. The average width of the switching transistors is 24 μm and the average size of the saturated resistors is 13 μm. The total chip area including the bonding pads is 400 μm × 400 μm.

The threshold voltage of the enhancement-mode switching transistors was controlled to be around + 0.1 V. For the first batch of wafers, the gate lengths of the transistors were made 1.4 μm. Very high yield was obtained for the divider circuits on these wafers. Most of the circuits were operational up to 2 GHz or higher. As an example at V_{DD} = 1 V, a typical device would operate with a maximum clock input frequency of 2.2 GHz with a total power consumption of 8.8 mW (excluding the power consumed by buffer stages). Most of the devices could be operated at very low V_{DD} without sacrificing the speed performance. This is probably due to the characteristics of HEMTs, which have high transconductance even at very low gate voltages. So the dividers were capable of operating at very high frequency yet with very small power consumption. The best result for the 1.4 μm gate devices achieved at room temperature was 2.6 GHz at V_{DD} = 0.7 V with a total chip power consumption of 6.7 mW. Subtracting the power con-

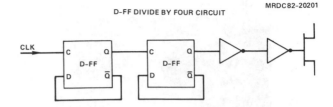

D-FF DIVIDE BY FOUR CIRCUIT　　　MRDC82-20201

MRDC 83-22759

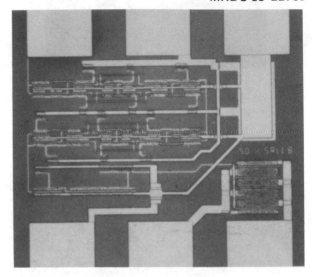

Fig. 6　Circuit diagram and photograph of a divide-by-four circuit.

sumed by the buffer stages the total power dissipation was 3.2 mW or 0.27 mW per gate. When the device was cooled down to - 70°C, it operated up to 3.54 GHz with a power consumption of only 0.21 mW per gate. At such frequency the average gate delay was 56.5 ps and the speed-power product was 12 fJ, which are the best divider results ever reported. When the gate lengths of the switching transistors were reduced to 1.1 μm, higher frequency operation was achieved. At room temperature the maximum input frequency obtained was 3.48 GHz for a power dissipation of 0.47 mw per gate. The corresponding propagation delay was 57 ps and the speed-power product was 27 fJ. The output waveform at this frequency is shown in Fig. 7.

The highest speed of GaAs DCFL frequency dividers reported previously corresponded to a propagation delay of 66 ps at a power dissipation of 1.2 mW per gate (7). These circuits had 0.6 μm gates and their power consumption was substantially higher than those of our HEMT dividers.

CONCLUSION

The high electron mobility and the structural advantage of HEMTs make them the most attractive candidates for many high speed integrated circuit applications such as next generation super computers and high speed real time signal processors. The performance of HEMTs demonstrated to date has already surpassed that demonstrated by all other

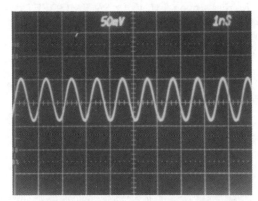

Fig. 7 Output waveform of a divide-by-four
        result at input frequency of 3.48 GHz.

technologies. Because of the similarity between
HEMTs and MESFETs most of the process techniques
used for conventional MESFETs can be used easily
for HEMTs. With the process technology and the
performance of HEMT circuits demonstrated by this
work, more complex ultra high speed HEMT integrated circuits should be achievable in the near
future.

ACKNOWLEDGMENT

The authors would like to thank F. Eisen, C.
Kirkpatrick and J. Donovan for their support concerning this work.

REFERENCES

1. T. Mimura, S. Hiyamizu, T. Fujii and K. Nanbu,
   "A New Field-Effect Transistor With Selec-
   tively Doped GaAs/n-Al$_x$Ga$_{1-x}$As Heterojunc-
   tions," Japan J. Appl. Phys. 19, pp. L225-227,
   1980.
2. S.L. Su, P. Fischer, T.J. Drummond, W.G.
   Lyons, R.E. Thorne, W. Kopp and H. Morkoc,
   "Modulation-Doped AlGaAs/GaAs FETs With High
   Transconductance and Electron Velocity,"
   Electron Lett. 18, pp. 794-796, 1982.
3. M. Abe, T. Mimura, N. Yokoyama and
   H. Ishikawa, "New Technology Towards GaAs
   LSI/VLSI for Computer Applications," IEEE
   Trans. Microwave Theory Tech. MTT-30, pp. 992-
   998, 1982.
4. P.N. Tung, P. Delescluse, D. Delagebeaudeuf,
   M. Laviron, J. Chaplart and N.T. Linh, "High-
   Speed Low-Power DCFL Using Planar Two-
   Dimensional Electron Gas FET Technology,"
   Electron Let. 18, pp. 517-519, 1982.
5. C.P. Lee and W.I. Wang, "High-Performance
   Modulation-Doped GaAs Integrated Circuits With
   Planar Structures," Electron Lett. 19, pp.
   155-157, 1983.
6. C.P. Lee, D.L. Miller, D. Hou and R.J.
   Anderson,"Ultra-High Speed Integrated Cirucits
   Using GaAs/GaAlAs High Electron Mobility
   Transistors," Device Research Conference,
   Burlington, Vermont, 1983.
7. M. Ohmori, T. M izutani and N. Kato, "Very Low
   Power Gigabit Logic Circuits With Enhancement-
   Mode GaAs MESFETs," IEEE MTT-S Internat.
   Microwave Symposium, Proc. pp. 188-190, 1981.

ULTRA-HIGH-SPEED RING OSCILLATORS BASED ON SELF-ALIGNED-GATE MODULATION-DOPED n^+-(Al, Ga)As/GaAs FETs

Indexing terms: Semiconductor devices and materials, Oscillators

Ultra-high-speed ring oscillator test circuits based on modulation-doped n^+-(Al, Ga)As/GaAs field-effect transistors have been fabricated using a completely planar, self-aligned gate by an ion-implantation process. A gate propagation delay of 11·6 ps/gate at a power dissipation of 1·56 mW/gate was measured at room temperature. On cooling to 77 K the gate delay decreased to 8·5 ps/gate at a power dissipation of 2·59 mW/gate. These ring oscillator gate delay times are the fastest ever reported for any semiconductor digital circuit technology.

Introduction: The ultra-high-speed operation of ring oscillator test circuits based on modulation-doped (Al, Ga)As/GaAs field-effect transistors (MODFETs), also called HEMTs and SDHTs, have been demonstrated by a number of research laboratories in recent years. Lee *et al.* reported a ring oscillator gate delay of 12·2 ps/gate at 1·1 mW/gate at room temperature.[1] The delay-power performance of this circuit at 77 K was not reported. A ring oscillator gate delay of 17·3 ps/gate at 1·33 mW/gate at room temperature was recently reported by Hendel *et al.*, which when cooled to 77 K had a gate delay of 9·4 ps/gate at 4·53 mW/gate.[2] These circuits were fabricated using the standard MODFET process based on a recessed gate structure to reduce the source resistance.

Cirillo *et al.* have reported the first fabrication of MODFETs[3] and ring oscillators with gate delays of 17·6 ps/gate at 2·56 mW/gate at room temperature,[4] using a completely planar, self-aligned gate by an ion-implantation process. The objective of the self-aligned-gate process is to reduce the high parasitic source resistance and improve the threshold voltage uniformity obtained using the standard recessed-gate process.

In this letter we report further improvements in ring oscillator gate propagation delay times, both at room temperature and 77 K, using the self-aligned-gate MODFET process.

Fabrication: The modulation-doped n^+-(Al, Ga)As/GaAs heterostructures were grown on semi-insulating LEC GaAs substrates by molecular beam epitaxy (MBE). Starting from the substrate, the layers were a thick undoped GaAs buffer layer, a thin undoped (Al, Ga)As spacer layer and an Si-doped n^+-(Al, Ga)As charge-control layer. The doping and thickness of the charge-control layer. The doping and thickness of the charge-control layer are selected to obtain normally off devices. Van der Pauw measurements made on similar structures having the same undoped (Al, Ga)As spacer layer thickness, but with the n^+-(Al, Ga)As charge-control layer thickness appropriate for normally on devices, yield a typical electron mobility of 5000–8000 cm²/V s at room temperature and 60 000–120 000 cm²/V s at 77 K. The sheet electron concentration is typically $0·5–1·0 \times 10^{12}$ cm⁻² at both room temperature and 77 K. The MODFET devices and ring oscillator test circuits were fabricated using a self-aligned gate by an ion-implantation process similar to that used to fabricate conventional, self-aligned-gate GaAs MESFET integrated circuits.[5] All the lithography steps were done using a 10:1 projection die-by-die aligner. The process is completely

planar, and no recess gate etch or device isolation etch is required. A cross-sectional view of a SAG MODFET device is shown in Fig. 1. Si ion implantation is used to form the

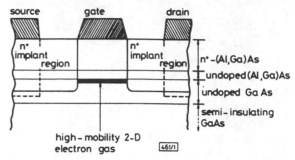

Fig. 1 *Cross-sectional view of a planar self-aligned-gate MODFET structure*

heavily doped n^+ source and drain regions, and the saturated resistor loads. A rapid thermal anneal using an argon arc lamp is used to activate the implants. The rapid thermal anneal does not degrade the quality of the (Al, Ga)As/GaAs as does a conventional furnace anneal.[4] The devices are isolated by a proton ion implant. The integrated circuits are completed using a two-level upper metal interconnect scheme.

Results: The DC characteristics of the self-aligned-gate MODFETs and the saturated resistor loads were determined at both room temperature and 77 K. The extrinsic transconductance g_m of nominally 1 μm-gate-length devices was typically 170–210 mS/mm at room temperature, with a threshold voltage of +0·42 V. On cooling to 77 K, the extrinsic transconductance g_m increased to 250–300 mS/mm. A threshold voltage shift of ~ +0·2 V was observed on cooling, indicating the presence of electron traps.[6] Fig. 2 shows the drain I/V characteristics of a 0·8 μm-gate-length self-aligned-gate MODFET and a saturated resistor load at 77 K. The sheet resistance of the n^+-implanted region and the contact resistance at room temperature were $R_s = 320\ \Omega/\square$ and $R_c = 0·45\ \Omega\,\text{mm}$, respectively, as measured by the transmission line method (TLM) on gateless structures. The source resistance R_s for the self-aligned-gate MODFETs having a gate-to-source spacing of 1 μm is then calculated to be $R_s = 0·77\ \Omega\,\text{mm}$.

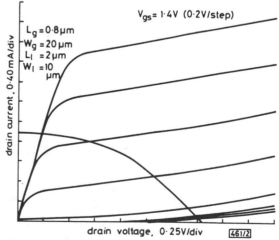

Fig. 2 *Drain I/V characteristics of a 0·8 μm-gate-length MODFET and saturated resistor load at 77 K*

25-stage direct-coupled FET logic (DCFL) ring oscillators were characterised to evaluate logic performance under ideal loading conditions at both room temperature and 77 K. The driver FETs have a nominal 1 μm gate length and were 20 μm wide. The saturated resistor loads were 2 μm long and 10 μm wide. Each inverter stage had a fan-in and fan-out equal to one. The oscillator waveforms were measured through three buffer inverter stages. The gate propagation delay τ_d and power dissipation P_d were measured for a number of ring oscillators as a function of the power supply voltage V_{dd} at room temperature. Two ring oscillators separated by ~ 1 cm were found to have the same minimum gate propagation delay of 11·6 ps/gate. The power dissipation was 1·55 mW/ gate at $V_{dd} = 1·63$ V for the ring oscillator with the best delay-power product at the minimum gate delay. Several other neighbouring ring oscillators were found with gate delays below 15 ps/gate. The two fastest ring oscillators were tested at 77 K. One has a minimum gate delay of 10·7 ps/gate at 2·04 mW/gate. The other had a minimum gate delay of 8·5 ps/gate at 2·59 mW/gate for $V_{dd} = 1·92$ V. This was also the ring oscillator with the better delay-power product at room temperature. Fig. 3 shows both the gate propagation

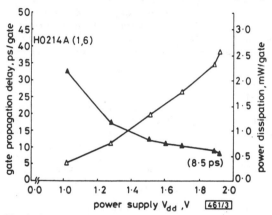

Fig. 3 *Gate propagation delay and power dissipation as a function of supply voltage for a 25-stage ring oscillator at 77 K*

delay and the power dissipation of this ring oscillator at 77 K as a function of the power supply voltage.

This is the fastest ring oscillator delay time ever reported for any semiconductor digital circuit technology. A comparison of delay-power performance of this ring oscillator fabricated using self-aligned-gate MODFETs with the best previously reported results of other recessed gate MODFET ring oscillators is shown in Fig. 4.

Conclusions: Ultra-high-speed ring oscillator test circuits were demonstrated using a self-aligned gate by an ion-implantation MODFET integrated circuit process. The fastest semiconductor gate propagation delay ever reported was obtained demonstrating the potential for very high-speed MSI/LSI

complexity integrated circuits using the completely planar self-aligned-gate MODFET process.

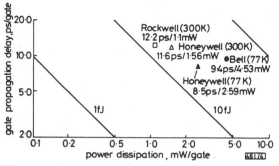

Fig. 4 *Comparison of propagation delay and power dissipation for MODFET ring oscillators*

Acknowledgments: This work was supported by the US Air Force, Avionics Laboratory, Wright Patterson AFB, OH 45433, USA. The authors would like to thank P. Joslyn for his assistance with the MBE growth, and to acknowledge S. A. Jamison and T. C. Lee for many helpful discussions.

N. C. CIRILLO   *14th June 1985*
J. K. ABROKWAH
A. M. FRAASCH
P. J. VOLD

Honeywell Inc.
Physics Sciences Center
Bloomington, MN 55420, USA

References

1 LEE, C. P., HOU, D., LEE, S. J., MILLER, D. L., and ANDERSON, R. J.: 'Ultra high speed digital integrated circuits using GaAs/AlGaAs high mobility transistors'. IEEE gallium arsenide integrated circuit symposium technical digest, 1983, pp. 162–165

2 HENDEL, R. H., PEI, S. S., TU, C. W., ROMAN, B. J., SHAH, N. J., and DINGLE, R.: 'Realization of sub-10 picosecond switching times in selectively doped (Al, Ga)As/GaAs heterostructure transistors'. IEEE international electron device meeting technical digest, 1985, pp. 857–858

3 CIRILLO, N. C., JUN., ABROKWAH, J. K., and SHUR, M. S.: 'Self-aligned modulation-doped (Al, Ga)As/GaAs field-effect transistors', *IEEE Electron Device Lett.*, 1984, **EDL-5**, pp. 129–131

4 CIRILLO, N. C., JUN., ABROKWAH, J. K., and JAMISON, S. A.: 'A self-aligned gate modulation-doped (Al, Ga)As/GaAs FET IC process'. IEEE gallium arsenide integrated circuit symposium technical digest, 1984, pp. 167–170

5 HELIX, M., HANKA, S., VOLD, P., and JAMISON, S.: 'A low power gigabit IC fabrication technology'. IEEE gallium arsenide integrated circuit symposium technical digest, 1984, pp. 163–169

6 VALOIS, A. J., ROBINSON, G. Y., LEE, K., and SHUR, M. S.: 'Temperature dependence of the *I-V* characteristics of modulation-doped FETs', *J. Vac. Sci. & Technol. B*, 1983, **1**, pp. 190–195

Realization of Sub-10 Picosecond Switching Times in Selectively Doped
(Al,Ga)As/GaAs Heterostructure Transistors

R.H. Hendel, S.S. Pei, C.W. Tu, B.J. Roman,
N.J. Shah, and R. Dingle*

AT&T Bell Laboratories
Murray Hill, New Jersey

ABSTRACT

We report gate delay times of less than 10 ps
for ring oscillators based on direct coupled FET
logic implemented with selectively doped (Al,Ga)
As/GaAs heterostructure transistors (SDHTs). The
minimum delay time observed was 9.4 ps at 77K with
a speed-power product of 42.4 fJ and 1.1 V bias.
The gate length the enhancement mode driver FETs
was measured to be 0.7   μm. Dual-clocked M/S
(master-slave) dividers on the same wafer using
dual gate SDHTs operated up to a maximum dividing
frequency of 6.3 (13.0) GHz at 300 (77)K.

I. INTRODUCTION

The high speed operation of integrated circuits based on selectively doped heterostructure
transistors (SDHTs) has been demonstrated. Lee et
al., for example, reported [1] a ring oscillator
delay time of 12.2 ps at 300K. Frequency dividers
using a gated master-slave flip-flop design and
0.5   μm single-gate selectively doped heterostructure transistors (SDHTs) were reported by
Abe et al. [2] and operated up to a maximum
dividing frequency of 5.5 (8.9) GHz at 300 (77) K.
Recently, Hendel et al. reported [3] a dual-
clocked frequency divider implemented with dual-
gate SDHTs. There, a maximum toggle frequency of
10.1 GHz was reported at 77K with a gate length of
1.0   μm. In this paper we report further
improvements in ring oscillator delay times and
maximum dividing frequencies by using sub-micron
gates.

II. TRANSISTOR PROPERTIES

The circuits were fabricated on selectively
doped (Al,Ga)As/GaAs heterostructures grown by
molecular beam epitaxy on 50 mm diameter semi-insulating GaAs substrates. Low field mobilities
for our structures typically are $6,000 cm^2/V-2$ at
300K and increase to $75,000 cm^2/V-s$ at 77K. All
features were patterned by optical contact lithography. The 50 m wide submicron driver-tran-
sistor gates (typical gate length 0.7 to 0.8 m)
were exposed using mid-range UV (310 nm). The
threshhold voltage was approximately +0.2 V with a
standard deviation of 70mV over an area of
$4.24 cm^2$. The average peak transconductance of all
driver transistors measured at room temperature
was 158 mS/mm. The average peak transconductance
of transistors measured at 77 K was 175 (302)
mS/mm at 300 (77) K. The average threshhold
voltage shifted from 0.21 V (300K) to 0.35 V
(77K). The loads were saturated resistors.

III. RING OSCILLATOR DELAY TIMES

We have fabricated 19-stage DCFL (direct
coupled FET Logic) ring oscillators. A histogram
of the speed distribution for all ring oscillators
measured in a $1.7 cm^2$ area is shown in Fig. 1. At
a fixed bias of 2.0 V, the average delay time is
22.9+-2.5 ps. The shortest gate delay measured
was 17.3 ps at 1.07 V bias with a 23.0 fJ speed-
power product. The minimum speed-power product
was 17.4 fJ at 0.86V bias and had a delay time of
18.1 ps.

For testing at 7K, we chose a subset of 18
circuits with average delay time of 21.1 + 1.1 ps
at 2.0 V bias. Optimizing the bias conditions for
each circuit to obtain maximum speed, we found an
average delay time of 10.6 ps at 77K. The minimum
propagation delay observed was 9.4 ps at a bias of
1.1 V and a speed-power product of 42.4 fJ. The
speed-power curve of a ring oscillator yielding a
mimimum delay time under 10 ps at 77K is shown in
Fig. 2. The circuit giving the shortest delay
times at 300K also was the fastest at 77K.

IV. DUAL GATE SDHT FREQUENCY DIVIDERS

The advantage of the dual-clocked master-
slave DCFL AND/NOR design in divide-by-two circuits for achieving maximum dividing frequencies
has been demonstrated [3]. A photomicrograph of
the basic AND/NOR gate is shown in Fig. 3. This
design flexibility offers a reduction of the
critical signal path from five to two loaded delay
times and allowed the implementation of divide-
by-two circuits operating up to 5.5 (10.1) GHz at
300 (77) K with 1 μm driver gates [3]. Using the
same design and technology but shorter (0.7 to
0.8 μm) driver gates we observed a maximum
dividing frequency of 6.3 GHz at 300 K. The total
power at this frequency, including the two buffered output gates, was 35.4 mW at 1.77 V bias.
Cooling to 77 K increased the maximum operating
frequency to 13.0 GHz at 1.24 V bias with a total
power dissipation of 33.7 mW.

V.   SUMMARY

We have presented gate delay times as short
as 9.4 ps and dividing frequencies up to 13 GHz
using SDHT technology with sub-micron driver
gates at 77K. This is the first time any
semiconductor technology has achieved sub-10 ps
switching times and confirms the potential of
SDHTs for high speed applications. The observed
performance improvement of approximately a factor
of two between 300 K and 77 K is significant.

Reprinted from *IEEE Int. Electron Devices Meeting, 1984, Tech. Dig.*, pp. 857-858, 1984.

The authors wish to acknowledge C.L. Allyn, P.F. Sciortino, L.D. Urbanek and T.M. Brennan for their support and assistance, and C.J. Mogab as well as J.V. DiLorenzo for helpful discussions.

This work was supported in part by the Air Force Wright Patterson Aeronautical Laboratories, Avionics Laboratory, Wright-Patterson Air Force Base, Ohio under contract F33614-83-C-1067.

*   Present address: Pivot III-V, P.O. Box 795, Cooper Station, New York, NY 10003

REFERENCES

[1] C.P. Lee, D. Hou, S.J. Lee, D.L. Miller, and R.J. Anderson, "Ultra High Speed Digital Integrated Circuits using GaAs/GaAlAs High Electron Mobility Transistors," GaAs IC Symposium Tech. Dig. 1983, pp, 162-165, Oct. 1983.

[2] M. Abe, T. Mimura, K. Nishiuchi, A. Shibatomi, and M. Kobayashi, "HEMT LSI Technology for high speed computers," GaAs IC Symposium Tech. Dig. 1983, pp. 158-161, Oct. 1983.

[3] R.H. Hendel, S.S. Pei, R.A. Kiehl, C.W. Tu, M.D. Feuer, and R. Dingle, "A 10 GHz Frequency Divider Using Selectively Doped Heterostructure Transistors", Electron Device Lett., Oct. 1984.

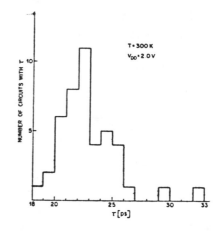

Fig. 2.

Histogram of gate delay times observed for 44 ring oscillators tested at 300 K in an area covering $1.7cm^2$.

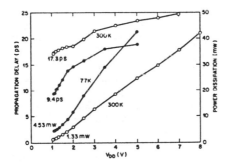

Fig. 1.

Propagation delay and power dissipation vs. bias voltage of a SDHT ring oscillator with 0.7 μm gates at 300 K (open symbols) and 77 K (solid symbols). The minimum values for propagation delay and total power dissipated are indicated.

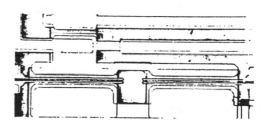

Fig. 3.

Photomicrograph of the dual gate AND/NOR logic gate, the basic building block of the M/S divide-by-two circuit.

n^+ SELF-ALIGNED-GATE AlGaAs/GaAs HETEROSTRUCTURE FET

Indexing terms: Semiconductor devices and materials, Field-effect transistors

The n^+ self-aligned-gate technology for high-performance AlGaAs/GaAs heterostructure FETs employing rapid lamp annealing have been studied. The large transconductance of 330 mS/mm at 300 K and 530 mS/mm at 83 K was obtained for the 0·7 μm gate length device, by reducing the source resistance to 0·6 Ω mm. The minimum delay time of 18·7 ps was obtained with a power dissipation of 9·1 mW at 300 K. The standard deviation of the delay time was as small as 1·1 ps at a fixed bias of 2·5 V.

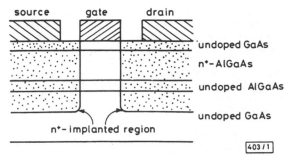

Fig. 1 *Schematic cross-sectional view of n^+ self-aligned-gate AlGaAs/ GaAs heterostructure FET*

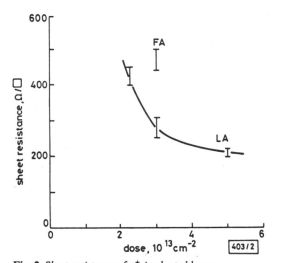

Fig. 2 *Sheet resistance of n^+ implanted layer*

LA and FA denote lamp annealing and furnace annealing, respectively

Introduction: In recent years, considerable interest has been focused on high-mobility two-dimensional electron-gas FETs (HEMTs or MODFETs) using AlGaAs/GaAs heterostructure to develop high-speed ICs and low-noise amplifiers.[1-4] For HEMTs to realise their full potential, it is necessary to reduce the parasitic source resistance because the sheet resistance of the two-dimensional electron gas is as large as 1 kΩ/$\square$ due to the small sheet carrier concentration. The introduction of a thick highly doped GaAs top layer proposed by Dämbkes *et al.*[5] needs a precise thickness control for the deep gate recess etching. The n^+ self-aligned-gate process using conventional furnace annealing is not suitable for HEMT fabrication because of the degradation of the AlGaAs/GaAs hetero-interface due to Si diffusion.[6] Recently, Cirillo *et al.*[7] have reported a self-aligned-gate HEMT process using lamp annealing to form the source and drain n^+ region and succeeded in high-speed operation. However, it seems that the implantation and annealing conditions are not optimised, judging from the large source resistance of 1 Ω mm.

This letter describes the detailed study of the n^+ self-aligned-gate HEMT process and shows that high transconductance of 330 mS/mm and high-speed operation below 20 ps at 300 K can be obtained.

Fabrication process and results: The modulation-doped AlGaAs/GaAs heterostructure was grown on $\langle 100 \rangle$-oriented semi-insulating GaAs substrates by molecular beam epitaxy at 630°C. A schematic cross-sectional view of the n^+ self-aligned-gate AlGaAs/GaAs heterostructure FET is shown in Fig. 1. The layers were undoped GaAs (6000 Å), undoped AlGaAs spacer (50 Å), Si-doped n^+ AlGaAs (300 Å, 1·5 × 10^{18} cm^{-3}) and undoped GaAs (50 Å) cap layer. The AlAs mole fraction in the AlGaAs layer was 0·3. The GaAs cap layer was grown to obtain good ohmic contacts.

Following the mesa etching for the electrical isolation, WSi$_x$ was sputter-deposited and patterned by reactive ion etching using SF$_6$ gas to form the gate electrodes. The etch rate ratio between WSi$_x$ and GaAs is sufficiently large as 140.

The source and drain n^+ regions were formed by Si ion implantation at 100 keV, followed by rapid lamp annealing at 950°C for 4 s in a N$_2$ atmosphere. The temperature increase rate was 100°C/s. For the annealing, the implanted wafer was set face-to-face on a GaAs wafer to prevent As dissociation. Prior to the annealing, preheating at 400°C for 10 s was performed to make the starting temperatures identical from run to run. The wafer surface was sufficiently smooth even with no encapsulant nor the As vapour pressure. The ohmic electrodes were made by evaporating AuGe/Ni, followed by the alloying at 450°C for 1 min. The gate width was 20 μm, and the gate-source and gate-drain spacing of the FET was 2 μm.

The sheet resistance $R_{\square}$ of the n^+ implanted layer is shown in Fig. 2 as a function of implantation dose, comparing them

with a furnace-annealed sample at 800°C for 20 min with P-CVD SiN$_x$ encapsulant. The lamp-annealed sample shows a smaller sheet resistance of 250–310 Ω/$\square$ than that obtained by furnace annealing ($R_{\square}$ = 440–500 Ω/$\square$) for the 3 × 10^{13} cm^{-2} dose. The sheet resistance is 200 Ω/$\square$ for the 5 × 10^{13} cm^{-2} dose. The ohmic-contact resistance measured by the transmission-line method was 0·2 Ω mm. Then, the source resistance is evaluated to be 0·6 Ω mm, which is smaller than that reported in Reference 7.

The transconductance for the typical enhancement-mode, self-aligned-gate HEMTs with a nominal 1·5 μm gate length is $\sim$180–240 mS/mm at 300 K. The best extrinsic transconductance of 330 mS/mm at 300 K and 530 mS/mm at 83 K have been measured on the 0·7 μm gate length HEMT as shown in Fig. 3. The present transconductance is much higher than that reported in Reference 7. The larger transconductance is thought to be attributed to the smaller source resistance. It should be noted that the HEMTs with a large transconductance can be fabricated without using a recessed-gate process. The saturation velocity evaluated using the formula derived by Delagebeaudeuf *et al.*[8] is 1·5 × 10^7 cm/s at 300 K and 2·8 × 10^7 cm/s at 83 K. No noticeable decrease in the saturation velocity is observed on the rapid-lamp-annealed HEMTs. The successful fabrication of the self-aligned-gate HEMTs with such high values of transconductance and saturation velocity promises routine fabrication of high-speed HEMT ICs.

The 15-stage DCFL ring oscillator was fabricated using the above-mentioned n^+ self-aligned-gate technology. The

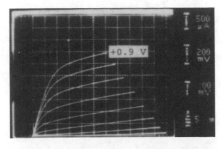

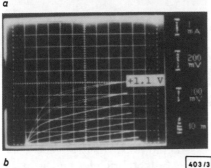

Fig. 3 *Drain I/V characteristics of n^+ self-aligned-gate HEMT at (a) 300 K and (b) 83 K*

Gate length and gate width of HEMT are $0.7 \mu m$ and $20 \mu m$, respectively

nominal gate length of the driver transistor was $1 \mu m$. As a load, an Si ion-implanted resistor was used, the dimensions of which were $8 \mu m$ wide and $6 \mu m$ long. The resistance was 570Ω for a 2.25×10^{13} cm^{-2} dose at 100 keV. The average delay time measured for nine chips is 21·5 ps and the standard deviation is as small as 1·1 ps at a fixed bias of 2·5 V at 300 K. The minimum delay time is 18·3 ps with a power dissipation of 9·1 mW at 3·4 V supply voltage. The small standard deviation of the delay time suggests the stability of the present n^+ self-aligned-gate process. Performance improvements are expected with further reduction of gate length and optimisation of the circuit design.

Conclusions: The n^+ self-aligned-gate technology employing rapid lamp annealing has been studied and successfully applied to the AlGaAs/GaAs HEMTs ICs. Large transconductances of 330 mS/mm at 300 K and 530 mS/mm at 77 K were obtained for the $0.7 \mu m$ gate length device by reducing the source resistance to $0.6 \Omega mm$. The minimum delay time of 18·7 ps was obtained with a power dissipation of 9·1 mW at 300 K. The small delay time standard deviation of 1·1 ps suggests that the n^+ self-aligned-gate process is one of the key processes for HEMT IC fabrication.

Acknowledgments: The authors would like to thank Dr. Y. Imamura for fruitful comments on the epitaxial growth and K. Maesawa for valuable discussions on lamp annealing.

T. MIZUTANI   *31st May 1985*
K. ARAI
K. OE
S. FUJITA
F. YANAGAWA

NTT Atsugi Electrical Communication Laboratories
3-1, Morinosato Wakamiya, Atsugi-shi
Kanagawa Pref. 243-01, Japan

References

1 KURODA, S., MIMURA, T., SUZUKI, M., KOBAYASHI, N., NISHIUCHI, K., SHIBATOMI, A., and ABE, M.: 'New device structure for 4 kb HEMT SRAM'. GaAs IC symposium, Technical digest, 1984, pp. 125–128
2 PEI, S. S., SHAH, N. J., HENDEL, R. H., YU, C. W., and DINGLE, R.: 'Ultra high speed integrated circuits with selectively doped heterostructure transistors', *ibid.*, 1984, pp. 129–132
3 LEE, C. P., HOU, D., LEE, S. J., MILLER, D. L., and ANDERSON, R. J.: 'Ultra high speed digital integrated circuits using GaAs/GaAlAs high electron mobility transistors', *ibid.*, 1983, pp. 162–165
4 MISHRA, U. K., PALMATEER, S. C., CHAO, P. C., SMITH, P. M., and HWANG, J. C. M.: 'Microwave performance of $0.25 \mu m$ gate length high electron mobility transistors', *IEEE Electron Device Lett.*, 1985, **EDL-6**, pp. 142–145
5 DÄMBKES, H., BROCKERHOFF, W., and HEIME, K.: 'Optimization of modulation-doped heterostructures for TEGFET operation at room temperature', *Electron. Lett.*, 1984, **20**, pp. 615–618
6 ISHIKAWA, T., HIYAMIZU, S., MIMURA, T., SAITO, J., and HASHIMOTO, H.: 'The effect of annealing on the electrical properties of selectively doped GaAs/n-AlGaAs heterojunction structures grown by MBE', *Jpn. J. Appl. Phys.*, 1981, **20**, pp. L814–L816
7 CIRILLO, N., JUN., ABROKWAH, J., and JAMISON, S. A.: 'A self-aligned gate modulation doped (Al, Ga)As/GaAs FET IC process'. GaAs IC symposium, Technical digest, 1984, pp. 167–170
8 DELAGEBEAUDEUF, D., and LINH, N. T.: 'Metal-(n)AlGaAs-GaAs two dimensional electron gas FET', *IEEE Trans.*, 1982, **ED-6**, pp. 955–960

Submicrometer Insulated-Gate Inverted-Structure HEMT for High-Speed Large-Logic-Swing DCFL Gate

HARUHISA KINOSHITA, TOSHIMASA ISHIDA, HIROKI INOMATA, MASAHIRO AKIYAMA, AND
KATSUZO KAMINISHI, MEMBER, IEEE

Abstract—Application of insulated-gate inverted-structure HEMT (I²-HEMT) to the enhancement/depletion (E/D) type direct-coupled FET logic circuits has been investigated. Superior electric characteristics were attained in a submicrometer-gate FET and ring oscillator. The threshold voltage shift with a reduction of gate length from 1.2 to 0.7 μm was as small as -0.05 V at 300 K. Drain conductances were very small and were 2.0 and 3.6 mS/mm at 300 and 77 K, respectively. Gate leakage current was small enough even at a gate voltage of $+1.4$ V both at 300 and 77 K, and a logic swing of larger than 1.2 V was achieved using a DCFL inverter. A 21-stage E/D-type DCFL ring oscillator with an 0.8-μm gate length showed a minimum gate delay of as small as 18.0 ps at a low power dissipation of 520 μW/gate at 77 K. High-speed and large logic-swing characteristics of the I²-HEMT DCFL circuits are accomplished by forming an undoped AlGaAs layer as a gate insulator on the inverted-structure HEMT structure.

I. INTRODUCTION

SINCE THE SUCCESSFUL fabrication of GaAs/AlGaAs superlattices, high electron mobility transistors (HEMT) with a modulation-doped GaAs/n-AlGaAs structure have been developed for an actual ultrahigh-speed-switching field-effect transistor (FET) operating at 77 K [1]. Through many efforts to improve the performance of the HEMT, a very high switching speed of 8.5 ps in a DCFL ring oscillator [2], a maximum input frequency of 10.1 GHz in a frequency divider [3], and a minimum access time of 2.0 ns in a 4-kbit static RAM [4] were achieved. Among many kinds of HEMT integrated circuits (IC's) fabricated by many researchers, almost all of the HEMT's had fundamentally the same layer structure, i.e., an n⁺-AlGaAs layer on a GaAs channel layer. On the other hand, another type of modulation-doped structure, i.e., a GaAs channel layer on an n⁺-AlGaAs layer, is also suited to HEMT's and should be applied to high-speed IC's. This kind of HEMT is called an inverted-structure HEMT (I-HEMT) and its name originates from its layer structure. The electron mobility of the I-HEMT structure was rather small at 77 K comparing

with that of the HEMT structure [6]. Because of the difficulty in obtaining high electron mobility, the I-HEMT did not attract much attention, and the development of the I-HEMT has been delayed.

Recently, due to improvements in epitaxial growth techniques, the optimization of layer structures, and the development of process techniques, an E/D-type DCFL ring oscillator was successfully fabricated and it has realized ultrahigh-speed and low-power characteristics at 77 K, i.e., a propagation delay time of 26.3 ps/gate at a power dissipation of 234 μW/gate [7]. In spite of high-speed operation, the I-HEMT had a weak point, i.e., the maximum voltage (V_{Gmax}) applicable to the gate without conspicuous leakage current was small (about $+0.6$ V) compared with that of the HEMT (about $+0.9$ V).

A small V_{Gmax} is not advantageous to the DCFL large-scale integrated circuit (LSI). This is because the logic swing of a DCFL inverter is proportional to the value of V_{Gmax}, and a small logic swing requires the severe control of the threshold voltage (V_{TH}) of a FET owing to the small margin of V_{TH} to obtain stable and high-speed operation of the DCFL circuit. By shortening the gate length to the submicrometer region, a large decrease of V_{TH} is usually observed, which is called the short-channel effect. The short-channel effect is influenced by the gate length and process conditions, and an inhomogenious gate length causes scattering of V_{TH} in LSI. The large scattering of V_{TH} and the small margin of adequate V_{TH} makes V_{TH} control difficult.

V_{Gmax} can be increased by the use of a p-n junction gate or a Schottky gate with high built-in potential or an insulated gate with a metal-insulator-semiconductor (MIS) structure. By adding an undoped AlGaAs layer on a GaAs channel layer of the I-HEMT, i.e., an MIS-type I-HEMT, the V_{Gmax} was increased to about 1.4 V from 0.6–0.7 V of I-HEMT [8]. The V_{Gmax} of the HEMT is about 0.8–0.9 V; therefore, the V_{Gmax} of an MIS-type I-HEMT, called an insulated-gate inverted-structure HEMT (I²-HEMT), is larger than that of the HEMT. This large V_{Gmax} leads to the predominancy of stable operation of an I²-HEMT DCFL LSI with high circuit reliability having a large logic margin.

In this paper we report the electrical characteristics of a submicrometer-gate I²-HEMT including short-channel

Manuscript received September 27, 1985; revised December 20, 1985. This work was supported by the Agency of Industrial Science and Technology, MITI of Japan, in the frame of the national research and development project, "Scientific Computing System."
The authors are with the Research Laboratory, Oki Electric Industry Co., Ltd., 550-5 Higashiasakawa, Hachioji, Tokyo 193, Japan.
IEEE Log Number 8607717.

Reprinted from *IEEE Transactions on Electron Devices*, vol. ED-33, no. 5, pp. 608–615, May 1986.

338

effects and high-speed operation of a DCFL ring oscillator with submicrometer-gate I^2-HEMT's [9], [10]. Gate leakage current and DCFL inverter characteristics are also discussed and compared with those of the I-HEMT.

II. Fabrication Process and Experimental Procedure

Modulation-doped structures of an I^2-HEMT and a I-HEMT were grown by molecular-beam epitaxy (MBE) on (100)-oriented Cr-doped semi-insulating GaAs substrates. The multilayer structure is composed of a top 300-Å-thick n^+-GaAs layer, a 200-Å-thick n^+-Al$_{0.5}$Ga$_{0.5}$As layer ($N_D \simeq 2 \times 10^{18}$ cm^{-3}), a 200-Å-thick insulated Al$_{0.5}$Ga$_{0.5}$As layer, a 400-Å-thick undoped GaAs layer (channel), an undoped Al$_{0.3}$Ga$_{0.7}$As spacer, an n^+-Al$_{0.3}$Ga$_{0.7}$As layer ($N_D \simeq 1 \times 10^{18}$ cm^{-3}) in the I^2-HEMT, and a top 400-Å-thick n^+-GaAs layer, a 400-Å-thick n-GaAs layer ($N_D \simeq 5 \times 10^{17}$ cm^{-3}), a 40-Å-thick undoped spacer layer, an n^+-Al$_{0.3}$Ga$_{0.7}$As layer ($N_D \simeq 1 \times 10^{18}$ cm^{-3}) in the I-HEMT. The fabrication process of the I^2-HEMT is essentially similar to that of the I-HEMT already reported [7], and a schematic cross section and energy-band diagram of the I^2-HEMT and I-HEMT are shown in Figs. 1 and 2.

The n-GaAs layer directly under the gate in the I-HEMT is formed to shorten the distance between the gate and the two-dimensional electron gas (2DEG) in the enhancement-mode (E-mode) I-HEMT. As a result, higher transconductance is obtained. The n-GaAs layer does not provide a parasitic conduction path under the normal operating conditions of an E-mode I-HEMT.

Devices were isolated by ^{16}O$^+$ ion double implanation at 30 keV with a dose of 2×10^{12} cm^{-2} and at 80 keV with a dose of 3×10^{12} cm^{-2} to form a planar structure. Under these conditions, a low leakage current of below 5 µA/mm was obtained between the 6-µm gap pattern at a supply voltage of 7 V [7]. Back-gate effect was evaluated applying a bias voltage (V_{BACK}) to the surrounding three ohmic electrodes placed a distance of 20 µm. A small V_{TH} shift of about 15 mV was observed for V_{BACK} of -5 to 5 V at 300 and 77 K, respecitvely. Ohmic electrodes are formed by the following procedure. Ohmic regions are recessed with low-energy Ar ion-beam etching [11] to a depth of about 500 Å and AuGe/Ni/Au are deposited to the recessed region and annealed. The buried ohmic structure was formed to lessen the thickness of the alloyed AuGe/Al$_x$Ga$_{1-x}$As layer, and was effective in reducing the contact resistance to lower than 2 Ω · mm. Gates are recessed with low-energy Ar ion-beam etching, which produces low damage at the AlGaAs surface [11]. Ti/Pt/Au metals are deposited into the gate recess using the lift-off technique. After metallization and passivation, the electric characteristics of the FET's on a wafer were measured at room temperature. Ring oscillators and FET's were mounted on TO-5 headers to measure the electric characteristics at 77 K.

The experimental data for the analysis of the short-channel effects and ring oscillator performance were taken

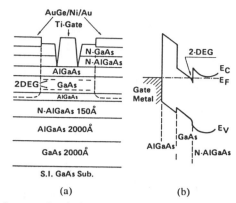

Fig. 1. (a) Cross-sectional view and (b) energy-band diagram of I^2-HEMT.

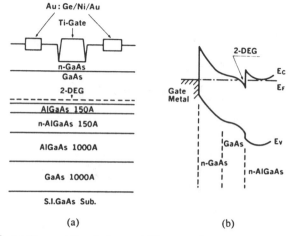

Fig. 2. (a) Cross-sectional view and (b) energy-band diagram of I-HEMT.

from a single wafer. The number of measured FET's and ring oscillators was about ten. Electric characteristics such as threshold voltage, K-value, and transconductance were evaluated by averaging the values measured. The performance of the ring oscillator reported in this paper concerns the best data. To average the data, uniformity of electric characteristics is important. In the wafer measured, the uniformity of the threshold voltage was not very good, but a standard deviation of 23 mV ($L_G = 0.7$ µm) was obtained by limiting the number of FET's used for analysis to about ten. The excellent value of 11.8 mV ($L_G = 1.5$ µm) already has been obtained in a 2-in wafer of an I-HEMT. The layer structures of the I^2-HEMT and I-HEMT are fundamentally similar. Therefore, good uniformity of threshold voltage is also expected to be obtained in an I^2-HEMT wafer by improving the process technology.

III. Device Characteristics

In the I^2-HEMT, quasi-two-dimensional electrons at the source region are accumulated in the GaAs channel layer by accepting electrons from the upper n^+-Al$_{0.5}$Ga$_{0.5}$As layer and the lower n^+-Al$_{0.3}$Ga$_{0.7}$As layer as shown in Fig. 3. The depth profile of the electron concentration was estimated by C–V measurement and a well-confined distribution of electrons is observed. In the GaAs layer under

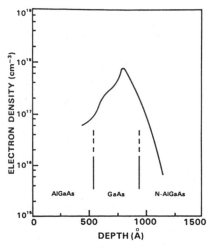

Fig. 3. Carrier depth profile of I²-HEMT measured by C–V method.

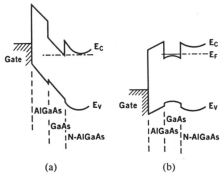

Fig. 4. Energy-band diagram of I²-HEMT at (a) closed-channel and (b) opened-channel states.

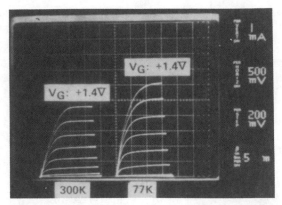

Fig. 5. Drain I–V characteristics of I²-HEMT at 300 and 77 K.

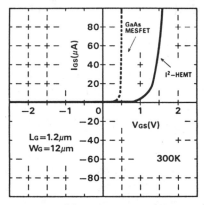

Fig. 6. I–V curve measured between gate and source electrodes of I²-HEMT and GaAs MESFET at 300 K.

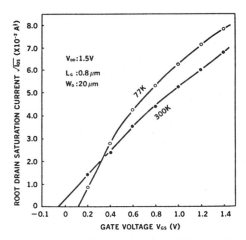

Fig. 7. Transfer characteristics of square root of drain saturation current $\sqrt{I_{DS}}$ versus gate voltage V_{GS} of E-mode I²-HEMT at 300 and 77 K.

the gate of the I²-HEMT, electrons are accumulated from the lower n⁺-AlGaAs layer and induced electrostatically owing to the positive bias applied to the undoped AlGaAs gate. Energy-band diagrams at the negatively and positively biased conditions of the gate are shown in Fig. 4.

The drain characteristics of E-mode I²-HEMT with 0.8-μm gate length and 20-μm gate width measured at 300 and 77 K are shown in Fig. 5. By the use of an insulated $Al_{0.5}Ga_{0.5}As$ layer, a high gate voltage of 1.4 V can be applied without serious gate leakage current. Fig. 6 shows the I–V curve between gate and source electrodes measured at 300 K. The applied gate voltages at a gate current of 20 μA were 1.3 V in the I²-HEMT with 1.2-μm gate length and 12-μm gate width, and only 0.5 V in a GaAs MESFET with the same gate size. The gate leakage current was further reduced by lowering the temperature to 77 K. The maximum transconductances observed were 230 mS/mm at 300 K and 280 mS/mm at 77 K.

Fig. 7 shows transfer characteristics of the square root of the drain saturation current versus gate voltage, i.e., $\sqrt{I_{DS}}$ versus V_{GS}, in the E-mode I²-HEMT. The K-values measured at the low-field region were 3.6 and 7.2 mA/V² at 300 and 77 K, respectively. The deflection of the $\sqrt{I_{DS}}$–V_{GS} curve from the square low $I_{DS} = K \cdot (V_{GS} - V_{TH})^2$ was observed at the high-field region. This phenomenon is caused by high source resistance (about 2.5–3 Ω · mm) including a high contact resistance of about 2 Ω ·

mm. In the I-HEMT, on the other hand, the source resistance was as small as 0.4–0.6 Ω · mm owing to a small contact resistance of about 0.2 Ω · mm [7]. The small transconductance increment of the I²-HEMT from 230 mS/mm (300 K) to 280 mS/mm (77 K) is caused by the high source resistance. A remarkable performance improvement by cooling the I²-HEMT is shown in the increment of the K-value by a factor of 2.0 rather than in the transconductance increment. The contact resistance can be reduced by optimizing the buried ohmic structure.

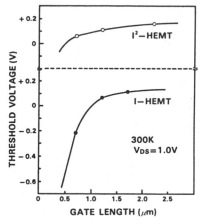

Fig. 8. Gate-length dependences of threshold voltage of I²-HEMT and I-HEMT.

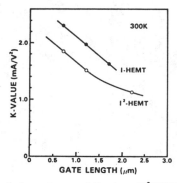

Fig. 9. Gate-length dependences of K-value of I²-HEMT and I-HEMT.

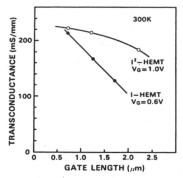

Fig. 10. Gate-length dependences of transconductance of I²-HEMT and I-HEMT.

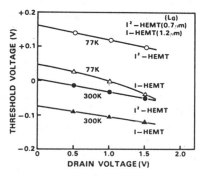

Fig. 11. Drain-voltage dependences of threshold voltage of I²-HEMT and I-HEMT.

The threshold voltage shift was +0.18 V on lowering the operating temperature from 300 to 77 K. The maximum saturation drain current was as high as 300 mA/mm.

IV. SHORT-CHANNEL EFFECTS

The threshold voltages (V_{TH}s) of the I²-HEMT and I-HEMT were measured at 300 K using FET's with a gate length longer than 0.7 μm. As shown in Fig. 8, the decrease of V_{TH} (ΔV_{TH}) due to a reduction of the gate length from 1.2 to 0.7 μm was only -0.05 V in the I²-HEMT and was -0.25 V in the I-HEMT. The above value of ΔV_{TH} for the I²-HEMT is very small compared to those of the I-HEMT and GaAs MESFET's. The ΔV_{TH} of GaAs MESFET's is typically more than -0.25 V [12], [13]. The K-value versus the gate length of each FET was measured at 300 K, and are shown in Fig. 9. The gate widths of the FET's measured were all 10 μm. The K-value of the I²-HEMT and I-HEMT increased monotonically with decreasing gate lengths from 2.2 to 0.7 μm. These effects suggest that the short-channel effects of both the I²-HEMT and I-HEMT are small. Fig. 10 shows the gate length dependences of transconductance of each FET measured at 300 K. The FET's used were all E-mode FET's and the gate voltages were 1.0 V for the I²-HEMT and 0.6 V for the I-HEMT. The smaller K-value of the I²-HEMT and the lower increment ratio of transconductance of the I²-HEMT compared with that of the I-HEMT is deeply influenced by the higher source resistance of the I²-HEMT, i.e., higher contact resistance. Therefore, by reducing the contact resistance, the estimated value of the extrinsic transconductance will become more than 400 mS/mm for the 0.5-μm gate I²-HEMT.

The drain voltage dependences of the threshold voltage in 0.7-μm gate length I²-HEMT and the 1.2-μm gate length I-HEMT were measured at 300 and 77 K, and are shown in Fig. 11. $\delta V_{TH}/\delta V_{DS}$ ($=\gamma$), i.e., gradient of a curve, is a parameter representing the degree of short-channel effects. In all cases at 300 and 77 K, γ was about 0.03–0.06. This value was very small compared to the values of GaAs MESFET's, which are typically 0.05–0.5 [12]. The drain conductance g_D of the I²-HEMT and I-HEMT were measured at 300 and 77 K. The values for an I²-HEMT with 0.7-μm gate length were 2.0 and 3.6 mS/mm at 300 and 77 K, respectively. On the contrary, the values for an I-HEMT with a gate length of as long as 1.2 μm were 7.4 and 15.0 mS/mm at 300 and 77 K, respectively. The drain conductance g_D of the I²-HEMT was much smaller than that of the I-HEMT in spite of the shorter gate length of the I²-HEMT. The γ-value and drain conductance of the I²-HEMT are summarized in Table I compared to those of the I-HEMT and GaAs MESFET's. The drain conductance of a standard HEMT is small and similar to that of the I²-HEMT. Such a small drain conductance value for the I²-HEMT indicates that the Al-GaAs/GaAs/AlGaAs layer structure of the I²-HEMT is effective in suppressing the influence of short-channel effects.

TABLE I
SUMMARY OF γ-VALUE AND DRAIN CONDUCTANCE g_D

	γ-value	g_D (mS/mm)	
I²-HEMT (Lɢ:0.7μm)	0.03–0.06	2.0 (300K)	3.6 (77K)
I-HEMT (Lɢ:1.2μm)		7.4	15.0
GaAs MESFET (Lɢ:0.7–1.2μm)	0.05–0.5	7–30 (300K)	

V. RING OSCILLATOR PERFORMANCE

Ring oscillators with DCFL gates were fabricated to evaluate high-speed characteristics. The gates were composed of 21-stage inverters including a 0.8-μm-length 24-μm-width E-mode FET, and a 0.8-μm-length 12-μm-width depletion-mode (D-mode) FET for each ones. The drain characteristics of an E-mode FET are shown in Fig. 5. The load current of a D-mode FET with 20-μm gate width was about 1.5 mA at 300 K and about 1.9 mA at 77 K.

Fig. 12 shows the voltage transfer curve of a DCFL inverter. The output terminal of the measured inverter was linked to an input terminal of the following inverter. Three curves show the transfer curves observed under bias voltages of 0.6, 1.0, and 1.5 V. Owing to the very small gate leakage current, i.e., large V_{Gmax}, a large logic swing of more than 1.2 V was obtained at a supply voltage of 1.5 V. This logic swing is larger than 0.6 V for the I-HEMT and 1.0 V for a standard HEMT. By optimizing the gate structure, the gate leakage current will be reduced and a larger logic swing is expected to be achieved. A noise margin larger than about 300 mV will be also obtained by adjusting the load current of the D-mode FET and the threshold voltage of the E-mode FET. In the case shown in Fig. 12, a higher threshold voltage for the E-mode FET, such as 0.25–0.35 V, will be effective in enlarging the noise margin.

A ring oscillator mounted on a TO-5 header was measured at 300 and 77 K in ambient light. The oscillation frequencies were measured using a spectrum analyzer and the wave form was monitored by a sampling oscilloscope. Fig. 13 shows an oscillation wave form of a ring oscillator at 77 K corresponding to a delay time of 19.8 ps/gate. The speed–power relations of a ring oscillator measured at 300 and 77 K are shown in Fig. 14. Minimum propagation delays of 33.5 and 18.0 ps/gate were obtained at 300 and 77 K, respectively, at the same power dissipation of as low as 520 μW/gate. A large speed improvement from 300 to 77 K was obtained. This improvement would be caused by the large increment of K-value by a factor of 2.0 by cooling from 300 to 77 K. These speed-power relations are superior to those of the 1.2-μm I-HEMT already reported, i.e., 60 and 26.3 ps/gate at 300 and 77 K [7], respectively, and comparable to those of standard HEMT's.

VI. DISCUSSION

A high saturation drain current of 300 mA/mm was obtained at 77 K. This current is higher than those of stan-

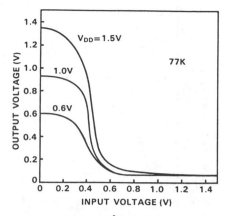

Fig. 12. Voltage-transfer curve of I²-HEMT inverter observed at 77 K.

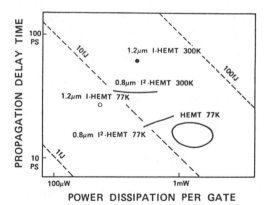

Fig. 13. Photograph of an oscillation waveform of 21-stage DCFL ring oscillator operated at 77 K showing propagation delay time of 19.8 ps/gate.

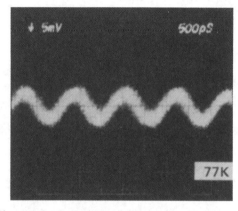

Fig. 14. Propagation delay versus power dissipation for I²-HEMT ring oscillator.

dard HEMT's (typically 100–200 mA/mm) and the I-HEMT (80–160 mA/mm). Such a high drain current was achieved because of electron accumulation in a channel by two mechanisms. One is the electron supply from a lower n^+-$Al_{0.3}Ga_{0.7}As$ layer, and the other is the electrostatic induction by positive voltage biased to the gate.

From the electric characteristics of the I²-HEMT and I-HEMT shown in Figs. 9–11, we can see that the short-channel effects are small in both the I²-HEMT and I-HEMT. The results shown in Fig. 8 and those for drain conductance indicate that the short-channel effect in the

TABLE II
SUMMARY OF SHORT-CHANNEL EFFECTS

	I^2-HEMT	I-HEMT
$\Delta V_{TH}/\Delta L_G$	Small	Large
g_D	Small	Large
ΔK-Value$/\Delta L_G$	Small	Small
$\Delta gm/\Delta L_G$	Small	Small
γ-Value	Small	Small

I^2-HEMT is smaller than that in the I-HEMT. These results on short-channel effects are summarized in Table II. Short-channel effects are strongly dependent on the mechanism of electron conduction in the channel layer. In both the I^2-HEMT and I-HEMT electrons are accumulated in the GaAs layer grown on the n^+-AlGaAs layer as shown in Figs. 1 and 2, and these electrons cannot contribute to the conduction in the n^+-AlGaAs layer with obstruction due to the potential barrier of n^+-AlGaAs. Therefore, electrons are confined to a narrow quantum well. The K-value, transconductance, and γ-value are strongly dependent on the conduction mechanism of electrons confined in the narrow quantum well. On the other hand, the I^2-HEMT has another AlGaAs layer on the GaAs channel layer. This AlGaAs layer also plays the role of a potential barrier confining the electrons, i.e., the I^2-HEMT has two potential barriers on both sides of the GaAs channel layer. Therefore, the conducting electrons in an I^2-HEMT are more strongly confined in the channel layer than in the case of an I-HEMT.

The differences in the short-channel effects on the threshold voltage shift and drain conductance observed between the I^2-HEMT and the I-HEMT were analyzed. To investigate the small threshold voltage shift with the reduction of the gate length to the submicrometer region, the electron conduction mechanism in the I^2-HEMT under three gate voltage biased conditions are investigated using the illustration in Fig. 15(a) relative to the case of the I-HEMT. Fig. 15(b) shows the gate bias conditions of the E-mode FET labeled as I, II, and III.

In the I-HEMT, the depletion region spreads below the gate. By biasing the positive voltage to the gate, the depletion region is shrunk and the electron channel is opened. The saturation electric current is regulated at a pinchoff point under the gate near drain region. At a high gate bias (condition III) in the I-HEMT, the channel depletion region width is limited to a narrow region under the gate. Therefore, electrons flow easily from the n^+-GaAs source to the n^+-GaAs drain through a narrow-channel depletion layer. This produces a higher drain conductance and a large V_{TH} shift in the I-HEMT.

In the I^2-HEMT, however, the channel depletion region spreads widely toward the source and drain regions because the upper insulating AlGaAs layer helps the depletion layer to spread in both directions and interrupts the electron flow from the n^+-GaAs surface layer to the GaAs channel layer. At a high gate bias (condition III), electrons are also accumulated near the upper GaAs–AlGaAs interface, the electric current is regulated at two pinchoff points near the drain region, and the depletion region be-

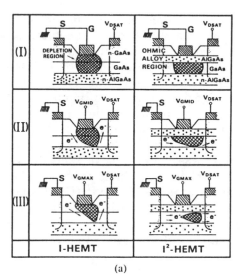

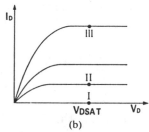

Fig. 15. (a) Channel states of I-HEMT and I^2-HEMT at three gate-bias condition of E-FET and (b) drain I–V characteristics indicating three gate-bias condition of (I), (II), and (III).

comes bullet-like in shape. The bullet-shaped depletion region is effective in reducing the drain conductance because it easily expands toward the drain region with increasing drain voltage. In the submicrometer-gate I^2-HEMT, the depletion region extends very close to the source and drain regions, and hence a further increase of drain current with drain voltage is suppressed. As a result, the change in V_{TH} with the reduction of the gate length is reduced.

In Fig. 14, which shows the propagation delay versus power dissipation of an I^2-HEMT DCFL ring oscillator, the reduction of oscillation frequency with the increase of supply voltage was observed at 77 K. This phenomenon also was observed in both a standard HEMT and I-HEMT, and is thought to be caused by the reduction of transconductance of the E-FET at the high drain current region [7], [14] and/or the reduction of drain current both of the E-mode and D-mode FET at the higher drain voltage region by hot-electron trapping effect in an n^+-AlGaAs layer [15], [16]. The hot-electron trapping effect was found at 77 K and causes capture of conducting electrons to deep levels such as DX centers in an n^+-AlGaAs layer. Trapped electrons cannot be emitted from DX centers due to large energy barrier heights of about 0.45 eV for electron emission. By the capture of them, the transfer speed is thought to be delayed, i.e., by reduction of the conducting electron number with increase of applied drain voltage, the propagation delay will be observed.

In the measurement of the I^2-HEMT ring oscillator, the

343

experimental data were measured in ambient light. By measuring this ring oscillator in complete darkness, a larger reduction of oscillation frequency with an increase of supply voltage was observed at 77 K. On the other hand, the speed–power relations measured at 300 K changed little by the measurement in complete darkness. The I^2-HEMT measured at 77 K in complete darkness did not show $I-V$ collapse, but a persistent decrease of drain current was observed [15], [16]. This persistent decrease of drain current was quickly recovered by the illumination of light. These phenomena indicate that the hot-electron trapping effect is generated in the I^2-HEMT ring oscillator operating at higher supply voltage. To obtain higher speed characteristics, a greater reduction of contact resistance and improvement of the n^+-AlGaAs layer are important.

VII. Conclusion

Very small short-channel effects and large logic swing of DCFL inverter were observed in a submicrometer-gate I^2-HEMT. Small short-channel effects are effective in keeping the high reproducibility of electric characteristics of FET's such as threshold voltage, K-value, transconductance, drain conductance, and so on. A large logic swing is adequate for ensuring stable operation, especially in the case of LSI and VLSI.

High-speed operation is easily attained by using a submicrometer-gate FET. In a DCFL ring oscillator with 0.8-μm I^2-HEMT, a high speed of 18.0 ps was obtained at a power dissipation of 520 μW/gate at 77 K. Such high-speed and low-power characteristics were obtained by the superior electric characteristics of small short-channel effects and low gate leakage current of the I^2-HEMT. The optimized I^2-HEMT will be one of the most promizing FET's among the HEMT groups for very high-speed LSI and VLSI.

Acknowledgment

The authors wish to thank Y. Sano for his helpful discussion. They would also like to thank Dr. S. Nishi and Dr. T. Itoh for preparing the MBE films.

References

[1] M. Abe, T. Mimura, N. Yokoyama, and H. Ishikawa, "New technology toward GaAs LSI/VLSI for computer applications," *IEEE Trans. Electron Devices*, vol. ED-29, pp. 1088–1094, July 1982.

[2] N. C. Cirillo, Jr. and J. K. Abrokwah, "8.5-picosecond ring oscillator gate delay with self-aligned gate modulation-doped n^+-(Al, Ga)As/GaAs FETs" in *Proc. 43rd Dev. Res. Conf.*, p. IIA-7, June 1985.

[3] R. H. Hendel, S. S. Pei, R. A. Kiehl, C. W. Tu, M. D. Feuer, and R. Dingle, "A 10-GHz frequency divider using selectively doped heterostructure transistors," *IEEE Electron Device Lett.*, vol. EDL-5, pp. 406–408, Oct. 1984.

[4] S. Kuroda, T. Mimura, M. Suzuki, N. Kobayashi, K. Nishiuchi, A. Shibatomi, and M. Abe, "New device structure for 4kb HEMT SRAM," in *Proc. GaAs IC Symp.*, pp. 125–128, Nov. 1984.

[5] R. E. Thorne, R. Fischer, S. L. Su, W. Kopp, T. J. Drummond, and H. Morkoç, "Performance of inverted structure modulation doped Schottky barrier field effect transistors," *Japan. J. Appl. Phys.*, vol. 21, pp. L223–L224, Apr. 1982.

[6] T. J. Drummond, J. Klem, D. Arnold, R. Fischer, R. E. Thorne, W. G. Lyons, and H. Morkoç, "Use of a superlattice to enhance the interface properties between two bulk heterolayers," *Appl. Phys. Lett.*, vol. 42, pp. 615–617, Apr. 1983.

[7] H. Kinoshita, S. Nishi, M. Akiyama, and K. Kaminishi, "High-speed low-power ring oscillator using inverted structure high electron mobility transistors," in *3rd Int. Conf. MBE Abstracts*, p. 53, Aug. 1984; also *Japan. J. Appl. Phys.*, vol. 24, pp. 1061–1064, Aug. 1985.

[8] H. Kinoshita, Y. Sano, T. Ishida, S. Nishi, M. Akiyama, and K. Kaminishi, "A new insulated-gate inverted-structure modulation-doped AlGaAs/GaAs/N-AlGaAs field-effect-transistor," *Japan. J. Appl. Phys.*, vol. 23, pp. L836–L838, Nov. 1984.

[9] H. Kinoshita, Y. Sano, S. Nishi, T. Ishida, M. Akiyama, and K. Kaminishi, "High performance AlGaAs/GaAs/N-AlGaAs insulated-gate inverted-structure HEMT ring oscillator," in *Proc. 43rd Dev. Res. Conf.*, p. IIA-6, June 1985.

[10] H. Kinoshita, T. Ishida, M. Akiyama, H. Inomata, Y. Sano, S. Nishi, and K. Kaminishi, "New insulated-gate inverted-structure AlGaAs/GaAs/N-AlGaAs HEMT ring oscillator," *Electron. Lett.*, vol. 21, pp. 1062–1064, Nov. 1985.

[11] H. Kinoshita, Y. Sano, T. Nonaka, T. Ishida, and K. Kaminishi, "Ion beam etching for GaAs application," in *Proc. Int. Ion Engineering Congress—ISIAT'83 and IPAT'83*, pp. 1629–1634, Sept. 1983.

[12] T. Ohnishi, Y. Yamaguchi, T. Onodera, N. Yokoyama, and H. Nishi, "Experimental and theoretical studies on short channel effects in lamp-annealed WSi$_x$-gate self-aligned GaAs MESFETs," in *Proc. 16th Conf. Solid State Devices Mater.*, pp. 391–394, Aug. 1984.

[13] K. Yamasaki, N. Kato, and M. Hirayama, "Below 10ps/gate operation with buried p-layer SAINT FETs," *Electron. Lett.*, vol. 20, pp. 1029–1031, Dec. 1984.

[14] A. Ketterson, M. Moloney, and H. Morkoç, "Modeling of GaAs/AlGaAs MODFET inverters and ring oscillators," *IEEE Electron Device Lett.*, vol. EDL-6, pp. 359–362, July 1985.

[15] H. Kinoshita, S. Nishi, M. Akiyama, T. Ishida, and K. Kaminishi, "Persistent channel depletion caused by hot electron trapping effect in selectively doped n-AlGaAs/GaAs structures," *Japan. J. Appl. Phys.*, vol. 24, pp. 377–378, Mar. 1985.

[16] H. Kinoshita, M. Akiyama, T. Ishida, S. Nishi, Y. Sano, and K. Kaminishi," Analysis of electron trapping location in gated and ungated inverted-structure HEMT's," *IEEE Electron Device Lett.*, vol. EDL-9, pp. 473–475, Sept. 1985.

AlGaAs/GaAs MESFET IC with Ni Buried Gate Technology

T. MIZUTANI, K. ARAI, K. OE, S. FUJITA, Y. IMAMURA, AND F. YANAGAWA, MEMBER, IEEE

Abstract—Ni buried gate technology for threshold voltage control using a Ni-GaAs reaction by a heat treatment is developed and successfully applied to AlGaAs/GaAs heterostructure MESFET IC's. Switching delay time of 36.7 ps with the power-delay product of 10 fJ (1-V supply voltage) was obtained at 83 K for a ring oscillator with 1.5-μm gate FET's. This technology, together with the saturated resistor loads, promises to simplify the process for AlGaAs/GaAs MESFET LSI's by not requiring active-layer etching.

I. INTRODUCTION

IN recent years, considerable interest has been focused on high-mobility two-dimensional electron gas (2-DEG) FET's using AlGaAs/GaAs heterostructure to develop high-speed IC's [1]-[4]. One of the 2-DEG FET inverters, direct-coupled FET logic (DCFL) is extensively studied because its circuit configuration is the simplest. However, it is difficult for DCFL to fabricate driver enhancement-mode FET's, because the thickness and the doping concentration of n-AlGaAs need to be tightly controlled.

Widely used recessed gate technology is not suitable for large-scale integrated circuits because etching of the epitaxial layer in the gate region with precise thickness control is very difficult and is apt to bring about threshold voltage scattering.

Recently, Takanashi *et al.* [5] have proposed a Ni buried gate technology for the threshold voltage control employing Ni-GaAs reaction by a heat treatment. This letter demonstrates the validity of the Ni buried gate technology for 2-DEG FET IC's. The fabrication process of 2-DEG FET IC's with Ni/Ti/Au gate and the ring oscillator characteristics are described. A saturated resistor [6] is used for a load to simplify the fabrication process. The simplicity of the process, in which no etching of the active layer for driver enhancement-mode FET's and loads is necessary, opens an interesting approach in high-speed 2-DEG FET IC's.

II. FABRICATION PROCESS AND RESULTS

The modulation-doped AlGaAs/GaAs heterostructure was grown on ⟨100⟩-oriented semi-insulating GaAs substrates by molecular beam epitaxy at 630°C. Typical electron mobility and sheet carrier concentration of the modulation-doped MBE wafer are 6460 cm²/V·s and 6.3×10^{11} cm⁻² at 300 K, and 23860 cm²/V·s and 4.8×10^{11} cm⁻² at 77 K. A

Manuscript received January 2, 1985; revised February 25, 1985.

The authors are with the Atsugi Electrical Communication Laboratory, Nippon Telegraph and Telephone Public Corporation, 1839 ONO, ATSUGI-SHI, Kanagawa, 243-01, Japan.

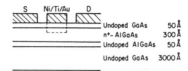

Fig. 1. Schematic cross-sectional view of an AlGaAs/GaAs MESFET.

schematic cross-sectional view of a modulation-doped AlGaAs/GaAs FET used in the present study is shown in Fig. 1. The layers were undoped GaAs (3000 Å), undoped AlGaAs spacer (50 Å), Si-doped n⁺ AlGaAs (300 Å, 1 ~ 2 $\times 10^{18}$ cm⁻³) and undoped GaAs (50-Å) cap layer. The AlAs mole fraction in the AlGaAs layer was 0.3. The GaAs cap layer was grown to obtain good ohmic contacts. The layer was not doped with Si donor impurity to improve threshold voltage reproducibility because the doping fluctuation in the layer produces threshold voltage scattering. No problem occurred in the ohmic contacts even though the GaAs cap layer was not doped. The ohmic contact resistance measured by the transmission-line method was 0.3 Ω·mm.

Ni buried gate technology that utilize a Ni-GaAs reaction by a heat treatment was used for threshold voltage control instead of recessed gate technology. Tung *et al.* have reported on DCFL circuits without recessed gate, utilizing a benefit of a low surface potential of 0.33 V [6]. In the present study, however, the planar FET's without recessed gate showed a large gate–source resistance and a small transconductance, less than 20 mS/mm for enhancement-mode FET's. Moreover, the drain current before gate electrode formation was almost zero for FET's with threshold voltage of 0.15 V. These facts suggests that the surface potential is larger than 0.33 V. This surface potential difference is presumably explained by the existance of the GaAs cap layer that was grown to improve the ohmic contact reproducibility in the present study. Ni buried gate technology was effective in reducing the influence of surface potential and decreasing the gate–source resistance. As gate metals, Ni (100 Å), Ti (1500 Å), and Au (3000 Å) were successively evaporated and lifted off, followed by annealing in Ar atmosphere for 2 min.

An example of the threshold voltage shift caused by the Ni–GaAs reaction is shown in Fig. 2 for two different wafers. As the threshold voltage shifts positively by the heat treatment, the transconductance increases. This result can be explained by the decrease of AlGaAs thickness. Roughly speaking, an intrinsic transconductance is proportional to the inverse of the AlGaAs thickness. It was possible to get FET's

Reprinted from *IEEE Electron Device Letters*, vol. EDL-6, no. 5, pp. 232–233, May 1985.

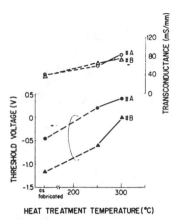

Fig. 2. Dependence on heat-treatment temperature of threshold voltage and transconductance.

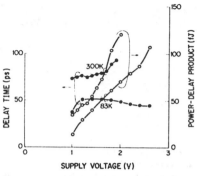

Fig. 3. Dependence on supply voltage of delay time and power-delay product.

with required threshold voltage by examining the relation between the threshold voltage and the heat treatment temperature prior to IC fabrication for a given wafer. The threshold voltage standard deviation σV_T before and after the heat treatment, for 36 FET's in the area of 3×6 mm^2, was 47 and 22 mV, respectively. The σV_T improvement can be explained by the elimination of surface contamination effect by Ni–GaAs reaction.

Good Schottky contacts were obtained even after heat treatment at 300°C for 5 min. The ideality factor n and the barrier height of the Ni Schottky contacts measured by foward I–V characteristics were 1.15 and 0.72 V after gate metal evaporation, and 1.11 and 0.91 V after heat treatment, respectively. As regards to the long term stability, it might be necessary that Ni completely reacts with GaAs in order to assure the realiability of the Ni Schottky contacts. Further study is necessary to clarify it.

The transconductance and the threshold voltage of the switching transistor were 100 mS/mm and 0.25 V at room temperature, and 280 mS/mm and 0.5 V at 83 K. The gate length and the gate width were 1.5 and 20 μm, and the source–drain spacing was 5 μm. The source resistance was 3.5 $\Omega \cdot$mm. A 20-μm-wide saturated resistor with 2-μm electrode spacing was used as a load. It was possible to create a very simple fabrication process with this saturated resistor load because one n-AlGaAs layer sufficed and it was not necessary to etch the active layer of the driver transistor and of the depletion FET load. Only five masks were used for the 2-DEG FET IC fabrication. Air-gapped crossover [7] was employed for the second-level interconnect.

The switching delay time and the power-delay product measured by a 15-stage E/R ring oscillator are shown in Fig. 3 as a function of supply voltage. High-speed operation was possible at a small supply voltage of 1 V. The delay time was 73 ps with the power-delay product of 34 fJ at room temperature, and 36.7 ps with the power-delay product of 10

fJ at 83 K. The delay time of 36.7 ps is reasonable taking the fairly long gate length of 1.5 μm into consideration. According to a simulation, it is possible to attain a smaller delay time below 10 ps by reducing gate length from 1.5 to 0.5 μm.

III. CONCLUSION

Ni buried gate technology which uses Ni–GaAs reaction by a heat treatment for the threshold voltage control has been developed and successfully applied to AlGaAs/GaAs 2-DEG FET IC's. Switching delay time of 36.7 ps with the power-delay product of 10 fJ was obtained at 83 K for a ring oscillator with 1.5-μm gate FET's. This technology, together with saturated resistor loads, promises to simplify the process for 2-DEG FET LSI's by not requiring active-layer etching.

REFERENCES

[1] S. Kuroda, T. Mimura, M. Suzuki, N. Kobayashi, K. Nishiuchi, A. Shibatomi, and M. Abe, "New device structure for 4 Kb HEMT SRAM," in *GaAs IC Symp. Tech. Dig.*, 1984, pp. 125–128.

[2] S. S. Pei, N. J. Shah, R. H. Hendel, C. W. Yu, and R. Dingle, "Ultra high speed integrated circuits with selectivity doped heterostructure transistors," in *GaAs IC Symp. Tech. Dig.*, 1984, pp. 129–132.

[3] N. Cirillo, Jr., J. Abrokwah, and S. A. Jamison, "A self-aligned gate modulation doped (Al,Ga)As/GaAs FET IC process," in *GaAs IC Symp. Tech. Dig.*, 1984, pp. 167–170.

[4] C. P. Lee, D. Hou, S. J. Lee, D. L. Miller, and R. J. Anderson, "Ultra high speed digital integrated circuits using GaAs/GaAlAs high electron mobility transistors," in *GaAs IC Symp. Tech. Dig.*, 1983, pp. 162–165.

[5] Y. Takanashi, M. Hirano, and T. Sugeta, "Control of threshold voltage of AlGaAs/GaAs 2 DEG FET's through heat treatment," *IEEE Electron Device Lett.*, vol. EDL-5, pp. 241–243, 1984.

[6] P. N. Tung, P. Delescluse, D. Delagebeaudeuf, M. Laviron, J. Chaplart, and N. T. Linh, "High-speed low-power DCFL using planar two-dimensional electron gas FET technology," *Electron. Lett.*, vol. 18, pp. 517–519, 1982.

[7] T. Mizutani, N. Kato, K. Osafune, and M. Ohmuri, "Gigabit logic operation with enhancement-mode GaAs MESFET IC's," *IEEE Trans. Electron. Devices*, vol. ED-29, pp. 199–204, 1982.

HIGH-SPEED RING OSCILLATORS USING PLANAR p^+-GATE n-AlGaAs/GaAs 2DEG FETs

Indexing terms: Semiconductor devices and materials, Logic circuits

An E/R DCFL ring oscillator has been fabricated using a planar n-AlGaAs/GaAs 2DEG enhancement FET with p^+-gate structure. For 1 μm gate length E-FETs, the standard deviation of threshold voltage was as small as 17·6 mV at 0·016 V average threshold voltage. At room temperature, a gate propagation delay of 15 ps/gate at a 5·25 mW/gate power dissipation was obtained from a 25-stage ring oscillator. A small minimum power-delay product of 14 fJ/gate was attained at 0·65 V supply voltage.

Introduction: N-AlGaAs/GaAs two-dimensional electron gas FETs (2DEG FETs) have attracted much attention for applications to ultra-high-speed ICs and low-noise amplifiers, due to the high electron mobility and velocity at the heterointerface.[1-6] However, since the FETs previously reported usually have a recessed gate structure,[1-4] precise etching in the n-layer beneath the gate is required to control threshold voltage, which degrades threshold voltage uniformity, surface flatness and so on.

The authors proposed and demonstrated a new planar enhancement/depletion FET technology without gate recessing, which could give high device performance.[6] E-FETs, called p^+-gate 2DEG E-FETs, consisted of a p^+-GaAs gate on the n-AlGaAs/GaAs structure.

This letter presents the first successful planar p^+-gate 2DEG E-FET application to digital ICs. Moreover, fast ring oscillator performance and good threshold voltage uniformity have also been demonstrated.

DCFL circuit structure: E/R direct-couple FET logic (DCFL) ring oscillators were fabricated to evaluate logic performance. A basic structure for planar DCFL circuits is shown in Fig. 1.

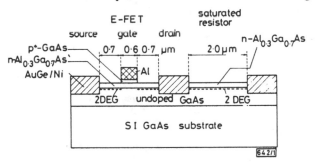

Fig. 1 *Basic structure for a planar p^+-gate DCFL circuit*

The driver is a p^+-gate 2DEG E-FET. The load is a saturated resistor using 2DEG. The present E-FET differs from the conventional Schottky gate structure in regard to the highly doped p^+-GaAs layer just under gate metal on an n-AlGaAs/GaAs structure. Therefore, owing to a high diffusion potential of the p^+-n junction, there is no 2DEG under the gate in thermal equilibrium. On the other hand, outside the gate and on the load, there is much 2DEG. The surface potential for saturated resistor is smaller, by about 0·6 V, than the diffusion potential for the present E-FET.

The main features for DCFL circuits using the p^+-gate structure are as follows:

(i) The gate potential barrier is so high that the present struc-

ture has an additional advantage of higher logic swing than the conventional Schottky gate structure.

(ii) The threshold voltage for E-FET is determined without any etching, so that the threshold voltage uniformity is guaranteed by the good uniformity that MBE layers offer. Also, it is hardly degraded by processing.

(iii) A DCFL circuit can be formed by a simple process with selective p^+-layer etching.

Modulation-doped AlGaAs/GaAs heterostructures, used for E/R circuit fabrication, were grown on a semi-insulating GaAs substrate by MBE at 600°C. The layers were undoped GaAs (8000 Å) buffer layer, Si-doped n-Al$_{0.3}$Ga$_{0.7}$As (270 Å, 2.5×10^{18} cm^{-3}) and Be-doped p^+-GaAs (80 Å, 3×10^{19} cm^{-3}) gate layer. The p^+-GaAs layer was so highly doped that good ohmic contact can be obtained with Al gate metal. The gate to source and gate to drain spacings for the E-FET were 0·7 μm. The electrode spacing for the load was 2 μm.

The fabrication process for the DCFL circuit was as follows. First, the active region was isolated by mesa etching. Then, an Al film for the gate was evaporated and a gate electrode was formed by side etching. Next, the source and the drain ohmic contacts both for the driver and for the load were formed by evaporating AuGe/Ni/Au, followed by alloying at 450°C. Finally, the p^+-GaAs layer outside the gate and on the load was etched. An SiO$_2$ film was used for an interlayer insulating film and Ti/Pt/Au were used for second-level interconnection.

DC characteristics: Characteristics for a typical E-FET, with 0·6 μm gate length and 20 μm gate width, and a saturated resistor of 20 μm width at room temperature, are shown in Fig. 2. The driver saturation current to load saturation

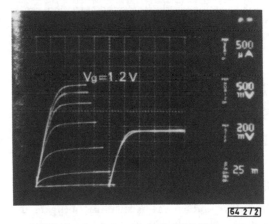

Fig. 2 *I/V characteristics of an E-FET (left) with $L_g = 0.6$ μm and $W_g = 20$ μm and a saturated resistor (right) at room temperature*

Horizontal axis: 0·2 V/div; vertical axis: 0·5 mA/div; gate voltage: 0·2 V/step from 0 V to 1·2 V

current ratio is about 2 for a 1·2 V logic swing with the same gate width, which is adequate for high-speed operation. For a typical E-FET, the extrinsic transconductance g_m was 210 mS/mm at room temperature, with a threshold voltage of 0·08 V. On cooling to 77 K, g_m increased to 365 mS/mm and the threshold voltage shifted 0·28 V. The source resistance for this present driver was high (2·5 Ωmm at room temperature and 1·1 Ωmm at 77 K). High performance is expected by reducing the source resistance.

A histogram of the threshold voltage for 1 μm gate length

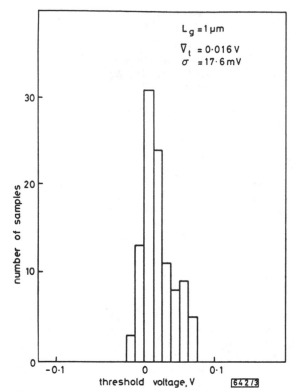

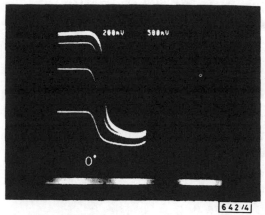

Fig. 4 *Transfer characteristics of a 2-stage inverter*

Supply voltages are 0·5, 1·0, 1·5, 2·0 and 2·5 V. Horizontal axis (input voltage): 0·5 V/div; vertical axis (output voltage): 0·2 V/div

Fig. 3 *Histogram of threshold voltage for 1 μm gate length E-FETs*

E-FETs on a 2 in (50·8 mm)-diameter wafer is shown in Fig. 3. An average threshold voltage was 0·016 V. A small standard deviation of 17·6 mV for the threshold voltage was obtained. Although variations in thickness and carrier concentration for this wafer were relatively large, ±3% at present, good uniformity of threshold voltage was obtained, proving the desired p^+-gate structure features.

Ring oscillator performance: Switching performance for the developed DCFL circuit was evaluated from 25-stage ring oscillators. The driver FETs had a 0·6 μm gate length and 40 μm gate width. The saturated load resistors were also of 40 μm width.

The propagation delays of 19 ps/gate and 15 ps/gate were typically obtained at 5·7 mW/gate and 3·7 mW/gate power dissipations at room temperature and at 77 K, respectively. A minimum propagation delay of 15 ps/gate was measured at a 5·25 mW/gate power dissipation at room temperature. The minimum power-delay product was as small as 14 fJ/gate at 0·65 V supply voltage at room temperature.

Fig. 4 shows the transfer characteristics for a 2-stage inverter. A higher logic swing than 1·1 V was obtained at room temperature. This high logic swing is due to the higher gate potential barrier, as described before, which allows an improvement in the noise margin.

It really has been confirmed that the DCFL circuit, using planar p^+-gate 2DEG FETs, had a large logic swing, good threshold voltage uniformity, high speed and small power-delay products. Further performance improvement can be accomplished by reducing parasitic source resistance and by optimising the current ratio between driver and load.

Conclusions: A planar 2DEG enhancement FET with p^+-gate structure has been applied to digital ICs. Good threshold voltage uniformity where standard deviation was 17·6 mV for 1 μm gate length was demonstrated. Ring oscillators with E/R DCFL circuits were fabricated. A propagation delay of 15 ps/gate was obtained at a 5·25 mW/gate power dissipation at room temperature. A high logic swing, more than 1·1 V, was obtained. Good threshold voltage uniformity, high speed and high logic swing characteristics indicate that the present structure provides high potentiality for high-speed ICs.

Acknowledgment: The authors would like to thank H. Miyamoto, Dr. N. Goto, T. Maeda and M. Ogawa for their valuable discussions and support. They also thank Dr. Y. Takayama and Dr. N. Kawamura for their helpful suggestions and encouragement.

Y. SUZUKI   *1st April 1986*
H. HIDA
H. TOYOSHIMA
K. OHATA

Microelectronics Research Laboratories
NEC Corporation
4-1-1 Miyazaki, Miyamae-ku, Kawasaki 213, Japan

References

1 MIMURA, T., HIYAMIZU, S., FUJI, T., and NANBU, K.: 'A new field-effect transistor with selectively-doped GaAs/n-Al$_x$Ga$_{1-x}$As heterojunctions', *Jpn. J. Appl. Phys.*, 1980, **19**, pp. L225–L227
2 LEE, C. P., MILLER, D. L., HOU, D., and ANDERSON, R. J.: 'Ultra high speed digital integrated circuits using GaAs/AlGaAs high electron mobility transistors'. IEEE GaAs IC symposium, Technical digest, 1983, pp. 162–165
3 PEI, S. S., SHAH, N. J., HENDEL, R. H., TU, C. W., and DINGLE, R.: 'Ultra high speed integrated circuits with selectively doped heterostructure transistors'. IEEE GaAs IC symposium, Technical digest, 1984, pp. 129–131
4 KURODA, S., MIMURA, T., SUZUKI, M., KOBAYASHI, N., NISHIUCHI, K., SHIBATOMI, A., and ABE, M.: 'New device structure for 4 Kb HEMT SRAM'. IEEE GaAs IC symposium, Technical digest, 1984, pp. 125–128
5 CIRILLO, N. C., ABROKWAH, J. K., FRAASCH, A. M., and VOLD, P. J.: 'Ultra-high-speed ring oscillators based on self-aligned-gate modulation-doped n^+-(Al, Ga)As/GaAs FETs', *Electron. Lett.*, 1985, **21**, pp. 772–773
6 OHATA, K., OGAWA, M., HIDA, H., and MIYAMOTO, H.: 'Planar p^+-gate E/D technology for n-AlGaAs/GaAs selectively doped high speed ICs'. Proc. of 11th int. symp. on GaAs and related compounds, Technical digest, 1984, pp. 653–658

p-Channel (Al,Ga)As/GaAs Modulation-Doped Logic Gates

R. A. KIEHL, MEMBER, IEEE, AND A. C. GOSSARD

Abstract—The operation of logic gates composed of modulation-doped field-effect transistors based on two-dimensional hole-gas conduction is reported for the first time. Direct coupled inverters fabricated on an MBE grown Be-doped (Al,Ga)As/GaAs wafer having a sheet density of 1.5×10^{12} cm^{-2} and a 77K mobility of 1800 cm^2/V·s exhibit logic states of -0.25 and -0.98 V at 77K for a -1-V bias. Propagation delays of 233.0 ps/gate are obtained at 77K in ring-oscillator circuits with a power dissipation of 0.31 mW/gate. Power-delay products as low as 9.1 fJ are also obtained.

THE ENHANCED low-temperature mobility of electrons confined near the interface of a modulation-doped heterojunction has been exploited over the past several years to realize high performance n-channel field-effect transistors. The results for logic circuits based on this approach in the (Al,Ga)As system have been impressive and include binary counters [1] operating above 10 Gb/s and 1-kB RAM's [2] with access times under 1 ns. The n-channel FET is generally employed in high-speed circuits due to the lower effective-mass of electrons compared to holes. However, for complementary circuits, high-speed p-channel FET's are also a necessity.

Two-dimensional hole-gas mobilities as high as 5000 cm^2/V·s at 77K and 43,000 cm^2/V·s at 4.2K have now been demonstrated [3] in a p-(Al,Ga)As/GaAs modulation-doped heterostructure analogous to the familiar n-(Al,Ga)As/GaAs structure. These values are to be compared with a mobility of approximately 400 cm^2/V·s for holes in bulk GaAs at a comparable free carrier density. Recently, p-channel (Al,Ga)As/GaAs modulation-doped FET's having transconductance values as high as 28 mS/mm at 77K and 44 mS/mm at 4.2 K have been demonstrated [4]. In this letter, we report the first results on p-channel logic gates based on this approach.

The direct coupled FET inverter gates are fabricated on a four layer Be-doped structure grown by MBE on a Cr-doped GaAs substrate. In their order of growth, the layers are 1 μm

of undoped GaAs, 75 Å of undoped Al$_{.6}$Ga$_{.4}$As, 450 Å of 2×10^{18} cm^{-3} p-Al$_{.6}$Ga$_{.4}$As, and 50 Å of 4×10^{18} cm^{-3} GaAs. The sheet density for holes in this structure is 1.5×10^{12} cm^{-2} and the Hall mobility is 190 and 1800 cm^2/V·s at 300 and 77K, respectively. A Au-Be-based ohmic contact metallization is employed. Transmission line measurements give a contact resistance of 19 Ω·mm and 18 000 $\Omega/\square$ at 300K. The values for these parameters are 3.3 Ω·mm and 2 500 $\Omega/\square$ at 77 K. The Ti/Au gates, which were 1.5 μm in length, were chemically recessed into the p-Al$_{.6}$Ga$_{.4}$As layer for threshold control of enhancement-mode driver FET's and current control of depletion-mode load FET's. The turn-on voltage measured for the Schottky barrier gates is -1.5 V, as shown in Fig. 1.

Typical current–voltage characteristics for a 25-μm-wide driver FET and a 12-μm-wide load FET are shown in Fig. 2 for both 300 and 77K. As can be seen, the threshold for the driver FET is approximately 0 V. At room temperature the driver characteristic exhibits a peak transconductance g_m^{Max} of 5 mS/mm, a source–drain resistance R_{sd} of 154 Ω·mm, and an output resistance R_o (at $V_{ds} = -2$, $V_g = -0.75$) of 3400 Ω·mm. At 77 K, a g_m^{Max} of 27 mS/mm is observed together with R_{sd} and R_o values equal to 22 Ω·mm and 12 000 Ω·mm, respectively. The best value of g_m^{Max} obtained on the present wafer at 77K is 35 mS/mm.

The dramatic improvement in the driver characteristics upon cooling is primarily due to the greatly enhanced hole mobility at 77K. While the gate voltage range for high transconductance in Fig. 2 (≈ 0.6 V) is adequate for providing a useful logic swing, we mention that considerably larger ranges have been observed for other samples. Due to the low hole mobility at room temperature, the load characteristic of the inverter is nearly linear at 300K, as shown in Fig. 2. The enhanced mobility at low temperature results, however, in a nonlinear characteristic which is more desirable for achieving a good transfer characteristic.

A typical transfer characteristic is shown in Fig. 3. This data was obtained by directly measuring the voltage transfer on an

Manuscript received June 25, 1984; revised August 2, 1984.
The authors are with AT&T Bell Laboratories, Murray Hill, NJ 07974.

Reprinted from *IEEE Electron Device Letters,* vol. EDL-5, no. 10, pp. 420–422, October 1984.

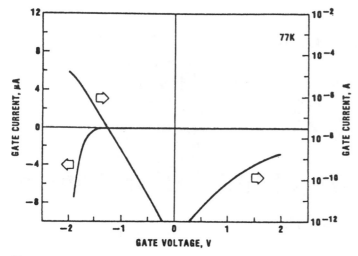

Fig. 1. Gate characteristic measured for a driver FET at 77K. The characteristic shows a turn-on voltage of approximately −1.5 V, which is the same value as seen at 300K.

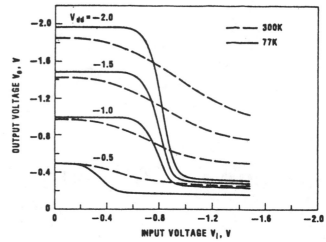

Fig. 3. Transfer characteristic at 300K (dashed curves) and 77K (solid curves) for various levels of the supply voltage V_{dd}.

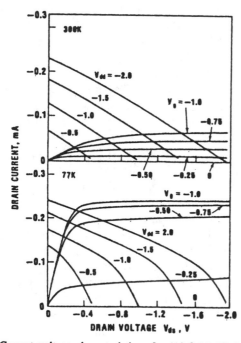

Fig. 2. Current–voltage characteristics of p-(Al,Ga)As/GaAs inverters. The parameters for the drive and load characteristics are gate voltage V_g and supply voltage V_{dd}, respectively.

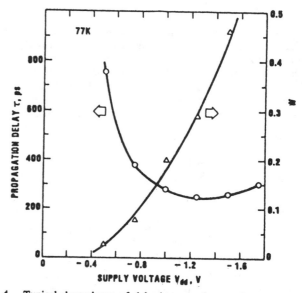

Fig. 4. Typical dependence of delay/stage and power/stage on supply voltage at 77K.

inverter having transistor characteristics nearly identical to those in Fig. 2, but a driver threshold equal to −0.3 V. At 300K the gain is seen to be too small to produce the stable logic states needed for inverter operation. At 77K, however, stable logic states are achieved for V_{dd} as low at −0.4 V. At a supply voltage of −0.1 V, the voltage gain of this inverter is 4.0 and the logic levels are −0.25 and −0.98 V. At V_{dd} = 2.0 V, the gain reaches 8.

Measurements were made on 11-stage ring-oscillator circuits located in a region of the wafer near the inverter of Figs. 2 and 3. The driver and load gate widths of the oscillator stages were 25 and 12 μm, respectively. The range in propagation delay seen for various circuits was 233–379 ps/stage at 77K. The 233-ps delay was obtained with a power dissipation of

0.31 mW/stage at a supply voltage of −2.0 V. The typical bias dependence of delay and power is shown in Fig. 4. The lowest value of power-delay product achieved in our experiments is 9.1 fJ and was achieved at −0.4-V bias with a delay of 910 ps/ stage. No oscillations have been obtained in our circuits at 300K.

In conclusion, the results presented here represent the first demonstration of high-speed logic gates based on a p-channel modulation-doped structure. To our knowledge this is also the first report on ring-oscillator performance in any p-channel structure not based on silicon. The results demonstrate the attractiveness of p-channel modulation-doped structures for obtaining the transfer characteristics and high-speed switching operation needed for the development of high-speed logic circuits. Owing to the excellent current saturation in the driver FET, considerable improvement in the transfer characteristic will be possible with improvements in the load design. In particular, our present results give a $g_m{}^{Max} R_0$ limit (ideal load limit) for the inverter gain of more than 100. The general performance will also improve as 2-D hole mobilities increase.

Already, a 2-D hole mobility almost three times greater than that in our present structures has been achieved [3]. The ultimate limit in switching speed for this approach will be determined by the details of carrier heating in the 2-D hole gas, which is currently under investigation.

We feel that a p-channel modulation-doped structure provides an attractive p-type element for use in ultra-low-power complementary circuits analogous to Si-CMOS and, therefore, is of great interest for the development of future ultra-high-speed VLSI integrated circuits.

ACKNOWLEDGMENT

The authors would like to thank H. L. Stormer for many useful discussions concerning this work. W. Wiegmann is gratefully acknowledged for his work on the MBE growth. We also thank K. Baldwin, R. J. Malik and V. G. Keramidas for useful discussions regarding the device fabrication.

REFERENCES

[1] S. S. Pei, R. H. Hendel, R. A. Kiehl, C. W. Tu, M. D. Feuer, and R. Dingle, "Selectively doped heterostructure transistors for ultrahigh speed integrated circuits," presented at 1984 Device Research Conf., Santa Barbara, CA, June 18–20, 1984.

[2] K. Nishiuchi, N. Kobayashi, S. Kuroda, S. Notomi, T. Nimura, M. Abe, and M. Kobayashi, "A subnanosecond HEMT 1 Kb SRAM," in *IEEE Intl. Solid State Circ. Conf., Tech. Dig.,* 1984, pp. 48–49.

[3] H. L. Stormer, A. C. Gossard, W. Wiegmann, R. Blondel, and K. Baldwin, "Temperature dependence of the mobility of two-dimensional hole systems in modulation-doped GaAs–(AlGa)As," *Appl. Phys. Lett.,* vol. 44, pp. 193–141, Jan. 1984.

[4] H. L. Stormer, K. Baldwin, A. C. Gossard, and W. Wiegmann, "Modulation-doped field effect transistor based on a two dimensional hole gas," *Appl. Phys. Lett.,* vol. 44, pp. 1062–1064, June 1984.

CIRCUIT PERFORMANCE OF COMPLEMENTARY HETEROSTRUCTURE MISFET INVERTER USING HIGH MOBILITY 2DEG AND 2DHG

T. Mizutani, S. Fujita, M. Hirano, and N. Kondo

NTT Electrical Communications Laboratories
Morinosato Wakamiya, Atsugi-shi, Kanagawa, 243-01 Japan

ABSTRACT

A complementary circuit employing high mobility two-dimensional electron and hole gases(2DEG and 2DHG) induced at the AlGaAs/GaAs heterointerface has been successfully fabricated on an n^+-Ge/undoped AlGaAs/undoped GaAs heterostructure. Transfer characteristics of a two-stage inverter show sufficiently high and low output levels. A 15-stage ring oscillator shows a minimum delay time of 94 ps at 300 K and 64 ps at 77 K. The circuit simulation using a new gate current model shows that, for multi-input logic gate, a NAND gate configuration is superior to a NOR gate configuration, and that a high speed operation below 20 ps at 77 K is possible by reducing the gate length to 0.5 μm.

INTRODUCTION

There has been much interest in a complementary circuit using high mobility two dimensional electron and hole gases(2DEG and 2DHG)(1)-(3) because of its potential for high speed, low power dissipation, and large noise margin. The authors have reported preliminary results of the first demonstration of a complementary circuit(1) using AlGaAs/GaAs heterostructure MISFETs. In order to apply the complementary circuit to LSIs, it is necessary to investigate the circuit performance of MISFET complementary inverters. In this paper, we report further improvement in ring oscillator performance, and discuss the observed transfer characteristics based on a new gate current model. Finally, we present results of circuit simulation for scaled-down devices.

DEVICE STRUCTURE

The complementary circuit is fabricated on an n^+-Ge(200 nm)/undoped GaAs(5 nm)/undoped AlGaAs (30 nm)/undoped GaAs(300 nm) heterostructure grown by molecular beam epitaxy (MBE). The schematic cross section of the complementary n- and p- channel MISFETs is shown in Fig. 1. By employing this structure, it becomes very simple to integrate n- and p- channel MISFETs on the same wafer. This is because doping and thickness control is not necessary for the threshold voltage control(4), (5), and because 2DEG and 2DHG can be induced at the same AlGaAs/GaAs

heterointerface.

An optimum AlAs mole fraction for AlGaAs gate barrier insulator was determined by measuring GaAs/AlGaAs/GaAs SIS diode I-V characteristics (6). Measured ΔE_c and ΔE_v are shown in Fig. 2 by open and closed circles, respectively. Solid and broken lines are calculated ΔE_c and ΔE_v, respectively. Here, the valence band discontinuity ΔE_v is assumed to be equal to $\Delta E_v = 0.35x \, \Delta Eg$. ΔE_c, which dominates the electron current through the AlGaAs /GaAs heterobarrier,

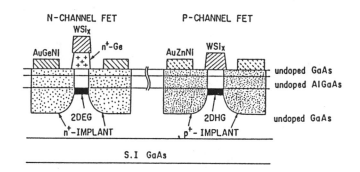

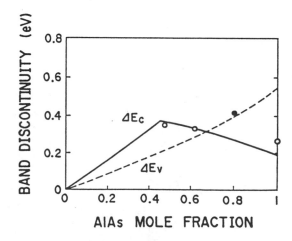

Fig. 1 Schematic cross-section of the complementary n- and p-channel MISFETs.

Fig. 2 Band discontinuity as a function of AlAs mole fraction.

Reprinted from *Rec. of the IEEE GaAs Integrated Circuits Symp.*, pp. 107–110, 1986.

is given by E_C^Γ (GaAs)-E_C^X (AlGaAs) in the indirect band gap region. Then, the optimum AlAs mole fraction to give a same barrier height for electrons and holes is determined to be about 0.7.

DEVICE FABRICATION

The complementary circuit was fabricated as follows. First, the n^+-Ge layer is removed except for the regions of n-channel FETs by reactive ion etching(RIE), followed by mesa etching for the electrical isolation. WSi$_x$ is then deposited by RF sputtering. The gate patterns of n^+-Ge/WSi$_x$ and WSi$_x$ for the n- and p-channel FETs are delineated by RIE. Source and drain regions self aligned to the gates are then formed by Si (n-channel FETs) and Be(p-channel FETs) ion implantation and successive lamp annealing. Ohmic contacts are formed by AuGeNi(n-channel FETs) and AuZnNi(p-channel FETs) metallization and the subsequent alloying. The complementary circuit is completed by the interconnect using the Ti-Au lift-off method.

DEVICE PERFORMANCE

Drain current-voltage characteristics at 300 K for the n- and p-channel FETs with 1.5 μm gate length are shown in Fig. 3. The maximum transconductances are 250 mS/mm for the n-channel and 50 mS/mm for the p-channel FETs. In a separate experiment, an n-channel MISFET with 10 nm AlGaAs thickness and 0.8 μm gate length showed an extrinsic transconductance of 430 mS/mm(7). This result makes evident the considerable room for improvement over the performance reported in this paper.

The two-stage inverter shows good transfer characteristics with a 1 V high-output level and a 0.02 V low-output level for 1 V supply voltage (V_{DD}) at 77 K, as shown in Fig. 4 by closed circles. These sufficiently high and low logic levels imply a larger noise margin than that of a DCFL gate.

A 15-stage ring oscillator(FI/FO=1/1) is successfully operated both at 300 K and 77 K. Fig. 5 shows the minimum delay time at 300 K as a function of the gate-width ratio of the p-channel and n-channel FETs with 1.5 μm gate length. The gate-width ratio of 1 gives the smallest delay time. This can be qualitatively explained by the input capacitance increase of the p-channel FET and by a logic threshold change in accordance with the gate-width ratio. Fig. 6 shows the supply voltage dependence of the delay time and the power dissipation at 300 K. The minimum delay time is 94 ps at V_{DD}=1.5 V. Power dissipation can be estimated to be 580 μW/gate if the logic circuit is operated at clock frequency(f_c) of 350 MHz. When the ring oscillator is operated at V_{DD}=1 V, the delay time is 217 ps and the associated power dissipation is 129 μW/gate at f_c=150 MHz. The minimum delay time at 77 K was 64 ps with a power dissipation of 1.8 mW/gate at V_{DD}=2.1 V.

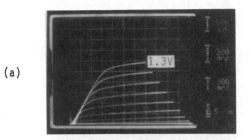
(a)

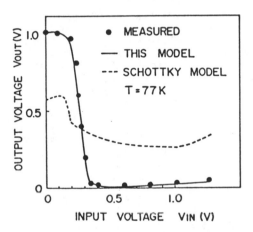
(b)

Fig. 3 Drain Current-voltage characteristics of MISFETs at 300 K;
(a)n-channel, (b)p-channel.

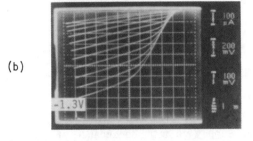

Fig. 4 Transfer characteristics of the two-stage inverter at 77 K.

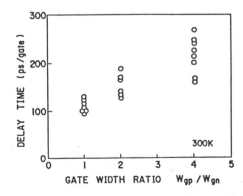

Fig. 5 Gate width ratio dependence of the minimum delay time at 300 K.

Ga As IC Symposium

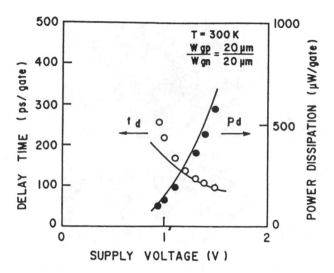

Fig. 6 Supply voltage dependence of a delay time and the power dissipation at 300 K. Solid lines are the simulated results.

GATE CURRENT MODEL

In order to fully apply the complementary MISFETs to LSIs, a circuit simulation is important for predicting LSI performance. In this case, a precise gate current model must be developed in addition to a drain current model, because the gate current causes extra power consumption, and because both the high and low logic levels are affected by the gate current through the source resistance. A Schottky diode model, which is widely used in a GaAs MESFET circuit design, is not suitable for AlGaAs/GaAs heterostructure MISFET LSIs. This is because the gate voltage(V_g) dependence of the barrier height is different between MISFETs and MESFETs.

Fig. 7 shows an energy band diagram of an n^+-Ge gate MISFET. The surface potential(ϕ_s) is almost proportional to V_g for gate voltages smaller than the threshold voltage(V_T). However, when V_g becomes larger than V_T and a two-dimensional electron gas is induced at the heterointerface, most of the gate voltage increase is applied across the undoped AlGaAs barrier layer. Therefore, the surface potential change caused by the gate bias is very small for V_g values larger than V_T.

A new gate current model(8) takes into account the small V_g dependence of ϕ_s. Using the effective barrier height for electrons measured from the Fermi level, the gate current can be expressed as folllows assuming a thermionic emission model:

$$Ig = A^* T^2 exp(-q\phi_b/kT)(exp(q\phi_s/kT)-1)$$

$$\phi_s = \begin{cases} V_g & \cdots\cdots\cdots (V_g < V_t) \\ V_t + S(V_g - V_t)^R & \cdots\cdots (V_g > V_t) \end{cases}$$

$$\phi_b = V_t + \Delta E_c/q$$

Here, ΔE_c is the conduction band discontinuity, R and S are fitting parameters, and the other parameters have the usual meanings. To achieve the small V_g dependence of ϕ_s, R and S should be less than 1[9].

Fig. 8 shows a simulated gate current(solid line) comparing it with the experimental results. Here, R=0.5 and S=0.21 are used, and ΔE_c is assumed to be 0.4 eV for an $Al_{0.5}Ga_{0.5}As/GaAs$ heterostructure. The calculated results using the present model agree well with the measured data as shown in the figure. For a p-channel FET, the gate current can be expressed by using valence band discontinuity ΔE_v instead of ΔE_c and using A^* and ϕ_b for holes.

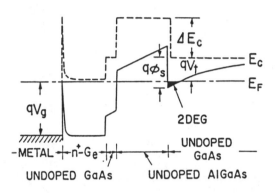

Fig. 7 Energy band diagram of an n^+-Ge gate MISFETs. Dashed and Solid lines show the conduction band edges for zero and positive gate voltages, respectively.

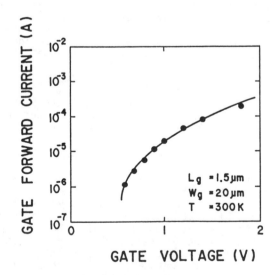

Fig 8, Gate current-voltage characteristics for the n-channel MISFET. Closed circles are experimental results and the solid line is a simulated one.

CIRCUIT SIMULATION

The simulated transfer characteristics of a two-stage complementary inverter show a good agreement with the experimental results as shown in Fig. 4 by a solid line. Here, an electron velocity saturation is taken into account for the FET drain current model. The conventional Schottky model does not predict sufficiently high and low logic levels, as shown by a broken line. This is because the Schottky model overestimates the gate current, which results in voltage drops across the source resistances of n- and p-channel FETs. The satisfactorily high and low output levels for the MISFETs indicate the superiority of the MISFET over the MESFET for the complementary circuit.

The simulated delay time and the power dissipation are in good agreement with the experimental results as shown in Fig.6 by the solid lines. Fig. 9 shows a simulated delay time as a function of fan-in/fan-out(FI/FO) for NAND and NOR logic gates at 77 K. Here, the supply voltage is 1 V and the gate length is 0.5 μm. A high speed operation below 20 ps is possible for FI/FO=1/1 with zero load capacitance(C_L). The delay time increase due to the FI/FO increase is 23 ps/FI=FO for the NAND gate and 33 ps/FI=FO for the NOR gate, respectively. The delay time increase due to the load capacitance is 13 ps/100fF for the NAND gate and 14 ps/100fF for the NOR gate, respectively. Consequently, the NAND gate is superior to the NOR gate when the FI/FO is large. This is because the n-channel FETs with large current driving capability are connected in series in the NAND gate. For the NOR gate, on the other hand, p-channel FETs with a smaller current driving capability are connected in series. For a FI/FO=3/3 and 100 fF load capacitance condition, a 54 ps delay time and a 57 μW power dissipation/gate are possible for a NAND gate.

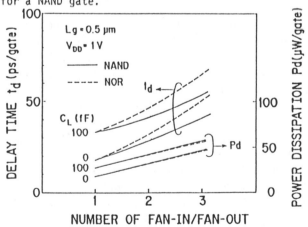

Fig. 9 Delay time and power dissipation as a function of fan-in/fan-out.

SUMMARY

A complementary circuit employing high mobility two-dimensional electron and hole gases(2DEG and 2DHG) induced at the AlGaAs/GaAs heterointerface has been successfully fabricated on an n$^+$-Ge/undoped AlGaAs/undoped GaAs heterostructure grown by MBE. Transfer characteristics of a two-stage inverter show sufficiently high and low output levels, 1V and 0.02 V, respectively. A 15-stage ring oscillator shows a minimum delay time of 94 ps at 300 K and 64 ps at 77 K. A new gate current model taking into account the small gate voltage dependence of a surface potential explains well the transfer characteristics of a two-stage inverter. The circuit simulation predicts that, for multi-input logic gates, a NAND gate configuration is superior to a NOR gate configuration, and that a high speed operation below 20 ps at 77 K is possible by reducing the gate length to 0.5 μm.

ACKNOWLEDGMENTS

The authors would like to express their thank to K. Maezawa for the measurement of band discontinuity and K. Arai, F. Yanagawa, and K. Oe for their fruitful discussions.

REFERENCES

(1) T. Mizutani, S. Fujita, and F. Yanagawa, "Complementary Circuit with AlGaAs/GaAs Heterostructure MISFETs Using High Mobility Two Dimensional Electron and Hole Gases," Int. Symp. GaAs and Related Compounds, Karuizawa, 1985 Int. Phys. Conf. Series No.79, p.733-734, Sep. 1985

(2) K.Matsumoto, M.Ogura, T.Wada, T.Yao, Y.Hayashi, N.Hashizume, M.Kato, N.Fukuhara, T.Miyashita, and H.Hirashima, "p$^+$-GaAs Gate p-Channel GaAs SIS FET Self-Aligned by Ion-Implantation," ibid., p.625-630, Sep. 1985

(3) N.C.Cirillo,Jr., M.S.Shur, P.J.Vold, J.K.Abrokwah, R.R.Daniels, and O.N.Tufte, "Complementary Heterostructure Insulated Gate Field Effect Transistors(HIGFETs)," IEDM Tech. Dig., p.317-320, Dec. 1985

(4) K.Arai, T.Mizutani, and F.Yanagawa, "An n$^+$-Ge Gate MIS-Like Heterostructure FET," Int. Symp. GaAs and Related Compounds, Karuizawa, 1985 Int. Phys. Conf. Series No.79,p.631-636, Sep. 1985

(5) M.Hirano and N.Kondo, "Threshold-Voltage Stability of P-Channel AlGaAs/GaAs MIS-Like Heterostructure FETs," Jpn. J. Appl. Phys., vol.25, to be published, 1986

(6) K.Maezawa, T.Mizutani, and F.Yanagawa, "Barrier Height in Indirect Bandgap AlGaAs/ GaAs Hetero-Junction Determined with n-Semiconductor/Insulator/Semiconductor Diodes," Jpn. J. Appl. Phys., vol.25, p.L557-L559, July 1986

(7) K.Maezawa, T.Mizutani, K.Arai, and F. Yanagawa, "Large Transconductance n$^+$-Ge Gate AlGaAs/GaAs MISFET with Thin Gate Insulator," IEEE Electron Device Letters, vol.EDL-7, p.454-456, July 1986

(8) S.Fujita, T.Mizutani, "Gate Current Model for Heterostructure MISFETs," Trans. IECE Japan, vol.288-290, April 1986

A High-Speed HEMT 1.5K Gate Array

YUU WATANABE, KIYOSHI KAJII, YOSHIMI ASADA, KOUICHIRO ODANI,
TAKASHI MIMURA, MEMBER, IEEE, AND MASAYUKI ABE, MEMBER, IEEE

Abstract—A 1.5K-gate HEMT gate array has been developed, using a direct-coupled FET logic (DCFL) circuit. The chip, containing 1520 basic cells and 72 I/O cells, was 5.5 mm × 5.6 mm. The basic circuit was designed for two different threshold voltages for D-HEMT, in order to obtain high-speed performance both at room temperature and low temperature.

Fully functional 8 × 8 bit parallel multipliers were fabricated on the gate-array chip. At room temperature a multiplication time of 3.7 ns including I/O buffer delay was achieved with power dissipation of 6.0 W at a supply voltage of 1.6 V, and at liquid-nitrogen temperature multiplication time was 3.1 ns where the supply voltage was 0.95 V and the power dissipation was 3.2 W.

I. INTRODUCTION

SINCE THE announcement of the high electron mobility transistor (HEMT) [1], a great deal of interest in HEMT technology has been shown. In the field of digital circuits, the HEMT is one of the most promising devices to realize very-high-speed systems. A switching time of 5.8 ps per gate at 77 K has been achieved for 0.35-μm gate devices [2] and several HEMT IC's have also shown high-speed performamce [3]–[9].

To construct the next generation system, ultra-high-speed LSI's are required. To meet these requirements, compound semiconductor device technology has been aggressively developed, as well as Si bipolar technology. Particularly in gate-array applications, a DCFL GaAs 2K gate array [10], a BFL GaAs 3K gate array [11], and a HBT 4K gate array [12] have been reported to date. In order to realize such LSI level complexity with HEMT's, it is necessary to control device parameters uniformly and precisely, and to reduce surface defects on MBE-grown wafers. We have overcome these problems by adopting selective dry-etching technology and optimizing the MBE growth conditions. Using these technologies in a DCFL circuit, we designed a 1.5K gate array as an adoption of HEMT devices to logic LSI, and applied it in an 8 × 8 bit parallel multiplier. In the basic circuit design, the differences in device parameters at room temperature and at low temperature were taken into account.

This paper describes the design of the gate-array fabrication process, and its application to an 8 × 8 bit parallel multiplier.

II. HEMT DEVICE CHARACTERISTICS AND DCFL CIRCUIT

To design a HEMT logic circuit, we must pay attention to special characteristics of the HEMT and their variation at low temperature. Fig. 1 shows the drain current and the transconductance of a HEMT with 1.2-μm gate length as a function of gate voltage at room temperature. It is noted that at gate voltages higher than about 0.6 V, transconductance declines. This is quite different from other devices such as the MESFET. The most effective way to obtain high-speed performance in HEMT IC's is to use this high-transconductance voltage region in circuit operation. This means that the desirable logic swing in a HEMT logic circuit is around 0.7 V.

At low-temperature, HEMT dc characteristics change remarkably. As shown in Fig. 2, at 77 K the current driving capability is enhanced and threshold voltage shifts about 0.2 V. Though the high performance at 77 K is especially attractive for high-speed IC applications, we must take account of these variations in device parameters to design HEMT logic circuits for low-temperatare operation.

Direct-coupled FET logic (DCFL) is one of the best circuits to make use of the high-transconductance region of the HEMT, in which the logic swing is about 0.8 V. It also has advantages over other circuits in terms of lower power dissipation and simplicity. However, there are some disadvantages in that precise control of device parameters is required, and their variation seriously affects circuit performance. Particularly in HEMT DCFL circuits, a variation of device parameters at low temperature makes it difficult to design a higher speed gate with good noise margin at both room temperature and low temperature. In order to make use of enhanced current drivability at low temperature in the DCFL circuit, it is necessary to increase the load current at low temperature in spite of the threshold voltage shift of the D-HEMT. In Fig. 2, the threshold voltage becomes higher than −0.3 V at 77 K, then the drain–source current at $V_{GS} = 0$ V, i.e., the load current of the DCFL circuit decreases. To increase the load current at low temperature, the threshold voltage of the D-HEMT (V_{THD}) at room temperature should be lower than −0.6 V, although it reduces the noise margin.

Therefore, we designed a basic circuit with a V_{THD} of −1 V for low-temperature operation (first design) and a

Manuscript received October 21, 1986; revised February 4, 1987. This work is part of the National Research and Development Program on "Scientific Computing System," conducted under a program set up by the Agency of Industrial Science and Technology, Ministry of International Trade and Industry.

The authors are with Fujitsu Laboratories Ltd., Fujitsu Limited, 10-1 Morinosato-Wakamiya, Atsugi 243-01, Japan.

IEEE Log Number 8714139.

Reprinted from *IEEE Transactions on Electron Devices*, vol. ED-34, no. 6, pp. 1253–1258, June 1987.

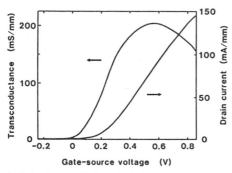

Fig. 1. Typical dc characteristics of a HEMT with 1.2-μm gate length.

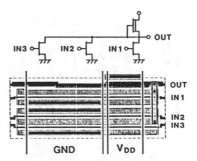

Fig. 4. Equivalent circuit and pattern layout of a basic cell, programmed as a three-input NOR circuit. Cell size is 132 μm $\times$ 54 μm.

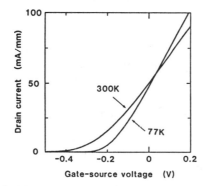

Fig. 2. Typical I_{DS}-V_{GS} characteristics of a HEMT at 300 and 77 K.

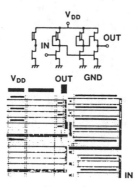

Fig. 5. Equivalent circuit and pattern layout of an I/O cell, programmed as an output buffer.

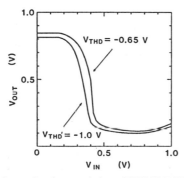

Fig. 3. Simulated transfer characteristics of HEMT DCFL inverters with the same E-HEMT and the same power consumption but different D-HEMT threshold voltages (V_{THD}).

V_{THD} of -0.65 V for room-temperature operation (second design). Fig. 3 shows simulated transfer characteristics of basic inverters at room temperature, using the SPICE II program with the HEMT model implemented. In this simulation, the driver FET's for both inverters are the same, and the load FET's have different sizes to make the load current the same in spite of the V_{THD} difference. The simulated delay times of the inverters at room temperature are 81 ps for the first design and 63 ps for the second design.

III. GATE-ARRAY DESIGN

Fig. 4 shows the pattern layout and the equivalent circuit of a basic cell in the first design. It measures 132 μm $\times$ 54 μm, and includes one D-HEMT and three E-HEMT's, where the gate lengths of the E-HEMT and D-

HEMT are 1 and 5 μm, respectively. A three-input NOR gate can be programmed with this basic cell by interconnection layers. Power supply and ground lines run over the cell, formed with second interconnection metallization.

The I/O cell is 200 μm $\times$ 165 μm, including the 3 D-HEMT's and five E-HEMT's. Fig. 5 shows the pattern layout of an I/O cell programmed as an output buffer, in which an E/E-type push-pull circuit was used. The power supply lines of the I/O cells were independent of those for the basic cells.

Fig. 6 is a photomicrograph of the HEMT 1.5K gate-array chip. It contains 1520 basic cells and 72 I/O cells. There are 20 columns, each containing 76 basic cells. Between the columns, there are 15 interconnection tracks. The interconnection lines are 2 μm wide and the spacing between them is at least 2 μm for both the first and second levels. The contact holes are 2 μm $\times$ 2 μm. To keep a sufficient noise margin in DCFL circuits with small logic swing, the supply-voltage drop and ground-voltage rise must be minimized. So we carefully designed the power-line layout, taking into account the resistivity of the power-line metal. In the worst case, the ground-voltage rise is less than 30 mV. There are 24 power pads, and 64 pads are available for I/O signals. The chip is 5.5 mm $\times$ 5.6 mm and incorporates 6656 HEMT's.

As a circuit to test the gate array, we designed an 8 $\times$ 8 bit parallel multiplier. Fig. 7 shows the schematic logic diagram of an 8 $\times$ 8 bit parallel multiplier with a carry-look-ahead circuit (CLA). It consists of half adders, full

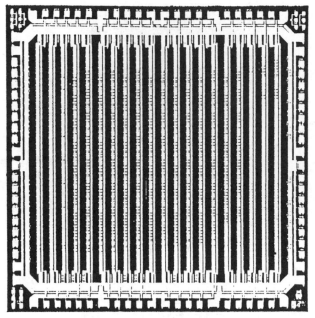

Fig. 6. Photomicrograph of a HEMT 1.5K gate array with chip size of 5.5 mm × 5.6 mm.

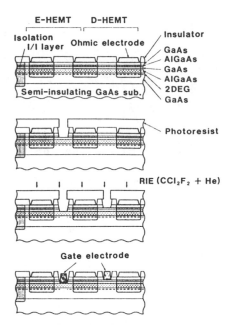

Fig. 8. Process sequence for E/D-HEMT's in DCFL circuit.

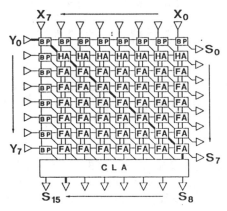

Fig. 7. Logic diagram of an 8 × 8 bit parallel multiplier. BP: bit product. FA: full adder and bit product. HA: half adder and bit product. CLA: carry look ahead. Bold lines show the critical path.

adders, bit products, and a CLA. The bold lines in Fig. 7 show the critical path, where the signal runs from Y_0 to S_{14}. The signal goes through 28 stages of gates along the critical path, including I/O buffers. 888 basic cells were used for the multiplier. Several ring oscillators were designed in the same chip to evaluate gate-delay performance. Including other test circuits, 74 percent of the cell array was used.

IV. FABRICATION PROCESS

Fig. 8 shows the process sequence for E/D-HEMT's in a DCFL circuit. An MBE-grown wafer with a 2-in diameter was used. In MBE technology, it is very important to reduce surface defects (oval defects) that affect FET characteristics and degrade IC yield. We used a wafer with a defect density of less than 100 cm^{-2}, which is necessary to obtain reasonable chip yield. Epitaxial layers are very

uniform, and the deviations of thickness and carrier concentration are less than 1 percent. The doping density in the cap layer and n-AlGaAs layer is $1.4 \times 10^{18} \text{ cm}^{-3}$. The AlGaAs layer embedded in the cap GaAs layer acts as an etching stopper against selective dry etching.

In fabrication, first the active region is isolated by 130-keV O^+ ion implantation. The source and drain for E- and D-HEMT's are metallized with AuGe/Au alloy to form ohmic contacts with the electron layer. Next, the very top GaAs layer and thin AlGaAs etching stopper for E-HEMT's are etched off by nonselective wet chemical etching. Then, selective dry etching is performed, which is a key technology for realizing HEMT LSI circuits. In CCl_2F_2 + He discharge there is a great difference in the etching rates between GaAs and AlGaAs. Therefore, applying this technology to the HEMT DCFL structure after pre-etching for an E-HEMT, the very top GaAs layer for D-HEMT's and the GaAs layer under the thin AlGaAs etching stopper for E-HEMT's can be removed in the same RIE run. We can control the gate-recessing process for E- and D-HEMT's uniformly with this technique. Schottky contacts for the E- and D-HEMT's are provided by depositing Al, which is also used as a first interconnection metal layer. Then silicon oxynitride film is deposited as an insulating layer. Finally the second interconnecting metal layer composed of Ti/Pt/Au is formed.

In the first design chip, the average threshold voltage was 0.12 V for the E-HEMT and −1.11 V for the D-HEMT. In the second design chip, the average threshold voltage was 0.10 V for the E-HEMT and −0.58 V for the D-HEMT. The threshold voltage is excellently uniform reflecting the uniformity of MBE-grown epitaxial film and the good controllability of processing. The standard deviation in the threshold voltage was around 10 and 20 mV for E- and D-HEMT's, respectively. The average transconductance was 230 mS/mm, where the gate length was

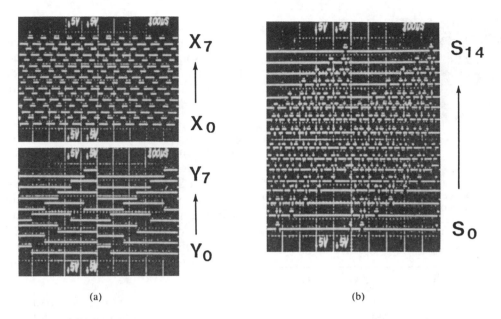

Fig. 9. (a) Input and (b) output signal of the 8 × 8 bit parallel multiplier.

1.2 μm. The 0.2-μm difference from the design parameter is due to photolithography.

V. Gate-Array Performance

Measurement at room temperature was done using a wafer probe card, and in low-temperature measurement a chip mounted on the ceramic flat package was dipped into liquid nitrogen.

To verify proper functioning of the 8 × 8 bit parallel multiplier, a low-frequency function test consisting of 2^{16}(65356) test patterns was performed. Several chips were fully functional, not only at room temperature, but also at liquid-nitrogen temperature. They operated stably over the wide range of supply voltage from 0.7 to 3.0 V. Fig. 9 shows some of the input patterns and successful output waveforms in the function test. These patterns were used as a preliminary function test.

High-speed performance was measured by the delay time through the critical path using the following multiplication:

$$11111111 \times 1000000P = P\overline{P}\overline{P}\overline{P}\overline{P}\overline{P}\overline{P}\overline{P}PPPPPPP.$$

Here P is the signal input into Y_0. This evaluation was performed for the chips that passed the low-frequency function test. Fig. 10 shows Y_0 input and S_{14} output waveforms, which indicate the multiplication time of a typical chip at 77 K including five stages of I/O buffer delay.

In the chips that were designed first, a multiplication time of 4.9 ns at 300 K and 3.1 ns at 77 K was achieved, where supply voltage and total chip power consumption were 2.2 V, 5.8 W and 0.95 V, 3.2 W, respectively. The supply-voltage dependence of the multiplication time and total power dissipation of the chip are shown in Fig. 11. In the second design, a multiplication time of 3.7 ns was achieved at 300 K, where supply voltage and total chip

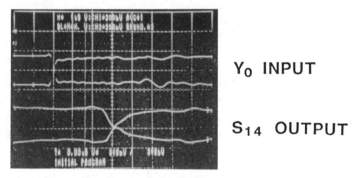

Y$_0$ INPUT

S$_{14}$ OUTPUT

Fig. 10. Example of switching waveforms of the multiplier at 77 K.

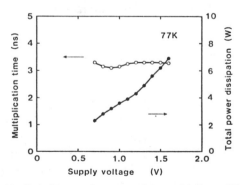

Fig. 11. Switching performance of the multiplier at 77 K.

power consumption were 1.6 V and 6.0 W. But at 77 K the multiplication time was not improved because of the threshold voltage shift problem. Fig. 12 shows Y_0 input and S_{14} output waveforms, indicating the multiplication time of the second design chip at 300 K. A fourth of the chip power was dissipated in the I/O cells.

From the simulated results we found that the I/O buffer delay is 15 percent of the multiplication time. So, the in-

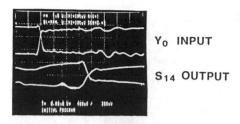

Y_0 INPUT

S_{14} OUTPUT

Fig. 12. Example of switching waveforms of the multiplier at 300 K.

TABLE I
BASIC DELAY TIME AND DELAY TIME INCREMENTS DUE TO FAN-OUT AND
INTERCONNECTION LINES

	1st design		2nd design
	300K	77K	300K
T_d [ps/gate]	89	54	66
t_{FO} [ps/fan-out]	56	22	20
t_{IL} [ps/mm]	31	18	36

TABLE II
MAIN FEATURES OF THE HEMT 1.5K GATE ARRAY

Basic cell	DCFL 3-input NOR
Number of cells	1520
Number of I/Os	72
Element count	6656
Gate length	1.2 μm
Line	2 μm (width & space)
Contact hole	2 μm x 2 μm
Chip size	5.5 mm x 5.6 mm
Test circuit	8 x 8 bit parallel multiplier

Multiplication time
3.7 ns   6.0 W   at 300K
3.1 ns   3.2 W   at 77K

trinsic multiplication time is evaluated to be 3.1 ns at 300 K or 2.6 ns at 77 K.

The dynamic performance of the basic gate was measured in several types of ring oscillators using basic cells with various loading conditions. These ring oscillators were designed on the gate array together with the 8 × 8-bit parallel multiplier. From the results, the basic delay time T_d, the increase in delay time due to fan-out t_{F0}, and the interconnection line length t_{IL} can be calculated. They are summarized in Table I. By rearranging these values we evaluated the loaded delay time, which is more important than the unloaded delay in an actual integrated circuit. With loading of three fan-outs and a 2-mm interconnection line length without any crossover, the delay time becomes 178 ps at 300 K with 2.1-mW dissipation or 134 ps at 77 K with 2.6-mW dissipation. In actual IC's using the gate array, we should take account of additional crossover capacitance between interconnection lines and power lines. Adding several crossovers, the loaded delay becomes a little longer.

Table II summarizes the characteristics of the HEMT 1.5K high-speed gate array.

VI. SUMMARY

A successful HEMT 1.5K gate array has been developed using E/D DCFL circuitry. It contains 1520 basic cells and 72 I/O cells. A fully functional 8 × 8 bit parallel multiplier designed on the gate array achieved a multiplication time of 3.7 ns at 300 K, and 3.1 ns at 77 K including I/O buffer delay, with a total chip power consumption of 6.0 and 3.2 W, respectively. This is the shortest multiplication time of an 8 × 8 bit multiplier to date. These results demonstrate the feasibility of the HEMT LSI for high-speed computer systems.

We believe that further improvements in high-speed performance can be achieved by optimizing the circuit design and/or reducing the gate length. The experimental data for the HEMT submicrometer gate device indicates that a loaded delay of less than 70 ps per gate can be expected.

ACKNOWLEDGMENT

The authors would like to thank T. Misugi and M. Kobayashi for their encouragement and support. They are also thankful to K. Kondo, N. Kobayashi, S. Notomi, and M. Suzuki for their valuable discussions, and to M. Nakayama and N. Ikeda for their technical support.

REFERENCES

[1] T. Mimura, S. Hiyamizu, T. Fujii, and K. Nanbu, "A new field-effect transistor with selectively doped GaAs/n-Al$_x$Ga$_{1-x}$As heterojunctions," *Japan. J. Appl. Phys.*, vol. 19, pp. L225-227, 1980.
[2] N. J. Shah, S.-S. Pei, C. W. Tu, and R. C. Tiberio, "Gate-length dependence of the speed of SSI circuits using submicrometer selectively doped heterostructure transistor technology," *IEEE Trans. Electron Devices*, vol. ED-33, pp. 543-547, 1986.
[3] M. Abe, T. Mimura, K. Nishiuchi, A. Shibatomi, and M. Kobayashi, "HEMT LSI technology for high speed computers," in *GaAs IC Symp. Tech. Dig.*, pp. 158-151, 1983.
[4] K. Nishiuchi, N. Kobayashi, S. Kuroda, S. Notomi, T. Mimura, M. Abe, and M. Kobayashi, "A subnanosecond HEMT 1Kb SRAM," in *ISSCC Dig. Tech. Papers*, pp. 48-49, 314, Feb. 1984.
[5] S. Kuroda, T. Mimura, M. Suzuki, N. Kobayashi, K. Nishiuchi, A. Shibatomi, and M. Abe, "New device structure for 4Kb HEMT SRAM," in *GaAs IC Symp. Tech. Dig.*, pp. 125-128, Oct. 1984.
[6] A. R. Shlier, S. S. Pei, N. J. Shah, C. W. Tu, and G. E. Mahoney, "A high-speed 4 × 4 bit parallel multiplier using selectively doped heterostructure transistors," in *GaAs IC Symp. Tech. Dig.*, pp. 91-93, Oct. 1985.
[7] C. P. Lee, N. H. Sheng, H. F. Lewis, H. T. Wang, D. L. Miller, and J. Donovan, "GaAs/GaAlAs high electron mobility transistors for analog-to-digital converter applications," in *IEDM Tech. Dig.*, pp. 324-327, 1985.
[8] N. Kobayashi, S. Notomi, M. Suzuki, T. Tsuchiya, K. Nishiuchi, K. Odani, A. Shibatomi, T. Mimura, and M. Abe, "A fully operational 1-kbit HEMT static RAM," *IEEE Trans. Electron Devices*, vol. ED-33, pp. 548-553, 1986.
[9] Y. Watanabe, K. Kajii, K. Nishiuchi, M. Suzuki, T. Hanyu, M. Kosugi, K. Odani, A. Shibatomi, T. Mimura, M. Abe, and M. Kobayashi, "A high electron mobility transistor 1.5K gate array," in *ISSCC Dig. Tech. Papers*, pp. 80-81, 314, Feb. 1986.
[10] N. Toyoda, N. Uchitomi, Y. Kitaura, M. Mochizuki, K. Kanazawa, T. Terada, Y. Ikawa, and A. Hojo, "A 2K-gate GaAs gate array with a WN gate self-alignment FET process," *IEEE J. Solid-State Circuits*, vol. SC-20, no. 5, pp. 1043-1049, Oct. 1985.
[11] H. Hirayama, T. Furutsuka, Y. Tanaka, M. Kaga, M. Kanamori, K. Takahashi, H. Kohzu, and A. Higashisaka, "A CML compatible GaAs gate array," in *ISSCC Dig. Tech. Papers*, pp. 72-73, Feb. 1986.
[12] H-T. Yuan, J. B. Delaney, H-D. Shih, and L. T. Tran, "A 4K GaAs bipolar gate array," in *ISSCC Dig. Tech. Papers*, pp. 74-75, 312, Feb. 1986.

A 40 ps High Electron Mobility Transistor 4.1K Gate Array

Kiyoshi Kajii, Yuu Watanabe, Masahisa Suzuki, Isamu Hanyu, Makoto Kosugi,
Kouichiro Odani, Takashi Mimura, and Masayuki Abe

Fujitsu Laboratories Ltd., Fujitsu Limited
10-1 Morinosato-Wakamiya, Atsugi, 243-01, Japan

ABSTRACT

We designed and fabricated a HEMT 4.1K gate array as the largest scale logic LSI using HEMT devices. This gate array realized a 16 x 16 bit parallel multiplier. We obtained the basic gate delay time of 40 ps and the multiplication time of 4.1 ns at 300 K. In this paper we will describe this gate array, which uses DCFL circuits, featuring 0.8 μm gate devices.

1. Introduction

Development in HEMT technology has intensified, because of requirements for the next generation of mainframes and supercomputers. Recently a 1.5K gate array configured as an 8 x 8 bit parallel multiplier was demonstrated by Watanabe et al.[1] Multiplication time of 4.9 ns and 3.1 ns were measured at 300 K and 77 K respectively, showing the high speed performance of HEMT LSI.

We have newly developed a HEMT 4.1K gate array. To improve high speed operation, we reduced the gate length of HEMTs and the cell size. The integration of the device was increased and the wiring load was reduced. In the following, we report the design, fabrication and performance of a 4.1K

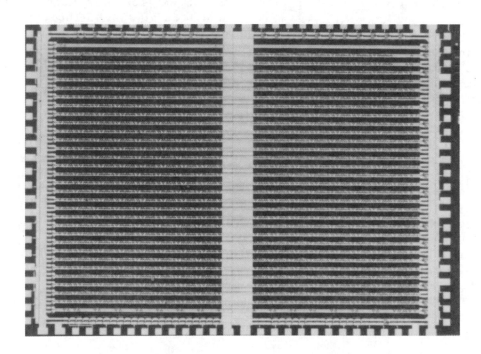

Fig. 1 Photomicrograph of 4.1K gate array

Reprinted from *Proc. IEEE 1987 Custom Integrated Circuits Conf.*, pp. 199–202, 1987.

gate array.

2. Design and Fabrication

Figure 1 is a photomicrograph of a 4.1K gate array. This gate array consists of 156 I/O cells and 4096 basic cells. The basic cell includes one depletion type (D-) HEMT and 3 enhancement type (E-) HEMTs. It can be programmed as a 3-input NOR gate. The cell size is 37.5 µm x 45 µm. Figure 2 shows the basic cell layout. An I/O cell, which includes 3 D-HEMTs and 5 E-HEMTs, can be made into an input buffer or an output buffer by using an overlay pattern. A push-pull type circuit is used as a final-stage output buffer amplifier for driving a large capacitance with low power consumption. The cell size is 100 µm x 125 µm. The basic cell array consists of 32 columns with 128 cells each. Between the columns there are 15 interconnection tracks , each line being 2 µm wide with 3 µm spacing. The chip of this gate array has 100 pads, including 72 for I/O signals and 28 for power supply. To obtain sufficient noise margin , this chip is designed to minimize the V_{DD} voltage drop and GND voltage rise by careful arrangement of the power supply pads. Therefore the chip has a relatively large number of power supply pads. The chip measures 6.3 mm x 4.8 mm.

The average values of standard deviation for threshold voltage (V_{TH}) are 17 mV for an E-HEMT and 59 mV for a D-HEMT. These HEMTs are fabricated by using a standard process based on MBE material growth and selective dry etching. The average of the transconductance over the wafer is 240 mS/mm. We have obtained high uniformity of device parameters.

3. Performances

The dynamic performance of the basic gate is measured with several ring oscillators arranged on the test element groups. These test element groups include inverters and basic gates with various loads. The top data of the delay time of the inverter without load is 21.7 ps. The average value of the delay time of the inverter and the basic gate are 26.7 ps and 40.3 ps respectively. The difference between 26.7 ps and 40.3 ps results from crossover capacitance. The basic gates are covered with power supply lines in order to make the array more compact. Therefore the basic gate has crossover capacitance. The dependence of the delay time on fan-out and the interconnection lines were 21.9 ps/fan-out and 12.3 ps/mm at 300 K.

A 4.1 K gate array uses a 16 x 16 bit parallel multiplier as a test vehicle. A 16 x 16 bit multiplier constructed of 93% of this array, employs a carry save algorithm with a carry look ahead circuit, which is well-suited to a gate array. Figure 3 is the logic diagram of the multiplier. The critical path is that from Y_0 to S_{30}, which consists of 49 stages, including a 5-stage I/O buffer. To evaluate the performance of the multiplier, the multiplication time was measured. Figure 4 shows the results. When the multiplication X(=1111111111111111) x Y(=100000000000000Y_0) is performed, the output signal of each bit is inverted by inputting a low level (0) and high level (1) to Y_0 alternately. This photograph shows the Y_0 input and S_{30} output waveforms. A multiplication time of 4.1 ns at 300 K, including a 5-stage I/O buffer delay, was achieved, where the supply voltage V_{DD} was 1.1 V and total chip power consumption was 6.2 W. This is the fastest multiplication time ever reported.[2,3]

From a simulation using the SPICE II program, we confirmed that the multiplication time was about 49 times the typical gate delay of 80 ps with a loading of 2.6 fan-outs and a 363 µm interconnection line. Therefore, 4.1 ns is a reasonable value for the multiplication time. This simulation gave an I/O buffer delay of about 8% of the multiplication time. So the intrinsic multiplication time was found to be 3.8 ns.

Table 1 summarizes the characteristics of the 16 x 16 bit parallel multiplier.

4. Conclusion

As the largest scale logic IC using HEMT devices, we designed and fabricated a 4.1K gate array. This gate array realized the basic gate delay of 40 ps and the fastest multiplication time of 4.1 ns at 300 K. These successful results showed that the HEMT devices can be applied to high speed logic LSIs.

Acknowledgments

The authors are deeply indebted to Y. Asada, N. Kobayashi and S. Notomi for valuable discussions. They are also grateful to M. Nakayama and N. Ikeda for technical support.

The present research effort is part of the National Research and Development Program on the "Scientific Computing System," conducted under a program set by the Agency of Industrial Science and Technology, Ministry of International Trade and Industry, Japan.

References

1. Y. Watanabe, K. Kajii, K. Nishiuchi, M. Suzuki, I. Hanyu, M. Kosugi, K. Odani, A. Shibatomi, T. Mimura, M. Abe and M. Kobayashi, "A High Electron Mobility Transistor 1.5K Gate Array," ISSCC DIGEST OF TECHNICAL PAPERS, pp.80-81, 1986.
2. T. Sakai, S. Konaka, Y. Yamamoto, and M. Suzuki, "PROSPECTS OF SST TECHNOLOGY FOR HIGH SPEED LSI," IEDM Technical Digest, pp.18-21, 1985.
3. B.E. Miller and R.E. Owen, "A SUB 10 ns LOW POWER BIPOLAR 16 x 16 BIT MULTIPLIER," the proceedings of CICC, pp.97-99, 1986.

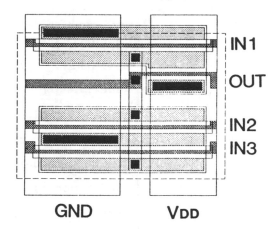

Fig. 2 Basic cell layout

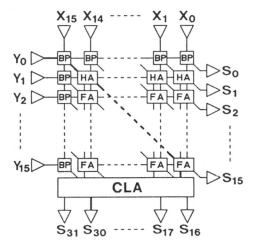

Fig. 3 Logic diagram of 16 x 16 bit parallel multiplier

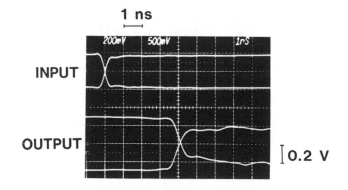

Fig. 4 Waveforms of input to Y_0 and output from S_{30}

Table 1 Characteristics of 16 x 16 bit parallel multiplier

Algorithm	Carry save
Chip size	6.3 mm x 4.8 mm
Gate length	0.8 μm
Multiplication time	4.1 ns
Power dissipation	6.2 W
Basic gate delay (fan-out=1)	40.3 ps/gate
Typical gate delay (fan-out=2.6, line length=363 μm)	80 ps/gate

HIGH-SPEED FREQUENCY DIVIDERS USING GaAs/GaAlAs HIGH-ELECTRON-MOBILITY TRANSISTORS

Indexing terms: Semiconductor devices and materials, Frequency dividers

Very-high-speed divide-by-four circuits have been fabricated by using modulation-doped GaAs/GaAlAs high-electron-mobility transistors. The circuits consist of two T-connected D-flip-flops and are capable of operating at 3·6 GHz with a power dissipation of 0·46 mW per gate at room temperature, and at 5·2 GHz with a power dissipation of 0·78 mW per gate at 77 K. The speed-power products achieved are the lowest ever reported.

Modulation-doped GaAs/GaAlAs high-electron-mobility transistors (HEMTs) have attracted much attention recently for their potential applications in ultra-high-speed integrated circuits. The high electron mobility and the MOSFET-like structure have resulted in HEMT devices with transconductances two to three times higher than those of conventional GaAs MESFETs.[1,2] Very fast HEMT ring oscillators have already been reported by several laboratories.[3-5] Recently we have achieved ring oscillators with propagation delays as low as 12·2 ps at room temperature.[6] In this letter the fabrication and performance of divide-by-four circuits HEMT devices are reported. These are the most complex HEMT circuits reported so far.

The modulation-doped structures used for device fabrication were grown by MBE on one-inch-square LEC semiinsulating GaAs substrates. The layers were, starting from the substrate, undoped GaAs (~ 1 μm), undoped Ga$_{0.7}$Al$_{0.3}$As spacer (~ 90 Å), N-type Ga$_{0.7}$Al$_{0.3}$As (~ 500 Å, 1×10^{18} cm^{-3}) and undoped GaAs (~ 40 Å). The low field electron mobility of these structures was typically around 5000 cm^2-V^{-1}s^{-1} at 300 K and 100 000 cm^{2}V^{-1}s^{-1} at 77 K. The fabrication procedure for the devices consisted of: (i) ohmic-contact formation; (ii) proton implantation for device isolation; (iii) delineation of Schottky gates and first-level interconnect; (iv) isolation dielectric deposition and second-level interconnect formation. The ohmic contact was AuGe/Ni, which typically had a contact resistance of 0·5 Ω mm. The Schottky metal and the first-level interconnect were Ti/Pt/Au, and the second level metal was Ti/Au. All the lithographic steps were done with a Sensor 10 × wafer stepper. Because of the use of proton implantation for device isolation, the process results in a planar structure which is desirable for high-density large-circuit fabrication.

The divide-by-four circuits use two T-connected D-flip-flops as binary counters to perform the frequency division. Direct-coupled FET logic (DCFL) was used for the circuit design. A basic NOR gate consists of several enhancement-mode switching transistors and a saturated resistor pull-up. The circuit diagram and the photograph of a fabricated circuit are shown in Fig. 1. The output signal goes through two buffer inverter stages and a 300 μm wide buffer FET, and is detected with the buffer FET connected as a source-follower with a 50 Ω load. Besides the buffer stages there are 12 gates in total to perform the logic functions. The total signal delay time in this circuit configuration is $5\tau_D$, where τ_D is the propagation delay per gate. The average fan-in and fan-out are 2·2 and 2, respectively. There are other configurations of flip-flops, such as master-slave flip-flops, which theoretically can operate faster than the configuration used here. However, they usually require complimentary clock inputs and more complex design. Therefore, the simpler D-type flip-flop was chosen for this

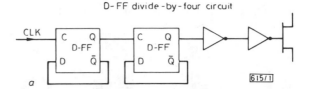

D-FF divide-by-four circuit

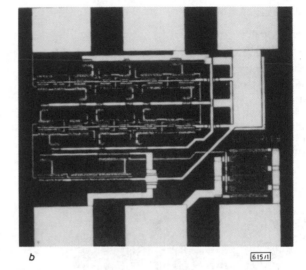

Fig. 1

a Circuit diagram of divide-by-four circuit
b Photograph of a fabricated divide-by-four circuit

divider circuit, with the understanding that overall circuit speed was being traded for design and test simplicity.

The transistor sizes vary with the gate loading conditions. The average width of the switching transistors was 24 μm and the average size of the saturated resistors was 13 μm. The total chip area including the bonding pads was 400 × 400 μm.

Two batches of wafers have been processed and tested. The devices on the first batch had 1·4 μm-long gates, and those on the second batch had 1 μm gates. The threshold voltages of all the enhancement-mode switching transistors were controlled to be around +0·1 V. The transconductance was about 150 mS/mm or higher. For the dividers on the first batch the typical maximum operation frequency was 2·3 GHz. The best result obtained at room temperature was 2·6 GHz at $V_{DD} = 0·7$ V with a total chip power consumption of 6·7 mW. Subtracting the power consumed by the buffer stages the total power dissipation was 3·2 or 0·27 mW per gate. When the device was cooled down to 200 K, it operated up to 3·54 GHz with a power consumption of 0·21 mW per gate. At 77 K, the device operated up to 4·6 GHz with a 0·57 mW/gate power dissipation. The devices on the second wafer batch were capable of being operated at higher frequencies. At room temperature the maximum operating frequency was 3·6 GHz with a 0·46 mW/gate power consumption, and at 77 K the maximum operating frequency was 5·2 GHz with a 0·78 mW/gate power consumption. A summary of the performance is shown in Table 1. The speed-power products obtained were between 12 and 30 fJ. These are the lowest values ever reported for high-speed GaAs frequency dividers. Nishiuchi et al.[7] and Kiehl et al.[8] have also recently reported high-speed HEMT frequency dividers, but those were divide-by-two circuits with shorter gate lengths (Nishiuchi et al.), and they consumed much higher power.

Unlike the MESFET circuits, most of the HEMT devices

Table 1 SUMMARY OF DIVIDE-BY-FOUR
PERFORMANCE

L_g	T	f_{in}	P	τ	$p\tau$
μm	K	GHz	mW/gate	ps	fJ
1·4	300	2·6	0·27	77	21
1·4	200	3·54	0·21	56·6	12
1·4	77	4·6	0·57	43·5	25
1	300	3·6	0·46	58	25
1	77	5·2	0·78	38	30

tested could be operated at very low V_{DD} without sacrificing the speed. This is probably due to the characteristics of HEMTs, which have high transconductances even at very low gate voltages; so the dividers were capable of operating at very high frequency yet with very small power consumption. The measured output voltage swing was higher than that of the regular normally-off GaAs MESFET devices. Fig. 2 shows the output waveforms of a HEMT divide-by-four circuit operating at an input frequency of 2·5 GHz. The voltage swing

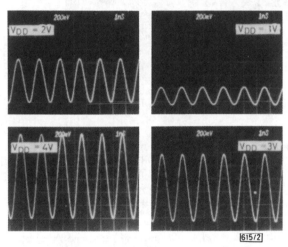

Fig. 2 *Waveform of divide-by-four output with different V_{DD}s at clock input of 2·5 GHz*

increased with the supply voltage V_{DD}. At $V_{DD} = 4$ V the amplitude of the output waveform was 1·3 V, which is substantially higher than the typical 0·7 V obtained for regular enhancement-mode GaAs MESFET. This large voltage swing, which is due to the high turn-on voltage of Schottky gate on GaAlAs, can greatly increase the noise margin and ease the

requirement on the control of the device parameters. For large digital circuits this will be a significant advantage for HEMT devices over regular MESFETs.

In conclusion, very-high-speed low-power HEMT divide-by-four circuits have been achieved. With a conservative $f_{max} = 1/5\tau_D$ design, devices operating at 5·2 GHz were obtained. The power consumption was very low and the speed-power product was below 30 fJ.

Acknowledgment: The authors would like to thank F. Eisen for his encouragement and support concerning this work.

C. P. LEE   *13th January 1984*
S. J. LEE
D. HOU
D. L. MILLER
R. J. ANDERSON
N. H. SHENG

Rockwell International
Microelectronics Research & Development Center
Thousand Oaks, CA 91360, USA

References

1 MIMURA, T., HIYAMIZU, S., FUJII, T., and NANBU, K.: 'A new field-effect transistor with selectively doped GaAs/n-Al$_x$Ga$_{1-x}$As heterojunctions', *Jpn. J. Appl. Phys.*, 1980, **19**, pp. L225–L227
2 SU, S. L., FISCHER, P., DRUMMOND, T. J., LYONS, W. G., THORNE, R. E., KOPP, W., and MORKOÇ, H.: 'Modulation-doped AlGaAs/GaAs FETs with high transconductance and electron velocity', *Electron. Lett.*, 1982, **18**, pp. 794–796
3 ABE, M., MIMURA, T., YOKOYAMA, N., and ISHIKAWA, H.: 'New technology towards GaAs LSI/VLSI for computer applications', *IEEE Trans.*, 1982, **MTT-30**, pp. 992–998
4 TUNG, P. N., DELESCLUSE, P., DELAGEBEAUDEUF, D., LAVIRON, M., CHAPLART, J., and LINH, N. T.: 'High-speed low-power DCFL using planar two-dimensional electron gas FET technology', *Electron. Lett.*, 1982, **18**, pp. 517–519
5 LEE, C. P., and WANG, W. I.: 'High-performance modulation-doped GaAs integrated circuits with planar structures', *ibid.*, 1983, **19**, pp. 155–157
6 LEE, C. P., MILLER, M. L., HOU, D., and ANDERSON, R. J.: 'Ultra high speed integrated circuits using GaAs/GaAlAs high electron mobility transistors'. 41st Device Research Conference, Burlington, Vermont, 1983
7 NISHIUCHI, K., MIMURA, T., KURODA, S., HIYAMIZU, S., NISHI, H., and ABE, M.: 'Device characteristics of short channel high electron mobility transistor'. *Ibid.*, 1983
8 KIEHL, R. A., FEUER, M. D., HENDEL, R. H., HWANG, J. C. M., KERAMIDAS, V. G., ALLYN, C. L., and DINGLE, R.: 'Selectively doped heterostructure frequency dividers', *IEEE Electron Device Lett.*, 1983, **EDL-4**, p. 377

Selectively Doped Heterostructure Frequency Dividers

R. A. KIEHL, MEMBER, IEEE, M. D. FEUER, R. H. HENDEL, J. C. M. HWANG, SENIOR MEMBER, IEEE, V. G. KERAMIDAS, C. L. ALLYN, AND R. DINGLE

Abstract—The operation of high-speed divide-by-two circuit (binary counter) composed of selectively doped heterostructure logic gates is reported for the first time. These field-effect transistor circuits utilize the enhanced transport properties of high-mobility electrons confined near a heterojunction interface in a selectively doped AlGaAs/GaAs structure. The dividers are based on a Type-D flip-flop composed of six direct-coupled NOR-gates having 1-μm gate lengths and 4-μm source–drain spacings. They are fabricated by conventional optical contact lithography on a four-layer Al$_{.3}$Ga$_{.7}$As/GaAs structure grown by molecular-beam epitaxy.

Successful operation is demonstrated at 5.9 GHz at 77 K for 1.3-V bias and 30-mW total power dissipation (including output buffers) and 3.7 GHz at 300 K for 1.4-V bias and 19-mW total power dissipation. Total power dissipation values as low as 3.9 mW at 0.65-V bias were also obtained for 2.85-GHz operation at 300 K. These preliminary results illustrate the promise of SDHT logic for ultrahigh-speed low-power applications.

SELECTIVELY doped heterostructure transistors (SDHT's) utilize the enhanced transport properties of high-mobility electrons confined near a heterojunction interface [1] to provide a high-performance field-effect device characterized by low parasitic resistance, high transconductance, and high output resistance. Due to a dramatic increase in electron mobility with decreasing temperature, SDHT's are particularly attractive for low-temperature operation, and transconductances as high as 409 mS/mm have been reported [2] at 77 K with an AlGaAs/GaAs structure similar to that employed here. Propagation delays as low as 16.7 and 12.8 ps have been reported [2] for SDHT ring oscillators at 300 and 77 K, respectively, representing the fastest operation yet achieved for any semiconductor device. Recently, the operation of selectively doped heterostructure frequency dividers was reported by us for 1.0-μm gate Type-D flip-flop circuits [3] and by others for 0.5-μm gate master-slave flip-flop circuits [4]. In this letter we present the details of our preliminary divider circuit results.

The divider circuits are composed of enhancement-mode driver and depletion-mode load transistors fabricated on a four-layer Al$_{0.3}$Ga$_{0.7}$Ga/GaAs structure grown by molecular-

Manuscript received May 26, 1983; revised August 4, 1983.

R. A. Kiehl, M. D. Feuer, R. H. Hendel, V. G. Keramidas, C. L. Allyn, and R. Dingle are with Bell Laboratories, Murray Hill, NJ 07974.

J. C. M. Hwang was with Bell Laboratories, Murray Hill, NJ 07974. He is now with GE Electronics Laboratory, Syracuse, NY 13221.

beam epitaxy on a Cr-doped semi-insulating substrate. The structure begins with the growth of a 1-μm undoped GaAs channel layer followed by a 60-Å undoped Al$_{0.3}$Ga$_{0.7}$As spacer layer, a 350-Å n-Al$_{0.3}$Ga$_{0.7}$As donor layer, and a 500-Å n-GaAs cap layer. The silicon doping in both the donor and cap layer is approximately 2 $\times$ 10^{18} cm^{-3}. The undoped Al$_{0.3}$Ga$_{0.7}$As spacer layer is included to increase the degree of mobility enhancement for the two-dimensional electron gas confined near the heterojunction interface in the GaAs channel layer. Mobilities as high as 8000 and 120 000 cm^2/(V·s) have been achieved at 300 and 77 K, respectively. The GaAs cap layer serves to reduce the source resistance in transistor structures.

The circuit fabrication involved six lithographic steps using optical contact lithography and is described elsewhere [5], [6]. Gates for the driver and load transistors were recessed by chemically etching through the n-GaAs cap layer and into the underlying n-Al$_{0.3}$Ga$_{0.7}$As layer. Driver and load gates were defined by different mask levels, thereby allowing the driver and load thresholds to be individually controlled. The length of the Al gates was nominally 1.0 μm and the spacing between the Au/Ge based ohmic source and drain metallizations was 4.0 μm. A 4500-Å-thick Au layer was used as a second-level interconnect and the crossover insulator was 4000 Å of plasma-enhanced deposited silicon nitride.

The divider circuit is based on a Type-D flip-flop consisting of six direct-coupled SDHT NOR-gates and two SDHT output buffer gates, as shown in Fig. 1. Such an arrangement represents a functional logic circuit having both two- and three-input NOR-gates, fan-outs of up to three, and realistically long interconnects. Although considerably faster operation should be possible with circuit designs based on gated master-slave flip-flops [7], the Type-D flip-flop circuit employed here is a conventional design which permits direct comparison with the results of other device technologies.

Each gate in the circuit is composed of one depletion-mode load transistor with its gate shorted to its source and one or more enhancement-mode driver transistors. Divider circuits with driver to load gate-widths of 25:12, 50:25, and 50:12 μm were included on the mask set together with direct-coupled ring oscillators and discrete transistors of identical dimensions.

Divider circuit performance was measured on-wafer using a GGB Model 7A 50-Ω coaxial input probe which was termi-

Reprinted from *IEEE Electron Device Letters,* vol. EDL-4, no. 10, pp. 377–379, October 1983.

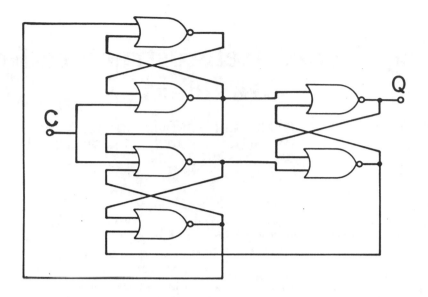

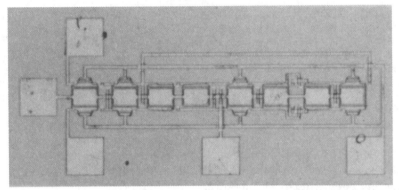

Fig. 1.   Logic diagram and photograph of SDHT divide-by-two circuit.

nated approximately 3 mm from the probe tip with a 50-Ω chip resistor. A short length of wire connected between the outer conductor of the input probe near its termination and the ground probe for the circuit was used to minimize parasitic reactance at the input. The microwave output was monitored on an HP 8569A spectrum analyzer which allowed the divider to be carefully analyzed for correct divide-by-two operation with no spurious responses within the entire range from 10 MHz to 18 GHz. Cooling of the wafer to 77 K during proving was provided by immersing the wafer in a liquid nitrogen bath. All measurements were made under the light of the probe station optical illuminator.

Room-temperature operation of a circuit with a 2.6-GHz input signal is illustrated in Fig. 2, where sampling oscilloscope traces for both the input and output signals are shown. The driver and load gate widths of this circuit were 50 and 25 μm, respectively. The dc bias in this case was 1.80 V and the dc offset voltage at the input was 0.4 V. Fig. 3 shows the maximum operating frequency as a function of bias voltage for a circuit at 300 K. In this experiment the offset voltage was held fixed at 1.0 V and the input signal frequency and amplitude were adjusted to obtain divide-by-two operation at the highest possible rate. The data reveal that high-speed operation with little bias sensitivity is achieved over the range from 0.68 to 3.25 V (the upper limit of the measurement range).

The highest operating frequency observed at room tempera-

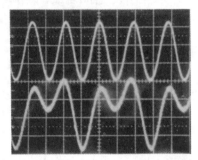

Fig. 2.   Sampling oscilloscope display of input (top) and output (bottom) signals for circuit at 300 K. Vert: uncal, Hor: 200 ps/div.

ture was 3.7 GHz with a total power dissipation, including output buffers, of 19 mW at 1.4-V bias. Total power dissipation values as low as 3.9 mW were obtained for 2.85-GHz operation at 300 K with 0.65-V bias in a circuit with driver:load gate widths of 25:12 μm. Considerably faster operation was obtained for cooled circuits. In particular, operation to frequencies as high as 5.9 GHz was possible at 77 K with a 1.3-V bias and a total power dissipation of 30 mW. The propagation delay in 19-stage ring oscillators on the same wafer was 38 ps at 300 K and 18 ps at 77 K. The transconductance of E-mode driver SDHT's on this wafer was 110 and 230 mS/mm at 300 and 77 K, respectively. More recently we have achieved considerably higher transconductance values

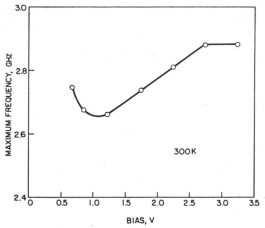

Fig. 3.   Bias dependence of maximum operation frequency at 300 K.

in modified SDHT wafers and improvement in the frequency divider performance is anticipated.

In conclusion, these preliminary results demonstrate the high-speed performance of SDHT divider circuits. While the performance of other technologies at low temperatures has not yet been reported, we note that the 6-GHz performance level achieved with 1-μm gate lengths in our cooled experiments is 2.4 times as fast as that reported for 0.8-μm gate (0.40-μm channel) silicon NMOS technology [8] and 1.6 times as fast as that reported for 0.6-μm gate GaAs MESFET technology with similar Type-D flip-flop circuitry [9]. Frequency division at nearly 6 GHz has been demonstrated [7] in 0.6-μm gate GaAs MESFET's using buffered FET logic and intrinsically faster master/slave circuitry; however, it should be noted that the power consumption of these circuits was about 5 times greater than that achieved here. These results demonstrate the promise of SDHT logic for ultrahigh-speed, low-power applications.

ACKNOWLEDGMENT

The authors wish to acknowledge S. H. Wemple and P. G. Flahive for their contributions to the circuit layout. We also thank A. Kastalsky for his work in materials characterization and A. S. Williams and P. F. Sciortino for their contribution to device fabrication.

REFERENCES

[1] R. Dingle, H. L. Stormer, A. C. Gossard, and W. Wiegmann, "Electron mobilities in modulation doped semiconductor heterojunction superlattices," *Appl. Phys. Lett.*, vol. 33, pp. 665–667, Oct. 1978.

[2] M. Abe, T. Mimura, N. Yokoyama, and H. Ishikawa, "New technology towards GaAs LSI/VLSI for computer applications," *IEEE Trans. Electron Devices*, vol. ED-29, pp. 1088–1093, July 1982.

[3] R. A. Kiehl, M. D. Feuer, R. H. Hendel, J. C. M. Hwang, V. G. Keramidas, C. L. Allyn, and R. Dingle, "Selectively doped heterostructure divide-by-two circuit," presented at the 41st Annual Device Research Conf., Burlington, VT, June 20–22, 1983.

[4] N. Nishiuchi, T. Mimura, S. Kuroda, S. Hiyamizu, H. Nishi, and M. Abe, "Device characteristics of short channel high electron mobility transistor (HEMT)," presented at the 41st Annual Device Research Conf., Burlington, VT., June 20–22, 1983.

[5] J. V. DiLorenzo, R. Dingle, M. Feuer, A. C. Gossard, R. Hendel, J. C. M. Hwang, A. Kastalsky, V. G. Keramidas, R. A. Kiehl, and P. O'Connor, "Material and device considerations for selectively doped heterojunction transistors," in *IEDM Tech. Dig.*, pp. 578–581, Dec. 1982.

[6] M. D. Feuer, R. H. Hendel, R. A. Kiehl, J. C. M. Hwang, V. G. Keramidas, C. L. Allyn, and R. Dingle, "High-speed low-voltage ring oscillators based on selectively doped heterostructure transistors," accepted for publication in *IEEE Electron Device Letters*.

[7] M. Cathelin, M. Gavant, and M. Rocchi, "A 3.5 GHz self-aligned single-clocked binary frequency divider on GaAs," *Proc. IEEE*, vol. 127, pp. 270–277, Oct. 1980.

[8] D. L. Fraser, Jr., H. J. Boll, R. J. Bayruns, N. C. Wittwer, and E. N. Fuls, "Gigabit logic circuits with scaled NMOS," *1981 European Solid State Circuits Conf. Tech. Dig.*, pp. 202–204, 1981.

[9] M. Ohmori, T. Mizutani, and N. Kato, "Very low power gigabit logic circuits with enhancement-mode GaAs MESFETs," in *IEDM Tech. Dig.*, pp. 188–190, 1981.

A 10-GHz Frequency Divider Using Selectively Doped Heterostructure Transistors

R. H. HENDEL, MEMBER, IEEE, S. S. PEI, R. A. KIEHL, MEMBER, IEEE, C. W. TU, M. D. FEUER, AND R. DINGLE

Abstract—We report the first complementary clocked frequency divider using dual gate selectively doped heterostructure transistors (SDHT's). The circuit employs a master-slave flip-flop design which consists of four direct coupled AND-NOR gates. The nominal gate length and the gate-gate, separation in the dual gate SDHT's are 1 μm.

A maximum dividing frequency of 10.1 GHz at 77 K was achieved; at this frequency the circuit dissipated 49.9 mW at 1.67-V bias. This is the highest operating frequency reported for static frequency dividers at any temperature. At room temperature the dividers were operated successfully at frequencies up to 5.5 GHz with a total power dissipation of 34.8 mW at 1.97-V bias. The lowest speed-power product at room temperature was obtained at 5 GHz with 14.9-mW power dissipation at 1.45-V bias.

I. INTRODUCTION

THE selectively doped heterostructure transistor has attractive potential for very high-speed and low power integrated circuit applications. SDHT circuits have demonstrated performance competitive with or better than the comparable GaAs or Si circuits at room temperature. Furthermore, notable improvements in performance have been achieved for SDHT circuits operated at 77 K. For example, an SDHT ring oscillator with a minimum propagation delay of 12.2 ps has been reported [1]. This represents the fastest operation of any room temperature device.

Although ring oscillators are useful test vehicles for demonstrating the potential of a device technology, it is essential to evaluate performance in more complex circuits under more realistic loading conditions. A variety of frequency divider demonstration circuits has been reported in the literature [2]–[6]. A SIX NOR gate edge-triggered D-flip-flop divider using SDHT's with 1-μm gates was operated successfully at frequencies up to 3.7 GHz at room temperature and 5.9 GHz at 77 K [2]. Recently a maximum toggle frequency of 3.8 GHz has also been observed for GaAs type-D divider operating at room temperature [3]. Another design uses 8 NOR gates to form a gated master-slave flip-flop. This design has resulted in a maximum toggle frequency of 5.5 and 8.9 GHz at room temperature and 77 K, respectively using SDHT's with 0.5-μm gates [4]. All SDHT dividers reported in the literature have used NOR gates as the basic logic element.

A master-slave flip-flop design using AND/NOR gates rather than NOR gates was described in [5]. The highest operating frequency reported for this design in a complementary clocked

Manuscript received June 15, 1984; revised July 17, 1984. This work was supported in part by the Air Force Wright Aeronautical Laboratories, Avionics Laboratory, Wright-Patterson Air Force Base, OH.

The authors are with AT&T Bell Laboratories, Murray Hill, NJ 07974.

frequency divider is 5.7 GHz at room temperature with a power dissipation of 60 mW using buffered FET logic, depletion-mode GaAs MESFET's and 0.8-μm gates [6].

In this letter we report the results of complementary clocked frequency dividers using dual gate SDHT's.

II. CIRCUIT DESIGN

The circuit uses enhancement/depletion (E/D) type direct coupled SDHT logic and a master-slave flip-flop design [5]. The set–reset (RS) flip-flop is formed by two cross-coupled AND/NOR gates as shown in Fig. 1. The divide-by-2 operation is accomplished by feeding the outputs of the first RS flip-flop (master) to the inputs of the second RS flip-flop (slave), and the inverted outputs of the slave to the inputs of the master. Two inverters are used as the output buffer gates. Since the storage latch is disabled during the data-entry cycle (i.e., when C is high), it takes only one AND/NOR gate propagation delay for the flip-flop to reach the equilibrium state after the data-enter cycle is activated. The data storage cycle is also completed in one gate delay. The divider is therefore capable of operating at frequencies as high as $1/2\tau$ where τ is the loaded gate delay. In comparison, for other divider designs where the latch is active all the time, the typical maximum operating frequency is $1/4.85\tau$ for the type-D divider and $1/4\tau$ for the master-slave divider [7].

III. PROCESSING

The processing sequence is briefly outlined as follows. The MBE grown active layers are electrically isolated from each other by mesa isolation. The ohmic contact is Au/Ge based, and the crossover dielectric is silicon nitride deposited by plasma enhanced CVD. The threshold voltages of the enhancement and depletion mode SDHT's are individually adjusted by two stages of wet chemical etching to approximately 0.2 and −0.7 V, respectively. The gate metal is Al. All features are patterned by conventional optical contact lithography, and all levels of metallization are defined by liftoff. Although the AND function can be implemented with two FET's connected in series, a single dual gate FET is used to reduce the source-drain resistance as well as the area of the circuit. Fig. 2 shows an SEM micrograph of a NAND gate using dual gate SDHT and the E/D direct coupled SDHT logic. The nominal gate length is 1 μm with a 6-μm source-drain spacing and 1-μm gate separation.

IV. TESTING AND RESULTS

Individual dual gate SDHT's were evaluated with test circuits close to the dividers. The drain–gate and the source–gate re-

Reprinted from *IEEE Electron Device Letters*, vol. EDL-5, no. 10, pp. 406–408, October 1984.

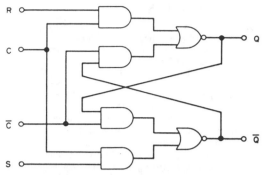

Fig. 1. Complementary clocked R/S flip-flop implemented with AND/NOR gates.

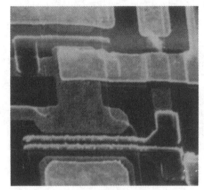

Fig. 2. SEM micrograph of a SDHT NAND gate.

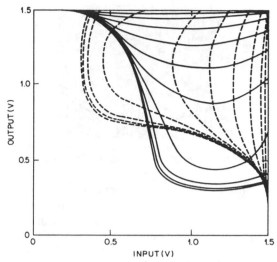

Fig. 3. Voltage transfer curves of a SDHT NAND gate with the supply voltage $V_{dd} = 1.5$ V. The output voltage is plotted against V_{in} on the upstream gate (the one on the source side) with voltage applied to the downstream gate (the one on the drain side) from 0.2 to 1.1 V in steps of 0.1 V. Dashed curves are the same data with axes interchanged to show the operational region of the NAND gate.

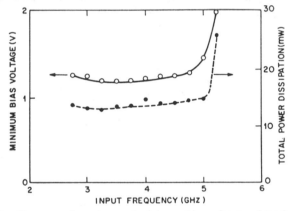

Fig. 4. Frequency dependence of minimum bias voltage and total power dissipation of a complementary clocked frequency divider at room temperature.

sistances of the dual gate SDHT were about the same as in the single gate SDHT (~ 2 Ω-mm), while the resistance between the two gates was usually slightly higher. Typical transfer characteristics of a dual gate SDHT NAND gate are shown in Fig. 3. Due to the high turn-on voltage of the Schottky gate, the noise margin of the SDHT NAND gate is expected to be better than that of a comparable GaAs NAND gate.

The dividers were tested on-wafer using probe cards or individual probes. All input probes were terminated with 50-Ω resistors close to the probe tip in order to maximize the coupling to the circuit and to minimize crosstalk. The procedure is similar to the testing of other dividers. The difference is that this circuit requires complementary inputs. They are generated by splitting a single clock signal with a power splitter and delaying one leg by an adjustable delay line. The same dc offset voltage was applied to both clock inputs to provide the optimum operating condition for the divider. The complementary clocked divider becomes unstable if both clock inputs are high. Under proper bias and offset conditions the divider will then start oscillating at a characteristic frequency, providing a simple way to evaluate the yield of functioning circuits, and giving an indication of the expected performance of the divider circuit.

Dividers with three different driver and/or load gate widths (25:12, 50:25, 50:12 μm) were implemented. The saturation current ratio of the driver and load transistors, measured at $V_{ds} = 1.2$ V and $V_{gs} = 1.2$ V (driver SDHT) or $V_{gs} = 0$ V (load SDHT) is adjusted to be about 2 to 4 by the load gate recess etching. Dividers with 50-μm drivers and 25-μm loads showed the best performance. The typical operating frequency was about 4 GHz. The maximum dividing frequency

achieved at room temperature was 5.5 GHz. At this frequency, the circuit including the two buffered output gates dissipated 34.8 mW in total at 1.97 V bias. The lowest speed–power product was obtained at 5 GHz with a total power dissipation of 14.9 mW at 1.45-V bias. The minimum bias voltage and the corresponding power dissipation as a function of input frequency of the divider is shown in Fig. 4.

When the circuit was cooled to 77 K, a maximum dividing frequency of 10.1 GHz was achieved with 49.9 mW power dissipation at 1.67-V bias. This is the highest operating frequency ever reported for static frequency dividers using any technology. The corresponding propagation delay of the 4-input AND-NOR gate is 49.5 ps/gate for an average fan-out of 2.5 (see Fig. 5).

V. SUMMARY

We have demonstrated the feasibility of direct coupled FET logic using dual gate SDHT's. The high turn-on voltage of the Schottky gate in SDHT's greatly improves the noise margin of the NAND gate and makes the dual gate SDHT an attractive alternative in circuit designs. The advantage of this design flexi-

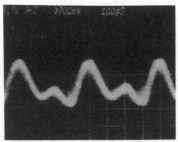

Fig. 5. The output waveform of the divider operating at 4.96 GHz and room temperature. Vert : uncal; Hor : 100 ps.

bility is demonstrated by the performance of a complementary clocked divider circuit. With 1-μm gates a maximum dividing frequency of 10.1 and 5.5 GHz was achieved at 77 K and room temperature, respectively.

ACKNOWLEDGMENT

The authors wish to acknowledge P. G. Flahive and S. H. Wemple for their contributions to circuit layout and design. We would also like to thank C. L. Allyn and P. F. Sciortino for their support and assistance in the device fabrication and T. M. Brennan for his help in MBE growth.

REFERENCES

[1] C. P. Lee, D. Hou, S. J. Lee, D. L. Miller, and R. J. Anderson, "Ultra high-speed digital integrated circuits using GaAs/GaAlAs high electron mobility transistors," in *GaAsIC Symp. Tech. Dig. 1983*, pp. 162–165, Oct. 1983.

[2] R. A. Kiehl, M. D. Feuer, R. H. Hendel, J. C. M. Hwang, V. G. Keramidas, C. L. Allyn, and R. Dingle, "Selectively doped heterostructure frequency dividers," *IEEE Electron Device Lett.*, vol. EDL-4, pp. 377–379, Oct. 1983.

[3] P. O'Connor, P. G. Flahive, W. Clemetson, R. L. Panock, S. H. Wemple, S. C. Shunk, and D. P. Takahashi, "A monolithic multigigabit/sec DCFL GaAs decision circuit," to be published.

[4] M. Abe, T. Mimura, K. Nisiuchi, A. Shibatomi, and M. Kobayashi, "HEMT LSI technology for high speed computers," in *GaAs IC Symp. Tech. Dig. 1983*. pp. 158–161, Oct. 1983.

[5] R. L. Van Tuyl, C. A. Liechti, R. E. Lee, and E. Gowen, "GaAs MESFET logic with 4-GHz clock rate," *IEEE J. Solid-State Circuits*, vol. SC-12, pp. 485–496, Oct. 1977.

[6] M. Gloanec, J. Jarry, and G. Nuzillat, "GaAs digital integrated circuits for very high-speed frequency division," *Electron. Lett.*, vol. 17, pp. 763–765, Oct. 1981.

[7] R. C. Eden, B. M. Welch, R. Zucca, and S. I. Long, "The prospects for ultra high-speed VLSI GaAs digital logic," *IEEE Trans. Electron Devices*, vol. ED-26, pp. 299–317, Apr. 1979.

[8] M. D. Feuer, R. H. Hendel, R. A. Kiehl, J. C. M. Hwang, V. G. Keramidas, C. L. Allyn, and R. Dingle, "High-speed low-voltage ring oscillators based on selectively doped heterostructure transistors," *IEEE Electron Device Lett.*, vol. EDL-4, pp. 306–307, Sept. 1983.

A High-Speed Frequency Divider Using n$^+$-Ge Gate AlGaAs/GaAs MISFET's

SHUICHI FUJITA, MAKOTO HIRANO, KOICHI MAEZAWA, AND TAKASHI MIZUTANI, MEMBER, IEEE

Abstract—A high-speed divide-by-four static frequency divider is fabricated using n$^+$-Ge gate AlGaAs/GaAs heterostructure MISFET's. The divider circuit consists of two master–slave T-type flip-flops (T-FF's) and an output buffer based on source-coupled FET logic (SCFL). A maximum toggle frequency of 11.3 GHz with a power dissipation of 219 mW per T-F/F is obtained at 300 K using 1.0-μm gate FET's.

I. INTRODUCTION

IN RECENT YEARS, undoped AlGaAs/undoped GaAs MIS-like heterostructure FET's [1]–[6] using high-mobility two-dimensional electron gas have been extensively studied. This is because these FET's have superior potential as high-speed LSI's using two-dimensional electron gas induced at the AlGaAs/GaAs heterointerface. In addition to that, the FET's have the advantages of good reproducibility and high threshold voltage uniformity because the threshold voltage is principally determined by physical parameters such as electron affinity and it is not affected by the thickness of the AlGaAs. Though good dc characteristics have been shown for MISFET's [7], there are few reports [6], [8] concerning RF performance and high-speed operation.

In this work, a divide-by-four frequency divider using n$^+$-Ge gate AlGaAs/GaAs MISFET's has been studied in order to realize the high potential of the AlGaAs/GaAs MISFET's for high-speed digital IC's. The divider circuit was based on source-coupled FET logic (SCFL) [9] which is expected to realize high speed and a large noise margin.

The following sections describe the device structure, fabrication process, circuit configuration, and performance of the divide-by-four frequency divider using n$^+$-Ge gate AlGaAs/GaAs MISFET's.

II. DEVICE STRUCTURE AND FABRICATION

A schematic cross-sectional view of the n$^+$-Ge gate AlGaAs/GaAs MISFET is shown in Fig. 1. The FET was fabricated using an MBE-grown structure composed of n$^+$-Ge (150 nm), undoped GaAs (5 nm), undoped Al$_x$Ga$_{1-x}$As (15 nm), and undoped GaAs (300 nm). The AlAs mole fraction x was chosen to be about 0.45 which gives a maximum band offset [10].

For the gate electrode formation, 200-nm WSi$_x$ was deposited and the gate pattern of n$^+$-Ge/WSi$_x$ was delineated

Manuscript received January 19, 1987.
The authors are with NTT Electrical Communications Laboratories, 3-1 Morinosato Wakamiya, Atsugi-shi, Kanagawa 243-01, Japan.
IEEE Log Number 8714598.

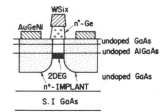

Fig. 1. Schematic cross-sectional view of the n$^+$-Ge gate AlGaAs/GaAs heterostructure MISFET.

by CF$_4$ RIE. Source and drain n$^+$ regions self-aligned to the gate were then implanted with Si at 100 keV to a dose of 1 $\times$ 10^{14} cm^{-2}. Load resistors were formed by an n-type conductive layer prepared by Si implantation. The implanted Si atoms were activated by lamp annealing at 800°C for 4 s in an N$_2$ atmosphere with the wafers in face-to-face contact. Source and drain ohmic contacts were formed by AuGeNi metallization and subsequent alloying. Finally, the frequency-divider circuit was completed by interconnection using air–bridge technology [11] in order to minimize the wiring capacitance.

III. CIRCUIT CONFIGURATION

The block diagram of the divide-by-four frequency divider is shown in Fig. 2(a). The divider circuit consists of two biphase-clock master–slave T-type flip-flops (T-F/F's) and an output buffer. The master–slave T-F/F is based on SCFL using a two-level series gate, as shown in Fig. 2(b). Each master and slave flip-flop consists of a switching stage and a source-follower stage. The switching state is composed of 1 $\times$ 20-μm^2 gate FET's (Q_1–Q_7) and 650-Ω load resistors (R_1, R_2), and the source-follower stage is composed of four FET's (Q_8–Q_{11}) with the same gate size. For high-speed operation, the supply voltage is important. As an example, when the gate bias voltage V_{cs} of the current-source FET's Q_7, Q_9, and Q_{11} is chosen to be 1.2 V to obtain a large transconductance at a large-signal operation, and the inverter gain is set to be 1.4, the supply voltage $|V_{ss}|$ should be higher than 4.8 V in order to operate the circuits in a bias region where the gate–drain capacitances of all the FET's are small.

IV. PERFORMANCE

The threshold voltage of the MISFET's was 0.3 V and the standard deviation was as small as 16 mV throughout the 2-in wafer. The fabricated MISFET's had transconductances of 150–220 mS/mm at room temperature with 1.0-μm gate lengths. From S-parameter measurements up to 26.5 GHz,

Reprinted from *IEEE Electron Device Letters*, vol. EDL-8, no. 5, pp. 226–227, May 1987.

373

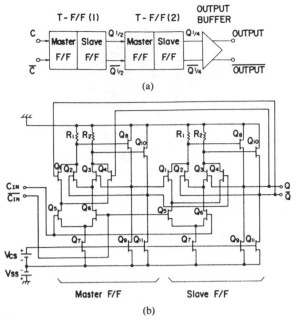

(a)

(b)

Fig. 2. (a) Block diagram of divide-by-four frequency divider. (b) Circuit configuration of master–slave *T*-F/F.

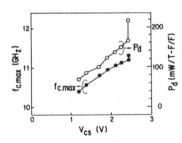

Fig. 3. V_{cs} dependence of maximum toggle frequency and power dissipation per *T*-F/F.

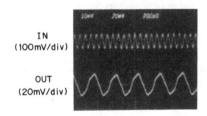

IN
(100mV/div)

OUT
(20mV/div)

Fig. 4. Operating waveforms of the divide-by-four frequency divider at input frequency of 11.3 GHz. Input and output losses of the measurement system were 7.5 dB, so the actual input and output voltages were 240 and 80 mV_{p-p}, respectively.

high cutoff frequencies f_T of 33 and 20 GHz were attained for 0.5- and 1.0-μm gate-length MISFET's, respectively.

The divide-by-four frequency dividers were measured using a high-frequency on-wafer probing system with a 50-Ω characteristic impedance. Fig. 3 shows the V_{cs} dependence of maximum toggle frequency ($f_{c\ max}$) and power dissipation per *T*-F/F (P_d). At the aforementioned bias condition ($V_{ss} = -4.8$ V, $V_{cs} = +1.2$ V), a high-frequency dividing operation at 10.4 GHz was obtained with a small power dissipation of 73 mW per *T*-F/F. When $|V_{ss}|$ and V_{cs} were increased to 5.81 and 2.45 V, respectively, the $f_{c\ max}$ of 11.3 GHz was achieved with a power dissipation of 219 mW per *T*-F/F. Fig. 4 shows the

operating waveforms of the divide-by-four frequency divider at an input frequency of 11.3 GHz. The divider operation was very stable and high-speed rise and fall times of 90 and 80 ps were obtained. The toggle frequency of 11.3 GHz is the highest for any reported static frequency dividers using FET's with gate lengths longer than 0.5 μm and operating at room temperature. Consequently, n$^+$-Ge gate AlGaAs/GaAs MISFET's using high-mobility two-dimensional electron gas have shown the possibility for high-speed digital LSI applications.

V. Conclusion

A divide-by-four static frequency divider has been fabricated using n$^+$-Ge gate AlGaAs/GaAs heterostructure MISFET's for the first time. The divider circuit was based on SCFL with 1.0-μm gate length MISFET's. A maximum toggle frequency of 11.3 GHz with a power dissipation of 219 mW per *T*-F/F has been achieved at room temperature. The results obtained here show the high potential of n$^+$-Ge gate AlGaAs/GaAs MISFET technology for very high-speed digital LSI applications.

Acknowledgment

The authors would like to thank Dr. T. Takada for his helpful suggestions on SCFL circuit design. They are also grateful to H. Yamazaki for the ion implantation and useful discussions.

References

[1] T. J. Drummond, W. Kopp, D. Arnold, R. Fischer, H. Morkoç, L. P. Erickson, and P. W. Palmberg, "Enhancement-mode metal/(Al, Ga)As/GaAs buried-interface field-effect transistor (BIFET)," *Electron. Lett.*, vol. 19, no. 23, pp. 986–988, Nov. 1985.

[2] Y. Katayama, M. Morioka, Y. Sawada, K. Ueyanagi, T. Mishima, Y. Ono, T. Usagawa, and Y. Shiraki, "A new two-dimensional electron gas field-effect transistor fabricated on undoped AlGaAs-GaAs heterostructure," *Japan. J. Appl. Phys.*, vol. 23, no. 3, pp. L150–L152, Mar. 1984.

[3] K. Matsumoto, M. Ogura, T. Wada, N. Hashizume, T. Yao, and Y. Hayashi, "n$^+$-GaAs/undoped GaAlAs/undoped GaAs field-effect transistor," *Electron. Lett.*, vol. 20, no. 11, pp. 462–463, May 1984.

[4] P. M. Solomon, C. M. Knoedler, and S. L. Wright, "A GaAs gate heterojunction FET," *IEEE Electron Device Lett.*, vol. EDL-5, no. 9, pp. 379–381, Sept. 1984.

[5] K. Arai, T. Mizutani, and F. Yanagawa, "Threshold voltage behavior for WSi/Al$_x$Ga$_{1-x}$As/GaAs MIS-like heterostructure FET," *Japan. J. Appl. Phys.*, vol. 24, no. 8, pp. L623–L625, Aug. 1985.

[6] K. Arai, T. Mizutani, and F. Yanagawa, "An n$^+$-Ge gate MIS-like heterostructure FET," in *Proc. Int. Symp. GaAs and Related Compounds* (Int. Phys. Conf. Series No. 79), Sept. 1985, pp. 631–636.

[7] K. Maezawa, T. Mizutani, K. Arai, and F. Yanagawa, "Large transconductance n$^+$-Ge Gate AlGaAs/GaAs MISFET with thin gate insulator," *IEEE Electron Device Lett.*, vol. EDL-7, pp. 454–456, July 1986.

[8] T. Mizutani, S. Fujita, and F. Yanagawa, "Complementary circuit with AlGaAs/GaAs heterostructure MISFETs using high mobility two dimensional electron and hole gases," in *Proc. Int. Symp. GaAs and Related Compounds* (Int. Phys. Conf. Series No. 79), Sept. 1985, pp. 733–734.

[9] T. Takada, M. Idda, and T. Sudo, "High-speed GaAs MESFET logic (source coupled FET logic)," in *Nat. Conv. Rec. Semiconductor and Mater. IECE Japan*, 1981, p. 123.

[10] K. Maezawa, T. Mizutani, and F. Yanagawa, "Barrier height in indirect bandgap AlGaAs/GaAs hetero-junction determined with n-semiconductor/insulator/semiconductor diodes," *Japan. J. Appl. Phys.*, vol. 25, no. 7, pp. L557–L559, July 1986.

[11] T. Mizutani, N. Kato, K. Osafune, and M. Ohmori, "Gigabit logic operation with enhancement-mode GaAs MESFET IC's," *IEEE Trans. Electron Devices*, vol. ED-29, no. 2, pp. 199–204, Feb. 1982.

A Self-Aligned Gate Superlattice (Al,Ga)As/n$^+$-GaAs MODFET 5 × 5-bit Parallel Multiplier

D. K. ARCH, MEMBER, IEEE, B. K. BETZ, P. J. VOLD, STUDENT MEMBER, IEEE, J. K. ABROKWAH, MEMBER, IEEE, AND N. C. CIRILLO, JR., MEMBER, IEEE

Abstract—A 5 × 5-bit parallel multiplier circuit has been demonstrated with self-aligned gate superlattice (Al,Ga)As/n$^+$-GaAs modulation-doped FET's (MODFET's). Multiplication times (gate delays) and corresponding power dissipations of 1.80 ns (73 ps/gate) at 0.43 mW/gate and 1.08 ns (43 ps/gate) at 0.75 mW/gate were measured at room temperature and 77 K, respectively. These are the shortest gate propagation delays ever reported for parallel multiplier circuits at room temperature or 77 K using any semiconductor IC technology.

I. INTRODUCTION

INTEREST in GaAs technologies has intensified because of future military and commercial system requirements for ultra-high-speed, low-power radiation-hard signal processing and memory IC's. n$^+$-(Al,Ga)As/GaAs modulation-doped FET's (MODFET's) and GaAs MESFET's have demonstrated the potential for meeting these performance levels. Cirillo *et al.* have demonstrated self-aligned gate n$^+$-(Al,Ga)As/GaAs MODFET ring oscillators with gate delays of 11.6 ps/gate at room temperature and 8.5 ps/gate at 77 K, at a nominal gate length of 1.0 μm [1]. At a gate length of 0.35 μm, n$^+$-(Al,Ga)As/GaAs MODFET ring oscillators fabricated using a recessed gate process have demonstrated gate propagation delays as low as 10.2 ps/gate at room temperature and 5.8 ps/gate at 77 K [2]. Many recent investigations into signal processing IC performance have used multiplier circuits as demonstration vehicles. Nakayama *et al.* have demonstrated GaAs MESFET 16 × 16-bit parallel multipliers with 10.2-ns multiply time (162 ps/gate) at 952-mW total power dissipation (0.3 mW/gate) [3]. Schlier *et al.* have fabricated a recessed-gate n$^+$-(Al,Ga)As/GaAs MODFET 4 × 4-bit parallel multiplier showing a 1.6-ns multiply time (114 ps/gate) at 0.34 mW/gate at room temperature [4]. No 77 K test results were reported. A 1.5 K recessed-gate n$^+$-(Al,Ga)As/GaAs E/D-MODFET (HEMT) gate array configured as an 8 × 8-bit parallel multiplier has been demonstrated by Watanabe *et al.* [5]. Multiplication times of 4.9 and 3.1 ns (158 ps/gate) were measured at room temperature and 77 K, respectively. A major problem with conventional n$^+$-(Al,Ga)As/GaAs MOD-FET's is drain *I–V* degradation or collapse at cryogenic temperature, due to Si doping induced "DX" center traps in

the (Al,Ga)As charge control layer [6]. The use of superlattice (Al,Ga)As/n$^+$-GaAs charge-control layers, in which doping is confined to the GaAs layer, has demonstrated significantly reduced low-temperature drain *I–V* degradation [7]. In this paper we report for the first time the ultra-high-speed performance at both room temperature and 77 K of a superlattice MODFET parallel multiplier. The self-aligned gate superlattice (Al,Ga)As/n$^+$-GaAs MODFET 5 × 5-bit parallel multiplier circuit demonstrated the shortest gate propagation delays of any semiconductor parallel multiplier circuit at either room temperature or 77 K.

II. CIRCUIT DESIGN

The self-aligned gate MODFET 5 × 5-bit parallel multiplier design employs approximately 350 direct-coupled FET logic (DCFL) gates for the logic portion of the circuit, and two cascaded differential amplifiers feeding into a large push–pull driver FET on each of the nine outputs. A photomicrograph of the MODFET multiplier circuit (area = 1.86 × 1.54 mm^2) is shown in Fig. 1. The circuit multiplies two 5-bit two's complement input numbers and produces a 9-bit two's complement product. The circuit architecture has a product term generator feeding into 16 full adders and four half adders, although the circuit uses full adder cells to provide the half-adder function. A schematic showing the adder cell array with the product term inputs and the multiplier outputs is shown in Fig. 2. The adder has two gate delays for computing a carry and four gate delays to compute a sum. The average fan-out and fan-in is 2.2. The standard dimensions used for the layout of the logic portion of the circuit were 1.0 × 10 μm^2 for the FET's and 2.0 × 5 μm^2 for the saturated resistors. Larger width FET's and resistors are used to drive the longer signal lines.

III. MATERIAL GROWTH AND FABRICATION

Epitaxial growth of the superlattice (Al,Ga)As/n$^+$-GaAs MODFET's was performed using a PHI 425A MBE machine on 3-in LEC GaAs substrates. Starting from the substrate the layers were an undoped GaAs buffer, an undoped (Al,Ga)As spacer, a five-period (Al,Ga)As/n$^+$-GaAs superlattice, and a thin undoped (Al,Ga)As cap. Details of the MBE material growth have been reported elsewhere [8]. Van der Pauw measurements on depletion MODFET's with similar superlattice (Al,Ga)As/n$^+$-GaAs structures show sheet electron con-

Manuscript received August 22, 1986; revised October 11, 1986.
The authors are with the Honeywell Physical Sciences Center, Bloomington, MN 55431.
IEEE Log Number 8611987.

Reprinted from *IEEE Electron Device Letters*, vol. EDL-7, no. 12, pp. 700–702, December 1986.

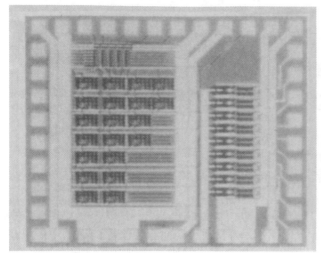

Fig. 1. Photomicrograph of a superlattice (Al,Ga)As/n$^+$-GaAs MODFET 5 × 5-bit parallel multiplier circuit.

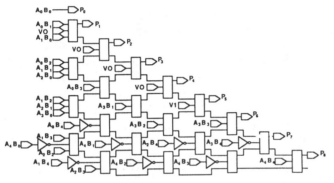

Fig. 2. Adder array schematic showing the product term inputs and the multiplier outputs.

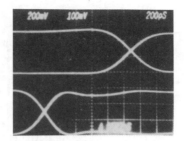

P$_6$ OUTPUT
(MSB, excluding sign bit)

B$_0$ INPUT
(LSB)

Fig. 3. Oscillograph showing a 1.08-ns multiply time at 77 K. (Upper trace = 100 mV/div., lower trace = 200 mV/div.)

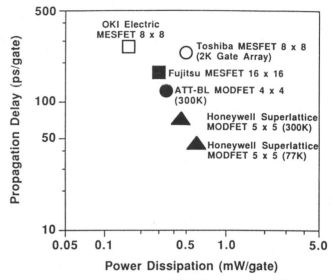

Fig. 4. Comparison of recent GaAs MESFET and (Al,Ga)As/GaAs MODFET multiplier gate propagation delays as a function of gate power dissipation.

centrations of 0.5-1.4 × 10^{12} cm^{-2} and mobilities of 40 000–120 000 cm^2/V·s at 77 K. The multiplier circuits were fabricated using a self-aligned gate MODFET IC process detailed elsewhere [9]. The self-aligned ion implants were annealed in a rapid optical anneal system in order to minimize dopant diffusion and heterostructure degradation.

IV. Device and Circuit Results

The self-aligned gate superlattice (Al,Ga)As/n$^+$-GaAs MODFET's with 1-μm gate lengths demonstrated extrinsic transconductances g_m of 254 mS/mm and threshold voltages V_t of 0.193 V at room temperature. Upon cooling to 77 K, a g_m of 326 mS/mm and V_t of 0.296 V were measured. The increase in V_t of 103 mV, from room temperature to 77 K, is significantly less than the 150–200-mV shift observed in conventional n$^+$-(Al,Ga)As/GaAs MODFET's. This is due to a significant reduction in the number of DX-center traps generated in the superlattice (Al,Ga)As/n$^+$-GaAs MODFET device [7]. Drain I–V collapse was not observed in the superlattice (Al,Ga)As/n$^+$-GaAs MODFET's. The source resistance R_s was 0.68 Ω·mm, as measured from a transmission-line method (TLM) structure.

The 5 × 5-bit parallel multiplier circuits were first characterized on-wafer to determine the yield of fully functional circuits. All 1024 input/output combinations were verified at room temperature using a logic analyzer with a 100-kHz cycle time. Typical wafer yields of about 40 percent are obtained for fully functional superlattice 5 × 5 multipliers. The fully functional multipliers were then tested on-wafer using a high-frequency probe card to determine the multiply times at room temperature and 77 K. The maximum multiply time was determined by measuring the delay through the largest number of gates, excluding the sign bit. The least-significant bit (LSB) of input B was toggled, with all of the other A and B inputs held at constant high or low logic levels, and the most-significant bit (MSB) of the output, excluding the sign bit, was observed to determine the delay time through the circuit and hence the multiply time. The superlattice (Al, Ga)As/n$^+$-GaAs MODFET 5 × 5-bit parallel multiplier demonstrated a multiply time of 1.80 ns at 0.43 mW/gate at room temperature. This yields an average propagation delay of 73 ps/gate at a switching energy of 31 fJ/gate. The 5 × 5-bit multiplier was then cooled to 77 K and tested using the same procedure. Multiply times of 1.18 ns at 0.44 mW/gate and as low as 1.08 ns at 0.75 mW/gate were observed. This yields average propagation delays of 47 and 43 ps/gate, at switching energies of 21 and 32 fJ/gate, respectively. Fig. 3 shows the LSB input and the MSB output of the multiplier circuit at 77 K, demonstrating a 1.08-ns multiply time. A comparison of recent GaAs MESFET and AlGaAs/GaAs MODFET multiplier gate propagation delays as a function of gate power dissipation is shown in Fig. 4.

V. Conclusion

A superlattice (Al,Ga)As/n$^+$-GaAs MODFET 5 × 5-bit parallel multiplier has been designed and fabricated demonstrating the shortest gate propagation delays ever reported for multiplier circuits at either room temperature or 77 K. This result illustrates the promise of superlattice (Al,Ga)As/n$^+$-GaAs MODFET IC's for ultra-high-speed logic at both room and cyrogenic operating temperatures.

Acknowledgment

The authors would like to thank P. Joslyn and T. Nohava for the MBE growth, and A. M. Fraasch for the wafer processing.

References

[1] N. Cirillo, Jr., J. Abrokwah, A. Fraasch, and P. Vold, "Ultra-high-speed ring oscillators based on self-aligned-gate modulation-doped n$^+$-(Al,Ga)As/GaAs FET's," *Electron. Lett.*, vol. 21, no. 17, 1985.

[2] N. J. Shah, S. S. Pei, C. W. Tu, and R. C. Tiberio, "Gate-length dependence of the speed of SSI circuits using submicrometer selectively doped heterostructure transistor technology," *IEEE Trans. Electron Devices*, vol. ED-33, no. 5, pp. 543–547, 1986.

[3] Y. Nakayama, K. Suyama, H. Shimizu, N. Yokoyama, A. Shibatomi, and H. Ishikawa, "A GaAs 16 × 16b parallel multiplier using self-alignment technology," in *Proc. IEEE Int. Solid-State Circuits Conf.*, 1983, pp. 48–49.

[4] A. R. Schlier, S. S. Pei, N. J. Shah, C. W. Tu, and G. E. Mahoney, "A high-speed 4 × 4 bit parallel mutliplier using selectively doped heterostructure transistors," in *Proc. IEEE 1985 GaAs IC Symp.*, pp. 91–93.

[5] Y. Watanabe *et al.*, "A high electron mobility transistor 1.5 K gate array," in *IEEE Int. Solid-State Circuits Conf.*, 1986, pp. 80–81.

[6] A. Katalsky and R. A. Kiehl, "On the low-temperature degradation of (Al,Ga)As/GaAs modulation-doped field-effect transistors," *IEEE Trans. Electron Devices*, vol. ED-33, no. 3, pp. 414–423, 1986.

[7] T. Baba, T. Mizutani, M. Ogawa, and K. Ohata, "High performance (AlAs)/n-GaAs superlattice 2DEGFETs with stabilized threshold voltage," *Japan. Appl. Phys.*, vol. 23, no,. 8, 1984.

[8] J. K. Abrokwah *et al.*, "Novel self-aligned gate Al$_x$Ga$_{1-x}$As/n-GaAs superlattice modulation-doped FETs," *J. Vac. Sci. Technol. B*, vol. 4, no. 2, 1986.

[9] N. C. Cirillo, Jr., J. K. Abrokwah, and S. A. Jamison, "A self-aligned gate modulation-doped (Al,Ga)As/GaAs FET IC Process," in *Proc. 1984 IEEE GaAs IC Symp.*, pp. 167–170.

A High-Speed 4x4 Bit Parallel Multiplier Using Selectively Doped Heterostructure Transistors

A. R. Schlier
S. S. Pei
N. J. Shah
C. W. Tu
G. E. Mahoney

AT&T Bell Laboratories
Murray Hill, New Jersey 07974

ABSTRACT

A 4x4 bit parallel multiplier using selectively doped heterostructure transistors (SDHT) has been demonstrated. The circuit was implemented with NOR gates using direct coupled FET logic (DCFL). The shortest propagation delay was measured to be 114ps/gate which corresponds to a multiplication time of 1.6ns. At this speed the circuit dissipated .34mW/gate. This is the best result ever reported for a parallel multiplier at room temperature.

I. INTRODUCTION

Selectively doped GaAs/AlGaAs heterostructure transistors (SDHT) have received considerable attention recently for their potential applications in ultrahigh-speed memory and logic circuits (1-2). The potential for the uniformity and the control of the threshold voltage (2) and the extremely high transconductance (3) make the SDHT technology attractive for large scale integration.

Memories as large as a 4K static random access memory (SRAM) have been reported, but these circuits have not yet been shown to be fully functional (2). For logic circuits, only small scale integrated circuits, such as ring oscillators and frequency dividers, have been reported up to now (4). It is necessary to go to a higher level of integration, where the parasitic capacitance of interconnects becomes important, to demonstrate the advantage of the high transconductance of SDHTs.

To obtain higher levels of integration, the direct coupled FET logic (DCFL) design is the most promising since it requires the smallest number of circuit elements, has low power consumption and requires only one power supply. We report in this paper a 4x4 multiplier implemented with DCFL gates and fabricated with a recessed-gate SDHT technology.

II. CIRCUIT DESIGN

The 4x4 multiplier circuit is comprised of 8 full adders and 4 half adders. The adders use the conventional NOR gate designs. The maximum number of fan-ins and fan-outs of the NOR gates are 4 and 6, respectively. A schematic of the multiplier design is shown in Fig. 1. With this design, the longest delay path has been reduced from $16\tau_{P.D.}$ for the conventional parallel multiplier design to $14\tau_{P.D.}$. The design has an average fan-out of 2.6.

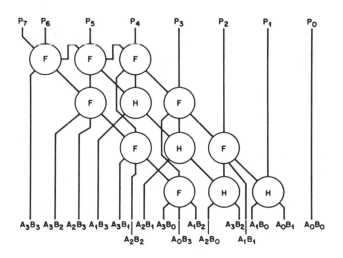

Fig. 1   4x4 multiplier connection scheme.

The circuit uses 141 DCFL gates with a two-stage source follower output buffer. All of the logic gates uses 5μm wide saturated resistor loads with 10μm wide enhancement-mode SDHT drivers. The eight output buffers are comprised of a 30μm inverter followed by a 75μm follower to drive a 50Ω load. The total device count is 315 drivers with 149 loads.

To aid in testing the multiplier, an electrically controllable self-oscillation feedback loop is designed into the circuit. The feedback path routes the most significant product bit (P_7) to the least significant input bit (A_0).

III. FABRICATION AND LAYOUT

The circuits were fabricated on semi-insulating GaAs wafers with the following MBE-grown epitaxial layers : ≈1μm undoped GaAs buffer, an undoped AlGaAs spacer layer, an n^+-AlGaAs donor layer and an n^+-GaAs cap layer. The process was a recess-gate technology using a wet chemical etch to adjust the threshold voltage of the enhancement-mode SDHT drivers and the current of the saturated resistor loads. Liftoff was used for all metallization levels and the ohmic contact to source and drain regions was made with a Au-Ge based alloy.

The circuits used SiN dielectric crossover technology for second level metallization to minimize interconnect line lengths. The design rules specify 3μm line widths with a 4.5μm spacing between interconnect lines. The gate length

Reprinted from *Rec. of the IEEE GaAs Integrated Circuits Symp.*, pp. 91–93, 1985.

of the E-mode devices was nominally 1μm. A micrograph of a fully processed 4x4 multiplier circuit is shown in Fig.2. The die size is 1.1mm x 1.2mm with 75μm square pads for probing. The output buffers are at the top of the figure with a center buss for V_{ss} and a split V_{dd} buss on either side of the circuit.

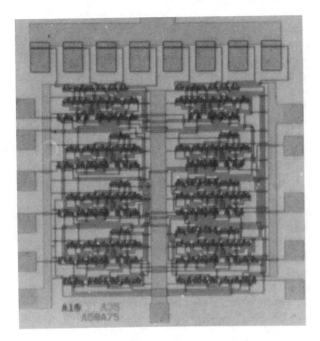

Fig. 2   Picture of fully processed 4x4 multiplier.

IV. DEVICE AND CIRCUIT RESULTS

The characteristics of discrete devices were measured on nearby testers. The average transconductance at 300K was 167mS/mm and the average threshold voltage was +.06V. The high speed performance of the devices was evaluated with a 19-stage ring oscillator. The results showed a very tight distribution of propagation delays across the wafer (see Fig.3). The mean gate delay was 29ps with a standard deviation of 4ps for a bias voltage of 1V. The average speed-power product for the ring oscillators was 8.3fJ. At 1V bias the average power dissipation was .28mW/gate.

The multiplier was evaluated by both D.C. and high speed testing. The D.C. testing was used to determine the full functionality of the circuit. The high-speed testing involved setting the multiplier in a self-oscillation mode. The oscillation frequency was used to measure the minimum propagation delay per stage which in turn would be used to predict the fastest complete multiplication time. The high speed testing was done using a probe card that has ceramic stripline probe blades with a semi-rigid coaxial cable attached to the blades. The V_{dd} lines on the probe card were terminated with 200pF capacitors and the input lines were terminated with 50Ω resistors on the probe blades.

The D.C. testing involved sending random 1's and 0's to the inputs of the multiplier and monitoring the output on a D.V.M. to verify the correct arithmetic operation. Many multipliers were found fully functional from a V_{dd} voltage of

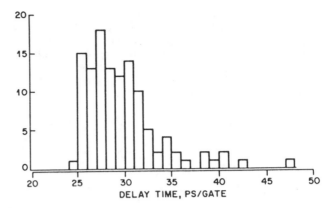

Fig. 3   Histogram of ring oscillator delay at 1V bias vs. number of devices.

.8V to 1.5V. The source follower provided an output voltage swing ranging from .3V for 1V bias to .5V for 1.5V.

To determine the performance of the multiplier, the self-oscillation feedback loop that was designed into the circuit was activated and the inputs were set to 1111 x 100S=SSSSSSSS. The S input corresponds to the signal fed back from P_7. Since the feedback loop has a propagation delay of $14\tau_{P.D.}$, an oscillating frequency of $1/28\tau_{P.D.}$ is expected. The results of the best multiplier obtained are shown in Table 1. The highest oscillation frequency observed was 318MHz which corresponds to a propagation delay of 114ps/gate. It would therefore yield a complete 4x4 multiplication in 1.6ns. The circuit dissipated 52.5mW or .34mW/gate at a bias voltage of 1.1V. Many fully functional multipliers were found on the wafer and a majority of them oscillated above 240MHz with similar bias voltage and power dissipation.

4X4 MULTIPLIER RESULTS

- MULTIPLICATION   TIME : 1.6nS

- PROPAGATION   DELAY : 114pS

- POWER DISSIPATION : 55mW (0.34mW/GATE)

- BIAS VOLTAGE : 1.1V

Table 1 Summary of 4x4 multiplier results.

V. SUMMARY

The results of the 4x4 multiplier are very encouraging. Fig.4 plots propagation delay vs. multiplier size for some of the fastest GaAs MESFET multipliers. It can be seen that the SDHT multiplier shows a considerable edge in terms of propagation delay than other multipliers. Also, by looking at the performance summary in Fig.4, it shows that the SDHT multiplier operates at low power and with good performance at room temperature. The results indicate that the advantage of the SDHT technology becomes more apparent at a higher level of integration. With the good results of the multiplier,the SDHT technology is a very good candidate for large scale integrated circuit applications.

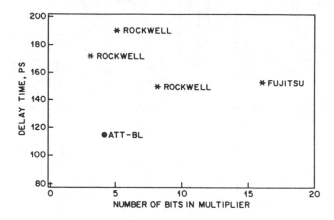

Fig. 4   Plot of propagation delay vs. multiplier size for fastest room temperature multipliers. (*=MESFET,o=SDHT)

The authors wish to thank L.D.Urbanek, P.F.Sciortino

and R.F.Kopf for their technical support and contributions toward this project.

This work was supported in part by the Air Force Wright Aeronautical Avionics Laboratory, Wright Patterson AFB,Ohio under contract F33615-84-C-1570.

REFERENCES

1. R.H.Hendel, S.S.Pei, C.W.Tu, B.J.Roman, N.J.Shah and R.Dingle,"Realization of sub-10 picosecond switching time with SDHTs, "Tech.Dig.Int. Electron Device Meet., San Francisco, California(1984).

2. S.Kuroda, T.Mimura, M.Suzuki, N.Kobayashi, K.Nishiuchi, A.Shibatomi and M.Abe, "New device structure for 4Kb HEMT SRAM, "IEEE GaAs IC Symposium, pp. 125-128, 1984.

3. L.H.Camnitz, P.J.Tasker, H.Lee, D.VanDerMerwe and L.F.Eastman, "Microwave Characterization of very high transconductance MODFET," Tech.Dig.Int.Electron Device Meet., pp. 360-363, 1984.

4. S.Pei, N.Shah, R.Hendel, C.Tu,and R.Dingle,"Ultra high speed integrated circuits with selectively doped heterostructure transistors, "IEEE GaAs IC Symposium, pp. 129-132,1984.

High performance inverted HEMT and its application to LSI

S.Nishi, T.Saito, S.Seki, Y.Sano, H.Inomata, T.Itoh, M.Akiyama, K.Kaminishi

Research Laboratory, Oki Electric Industry Co., Ltd.
550-5, Higashiasakawa, Hachioji, Tokyo 193, Japan

Abstract The growth condition of MBE and processing technique for inverted HEMT were studied. A high electron mobility ($1.1 \times 10^5 cm^2/Vs$), a high transconductance and a K-value ($gm=400mS/mm$ $K=480mA/V^2mm$ at R.T. and $gm=550mS/mm$ $K=860mA/V^2mm$ at 77K) were obtained. The minimum ring-oscillator propagation delay of 19.7ps was obtained at R.T. with the gate length of 0.5μm. A fully operational 6x6 parallel multiplier was fabricated on a 1K gate array using the inverted HEMTs with a gate length of 0.8μm. Multiply times of 6ns(0.7W) and 5.1ns(1W) were measured at R.T. and 77K, respectively.

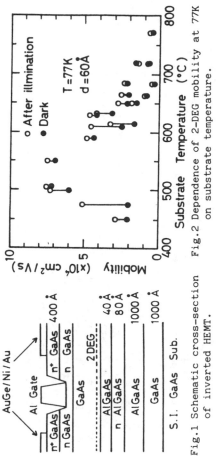

AuGe/Ni/Au — Al Gate — n⁺ GaAs — n GaAs

n⁺ GaAs	400 Å
n GaAs	
2DEG	
GaAs	40 Å
n AlGaAs	80 Å
n AlGaAs	1000 Å
AlGaAs	1000 Å
GaAs	
S.I. GaAs Sub.	

Fig.1 Schematic cross-section of inverted HEMT.

Fig.2 Dependence of 2-DEG mobility at 77K on substrate temperature.

(Mobility ($\times 10^4 cm^2/Vs$) vs Substrate Temperature (°C); ○ After illumination, ● Dark; T=77K, d=60Å)

1. Introduction

The selectively doped GaAs/N-AlGaAs single heterostructures was widely used for high speed devices (Kobayashi et al. 1985, Watanabe et al. 1986, Kinoshita et al. 1985). A high transconductance (Inomata et al. 1986) and an extremely short propagation delay time (Shah et al. 1986) were reported by this time. But these reports were using the heterostructure of a ternary on top of a binary (conventional HEMT) because of the difficulty in the growth of inverted heterostructure (binary on top of ternary) (Morkoç et al. 1982a,b). In the conventional HEMT, as the sheet resistance of 2 dimensional electron gas(2-DEG) is rather high, some processing techniques are required for the reduction of the source resistance to improve the FET performance at R.T. On the other hand, in the inverted heterostructure, the source resistance can be easily lowered at R.T. by the top n⁺GaAs layer and recessed gate process. Then a high performance is expected at room temperature in an inverted HEMT.

In this report, we demonstrate the growth condition of MBE and the processing technique for inverted HEMT. A high electron mobility was obtained in the inverted heterostructure. The optimum layer structure was studied and an inverted HEMT with high performance was fabricated. A 1K gate array was successfully fabricated using the inverted HEMTs with the gate length of 0.8μm.

2. Crystal Growth

The inverted heterostructures were grown by conventional III-V MBE system. The growth rate was about 0.5μm/h. A 2-inch HB semi-insulating GaAs was used as a substrate. After chemical etching, a substrate was immediately fixed to a substrate holder by In solder. A direct heating system without In solder was also used. Schematic cross section of an inverted

HEMT is shown in Fig.1. An epitaxial layer consists of undoped GaAs and AlGaAs with both thicknesses of 1000Å, n-AlGaAs ($1.1 \times 10^{18}cm^{-3}$, 80Å) AlGaAs separation layer (40Å), undoped GaAs (200Å or 400Å), n-GaAs ($5 \times 10^{17}cm^{-3}$) and n⁺-GaAs ($4 \times 10^{18}cm^{-3}$, 400Å). With this structure, the maximum carrier concentration of 2-DEG is $8 \times 10^{11}cm^{-2}$.

To fabricate a high performance inverted HEMT, it is important to grow a heterostructure with 2-DEG of high electron mobility. We studied the substrate temperature dependence of the mobility of 2-DEG at the inverted hetero-interface. In Fig.2, mobility of 2-DEG at 77K is shown as a function of a substrate temperature. In the experiment, the substrate temperature was monitored by a thermocouple behind a Mo substrate holder. The thickness of an AlGaAs separation layer was fixed to be 60Å. To suppress both a depletion of 2-DEG by surface potential and a parallel conduction in n-GaAs, thick undoped GaAs channel layer of 2000Å with thin top n-GaAs layer was grown on AlGaAs layer. The electron mobility was measured by van der Pauw method. The electron mobility increased gradually with decreasing the substrate temperature and it decreased rapidly with the substrate temperature below 500°C. The sheet carrier concentrations of 2-DEG were ranging between $5 \times 10^{11}cm^{-2}$ and $1.5 \times 10^{12}cm^{-2}$ above 500°C and decreased to about $2 \times 10^{11}cm^{-2}$ below 500°C. The decrements of mobility and sheet carrier concentration below 500°C were due to the increase of the deep level concentration in the grown layer. The higher mobilities at fairly low substrate temperatures may be due to the improvement of AlGaAs surface morphology and the reduction of surface segregation of doped Si. The maximum mobility of $1.1 \times 10^5 cm^2/Vs$ at 77K was obtained with a sheet carrier concentration of $3.5 \times 10^{11}cm^{-2}$ when the separation layer thickness was increased to 250Å. As the sheet carrier concentrations were extremely low at low substrate temperatures and a control of substrate temperature is rather difficult, a little high substrate temperature (600-650°C) was used for inverted HEMT. In this temperature range, a sufficiently high electron mobility was obtained at R.T. ($>6000cm^2/Vs$).

For the inverted HEMT, the thickness of n-AlGaAs must be chosen to the proper value because the thicker n-AlGaAs causes a parallel conduction to a channel and degrades a FET performance. Fig.3 shows the relation between the sheet carrier concentration and the thickness of n-AlGaAs.

Reprinted with permission from *Inst. Phys. Conf. Ser. No. 83*, Chapter 8, pp. 515–520 (*1986 Int. Symp. GaAs and Related Compounds*).

High-speed devices

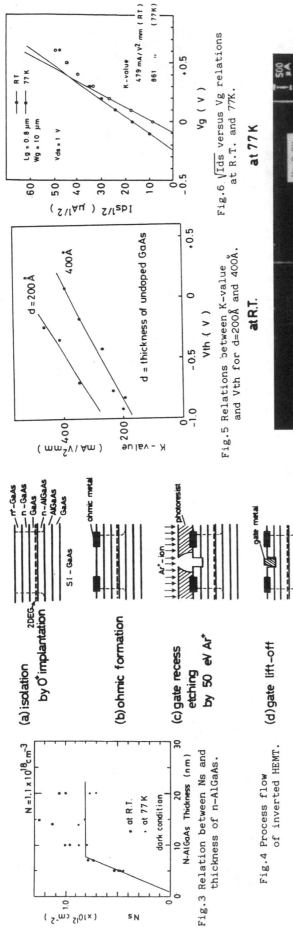

Fig.3 Relation between Ns and thickness of n-AlGaAs.

Fig.4 Process flow of inverted HEMT.

Fig.5 Relations between K-value and Vth for d=200Å and 400Å.
at R.T.

Fig.6 $\sqrt{Ids}$ versus Vg relations at R.T. and 77K.
at 77K

Fig.7 I-V characteristics of inverted HEMT. Lg=0.8μm and Wg=10μm.

The thickness of AlGaAs separation layer was 40Å. The sheet carrier concentration increased to $8\times10^{11}cm^{-2}$ with increasing the thickness of n-AlGaAs to 80Å. Above the thickness of 80Å, the sheet carrier concentration saturated to the same value at 77K, indicating the parallel conduction at R.T. occured in n-AlGaAs layer.

3. Device Fabrication and Characteristics

The fabrication process flow is shown in Fig.4. At first, devices were isolated by ion implantation of oxygen to form a planar structure (Fig.4(a)). Source and drain contacts were formed by AuGe/Ni/Au (Fig.4(b)). Gate regions were recessed by 50eV Ar ion-beam (Fig.4(c)). Al metal was deposited into the recessed region by a self-alignment technique using LMR deep-UV photoresist (Fig.4(d))(Yamashita et al. 1985). The threshold voltage was controlled by the recessed depth.

To improve the performance of an inverted HEMT, a short distance from the gate electrode to 2-DEG was desirable. Two different thicknesses (200Å and 400Å) of the undoped GaAs channel layers were compared in FET performance. In Fig.5, K-values at R.T. are shown as a function of threshold voltage for two different thicknesses. The gate length and the width were 0.8μm and 10μm, respectively. K-values were much improved by reducing the thickness of undoped GaAs. At the threshold voltage of 0V, the distances from the gate elctrode to the hetero-interface are calculated to be 460Å and 590Å for the undoped GaAs thickness of 200Å and 400Å. K-value at the threshold voltage of 0V was improved about 40%. This larger improvement may be explained if the distribution of 2-DEG in GaAs layer is considered. In Fig.6, square roots of saturation current at R.T. and 77K are shown as a function of a gate voltage. The thickness of an undoped GaAs was 200Å. Extremely high K-values of $480mA/V^2mm$ at R.T.

and $860mA/V^2mm$ at 77K were obtained. I-V characteristics of the inverted HEMT at R.T. and 77K are shown in Fig.7. The maximum transconductances at R.T. and 77K were 400mS/mm and 550mS/mm, respectively. The source resistances were as low as 0.7Ωmm and 0.54Ωmm at R.T. and 77K. Drain current saturation characteristics were fairly good, showing the small short channel effect.

To fabricate LSI using inverted HEMTs, the threshold voltage must be uniform on the microscopic scale as well as on the full wafer. On a whole area of a 2-inch wafer, the very uniform threshold voltage distribution was obtained with the gate length of 1.5μm at R.T. The standard deviation was 11.8mV at the threshold voltage of 0.1V. The microscopic uniformity was studied at R.T. using the 50μm×50μm pitch FET array (Nakamura et al. 1985). The histogram of the threshold voltage is shown in Fig.8. The gate length and width were 0.8μm and 10μm, respectively. The threshold voltages of 780 samples, in the area of 0.5mm × 3.9mm, were included in the figure. The standard deviation of 19mV at the mean threshold voltage of -0.38 V was obtained, which was sufficiently small to fabricate LSI.

The propagation delay time of an inverted HEMT E/D DCFL circuit was measured at R.T. using the 21-stage ring-oscillator (R/O). The gate length was reduced to 0.5μm and the minimum propagation delay times of 19.7ps/gate with the power dissipation of 0.2mW/gate and 20ps/gate with 2mW/gate were obtained. No dependence of a propagation delay time on a power dissipation may be due to the good saturation characteristics of drain current.

4. 1K Gate Array

To confirm the ability of the inverted HEMT for LSI, we fabricated 1K gate array using the inverted HEMTs with the gate length of 0.8μm. 1K gate array consisted of 1000 basic cells and 52 I/Os. The basic cell was 3-input NOR gate by E/D DCFL circuit and I/O was constructed by SBFL circuit (Tanaka et al. 1984). The chip size was 3.8mm x 4.2mm. On this masterchip a 6x6 parallel multiplier using the carry save algorithm and the R/Os under various load conditions were fabricated. Fig.9 shows the photograph of a 1K gate array.

Functional check was performed on a 6x6 parallel multiplier. All 4096 bit patterns were tested at 10kHz and full operation was observed. The multiply time was measured from the delay time through critical path (26-stage internal gates and 5-stage I/O gates). The multiply times at R.T. and 77K were 6.0ns (0.7W), and 5.1ns (1W), respectively. Table 1 shows increments of delay times due to various loads calculated from the results of R/Os under various loads at R.T. The multiply time of 26-stage internal gates could be calculated from the actual fan in, fan out, line length and crossovers, using the data of Table 1, which was 4ns at R.T. The difference between the measured and calculated ones is due to the delay at I/O. The transconductance of inverted HEMT used to fabricate the gate array was 240mS/mm, which was not improved yet. The multiply time will be shortened if the improved inverted HEMTs are used.

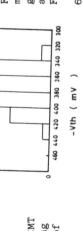

N=780
$V_{th} = -382mV$
$\sigma V_{th} = 19mV$

Fig.8 Histogram of threshold voltages of 50μmx50μm pitch inverted HEMTs. Lg=0.8μm.

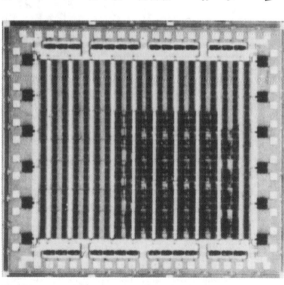

Fig.9 Photograph of 6x6 parallel multiplier on 1K gate array. Lg=0.8μm.

delay time	τ0 (ps/gate)	70.6	-
fan-in delay	τfo (ps/f.o.)	37.2	(0.528)
fan-out delay	τfl (ps/f.l.)	12.5	(0.177)
line load delay	τl (ps/mm)	56.8	(0.804)
crossover delay	τc (ps/c.o.)	0.90	(0.013)
power dissipation P	(mW/gate)	0.86	

Table 1 The increments of delay time due to various load conditions.

5. Summary

MBE growth condition and processing technique for inverted HEMT were optimized. High performance inverted HEMT was fabricated and short propagation delay time was obtained at R.T. Fully operating 6x6 parallel multiplier was fabricated on 1K gate array. These results confirm an inverted HEMT is a promising FET for high speed LSI at R.T.

6. Acknowledgement

The present research effort is part of the National Research and Development program on "Scientific Computing System", conducted under a program set by the Agency of Industrial Science and Technology, Ministry of International Trade and Industry.

7. References

Inomata H, Nishi S, Takahashi S and Kaminishi K 1986 Jpn. J. Appl. Phys. 25 L731

Kobayashi N, Notomi S, Suzuki M, Tsuchiya T, Nishiuchi K, Odani K, Shibatomi A, Mimura T and Abe M 1985 GaAs IC Symp. Tech. Dig. 207

Kinoshita H, Nishi S, Akiyama M and Kaminishi K 1985 Jpn. J. Appl. Phys. 24 1061

Morkoc H, Drummond T J and Fischer R 1982a J. Appl. Phys. 53 1030

Morkoc H, Drummond T J, Fischer R and Cho A Y 1982b J. Appl. Phys. 53 3321

Nakamura H, Matuura H, Egawa T, Sano T, Ishida T and Kaminishi K 1985 Extended Abstracts 17th Conf. Solid State Devices and Materials, Tokyo, 429

Shah N J, Pei S S, Tu C W, Tiberio R C 1986 IEEE Trans. Electron Devices ED-33 543

Tanaka K, Nakamura H, Kawakami Y, Akiyama M, Ishida T and Kaminishi K 1984 Extended Abstracts 16th Conf. Solid State Devices and Materials, Kobe. 399

Watanabe Y, Kajii K, Nishiuchi K, Suzuki M, Hanyu I, Kosugi M, Odani K, Shibatomi A, Mimura T, Abe M and Kobayashi M 1986 IEEE ISSCC Tech. Dig. 80

Yamashita Y, Kawazu R, Kawamura K, Ohno S, Asano T, Kobayashi K and Nagamatsu G 1985 J. Vac. Sci. Technol. 3 314

Section 2.5: Converters

GaAs/GaAlAs HIGH ELECTRON MOBILITY TRANSISTORS FOR ANALOG-TO-DIGITAL
CONVERTER APPLICATIONS

C.P. Lee, N.H. Sheng, H.F. Lewis, H.T. Wang, D.L. Miller and J. Donovan

Microelectronics Research and Development Center, Rockwell International
Thousand Oaks, CA 91360

ABSTRACT

The use of GaAs/GaAlAs high electron mobility
transistors (HEMTs) for analog-to-digital converter applications has been studied. HEMTs have
shown superior device stability and uniformity to
those of GaAs MESFETs. Because of the buried
channel conduction and the control of MBE layers,
HEMTs are suitable for high speed and high resolution ADCs. 5-bit ADCs, 5-bit DACs, and timing and
control circuits, all built with HEMTs, have been
demonstrated. A waveform reconstruction using
these three chips has been performed at 140 Mega
samples per second with 4-bit accuracy.

1.0 INTRODUCTION

High-speed high-resolution analog-to-digital
converters (ADCs) are one of the most important
components in many electronic systems. However,
it is not an easy task to build such ADCs with
combined requirements on speed and resolution.
Several attempts have been made to use GaAs
MESFETs to build high speed ADCs, but success has
been limited. Since the MESFETs technology is
based on direct ion-implantation in semi-
insulating GaAs, the device performance is affected by substrate related problems. The distribution and the concentration of residual impurities, deep traps and dislocations can all cause
uniformity and reproducibility problems for the
devices. Furthermore, the existence of the
channel-substrate interface backjunction causes
the so-called backgating effect, resulting in
instability in the device characteristics (1).
The various backgating-caused phenomena such as
crosstalk, frequency dependent in I-V, I-V
hysteresis, and temperature dependence, which are
normally tolerated by digital circuits, can be
detrimental to ADC operation.

In this paper we report the use of GaAs/GaAlAs
high electron mobility transistors (HEMTs) for
high-speed and high-resolution ADCs. The HEMT
technology which combines the advantages of FET
planar processing, high velocity transport and the
control of MBE growth is ideal for such applications. Besides its superior speed, it is not sensitive to the substrate effects which affect
MESFET performance.

2.0 DEVICE STABILITY

In high precision circuits such as ADCs, the
stability of the devices is very important. Any
drift in device characteristics with time, temperature and frequency can cause inaccuracies in
the result. In regular ion-implanted MESFETs,
some of these problems do exist because of the
trapping effect from the channel-substrate back
junction (2). For HEMTs, since the conduction
electrons are separated from the substrate by a
thick GaAs buffer layer, the substrate related
phenomena are much less pronounced.

Temperature Sensitivity

Stability of HEMT device characteristics over
the mil-spec temperature range, from -55°C to
125°C, has been evaluated and compared with that
of MESFETs. A HEMT and a regular ion-implanted
MESFET, both of which were enhancement mode with
dimensions of 1 μm × 50 μm, were tested in an en-
vironmental chamber over the whole mil-spec temperature range. Figure 1 shows the drain current
versus gate voltage plot of these devices measured
at -55°C, 24°C, and 125°C. For the HEMT, only
very slight change in the characteristics was observed. For the MESFET, a much higher change in
device characteristics is observed; the drain current drops by 41%, and the threshold voltage
changes more than 0.2 V when the temperature is
changed from 125°C to -55°C. The major contribution to the temperature sensitivity of MESFETs is
the change in built-in potential at the Schottky
junction and the channel-substrate interface junction due to Fermi level shift (3). For HEMTs, the
channel-substrate interface junction does not
exist and the only temperature dependent factors
which may influence the device performance are the
mobility and the build-in potential of the
Schottky junction. These two factors actually
compensate each other as the temperature changes,
which is why the net change of the drain current
is very small.

Output Conductance vs Frequency

The output conductance of a device is one of
the determining factors for the gain of an amplifier. Therefore its stability is very important
for analog devices. We have measured the output
conductance of a HEMT at different frequencies and
compared it with that of a MESFET. Figure 2 shows

Reprinted from *IEEE Int. Electron Devices Meeting, 1985, Tech. Dig.*, pp. 324–327, 1985.

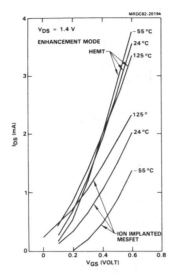

Fig. 1 Transfer characteristics (I_{ds} vs V_{gs}) of a MESFET and a HEMT measured at -55°C, 24°C and 125°C.

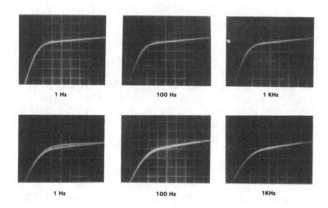

Fig. 2 I_{dss} vs V_{ds} curves of a depletion mode HEMT (upper row) and a depletion mode MESFET (lower row) measured at scanning rates of 1 Hz, 100 Hz and 1 kHz.

the I-V curves (at V_{gs} = 0 V) of a depletion-mode HEMT and a depletion-mode MESFET measured at 1 Hz, 100 Hz, and 1 kHz. For the HEMT the output conductance is the same at these two different frequencies, but for the MESFET the output conductance is smaller when the frequency is higher. The degradation of output conductance for MESFETs is the direct result of the trapping effect at the channel-substrate interface. The loops in the I-V characteristics of the MESFET also indicate the existence of such traps.

Current Lag Effect

When a square voltage pulse is applied to the gate of a FET, a corresponding current pulse should be detected flowing from the drain to the source. However for GaAs MESFETs, due to traps at

the surface of the device, the current pulse is often not square; there is a time lag in reaching the maximum of the current value. The time constant for the recovery of the current is of the order of microseconds to milliseconds. These current lag effects could result in inaccuracies in ADC operations. In HEMTs, however, no such effect has been observed. Figure 3 shows the measured results. Voltage pulses with widths varied from 10 μs to 1 ms were applied to the gate of an enhancement-mode HEMT. The low level of the pulses was -0.5 V, an off state for the HEMT, and the high level was 1 V, an on state for the HEMT. The measured drain current pulses show perfect square shapes, indicating no lag effect. The absence of such an effect in HEMTs is probably because of the separation of the 2-DEG conduction channel from the surface which is protected by a N-GaAs layer.

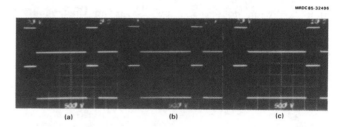

LOWER TRACE = GATE VOLTAGE = -0.5V → +0.9V
UPPER TRACE = DRAIN CURRENT = 0 → 4.3mA

Fig. 3 HEMT drain current response to a square gate voltage pulse with width of (a) 1 ms, (b) 100 μs and (c) 10 μs.

3.0 MATERIAL UNIFORMITY

For high resolution ADCs, uniformity of the device characteristics on a wafer is probably the most important factor in determining the success of the circuit. The offset voltage of a comparator in an ADC needs to be controlled in the range of millivolts. In order to study the device uniformity in a circuit area we have fabricated HEMT arrays each containing 100 1 μm × 20 μm HEMTs in an area of about 1 mm × 1 mm. Figure 4 shows the measured I-V characteristics of one HEMT array. The drain currents (log scale) of the devices are plotted as functions of the gate voltage. The threshold voltage, defined as the gate voltage corresponding to a drain current of 10 μA, shows a standard deviation of only 4 mV. This result is much better than what has been reported for conventional MESFETs. The 4 mV is less than half a LSB of a 8-bit ADC with 2.56 V analog input.

The device uniformity demonstrated above is a direct result of the good control of layer properties achieved with MBE growth. We have investigated using MBE layers as the resistors for ADCs' resistor ladders. Resistor strings with 32 contact nodes have been fabricated for 5-bit flash converters. The widths of the resistors were defined by oxygen implantation. The error voltage at each

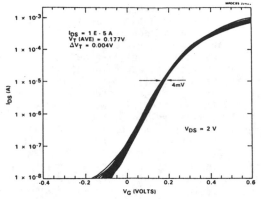

Fig. 4 Superimposed I_{ds} vs V_{gs} curves (semi-log plot) of a HEMT array.

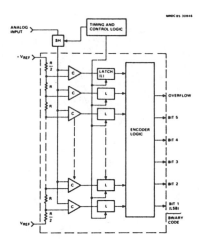

Fig. 6 5-bit flash ADC with binary output.

node has been determined by comparing the measured voltage value and the theoretical value. Figure 5 shows the measured error voltages of the 32 nodes of a resistor string. The errors are all within 5 mV. This value, which exceeds the requirement of an 8-bit ADC, is a direct result of the uniformity of the MBE layers.

5 binary output bits plus one overflow bit. The whole circuit consists of more than 1500 transistors.

In the 5-bit DAC design, a capacitor network was used to perform the voltage summing operation instead of a resistor network. The schematic diagram of a 5-bit capacitive DAC is shown in Fig. 7. The circuit utilizes five binary weighted capicitors. The top plates of these capacitors are connected to form an output node. The bottom plate of these capacitors is driven from the output of a precision DAC switch. The capacitor array output node is first switched to ground prior to a conversion. After the reset interval, the digital input code to be converted is switched into the digital input lines which switch the DAC switch outputs to their respective states depending upon the digital input codes, thus realizing, at the DAC output voltage equivalent to the digital input quantity.

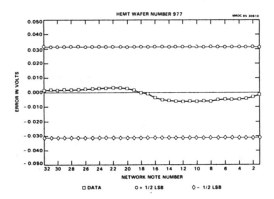

Fig. 5 Error voltage of each of the 32 nodes of a resistor string.

4.0 ADC CIRCUIT RESULT

5-bit flash ADCs, 5-bit DACs and timing and control circuits have been fabricated using our planar HEMT technology. Enhancement-mode and depletion-mode transistors were each used in both the digital and the analog circuits.

A functional diagram of a 5-bit ADC is shown in Fig. 6. The resistor string provides 32 node voltages as inputs to one side of the comparators. The other side of the comparators receives the analog input voltage. The output of each comparator is connected to a latch which is clocked by a clock pulse obtained from the timing and control logic which encodes the signal from 32 lines to

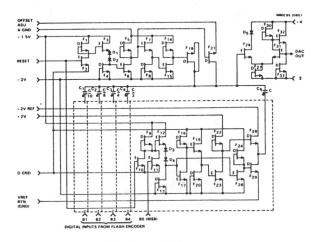

Fig. 7 5-bit capacitive digital-to-analog converter.

The capacitor array for this circuit is realized by utilizing a MOM building block capacitor, which is replicated twice, four times, eight times, and sixteen times, to form the array. The capacitors were fabricated using the two levels of metals and the interlevel dielectric used in our process. The capacitance value is determined by the size of the metal, which is accurately defined by the photolithography.

The timing and control logic was designed to generate the timing pulses for the ADCs and DACs. It consists of six D flip-flops with proper feedback loops to generate the necessary logic signals required to generate the timing terms. The output of the flip-flops is connected to six latches, which are then connected to digital drivers for the outside world.

The ADCs, DACs, and timing and control circuits have all been tested. For the ADCs we have found circuits with four bits working. For the DACs, all five bits worked properly. A photograph of the analog outputs corresponding to all the digital inputs from 00000 to 11111 is shown in Fig. 8. The test was carried out with the digital input increasing at a rate of 1 bit per microsecond. The timing and control circuits also worked properly. The six timing signals worked correctly up to more than 1 GHz.

frequency of 700 MHz, which corresponds to a conversion rate of 140 Mega samples per second. The photograph of the reconstructed sine wave is shown in Fig. 9. Because of the hybrid set up, in which the three packages were connected by long coaxial cables, the real speed of the devices could not be exploited.

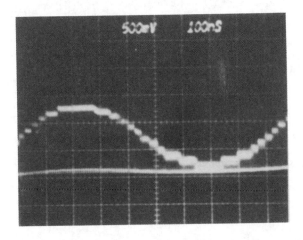

Fig. 9 Sine wave reconstruction using a ADC and a DAC. The sampling rate is 140 MHz.

5.0 CONCLUSION

The GaAs/GaAlAs HEMT has shown great promise for high speed and high resolution ADC applications. Besides its high speed capabilities, it has the advantages of low substrate effect, low temperature sensistivity, and good threshold control by MBE. Because it is still basically a FET technology with planar process, it is suitable for high density and low power operation. 5-bit ADCs, 5-bit DACs, and timing and control circuits, all built with HEMTs, have been demonstrated. A waveform reconstruction using these three chips has been achieved with a sample rate of 140 Mega samples per second with 4-bit accuracy.

6.0 REFERENCES

1. C.P. Lee, "Influence of Substrates on the Electrical Properties of GaAs FET Devices and Integrated Circuits," in Semi-Insulating III-V Materials, p. 324, Shiva Publishing Limited, 1982.

2. C. Kocot and C.A. Stolte, "Backgating in GaAs MESFETs," IEEE Trans. Electron Devices ED-29, p. 1059, 1982.

3. S.J. Lee and C.P. Lee, "Temperature Effect on Low Threshold Voltage Ion-implanted GaAs MESFETs," Electronics Lett. 17, p. 760, 1981.

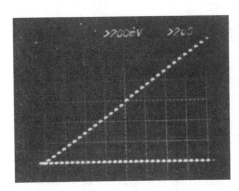

Fig. 8 5-bit DAC output with digital inputs from 00000 to 11111.

It is a difficult task to test the accuracy of A to D conversion at high speed. One way to do it is to connect the ADC circuit with a high speed DAC to reconstruct the input analog signal. We have carried out such experiments using three separate packaged chips, a 5-bit flash ADC, a 5-bit DAC and a timing and control logic. The 5-bit ADC had only four bit resolution because the least significant bit did not function. The timing and control circuit provided strobe pulses for the ADC and the reset pulses for the DAC. A sine wave was sent to the ADC and the parallel digital output was fed into the 5-bit DAC. The reconstructed sine wave was monitored by an oscilloscope. We have operated these circuits with a maximum clock

MODULATION-DOPED FET DCML COMPARATOR

Indexing terms: Semiconductor devices and materials, Comparators, Analogue-digital conversion

A differential current mode logic (DCML) comparator has been inplemented with modulation-doped FET technology. The dynamic input/output hysteresis is low enough to realise a flash quantiser with up to 5-bit accuracy.

Modulation-doped FET technology has been used to fabricate a number of functional digital integrated circuits[1] and low-noise microwave amplifiers.[2] However, little has been published to date on the application of this technology to the realisation of analogue/digital signal convertors. We believe that ultra-high-speed A/D convertors will greatly benefit from this new technology. The combination of faster switching speed, lower power dissipation and high-g_m/low-hysteresis enhancement-mode devices will improve clock rate and resolution, while reducing power consumption, circuit complexity and size.

The key building blocks of an A/D convertor are the flash quantiser and sample-and-hold circuits. In this letter we report on the first one-bit quantiser, i.e. comparator, made using modulation-doped FET technology. This comparator uses differential current mode logic (DCML), a form of balanced coalesced logic which is particularly well suited to A/D convertors.

The basic DCML comparator circuit is illustrated in Fig. 1. It consists of a differential input stage with latched positive feedback. This circuit approach has several inherent advantages, including faster effective system speed and improved noise immunity. In practice, source-follower buffers are used at the outputs to increase the current drive capability, and a clock buffer is employed at the input. Another advantage of this type of comparator circuit is that it requires only two types of circuit elements: an enhancement-mode transistor and a thin-film resistor.

An important constraint for any comparator design is that the characteristics of the constituent devices must be well controlled. In an enhancement-mode MODFET the drain-source current is a very sensitive function of the doping charge of the AlGaAs layer beneath the gate. As a result, a major source of nonuniformity in enhancement-mode FET characteristics is due to lack of precise control over the depth of the gate recess. We have successfully achieved uniform gate recess etching using an aluminium arsenide (AlAs) 'etch-stop' layer, as shown in Fig. 2. With this approach we have achieved less than 7% absolute variation in drain-source current after gate recess etching over an area of 5 cm².

Enhancement-mode FETs were fabricated with 1 μm gate length centred in a 4 μm channel using conventional optical lithography. Both mesa etching and oxygen implantation were used for device isolation. Alloyed NiGeAu ohmic contacts provided a source-gate resistance of 1·4 Ωmm. Transconductance of 200 mS/mm was typically obtained with a saturation voltage of less than 0·4 V. Thin-film resistors were fabricated by electron beam evaporation of nickel-chromium followed by photoresist lift-off for definition. The mean value of sheet resistance was 100 Ω/□ with a standard deviation of only 2 Ω/□.

15-stage ring oscillators composed of the same circuit elements were fabricated along with the DCML comparator circuits on the same wafers. The best measured performance of the ring oscillators was a propagation delay of 20 ps and a

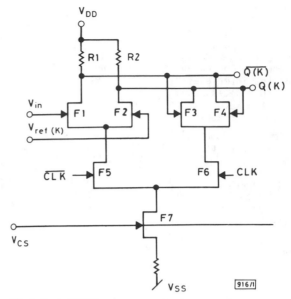

Fig. 1 *Basic DCML comparator circuit*

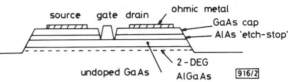

Fig. 2 *AlAs 'etch-stop' layer approach to control of gate recess depth*

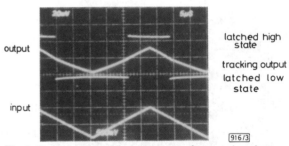

Fig. 3 *Comparator input and output waveforms against time*

speed-power product of 3·4 fJ at room temperature. This result compares favourably with other state-of-the-art published results for ring oscillators optimised for minimum power dissipation.[3]

DCML comparator circuit functionality was verified by on-wafer probe testing at clock rates up to 50 MHz. Fig. 3 illustrates typical input and output voltage waveforms as a function of time. The input was a 30 kHz triangular wave with adjustable offset bias. The reference was grounded, and the clock signal was a 50 MHz square wave from a word generator. As shown in Fig. 3, whenever the input signal passes the grounded reference level, the output latch switches states.

The low-frequency resolution or hysteresis was evaluated by displaying output against input voltage waveforms, as shown in Fig. 4. The ramp though the centre of the display is a closely spaced series of dots representing the output/input during the 'track' time which occurs each time the high-frequency clock is 'low'. Each time the clock goes 'high' the output voltage latches, either up to the logic '1' level or down to the logic '0' level, depending on whether the input is above or below the input decision threshold (0 V). For proper operation the comparator must correctly and monotonically

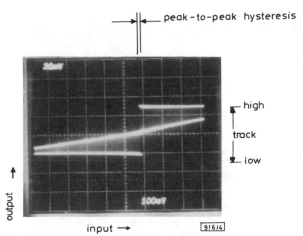

Fig. 4 *Low-frequency comparator hysteresis evaluation*

switch logic states as the input sweeps through 0 V, regardless of which direction the input is 'coming from'. The indicated input voltage decision level varies with direction by less than 15 mV. The magnitude of this hysteresis is important because it determines the accuracy of the A/D convertor which can be implemented with this comparator.

The sensitivity of this circuit performance to changes in ambient illumination has been evaluated at room temperature and found to be dependent on whether the FET devices were mesa-etch or oxygen-implant isolated. Circuits with oxygen-implant isolation were insensitive to changes in ambient illumination. One plausible explanation for the observed sensitivity of mesa isolated circuits is that the AlGaAs layer is exposed by mesa etching and acts as a 'light pipe' to the active region of the devices.

The resolution which an A/D convertor can achieve depends on the type of architecture and the accuracy of the

elements. In a high-speed A/D with parallel-processing implementation, $2^N - 1$ comparators quantise N bits simultaneously. For this architecture, the peak-to-peak decision level errors as described should be less than one quarter of the least significant bit. Using this criterion the MODFET DCML comparator will support a minimum quantum step size of 60 mV. Since a practical A/D input voltage range may be limited to about 2 V, this implies that the uncalibrated A/D resolution which can be realised with this comparator is 5 bits.

There are numerous techniques for improving this performance. However, we believe these preliminary results adequately demonstrate the suitability of the MODFET DCML comparator for use in high-speed A/D convertors.

Acknowledgments: This work was performed with the support of the Office of Naval Research, under contract no. N00014-83-C-0389, and Mr. Max Yoder, Technical Program Monitor.

J. BERENZ   *21st January 1985*
K. NAKANO
T. SATO
K. FAWCETT

TRW Electronic Systems Group
One Space Park, Redondo Beach, CA 90278, USA

References

1 KURODA, S., MIMURA, T., SUZUKI, M., KOBAYASHI, N., SHIBATOMI, A., and ABE, M.: 'New device structure for 4 kbit HEMT SRAM'. IEEE GaAs IC symposium, 1984, pp. 125–128
2 BERENZ, J.: 'High electron mobility transistors (HEMT)'. IEEE MTT newsletter, Summer 1984, pp. 43A52
3 ABE, M., MIMURA, T., NISHIUCHI, K., SHIBATOMI, A., and KOBAYASHI, M.: 'HEMT LSI technology for high speed computers'. IEEE GaAs IC symposium, 1983, pp. 158–161

High-Performance Self-Aligned Gate AlGaAs/GaAs MODFET Voltage Comparator

P. J. VOLD, STUDENT MEMBER, IEEE, DAVID K. ARCH, MEMBER, IEEE, K. L. TAN, A. I. AKINWANDE, MEMBER, IEEE, AND NICHOLAS C. CIRILLO, JR., MEMBER, IEEE

Abstract—A very high-performance voltage comparator circuit has been demonstrated using self-aligned gate AlGaAs/GaAs modulation-doped FET's (MODFET's) and laser-trimmable CrSi-based thin-film resistors. The MODFET master/slave comparator circuits demonstrated analog input resolutions of <1 and 2.5 mV at sampling rates of 0.5 and 1 GHz, respectively, at Nyquist analog input rates at room temperature. The MODFET comparators operated to sampling rates greater than 2.5 GHz at Nyquist analog input rates. Static hysteresis of less than 1 mV was observed for some comparators at room temperature. The self-aligned gate MODFET's demonstrated average threshold-voltage offsets for closely spaced FET pairs of 2.53 ± 1.15 mV, and typical static hysteresis levels of <1 to 3 mV. These MODFET comparators demonstrated the highest analog input resolution at gigahertz sampling frequencies ever reported, including comparators fabricated using AlGaAs/GaAs heterojunction bipolar transistors (HBT's).

I. INTRODUCTION

CONSIDERABLE attention has beed focused recently on the development of very high-speed high-resolution analog-to-digital converters (ADC's) for future high-speed signal processing applications. GaAs-based technologies have received significant interest, because of their high analog cutoff frequencies f_t and very fast digital switching speeds. One of the most critical components of high-speed ADC's is the voltage comparator. Low device hysteresis, high transconductance, and low input offsets are required for very high-speed high-resolution voltage comparator circuits. AlGaAs/GaAs modulation-doped FET (MODFET) and heterojunction bipolar transistor (HBT) technologies appear to be ideal for high-speed high-resolution comparators, because of their high gains, low offsets, and low hysteresis. GaAs MESFET technologies must employ extra calibration circuitry to compensate for device instabilities and hence are not as attractive for high-resolution ADC's [1]. Berenz *et al.* have demonstrated recessed-gate AlGaAs/GaAs MODFET comparator circuits showing 15-mV hysteresis at low sampling frequencies [2]; high-frequency characterization results were not reported. Wang *et al.* have fabricated AlGaAs/GaAs HBT comparator circuits demonstrating input offsets of about 4 mV and hysteresis of 1 mV at low sampling frequencies [3]. At sampling rates of 1 GHz, an analog input resolution of about 10 mV was obtained at Nyquist analog input rates. This HBT comparator operated to sampling rates greater than 2.5 GHz at Nyquist analog input rates.

In this paper we report for the first time the material growth,

fabrication, and high-speed test results of a self-aligned gate AlGaAs/GaAs MODFET voltage comparator circuit. The MODFET comparator demonstrated the highest input resolution ever reported for comparators or ADC's at gigahertz sampling frequencies.

II. CIRCUIT DESIGN

The voltage comparator was implemented using the current-mode-logic (CML) bi-level latch architecture shown in Fig. 1. The design uses two latches in series, with a source-follower buffer between them, to obtain master/slave operation. The CML bi-level latch operates in either a latch or track mode based on the polarity of the clock; it tracks the input when the clock is high and latches when the clock is low. The master and slave clocks are complementary, yielding a fully latched output from the slave. A source follower is utilized to buffer the output of the circuit. The comparator was designed with FET gate lengths of 1.0 μm and nominal gate widths of 20 μm. The resistors were implemented using a CrSi-based thin-film resistor technology, which allows for small input offsets to be nulled out by laser trimming/annealing.

III. MATERIAL GROWTH AND FABRICATION

The epitaxial growth of the AlGaAs/GaAs modulation-doped heterostructures was done by molecular beam epitaxy (MBE) on 3-in LEC GaAs wafers. Starting from the substrate the layers were a thick undoped GaAs buffer layer, a thin undoped (Al, Ga)As spacer layer, and a Si-doped n^+-(Al, Ga)As charge control layer. Details of the growth procedure have been reported previously [4].

The MODFET comparator circuits were fabricated using a completely planar self-aligned gate by the ion-implantation process, with no gate recess etch or isolation etch required [5]. A rapid thermal anneal is used to activate the self-aligned source and drain implants in order to avoid heterointerface damage [6].

IV. RESULTS AND DISCUSSION

The MODFET devices demonstrated threshold voltages V_t of +0.11 V, extrinsic transconductances g_m of 210 mS/mm, and output conductances g_{ds} of 7 mS/mm at room temperature. The source resistance R_s was 0.7 $\Omega \cdot$mm, as measured from a TLM structure. Closely spaced FET pairs, which are similar in structure to the differential input stage of the comparator, demonstrated average V_t offsets of 2.53 ± 1.15 mV for eight die across a 3-in wafer at room temperature. Typical FET pair V_t offsets for self-aligned gate MODFET's range from <1 to

Manuscript received May 11, 1987; revised July 9, 1987.

The authors are with Honeywell Physical Sciences Center, Bloomington, MN 55420.

IEEE Log Number 8716691.

Reprinted from *IEEE Electron Device Letters*, vol. EDL-8, no. 9, pp. 431–433, September 1987.

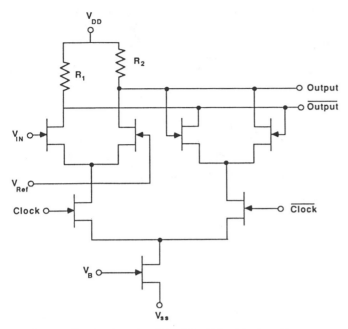

Fig. 1. Circuit schematic of the CML bi-level latch architecture.

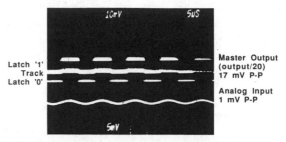

Fig. 2. Oscillograph of comparator output demonstrating correct latch operation with a 1-mV analog input at low input and clock frequencies.

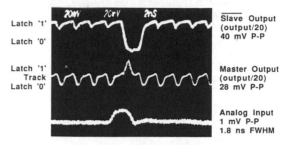

Fig. 3. Oscillograph of high-speed comparator test showing correct operation with a 1-mV analog input pulse at a sampling rate of 0.5 Gsps.

6 mV over the 3-in wafer. The CrSi-based thin-film resistor material had a sheet resistance of 495 $\Omega/\square$. The typical temperature coefficient of resistivity for the CrSi-based resistors is less than ± 300 ppm/°C over the military temperature range. Laser trimming/annealing was not performed on these comparator circuits, but results on similarly processed wafers show that resistance can be reliably adjusted up by trimming or down by annealing.

The master/slave comparator circuits were tested first at low sampling rates to identify functional parts and to measure static hysteresis. Fig. 2 shows the output of the master latch of the comparator demonstrating correct operation with a 1-mV analog input at low input and clock frequencies. This demonstrates that static hysteresis is less than 1 mV and it corroborates earlier discrete device testing which showed typical gate referred hysteresis in the range of <1 to 3 mV. The high-speed testing was performed on wafer, using a high-frequency probe card. The clock and analog input waveforms were phase locked so that accurate measurements could be made. The analog input consists of a pulse whose full width at half maximum (FWHM) is of shorter duration than the clock period and is repeated only once every 20 clock cycles. This is a very stringent test since the comparator must latch to a pulse during one clock period and then return to its original state during the very next clock period, hence testing the comparator at Nyquist analog input rates. An oscillograph of a comparator latching to a 1-mV analog input pulse at a sampling rate of 0.5 gigasamples/s (Gsps) at room temperature is shown in Fig. 3. The master latch of the comparator tracks the analog input when the clock is high and latches when the clock goes low. The slave output is always latched, because when the clock is low it tracks a latched master output and when the clock is high it latches. The bias current used for these tests was about 2.5 mA per latch, which yields a power dissipation of about 30 mW for the master/slave comparator, excluding the output drivers.

TABLE I

Analog Input Resolution	Sampling Rate	ADC Accuracy*
< 1.0 mV	0.5 Gsps	9-10 bits
2.5 mV	1.0 Gsps	8-9 bits
24 mV	1.5 Gsps	5 bits
75 mV	2.0 Gsps	4 bits
105 mV	2.5 Gsps	3 bits

* Based on present MODFET technology

The resolution of an ADC circuit often will be limited by the comparator resolution and the analog input voltage range. Present MODFET technology has device breakdowns sufficient for a 2–3-V analog input range. Further device enhancements could increase breakdowns to expand the useful analog input range. For accurate reliable ADC operation the magnitude of the comparator resolution should be below one-third of the ADC least significant bit (LSB). Thus a flash ADC resolution capability of 9–10 bits at 0.5 Gsps and 8–9 bits at 1.0 Gsps is predicted, assuming V_t offsets are eliminated by laser trimming of the CrSi resistors. The analog input resolution for the self-aligned gate MODFET comparator and calculated ADC accuracy as a function of sampling rate is shown in Table I.

V. CONCLUSION

A self-aligned gate (Al,Ga)As/GaAs MODFET voltage comparator circuit has been fabricated utilizing laser trimmable/annealable CrSi-based thin-film resistors. The MODFET comparator demonstrated the highest analog input resolution at gigahertz sampling rates ever reported for any semiconductor IC technology, including AlGaAs/GaAs HBT. The master/slave comparator demonstrated input resolutions (ADC accuracies) of <1 mV (9–10 bits) at 0.5 Gsps and 2.5 mV (8–9 bits) at 1.0 Gsps at Nyquist analog input rates at room temperature. These results demonstrate the suitability of self-aligned gate MODFET technology for very high-sampling-rate/high-resolution ADC's.

ACKNOWLEDGMENT

The authors would like to recognize the contributions of T. Nohava and P. Joslyn for the MBE growths, A. Fraasch and B. Fure for the circuit fabrication, and M. Listvan for technical consultations.

REFERENCES

[1] K. Fawcett *et al.,* "High-speed, high-accuracy, self-calibrating GaAs MESFET voltage comparator for A/D converters," in *Proc. 1986 IEEE GaAs IC Symp.*, pp. 213–216.

[2] J. Berenz, K. Nakano, T. Sato, and K. Fawcett, "Modulation-doped FET DCML comparator," *Electron. Lett.*, vol. 21, no. 6, pp. 242–243, 1985.

[3] K. C. Wang, P. M. Asbeck, D. L. Miller, and F. H. Eisen, "High-speed, high accuracy voltage comparators implemented with GaAs/(GaAl)As heterojunction bipolar transistors," in *Proc. 1985 IEEE GaAs IC Symp.*, pp. 99–102.

[4] J. K. Abrokwah, N. C. Cirillo, Jr., M. J. Helix, and M. Longerbone, "Modulation-doped FET threshold voltage uniformity of a high throughput 3 inch MBE system," *J. Vac. Sci. Technol. B*, vol. 2, no. 2, pp. 252–255, 1984.

[5] N. C. Cirillo, Jr., J. K. Abrokwah, and S. A. Jamison, "A self-aligned gate modulation-doped (Al, Ga)As/GaAs FET IC process," in *Proc. 1984 IEEE GaAs IC Symp.*, pp. 167–170.

[6] T. Ishikawa, S. Hiyamizu, T. Mimura, J. Saito, and H. Hashimoto, "The effects of annealing on the electrical properties of seletively doped GaAs/n-AlGaAs heterojunction structures grown by MBE," *Japan. J. Appl. Phys.*, vol. 20, no. 11, 1981.

A 1-μm MODFET Process Yielding
MUX and DMUX Circuits Operating at 4.5 Gb/s

B.J.F. Lin, H. Luechinger, C.P. Kocot, E. Littau, C. Stout, B. McFarland,
H. Rohdin, J.S. Kofol, R.P. Jaeger, and D.E. Mars

Hewlett-Packard Labs. 3500 Deer Creek Rd.,
Palo Alto, CA 94304-1317

ABSTRACT

We report a high yield 1-μm enhancement-and depletion-mode MODFET (E/D MODFET) process capable of producing medium scale integrated circuits. Special care has been taken to reduce anomalous transients. Reproducible MBE growth of epi-layers and a uniform and reproducible RIE process have resulted in excellent threshold voltage (V_{TH}) control. The standard deviation of V_{TH} for EFETs with a median V_{TH} value of 0.27 V was 42 mV from wafer to wafer and 21 mV across a wafer. With this process, we have demonstrated a stand-alone, general purpose 8-bit multiplexer (MUX) and demultiplexer (DMUX) pair operating at 4.5 Gb/s with a power consumption of 1.4 W and 2.2 W, respectively. This is presently the fastest GaAs MUX/DMUX reported in the literature.

INTRODUCTION

The modulation-doped field-effect transistor (MODFET) has demonstrated promise as a candidate for ultra-high switching speed, low power and radiation-hard integrated circuits. Excellent results have been achieved in ring oscillator propagation delay [1] and several application circuits, such as an 8×8 multiplier [2] and a 4K SRAM [3]. However, the application of MODFET technology for MUX- and DMUX-circuits has never been reported. In recent years, several groups [4,5,6] have reported MUX- and DMUX-circuits using GaAs MESFET technology. Takada et. al. [5] fabricated an 8-bit MUX/DMUX pair operating up to 2.8 Gb/s. Liechti et al. [4] reported a 5 Gb/s, 2W Word Generator that possessed some but not all of the features of the MUX reported here.

In a recessed gate MODFET structure, the Si doping density and the epi-layer thickness under the gate determine the threshold voltage. A highly selective reactive ion etch (RIE) is needed to achieve high threshold voltage uniformity across a wafer [7]. However, due to drifts of the doping density during MBE growth and the instability of the RIE, process control of V_{TH} is not trivial. In this work, we resolved this issue with a reproducible low-voltage RIE process and a wafer screening procedure. A simple 1-μm non-self-aligned process was chosen for high yield. In addition, anomalous transients, as reported by Kaneshiro et al. [8], were addressed. Three major contributors (DX-center, RIE, and surface-state induced traps) were reduced

sufficiently to yield functional circuits.

FABRICATION PROCESS

1. MBE Structure Design

The MBE epi-layer structure used is shown in Table I. A pulse-doped scheme [9] was implemented with a heavily doped pulse layer of 60 $\overset{\circ}{A}$. Pulse doping increases channel carrier density without sacrificing the Schottky barrier height (V_b), and reduces sensitivity of V_{TH} to process variations. The capping layer was heavily doped to decrease the source resistance (R_s) and to provide a 285 Ω/□ sheet resistance for resistor. 20% aluminum mole fraction was used in the pulse-doped layer to minimize DX centers. Two $Al_{.2}Ga_{.8}As$ spacer layers were used to minimize the chance of DX-center formation due to Si diffusion.

TABLE I. PULSE-DOPED MODFET EPILAYERS

Layer	Thickness (A)	Al Mole Fraction	Doping (x10^{18}cm^{-3})
N+ Cap	330	0	3.0
DFET Stopping Layer	30	0.3	0
ϕ-GaAs	110	0	0
EFET Stopping Layer	140	0.3	0
Si Diffusion Spacer	60	0.2	0
Pulse-doped Layer	60	0.2	5.0
Spacer	20	0.2	0
Buffer	1μm	0	0

2. Process Sequence

The basic device structure is a 1μm gate MODFET with a 3μm source-drain spacing (L_{SD}). The process sequence is shown in Fig. 1. Electron-beam SiO_2 was first evaporated to protect the active region and eventually served as the passivation layer for the gate access region. Oxygen-implant isolation and Au-Ge-Ni based alloyed contacts were formed in sequence. A typical isolation leakage current at -10 V was 100 nA/mm across 3μm at room temperature. Nominal ohmic contact resistance was 0.1 Ω·mm. Next, RIE-recessed D- and E-gates were formed. Details of the gate process were described previously [7]. To reduce RIE-induced traps, a selective and reproducible low-voltage RIE process was used. The E-gate metal also served as first interconnect. Sequential deposition of e-beam SiO_2 and PECVD SiO_2 served as passivation. 1 μm thick polyimide cured at 200°C served as the planarizing dielectric layer. Via formation followed by second metal deposi-

Reprinted from *Rec. of the IEEE GaAs Integrated Circuits Symp.*, pp. 143–146, 1988.

tion and etching complete the process. Side wall coated vias, as shown in Fig.2, were used to improve via yield and resistance. Side walls were coated with a back-etched sacrificial Au layer.

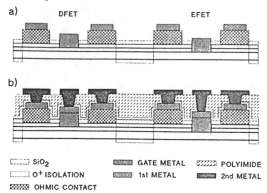

FIGURE 1. (a) front end process and (b) back end process.

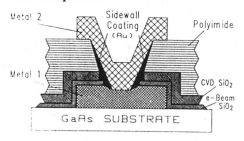

FIGURE 2. Sidewall coated via.

3. Transient Control

In Fig. 3, the Conductance DLTS (CDLTS) data measured on wafers etched with three different RIE self-bias voltages are shown. A positive going signature with an activation energy $E_a = E_v + 0.4\ eV$ was observed for the first time to be associated with the RIE. The time constant of this trap is on the order of 100 μs. For a 300 V RIE, the amplitude of this "hole-like" DLTS spectrum is comparable to that of the input signal. It decreases with decreasing RIE voltage (or RIE power). With a 45 V RIE, the concentration of this trap is reduced by an order of magnitude. After a 300°C, 2 hr annealing, this signal is further reduced by a factor of five. This result strongly suggests that this trap is induced by the RIE. As a reference, Fig. 3 also contains CDLTS data typical of wafers with 30% Al mole fraction to indicate DX-center induced traps.

The surface-state induced transient, as reported in [8], was reduced by decreasing the spacing of the access region. In Fig. 4, we show data taken from an inverter chain similar to the one described in Ref. [8]. The upper panel shows the gated burst of input pulses. The middle panel shows the output response to a 1% duty cycle, 200 MHz burst input signal after 96 inverter stages for a device with $L_{SD} = 5\ \mu m$. The loss of the initial and final pulses was due to a nanosecond transient most likely caused by the surface states in the access region. For a device with $L_{SD} = 3\ \mu m$ all the pulses were passed.

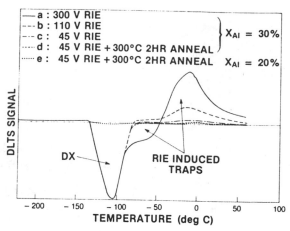

FIGURE 3. CDLTS spectra of DX-center and RIE induced traps.

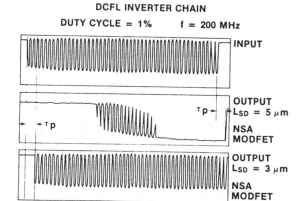

FIGURE 4. Gated burst measurement on a DCFL inverter chain. (a) input signal, (b) and (c) output reponse after 96 stages for non-self-aligned (NSA) process.

4. Process Control

A reproducible MBE growth was achieved using three techniques: daily flux measurements of the group III sources, calibration wafer growths before MODFET IC wafer lots, and individual screening of wafers. The fluxes of the gallium and the two aluminum sources were measured by an ion gauge rotated into the wafer growth position. Deviations from the target values were corrected by changes in the corresponding source temperatures. The target values were periodically updated to account for long-term drift of the ion gauge. This daily flux monitoring allowed an initial settability of ± 2%. Three calibration wafers were grown before the growth of a MODFET IC wafer lot: two test MODFET structures (one with 20% and the other 30% Al mole fraction) and a 1μm thick Si-doped GaAs layer. From quick turnaround optical and electrical measurements, the aluminum concentrations, the GaAs growth rate, and the material quality are determined. Wafer screening is accomplished with a non-contact resistivity monitor which allows the overall sheet resistance of the as-grown material to be measured with a repeatability of ± 2%. Changes in the Si-doping calibration which shift V_{TH} also affect the overall sheet resistance of the structure.

With a fixed D-gate RIE time, a correlation of the V_{TH} vs. the sheet resistance was observed, as shown in Fig. 5. The sensitivity is $7.7 mV/\Omega$. In order to meet the circuit requirement for the wafer-to-wafer variation of V_{TH}, only wafers with the sheet resistance within a specified range were used in the processing. To improve the reproducibility of the RIE, the etching chamber was carefully pretreated and a load-lock was used. These steps helped provide reproducible initiation of the etch. A 300-500% overetch is essential to achieve high uniformity. Details of EFET threshold voltage control have been described in [7].

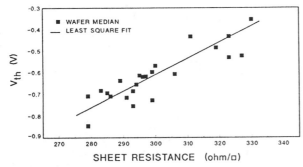

FIGURE 5. The relation of DFET V_{TH} to the sheet resistance.

MODFET PERFORMANCE AND DISCUSSION

The MODFET performance is listed in Table II. The averaged median value, averaged standard deviation across wafer (σ_w), and the standard deviation of the median value from wafer to wafer ($\sigma_{w\text{-}w}$) are evaluated over eight wafers in three runs. The design values of V_{TH} for the DFET and EFET are -0.60 V and 0.28 V respectively. The low values of σ_w and $\sigma_{w\text{-}w}$ for both DFET and EFET V_{TH} greatly improve the circuit yield. The relatively high values of g_{mmax} and I_{dssmax} are due to our pulse-doped structure in which a very high doping level is used. The extraordinarily high value of I_{dssmax} for the DFET is possibly due to a parasitic conduction in the top layer because it was measured with $V_G > 1$ V. The source resistances (R_s) are higher than expected. We believe that this is due to a relatively high tunneling resistance between the top N^+ cap layer and the two-dimensional-electron-gas layer caused by the high number of potential barriers and the thicker surface depletion region in our structures. The cut-off frequencies (f_T) of the DFETs and EFETs were measured on

devices with ten interdigitated 60 μm fingers. The interfinger parasitic capacitance may explain the lower values of (f_T). The speed performance of these FETs was benchmarked with a Source Coupled FET Logic (SCFL) divide-by-two circuit operating at a maximum frequency of 7.5 GHz.

CIRCUIT PERFORMANCE

An 8:1 multiplexer and a 1:8 demultiplexer were designed and fabricated in this process. The application for these circuits is in high speed fiber optic communication systems. The multiplexer and demultiplexer operate up to bit rates of 4.5 Gb/s, NRZ. The effects of process variation were minimized by careful circuit design resulting in good timing margin at all points in the circuits. Only one single-ended clock input is required for operation. The input clock frequency is half of the output data rate in both the multiplexer and the demultiplexer because the final selector stage of the multiplexer and the input of the demultiplexer use both halves of the clock cycle. Both circuits allow the user to digitally select one of four phases of the low speed output clock to insure proper timing with the rest of the system.

Differential SCFL was employed in both circuits. SPICE simulations using a three-terminal MODFET model [10] showed that the highest operating speed could be achieved with this logic family. Also, the effect of threshold voltage variations is reduced by using source coupled pairs. Disadvantages of this logic family are higher power dissipation and the extra area required for differential signal paths. In this case, these disadvantages were not significant and SCFL was chosen because of its speed advantage. Enhancement devices were used for the switching pairs and source followers while depletion devices were used as current sources and level shift diodes.

The multiplexer inputs are single-ended and ECL compatible, but can be referenced to ground due to an adjustable input threshold. The multiplexer produces complementary outputs with adjustable output amplitudes. The demultiplexer input is differential but can be used with a single input and an external reference voltage. Demultiplexer outputs are single ended and ECL compatible. Both chips are 1.5mm by 3mm. The multiplexer contains 770 active devices and the demultiplexer contains 940. The average functional yield for both circuits was 22%.

The chips were mounted on sapphire substrates for testing in a high frequency test fixture. Correct operation of the multiplexer was verified by applying 8 pseudo-

TABLE II. MODFET PERFORMANCE

Parameter	Units	DFET			EFET		
		Average Median	σ_ν	$\sigma_{\nu\text{-}\nu}$	Average Median	σ_ν	$\sigma_{\nu\text{-}\nu}$
V_{th}	V	-.638	.024	.057	.272	.021	.042
g_{mmax}	mS/mm	198	2.4	7	257	3.6	11
$I_{ds}(V_g=0)$	mA/mm	133	3.9	15	--	--	--
I_{dssmax}	mA/mm	343	4.8	15	249	9.3	31*
R_s	Ohm.mm	1.25	.04	.02	1.29	.04	.06
V_b	eV	.61	.003	.02	.65	.004	.02
n		1.48	.01	.13	1.68	.01	.06
f_T	GHz	17.3*			16.5*		

*Best Value

random sequences to the inputs. These sequences were the same pattern but delayed by 1/8th of the sequence length with respect to each other. The correct pattern was observed at the multiplexer output at 8 times the input data rate. The demultiplexer was tested by applying the pseudo-random sequence produced by the multiplexer to its input, observing the 8 outputs and comparing them with the original inputs to the multiplexer.

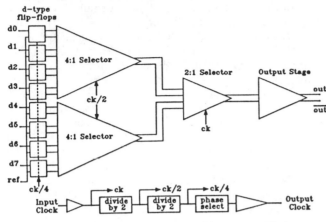

FIGURE 6. 8:1 Multiplexer simplified block diagram.

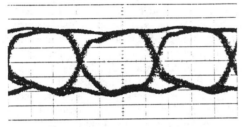

Output Data Rate = 3.0 Gb/s
Amplitude: 200 mV/div, Time: 100 ps/div.

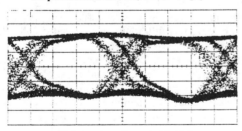

Output Data Rate = 4.5 Gb/s
Amplitude: 200 mV/div, Time: 50 ps/div.
FIGURE 7. Multiplexer eye diagrams.

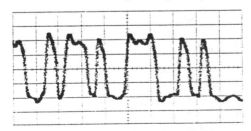

Amplitude: 200 mV/div, Time: 500 ps/div.
FIGURE 8. 4.5 Gb/s multiplexer output waveform.

TABLE III. MUX/DMUX PERFORMANCE SUMMARY		
	Multiplexer	Demultiplexer
Maximum Data Rate	4.5 Gb/s	4.5 Gb/s
10-90% Transition Time	135 ps	600 ps
Output Amplitude	0.4 - 1.0 Vpp	0.8 Vpp
Power Dissipation	1.4 W	2.2 W

CONCLUSION

We report a high-yield 1-μm E/D MODFET process capable of producing MSI circuits. Transient control and process control are emphasized, and we have demonstrated the first E/D MODFET general-purpose, 8-bit MUX/DMUX pair. Both circuits operate at 4.5 Gb/s with a power consumption of 1.4 W (MUX) and 2.2 W (DMUX). The best functional yield for both circuits is 38% on one wafer, and the average yield over four wafers is 22%. The speed/power performance of these circuits exceeds previously-reported MESFET technology counterparts [5] by 140%.

ACKNOWLEDGEMENT

We acknowledge H. R. Yeager, C. Yen, G. Baldwin, F. S. Fei, J. N. Miller, R. Kaneshiro, N. Moll, and K. Seaward for many helpful discussions. We also thank all the technicians in the high speed devices lab. at HPL for their technical support.

REFERENCES

1. N.J.Shah, IEEE Trans. Electron Devices, vol. ED-33, pp543 (1986).

2. N.Cirillo Jr., D.Arch, P.Vold, B.Betz, I.Mactaggart, and B.Grung, IEEE GaAs IC Symp. Tech. Digest, pp257 (1987).

3. S.Notomi, Y.Awano, M.Kosugi, T.Nagata, K.Kosemura, M.Ono, N.Kobayashi, H.Ishiwari, K.Odani, T.Mimura, and M.Abe, IEEE GaAs IC Symp. Tech. Digest, pp171 (1987).

4. C.A.Liechti, G.L.Baldwin, E.Gowen, R.Joly, M.Namjoo, and A.F.Podell, IEEE Trans. Electron Devices, vol. ED-29, pp1094 (1982).

5. T.Takada, K.Nozawa, M.Ida, and K.Asai, IEEE GaAs IC Symp. Tech. Digest, pp7 (1986).

6. A.Kameyama, K.Kawakyu, K.Ishida, Y.Kitaura, M.Mochizuki, T.Terada, Y.Ikawa, and N.Toyoda, IEEE GaAs IC Symp. Tech. Digest, pp265 (1987).

7. B.J.F.Lin, S.Kofol, C.Kocot, H.Luechinger, J.N.Miller, D.E.Mars, B.White, and E. Littau, IEEE GaAs IC Symp. Tech. Digest, pp51 (1986).

8. R. Kaneshiro, C.P.Kocot, R.P.Jaeger, J.S.Kofol, B.J.F.Lin, E.Littau, H.Luechinger, and H. Rohdin, IEEE Electron Devices Lett., vol. EDL-9, pp250 (1988).

9. M.Hueschen, N.Moll, E.Gowen, and J. Miller, IEDM Tech. Digest, pp348 (1984).

10. H.R. Yeager, and R.W. Dutton, IEEE Trans. Electron Device , vol. ED-33, pp682 (1986).

Static Random Access Memory Using High Electron Mobility Transistors

S. J. LEE, C. P. LEE, MEMBER, IEEE D. L. HOU, R. J. ANDERSON, AND D. L. MILLER

Abstract—A 4-bit fully decoded static random access memory (RAM) has been designed and fabricated using high electron mobility transistors (HEMT's) with a direct-coupled FET logic approach. The circuit incorporates approximately 50 logic gates. A fully operating memory circuit was demonstrated with an access time of 1.1 ns and a minimum WRITE-enable pulse of less than 2-ns duration at room temperature. This memory consumes a total power of 14.89 mW and 87.8 μW per memory cell.

IN future high-performance digital systems, such as signal processing systems and high-speed computers, high-speed memory circuits with very short access time and low power consumption will be required. The development of these memory circuits in GaAs has been reported either by using depletion-mode GaAs MESFET's [1] or by using enhancement-mode GaAs MESFET's [2]. The depletion-mode MESFET approach represents an easier fabrication process because of its larger noise margin, but leads to a complicated circuit implementation. The enhancement-mode MESFET approach has a simpler circuit implementation, but results in a tighter fabrication process because of its small noise margin. However, with the GaAs–GaAlAs high electron mobility transistors (HEMT's) [3], one can use the simpler enhancement-mode circuit approach but yet obtain larger noise margin, because of the high turn-on voltage for the metal–GaAlAs Schottky junction [4]. Furthermore, the higher transconductance and shorter gate delay of HEMT's over conventional GaAs MESFET's can greatly improve the circuit performance [5]–[7]. In this letter we report, for the first time, the design, fabrication, and test results of a static random access memory circuit (RAM) using HEMT's.

The memory circuits consist of enhancement-mode switching FET's and depletion-mode pullup FET's fabricated on a modulation-doped structure grown by MBE on a 1-in-sq LEC semi-insulating GaAs substrate. The layers were, starting from the substrate, undoped GaAs ($\sim$1 μm), an undoped $Ga_{0.7}Al_{0.3}As$ spacer ($\sim$90 Å), n-type $Ga_{0.7}Al_{0.3}As$ ($\sim$500 Å, 1×10^{18} cm^{-3}), and undoped GaAs ($\sim$40 Å). The fabrication procedure for the devices consisted of: 1) ohmic contact formation; 2) proton implantation for device isolation; 3) delineation of Schottky gates and a first-level interconnect;

and 4) isolation dielectric deposition and second-level interconnect formation. The ohmic contact was AuGe–Ni, which typically had a contact resistance of 0.5 $\Omega \cdot$mm. The Schottky gate metal and first-level interconnect were Ti–Pt–Au, and the second-level metal was Ti–Au. All the lithographic steps were done with a Censor 10X wafer stepper. The enhancement-mode FET's typically had a threshold voltage of 0.2 V and were obtained by recessing the n-type GaAlAs layer, while the depletion-mode FET's obtained without gate recessing typically had a threshold voltage of −0.6 V. Due to the use of proton implantation for device isolation, the process results in a planar structure, which is desirable for VLSI fabrication.

The memory cell was designed using the *E/D*-type crosscoupled flip–flop with two enhancement FET's as transfer gates. The total number of FET's for a memory cell is six. The transistor sizes were 1 μm X 20 μm for the driver enhancement FET's and 2 μm X 4 μm for the depletion-load FET's. The memory cell size is 60 μm X 75 μm. The 20-μm-wide drivers (enhancement FET's) of the memory cell were designed to sink 1.5 mA when the bit line was discharged to ground. The organization of this memory chip with peripheral circuit is shown in Fig. 1. This organization allows expansion of the present design to a larger size memory by simply increasing the size of the memory array and the number of decoders. The external address signal was applied to an address buffer to generate complementary address signals. This buffer circuit consisted of a cross-coupled flip–flop followed by a larger inverter driver which can supply the necessary current to charge or discharge the address line. NOR gates were used as column decoders and row decoders. In the column selection circuits the decoder was used directly to control the data bus transfer gate while in the row selection circuit a complementary driver was used after the decoder because of large loading. In the present prototype circuit, only one row was utilized. The bit line in the memory was precharged to the high state before either READ or WRITE. The WRITE circuit consisted of two tri-state drivers which are in the high impedance state if READ/WRITE (R/W) is high, and are activated if R/W is low. The sense amplifier is a simple crosscoupled flip-flop which is activated if R/W is high. The total number of logic gates for this memory circuit is approximately 50. The access time is equivalent to nine logic gate delays, in addition to the line delay due to the data bus. A fabricated 4-bit memory circuit is shown in Fig. 2. The chip size is 340 μm X 750 μm. The memory cells are shown in the upper left part of Fig. 2.

Manuscript received December 16, 1983.

S. J. Lee is with Rockwell International, Microelectronics Research and Development Center, Thousand Oaks, CA 91360. He is also with the University of Southern California, Department of Materials Science, Los Angeles, CA 90007.

C. P. Lee, D. L. Hou, R. J. Anderson, and D. L. Miller are with Rockwell International, Microelectronics Research and Development Center, Thousand Oaks, CA 91360.

Reprinted from *IEEE Electron Device Letters*, vol. EDL-5, no. 4, pp. 115–117, April 1984.

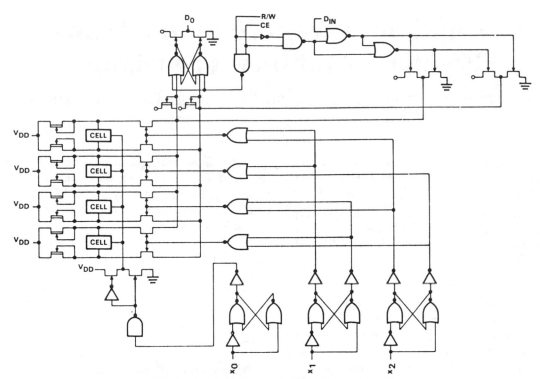

Fig. 1.   The overall organization of a RAM circuit.

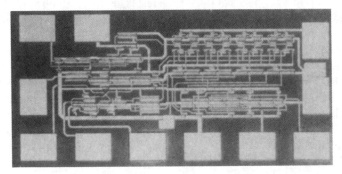

Fig. 2.   Photomicrograph of a fabricated RAM chip.

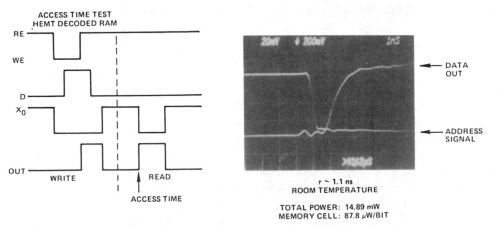

Fig. 3.   Access time measurement of a RAM circuit.

Testing of the memory circuits was performed at room temperature. Two supply voltages were used, one for the peripheral circuitry with a typical value of 1.2 V and the other for the memory cell array with a typical value of 1.0 V. The measurement of the memory access time is shown in Fig.

3(a). The address signal was pulled to ground to access a memory cell, then data "1" was written into the cell as a "low" signal was applied to R/W. The address signal then returned to "high" to disable the cell. The same memory cell was accessed to be read as the address signal returned to ground

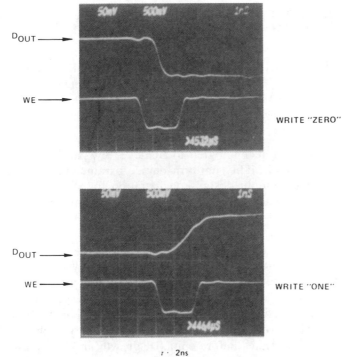

Fig. 4.   WRITE pulsewidth measurement of a RAM circuit.

again. The time delay between the address signal and the data output was measured to be 1.1 ns, as shown in Fig. 3(b). This access time is about a factor of two better than the best result obtained by conventional GaAs DCFL RAMS [8].

Since the present circuit is based on a very conservative design and layout, a better design and a more compact layout of the memory cell can further improve the device performance. For example, in the sense amplifiers, differential amplifiers can be used instead of the flip–flops used in the current design to improve the input sensitivity and the speed. Although the present HEMT RAM has only four bits and the device performance will probably degrade as the size of the memory increases, access time of 1 ns or lower should be reachable for 1K HEMT RAM with the improved design and layout. The minimum width of the WRITE pulse was also measured by varying the pulsewidth of the R/W signal. Observed results for writing "0" are shown in Fig. 4(a), which indicates that the writing pulsewidth can be as small as 2 ns. Investigation of narrower pulsewidth was not possible due to test equip-

ment limitations. The real response of the device should be much faster. Fig. 4(b) also shows the measurement of the minimum pulsewidth required to write "1"; it was also 2 ns. It was found that the D_{out} transient for writing "0" is faster than writing "1". It is generally observed in direct-coupled FET logic that the rising transient is slower than the falling transient. The total power consumption for the memory chip was 14.89 mW and the power consumption for the memory cell was 87.8 μW per bit.

In conclusion, the feasibility of using high electron mobility transistors for static random access memories was demonstrated. A memory test circuit was fabricated having a 50-gate integration level. The measured access time was 1.1 ns. The low power operation of the memory cell shows the potential for its use as a building block for large memory circuits.

ACKNOWLEDGMENT

The authors would like to acknowledge the technical support of all members of the Rockwell International GaAs Integrated Circuit Group.

REFERENCES

[1] S. J. Lee, R. P. Vahrenkamp, G. R. Kalen, L. D. Hou, R. Zucca, C. P. Lee, and C. G. Kirkpatrick, "Ultra-low power, high speed GaAs 256-bit static RAM," in *GaAs IC Symp. Proc.*, 1983, p. 74.

[2] N. Yokoyama, T. Ohnishi, H. Onodera, T. Shinoki, A. Shibatomi, H. Ishikawa, "A GaAs 1K static RAM using tungsten–silicide gate self-alignment technology," in *ISSCC Proc..*, 1983, pp. 44–45.

[3] T. Mimura, S. Hiyamizu, T. Fujii, and K. Nanbu, "A new field-effect transistor with selectively doped GaAs/n-Al$_x$Ga$_{1-x}$As heterojunctions," *Japan J. Appl. Phys.*, vol. 1–9, pp. L225–227, 1980.

[4] D. Delagebeaudeuf, M. Laviron, P. Delescluse, P. N. Tung, J. Chaplart, and N. T. Linh, "Planar enhancement mode two-dimensional electron Gas GET associated with a low AlGaAs surface potential," *Electron. Lett.*, vol. 18, pp. 103–105, 1982.

[5] C. P. Lee, D. Hou, S. J. Lee, D. L. Miller, and R. J. Anderson, "Ultra high speed digital integrated circuits using GaAs/GaAlAs high electron mobility transistors," in *GaAs IC Symp. Proc.*, 1983, pp. 162–165.

[6] R. A. Kiehl, M. D. Feuer, R. H. Hendel, J. C. M. Hwang, V. G. Keramidas, C. L. Allyn, and R. Dingle, "Selectively doped heterostructure frequency dividers," *IEEE Electron. Device Lett.*, vol. EDL-4, pp. 377–379, 1983.

[7] N. Nishiuchi, T. Mimura, S. Kuroda, S. Hiyamizu, H. Nishi, and M. Abe, "Device characteristics of short channel high electron mobility transistor (HEMT)," presented at 41st Annual Device Res. Conf., Burlington, VT, June 20–22, 1983.

[8] K. Asai, K. Kurumad, M. Hirayama, and M. Ohmori, "1 Kb static RAM using self-aligned FET technology," in *ISSCC Tech. Dig.*, 1983, p. 46–47.

A Fully Operational 1-kbit HEMT Static RAM

NAOKI KOBAYASHI, SEISHI NOTOMI, MASAHISA SUZUKI, TAKUMA TSUCHIYA, KOICHI NISHIUCHI, MEMBER, IEEE, KOUICHIRO ODANI, AKIHIRO SHIBATOMI, MEMBER, IEEE, TAKASHI MIMURA, MEMBER, IEEE, AND MASAYUKI ABE, MEMBER, IEEE

Abstract—In this paper we describe the current status of materials and fabrication technologies, and optimal design of a memory cell, and the performance of fully functional 1-kbit HEMT SRAM's. The surface defect density on MBE-grown wafers has been reduced to less than 100 cm^{-2} by improving MBE technology. Standard deviations of threshold voltages are 6.7 and 11.8 mV for enhancement-type and depletion-type HEMT's, respectively, measured in a 10 mm × 10 mm area. These deviations are sufficiently small for DCFL circuits. Memory cell design parameters have been optimized by circuit simulation, where the effects of variations in threshold voltages are taken into account. Full function of 1-kbit SRAM's has been confirmed by marching tests and partial galloping tests. The RAM chips have also shown excellent uniformity in access time. The difference between maximum and average values on the RAM chip is 4 percent.

I. INTRODUCTION

HIGH ELECTRON mobility transistors (HEMT's) are promising for very high-speed LSI's [1]–[3] as demonstrated with 1- and 4-kbit SRAM's in previous papers [4], [5]. Currently, it has become important to establish HEMT LSI technology to realize fully functional LSI circuits.

As reported in previous papers [5], [6], an intensive effort is required to improve materials and fabrication technology. A low density of surface defects on MBE-grown wafers and high uniformity of device parameters are necessary to obtain fully functional chips. Variations in threshold voltages must be sufficiently less than logic swings in LSI circuits. The logic swing in DCFL circuits, adopted in HEMT LSI's, is smaller than that found in normally-ON type logic circuits; however, DCFL circuits have the advantages of simple circuit configuration and low power dissipation [7]. Therefore, a DCFL circuit is preferable for the design of SRAM's if high uniformity of threshold voltages can be obtained.

Optimal design of a memory cell is also important. Stable memory operation depends greatly on the memory cell stability, which is affected by variations in threshold voltages. It is necessary to take this effect into account for memory cell design.

This paper reports the current status of materials and fabrication technologies to achieve low surface defect density and highly uniform device parameters, and also discusses circuit design consideration for static RAM's, mainly memory cell stability, and the performance of fully functional 1-kbit SRAM's.

II. MATERIALS

An important problem in MBE technology is to reduce surface defect (oval defect) density. If there are defects in the gate region, the drain current cannot be pinched off. We have reduced the defect density to less than 100 cm^{-2} by using wet-chemical and Ar^+ plasma dry etching of GaAs substrates before epitaxial growth and by optimizing growth conditions. Assuming a defect density of 100 cm^{-2}, the number of defects in the total gate area of the chip is calculated to be less than 0.31 for 1-kbit and 0.96 for 4-kbit HEMT SRAM's.

Another important problem is to reduce variations in device parameters. We have developed the MBE growth conditions for highly uniform epitaxial layers. As shown in Fig. 1, the thickness variations ($\Delta t/t$) and the carrier concentration variations ($\Delta n/n$) are less than ± 1 percent within 60 mm.

III. THE FABRICATION PROCESS

We have obtained excellent uniformity in threshold voltages for both enhancement-type (E-HEMT's) and depletion-type HEMT's (D-HEMT's) by using a self-terminating selective dry-etching technology. This technology is based on the large difference in the etching rate between GaAs and AlGaAs, and is very suitable for HEMT's because the excellent uniformity of MBE-grown epitaxial layers guarantees uniformity of the threshold voltages.

Fig. 2 shows etching characteristics, using CCl_2F_2 + He gas, of GaAs (60-nm-thick)–AlGaAs heterojunction material. A high-selectivity ratio of more than 260 is achieved, with the etching rate of AlGaAs as low as 2 nm/min and that of GaAs about 520 nm/min at 140 V in self-generated bias voltage.

Fig. 3 shows threshold voltages at 300 K versus thickness of the cap layer and the AlGaAs layer between surface and heterointerface. The basic epilayer structure consists of a 600-nm undoped GaAs layer, a 30-nm $Al_{0.3}Ga_{0.7}As$ layer doped to $2 \times 10^{18} \text{ cm}^{-3}$ with Si, and a 70-nm GaAs cap layer. They are successively grown on a semi-insulating GaAs substrate by MBE. A very thin

Manuscript received October 7, 1985; revised January 7, 1986. The present research effort is part of the National Research and Development Program on "Scientific Computing System," conducted under a program set by the Agency of Industrial Science and Technology, Ministry of International Trade and Industry.

The authors are with Fujitsu Laboratories, Ltd., Fujitsu Limited, 10-1 Morinosato-Wakamiya, Atsugi 243-01, Japan.

IEEE Log Number 8607885.

Reprinted from *IEEE Transactions on Electron Devices*, vol. ED-33, no. 5, pp. 548–553, May 1986.

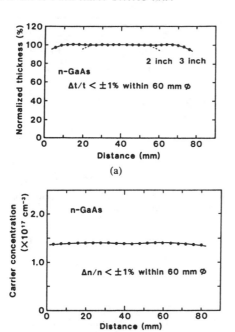

Fig. 1. n-GaAs (a) thickness and (b) carrier concentration variations on a 3-in φ wafer.

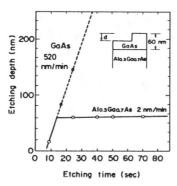

Fig. 2. Characteristics of selective dry etching.

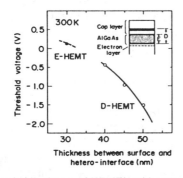

Fig. 3. Threshold voltages of HEMT's with new epi-structure.

AlGaAs layer, embedded in the cap layer, acts as an etching stopper against the selective dry etching in the gate recess process. Selective dry etching is carried out to remove the cap layer and expose the top surface of the thin AlGaAs stopper for D-HEMT's. To fabricate E-HEMT's, nonselective wet-chemical etching is carried out to remove the stopper before the selective dry etching. When the temperature is lowered from 300 to 77 K, threshold voltages shift to positive by around 0.1 V.

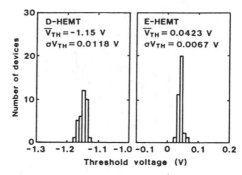

Fig. 4. Histogram of threshold voltages of HEMT's with 1.2-μm gate lengths.

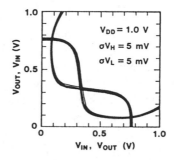

Fig. 5. Superimposed transfer curves of 20 HEMT DCFL inverters on a 2-in φ wafer.

Fig. 4 shows histograms of threshold voltages of HEMT's with gate lengths of 1.2 μm, measured on a 10 mm × 10 mm area at room temperature. The standard deviations are 6.7 mV for E-HEMT's and 11.8 mV for D-HEMT's, about $\frac{1}{2}$ of those for GaAs MESFET's. Moreover, for transconductance, a standard deviation of 3 percent has been obtained for E-HEMT's, having an average value of 180 mS/mm.

Fig. 5 shows superimposed transfer curves of 20 HEMT E/D type DCFL inverters on a 2-in φ wafer, measured at a supply voltage of 1.0 V at room temperature. Standard deviations of 5 mV are obtained for both the high and low logic levels. These deviations are less than 1 percent of the logic swing, which is approximately 0.7 V. Therefore, the high uniformity of the device parameters satisfies the requirements for stable operation in a DCFL circuit.

IV. CIRCUIT DESIGN

A. Memory Cell Stability

The 1-kbit SRAM has the same circuit configuration as reported in a previous paper [4]. There are two requirements for the design of HEMT SRAM's; that is, high-speed performance and high memory density. Because the memory density depends entirely on a memory cell size, the cell size should be small enough although the cell may have a contribution to the dynamic performance. To achieve high-speed performance, it is better to increase the driving ability of the peripheral circuits. Therefore, the driver devices have very wide gates in the peripheral circuits of HEMT SRAM's. A threshold voltage of 0.15 V is adopted for E-HEMT's to guarantee high-speed per-

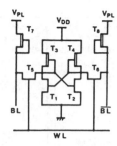

Fig. 6. Schematic of the memory cell.

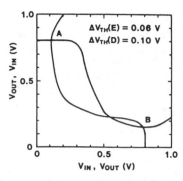

Fig. 7. Simulated dc characteristics of a memory cell in the worst case for $I_A/I_D = I_{PL}/I_D = 0.5$.

formance and sufficient noise margin in the peripheral circuits. The threshold voltage for D-HEMT's is -1.00 V.

Stable memory operation depends greatly on memory cell stability, which means that the input data are surely written and that the stored data are retained until the next WRITE operation. It is impossible to obtain fully functional SRAM's without considering cell stability. Because cell stability may be affected by variations in device parameters, detailed circuit analysis for optimal cell design must take this effect into account.

B. Memory Cell Design Parameters

Fig. 6 shows a schematic of the memory cell, which is a usual flip-flop circuit, consisting of 6 HEMT's. The current ratios I_A/I_D, I_{PL}/I_D, and I_D/I_L are cell-design parameters. I_D is the drain current of the drivers T_1 and T_2, and I_L is the drain current of the cell loads T_3 and T_4. I_A is the drain current of the access transistors T_5 and T_6, and I_{PL} is the drain current of the bit-line-pull-up loads T_7 and T_8. I_D and I_A are measured at $V_{DS} = 1$ V and $V_{GS} = 0.8$ V, and I_L and I_{PL} at $V_{DS} = 1$ V and $V_{GS} = 0$ V.

I_A/I_D and I_{PL}/I_D have greater effect on cell stability. The READ error may be caused by the transient current from the high-voltage storage node to the low-voltage bit line when the cell is selected. If I_A/I_D is larger and I_{PL}/I_D is smaller, the data destruction becomes more likely because the voltage drop at the storage node becomes larger. The WRITE condition is the opposite of the READ condition. The WRITE error may be caused by an insufficient voltage drop at the storage node in the WRITE operation. If I_A/I_D is smaller and I_{PL}/I_D is larger, it becomes harder to write in the cell. It is necessary to optimize these parameters by circuit simulation, taking into account the effects of variations in threshold voltages.

The current ratio of the cell inverter I_D/I_L has less effect on cell stability because the bit lines are pulled up. Although the cell noise margin decreases with the I_D/I_L ratio for $I_D/I_L < 5$, the margin is less dependent on the ratio for $I_D/I_L > 5$. This ratio also concerns power dissipation in a memory cell. Although a larger ratio is preferable for lower power dissipation, it is difficult to realize a very large ratio because the gate width of the load becomes very narrow and the gate length becomes very long. Based on these, I_D/I_L is chosen to be 10.

In the 1-kbit chips actually fabricated, the gate of the bit-line-pull-up load is not connected with the source, but connected to a pad. The gate bias can be controlled by an external voltage source.

C. Circuit Simulation

To investigate the effect of the variation in threshold voltages, we executed worst case simulation as follows. For a given variation ΔV_{TH}, we consider the worst case of many possible configurations in threshold voltages of the 8 HEMT's (T_1-T_8). That is, a threshold voltage of each HEMT is shifted by + or $-\Delta V_{TH}$ to maximize the difference in the noise margin between two stable states in the cell. The threshold voltage shifts used in the worst case simulation are as follows. A shift by $+\Delta V_{TH}$ is for T_1, T_4, T_6, and T_8, and a shift by $-\Delta V_{TH}$ for T_2, T_3, T_5, and T_7.

Fig. 7 shows the simulated dc characteristics of the memory cell for $I_A/I_D = I_{PL}/I_D = 0.5$ in the worst case, with a variation of 0.06 V for E-HEMT's, and 0.10 V for D-HEMT's. As shown in the figure, the difference in the noise margin between the states, A and B, is maximized. A corresponds to the worst case for writing, and B for reading.

By the worst case simulation discussed above, we can obtain cell design conditions to achieve a fully functional chip if actual variations in threshold voltages are within a shift of $\pm\Delta V_{TH}$. It should be noted that transient analysis is necessary to examine cell stability for both reading and writing in the worst case. As discussed before, the transient currents act in an important role in the READ and WRITE operations.

We performed the circuit simulation with the SPICE II JFET model. This model can describe HEMT dc characteristics for gate bias in DCFL circuits. The device model parameters, for 1-μm gate length and 50-μm gate width at room temperature, are as follows: $\beta = 16$ mA/V^2, $\lambda = 0.1$, $R_S = R_D = 15$ Ω, $C_{GS} = 150$ fF, $C_{GD} = 15$ fF, $I_S = 1.4 \times 10^{-17}$ A.

D. Optimization of the Parameters

Two simulation results are shown in Fig. 8. For E-HEMT's, Fig. 8(a) is for a variation of 0.03 V and (b) for 0.06 V. For D-HEMT's, Fig. 8(a) is for a variation of 0.05 V and (b) for 0.10 V. The stable operating region and the error regions are shown in the figure. The WRITE error boundary is calculated for a WRITE-enable pulse

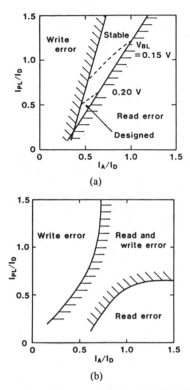

(a)

(b)

Fig. 8. Stable operating and error regions for (a) $\Delta V_{TH}(E) = 0.03$ V and $\Delta V_{TH}(D) = 0.05$ V, and (b) $\Delta V_{TH}(E) = 0.06$ V and $\Delta V_{TH}(D) = 0.10$ V. The broken lines show the cell design parameters where logic swings on the bit lines, V_{BL}, are 0.15 and 0.20 V.

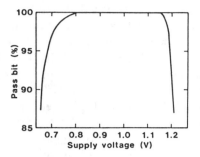

Fig. 9. Normalized pass-bit number for several supply voltages.

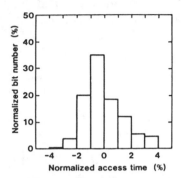

Fig. 10. Histogram of normalized X-address access time.

width of 500 ps. The stable operating region is significantly narrow in the case shown in Fig. 8(a), and disappears in the case shown in Fig. 8(b). These results indicate that the variation in threshold voltage affects the cell stability considerably. As discussed before, high uniformity of threshold voltages has been obtained for HEMT's. Three times the standard deviation for E-HEMT's is 20 mV, and for D-HEMT's is 35 mV, which is less than ΔV_{TH} in the case of Fig. 8(a). It can be considered that the measured variations in threshold voltages are within ΔV_{TH} in the case of Fig. 8(a). Therefore, the cell design conditions to achieve fully functional chips are found to be satisfied in the case shown in Fig. 8(a).

The logic swing on the bit lines also should be considered to optimize the design parameters because the logic swing depends on I_A/I_D and I_{PL}/I_D. The broken lines in the figure show the design parameters where logic swings on the bit lines V_{BL} are 0.15 and 0.20 V. The logic swing on the bit lines must be at least 0.2 V for driving the sensing amplifiers. Based on these, the optimal design parameters I_A/I_D and I_{PL}/I_D have been chosen to be 0.5.

V. PERFORMANCE OF 1-kbit SRAM's

We have confirmed that the 1-kbit SRAM's are fully functional by marching tests with a 1-kHz clock at room temperature. The high-level voltage of input signals was 0.7 V, and the low-level voltage was 0 V. The tests were executed on wafers. The chip size was 3.0 mm × 2.9 mm, consisting of 7244 HEMT's, and 100 chips were fabricated on a wafer. Fig. 9 shows the normalized pass

bit number for several supply voltages. The chip passed the marching test for supply voltages between 0.8 and 1.15 V.

The marching test is not sufficient for detecting interference between memory cells. An additional test (partial galloping test) was performed to detect the interference. We examined the interference between a test cell, which was selected in a cell array, and any other cells in the READ cycle. Several test cells were selected on each chip. This test was similar to a galloping test, although all the combinations of cells were not tested. The chips which passed the marching test also passed this additional test.

The SRAM's have also shown excellent uniformity in access time. Fig. 10 shows a histogram of the normalized X-address access time for 512 pairs of adjacent cells, measured at room temperature. The difference between maximum and average values is only 4 percent, which is much smaller than that for GaAs MESFET SRAM's. The average value is 9 ns, not fast, because the load current in the whole circuit is about $\frac{1}{3}$ of the designed value. It has been demonstrated, however, that the access time can be as fast as 0.9 ns at 77 K if the load current is just the designed value [4]. The access time scattering is caused by variations in load current. Even if the access time becomes faster, the normalized scattering is constant because the relative variation in load current is considered to be constant.

VI. SUMMARY

HEMT LSI technology to achieve fully functional chips is presented. Highly uniform MBE-grown wafers and selective dry-etching technology enable excellent control of the device parameters. The standard deviations of threshold voltages are 6.7 mV for E-HEMT's and 11.8 mV for D-HEMT's. Surface defect densities have been reduced to less than 100 cm^{-2}. The cell design parameters have been optimized by worst case simulation, taking into account the effects of variations in threshold voltage. I_A/I_D and I_{PL}/I_D are chosen to be 0.5. Fully functional 1-kbit SRAM's with excellent uniformity in access time have been fabricated. These indicate that HEMT technology is promising for application to high-speed LSI's, and that higher density HEMT static RAM's can be projected.

ACKNOWLEDGMENT

The authors wish to thank Dr. T. Misugi and M. Kobayashi for their encouragement and support, and also wish to thank K. Kondo and M. Nakayama for their support in materials and fabrication.

REFERENCES

[1] T. Mimura, S. Hiyamizu, T. Fujii, and K. Nanbu, "A new field-effect transistor with selectively doped GaAs/n-AlGaAs heterojunctions," *Japan J. Appl. Phys.*, vol. 19, pp. L225–L227, 1980.

[2] T. Mimura, K. Joshin, S. Hiyamizu, K. Hikosaka, and M. Abe, "High electron mobility transistor logic," *Japan J. Appl. Phys.* vol. 20, pp. L598–L600, 1981.

[3] K. Nishiuchi, T. Mimura, S. Kuroda, S. Hiyamizu, H. Nishi, and M. Abe, "Device characteristics of short channel high electron mobility transistor (HEMT)," in *Proc. Device Res. Conf.* (Vermont), IIA-8, June 1983.

[4] K. Nishiuchi, N. Kobayashi, S. Kuroda, S. Notomi, T. Mimura, M. Abe, and M. Kobayashi, "A subnanosecond HEMT 1 Kb static RAM," in *ISSCC Dig. Tech. Papers*, pp. 48–49, 314, 1984.

[5] S. Kuroda, T. Mimura, M. Suzuki, N. Kobayashi, K. Nishiuchi, A. Shibatomi, and M. Abe, "New device structure for 4 Kb HEMT SRAM," in *GaAS IC Symp. Tech. Dig.*, pp. 125–128, 1984.

[6] A. Shibatomi, J. Saito, M. Abe, T. Mimura, K. Nishiuchi, and M. Kobayashi, "Material and device considerations for HEMT LSI," in *IEDM Tech. Dig.*, pp. 340–343, 1984.

[7] R. C. Eden, B. M. Welch, R. Zucca, and S. I. Long, "The prospects for ultrahigh-speed VLSI GaAs digital logic," *IEEE J. Solid-State Circuits*, vol. SC-14, pp. 221–239, 1979.

A Subnanosecond HEMT 1-kbit Static RAM

KOICHI NISHIUCHI, MEMBER, IEEE, NAOKI KOBAYASHI, SHIGERU KURODA, MEMBER, IEEE,
SEISHI NOTOMI, TAKASHI MIMURA, MEMBER, IEEE, MASAYUKI ABE, MEMBER, IEEE,
AND MASAAKI KOBAYASHI

Abstract —A static RAM LSI using high electron mobility transistor (HEMT) technology has been realized for the first time. The RAM has a memory capacity of 1024 bits and integrates 7244 HEMT devices into 1024-words × 1-bit organization. The RAM uses enhancement/depletion-(E/D) type direct-coupled FET logic (DCFL) circuitry as a basic circuit and can operate fully statically.

The design rules used are 1.5-μm minimum gate length, 2×2-μm^2 contact hole, and 3-μm linewidth and spacing of the wiring electrodes. The memory cell is 55×39 μm and the chip is 3.0×2.9 mm. The RAM is fabricated on an AlGaAs/GaAs heterojunction epi-structure grown by molecular beam epitaxy (MBE) on a Cr-doped 2-in LEC GaAs substrate wafer.

A subnanosecond access time of 0.87 ns with a 1.60-V supply and 360-mW dissipation has been attained at liquid nitrogen temperature.

I. INTRODUCTION

THERE have been strong demands for faster high-performance LSI's required to develop high-performance digital systems such as mainframe computers. In recent years, GaAs LSI technology has been attracting much interest for its high-speed possibility [1], [2]—faster than the bipolar ECL which is the fastest of the conventional Si LSI technologies.

High-speed random access memory (RAM) LSI is one of the most important LSI devices in high-performance digital systems. The bipolar ECL technology has been offering the fastest static RAM LSI's to the industry. MOS static RAM LSI's have also been advancing their operation speed. However, there are strong demands for still faster static RAM LSI.

High electron mobility transistor (HEMT) was introduced as a new high-speed FET device in the GaAs LSI technology [3]. This device, which uses the selectively doped GaAs/n^+–AlGaAs heterojunction structure [4], has shown much higher electron mobility than the conventional GaAs FET's especially at very low temperatures. The electron mobility is typically 7–8×10^3 cm^2/V·s at 300 K and 4–5×10^4 cm^2/V·s at liquid nitrogen temperature (77 K).

Manuscript received August 15, 1984; revised February 12, 1986. This work was supported by the Agency of Industrial Science and Technology, MITI of Japan in the frame of the National Research and Development Project "Scientific Computing System."

The authors are with Fujitsu Laboratories Ltd., Fujitsu Ltd., 10-1 Morinosato-Wakamiya, Atsugi 243-01, Japan.

IEEE Log Number 8609688.

Application of this device in digital circuits has attracted much attention as a high-speed switching element [5] and has shown excellent switching speeds. A 27-stage ring-oscillator circuit constructed with the enhancement- and depletion-type HEMT's with gate length of 1.7 μm demonstrated propagation delay time of 17.1 ps at 77 K [6]. Recently, 1-μm gate ring oscillators showed propagation delay of 12.2 ps at 300 K [7]. Furthermore, frequency dividers showed excellent operational frequencies of 5–10 GHz [5], [8], [9].

This paper describes the design and fabrication of a high-speed static RAM LSI constructed with HEMT technology. In order to assess the feasibility of this technology at the large-scale integration level, the static RAM with 1-kbit capacity was designed and evaluated. The RAM LSI, integrating over 7000 HEMT devices, is the first LSI implementation using this technology.

II. DESIGN OF 1-KBIT STATIC RAM

A. RAM Circuit Design

HEMT is a Schottky-gate FET of the same kind of GaAs MESFET's. Therefore basic circuit configurations applicable for HEMT technology are the same as MESFET circuits, such as the buffered FET logic (BFL), Schottky-diode FET logic (SDFL), and direct-coupled FET logic (DCFL) circuit [1]. In HEMT technology, the enhancement-type (E-type) and depletion-type (D-type) HEMT were both developed in the early stage, and introduced to enhancement/depletion-(E/D) type DCFL circuitry [6]. This circuit is especially suited for LSI implementation because of its simple circuit configuration and high packing density.

In the design of the first HEMT RAM LSI, the E/D-type DCFL circuit was chosen as the basic circuit. The memory cell is a conventional six-transistor cross-coupled flip-flop circuit. The transfer gate is constructed with an E-type device. In the memory cell design, a driver device having a gate length L_G of 2 μm was chosen for its process tolerance, and a 15-μm-long D-type device was used as the load device. Each cell was designed to consume the retaining power of 150 μW. The threshold voltages for E- and D-HEMT were designed to be 0.1 and -1.2 V, respectively.

Reprinted from *IEEE Journal of Solid-State Circuits*, vol. SC-21, no. 5, pp. 869–874, October 1986.

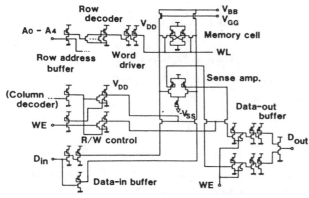

Fig. 1.   Circuit of the HEMT 1-kbit static RAM.

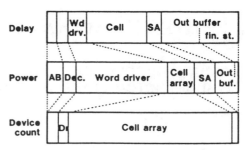

Fig. 2.   1-kbit RAM delay time, dissipation power, and device count broken down by circuit.

A HEMT static RAM having 1024 bits of memory capacity was designed. The RAM is organized into 1024 words × 1 bit, and arranged as a 32-row × 32-column memory plane. The circuit diagram of the 1-kbit RAM is shown in Fig. 1. The whole circuit is constructed with the E/D-type DCFL circuitry. The RAM does not require synchronous operation and can operate fully statically. For the peripheral circuits, a 1.5-μm gate E-HEMT was chosen for performance reasons and a 5-μm gate D-HEMT was chosen for the load device. The gate width of the driver device is from 30 to 200 μm corresponding to the current driving capability that is needed for a circuit stage in the RAM. Pull-up of bit lines and a differential-amplifier-type sensing circuit are adopted in order to fetch stored data in a short time from the low-powered memory cell. The data output circuit is designed to drive a heavy off-chip load of 50-Ω resistor and 15-pF capacitor, the same as in ECL LSI. The output buffer has four amplifier stages with a push–pull type final stage constructed of high-current E-type HEMT's. The output device has a gate width of 800 μm.

B. High-Speed Design

For the design of a high-speed static RAM LSI using HEMT technology, the following points are adopted. To obtain high-speed intrinsic delay time in the basic circuit, the gate length of the switching device is shrunk to 1.5 μm for the peripheral circuits which play an important role in signal propagation in the RAM LSI. In order to obtain a high-speed access time, it is important to design peripheral circuits by assigning high operating power, and to design the memory cell for low power. Therefore sufficiently high operating power is assigned to peripheral circuits, especially to the address buffer, word driver, and output buffer which have high load capacitances. These circuits have driver devices whose gate width is 100–800 μm.

Fig. 2 shows the delay time, dissipation power, and device count, broken down by circuit in the RAM design. These results were obtained from circuit simulations performed by SPICE-II circuit simulator. The word driver dissipates 47 percent of the chip power. The entire peripheral circuit dissipates 85 percent of the total power, whereas

the delay of this stage, from the opening of the transfer gate to the input of the sense amplifier, forms 31 percent of the total delay. The total dissipation power per chip is expected to be 1.0 W at 77 K in this design and the row address access time can be expected to be 0.45 ns.

The DCFL circuit with HEMT technology has a logic swing of 0.9 V at 77 K and 0.7 V at 300 K. In order to obtain enough of a noise margin and guarantee stable logic operation, power lines have to be designed carefully to avoid voltage drops due to the high operating current, especially in the ground GND lines. Linewidths of the GND power lines are from 50 to 200 μm, and the voltage drop is limited to less than 50 mV. Separate lines were used to supply power to each circuit block—the cell array, row decoder, and column decoder/output-buffer part. This resulted in many power pads (23) in the RAM.

C. Layout Design

The design rule adoped for the HEMT static RAM is a rather conservative one. In the switching HEMT devices, the gate lengths chosen were 1.5 μm for the driver of the peripheral circuits, 2 μm for the driver device and transfer gate of the memory cell, and 5 μm or more for the load devices.

The minimum contact hole is 2.0 × 2.0 μm. Alignment tolerance is 1.5 μm. The metal line width and spacing are 3 μm and 3 μm for both the first and second interconnect layers.

As the result of RAM layout design, the chip is 3.0 × 2.9 mm. The RAM cell is 55 × 39 μm (2145 μm²). The RAM has a total device count of 7244, including 2452 D-type load devices. As a result of the high-power design of the peripheral circuits, the total area of the peripheral circuits is just the same as the cell array, i.e., 2.2 mm².

III.   IC Fabrication Technology

A. Basic Device Structure

The basic structure of the HEMT device for IC application is shown in Fig. 3. Cross-sectional structure of both E- and D-type HEMT's for use in the E/D-type DCFL basic circuit is shown. The switching device is E-HEMT. It has a heterojunction epi-structure consisting of a 600-nm un-

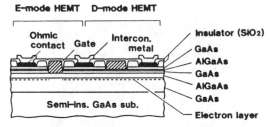

Fig. 3. Cross section of E- and D-mode HEMT for DCFL circuit.

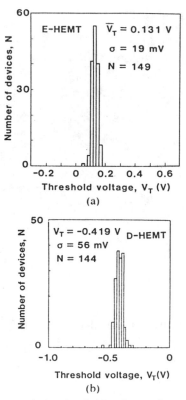

Fig. 4. Histograms of the threshold voltages for (a) E-HEMT and (b) D-HEMT.

doped GaAs layer and a 30-nm n-type AlGaAs layer, and a 70-nm n-type GaAs cap layer grown by molecular beam epitaxy (MBE) on a Cr-doped semi-insulating GaAs substrate. The n-AlGaAs and n-GaAs cap layer are doped to 2×10^{18} cm^{-3} with Si during the MBE growth. The sheet carrier electron density of the heterojunction structure is 1.0×10^{12} cm^{-2} at 300 K and 8.4×10^{11} cm^{-2} at 77 K. The low field electron mobility of the 2-DEG measured by Hall measurements is 7200 cm^2/V·s at 300 K and 38 000 cm^2/V·s at 77 K. The AlAs mole fraction, x of Al$_x$Ga$_{1-x}$As layer, is 0.3.

The gate electrode of Ti/Pt/Au is formed on the AlGaAs layer by recessing the n-GaAs cap layer forming the Schottky contact to the AlGaAs layer. As the source and drain electrode, AuGe and Au layers are formed on the n-GaAs layer and ohmic contact is made by thermal eutectic alloying at 450 °C. The space between the gate electrode and the source and drain regions is 1.5 μm.

The threshold voltage V_{TH} of HEMT is expressed as

$$V_{TH} = \Psi_M - \frac{qNd^2}{2\epsilon} - \frac{\Delta E_C}{q}$$

where Ψ_M is the metal–semiconductor potential difference, N is the dopant concentration of the AlGaAs layer, d is the thickness of the AlGaAs layer, and ΔE_C is the difference in energy between the AlGaAs and undoped GaAs layers. So, V_{TH} is determined mainly by the doping concentration N and the thickness of the AlGaAs layer d.

To form the gate recessed structure, reactive ion etching using CCl$_2$F$_2$ and He gas is used [10]. This etching method has a high selectivity ratio of more than 200 for the GaAs and AlGaAs. V_{TH} is determined mainly by the accuracy of MBE epilayer parameters. Fig. 4(a) shows a histogram of V_{TH} for E-HEMT measured within an area of 15×30 mm. The standard deviation of V_{TH} is 19 mV, lower than that of conventional GaAs MESFET using bulk crystal by a factor of 2 to 3.

The D-HEMT has a double heterojunction structure as shown in Fig. 3. The gate electrode of the D-HEMT is deposited on an upper AlGaAs layer, also using selective dry etching to recess it. Fig. 4(b) shows a typical histogram of V_{TH} for D-HEMT. The standard deviation of V_{TH} is 56 mV. This value is better than the case in which the gate recess structure is formed by wet chemical etching by a factor of 2. With the adoption of this device structure, V_{TH} for both E- and D-HEMT can be determined precisely by

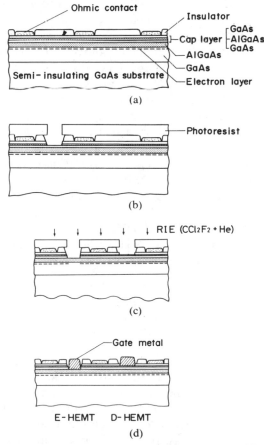

Fig. 5. Essential parts of the IC fabrication process. (a) Ohmic metallization. (b) Gate patterning for E-HEMT. (c) Reactive ion etching for gate recessing of E- and D-HEMT. (d) Schottky gate metallization.

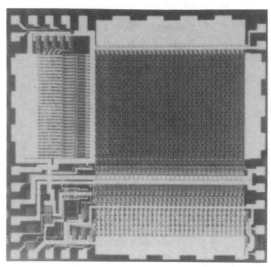

Fig. 6. Photomicrograph of a HEMT 1-kbit static RAM. Chip size is 3.0 × 2.9 mm.

selective dry etching and excellent uniformity of the device parameters can be realized.

B. IC Fabrication Process

The fabrication process for the HEMT static RAM is as follows. The essential parts of the sequence are shown schematically in Fig. 5(a)–(d). First, semi-insulating GaAs wafer, 2-in LEC wafer is used as the substrate on which the single heterojunction structure is grown by MBE. Mesa isolation is employed for interdevice isolation by etching a shallow mesa step of 200 nm. The source and drain ohmic contact regions are made by depositing AuGe and Au layer (300 nm) and alloying at 450 °C. Then, gate recess structures for E- and D-HEMT's are formed by etching the cap layer by selective dry etching with CCl_2F_2 and He gas at 140-V bias voltage. For the gate electrode, Ti/Pt/Au (100/60/240 nm) is evaporated and the gate pattern is formed by the lift-off method. This layer is used also as the first interconnection layer. Depositing a 700-nm SiO_2 film for an isolation layer and forming contact holes by dry etching are carried out. As the second interconnection layer, Ti/Pt/Au (50/100/800 nm) is formed and patterned by ion milling. Fig. 6 is a photomicrograph of a finished HEMT RAM chip.

IV. RAM PERFORMANCE

A. dc Characteristics

Fig. 8 shows basic READ/WRITE operation waveforms of the 1-kbit static RAM at 77 K. READ and WRITE operations are performed with ZERO and ONE data for two different address points. The measurement is done with 1.20-V supply voltage. The RAM consumes about 300 mA. Amplitude for input signals, the row address, data input D_{in}, and write enable WE are all 1.0 V. Test clock cycle is set to 10 kHz. The RAM can output a 0.55-V data output D_{out}

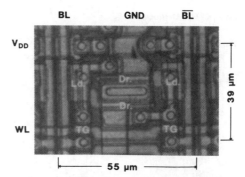

Fig. 7. Photomicrograph of the RAM cell.

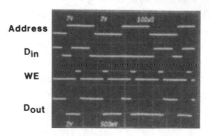

Fig. 8. READ/WRITE operation waveforms of the RAM at 77 K. Test cycle is 10 kHz and the supply voltage is 1.20 V.

Fig. 9. The RAM chip mounted in a 42-pin flat package.

signal. The D_{out} signal is measured with a 50-Ω output load. Normal READ/WRITE operation was also confirmed at room temperature.

The dc characteristics for the basic circuit and devices were evaluated at 77 K from the measurements of the inverter gate provided in the same chip. The high level was 1.0 V and the low level was 0.08 V. The K factor of the driver E-HEMT was typically 400 mA/V^2·mm, and the threshold voltage was 0.41 V. The threshold voltage of the D-HEMT was −0.43 V. Functional test of the RAM was also performed using a marching test pattern, and typical bit yield of 97 percent was evaluated with the RAM.

B. RAM Dynamic Performance

Dynamic performance such as the address access time of the HEMT 1-kbit RAM was evaluated at both room temperature and liquid-nitrogen temperature. At room temperature, the measurement was done on the wafer using probing pins on a card connected with coaxial cables for high-frequency signals. Measurements at liquid-nitrogen temperature were made with the RAM chip mounted in a flat package and immersed in liquid nitrogen. Fig. 9 is

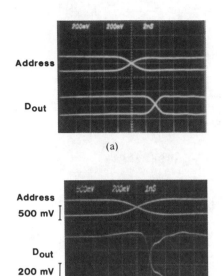

Fig. 10. Oscillograph of the access time measurements at (a) 300 K and (b) 77 K.

a photograph of the RAM chip mounted in a 42-pin flat package.

Measurements were done using a coaxial line setup. High-frequency signal inputs for the RAM were terminated with 50-Ω resistors at input points of the chip. The output signal from the chip was led to an oscilloscope whose input impedance was 50 Ω. The access time was measured by applying an address pulse having 2-ns rise time to a row address input. Calibration of the access time measurement was done using a package with a through line in it.

Fig. 10 shows an oscillograph of the access operations at 300 K and 77 K. At 300 K, the best data for the row address access time were 3.4 ns. The drain supply voltage was 1.30 V and the chip dissipation was 290 mW. The performance was greatly improved when RAM was cooled to 77 K. The access time of 0.87 ns was obtained for a chip at 77 K, as shown in Fig. 10(b), where the supply voltage was 1.60 V and the chip dissipation was 360 mW.

V. SUMMARY

Using HEMT technology, the first RAM LSI was designed and fabricated successfully. By employing the IC process which has good uniformity of device parameters for both the E- and D-HEMT's, a 1-kbit static RAM LSI integrating 7244 elements was successfully fabricated. Precise control of the epi-structure parameters at MBE growth and reactive ion etching for the device fabrication made the uniformity superior to the conventional GaAs MESFET IC process.

The RAM dynamic performance was evaluated and an address access time of 0.87 ns with chip dissipation of 360 mW at liquid-nitrogen temperature was recorded. The result is the first subnanosecond access operation with 1-kbit semiconductor RAM LSI, showing availability of high-speed HEMT LSI technology.

ACKNOWLEDGMENT

The authors wish to thank Dr. S. Hiyamizu, J. Saito, K. Odani, Y. Watanabe, and M. Suzuki for their support. They are also grateful to Dr. T. Misugi, Dr. A. Shibatomi, T. Kotani, and Dr. M. Fukuta for their encouragement.

REFERENCES

[1] R. C. Eden, B. M. Welch, R. Zucca, and S. I. Long, "The prospects for ultrahigh-speed VLSI GaAs digital logic," *IEEE J. Solid-State Circuits*, vol. SC-14, pp. 227–239, Apr. 1979.
[2] M. Abe, T. Mimura, N. Yokoyama, and H. Ishikawa, "New technology towards GaAs LSI/VLSI for computer applications," *IEEE Trans. Electron Devices*, vol. ED-29, pp. 1088–1093, July 1982.
[3] T. Mimura, S. Hiyamizu, T. Fujii, and K. Nanbu, "A new field-effect transistor with selectively doped GaAs/n-AlGaAs heterojunctions," *Japan. J. Appl. Phys.*, vol. 19, pp. L225–L227, May 1980.
[4] R. Dingle, H. L. Störmer, A. C. Gassard, and W. Wiegmann, "Electron mobilities in modulation-doped semiconductor heterojunction superlattices," *Appl. Phys. Lett.*, vol. 33, pp. 665–667, Oct. 1978.
[5] M. Abe, T. Mimura, K. Nishiuchi, A. Shibatomi, and M. Kobayashi, "HEMT LSI technology for high-speed computers," in *Proc. IEEE GaAs IC Symp.*, Oct. 1983, pp. 158–161.
[6] T. Mimura, K. Joshin, S. Hiyamizu, K. Hikosaka, and M. Abe, "High electron mobility transistor logic," *Japan. J. Appl. Phys.*, vol. 20, pp. L598–600, Aug. 1981.
[7] C. P. Lee, D. L. Miller, D. Hou, and R. J. Anderson, "Ultra high-speed ICs using GaAs/GaAlAs HEMT," in *Proc. Device Res. Conf.*, June 1983, Paper IIA-7.
[8] K. Nishiuchi, T. Mimura, S. Kuroda, S. Hiyamizu, H. Nishi, and M. Abe, "Device characteristics of short channel high-electron mobility transistor (HEMT)," in *Proc. Device Res. Conf.*, June 1983, Paper IIA-8.
[9] R. A. Kiehl, M. D. Feuer, R. H. Hendel, J. C. M. Hwang, V. G. Keramidas, C. L. Allyn, and R. Dingle, "Selectively doped heterojunction divide-by-two circuit," in *Proc. Device Res. Conf.*, June 1983, Paper IVA-3.
[10] K. Hikosaka, T. Mimura, and K. Joshin, "Selective dry-etching of AlGaAs–GaAs heterojunction," *Japan. J. Appl. Phys.*, vol. 20, pp. L847–L850, Nov. 1981.

409

NEW DEVICE STRUCTURE FOR 4Kb HEMT SRAM

S. Kuroda, T. Mimura, M. Suzuki, N. Kobayashi,

K. Nishiuchi, A. Shibatomi and M. Abe

Fujitsu Laboratories Ltd., Fujitsu Limited

1677, Ono, Atsugi 243-01, Japan

ABSTRACT

Self-aligned High Electron Mobility Transistor (HEMT) technology based on selectively doped GaAs/AlGaAs heterostructure for LSI circuits is presented. A unique epistructure, associated with gate dry recessing process which is one of the most important technologies for realizing E/D type HEMT DCFL circuits is described. Making the best use of refined gate recess technology, we have successfully fabricated and tested 4Kb HEMT SRAM, with minimum address access time of 2.0 ns at 77K. This is the highest speed ever reported for 4Kb RAMs.

INTRODUCTION

To meet the needs of information processing in the 1990's, the necessity of ultrahigh speed computers has been widely recognized. These computers will require high speed LSI circuits with sub-100 ps logic delays (1,2).

In 1980, Mimura et al. (3) developed HEMT which have realized field-effect control of the high mobility electrons in selectively doped GaAs/n-AlGaAs single heterojunction structures, offering new possibilities for high speed low power LSI circuits. In 1981, 1.7 μm-gate HEMT ring oscillators demonstrated 17.1 ps switching delay at 77K with 0.96 mW power dissipation per gate (4). Lee et al. (5) fabricated 1 μm-gate HEMT ring oscillators and obtained switching delay of 12.2 ps with 1.1 mW power dissipation per gate at room temperature. This is the fastest switching delay time in semiconductor digital circuits and also shows the dominant position of HEMT circuits at room temperature as well as at low temperature. Nishiuchi et al. (6) reported the successful operation of HEMT frequency divider circuits with direct coupled FET logic (DCFL) circuit configuration. Divide-by-two operation was demonstrated at up to 8.9 GHz at 77K and up to 5.5 GHz at 300K. The first HEMT LSI, a 1Kb SRAM, was developed and demonstrated minimum address access time of 0.87 ns with power dissipation of 360 mW at 77K (7). This paper presents the new device structure and self-alignment fabrication process using selective dry etching, optimization of dry etching, and performance of 4Kb HEMT SRAM.

HEMT TECHNOLOGY FOR LSI/VLSI CIRCUITS

Dry Recess Etching and Threshold Voltage

The selective dry etching technology (8), which is a key technology for realizing HEMT LSI/VLSI circuits gives excellent uniformity in control of device parameters such as threshold voltage. Fig. 1 shows etching characteristics in CCl_2F_2 + He discharges by using GaAs (60 nm thick)-$Al_{0.3}Ga_{0.7}As$ heterojunction material. High selectivity ratio of more than 260 is achieved, where the etching rate of $Al_{0.3}Ga_{0.7}As$ is as low as 2 nm/min and that of GaAs is about 520 nm/min at 140 V in self-generated bias voltage. The basic epilayer structure consists of a 600 nm undoped GaAs layer, a 30 nm $Al_{0.3}Ga_{0.7}As$ layer doped to 2×10^{18} cm^{-3} with Si, and a 70 nm GaAs cap layer successively grown on a semi-insulating GaAs substrate by molecular beam epitaxy. The low field electron mobility is found from Hall measurements to be 7200 $cm^2/V \cdot s$ at 300K and 38000 $cm^2/V \cdot s$ at 77K. The concentration of two Dimensional Electron Gas (2DEG) is 1.0×10^{12} cm^{-2} at 300K and 8.4×10^{11} cm^{-2} at 77K respectively. A very thin $Al_{0.3}Ga_{0.7}As$ layer to act as a etching stopper against selective dry recess etching is embedded in the cap GaAs layer to fabricate E- and D-HEMTs in the same wafer. Adopting this new device structure, we can apply the selective dry etching of GaAs to AlGaAs to achieve precise control of the gate recessing process for both E- and D-HEMTs.

Fig. 2 shows threshold voltage versus thickness between surface and hetero-interface at 300K. Selective dry etching is carried out to remove the GaAs cap layer, exposing the top surface of the thin $Al_{0.3}Ga_{0.7}As$ stopper for D-HEMTs. To fabricate E-HEMTs the thin $Al_{0.3}Ga_{0.7}As$ stopper is removed by non-selective wet-chemical etching followed by selective dry etching of the cap GaAs layer under the stopper. When the temperature is lowered to 77K, threshold voltage shifts to positive by around 0.1 V compared with that at 300K.

Reprinted from *Rec. of the IEEE GaAs Integrated Circuits Symp.*, pp. 125-128, 1984.

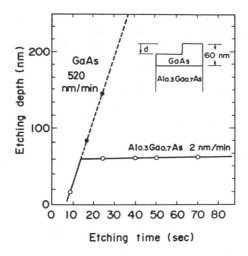

Fig. 1. Characteristics of selective dry etching.

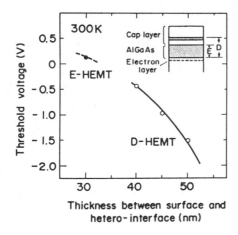

Fig. 2. Threshold voltages of HEMTs with new epistructure.

Fabrication of HEMT LSI

Fig. 3 indicates process sequence for self-aligned gate process in the fabrication of HEMT LSIs including enhancement-mode and depletion-mode HEMTs.

First of all, the active region is isolated by shallow mesa step (180 nm) which is produced by a very simple process and can be made nearly planar. The source and drain for E- and D-HEMTs are metallized with AuGe eutectic alloy and Au overlay alloying to form ohmic contacts with the electron layer. Then, fine gate patterns are formed for E-HEMTs, and the top GaAs layer and thin $Al_{0.3}Ga_{0.7}As$ stopper are etched off by non-selective chemical etching. Using the same resist, after formation of gate patterns for D-HEMTs, selective dry etching is performed to remove the top GaAs layer for D-HEMTs and also remove the GaAs layer under the thin $Al_{0.3}Ga_{0.7}As$ stopper for E-HEMTs. Next, Schottky contacts for the E- and D-HEMT gates are provided by depositing Al, the Schottky gate contacts and

GaAs cap layer for ohmic contacts being self-aligned to achieve high speed performance. Finally, electrical connections from the interconnecting metal, composed of Ti/Pt/Au, to the device terminals are provided through contact holes etched in a crossover insulator film. As described above, a unique epistructure in combination with self-terminating selective dry recess etching makes it possible to fabricate super-uniform E- and D-HEMTs simultaneously, reflecting the uniformity of MBE grown epitaxial film.

A histogram of threshold uniformity for E- and D-HEMTs is shown in Fig. 4. The standard deviations in threshold voltages, measured for over 100 E- and D-HEMTs, are 12 mV and 20 mV at threshold of 0.02 V and -1.11 V, respectively. It is clear that the key difficulty in the DCFL circuit approach is in controlling the threshold voltage of the FETs. But these results are sufficient to realize HEMT DCFL circuits with LSI/VLSI level complexities, because the ratio of standard deviation of threshold voltage to the logic swing (0.5 V for DCFL) is only 2.4%.

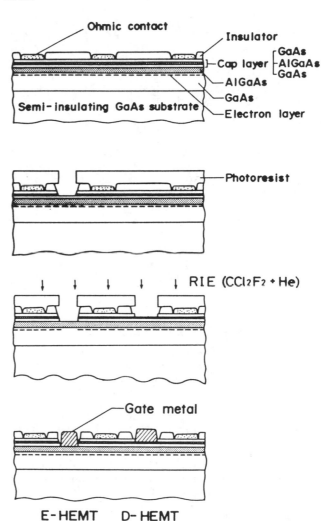

Fig. 3. Process sequence for self-aligned gate E/D-HEMTs.

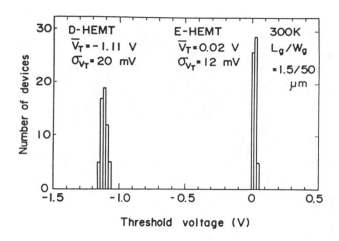

Fig. 4. Histogram of threshold voltages for E/D-HEMTs with 1.5-μm gate length.

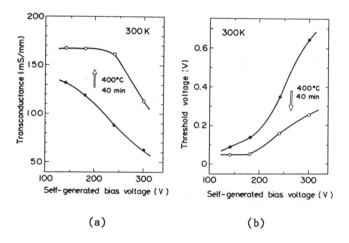

(a)   (b)

Fig. 5. Effect of annealing on (a) transconductance and (b) threshold voltage as a function of self-generated bias voltage.

Optimization of Dry Etching

The gate formation of HEMT is performed by selective dry recess etching, as described above. This dry etching may introduce damage and therefore may affect the electrical characteristics of HEMT such as transconductance and threshold voltage. Fig. 5 (a) shows that transconductance decreases with increasing self-generated bias voltage but it is recovered, by comparatively low temperature annealing at 400°C, up to self-generated bias voltage (V_b) of 240 V. This result apparently indicates that there still remains some damage caused by selective dry etching above V_b of 240 V. Fig. 5 (b) shows the relationship between threshold voltage and self-generated bias voltage. By annealing the devices, up to V_b of 180 V threshold voltage settles down at 0.05 V. In the V_b range above 180 V rather high threshold voltage is obtained. This is because there remains some

damage due to dry etching as described above, and/or because n-$Al_{0.3}Ga_{0.7}As$ layer is strongly sputtered and etched due to high self-generated bias voltage. The etching rate of $Al_{0.3}Ga_{0.7}As$ is estimated to be about 4 nm/min at 300 V in self-generated bias voltage condition.

Though damage due to selective dry recess etching is observed in the bias range previously described, comparatively low temperature (400°C) annealing is useful to recover the damage to equivalent performance of wet chemical recess gate HEMT up to V_b of 240 V.

PERFORMANCE OF 4Kb HEMT SRAM

A successful 4K x 1b static RAM has been developed with the unique epistructure making the best use of selective dry recess etching. A photomicrograph of the RAM is shown in Fig. 6. The RAM is organized into 4096 words x 1b, and arranged as a 64 x 64 matrix. Using depletion-mode HEMTs for load devices, E/D type DCFL circuits are employed as the basic circuit. The memory cell is a 6-transistor cross-coupled flipflop circuit with switching devices having gate lengths of 2 μm. For peripheral circuits, a 1.5-μm gate switching device is chosen for performance reasons. The memory cell is 55 μm x 39 μm, the chip is 4.76 mm x 4.35 mm and 26864 HEMTs are integrated in a 4Kb SRAM. The design rule adopted is a 3 μm line width and spacing at minimum. Minimum size of contact hole is 2 μm x 2 μm.

Normal read-write operation was confirmed both at 300K and 77K. Fig. 7 is a typical oscillograph of read-write operations at 77K. From top to bottom, these traces are X-address input, data-input pulses, write-enable pulses and data-output signals. Fig. 8 shows an oscillograph of X-address input and output waveforms at minimum memory access time at 77K. From this oscillograph, the minimum address access time was estimated to be 2.0 ns, with chip dissipation power of 1.6 W with a supply voltage of 1.54 V. At 300K, typical address access time was 4.4 ns with chip dissipation of 0.86 W.

Integration speed in HEMT memory is around fourfold each year up to 4Kb SRAM, as shown in Fig. 9. It is significantly higher than that of Si memory. In a sense, various Si and GaAs technologies, for example, lithography, microfabrication process, circuit design and so on, are available to HEMT LSI technology. In addition to this, the rapid growth in integration level is mainly due to original large flexibility in device parameter of HEMT such as this new epistructure which is adapted to the HEMT dry process.

Fig. 6. Microphotograph of 4Kb HEMT SRAM. The chip is 4.76 mm x 4.35 mm and contains 26864 HEMTs.

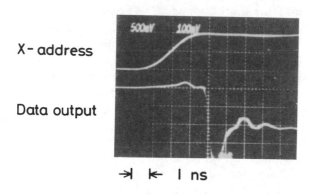

X-address

Data output

→ ← 1 ns

Fig. 8. Oscillograph of X-address input and data output waveforms at minimum access time of 2.0 ns.

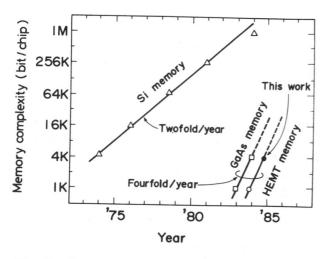

Fig. 9. Integration level as a function of year.

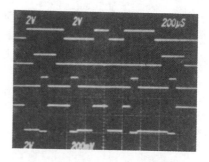

X-address
Data input
Write-enable

Data output

Fig. 7. Oscillograph of read-write operations.

SUMMARY

Recent advances in HEMT technology for high performance LSI circuits are presented, focusing on a new device structure for selective dry etching. A successful 4Kb HEMT SRAM has been developed, achieving address access time of 2.0 ns with power dissipation of 1.6 W at 77K, using 1.5-μm gate devices and 3-μm design rule. This is the highest speed ever reported for 4Kb RAMs. From the results above, address access time below 1 ns can be assessed for 1-μm gate technology and 2-μm design rule.

ACKNOWLEDGMENT

The authors wish to thank Dr. T. Misugi, Dr. O. Ryuzan, Dr. M. Fukuta, M. Kobayashi, T. Kotani, K. Odani, and S. Notomi for their encouragement and support.

This work is supported by the Agency of Industrial Science and Technology, MITI of Japan in the frame of National Research and Development Project "Scientific Computing System".

REFERENCES

(1) T. Misugi, K. Kurokawa and M.Mukai, High Speed Digital Tech. Conf., San-Diego. California, January (1981).
(2) M. Abe, T. Mimura, N. Yokoyama and H. Ishikawa, IEEE Trans. Electron Devices, ED-29 (1982) pp.1088-1094.
(3) T. Mimura, S. Hiyamizu, T. Fujii and K. Nanbu, Jpn. J. Appl. phys. Lett. 19 (1980) pp.L225-L227.
(4) T. Mimura, K. Joshin, S, Hiyamizu, K. Hikosaka and M. Abe, Jpn. J. Appl. Phys. Lett. 20 (1981) pp.L598-L600.
(5) C. P. Lee, D. L. Miller, D. Hou and R. J. Anderson, Device Res. Conf. Vermont, June (1983) IIA-7.
(6) K. Nishiuchi, T. Mimura, S. Kuroda, S. Hiyamizu, H. Nishi and M. Abe, Device Res. Conf. Vermont, June (1983) IIA-8.
(7) K. Nishiuchi, N. Kobayashi, S. Kuroda, S. Notomi, T. Mimura, M. Abe and M. Kobayashi, ISSCC Digest Tech. Papers (1984) pp. 48-49,314.
(8) K. Hikosaka, T. Mimura and K. Joshin, Jpn. J. Appl. Phys. Lett., 20 (1980) pp.L847-L850.

A High-Speed 1-kbit High Electron Mobility Transistor Static RAM

N. H. SHENG, MEMBER, IEEE, H. T. WANG, C. P. LEE, MEMBER, IEEE, G. J. SULLIVAN,
AND D. L. MILLER

Abstract—A 1-kbit static RAM with enhancement and depletion-mode devices was designed and fabricated using the high electron mobility transistor (HEMT) technology. The RAM circuit was optimized to achieve ultra-high-speed performance. A subnanosecond address access time of 0.6 ns was measured at room temperature for a total power dissipation of 450 mW. The minimum WRITE-ENABLE pulse width required to change the state of memory cell is less than 2 ns on probe testing. The best chip has 3 bits that failed to function, which corresponds to a bit yield of 99.7 percent. According to the simulation, variations of the threshold voltage inside the memory cell greatly reduce its stable functional range. High-speed operation requires more uniform threshold voltage control to achieve fully operational LSI memory circuits.

I. INTRODUCTION

THE SPEED advantage offered by GaAs/AlGaAs HEMT technology has placed it as a prime candidate for the device technology of next-generation high-speed systems. Digital HEMT circuits with propagation delays around 10 ps per gate and discrete HEMT's with current gain cutoff frequencies above 70 GHz have been demonstrated. These impressive performances are attributed to the superior electron transport properties of the HEMT's. In order to assess the speed advantage and process yield at large integration levels, we have chosen static RAM's as a test vehicle. Although 1-kbit and 4-kbit HEMT static RAM's have been demonstrated [1], [2], the speed performance of these reported results at room temperature is not superior to that demonstrated by GaAs MESFET self-aligned technology. However, for a fair comparison at large integration levels, all the device characteristics, the circuit structure, and the operating points have to be optimized. In this paper, we report the HEMT fabrication technology, device characteristics, circuit design, and measured results of a 1-bit RAM. In the design phase, the effect of a threshold voltage variation on memory cell stability has been investigated. In the high-speed operation region, the memory cells were found to require very uniform threshold voltage control to obtain a stable op-

eration point. A subnanosecond access time of 0.6 ns has been measured at room temperature from the fabricated 1K RAM circuits, with the best bit yield of 99.7 percent.

II. HEMT FABRICATION TECHNOLOGY AND DEVICE CHARACTERISTICS

The HEMT layer structure used for device fabrication is the conventional single heterostructure HEMT. As shown in Fig. 1, an AlAs/GaAs super-lattice buffer layer was first grown to improve the material quality of the subsequent 1-μm-thick undoped GaAs buffer layer. On top of the buffer layer is a 30-Å-thick undoped AlGaAs spacer layer and two layers of n-AlGaAs that are used as the etch-stop layers for RIE gate recess to control the threshold voltages of both depletion- and enhancement-mode devices. The gate lengths for all the devices in this circuit are 1 μm, except the depletion-load devices in memory cells, whose gate lengths are 10 μm to reduce power. The minimum interconnection line/space is 2 μm/2 μm for both first- and second-level metal. The minimum via hole size is 2 μm × 2 μm. The first step in the fabrication process was an O^+ implant to isolate the devices. The ohmic contacts were then put on and alloyed. The gates of enhancement devices were recessed by a dry-etching technique discussed below. Polyimide was used as an interlevel dielectric to isolate first- and second-level metal.

A. Dry Recess Process

The selective dry-etching process, which stops etching on the AlGaAs surface, is the best technology at present to control the threshold. The threshold voltage is then completely determined by the doping and thickness of the AlGaAs layer during MBE growth. Low power and bias voltage were used to minimize RIE radiation damage, while keeping the etching profile anisotropic. Therefore, no annealing procedure was required after dry etching, and the typical transconductance achieved by dry recessing was between 270 and 300 ms/mm. To study device uniformity in a circuit area, we have fabricated HEMT arrays, each containing 100 HEMT devices of 1 μm × 20 μm in an area of about 1 mm × 1 mm. Fig. 2 shows the measured *I–V* characteristics of one HEMT array. The drain currents (log scale) of the devices are plotted as functions of the gate voltages. The threshold voltages, defined as the gate voltage corresponding to a drain current of 10 μA, show a standard deviation of only 4 mV.

Manuscript received December 8, 1986; revised March 24, 1987. This work was supported by the Office of Naval Research under Contract N00014-83-C-0347.

N. H. Sheng, H. T. Wang, G. J. Sullivan, and D. L. Miller are with the Rockwell International Science Center, Thousand Oaks, CA 91360.

C. P. Lee was with the Rockwell International Science Center, Thousand Oaks, CA 91360. He is now on leave with Chiao-Tung University, Taiwan, Republic of China.

IEEE Log Number 8715021.

Reprinted from *IEEE Transactions on Electron Devices*, vol. ED-34, no. 8, pp. 1670–1675, August 1987.

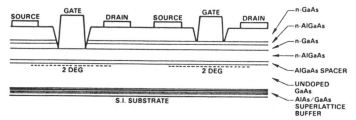

Fig. 1. HEMT device structure for IC fabrication.

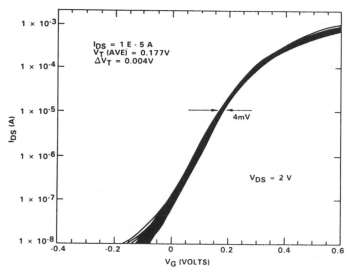

Fig. 2. *I–V* characteristics of a HEMT array.

B. Device Characteristics

Recent progress in self-aligned and submicrometer GaAs MESFET technology led to a peak transconductance comparable to 1-μm HEMT device technology. However, the electron transport mechanisms are quite different in these two devices. A comparison of nonself-aligned HEMT with self-aligned MESFET characteristics is shown in Fig. 3, in which the drain current (I_D) and the transconductance (G_m) are plotted against the gate voltage (V_{gs}) for 50-μm-wide devices. Although peak transconductances are comparable at room temperature, HEMT circuits are faster because the K value, which is the slope of the G_m curve of a HEMT (383 ms/mm $\cdot$ V) is about two times that of a MESFET (192 ms/mm $\cdot$ V). The drain current is proportional to the K value; therefore, HEMT's have higher current and faster speed than MESFET's.

III. Circuit Design

The E/D DCFL was used as the basic logic gate in the RAM circuit design. Push-pull drivers and E/D source followers were also used in places where long line driving is required. The HEMT 1-kbit memory circuit diagram is shown in Fig. 4. The memory cell is a conventional six-transistor cell that consists of an E/D-type cross-coupled flip-flop with two enhancement-mode transfer gates. Depletion-mode pull-up FET's were used to connect the bit lines to the supply voltage V_{DC}. The advantage of using these pull-up transistors is to precharge the bit lines, re-

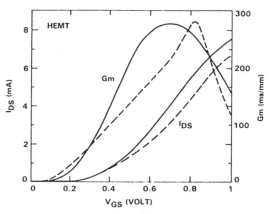

Fig. 3. Comparison of a nonself-aligned HEMT (solid lines) with a self-aligned MESFET.

duce the voltage swing on the bit lines, and speed up the access time. The sense amplifier consists of an E/D DCFL inverter amplifier with a series gate for column selection. In the row selection circuit, an E/D source follower was used after the decoder to reduce the high-level voltage on the word line and avoid dc gate current flowing through the transfer gate into the memory cells. This circuit architecture is quite similar to the conventional static RAM design, which has been widely used by both GaAs MESFET and HEMT technologies.

However, the FET sizes in memory cell and pull-up transistor sizes must be optimized with respect to speed and READ-WRITE stability. The memory cell is a bistable-state circuit with low-noise margin, especially for the low-power case. Variations in threshold voltage uniformity will make the cell favor one state over the other and reduce the operation margin. The operating point and the gain of the sense amplifier also have to be optimized to sense the minimum voltage on the bit lines. These optimizations involve several iterations of computer simulation, which was done with the SPICE 2 program by modifying the JFET model and fitting the appropriate parameters. An enhancement FET threshold voltage of 0.15 V and depletion FET threshold voltage of -0.6 V were used for circuit simulation. In the simulation, all the capacitance loading was included and analyzed to study the tradeoff between speed and cell stability. Line capacitances of 75 fF/mm were taken into consideration for address buffer outputs, writing signal lines, word lines, bit lines, and sense amplifier output lines.

A. Memory Cell Stability

Cell stability depends on the relative current flowing through pull-up, transfer gate, load, and driver FET's in the memory cell. A switching FET size of 14 μm was selected and the size of the other three were selected according to the stability, speed, and power considerations. Fig. 5(a) shows cell stability as a function of the transfer gate and bit line pull-up FET size. When the transfer gate is too small, the voltage drop across it is too large and results in a WRITE error. On the other hand, when the transfer gate size is too large, too much current will dump

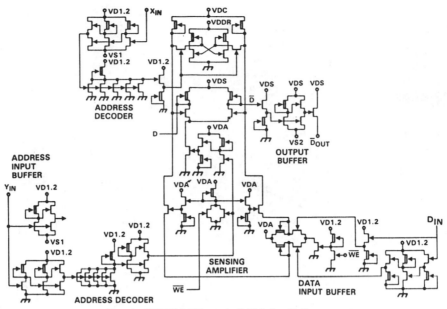

Fig. 4. HEMT 1-kbit static RAM circuit diagram.

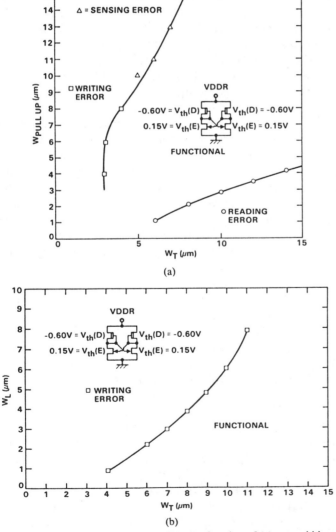

(a)

(b)

Fig. 5. Region of stable cell operation as a function of (a) gate widths of transfer gate FETs (W_T) and pull-up FET's (W_{pullup}), and (b) gate widths of transfer gate and load FET's (W_L). The switching FET size of memory cell is assumed to be 14 μm.

into the cell, upset the state inside the cell, and cause a READ error. An increase in the size of the pull-up transistor will reduce the bit line voltage swing and speed up the access time. When the pull-up FET's are too large, the bit line voltage swing will be too small to be sensed, as shown in Fig. 5(a). Fig. 5(b) shows the cell stability as a function of the transfer gate and FET load size. Typically, small-load transistors in the memory cell are desirable in order to reduce the cell power while keeping enough of a noise margin. The load FET's in our RAM cell were chosen to be 4 μm × 10 μm, which are equivalent to FET's with 1 μm × 1 μm sizes in terms of current values. A pull-up transistor size of 8 μm was selected for the consideration of high-speed operation and a transfer gate size of 7 μm was chosen to keep cell operation inside the stable regime. Simulated stability dependence on the threshold (V_{th}) of the FET is shown as solid lines in Fig. 6(a), where a wide tolerance of threshold variation is designed to improve the memory manufacturability with the present process technology.

In the above simulations, the threshold voltage has been assumed to be uniform across the whole circuit. Cell stability has also been studied under the worst case with an extremely nonuniform threshold distribution of ±50 mV inside the memory cell, as shown in the insert of Fig. 6(b), where the left side favors the high state and the right side favors the low state. The operation margin has been greatly reduced, especially in the high-speed operation regime, compared to Fig. 6(a) and (b). An enhancement-mode threshold voltage of +0.15 V and a depletion threshold of −0.60 V were chosen as the process target for speed, cell stability, and process yield considerations. Although the above nonuniformity assumpton is worse than what can be achieved by the present technology, the effect of the nonsymmetric memory cell on circuit yield should be carefully considered, especially for high-speed, low-power, and large-integration memory circuits.

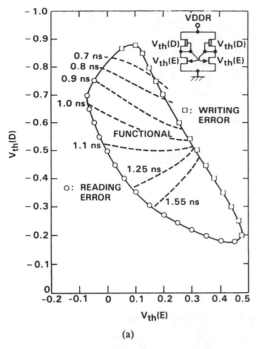

(a)

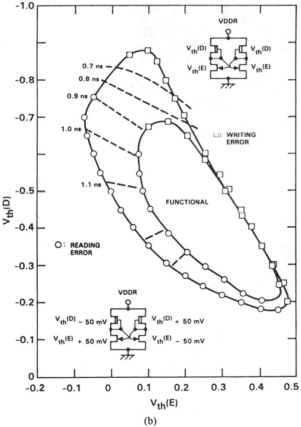

(b)

Fig. 6. Region of stable cell operation as a function of threshold voltages for enhancement-mode FET's ($V_{th}(E)$) and depletion-mode FET's ($V_{th}(D)$): (a) with uniform threshold distribution in memory cell; (b) with the extremely nonuniform threshold voltage distribution of ± 50 mV inside the memory cell. The dotted line shows the address access time as a function of threshold voltage.

B. Speed and Power Simulation

To achieve fast speed, an enhancement-mode threshold voltage of 0.1 V and a depletion-mode threshold voltage

TABLE I
SIMULATED ROW-ADDRESS ACCESS TIME

	Address Buffer	Word Line	Bit Line	Sensing Amp	Output Buffer	Total
Propagation Delay	75 ps	225 ps	200 ps	150 ps	100 ps	750 ps

TABLE II
SIMULATED POWER DISSIPATION FOR 1K RAM

	Voltage (V)	Current (mA)	Power
Data Input Buffer	1.5	0.59	0.885
	1.2	4.245	5.09
X-Address Buffer	1.2	19	22.8
X-Address Decoder	1.2	24.38	29.256
Y-Address Buffer	1.2	19	22.8
Y-Address Decoder	1.5	12.1	18.15
WE and Column Control	1.5	46.7	70
Bit Line Pull-Up	1.0	18	18
Memory Cell	1.2	156.77	188.1
Sensing Amplifier Output Buffer	1.5	24.9	37.4
Total	325.7 mA		412.5 mW

of -0.8 V were used to simulate the signal delay time on each stage in row-address access and are shown in Table I. The total address access time is 750 ps. In the memory circuit, the output of the word-line driver has the largest capacitance loading due to the transfer gate capacitance. A propagation delay of 225 ps was simulated on the word line, which is not much worse than delay contributions from other points for the 1-kbit size. The simulated distribution of power dissipation with a 50-Ω output load is shown in Table II with a total power of 413 mW. About 45 percent of the total power is dissipated in memory cells. The driver transistor size, and therefore the load transistor size, in the memory cell can be reduced to cut down the power dissipation without affecting the speed performance in the next design.

IV. TEST RESULTS

A 1-kbit RAM has been successfully fabricated on MBE-grown HEMT wafers. A photomicrograph of the fabricated 1-kbit RAM is shown in Fig. 7(a). The circuit is arranged as a 32 × 32 matrix. A photomicrograph of the fabricated memory cell is shown in Fig. 7(b). The cell size is 47 μm × 46 μm. The 1-kbit memory chip size is 2.5 mm × 2.35 mm with 2443 depletion-mode FET's and 4983 enhancement-mode FET's. Fig. 8(a) shows the functional READ-WRITE operation of a 2 × 2 neighboring cell out of the 1-kbit RAM (a reverse state was written and read from these four cells). The functionality tests include the walking test and the checkerboard test. The best 1-kbit RAM chip had only 3 bits that failed to function. This corresponds to a bit yield of 99.7 percent. Each chip on the same wafer has a bit yield of more than 95 percent, and the failed bit cells are randomly distributed from chip to chip. Fig. 8(b) shows a minimum address access time of 600 ps, which has been measured on one wafer with a total power consumption of 450 mW for the

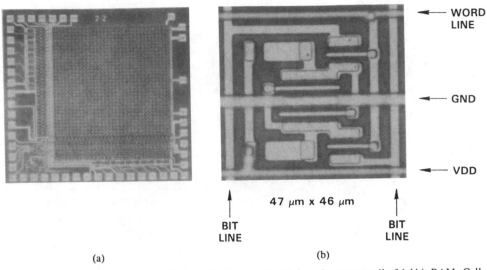

Fig. 7. (a) Fabricated 1-kbit HEMT RAM. Chip size is 2.5 mm × 2.35 mm. (b) Fabricated memory cell of 1-kbit RAM. Cell size is 47 μm × 46 μm.

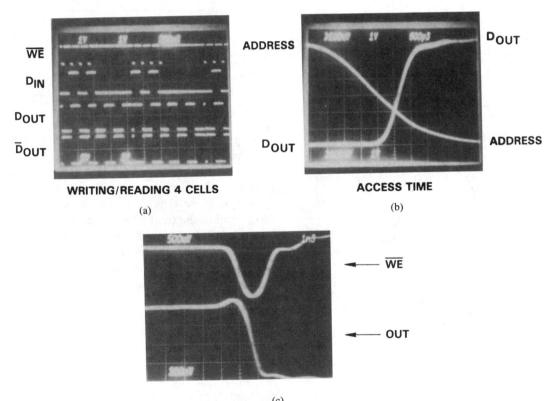

Fig. 8. (a) Functional READ-WRITE operation of 2 × 2 neighboring cells out of 1-kbit RAM. (b) Row-address input and data output waveforms. Access time is 0.6 ns. (c) WRITE access measurements with a WRITE-ENABLE pulse less than 2 ns.

1-kbit RAM at room temperature. The average threshold voltage on this wafer for the enhancement-mode devices is 0.05 V and −0.70 for the depletion-mode devices. This is the fastest access time ever reported for 1-kbit RAM's operated at room temperature. On the best bit yield chip, the access time varies randomly from 800 ps to 1 ns across the whole 1-kbit RAM. These results agree well with the simulated results. Fig. 8(c) shows the WRITE access measurement with a short WRITE-ENABLE pulse. The access time is 0.8 ns and the minimum WRITE-ENABLE pulse width is less than 2 ns, which is limited by probe-card loading.

V. CONCLUSION

In conclusion, we have designed and fabricated HEMT 1-kbit RAM's and have achieved a record high speed of 0.6-ns access time and obtained a bit yield of 99.7 percent (3 bits short of full functionality).

ACKNOWLEDGMENT

The authors would like to thank C. Hill for wafer processing and D. S. Deakin for testing.

References

[1] K. Nishiuchi, N. Kobayashi, S. Kuroda, S. Notomi, T. Mimura, M. Abe, and M. Kobayashi, in *ISSCC Dig. Tech. Papers*, pp. 48–49, 1984.
[2] S. Kuroda, T. Mimura, M. Suzuki, N. Kobayashi, K. Nishiuchi, A. Shibatomi, and M. Abe, in *Proc. GaAs IC Symp.*, pp. 125–128, 1984.

Part 3
Optoelectronic Applications

Monolithic integrated receiver front end consisting of a photoconductive detector and a GaAs selectively doped heterostructure transistor

C. Y. Chen, N. A. Olsson, C. W. Tu, and P. A. Garbinski

AT&T Bell Laboratories, Murray Hill, New Jersey 07974

(Received 6 December 1984; accepted for publication 23 January 1985)

We report the first demonstration of a monolithically integrated receiver front end consisting of a GaAs selectively doped heterostructure transistor (SDHT) and a photoconductive detector. Due to its simplicity in the required epitaxial layers and device fabrication, the photoconductive detector/SDHT integration scheme is one of the simplest ever reported. The detector has a measured gain-bandwidth product of 5 GHz and the SDHT has a transconductance of 100–140 mS/mm. For an error rate of 10^{-9} at 90 MHz, the receiver has sensitivities at 0.82 μm of -36.2 dBm and -42.2 dBm for pseudorandom patterns and fixed bit patterns of 101010..., respectively, for an uncoated device.

The monolithic integration of electronic components with optoelectronic devices is an attractive approach for making optoelectronic modules for lightwave communication applications. A monolithically integrated circuit not only eliminates elaborate wire connections between components but also reduces the parasitic reactances and capacitances, thereby increasing reliability and operation speed. Fabricating all relevant devices in one processing sequence also reduces the overall cost and increases the yield. Monolithic integration of laser diodes and field-effect transistors[1-4] (FET) as well as photodetectors and FET's[5-9] has been demonstrated by several workers. In the letter, we demonstrate the monolithic integration of a GaAs selectively doped heterostructure transistor (SDHT)[10-11] and a photoconductive detector (PCD),[12] and report its characteristics for use as a receiver front end for lightwave communication systems. This integration scheme has several advantages compared with the integration of a *pin* photodiode and a FET. Firstly the epitaxial layers used for FET's are also suited for photoconductive detectors, thus avoiding the complexity of ion implantation, growing epitaxial layers in an etched well, and planarizing the surface. Secondly, a selectively doped transistor is expected to have a high transconductance even at room temperature. Furthermore, a GaAs selectively doped PCD[12] can have lower dark current because the light absorbing layer is mostly lightly doped *p* type, which reduces the dark noise of the detector.

The epitaxial layers of this integrated circuit were grown by molecular beam epitaxy (MBE) on (100) oriented semi-insulating GaAs substrates. The structure consists of a 0.9-μm unintentionally doped GaAs layer (*p* type), a 25-Å spacer layer, a 380-Å Si-doped $Al_{0.3}Ga_{0.7}As$ layer, and a 410-Å GaAs cap layer for ohmic contacts. Both the SDHT and the PCD were fabricated by lift-off techniques. Isolation between the devices is provided by the semi-insulating substrate together with a mesa structure. Figure 1 shows a schematic cross-sectional view of the devices. The *n*-type ohmic metal is Ni/Ge/Au and the gate metal is Al.

The integrated circuit consists of two interdigitated photoconductive detectors directly coupled to the gate of an SDHT. An FET for testing purpose is also included. Figure 2(a) shows the photograph of a completed receiver front end. In this study, only one of the PCD's is used, and the other is a back-up. The large PCD has an electrode spacing of 4 μm and a mesa dimension of 36 $\mu \times$ 57 μ, while the small PCD has an electrode spacing of 6 μm and a mesa dimension of 30 $\mu \times$ 47 μ. Unless otherwise specified, the results will be referred to the large PCD. The SDHT used in the front-end circuit has a gate length of 1.2 μm and a gate width of 250 μm. Figure 2(b) shows the circuit diagram of the integrated receiver front end. No antireflection (AR) coating was applied to the detector in this study.

The PCD and the SDHT were characterized prior to the receiver sensitivity measurement. The experimental results showed that the detector had a saturated dark current of 1.5 mA and a saturation voltage of 0.6 V. When tested by a 0.82-μm AlGaAs laser, the detector showed an average gain of ~5000 at a power level of -36.2 dBm. We did observe, however, a decrease of gain with increasing incident power levels. Preliminary results showed that the decrease in gain was accompanied by an increase in speed. This study thus suggested the existence of saturable traps, presumably in the n^+-$Al_{0.28}Ga_{0.72}As$ layer, making the effective carrier lifetimes dependent upon the incident power levels.

The SDHT has a transconductance of 100–140 mS/mm and a reverse gate leakage current of 2 nA at 1-V bias at room temperature. The total zero-bias gate capacitance was ~0.63 pF at 1 MHz.

The completed chip together with a bias resistor for the PCD and a load resistor for the FET was then mounted in a microwave package for bit error rate measurements, which take into account not only the gain but also noise of the

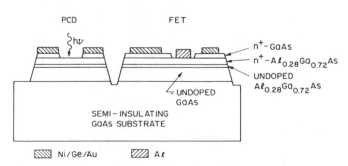

FIG. 1. Schematic cross section of the integrated selectively doped photoconductive detector FET front end.

Reprinted with permission from *Applied Physics Letters*, vol. 46, no. 7, pp. 681–683, (1985).

Appl. Phys. Lett., Vol. 46, No. 7, 1 April 1985

Chen *et al.*

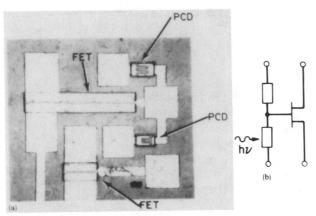
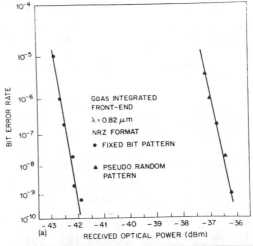

FIG. 2. Photomicrograph of the integrated PCD/FET. The gate width of the large FET is 250 μm. (b) Circuit diagram.

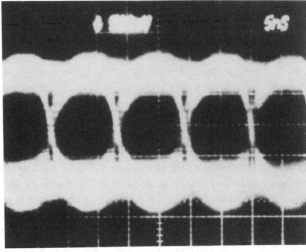

FIG. 3. (a) Bit error rate vs received optical power of the integrated device for pseudorandom pattern (solid triangles) and fixed bit pattern of 101010...(solid circles). The bit rate is 90 Mb/s. (b) Eye pattern at a bit error rate of 10^{-9}. Horizontal: 5 ns/div. vertical: 500 mV/div.

detector.[13] As a result of the aforementioned slow traps, the integrated front end reported here has a 3-dB bandwidth of 1 MHz. An *RC* equalizer was used to give a sufficient bandwidth for 90-Mb/s error rate measurement.[13] The integrated receiver front end was connected to a chain of amplifiers, an *RC* equalizer, and a 400-MHz low-pass filter. An (AlGa)As injection laser with $\lambda = 0.82$ μm was used as a light source. The signal level was measured by a broad-area Si *pin* photodiode.

Figure 3(a) shows the measured bit error rate as a function of received optical power for two different patterns: a pseudorandom pattern (solid triangles) and a fixed bit pattern of alternate 0 and 1 (solid circles). The detector was biased at a dark current of 0.28 mA and a voltage of 0.12 V. For an error rate of 10^{-9} at 90 Mb/s, the received signal levels were -36.2 dBm and -42.2 dBm for pseudorandom patterns and fixed patterns, respectively, with nonreturn to zero (NRZ) formats. Figure 3(b) shows the corresponding eye pattern at error rate of 10^{-9} for pseudorandom bit patterns. Correcting for the reflection from the semiconductor surface leads to expected receiver sensitivities of -37.7 dBm and -43.7 dBm for pseudorandom and fixed patterns, respectively. The 6-dB difference between pseudorandom patterns and fixed ones reflects the penalty as a result of poor equalization. In other words, the sensitivity for pseudorandom patterns can be improved by 6 dB, in principle, by improving the equalizer design. For the purpose of comparison, a $Ga_{0.47}In_{0.53}As$ *pin* photodiode integrated with an InP metal-insulator-semiconductor FET showed a receiver sensitivity of -34.5 dBm at 1.3 μm, 100 Mb/s.[14] Correction for the quantum efficiency, photon energy, and bit rates leads to an expected sensitivity ($\bar{p}$) of -32.3 dBm at 0.82 μm, 90 Mb/s, and an error rate of 10^{-9}. In addition, the results on a discrete $Ga_{0.47}In_{0.53}As$ *pin*/GaAs FET receiver[15] suggest a receiver sensitivity of -42.3 dBm at 0.82 μm, 90 Mb/s after correction for the quantum efficiency, photon energy, and bit rates. Therefore, the experimental results of our integrated front end, which is in the early stage of development, are very encouraging.

The theoretical gain-bandwidth product (GB) of a PCD is given by[16]

$$GB = 1/2\pi t_n = v_n/2\pi L, \tag{1}$$

where t_n is electron transit time, v_n is the saturated electron velocity, and L is the finger spacing. For a detector having 4-μm finger spacings, the theoretical gain-bandwidth product should be ~ 8 GHz. Correcting for the surface reflection leads to an experimental intrinsic gain-bandwidth product of ~ 7 GHz. We attribute the 0.6-dB difference to a thin light absorbing layer.

In conclusion, we have developed a monolithically integrated receiver front end consisting of a GaAs selectively doped FET and a photoconductive detector. As a result of the compatibility between the PCD and selectively doped FET, the fabrication processes were very simple, requiring no ion implantation or growth over a well. The detector showed an average gain of ~ 5000 and a 3-dB bandwidth of ~ 1 MHz (or an extrinsic gain-bandwidth product of 5 GHz) at an incident power level of -36.2 dBm. The SDHT's have a transconductance of 100–140 mS/mm and a gate leakage current of 2 nA at 1-V bias at room temperature. The error rate measurement at 0.82 μm showed that the required optical power to achieve an error rate of 10^{-9} was -36.2 dBm for an uncoated device. The study using fixed bit patterns of alternate 0 and 1 suggests that the receiver sensitivity can be improved approximately 6 dB by improving the equalizer

Appl. Phys. Lett., Vol. 46, No. 7, 1 April 1985

Chen *et al.*

design. Compared to the published results on receiver sensitivities, this integrated receiver looks very promising and may find applications in local area networks.

The authors would like to thank B. F. Levine and S. Wemple for their encouragement during the course of this work.

[1]I. Ury, S. Margalit, M. Yust, and A. Yariv, Appl. Phys. Lett. **34**, 430 (1979).

[2]T. Fukuzawa, M. Nakamura, M. Hirao, T. Kuroda, and J. Umeda, Appl. Phys. Lett. **36**, 181 (1980).

[3]U. Koren, K. L. Yu, T. R. Chen, N. Bar-Chaim, S. Margalit, and A. Yariv, Appl. Phys. Lett. **40**, 643 (1982).

[4]H. Matsueda, S. Sasaki, and M. Nakamura, IEEE J. Lightwave Tech. **LT-1**, 261 (1983).

[5]R. F. Leheny, R. E. Nahory, M. A. Pollack, A. A. Ballman, E. D. Beebe, J. C. DeWinter, and R. J. Martin, Electron. Lett. **16**, 353 (1980).

[6]J. Barnard, H. Ohno, E. C. Wood, and L. F. Eastman, Electron Dev. Lett. **EDL-2**, 7 (1981).

[7]O. Wada, S. Miura, M. Ito, T. Fujii, T. Sakurai, and S. Hiyamizu, Appl. Phys. Lett. **42**, 380 (1983).

[8]R. M. Kolbas, J. Abrokwah, J. K. Carney, D. H. Bradshaw, B. R. Elmer, and J. R. Biard, Appl. Phys. Lett. **43**, 821 (1983).

[9]Y. Yamada, M. Kawachi, M. Yasu, and M. Kabayashi, Electron. Lett. **20**, 314 (1984).

[10]R. Dingle, H. L. Stormer, A. C. Gossard, and W. Wiegmann, Appl. Phys. Lett. **33**, 665 (1978).

[11]T. Mimura, S. Hiyamizu, T. Fujii, and K. Nanbu, Jpn. J. Appl. Phys. **19**, L225 (1980).

[12]C. Y. Chen, A. Y. Cho, C. G. Bethea, P. A. Garbinski, Y. M. Pang, and B. F. Levine, Appl. Phys. lett. **42**, 1040 (1983).

[13]C. Y. Chen, B. L. Kasper, and H. M. Cox, Appl. Phys. Lett. **44**, 1142 (1984).

[14]K. Kasahara, J. Hayashi, K. Makita, K. Taguchi, A. Suzuki, H. Nomura, and S. Matushita, Electron. Lett. **20**, 314 (1984).

[15]D. J. Malyon, T. G. Hodgkinson, D. W. Smith, R. C. Booth, and B. E. Daymond-John, Electron. Lett. **19**, 144 (1983).

[16]H. Beneking, IEEE Trans. Electron Devices. **ED-29**, 1420 (1982).

OPERATION OF A HIGH-FREQUENCY PHOTODIODE-HEMT HYBRID PHOTORECEIVER AT 10 GHz

Indexing terms: Photoelectric devices, Photoreceivers, Microwave devices and components

A hybrid photoreceiver based on an ITO/GaAs photodiode integrated with a 0·5 μm gate length high electron mobility transistor (HEMT) is reported. The receiver gave an 8 dB associated gain at 10 GHz compared with a discrete detector.

The use of optical fibre technology for signal distribution and information transfer is well known at frequencies <1 GHz. However, with the advent of high-frequency light modulation techniques such as direct modulation of laser diodes[1] or the use of an external modulator,[2] and the development of high-frequency optical detectors,[3] the technology now exists for systems operating from 1 to 20 GHz. (Furthermore, the potential exists to extend this up to 100 GHz.) The ideal detector element to use for high-frequency applications is a PIN or Schottky barrier photodiode as it combines potentially high-frequency performance with reasonable sensitivity. For many system applications, however, the RF power output of a discrete photodiode is insufficient. Hence an amplification stage is required. In this letter we will present, to our best knowledge, the first results of an ITO/GaAs photodiode-HEMT hybrid photoreceiver operating at frequencies up to 10 GHz.

The detector used was a quasi-Schottky barrier photodiode which employs indium tin oxide (ITO) to form the rectifying junction.[4] It consists of 100 nm of sputtered ITO of composition $In_{0.9}Sn_{0.1}O$ on 1·2 μm-thick n-type GaAs ($\sim 10^{16}$ cm^{-3}) with an underlying highly doped ($\geq 10^{18}$ cm^{-3}) n^+ layer for the creation of a low resistance ohmic contact. The device active area is 40 μm square. The diode has a reverse breakdown voltage in excess of 25 V with an associated leakage current of typically 1 nA at -5 V. Owing to the wide bandgap nature of the ITO, and the chosen thickness of the layer also acting as an antireflection coating, the external quantum efficiency of the device has been measured to be as high as 60% at 840 nm. The impulse response has been measured to be 20 ps FWHM, which translates into a -3 dB bandwidth slightly in excess of 20 GHz. Hence, the detector response is uniform in the frequency region of interest here (10 GHz).

This photodiode was combined with a GaAs/AlGaAs HEMT in a simple hybrid receiver. The HEMT is an ideal device for use in such a system owing to its inherent low noise and high gain performance at microwave frequencies. Recently reported device results demonstrate a noise figure of 1·3 dB at 18 GHz with 12 dB associated gain for a 0·25 μm gate length device.[5]

This performance is due principally to the technique of modulation doping[6] in which the doping is introduced only into the wide-bandgap AlGaAs layer of the device. This leads to higher electron mobilities and saturated drift velocities as compared with GaAs MESFETs. The resulting improvement in device transconductance coupled with lower source resistance leads to the low noise and high gain performance. This device is, therefore, ideally suited for use as an amplifier in a high-frequency, low-noise hybrid receiver.

The material structure of the HEMT is shown in Fig. 1. The layers were grown by molecular beam epitaxy (MBE) on 2 in (50·8 mm)-diameter SI substrates. A superlattice buffer is used to impede the migration of impurities from the substrate into the device layers. The 500 Å-thick n^+ GaAs cap layer was used for ohmic contact formation and to reduce the source access resistance.

The fabrication process for the HEMT[7] was based on direct writing electron beam lithography for fine feature definition and registration. This allows accurate control of device dimensions and alignment tolerances. The gate was recessed, the etch and subsequent metallisation being performed in one lithographic step. Device isolation was achieved by boron implantation which results in a planar process. The gate

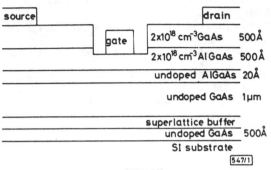

Fig. 1 *Material structure of HEMT*

length of the HEMT was 0·5 μm with a gate width of 300 μm. At 10 GHz, a noise figure of 1·3 dB was obtained with 8 dB associated gain.

To demonstrate the potential of an ITO/GaAs photodiode-HEMT combination a simple hybrid receiver was constructed using standard 50 Ω microstrip technology. To assess the performance of the receiver at 10 GHz, an Ortel Corporation GaAs/AlGaAs laser diode was used which has a -3 dB modulation bandwidth of approximately 5 GHz. An RF modulated signal was incident on the laser via an HP33180A bias network which also allowed the appropriate direct current level to be easily applied. The light output was then coupled to the detector via a series of collimating and refocusing lenses.

The photodiode was reverse-biased at -5 V, which resulted in a typical leakage current of less than 2 nA. This low-leakage behaviour suggests that the photodiode should be an intrinsically low-noise device. The HEMT was operated at $\frac{1}{2}I_{DSS}$ (30 mA) with a drain-source potential of 2 V and a gate-source potential of -0.88 V. The HEMT was impedance-matched at input and output for minimum noise at 10 GHz.

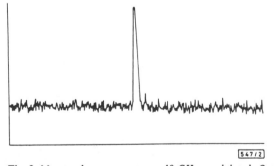

Fig. 2 *Measured response to a 10 GHz modulated Ortel laser as obtained from ITO/GaAs photodiode–HEMT combination*

Response of a discrete photodiode is included for reference

The output from the hybrid receiver, as obtained using an HP8569B high-frequency spectrum analyser, is shown in Fig. 2 for a 10 GHz modulated signal. For comparison the output from an identical discrete detector subjected to the same optical input (confirmed by establishing the same direct current conditions) demonstrates that a power gain of 8 dB, associated with the HEMT, has been achieved for the hybrid case. The relatively high noise floor is associated with noise from the laser diode which is transmitted on the optical carrier. This dominant effect makes an estimate of the overall receiver noise difficult. It should, however, be very low for the reasons outlined above.

In conclusion, a hybrid photoreceiver based on an ITO/GaAs photodiode and an HEMT is reported for the first time. The module exhibits an 8 dB associated gain at 10 GHz. Work is currently under way to increase the bandwidth capability of the receiver and to assess its true noise performance.

Acknowledgments: We would like to thank P. Say and C. Walker for assistance with device fabrication, T. Kerr and S. Jones for material growth, M. Harrison for electron-beam lithography, and C. Lau and U. Dhaliwal for assistance with HEMT measurements. The helpful discussions with Dr. J. A. Barnard, Dr. W. Sibbett and Dr. C. Snowden are also appreciated. The HEMT device work was partly funded by the Procurement Executive, UK Ministry of Defence (Directorate of Components, Valves & Devices) and sponsored from the Royal Signals & Radar Establishment.

W. A. HUGHES   *10th March 1986*
D. G. PARKER
GEC Research Limited
Hirst Research Centre
East Lane, Wembley, Middx. HA9 7PP, United Kingdom

References

1 BOWERS, J. E.: 'Millimetre wave response of InGaAsP lasers', *Electron. Lett.*, 1985, **21**, pp. 1195–1196
2 GEE, C. M., and THAMMOND, G. D.: '17 GHz bandwidth electro optic modulator', *Appl. Phys. Lett.*, 1983, **43**, pp. 998–1000
3 See for example WANG, S. Y., and BLOOM, D. M.: '100 GHz bandwidth planar GaAs Schottky photodiode', *Electron. Lett.*, 1983, **19**, pp. 554–555
4 PARKER, D. G.: 'Use of transparent indium tin oxide to form a highly efficient 20 GHz Schottky barrier photodiode', *ibid.*, 1985, **21**, pp. 778–779
5 MISHRA, U. K., *et al.*: 'Microwave performance of 0·25 μm gate length high electron mobility transistor', *IEEE Electron. Device Lett.*, 1985, **EDL-6**, pp. 142–145
6 DINGLE, R., *et al.*: 'Electron mobilities in modulation-doped semiconductor heterojunction superlattices', *Appl. Phys. Lett.*, 1978, **33**, pp. 665–667
7 HUGHES, W. A.: 'A fabrication process for high electron mobility transistors (HEMTs) based on electron beam lithography', *GEC J. Res.* (to be published)

A Monolithically Integrated AlGaAs/GaAs p-i-n/FET Photoreceiver by MOCVD

S. MIURA, O. WADA, H. HAMAGUCHI, M. ITO, M. MAKIUCHI, K. NAKAI, AND T. SAKURAI

Abstract—An AlGaAs/GaAs p-i-n photodiode and a GaAs FET have been monolithically integrated on a GaAs substrate by using the metal-organic chemical vapor deposition (MOCVD) technique and by applying a new interconnection technique. A current amplification characteristic consistent to the device parameters has been demonstrated. This result indicates a suitability of MOCVD to realize the monolithic integration of p-i-n/FET photoreceiver.

THE INTEGRATION of a p-i-n photodiode with a field-effect transistor (FET) preamplifier is attractive to obtain low-noise high-speed photoreceivers for the application in optical communication and data-processing systems. Several workers have shown p-i-n/FET hybrid photoreceiver to have a wide dynamic range and an increased sensitivity [1]–[3]. In this line, the monolithic integration of a p-i-n photodiode and a FET on the same semiconductor substrate is expected to further increase the sensitivity and the response speed primarily because of the reduction of the parasitic reactances. As for the fabrication of semiconductor structures, liquid phase epitaxy method has been used in most of the previous works [4]. However, superior uniformity over a wide area of the wafer is necessary for the near-future application of optoelectronic integrated circuits (OEIC's). The authors have previously reported the application of molecular beam epitaxy for this purpose [5]. The metal-organic chemical vapor deposition (MOCVD) technique is also considered to fulfill such a requirement. Besides, a high throughput at a low cost is expected for this growth technique. In this letter, the fabrication of a monolithically integrated p-i-n/FET photoreceiver by using MOCVD is described. An improved interconnection technique is developed and used in the present fabrication.

A cross-sectional view of the present integrated circuit is shown in Fig. 1. The AlGaAs/GaAs multilayer structure was grown on the (100) oriented semi-insulating GaAs substrate by a single growth run of MOCVD at the temperature of approximately 700°C. The n-type dopant used was Se. An n^+-GaAs contact layer with the carrier concentration of 1×10^{18} cm^{-3} was first grown. It works as an n-type contact layer of the p-i-n structure. The next n^--GaAs light absorption layer had the thickness of 3.5 μm and the carrier concentration as low as 5×10^{14} cm^{-3}, by which the bias voltage required for

Manuscript received July 14, 1983. This work was supported by the Agency of Industrial Science and Technology, MITI, of Japan in the framework of the National Research and Development Project "Optical Measurement and Control Systems."

The authors are with Fujitsu Limited, 1677 Ono, Atsugi, Kanagawa, 243-01 Japan.

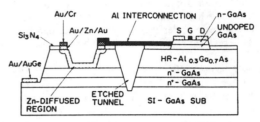

Fig. 1. Cross-sectional view of integrated p-i-n/FET photoreceiver.

the p-i-n photodiode operation can be reduced down to approximately 5 V. A high-resistivity (HR) AlGaAs layer, which works not only as the window layer of a p-i-n photodiode but also as the isolation layer for the FET channel, was successfully obtained by an MOCVD-grown undoped layer. Finally, an undoped GaAs buffer layer and then a 0.25-μm-thick FET channel layer ($n = 1.3 \times 10^{17}$ cm^{-3}) were grown. The p^+-region of the p-i-n photodiode was formed by selective Zn diffusion using a P-CVD Si$_3$N$_4$ mask. It was carried out by using ZnAs$_2$ source in a closed silica tube at the temperature of 600°C. The diffusion coefficient of Zn in Al$_{0.3}$Ga$_{0.7}$As is 4–5 times larger than in GaAs at this temperature, so that diffusion front was exactly controlled to be at the hetero interface. Ohmic contacts to p- and n-type layers composed of Au/Zn/Au [6] and Au/AuGe films, respectively, were formed by liftoff technique. The Al films were deposited both for the FET gate and the interconnection. Then an isolation groove was formed between the p-i-n photodiode and the FET by using an etchant 8H$_2$O$_2$ + 1H$_2$SO$_4$ + 1H$_2$O. In this etching procedure, the epitaxial layers just under the Al interconnection are also etched away by the lateral etching from both edges of the Al line [7], [8], so that the Al bridge crosses over a deep groove, as illustrated in Fig. 1.

Fig. 2 shows the photomicrograph of the device. The diameter of the photosensitive area is 100 μm. The FET channel width, the channel length, and the gate length are 80, 6, and 1.8 μm, respectively. The device characteristics were measured after it was bonded on a copper heat sink.

The reverse current–voltage characteristic of the p-i-n photodiode is shown in Fig. 3. The dark current at the reverse bias voltage of 5 V is as low as 7×10^{-10} A. At this voltage, the n^--GaAs light-absorption layer is fully depleted, consistent with the capacitance–voltage measurement. The low-leakage current characteristic is considered to be due to the low carrier concentration and the defect-free nature of the MOCVD-grown layers and also due to the surface leakage reduced by the P-CVD Si$_3$N$_4$ passivation. This result implies the usefullness

Reprinted from *IEEE Electron Device Letters*, vol. EDL-4, no. 10, pp. 375–376, October 1983.

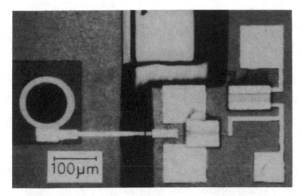

Fig. 2. Surface photomicrograph of integrated p-i-n/FET photoreceiver.

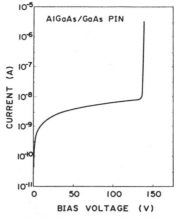

Fig. 4. Dependences of photocurrent induced in p-i-n photodiode ($\Delta I_{\text{p-i-n}}$) and photo-induced current in FET (ΔI_{FET}) on optical input power into the p-i-n photodiode. Current amplification ratio of 10 is obtained.

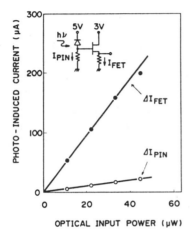

Fig. 3. Reverse current–voltage characteristic of p-i-n photodiode.

the p-i-n photodiode and the FET is achieved in this integration scheme.

In conclusion, a monolithic p-i-n/FET photoreceiver has been fabricated on a semi-insulating GaAs substrate by using an MOCVD-grown multilayer structure and by applying a new interconnection technique. The current amplification characteristic of the monolithic device was successfully achieved in the present device structure. This result is encouraging for the use of MOCVD and this device structure for the future integrated photoreceiver applications.

ACKNOWLEDGMENT

The authors wish to thank M. Ozeki and K. Dazai for their continuous encouragement.

of MOCVD for low-leakage current p-i-n photodiode fabrication. The current–voltage characteristic of the FET was measured under dc bias application. The pinchoff voltage of 2.5 V was observed in this case. By applying an HR (as high as 10^8 $\Omega \cdot cm$) AlGaAs layer to the isolation of the FET channel in the present structure, the leakage current between the source and the drain was negligibly small even at the pinchoff voltage.

The characteristics of the integrated p-i-n/FET device was examined using a 0.78-μm wavelength light emitting diode coupled with an optical fiber. For the measurement, a butt-ended fiber was adjusted to the light-sensitive area of the p-i-n photodiode. The circuit used in the measurement is illustrated in the inset of Fig. 4. The photocurrent of the p-i-n photodiode and the photoinduced current in the FET are shown versus the optical input power in Fig. 4. The quantum efficiency of the p-i-n photodiode can be estimated from the relation of the photocurrent and the input power. A high quantum efficiency of approximately 70 percent is obtained by this estimation. This value agrees well with our previous result on discrete AlGaAs/GaAs p-i-n photodiodes [9]. The current amplification ratio $\Delta I_{\text{FET}}/\Delta I_{\text{p-i-n}}$ of approximately 10 is obtained. This is consistent with the value expected from the transconductance multiplied with the load resistance of the photodiode, indicating that a sufficient electronic isolation between

REFERENCES

[1] B. Owen, "PIN-GaAs FET optical receiver with a wide dynamic range," Electron. Lett., vol. 18, pp. 626–627, 1982.

[2] K. Ogawa and E. L. Chinnock, "GaAs-FET transimpedance front-end design for a wideband optical receiver," Electron. Lett., vol. 15, pp. 650–653, 1979.

[3] D. R. Smith, R. C. Hooper, and I. Garret, "Receivers for optical communications: A comparison of avalanche photodiodes with PIN-FET hybrids," Opt. Quant. Electron., vol. 10, pp. 293–300, 1978.

[4] R. F. Leheny, R. E. Nahory, M. A. Pollack, A. A. Ballman, E. D. Beebe, J. C. Dewinter, and R. J. Martin, "Integrated $In_{0.53}$-$Ga_{0.47}As$ p-i-n F.E.T. photoreceiver," Electron. Lett., vol. 16, pp. 353–355, 1980.

[5] O. Wada, S. Miura, M. Ito, T. Fujii, T. Sakurai, and S. Hiyamizu, "Monolithic integration of a photodiode and a field-effect transistor on a GaAs substrate by molecular beam epitaxy," Appl. Phys. Lett., vol. 42, pp. 380–382, 1983.

[6] T. Sanada and O. Wada, "Ohmic contacts to P-GaAs with Au/Zn/Au structure," Jap. J. Appl. Phys., vol. 19, pp. L491–L494, 1980.

[7] O. Wada, S. Yanagisawa, and H. Takanashi, "Process of GaAs monolithic integration applied to Gunn-effect logic circuits," J. Electrochem. Soc., vol. 123, pp. 1546–1551, 1976.

[8] S. Yanagisawa, O. Wada, and Y. Toyama, "CW performances of planer Gunn-effect devices," IEEE Trans. Electron. Devices, vol. ED-26, pp. 1313–1319, 1979.

[9] M. Ito, O. Wada, S. Miura, K. Nakai, and T. Sakurai, "Planer structure AlGaAs/GaAs pin photodiode grown by MOCVD," to be published in Electron. Lett..

Ultrahigh speed modulation-doped heterostructure field-effect photodetectors

C. Y. Chen, A. Y. Cho, C. G. Bethea, P. A. Garbinski, Y. M. Pang,[a] and B. F. Levine

Bell Laboratories, Murray Hill, New Jersey 07974

K. Ogawa

Bell Laboratories, Holmdel, New Jersey 07733

(Received 18 October 1982; accepted for publication 22 March 1983)

We have developed a sensitive, ultrahigh speed photodetector which has a structure of a modulation-doped $Al_x Ga_{1-x} As$/GaAs field-effect transistor. In spite of a large gate-drain spacing of $> 8 \mu m$ and a gate length of $> 20 \mu m$, this detector exhibited a rise time of 12 ps and a full width at half-maximum of 27 ps. When tested by a 8200-Å GaAs injection laser, the detector showed an ac (> 20 MHz) external quantum efficiency of $> 630\%$, i.e., 9 times more sensitive than a *pin* photodiode. In view of its high sensitivity, ultrahigh speed, and compatibility with modulation-doped field-effect transistors, this detector has promise for a variety of high-speed optical applications.

PACS numbers: 85.60.Dw, 42.80.Sa, 73.60.Fw, 73.40.Lq

The use of metal-semiconductor field-effect transistors (MESFET's) as photodetectors has been reported by several workers.[1-10] Because of their high response speed and the possibility of achieving internal gain, they can be used in high-speed photodetection and optoelectronic integrated circuits. However, the MESFET detectors reported so far typically have a thin active layer ($\approx 0.2 \mu m$) and a small gate drain spacing (on the order of $2 \mu m$). A thin active layer can make the incident radiation be absorbed mostly in the semi-insulating layer, which is known to be filled with traps. And a small gate-drain spacing can reduce the coupling efficiency between the incident radiation and the detectors. Gammel and Ballantyne have overcome this problem by replacing the gate metal electrode with a notch.[5] In this letter we report a high-speed $Al_x Ga_{1-x} As$/GaAs modulation-doped field-effect photodetector (MDFEP) which has device dimensions much larger than those of a conventional high-speed FET. The impulse response with a full width at half-maximum (FWHM) of 27 ps, a relatively large gate-drain spacing ($> 8 \mu m$), and a thick absorption layer ($> 1 \mu m$) suggest that this detector may be useful in gigabit rate optical detection. This device also possesses the capability of monolithic integration with low noise modulation-doped FET's forming the basis for optoelectronic integrated circuits.

The MDFEP is made on a modulation-doped heterostructure grown by molecular beam epitaxy.[11,12] Figure 1(a) illustrates the layer compositions and device structure. An undoped GaAs (1 μm), an undoped $Al_{0.3} Ga_{0.7} As$ (75 Å), a Si-doped n^+-$Al_{0.3} Ga_{0.7} As$ (520 Å, 1×10^{18} cm^{-3}), and finally an n^+-GaAs layer (200 Å, 2×10^{18} cm^{-3}) are grown sequentially on a semi-insulating GaAs substrate. The undoped layer is normally p type with a carrier concentration of $\approx 1 \times 10^{14}$ cm^{-3}. Owing to a sufficient band bending, two-dimensional electron gas [2DEG, shown as dashed line in Fig. 1(a)] exists in the undoped GaAs layer within ≈ 100 Å from the $Al_{0.3} Ga_{0.7} As$/GaAs interface.[13] Reduced impurity scattering and enhanced electron screening make the 2DEG

have a higher mobility, and perhaps a higher drift velocity. Hall measurement shows an electron Hall mobility of 6500 cm^2 /Vs at 300 K and 66 000 cm^2/Vs at 77 K. This strong mobility enhancement indirectly indicates the existence of the 2DEG. As will be shown later, the 2DEG plays an important role in the device operation. The detector electrodes are made by using a lift-off technique. Source and drain oh-

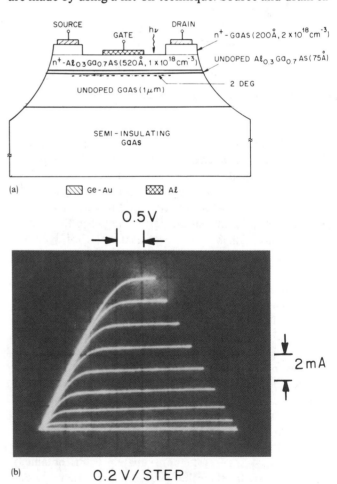

FIG. 1. (a) Schematic illustration of a modulation-doped field-effect photodetector; (b) drain current vs voltage characteristics of sample C0902 measured at 120 Hz.

[a] On leave from Chung-Shan Institute of Science and Technology, Taiwan, Republic of China.

Appl. Phys. Lett., Vol. 42, No. 12, 15 June 1983

Chen *et al.*

mic contacts are made of Ge/Au/Ni, which is alloyed at 450 °C for 15 s to ensure that it penetrates through both $Al_{0.3}Ga_{0.7}As$ layers. Unless otherwise specified, the device under study (sample C0902) has a 20-μm gate length, 450-μm gate width, 8-μm gate-drain spacing, and 8-μm gate-source spacing. The detector was mounted on a three-terminal 50-Ω microstrip line fixture for characterization.

Figure 1(b) illustrates the low-frequency (120 Hz) drain current-voltage characteristics of the transistor. It shows a transconductance (g_m) of 9 mmho, which corresponds to 20 mmho/mm, and a pinch-off voltage of -1.7 V. We also measured a source-gate capacitance (C_{gs}) of ≈ 0.3 pF and ≈ 0.15 pF at $V_{GS} = 0$ V and $V_{GS} < -1.7$ V, respectively. The source drain capacitance, which is mainly due to the stray capacitance of the package, is 0.05 pF. All the capacitance values are measured at 1 MHz.

The response speed of the detector is tested by a mode-locked dye laser ($\lambda \approx 6000$ Å) with a pulse duration of ≈ 6 ps, a repetition rate of 80 MHz, and an average power of 5 mW. The laser beam with a spot size of ≈ 30 μm is incident

through the gap between the gate and the drain. Figure 2(a) illustrates the detector response measured by an S-4 sampling head ($t_r < 25$ ps) with the detector biased at $V_{DS} = 5$ V and $V_{GS} = -2$V. We observed a rise time of ≈ 30 ps and a FWHM of 44 ps, which are largely limited by the sampling oscilloscope. The detector response was found to be independent of the drain bias within the oscilloscope limit. High-speed response similar to that shown in Fig. 2(a) was also observed at V_{DS} as low as 1.0 V. To determine the true response time of our detectors, we performed a cross-correlation measurement using a sampling optical gate which consists of two similar detectors. Details of this measurement will be discussed elsewhere. Results of this cross-correlation measurement are shown in Fig. 2(b). A rise time of 12 ps and a FWHM of ≈ 27 ps were observed. A longer fall time in Fig. 2(b) is probably due to reflection in the sampling gate. As far as we know, this device is among the fastest photodetectors reported on crystalline III-V compounds. It is worth mentioning that the detector showed no optical gain in this transient measurement. This is because the detector never reached a steady state during the period of incident radiation. The peak sensitivity under the high-speed testing is approximately 0.01 A/W, which is comparable to that of a Si *pin* detector tested under the same conditions.

Because of a thick GaAs absorption layer and a relatively short wavelength used in the experiment, absorption in the semi-insulating GaAs substrate is minute and can be neglected in the following discussion. The devices are operated in the low field mobility regime rather than the saturated velocity regime since the gate length is greater than or equal to 20 μm. In the present case, electrons have a mobility ≈ 30 times higher than that of holes. Consequently, the photocurrent is almost completely determined by the electrons. Photoelectrons (minority carriers) generated in the p^--GaAs layer will experience a vertical field associated with the band bending of the heterojunction as shown in Fig. 3, and a lateral field associated with the applied drain voltage.[14] As long as the electrons traverse through the 1-μm GaAs layer

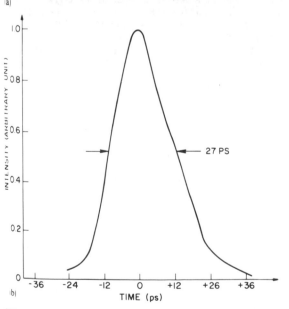

(a)

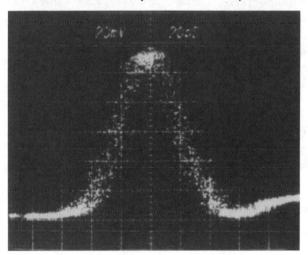

(b)

FIG. 2. (a) Impulse response of the detector tested by 6-ps dye laser pulses ($\lambda \approx 6000$ Å, average power ≈ 5 mW). (b) Impulse response of the detector measured by cross-correlation technique.

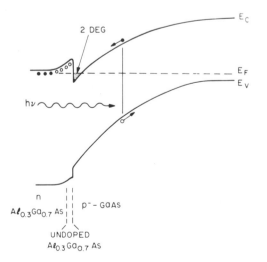

FIG. 3. Energy-band diagram in the direction perpendicular to the heterojunction.

in the vertical direction, they will be collected by the quasi-metallic 2DEG layer. This situation is similar to the collection of photogenerated electrons by n-type semiconductors in a conventional p-n junction photodetector. Thus, the response time of this detector is determined by the electron transit time across the 1-μm undoped GaAs layer rather than by the 8-μm gap. Taking ν_s to be 1.5×10^7 cm/s, we obtain a vertical transit time of 7 ps. We should point out that electrons collected by the drain electrode have to charge a small source-drain stray capacitance of 0.05 pF, resulting in a time constant ($2.2\,RC$) of 5.5 ps. Taking this into account, we obtained a total response time of 8.9 ps. A measured value of $\simeq 12$ ps is therefore in agreement with this calculation within the experimental error. In order to further confirm the mechanism that determines the response time, we tested a device (sample C0418) fabricated from the same wafer, which has a gate length of $280\,\mu$m, a gate-drain spacing of $15\,\mu$m, and a gate width of $450\,\mu$m. We observed a high-speed response similar to that shown in Fig. 2(a). It is worth noting that the response speed of the detector is not limited by the RC time constants associated with the gate-source capacitance. Although the gate-source capacitance is 0.15–0.3 pF, it plays no role in this case since photodetection does not require charging and discharging of this capacitance (notice that the gate electrode is connected to a dc bias only).

To measure the sensitivity of the detector, we utilize an $Al_x Ga_{1-x} As$ semiconductor laser ($\lambda \approx 8200$ Å), which is driven by 50-ns electrical pulses (repetition rate 12.5 kHz). This measurement avoids the difficulty associated with the response speed of the instruments and also allows the device to reach a steady state. A silicon photodiode, which has a sensitivity of 0.6 A/W at 8200 Å, is used for sensitivity calibration. At zero gate bias, the detector (sample C0902) shows an ac sensitivity of >4.2 A/W, which corresponds to an external quantum efficiency of $>630\%$. The observed optical gain is believed to be due to an increase in the number of carriers in the conduction channel. Following the guideline discussed above, we can show that the quantum efficiency of our detectors is given by $(\eta_i \tau_n L_g)/(t_n L_{gd})$, where η_i is the internal quantum efficiency, τ_n is the excess electron lifetime, t_n is the electron transit time across the gate, L_g is the gate length, and L_{gd} is the gate-drain spacing. Taking a gain of 6.3, a transit time of 133 ps, and η_i of 0.7, we calculated an excess electron lifetime of 0.48 ns. No comparison can be made at this time since excess electron lifetime in 2DEG is unknown. Furthermore, we also measured the gate current to see whether or not transconductance amplification is re-

sponsible for the optical gain. Our experimental results suggest that the observed gain is not due to $(1 + g_m R_g)$ as suggested in Ref. 6, where R_g is the gate resistance.

We have also performed a preliminary measurement on the noise spectrum of our detectors without incident radiation. At low frequency, the noise current is dominated by 1/f noise which has a corner frequency at 10 MHz. Beyond 10 MHz, FET channel noise which is given by $i_n^2 = 4kTg_m \Gamma \Delta f (\Gamma \approx 1)$ becomes important, where k is Boltzmann's constant, T is absolute temperature, and Δf is the bandwidth. This noise current has a value of 12 PA/$\sqrt{\text{Hz}}$ and has a flat spectrum up to the highest frequency used in the measurement (200 MHz).

In conclusion, we have reported a sensitive ultrahigh speed photodetector with a modulation-doped heterostructure FET. The built-in electric field and the two-dimensional electron gas existing in this system provide a mechanism for collecting the photogenerated carriers. This feature, not observed in a conventional MESFET detector, has led to the high response speed of the device. This detector shows a rise time of 12 ps, FWHM of 27 ps, an ac external quantum efficiency of 630% at 8200 Å, and a high-frequency noise current of 12 PA/$\sqrt{\text{Hz}}$. This new heterostructure photodetector should find applications such as in gigabit rate optical communication, optoelectronic monolithic integrated circuits, fiber dispersion measurement, and in studies of the dynamic response of semiconductor lasers.

The authors would like to thank F. R. Merritt for supplying us GaAs lasers used in this work, and D. Kahng for many valuable discussions.

[1] C. Baack, G. Elze, and G. Walf, Electron. Lett. **13**, 193 (1977).
[2] J. C. Gammel and J. M. Ballantyne, IEDM Technical Digest 120 (1978).
[3] J. J. Pan, 22nd SPIE International Technical Symposium, San Diego, CA 1978.
[4] J. M. Osterwalder and B. J. Rickett, Proc. IEEE **67**, 966 (1979).
[5] J. C. Gammel and J. M. Ballantyne, Appl. Phys. Lett. **36**, 149 (1980).
[6] T. Sugeta and Y. Mizushima, Jpn. J. Appl. Phys. **19**, L27 (1980).
[7] W. D. Edwards, IEEE Electron Device Lett. **1**, 149 (1980).
[8] J. Graffeuil, P. Rossel, and H. Martinot, Electron. Lett. **15**, 439 (1979).
[9] E. H. Hara and R. I. MacDonald, IEEE Trans. Microwave Theory and Tech. **28**, 662 (1980).
[10] R. I. MacDonald, Appl. Opt. **20**, 591 (1981).
[11] R. Dingle, H. L. Stormer, A. C. Gossard, and W. Wiegmann, Appl. Phys. Lett. **33**, 665 (1978).
[12] T. Mimura, S. Hiyamizu, T. Fujii, and K. Nanbu, Jpn. J. Appl. Phys. **19**, L225 (1980).
[13] H. L. Stormer, R. Dingle, A. C. Gossard, W. Wiegmann, and M. D. Sturge, Solid State Commun. **29**, 705 (1979).
[14] C. Y. Chen, A. Y. Cho, C. G. Bethea, and P. A. Garbinski, Appl. Phys. Lett. **41**, 282 (1982).

Modulation-Doped AlGaAs/GaAs Heterostructure Charge Coupled Devices

R. A. MILANO, MEMBER, IEEE, M. J. COHEN, MEMBER, IEEE, AND D. L. MILLER

Abstract—The advantages of the modulation-doped heterostructure over conventional materials structures for high speed CCD applications are outlined. In addition, the first demonstration of charge transfer in a modulation-doped AlGaAs/GaAs heterojunction is reported. A ten cell, three phase Schottky barrier gate CCD was fabricated using this structure and operated as a shift register. The details of the device fabrication and characterization are presented.

GaAs CHARGE coupled devices are interesting because of the high speed charge transfer that can be achieved [1]. High speed operation is possible owing primarily to the high electron mobility of GaAs and the use of a buried channel for electron transport. Recently, CCD operation at f_{cl} = 1 GHz has been demonstrated [2]. Modulation-doped AlGaAs/GaAs heterostructures have exhibited enhanced mobility at low electric fields [3-6], thus making this structure particularly attractive for CCD applications. In an optimized device, several advantages should result from the use of modulation-doped structures. An increased dynamic range is expected due to the larger charge storage capacity of the two-dimensional electron gas (2DEG). Enhanced high speed charge transfer from gate to gate results from the higher mobility and highly localized nature of the 2DEG, which permits the entire charge transfer to occur at the position of maximum electric field. Clocking rates in excess of those attainable with bulk GaAs devices can therefore be envisioned. In this article, the criteria required for the realization of an optimized high speed CCD using a modulation-doped heterostructure are presented. In addition, the first demonstration of charge transfer in a modulation-doped AlGaAs/GaAs heterostructure is reported.

This effort is motivated by three factors. An optimized modulation-doped heterostructure CCD has the potential for larger dynamic range, higher maximum clock frequency and more simplified integration with electronic drive circuitry than existing devices. For purposes of discussion consider the typical GaAs high speed CCD [1], which has 5 μm × 100 μm Schottky barrier gates separated by 1 μm gaps. The channel doping and layer thickness are N_d = 1 × $10^{16} cm^{-3}$ and t = 1.0 μm respectively. The full well-charge capacity is 5 × 10^6 electrons. If the same gate geometry is assumed, a high speed CCD fabricated on a single period modulation-doped heterostructure with N_s = 2 × $10^{12} cm^{-2}$ has a full well charge capacity of 1 × 10^7 electrons. This corresponds to

a dynamic range of ~100 db compared to ~90 db expected from a device fabricated on bulk GaAs, assuming equal noise sources. Further increase in dynamic range is possible if a multiple period structure is used. By appropriately choosing the thickness of the AlGaAs layer, the charge transport region can be located at the fringing field maximum, thereby improving the frequency response of the device. The gate-to-gate transit time in this case is $\tau \approx L/\mu E$, where L is the gate length and E is the magnitude of the fringing field. Modulation-doped heterostructures have $\mu_n(300K) \sim$ 6000–8000 cm^2/V-sec. In comparison, the μ_n for a bulk buried channel GaAs CCD is 3000–5000 cm^2/V-sec. Therefore, assuming L/E constant, the modulation-doped heterostructure CCD will have a maximum clock frequency approximately twice that of the conventional buried channel device.

The design of a complex chip requiring the integration of a high speed CCD with other electronic circuitry using conventional materials technology presents problems because of the conflicting material requirements. The CCD channel must be much thicker but more lightly doped than the channel of a high performance MESFET, that is, N_d(CCD) $\leqslant$ 1 × $10^{16} cm^{-3}$, d_{ch}(CCD) $\geqslant$ 1 μm, N_d(FET) $\geqslant$ 1 × $10^{17} cm^{-3}$, d_{ch}(FET) $\leqslant$ 0.3 μm. Recently, both enhancement mode and depletion mode MESFET's have been demonstrated using the modulation-doped heterostructure [7-8]. Therefore, both the CCD and the MESFET's can be fabricated from the same material structure, with isolation of the various active regions accomplished via proton bombardment. Considerable simplification of both design and fabrication can thus be achieved.

A modulation-doped heterostructure was grown by molecular beam epitaxy on a ⟨100⟩—oriented undoped semi-insulating GaAs substrate. The structure, shown schematically in Fig. 1(a), consists of a 1 μm thick undoped GaAs layer, an 80 Å thick updoped $Al_{0.3}Ga_{0.7}As$ spacer layer, and a 400 Å thick $n^+ Al_{0.3}Ga_{0.7}As$ layer. The top layer is doped to 1 × $10^{18} cm^{-3}$ with Si. Differential Hall mobility measurements made on this material show that $\mu n(300K)$ = 6100 cm^2/V-sec, $N(300K)$ = 6.4 × $10^{11} cm^{-2}$ and $\mu n(77K)$ = 93,000 cm^2/V-sec Schottky barrier gate charge coupled devices were fabricated using standard photolithographic techniques. The devices are ten cells in length and employ a three phase clocking scheme. The gates are designed to be 40 μm long by 400 μm wide and seperated by 2 μm gaps. Electron beam evaporated Cr (400 Å)/ Au (400 Å) was used for the gate metallization. The CCD channel was isolated by proton bombardment with a dose

Manuscript received May 3, 1982; revised June 4, 1982.

The authors are with Rockwell International, Microelectronics Research and Development Center, Thousand Oaks, CA 91360.

Reprinted from *IEEE Electron Device Letters*, vol. EDL-3, no. 8, pp. 194–196, August 1982.

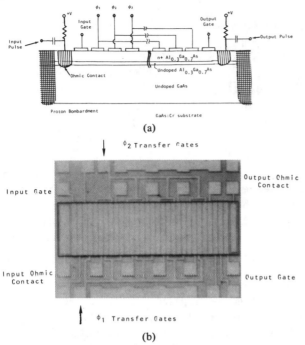

(a)

(b)

Fig. 1. (a) Schematic cross-section (not to scale) of a modulation-doped heterojunction charge coupled device. (b) Photograph of a modulation-doped AlGaAs/GaAs heterostructure CCD. The gate dimensions are 40 μm $\times$ 400 μm and the total active area is 1.4 mm $\times$ 0.4 mm.

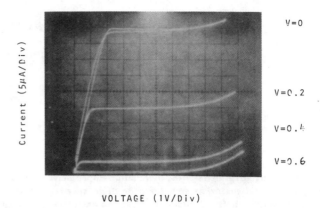

Fig. 2. *I–V* characteristics of the modulation-doped heterostructure CCD channel.

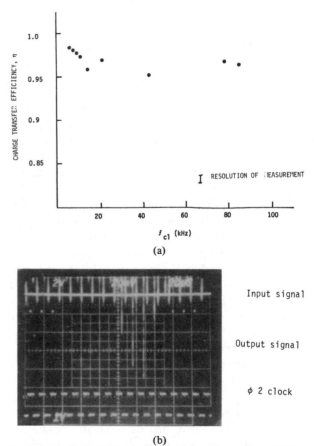

(a)

(b)

Fig. 3. (a) Frequency dependence of the charge transfer efficiency of the modulation-doped heterostructure CCD. (b) Charge transfer characteristics of the modulation-doped heterostructure CCD. The charge transfer efficiency is 0.97, with no fat zero, at f_{c1} = 76.9 kHz.

of 1 $\times$ 10^{12}cm^{-2} at 100 keV followed by a second implant at a dose of 1 $\times$ 10^{13}cm^{-2} at 30 keV. The second proton implant, which penetrates only 0.2-0.3 μm, was required to reduce the conductivity of the heavily doped Al$_{0.3}$Ga$_{0.7}$As. Input and output ohmic contacts were formed using alloyed AuGe/Ni films. A photograph of the device is shown in Fig. 1(b). It should be noted that CCD's fabricated from this wafer are not expected to exhibit optimum performance, due to the proximity of the 2DEG to the surface.

Prior to operating the CCD as a shift register, the *I–V* characteristics of the channel were measured to determine the pinchoff voltage. Figure 2 shows the channel characteristics when the device is operated as a FET, that is, all of the gates are held at the same potential. It is clear from this figure that the device operates in the depletion mode. A gate voltage of 0.5 V is required to pinch off the channel. Enchancement mode operation using a very thin n$^+$AlGaAs layer (~200 Å) has also been recently observed [9]. The reason for the upward bend in the channel *I–V* characteristics is not well-understood at present. A field dependent mechanism, such as avalanche breakdown or field-assisted emptying of traps, may be responsible for this behavior. Gate leakage current is probably not an important factor, since for reverse bias voltages $V < 7$ V the leakage current of a single Schottky CCD gate is < 10nA. Charge transfer was demonstrated by operating the device as a shift register and electrically injecting a signal at the input ohmic contact. A symmetric clocking scheme with $V_{cl} = \pm 0.5$ V yielded the best performance. The charge transfer efficiency was measured with no fat zero over the frequency range 5.9 kHz–83.3 kHz. Over this frequency range, CTE varies from 0.98 to 0.968 with the highest effi-

ciency occurring at the lowest clock frequency. This data along with an example of the transfer characteristics of the CCD are shown in Figs. 3(a) and (b), respectively.

An examination of Fig. 3(a) indicates that although the CTE of the device improves somewhat as the clocking frequency is reduced, a significant amount of loss is still present. The proximity of the charge transfer region to the surface (400 Å) is such that fringing fields do not significantly aid the transfer process. The maximum clock frequency is then limited by thermal diffusion, which is described by the time

434

constant $\tau_{th} = L^2/2.5D$ [10], where L is the gate length and D is the diffusion constant. For this device $\tau_{th} \approx 1 \times 10^{-7}$ sec and f_{max} ($= 1/4\tau_{th}$) ≈ 3.3 MHz. The frequency range over which CTE was determined is sufficiently below this limit implying that much of the loss may be frequency-independent. Measurement of the inter-electrode gap length revealed that for this device $1_g \approx 4$ μm, due to poor mask-to-wafer contact during photolithography. The rather large gaps cause significant perturbations in the channel potential in the direction of charge transfer. These potential variations are a major source of frequency-independent loss [10]. The uniformity of charge loss along the device was examined by injecting a burst of input pulses greater than the number of cells. A steady state output condition was achieved and the magnitude of the output signal from each cell was identical. The charge levels (full well signal) of all the cells are equal, therefore, and the transfer efficiency is uniform along the device [11]. The result is not unexpected if the primary limitation to charge transfer efficiency is associated with perturbations of the channel potential since the loss is determined by the length of the inter-electrode gap, which is constant. Further work is presently underway to better characterize the sources of charge transfer inefficiency.

Conclusion

The advantages of the modulation-doped heterostructure over the conventional buried channel structure are examined and it is concluded that this materials system presents some significant advantages for the design of high speed charge coupled devices. In addition, charge transfer has been demonstrated for the first time in a modulation-doped heterostructure. A Schottky barrier gate CCD has been fabricated which has a maximum charge transfer efficiency of 0.98 at $f_{c1} \sim$ 6 kHz. A preliminary study indicates that the CTE of this unoptimized device is limited primarily by the device geometry, and is not indicative of its ultimate performance.

Acknowledgment

The authors would like to acknowledge the technical assistance of P. Newman.

References

[1] I. Deyhimey, J. S. Harris, Jr., R. C. Eden, and R. J. Anderson, *Japan J. Appl. Phys.*, vol. 19-1, p. 269, 1980.

[2] I. Deyhimey, W. A. Hill, and R. J. Anderson, *Elec. Dev. Lett.*, EDL-2, 1981

[3] S. Hiyamizu, T. Mimura, T. Fujii, K. Nanbu, and H. Hashimoto, *Japan J. Appl. Phys.*, vol. 20, p. L245, 1981.

[4] T. J. Drummond, H. Morkoc, K. Hess, and A. Y. Cho, *J. Appl. Phys.*, vol. 52, p. 5231 1981.

[5] T. J. Drummond, H. Morkoc, and A. Y. Cho, *J. Appl. Phys.* vol. 52, p. 1380, 1981.

[6] J. J. Coleman, P. D. Dapkus, J. J. J. Yang, *Electron. Lett.*, vol. 17, p. 606, 1981.

[7] T. Mimura, S. Hiyamizu, K. Joshiu, and K. Hikosaka, *Japan J. Appl. Phys.* vol. 20, p. L317, 1981.

[8] T. Mimura, S. Hiyamizu, T. Fujii, and K. Nanbu, *Japan J. Appl. Phys.*, vol. 19, p. L225, 1980.

[9] Y. Z. Liu, private communication.

[10] W. F. Kosonosky and J. E. Carnes, *RCA Review* vol. 36, p. 566, 1975.

[11] R. M. Barsan, *Int. J. Electron.*, vol. 48, p. 149, 1980.

OPTICAL CONTROL OF MILLIMETER WAVE DEVICES

Harold R. Fetterman, Wei-Yu Wu and David Ni
Department of Electrical Engineering
University of California, Los Angeles
Los Angeles , CA 90024

Abstract

Coherent mixing of optical radiation from a tunable CW ring dye laser and a stablized HeNe laser was used to inject broadband microwave signals into GaAs MESFETs (FETs), AlGaAs-GaAs HEMTs and monolithic FET amplifiers up to 55 GHz. Using this technique, amplification of an optically injected signal at 32 GHz, direct injection locking of a 17 GHz GaAs oscillator, and frequency tuning of a 40 GHz oscillator by laser intensity modulation were demonstrated. Comparison of the millimeter wave frequency generation mechanism, noise sources and system performance with other techniques used for signal synchronization and distributed control application are discussed.

Introduction

Current interest in optical control of millimeter wave devices can be divided into two areas. Firstly, there is the in device physics area with detailed investigations of mechanisms of photoconductive and photovoltaic response under different conditions [1-5]. Of particular interest are the new device such as HEMTs which use layered structures and extremely short gate FETs with high electric fields. Secondly, in the application areas, considerable attention has been devoted toward dealing with overcoming the frequency constraints of existing devices and performing signal synchronization and distributed control for communication and radar systems [6-8]. Nonlinearities of semiconductor lasers, pin detectors as well as FETs have been utilized extensively to sideband lock [6] or to generate harmonics for multiplication of master oscillation signals [7-8]. In these approaches, reliability and FM noise degradation issues require significant engineering efforts since the devices operating in the nonlinear region are normally under significantly bias stress.

In these study, we explored laser mixing techniques to inject millimeter wave signals in GaAs FETs as well as related devices, such as HEMT [9-10] (High Electron Mobility Transistor). We also examined systems such as monolithic oscillators and amplifiers.

Experiment

A visible CW ring dye laser and stablized HeNe laser were selected as light sources based on their absorption coefficient for higher carrier generation [11], shallower light penetration for lower backgating and trap effects [4], and lower carrier concentration flunctuation (CW illumination) for lower capacitance effects [12]. The penetration depth of these lasers is about 0.3 μm, which is the same order as the thickness of active region of the FETs (0.2 μm thick n-GaAs on 2 μm thick undoped GaAs buffer layer) and sufficient to illuminate the GaAs and 2D electron gas region of HEMT structures. (0.04 μm thick n-AlGaAs on 1 μm thick undoped GaAs layer). Figure 1 shows the experimental arrangement used in all these measurements. The devices under investigation were illuminated with light obtained by coherent mixing of between two lasers: a frequency stablized He-Ne laser (632.8 nm) and a Coherent Model 699-21 ring dye laser. The wavelength of the ring dye laser, monitored by a Burleigh WA-20 wavemeter (accurate to 0.001 nm), can be continuously tuned to any value within the dye emission spectrum. For Kiton 620 dye (600 - 640 nm), the tuning range is more than 50 nm around the HeNe line. Consequently the beat frequency (difference of light frequency / optical i.f. signals) of two CW lasers can be easily varied from zero to several hundred GHz. The laser power density on the samples was kept to approximatedly 10 mW/cm^2 for the HeNe laser and 1 W/cm^2 for the dye laser with the linewidth and stability in the sub MHz range.

THE samples were FETs (commercial products from Dexcel, Hughes, NEC etc.) or HEMT'S (from Rockwell, TRW etc.) directly mounted onto a fixture with 50 ohm microstrip lines. The monolithic oscillators and multiple stage amplifiers used in this study were basically composed of GaAs FETs and impedance matching circuits, mounted in appropriated fixtures. HP 8569B and Tektronix 492P spectrum analyzers were used to monitor the outputs. Illuminating the active region of the samples with two coherent CW lasers resulted in the generation of rf electrical signals at the optical beat frequency in these devices. The experiments we examined involve three basic processes: amplification of the injected rf signal, optical injection locking, and frequency tuning by controlling the sample's carrier concentrations.

A. Amplification of the injected rf signals

Reprinted with permission from *Proc. SPIE—Optical Technology for Microwave Applications III (1987)*, vol. 789, pp. 50–53.

Although this CW method has considerable advantages over using diode lasers, it makes severe requirements on the lasers. In order to generate optical rf beat frequencies, the wavelength of the dye laser (the resolution of tuning is about the linewidth of laser) must be very close to the wavelength of HeNe laser. To generate a 7.5 GHz optical beat frequency, for example, corresponds to tuning a 0.01 nm wavelength difference between two lasers. Using the FET, we can directly observed the optical beat frequency up to 18 GHz , which was limited by the impedance matching between the device and fixture. Therefore to detect higher frequency optical beat signals, we have superimposed an electrical rf signal on the gate of the FET. As a result of the strong nonlinearity of GaAs FET, the electrical signal mixes with the rf signal produced by optical beating, and produes a secondary beat signal. A 30 dB gain rf amplifier (1-3 GHz) was inserted between the FET and spectrum analyzer and HP 8350A sweep generator supplied rf signal power to 40 GHz. Observing the secondary beat signal at 2.6 GHz, therefore permits accurate measurements of the mixing signal up to 42.6 GHz. Beyond this frequency, we used the secondary harmonics of electrical rf signal to obtain optical beat signals up to 52 GHz. Similar mixing techniques have used electrical rf signals to beat optically injected signal from directly modulated semiconductor lasers at lower frequencies [13].

This method provided the means of directly generating rf signals at millimeter frequencies by laser mixing and in addition permitted us to accurately monitor the frequency of our optically generated broadband rf signals. Therefore this system, with a commercial FET, was used in all our measurements giving us a resolution on the order of Kilocycles. As a example of how we used this system to directly observe optical beat signals in Ka band with a monolithic two stage GaAs FET amplifier having 16 dB gain at 32 GHz, fabricated by Hughes Aircraft Company. The active region of the FET at the first stage was illuminated with the mixing light at beat frequency around 32 GHz. The generated rf signal was further amplified through second stage and we observed, as shown in Fig. 2, a -60 dBm (S/N is around 12 - 15 dB) signal. We anticipated better output power with optimized optical access, however, this experiment shows the potential of this technique in distributing power optically to arrays etc. Furthermore, this technique can be used to measure the gain bandwidth of amplifier efficiently.

B. Injection Locking

The device used for injection locking was originally designed to be a monolithic three stage GaAs FET amplifier, by Rockwell, with 20 dB gain between 17 to 21 GHz. By introducing feedback to this common source configuration device, it was made into an broadband oscillator at 17.8 GHz with 15 dbm power output (Q factor was approximatedly 20). The dc bias levels were as follows: all three stages, drain to source voltage (V_d) were biased at 3.0 V, V_{g1} = 0 and V_{g2} = V_{g3} = -0.5 V, where V_{gn} represents the gate bias at nth stage. This oscillator can be electrically injection locked by applying a -10 dBm rf signal to the input (first gate) with a locking range was about 40 MHz. To offset the photovoltaic voltage introduced by laser illumination [2], we applied more negative bias to the illuminated gate, V_{g2} = -2.0 V, and reduced drain to source voltages V_{d1} = V_{d2} = 2.3 V to bias the oscillator below the threshold of oscillation, as shown in Fig. 3 (a). The oscillation was optically turned on, as shown in Fig. 3(b), when the second FET was illuminated. Finally, when the optical beat frequency approached the free run frequency of the oscillator, the signal was locked as shown in Fig. 3(c). The locking range was only about 2 MHz. We attribute the narrow locking range to the optical tuning discussed in the next section. Increasing the power actually tunes the devices and therefore reduce the efficiency of frequency lock.

C. Frequency tuning

Optical frequency tuning using a single laser has been related to capacitance changes, due to optically excited electron-hole pairs. Common-source mode oscillators have found to have an optical-frequency sensitivity approximately 5 times higher than common-drain mode configuration at oscillation frequencies between 4 - 9 GHz [12]. In our experiment, we used a Ka-band (at 40 GHz) GaAs FET oscillator with a common gate configuration. In this configuration, we examined the relative optical sensitivity of frequency tuning of Cgs (with the drain shorted to gate) vs Cds (with the source shorted to gate). A 480 MHz vs 80 MHZ (Fig. 4 (a)-(b)) frequency increase was observed with light intensities of about 1 W/cm^2 (only the dye laser was used with attenuators), which was in agreement with previous result in X band. Using higher illumination density, we have demonstrated 12% optical-frequency tuning in Ka band. Along with the oscillation frequency shift, we have also observed controllable variations in output power.

Result and Discussion

Due to the observation of acoustic noise introduced by light scattering below S band [14] and the difficulty to characterize secondary beating signal (beyond 18 GHz) by current setup, our major characterizations were in S band through Ku band (2 - 18 GHz). Fig. 5 shows the frequency response of optical mixing from 6 GHz to 18 GHz. The response of this 0.25 μm device is relative flat and indicates by its voltage

SPIE Vol. 789 Optical Technology for Microwave Applications III (1987)

dependance that the mechanism is primary photoconductive. From the above observations, we find that under intensive illumination and high photoexcited carrier concentration (photoexcited current was the same magnitude as intrinsic current) the Schottky gate might become unnecessary. This result is in agreement with non-gate high speed photodetectors which responding to picosecond light signals [15,16].

Since the photoconductive mechanism dominates, HEMT structures significantly improve the frequency performance only if properly designed and utilized. The undoped GaAs layer in such devices where the photoexcited carriers are generated reduce the impurity scattering and increase the mobility of carriers. However, the parasitic AlGaAs layer can reduce this advantage , especially if the illumination is absorbed heavily near the surface. The high doping concentration of AlGaAs results in a high trap concentration and low mobility [17]. To take full advantage of HEMT structure, AlGaAs layers with large energy gaps (such as obtained by increasing the Al content) should be fabricated. Then the carriers can be generated in the GaAs near the two dimensional electron gas (2 DEG) channel ("window effect") and shorter electron transit times will be obtained. Preliminary measurements on HEMTs with relatively large gate drain spacings have been relatively encouraging.

Conclusion

By using laser mixing techniques, we have extended direct optical control of FETs, related devices and systems from X band to Ka band. Infrared semiconductor lasers with wavelength stablization and tunability by temperature, cavity and current control) [6] can now be used to replace CW dye and HeNe lasers as light sources for realization of compactness. Specially design structures, such as HEMTs with tailored AlGaAs layers can be incorporated with the optimized lasers to form unique, optically controlled, systems working well into millimeter wave frequency.

Acknowledgement

We acknowledge this work is supported by the Air Force Office of Scientific Research.

References

1. J. P. Noad, E. H. Hara, R. H. Hum, R.I. MacDonald, IEEE Trans. Electron. Devices, ED-29, 1792 (1982).

2. A. A. de Salles, IEEE Trans. Microwave Theory Tech, MTT-31, 812 (1983)

3. J. L. Gautier, D. Pasquet, P. Pouvil, IEEE Trans. Microwave Tech, MTT-33, 819 (1985)

4. G. J. Papaionanou, J. R. Forrest, IEEE Trans. Electron. Devices, ED-33, 373 (1986).

5. R. B. Darling, J. P. Uyemura, IEEE J. Quantum Electron., QE-23, 1160 (1987)

6. L. Goldberg, H. F. Taylor, J. F. Weller, Electorn. Lett. Vol 19, No 13, 491 (1983)

7. A. S. Daryoush, P. R. Herczfeld, Z. Turski, P. Wahi, IEEE Trans. Microwave Theory Tech., MTT-34, 1363 (1986)

8. P. Herczfeld, A. S. Daryoush, A. Rosen, A. Sharma, V. M. Contarino, IEEE Trans. Microwave Theory Tech., MTT-34, 1371 (1986)

9. C. Y. Chen, A. Y. Cho, C. G. Bethea, P. A. Garbinski, Y. M. Pang, B. F. Levine, K. Ogawa, Appl. Phys. Lett., 42, 1040 (1983)

10. R. N. Simons, K. B. Bhasin, IEEE Trans. Microwave Theory Tech., MTT-34, 1349 (1986)

11. H. Mizuno, IEEE Trans Microwave Theory Tech., MTT-31, 576 (1983)

12. H. J. Sun, R. J. Gutmann, J. M. Borrego, Solid St. Electron., Vol 24, 935 (1981)

13. D. K. W. Lam, R. I. MacDonald, IEEE Trans. Electron. Device, ED-31, 1766 (1984)

14. David Ni, Harold R. Fetterman, submitted for publication

15. S. Y. Wang, D. M. Bloom, Electron. Lett. Vol. 19, No. 14, 554 (1983)

16. S. Y. Wang, D. M. Bloom, D. M. Collins, Appl. Phys. Lett. 42, 190 (1983)

17. A. Kastalsky, J. C. M. Huang, Solid St. Commumnication, Vol 15, No. 5, 317 (1984)

OPTICAL MIXING AT MILLIMETER IF's

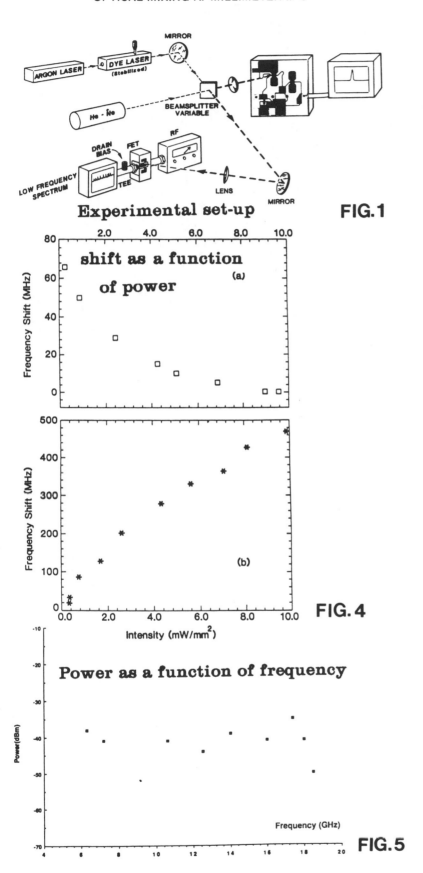

Experimental set-up **FIG. 1**

FIG. 4

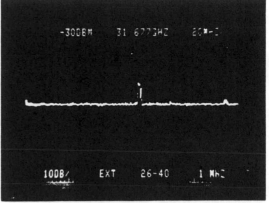

FIG. 2

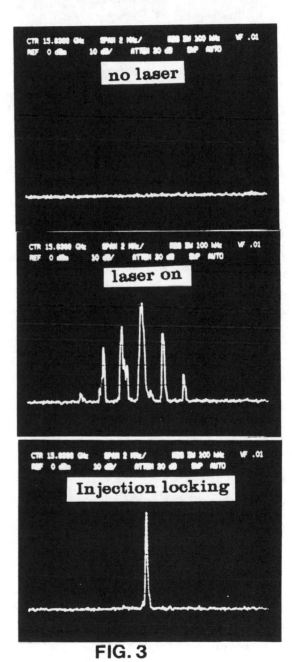

FIG. 3

FIG. 5

Author Index

Subject Index

453

Heinrich Daembkes (M'79) was born in Willich, Germany, on January 1, 1951. He received the Diplom-Ingenieur degree in electronics from the Technische Hochschule Aachen in 1976, and the Dr.-Ing. degree in electronics from the Duisburg University, in 1983.

After working on the design of receiver circuits at AEG Telefunken in Hannover, he joined the Solid State Electronics Department of the University in Duisburg. There he was engaged in the design, technology and characterization of GaAs MESFETs and GaAs/AlGaAs MOD-FETS. In 1984 he joined the AEG Research Center, Ulm, Germany (now Daimler-Benz AG Research Center Ulm), where he is now head of the Microwave Components Department. His main areas of interest are all kinds of heterostructure transistors, using InP, GaAs, and Si/SiGe based materials.